DATE DUE

	PRINTED IN U.S.A.

Physical Mathematics

Unique in its clarity, examples, and range, *Physical Mathematics* explains as simply as possible the mathematics that graduate students and professional physicists need in their courses and research. The author illustrates the mathematics with numerous physical examples drawn from contemporary research. In addition to basic subjects such as linear algebra, Fourier analysis, complex variables, differential equations, and Bessel functions, this textbook covers topics such as the singular-value decomposition, Lie algebras, the tensors and forms of general relativity, the central limit theorem and Kolmogorov test of statistics, the Monte Carlo methods of experimental and theoretical physics, the renormalization group of condensed-matter physics, and the functional derivatives and Feynman path integrals of quantum field theory. Solutions to exercises are available for instructors at www.cambridge.org/cahill.

KEVIN CAHILL is Professor of Physics and Astronomy at the University of New Mexico. He has done research at NIST, Saclay, Ecole Polytechnique, Orsay, Harvard, NIH, LBL, and SLAC, and has worked in quantum optics, quantum field theory, lattice gauge theory, and biophysics. *Physical Mathematics* is based on courses taught by the author at the University of New Mexico and at Fudan University in Shanghai.

Physical Mathematics

KEVIN CAHILL
University of New Mexico

CAMBRIDGE
UNIVERSITY PRESS

CAMBRIDGE
UNIVERSITY PRESS

University Printing House, Cambridge CB2 8BS, United Kingdom

Published in the United States of America by Cambridge University Press, New York

Cambridge University Press is part of the University of Cambridge.

It furthers the University's mission by disseminating knowledge in the pursuit of education, learning, and research at the highest international levels of excellence.

www.cambridge.org
Information on this title: www.cambridge.org/9781107005211

First published 2013
Reprinted with corrections 2014

Printed in the United Kingdom by Clays, St Ives plc.

A catalog record for this publication is available from the British Library

Library of Congress Cataloging in Publication data
Cahill, Kevin, 1941–, author.
Physical mathematics / Kevin Cahill, University of New Mexico.
pages cm
ISBN 978-1-107-00521-1 (hardback)
1. Mathematical physics. I. Title.
QC20.C24 2012
530.15–dc23
2012036027

ISBN 978-1-107-00521-1 Hardback

For Ginette, Mike, Sean, Peter, Mia, and James,
and in honor of Muntadhar al-Zaidi.

Contents

Preface		*page* xvii
1	**Linear algebra**	**1**
1.1	Numbers	1
1.2	Arrays	2
1.3	Matrices	4
1.4	Vectors	7
1.5	Linear operators	9
1.6	Inner products	11
1.7	The Cauchy–Schwarz inequality	14
1.8	Linear independence and completeness	15
1.9	Dimension of a vector space	16
1.10	Orthonormal vectors	16
1.11	Outer products	18
1.12	Dirac notation	19
1.13	The adjoint of an operator	22
1.14	Self-adjoint or hermitian linear operators	23
1.15	Real, symmetric linear operators	23
1.16	Unitary operators	24
1.17	Hilbert space	25
1.18	Antiunitary, antilinear operators	26
1.19	Symmetry in quantum mechanics	26
1.20	Determinants	27
1.21	Systems of linear equations	34
1.22	Linear least squares	34
1.23	Lagrange multipliers	35
1.24	Eigenvectors	37

1.25	Eigenvectors of a square matrix	38
1.26	A matrix obeys its characteristic equation	41
1.27	Functions of matrices	43
1.28	Hermitian matrices	45
1.29	Normal matrices	50
1.30	Compatible normal matrices	52
1.31	The singular-value decomposition	55
1.32	The Moore–Penrose pseudoinverse	63
1.33	The rank of a matrix	65
1.34	Software	66
1.35	The tensor/direct product	66
1.36	Density operators	69
1.37	Correlation functions	69
	Exercises	71
2	**Fourier series**	**75**
2.1	Complex Fourier series	75
2.2	The interval	77
2.3	Where to put the 2πs	77
2.4	Real Fourier series for real functions	79
2.5	Stretched intervals	83
2.6	Fourier series in several variables	84
2.7	How Fourier series converge	84
2.8	Quantum-mechanical examples	89
2.9	Dirac notation	96
2.10	Dirac's delta function	97
2.11	The harmonic oscillator	101
2.12	Nonrelativistic strings	103
2.13	Periodic boundary conditions	103
	Exercises	105
3	**Fourier and Laplace transforms**	**108**
3.1	The Fourier transform	108
3.2	The Fourier transform of a real function	111
3.3	Dirac, Parseval, and Poisson	112
3.4	Fourier derivatives and integrals	115
3.5	Fourier transforms in several dimensions	119
3.6	Convolutions	121
3.7	The Fourier transform of a convolution	123
3.8	Fourier transforms and Green's functions	124
3.9	Laplace transforms	125
3.10	Derivatives and integrals of Laplace transforms	127

3.11	Laplace transforms and differential equations	128
3.12	Inversion of Laplace transforms	129
3.13	Application to differential equations	129
	Exercises	134
4	**Infinite series**	**136**
4.1	Convergence	136
4.2	Tests of convergence	137
4.3	Convergent series of functions	138
4.4	Power series	139
4.5	Factorials and the gamma function	141
4.6	Taylor series	145
4.7	Fourier series as power series	146
4.8	The binomial series and theorem	147
4.9	Logarithmic series	148
4.10	Dirichlet series and the zeta function	149
4.11	Bernoulli numbers and polynomials	151
4.12	Asymptotic series	152
4.13	Some electrostatic problems	154
4.14	Infinite products	157
	Exercises	158
5	**Complex-variable theory**	**160**
5.1	Analytic functions	160
5.2	Cauchy's integral theorem	161
5.3	Cauchy's integral formula	165
5.4	The Cauchy–Riemann conditions	169
5.5	Harmonic functions	170
5.6	Taylor series for analytic functions	171
5.7	Cauchy's inequality	173
5.8	Liouville's theorem	173
5.9	The fundamental theorem of algebra	174
5.10	Laurent series	174
5.11	Singularities	177
5.12	Analytic continuation	179
5.13	The calculus of residues	180
5.14	Ghost contours	182
5.15	Logarithms and cuts	193
5.16	Powers and roots	194
5.17	Conformal mapping	197
5.18	Cauchy's principal value	198
5.19	Dispersion relations	205

5.20 Kramers–Kronig relations 207
5.21 Phase and group velocities 208
5.22 The method of steepest descent 210
5.23 The Abel–Plana formula and the Casimir effect 212
5.24 Applications to string theory 217
Exercises 219

6 Differential equations 223
6.1 Ordinary linear differential equations 223
6.2 Linear partial differential equations 225
6.3 Notation for derivatives 226
6.4 Gradient, divergence, and curl 228
6.5 Separable partial differential equations 230
6.6 Wave equations 233
6.7 First-order differential equations 235
6.8 Separable first-order differential equations 235
6.9 Hidden separability 238
6.10 Exact first-order differential equations 238
6.11 The meaning of exactness 240
6.12 Integrating factors 242
6.13 Homogeneous functions 243
6.14 The virial theorem 243
6.15 Homogeneous first-order ordinary differential equations 245
6.16 Linear first-order ordinary differential equations 246
6.17 Systems of differential equations 248
6.18 Singular points of second-order ordinary differential equations 250
6.19 Frobenius's series solutions 251
6.20 Fuch's theorem 253
6.21 Even and odd differential operators 254
6.22 Wronski's determinant 255
6.23 A second solution 255
6.24 Why not three solutions? 257
6.25 Boundary conditions 258
6.26 A variational problem 259
6.27 Self-adjoint differential operators 260
6.28 Self-adjoint differential systems 262
6.29 Making operators formally self adjoint 264
6.30 Wronskians of self-adjoint operators 265
6.31 First-order self-adjoint differential operators 266
6.32 A constrained variational problem 267

6.33 Eigenfunctions and eigenvalues of self-adjoint systems 273
6.34 Unboundedness of eigenvalues 275
6.35 Completeness of eigenfunctions 277
6.36 The inequalities of Bessel and Schwarz 284
6.37 Green's functions 284
6.38 Eigenfunctions and Green's functions 287
6.39 Green's functions in one dimension 288
6.40 Nonlinear differential equations 289
Exercises 293

7 Integral equations 296
7.1 Fredholm integral equations 297
7.2 Volterra integral equations 297
7.3 Implications of linearity 298
7.4 Numerical solutions 299
7.5 Integral transformations 301
Exercises 304

8 Legendre functions 305
8.1 The Legendre polynomials 305
8.2 The Rodrigues formula 306
8.3 The generating function 308
8.4 Legendre's differential equation 309
8.5 Recurrence relations 311
8.6 Special values of Legendre's polynomials 312
8.7 Schlaefli's integral 313
8.8 Orthogonal polynomials 313
8.9 The azimuthally symmetric Laplacian 315
8.10 Laplacian in two dimensions 316
8.11 The Laplacian in spherical coordinates 317
8.12 The associated Legendre functions/polynomials 317
8.13 Spherical harmonics 319
Exercises 323

9 Bessel functions 325
9.1 Bessel functions of the first kind 325
9.2 Spherical Bessel functions of the first kind 335
9.3 Bessel functions of the second kind 341
9.4 Spherical Bessel functions of the second kind 343
Further reading 345
Exercises 345

10	**Group theory**	**348**
10.1	What is a group?	348
10.2	Representations of groups	350
10.3	Representations acting in Hilbert space	351
10.4	Subgroups	353
10.5	Cosets	354
10.6	Morphisms	354
10.7	Schur's lemma	355
10.8	Characters	356
10.9	Tensor products	357
10.10	Finite groups	358
10.11	The regular representation	359
10.12	Properties of finite groups	360
10.13	Permutations	360
10.14	Compact and noncompact Lie groups	361
10.15	Lie algebra	361
10.16	The rotation group	366
10.17	The Lie algebra and representations of $SU(2)$	368
10.18	The defining representation of $SU(2)$	371
10.19	The Jacobi identity	374
10.20	The adjoint representation	374
10.21	Casimir operators	375
10.22	Tensor operators for the rotation group	376
10.23	Simple and semisimple Lie algebras	376
10.24	$SU(3)$	377
10.25	$SU(3)$ and quarks	378
10.26	Cartan subalgebra	379
10.27	Quaternions	379
10.28	The symplectic group $Sp(2n)$	381
10.29	Compact simple Lie groups	383
10.30	Group integration	384
10.31	The Lorentz group	386
10.32	Two-dimensional representations of the Lorentz group	389
10.33	The Dirac representation of the Lorentz group	393
10.34	The Poincaré group	395
	Further reading	396
	Exercises	397
11	**Tensors and local symmetries**	**400**
11.1	Points and coordinates	400
11.2	Scalars	401
11.3	Contravariant vectors	401

11.4 Covariant vectors 402
11.5 Euclidean space in euclidean coordinates 402
11.6 Summation conventions 404
11.7 Minkowski space 405
11.8 Lorentz transformations 407
11.9 Special relativity 408
11.10 Kinematics 410
11.11 Electrodynamics 411
11.12 Tensors 414
11.13 Differential forms 416
11.14 Tensor equations 419
11.15 The quotient theorem 420
11.16 The metric tensor 421
11.17 A basic axiom 422
11.18 The contravariant metric tensor 422
11.19 Raising and lowering indices 423
11.20 Orthogonal coordinates in euclidean n-space 423
11.21 Polar coordinates 424
11.22 Cylindrical coordinates 425
11.23 Spherical coordinates 425
11.24 The gradient of a scalar field 426
11.25 Levi-Civita's tensor 427
11.26 The Hodge star 428
11.27 Derivatives and affine connections 431
11.28 Parallel transport 433
11.29 Notations for derivatives 433
11.30 Covariant derivatives 434
11.31 The covariant curl 435
11.32 Covariant derivatives and antisymmetry 436
11.33 Affine connection and metric tensor 436
11.34 Covariant derivative of the metric tensor 437
11.35 Divergence of a contravariant vector 438
11.36 The covariant Laplacian 441
11.37 The principle of stationary action 443
11.38 A particle in a gravitational field 446
11.39 The principle of equivalence 447
11.40 Weak, static gravitational fields 449
11.41 Gravitational time dilation 449
11.42 Curvature 451
11.43 Einstein's equations 453
11.44 The action of general relativity 455
11.45 Standard form 455

11.46 Schwarzschild's solution 456
11.47 Black holes 456
11.48 Cosmology 457
11.49 Model cosmologies 463
11.50 Yang–Mills theory 469
11.51 Gauge theory and vectors 471
11.52 Geometry 474
Further reading 475
Exercises 475

12 Forms 479
12.1 Exterior forms 479
12.2 Differential forms 481
12.3 Exterior differentiation 486
12.4 Integration of forms 491
12.5 Are closed forms exact? 496
12.6 Complex differential forms 498
12.7 Frobenius's theorem 499
Further reading 500
Exercises 500

13 Probability and statistics 502
13.1 Probability and Thomas Bayes 502
13.2 Mean and variance 505
13.3 The binomial distribution 508
13.4 The Poisson distribution 511
13.5 The Gaussian distribution 512
13.6 The error function erf 515
13.7 The Maxwell–Boltzmann distribution 518
13.8 Diffusion 519
13.9 Langevin's theory of brownian motion 520
13.10 The Einstein–Nernst relation 523
13.11 Fluctuation and dissipation 524
13.12 Characteristic and moment-generating functions 528
13.13 Fat tails 530
13.14 The central limit theorem and Jarl Lindeberg 532
13.15 Random-number generators 537
13.16 Illustration of the central limit theorem 538
13.17 Measurements, estimators, and Friedrich Bessel 543
13.18 Information and Ronald Fisher 546
13.19 Maximum likelihood 550
13.20 Karl Pearson's chi-squared statistic 551

13.21	Kolmogorov's test	554
	Further reading	560
	Exercises	560
14	**Monte Carlo methods**	**563**
14.1	The Monte Carlo method	563
14.2	Numerical integration	563
14.3	Applications to experiments	566
14.4	Statistical mechanics	572
14.5	Solving arbitrary problems	575
14.6	Evolution	576
	Further reading	577
	Exercises	577
15	**Functional derivatives**	**578**
15.1	Functionals	578
15.2	Functional derivatives	578
15.3	Higher-order functional derivatives	581
15.4	Functional Taylor series	582
15.5	Functional differential equations	583
	Exercises	585
16	**Path integrals**	**586**
16.1	Path integrals and classical physics	586
16.2	Gaussian integrals	586
16.3	Path integrals in imaginary time	588
16.4	Path integrals in real time	590
16.5	Path integral for a free particle	593
16.6	Free particle in imaginary time	595
16.7	Harmonic oscillator in real time	595
16.8	Harmonic oscillator in imaginary time	597
16.9	Euclidean correlation functions	599
16.10	Finite-temperature field theory	600
16.11	Real-time field theory	603
16.12	Perturbation theory	605
16.13	Application to quantum electrodynamics	609
16.14	Fermionic path integrals	613
16.15	Application to nonabelian gauge theories	619
16.16	The Faddeev–Popov trick	620
16.17	Ghosts	622
	Further reading	624
	Exercises	624

17	**The renormalization group**	**626**
17.1	The renormalization group in quantum field theory	626
17.2	The renormalization group in lattice field theory	630
17.3	The renormalization group in condensed-matter physics	632
	Exercises	634
18	**Chaos and fractals**	**635**
18.1	Chaos	635
18.2	Attractors	639
18.3	Fractals	639
	Further reading	642
	Exercises	642
19	**Strings**	**643**
19.1	The infinities of quantum field theory	643
19.2	The Nambu–Goto string action	643
19.3	Regge trajectories	646
19.4	Quantized strings	647
19.5	D-branes	647
19.6	String–string scattering	648
19.7	Riemann surfaces and moduli	649
	Further reading	650
	Exercises	650

References 651

Index 656

Preface

To the students: you will find some physics crammed in amongst the mathematics. Don't let the physics bother you. As you study the math, you'll learn some physics without extra effort. The physics is a freebie. I have tried to explain the math you need for physics and have left out the rest.

To the professors: the book is for students who also are taking mechanics, electrodynamics, quantum mechanics, and statistical mechanics nearly simultaneously and who soon may use probability or path integrals in their research. Linear algebra and Fourier analysis are the keys to physics, so the book starts with them, but you may prefer to skip the algebra or postpone the Fourier analysis. The book is intended to support a one- or two-semester course for graduate students or advanced undergraduates. The first seven, eight, or nine chapters fit in one semester, the others in a second.

A list of errata is maintained at panda.unm.edu/cahill, and solutions to all the exercises are available for instructors at www.cambridge.org/cahill.

Several friends – Susan Atlas, Bernard Becker, Steven Boyd, Robert Burckel, Sean Cahill, Colston Chandler, Vageli Coutsias, David Dunlap, Daniel Finley, Franco Giuliani, Roy Glauber, Pablo Gondolo, Igor Gorelov, Jiaxing Hong, Fang Huang, Dinesh Loomba, Yin Luo, Lei Ma, Michael Malik, Kent Morrison, Sudhakar Prasad, Randy Reeder, Dmitri Sergatskov, and David Waxman – have given me valuable advice. Students have helped with questions, ideas, and corrections, especially Thomas Beechem, Marie Cahill, Chris Cesare, Yihong Cheng, Charles Cherqui, Robert Cordwell, Amo-Kwao Godwin, Aram Gragossian, Aaron Hankin, Kangbo Hao, Tiffany Hayes, Yiran Hu, Shanshan Huang, Tyler Keating, Joshua Koch, Zilong Li, Miao Lin, ZuMou Lin, Sheng Liu, Yue Liu, Ben Oliker, Boleszek Osinski, Ravi Raghunathan, Akash Rakholia, Xingyue Tian, Toby Tolley, Jiqun Tu, Christopher Vergien, Weizhen Wang, George Wendelberger, Xukun Xu, Huimin Yang, Zhou Yang, Daniel Young, Mengzhen Zhang, Lu Zheng, Lingjun Zhou, and Daniel Zirzow.

1

Linear algebra

1.1 Numbers

The **natural** numbers are the positive integers and zero. **Rational** numbers are ratios of integers. **Irrational** numbers have decimal digits d_n

$$x = \sum_{n=m_x}^{\infty} \frac{d_n}{10^n} \tag{1.1}$$

that do not repeat. Thus the repeating decimals $1/2 = 0.50000\ldots$ and $1/3 = 0.\bar{3} \equiv 0.33333\ldots$ are rational, while $\pi = 3.141592654\ldots$ is irrational. Decimal arithmetic was invented in India over 1500 years ago but was not widely adopted in the Europe until the seventeenth century.

The **real** numbers $\mathbb{R}$ include the rational numbers and the irrational numbers; they correspond to all the points on an infinite line called the **real line**.

The **complex** numbers $\mathbb{C}$ are the real numbers with one new number i whose square is -1. A complex number z is a linear combination of a real number x and a real multiple iy of i

$$z = x + iy. \tag{1.2}$$

Here $x = \mathrm{Re}z$ is the **real part** of z, and $y = \mathrm{Im}z$ is its **imaginary part**. One adds complex numbers by adding their real and imaginary parts

$$z_1 + z_2 = x_1 + iy_1 + x_2 + iy_2 = x_1 + x_2 + i(y_1 + y_2). \tag{1.3}$$

Since $i^2 = -1$, the product of two complex numbers is

$$z_1 z_2 = (x_1 + iy_1)(x_2 + iy_2) = x_1 x_2 - y_1 y_2 + i(x_1 y_2 + y_1 x_2). \tag{1.4}$$

The polar representation $z = r\exp(i\theta)$ of $z = x + iy$ is

$$z = x + iy = re^{i\theta} = r(\cos\theta + i\sin\theta) \tag{1.5}$$

in which r is the **modulus** or **absolute value** of z

$$r = |z| = \sqrt{x^2 + y^2} \tag{1.6}$$

and θ is its **phase** or **argument**

$$\theta = \arctan(y/x). \tag{1.7}$$

Since $\exp(2\pi i) = 1$, there is an inevitable ambiguity in the definition of the phase of any complex number: for any integer n, the phase $\theta + 2\pi n$ gives the same z as θ. In various computer languages, the function atan2(y, x) returns the angle θ in the interval $-\pi < \theta \leq \pi$ for which $(x, y) = r(\cos\theta, \sin\theta)$.

There are two common notations z^* and $\bar{z}$ for the **complex conjugate** of a complex number $z = x + iy$

$$z^* = \bar{z} = x - iy. \tag{1.8}$$

The square of the modulus of a complex number $z = x + iy$ is

$$|z|^2 = x^2 + y^2 = (x + iy)(x - iy) = \bar{z}z = z^*z. \tag{1.9}$$

The inverse of a complex number $z = x + iy$ is

$$z^{-1} = (x + iy)^{-1} = \frac{x - iy}{(x - iy)(x + iy)} = \frac{x - iy}{x^2 + y^2} = \frac{z^*}{z^*z} = \frac{z^*}{|z|^2}. \tag{1.10}$$

Grassmann numbers θ_i are **anticommuting** numbers, that is, the **anticommutator** of any two Grassmann numbers vanishes

$$\{\theta_i, \theta_j\} \equiv [\theta_i, \theta_j]_+ \equiv \theta_i\theta_j + \theta_j\theta_i = 0. \tag{1.11}$$

So the square of any Grassmann number is zero, $\theta_i^2 = 0$. We won't use these numbers until chapter 16, but they do have amusing properties. The highest monomial in N Grassmann numbers θ_i is the product $\theta_1\theta_2 \ldots \theta_N$. So the most complicated power series in two Grassmann numbers is just

$$f(\theta_1, \theta_2) = f_0 + f_1\,\theta_1 + f_2\,\theta_2 + f_{12}\,\theta_1\theta_2 \tag{1.12}$$

(Hermann Grassmann, 1809–1877).

1.2 Arrays

An **array** is an **ordered set** of numbers. Arrays play big roles in computer science, physics, and mathematics. They can be of any (integral) dimension.

A one-dimensional array $(a_1, a_2, \ldots, a_n)$ is variously called an ***n*-tuple**, a **row vector** when written horizontally, a **column vector** when written vertically, or an ***n*-vector**. The numbers a_k are its **entries** or **components**.

A two-dimensional array a_{ik} with i running from 1 to n and k from 1 to m is an $n \times m$ **matrix**. The numbers a_{ik} are its **entries**, **elements**, or **matrix elements**.

One can think of a matrix as a stack of row vectors or as a queue of column vectors. The entry a_{ik} is in the ith row and the kth column.

One can add together arrays of the same dimension and shape by adding their entries. Two n-tuples add as

$$(a_1, \ldots, a_n) + (b_1, \ldots, b_n) = (a_1 + b_1, \ldots, a_n + b_n) \tag{1.13}$$

and two $n \times m$ matrices a and b add as

$$(a + b)_{ik} = a_{ik} + b_{ik}. \tag{1.14}$$

One can multiply arrays by numbers. Thus z times the three-dimensional array a_{ijk} is the array with entries $z\, a_{ijk}$. One can multiply two arrays together no matter what their shapes and dimensions. The **outer product** of an n-tuple a and an m-tuple b is an $n \times m$ matrix with elements

$$(a\, b)_{ik} = a_i\, b_k \tag{1.15}$$

or an $m \times n$ matrix with entries $(ba)_{ki} = b_k a_i$. If a and b are complex, then one also can form the outer products $(\overline{a}\, b)_{ik} = \overline{a_i}\, b_k$, $(\overline{b}\, a)_{ki} = \overline{b_k}\, a_i$, and $(\overline{b}\, \overline{a})_{ki} = \overline{b_k}\, \overline{a_i}$. The outer product of a matrix a_{ik} and a three-dimensional array $b_{j\ell m}$ is a five-dimensional array

$$(a\, b)_{ikj\ell m} = a_{ik}\, b_{j\ell m}. \tag{1.16}$$

An **inner product** is possible when two arrays are of the same size in one of their dimensions. Thus the **inner product** $(a, b) \equiv \langle a|b\rangle$ or **dot-product** $a \cdot b$ of two real n-tuples a and b is

$$(a, b) = \langle a|b\rangle = a \cdot b = (a_1, \ldots, a_n) \cdot (b_1, \ldots, b_n) = a_1 b_1 + \cdots + a_n b_n. \tag{1.17}$$

The inner product of two complex n-tuples often is defined as

$$(a, b) = \langle a|b\rangle = \overline{a} \cdot b = (\overline{a_1}, \ldots, \overline{a_n}) \cdot (b_1, \ldots, b_n) = \overline{a_1}\, b_1 + \cdots + \overline{a_n}\, b_n \tag{1.18}$$

or as its complex conjugate

$$(a, b)^* = \langle a|b\rangle^* = (\overline{a} \cdot b)^* = (b, a) = \langle b|a\rangle = \overline{b} \cdot a \tag{1.19}$$

so that the inner product of a vector with itself is nonnegative $(a, a) \geq 0$.

The product of an $m \times n$ matrix a_{ik} times an n-tuple b_k is the m-tuple b' whose ith component is

$$b'_i = a_{i1} b_1 + a_{i2} b_2 + \cdots + a_{in} b_n = \sum_{k=1}^{n} a_{ik} b_k. \tag{1.20}$$

This product is $b' = a\, b$ in matrix notation.

If the size n of the second dimension of a matrix a matches that of the first dimension of a matrix b, then their product $a\, b$ is a matrix with entries

$$(a\, b)_{i\ell} = a_{i1}\, b_{1\ell} + \cdots + a_{in}\, b_{n\ell}. \tag{1.21}$$

1.3 Matrices

Apart from n-tuples, the most important arrays in linear algebra are the two-dimensional arrays called matrices.

The **trace** of an $n \times n$ matrix a is the sum of its diagonal elements

$$\operatorname{Tr} a = \operatorname{tr} a = a_{11} + a_{22} + \cdots + a_{nn} = \sum_{i=1}^{n} a_{ii}. \tag{1.22}$$

The trace of two matrices is independent of their order

$$\operatorname{Tr}(a\,b) = \sum_{i=1}^{n} \sum_{k=1}^{n} a_{ik} b_{ki} = \sum_{k=1}^{n} \sum_{i=1}^{n} b_{ki} a_{ik} = \operatorname{Tr}(ba) \tag{1.23}$$

as long as the matrix elements are numbers that commute with each other. It follows that the trace is **cyclic**

$$\operatorname{Tr}(a\,b \ldots z) = \operatorname{Tr}(b \ldots z\,a). \tag{1.24}$$

The **transpose** of an $n \times \ell$ matrix a is an $\ell \times n$ matrix a^{T} with entries

$$\left(a^{\mathsf{T}}\right)_{ij} = a_{ji}. \tag{1.25}$$

Some mathematicians use a prime to mean transpose, as in $a' = a^{\mathsf{T}}$, but physicists tend to use primes to label different objects or to indicate differentiation. One may show that

$$(a\,b)^{\mathsf{T}} = b^{\mathsf{T}}\,a^{\mathsf{T}}. \tag{1.26}$$

A matrix that is equal to its transpose

$$a = a^{\mathsf{T}} \tag{1.27}$$

is **symmetric**.

The (hermitian) **adjoint** of a matrix is the complex conjugate of its transpose (Charles Hermite, 1822–1901). That is, the (hermitian) adjoint $a^{\dagger}$ of an $N \times L$ complex matrix a is the $L \times N$ matrix with entries

$$(a^{\dagger})_{ij} = (a_{ji})^{*} = a_{ji}^{*}. \tag{1.28}$$

One may show that

$$(a\,b)^{\dagger} = b^{\dagger}\,a^{\dagger}. \tag{1.29}$$

A matrix that is equal to its adjoint

$$(a^{\dagger})_{ij} = (a_{ji})^{*} = a_{ji}^{*} = a_{ij} \tag{1.30}$$

(and which must be a square matrix) is **hermitian** or **self adjoint**

$$a = a^\dagger. \tag{1.31}$$

Example 1.1 (The Pauli matrices)

$$\sigma_1 = \begin{pmatrix} 0 & 1 \\ 1 & 0 \end{pmatrix}, \quad \sigma_2 = \begin{pmatrix} 0 & -i \\ i & 0 \end{pmatrix}, \quad \text{and} \quad \sigma_3 = \begin{pmatrix} 1 & 0 \\ 0 & -1 \end{pmatrix} \tag{1.32}$$

are all hermitian (Wolfgang Pauli, 1900–1958). □

A real hermitian matrix is symmetric. If a matrix a is hermitian, then the quadratic form

$$\langle v|a|v\rangle = \sum_{i=1}^{N} \sum_{j=1}^{N} v_i^* a_{ij} v_j \in \mathbb{R} \tag{1.33}$$

is real for all complex n-tuples v.

The **Kronecker delta** δ_{ik} is defined to be unity if $i = k$ and zero if $i \neq k$ (Leopold Kronecker, 1823–1891). The **identity matrix** I has entries $I_{ik} = \delta_{ik}$.

The **inverse** a^{-1} of an $n \times n$ matrix a is a square matrix that satisfies

$$a^{-1}\, a = a\, a^{-1} = I \tag{1.34}$$

in which I is the $n \times n$ identity matrix.

So far we have been writing n-tuples and matrices and their elements with lower-case letters. It is equally common to use capital letters, and we will do so for the rest of this section.

A matrix U whose adjoint $U^\dagger$ is its inverse

$$U^\dagger U = U U^\dagger = I \tag{1.35}$$

is **unitary**. Unitary matrices are square.

A real unitary matrix O is **orthogonal** and obeys the rule

$$O^{\mathsf{T}} O = O O^{\mathsf{T}} = I. \tag{1.36}$$

Orthogonal matrices are square.

An $N \times N$ hermitian matrix A is **nonnegative**

$$A \geq 0 \tag{1.37}$$

if for all complex vectors V the quadratic form

$$\langle V|A|V\rangle = \sum_{i=1}^{N} \sum_{j=1}^{N} V_i^* A_{ij} V_j \geq 0 \tag{1.38}$$

is nonnegative. It is **positive** or **positive definite** if

$$\langle V|A|V\rangle > 0 \tag{1.39}$$

for all nonzero vectors $|V\rangle$.

Example 1.2 (Kinds of positivity) The nonsymmetric, nonhermitian 2×2 matrix

$$\begin{pmatrix} 1 & 1 \\ -1 & 1 \end{pmatrix} \tag{1.40}$$

is positive on the space of all real 2-vectors but not on the space of all complex 2-vectors. □

Example 1.3 (Representations of imaginary and Grassmann numbers) The 2×2 matrix

$$\begin{pmatrix} 0 & -1 \\ 1 & 0 \end{pmatrix} \tag{1.41}$$

can represent the number i since

$$\begin{pmatrix} 0 & -1 \\ 1 & 0 \end{pmatrix}\begin{pmatrix} 0 & -1 \\ 1 & 0 \end{pmatrix} = \begin{pmatrix} -1 & 0 \\ 0 & -1 \end{pmatrix} = -I. \tag{1.42}$$

The 2×2 matrix

$$\begin{pmatrix} 0 & 0 \\ 1 & 0 \end{pmatrix} \tag{1.43}$$

can represent a Grassmann number since

$$\begin{pmatrix} 0 & 0 \\ 1 & 0 \end{pmatrix}\begin{pmatrix} 0 & 0 \\ 1 & 0 \end{pmatrix} = \begin{pmatrix} 0 & 0 \\ 0 & 0 \end{pmatrix} = 0. \tag{1.44}$$

To represent two Grassmann numbers, one needs 4×4 matrices, such as

$$\theta_1 = \begin{pmatrix} 0 & 0 & 1 & 0 \\ 0 & 0 & 0 & -1 \\ 0 & 0 & 0 & 0 \\ 0 & 0 & 0 & 0 \end{pmatrix} \quad \text{and} \quad \theta_2 = \begin{pmatrix} 0 & 1 & 0 & 0 \\ 0 & 0 & 0 & 0 \\ 0 & 0 & 0 & 1 \\ 0 & 0 & 0 & 0 \end{pmatrix}. \tag{1.45}$$

The matrices that represent n Grassmann numbers are $2^n \times 2^n$. □

Example 1.4 (Fermions) The matrices (1.45) also can represent lowering or annihilation operators for a system of two fermionic states. For $a_1 = \theta_1$ and $a_2 = \theta_2$ and their adjoints $a_1^\dagger$ and $a_2^\dagger$, the creation operators satisfy the anticommutation relations

$$\{a_i, a_k^\dagger\} = \delta_{ik} \quad \text{and} \quad \{a_i, a_k\} = \{a_i^\dagger, a_k^\dagger\} = 0 \tag{1.46}$$

where i and k take the values 1 or 2. In particular, the relation $(a_i^\dagger)^2 = 0$ implements **Pauli's exclusion principle**, the rule that no state of a fermion can be doubly occupied. □

1.4 Vectors

Vectors are things that can be multiplied by numbers and added together to form other vectors in the same **vector space**. So if U and V are vectors in a vector space S over a set F of numbers x and y and so forth, then

$$W = x\,U + y\,V \tag{1.47}$$

also is a vector in the vector space S.

A **basis** for a vector space S is a set of vectors B_k for $k = 1, \ldots, N$ in terms of which every vector U in S can be expressed as a linear combination

$$U = u_1 B_1 + u_2 B_2 + \cdots + u_N B_N \tag{1.48}$$

with numbers u_k in F. The numbers u_k are the **components** of the vector U in the basis B_k.

Example 1.5 (Hardware store) Suppose the vector W represents a certain kind of washer and the vector N represents a certain kind of nail. Then if n and m are natural numbers, the vector

$$H = nW + mN \tag{1.49}$$

would represent a possible inventory of a very simple hardware store. The vector space of all such vectors H would include all possible inventories of the store. That space is a two-dimensional vector space over the natural numbers, and the two vectors W and N form a basis for it. □

Example 1.6 (Complex numbers) The complex numbers are a vector space. Two of its vectors are the number 1 and the number i; the vector space of complex numbers is then the set of all linear combinations

$$z = x1 + yi = x + iy. \tag{1.50}$$

So the complex numbers are a two-dimensional vector space over the real numbers, and the vectors 1 and i are a basis for it.

The complex numbers also form a one-dimensional vector space over the complex numbers. Here any nonzero real or complex number, for instance the number 1, can be a basis consisting of the single vector 1. This one-dimensional vector space is the set of all $z = z1$ for arbitrary complex z. □

Example 1.7 (2-space) Ordinary flat two-dimensional space is the set of all linear combinations

$$\boldsymbol{r} = x\hat{\mathbf{x}} + y\hat{\mathbf{y}} \tag{1.51}$$

in which x and y are real numbers and $\hat{\mathbf{x}}$ and $\hat{\mathbf{y}}$ are perpendicular vectors of unit length (unit vectors). This vector space, called $\mathbb{R}^2$, is a 2-d space over the reals.

Note that the same vector $\boldsymbol{r}$ can be described either by the basis vectors $\hat{\mathbf{x}}$ and $\hat{\mathbf{y}}$ or by any other set of basis vectors, such as $-\hat{\mathbf{y}}$ and $\hat{\mathbf{x}}$

$$\boldsymbol{r} = x\hat{\mathbf{x}} + y\hat{\mathbf{y}} = -y(-\hat{\mathbf{y}}) + x\hat{\mathbf{x}}. \tag{1.52}$$

So the components of the vector $\boldsymbol{r}$ are (x, y) in the $\{\hat{\mathbf{x}}, \hat{\mathbf{y}}\}$ basis and $(-y, x)$ in the $\{-\hat{\mathbf{y}}, \hat{\mathbf{x}}\}$ basis. **Each vector is unique, but its components depend upon the basis.** □

Example 1.8 (3-space) Ordinary flat three-dimensional space is the set of all linear combinations

$$\boldsymbol{r} = x\hat{\mathbf{x}} + y\hat{\mathbf{y}} + z\hat{\mathbf{z}} \tag{1.53}$$

in which x, y, and z are real numbers. It is a 3-d space over the reals. □

Example 1.9 (Matrices) Arrays of a given dimension and size can be added and multiplied by numbers, and so they form a vector space. For instance, all complex three-dimensional arrays a_{ijk} in which $1 \le i \le 3$, $1 \le j \le 4$, and $1 \le k \le 5$ form a vector space over the complex numbers. □

Example 1.10 (Partial derivatives) Derivatives are vectors, so are partial derivatives. For instance, the linear combinations of x and y partial derivatives taken at $x = y = 0$

$$a \frac{\partial}{\partial x} + b \frac{\partial}{\partial y} \tag{1.54}$$

form a vector space. □

Example 1.11 (Functions) The space of all linear combinations of a set of functions $f_i(x)$ defined on an interval $[a, b]$

$$f(x) = \sum_i z_i f_i(x) \tag{1.55}$$

is a vector space over the natural, real, or complex numbers $\{z_i\}$. □

Example 1.12 (States) In quantum mechanics, a state is represented by a vector, often written as ψ or in Dirac's notation as $|\psi\rangle$. If c_1 and c_2 are complex numbers, and $|\psi_1\rangle$ and $|\psi_2\rangle$ are any two states, then the linear combination

$$|\psi\rangle = c_1|\psi_1\rangle + c_2|\psi_2\rangle \tag{1.56}$$

also is a possible state of the system. □

1.5 Linear operators

A **linear operator** A maps each vector U in its **domain** into a vector $U' = A(U) \equiv AU$ in its **range** in a way that is linear. So if U and V are two vectors in its domain and b and c are numbers, then

$$A(bU + cV) = bA(U) + cA(V) = bAU + cAV. \tag{1.57}$$

If the domain and the range are the same vector space S, then A maps each basis vector B_i of S into a linear combination of the basis vectors B_k

$$AB_i = a_{1i}B_1 + a_{2i}B_2 + \cdots + a_{Ni}B_N = \sum_{k=1}^{N} a_{ki}\, B_k. \tag{1.58}$$

The square matrix a_{ki} **represents** the linear operator A in the B_k basis. The effect of A on any vector $U = u_1 B_1 + u_2 B_2 + \cdots + u_N B_N$ in S then is

$$AU = A\left(\sum_{i=1}^{N} u_i B_i\right) = \sum_{i=1}^{N} u_i A B_i = \sum_{i=1}^{N} u_i \sum_{k=1}^{N} a_{ki}\, B_k$$
$$= \sum_{k=1}^{N}\left(\sum_{i=1}^{N} a_{ki} u_i\right) B_k. \tag{1.59}$$

So the kth component u'_k of the vector $U' = AU$ is

$$u'_k = a_{k1}u_1 + a_{k2}u_2 + \cdots + a_{kN}u_N = \sum_{i=1}^{N} a_{ki}\, u_i. \tag{1.60}$$

Thus the column vector u' of the components u'_k of the vector $U' = AU$ is the product $u' = a\,u$ of the matrix with elements a_{ki} that represents the linear operator A in the B_k basis and the column vector with components u_i that represents the vector U in that basis. So in each basis, vectors and linear operators are represented by column vectors and matrices.

Each linear operator is unique, but its matrix depends upon the basis. If we change from the B_k basis to another basis B'_k

$$B_k = \sum_{\ell=1}^{N} u_{\ell k}\, B'_\ell \tag{1.61}$$

in which the $N \times N$ matrix $u_{\ell k}$ has an inverse matrix u^{-1}_{ki} so that

$$\sum_{k=1}^{N} u^{-1}_{ki} B_k = \sum_{k=1}^{N} u^{-1}_{ki} \sum_{\ell=1}^{N} u_{\ell k} B'_\ell = \sum_{\ell=1}^{N}\left(\sum_{k=1}^{N} u_{\ell k} u^{-1}_{ki}\right) B'_\ell = \sum_{\ell=1}^{N} \delta_{\ell i} B'_\ell = B'_i, \tag{1.62}$$

then the new basis vectors B'_i are given by

$$B'_i = \sum_{k=1}^{N} u^{-1}_{ki} B_k. \tag{1.63}$$

Thus (exercise 1.9) the linear operator A maps the basis vector B'_i to

$$AB'_i = \sum_{k=1}^{N} u^{-1}_{ki} AB_k = \sum_{j,k=1}^{N} u^{-1}_{ki} a_{jk} B_j = \sum_{j,k,\ell=1}^{N} u_{\ell j} a_{jk} u^{-1}_{ki} B'_\ell. \tag{1.64}$$

So the matrix a' that represents A in the B' basis is related to the matrix a that represents it in the B basis by a **similarity transformation**

$$a'_{\ell i} = \sum_{jk=1}^{N} u_{\ell j} a_{jk} u^{-1}_{ki} \quad \text{or} \quad a' = u\, a\, u^{-1} \tag{1.65}$$

in matrix notation.

Example 1.13 (Change of basis) Let the action of the linear operator A on the basis vectors $\{B_1, B_2\}$ be $AB_1 = B_2$ and $AB_2 = 0$. If the column vectors

$$b_1 = \begin{pmatrix} 1 \\ 0 \end{pmatrix} \quad \text{and} \quad b_2 = \begin{pmatrix} 0 \\ 1 \end{pmatrix} \tag{1.66}$$

represent the basis vectors B_1 and B_2, then the matrix

$$a = \begin{pmatrix} 0 & 0 \\ 1 & 0 \end{pmatrix} \tag{1.67}$$

represents the linear operator A. But if we use the basis vectors

$$B'_1 = \frac{1}{\sqrt{2}}(B_1 + B_2) \quad \text{and} \quad B'_2 = \frac{1}{\sqrt{2}}(B_1 - B_2) \tag{1.68}$$

then the vectors

$$b'_1 = \frac{1}{\sqrt{2}} \begin{pmatrix} 1 \\ 1 \end{pmatrix} \quad \text{and} \quad b'_2 = \frac{1}{\sqrt{2}} \begin{pmatrix} 1 \\ -1 \end{pmatrix} \tag{1.69}$$

would represent B_1 and B_2, and the matrix

$$a' = \frac{1}{2} \begin{pmatrix} 1 & 1 \\ -1 & -1 \end{pmatrix} \tag{1.70}$$

would represent the linear operator A (exercise 1.10). □

A linear operator A also may map a vector space S with basis B_k into a different vector space T with its own basis C_k. In this case, A maps the basis vector B_i into a linear combination of the basis vectors C_k

$$AB_i = \sum_{k=1}^{M} a_{ki}\, C_k \tag{1.71}$$

and an arbitrary vector $U = u_1 B_1 + \cdots + u_N B_N$ in S into the vector

$$AU = \sum_{k=1}^{M} \left(\sum_{i=1}^{N} a_{ki}\, u_i \right) C_k \tag{1.72}$$

in T.

1.6 Inner products

Most of the vector spaces used by physicists have an inner product. A **positive-definite inner product** associates a number (f, g) with every ordered pair of vectors f and g in the vector space V and satisfies the rules

$$(f, g) = (g, f)^* \tag{1.73}$$

$$(f, z\, g + w\, h) = z\, (f, g) + w\, (f, h) \tag{1.74}$$

$$(f, f) \geq 0 \quad \text{and} \quad (f, f) = 0 \iff f = 0 \tag{1.75}$$

in which f, g, and h are vectors, and z and w are numbers. The first rule says that the inner product is **hermitian**; the second rule says that it is **linear** in the second vector $z\, g + w\, h$ of the pair; and the third rule says that it is **positive definite**. The first two rules imply that (exercise 1.11) the inner product is **antilinear** in the first vector of the pair

$$(z\, g + w\, h, f) = z^*(g, f) + w^*(h, f). \tag{1.76}$$

A **Schwarz inner product** satisfies the first two rules (1.73, 1.74) for an inner product and the fourth (1.76) but only the first part of the third (1.75)

$$(f, f) \geq 0. \tag{1.77}$$

This condition of **nonnegativity** implies (exercise 1.15) that a vector f of zero length must be orthogonal to all vectors g in the vector space V

$$(f, f) = 0 \implies (g, f) = 0 \ \text{for all} \ g \in V. \tag{1.78}$$

So a Schwarz inner product is *almost* positive definite.

Inner products of 4-vectors can be negative. To accommodate them we define an **indefinite** inner product without regard to positivity as one that satisfies the first two rules (1.73 & 1.74) and therefore also the fourth rule (1.76) and that instead of being positive definite is **nondegenerate**

$$(f, g) = 0 \ \text{for all} \ f \in V \implies g = 0. \tag{1.79}$$

This rule says that only the zero vector is orthogonal to all the vectors of the space. The positive-definite condition (1.75) is stronger than and implies nondegeneracy (1.79) (exercise 1.14).

Apart from the indefinite inner products of 4-vectors in special and general relativity, most of the inner products physicists use are Schwarz inner products or positive-definite inner products. For such inner products, we can define the **norm** $|f| = \| f \|$ of a vector f as the square-root of the nonnegative inner product (f,f)

$$\| f \| = \sqrt{(f,f)}. \tag{1.80}$$

The **distance** between two vectors f and g is the norm of their difference

$$\| f - g \| . \tag{1.81}$$

Example 1.14 (Euclidean space) The space of real vectors U, V with N components U_i, V_i forms an N-dimensional vector space over the real numbers with an inner product

$$(U, V) = \sum_{i=1}^{N} U_i V_i \tag{1.82}$$

that is nonnegative when the two vectors are the same

$$(U, U) = \sum_{i=1}^{N} U_i U_i = \sum_{i=1}^{N} U_i^2 \geq 0 \tag{1.83}$$

and vanishes only if all the components U_i are zero, that is, if the vector $U = 0$. Thus the inner product (1.82) is positive definite. When (U, V) is zero, the vectors U and V are **orthogonal**. □

Example 1.15 (Complex euclidean space) The space of complex vectors with N components U_i, V_i forms an N-dimensional vector space over the complex numbers with inner product

$$(U, V) = \sum_{i=1}^{N} U_i^* V_i = (V, U)^*. \tag{1.84}$$

The inner product (U, U) is nonnegative and vanishes

$$(U, U) = \sum_{i=1}^{N} U_i^* U_i = \sum_{i=1}^{N} |U_i|^2 \geq 0 \tag{1.85}$$

only if $U = 0$. So the inner product (1.84) is positive definite. If (U, V) is zero, then U and V are orthogonal. □

Example 1.16 (Complex matrices) For the vector space of $N \times L$ complex matrices A, B, ..., the trace of the adjoint (1.28) of A multiplied by B is an inner product

$$(A, B) = \mathrm{Tr} A^\dagger B = \sum_{i=1}^{N} \sum_{j=1}^{L} (A^\dagger)_{ji} B_{ij} = \sum_{i=1}^{N} \sum_{j=1}^{L} A^*_{ij} B_{ij} \tag{1.86}$$

that is nonnegative when the matrices are the same

$$(A, A) = \mathrm{Tr} A^\dagger A = \sum_{i=1}^{N} \sum_{j=1}^{L} A^*_{ij} A_{ij} = \sum_{i=1}^{N} \sum_{j=1}^{L} |A_{ij}|^2 \ge 0 \tag{1.87}$$

and zero only when $A = 0$. So this inner product is positive definite. □

A vector space with a positive-definite inner product (1.73–1.76) is called an **inner-product space**, a **metric space**, or a **pre-Hilbert space**.

A sequence of vectors f_n is a **Cauchy sequence** if for every $\epsilon > 0$ there is an integer $N(\epsilon)$ such that $\|f_n - f_m\| < \epsilon$ whenever both n and m exceed $N(\epsilon)$. A sequence of vectors f_n **converges** to a vector f if for every $\epsilon > 0$ there is an integer $N(\epsilon)$ such that $\|f - f_n\| < \epsilon$ whenever n exceeds $N(\epsilon)$. An inner-product space with a norm defined as in (1.80) is **complete** if each of its Cauchy sequences converges to a vector in that space. A **Hilbert space** is a complete inner-product space. Every finite-dimensional inner-product space is complete and so is a Hilbert space. But the term *Hilbert space* more often is used to describe infinite-dimensional complete inner-product spaces, such as the space of all square-integrable functions (David Hilbert, 1862–1943).

Example 1.17 (The Hilbert space of square-integrable functions) For the vector space of functions (1.55), a natural inner product is

$$(f, g) = \int_a^b dx\, f^*(x) g(x). \tag{1.88}$$

The squared norm $\| f \|$ of a function $f(x)$ is

$$\| f \|^2 = \int_a^b dx\, |f(x)|^2. \tag{1.89}$$

A function is **square integrable** if its norm is finite. The space of all square-integrable functions is an inner-product space; it also is complete and so is a Hilbert space. □

Example 1.18 (Minkowski inner product) The Minkowski or Lorentz inner product (p, x) of two 4-vectors $p = (E/c, p_1, p_2, p_3)$ and $x = (ct, x_1, x_2, x_3)$ is

$\boldsymbol{p} \cdot \boldsymbol{x} - Et$. It is indefinite, nondegenerate, and invariant under Lorentz transformations, and often is written as $p \cdot x$ or as px. If p is the 4-momentum of a freely moving physical particle of mass m, then

$$p \cdot p = \boldsymbol{p} \cdot \boldsymbol{p} - E^2/c^2 = -c^2 m^2 \leq 0. \tag{1.90}$$

The Minkowski inner product satisfies the rules (1.73, 1.75, and 1.79), but it is **not positive definite**, and it does not satisfy the Schwarz inequality (Hermann Minkowski, 1864–1909; Hendrik Lorentz, 1853–1928). □

1.7 The Cauchy–Schwarz inequality

For any two vectors f and g, the Schwarz inequality

$$(f,f)(g,g) \geq |(f,g)|^2 \tag{1.91}$$

holds for any Schwarz inner product (and so for any positive-definite inner product). The condition (1.77) of nonnegativity ensures that for any complex number λ the inner product of the vector $f - \lambda g$ with itself is nonnegative

$$(f - \lambda g, f - \lambda g) = (f,f) - \lambda^*(g,f) - \lambda(f,g) + |\lambda|^2(g,g) \geq 0. \tag{1.92}$$

Now if $(g,g) = 0$, then for $(f - \lambda g, f - \lambda g)$ to remain nonnegative for all complex values of λ it is necessary that $(f,g) = 0$ also vanish (exercise 1.15). Thus if $(g,g) = 0$, then the Schwarz inequality (1.91) is trivially true because both sides of it vanish. So we assume that $(g,g) > 0$ and set $\lambda = (g,f)/(g,g)$. The inequality (1.92) then gives us

$$(f - \lambda g, f - \lambda g) = \left(f - \frac{(g,f)}{(g,g)} g, f - \frac{(g,f)}{(g,g)} g\right) = (f,f) - \frac{(f,g)(g,f)}{(g,g)} \geq 0$$

which is the Schwarz inequality (1.91) (Hermann Schwarz, 1843–1921)

$$(f,f)(g,g) \geq |(f,g)|^2. \tag{1.93}$$

Taking the square-root of each side, we get

$$\| f \| \| g \| \geq |(f,g)|. \tag{1.94}$$

Example 1.19 (Some Schwarz inequalities) For the dot-product of two real 3-vectors $\boldsymbol{r}$ and $\boldsymbol{R}$, the Cauchy–Schwarz inequality is

$$(\boldsymbol{r} \cdot \boldsymbol{r})\ (\boldsymbol{R} \cdot \boldsymbol{R}) \geq (\boldsymbol{r} \cdot \boldsymbol{R})^2 = (\boldsymbol{r} \cdot \boldsymbol{r})\ (\boldsymbol{R} \cdot \boldsymbol{R}) \cos^2\theta \tag{1.95}$$

where θ is the angle between $\boldsymbol{r}$ and $\boldsymbol{R}$.

The Schwarz inequality for two real n-vectors $\boldsymbol{x}$ is

$$(\boldsymbol{x} \cdot \boldsymbol{x})\ (\boldsymbol{y} \cdot \boldsymbol{y}) \geq (\boldsymbol{x} \cdot \boldsymbol{y})^2 = (\boldsymbol{x} \cdot \boldsymbol{x})\ (\boldsymbol{y} \cdot \boldsymbol{y}) \cos^2\theta \tag{1.96}$$

and it implies (Exercise 1.16) that

$$\|x\| + \|y\| \geq \|x + y\|. \tag{1.97}$$

For two complex n-vectors $\boldsymbol{u}$ and $\boldsymbol{v}$, the Schwarz inequality is

$$\left(\boldsymbol{u}^* \cdot \boldsymbol{u}\right)\left(\boldsymbol{v}^* \cdot \boldsymbol{v}\right) \geq \left|\boldsymbol{u}^* \cdot \boldsymbol{v}\right|^2 = \left(\boldsymbol{u}^* \cdot \boldsymbol{u}\right)\left(\boldsymbol{v}^* \cdot \boldsymbol{v}\right) \cos^2\theta \tag{1.98}$$

and it implies (exercise 1.17) that

$$\|\boldsymbol{u}\| + \|\boldsymbol{v}\| \geq \|\boldsymbol{u} + \boldsymbol{v}\|. \tag{1.99}$$

The inner product (1.88) of two complex functions f and g provides a somewhat different instance

$$\int_a^b dx\, |f(x)|^2 \int_a^b dx\, |g(x)|^2 \geq \left|\int_a^b dx f^*(x) g(x)\right|^2 \tag{1.100}$$

of the Schwarz inequality. □

1.8 Linear independence and completeness

A set of N vectors $V_1, V_2, \ldots, V_N$ is **linearly dependent** if there exist numbers c_i, *not all zero*, such that the linear combination

$$c_1 V_1 + \cdots + c_N V_N = 0 \tag{1.101}$$

vanishes. A set of vectors is **linearly independent** if it is not linearly dependent.

A set $\{V_i\}$ of linearly independent vectors is **maximal** in a vector space S if the addition of any other vector U in S to the set $\{V_i\}$ makes the enlarged set $\{U, V_i\}$ linearly dependent.

A set of N linearly independent vectors $V_1, V_2, \ldots, V_N$ that is maximal in a vector space S can represent any vector U in the space S as a linear combination of its vectors, $U = u_1 V_1 + \cdots + u_N V_N$. For if we enlarge the maximal set $\{V_i\}$ by including in it any vector U not already in it, then the bigger set $\{U, V_i\}$ will be linearly dependent. Thus there will be numbers $c, c_1, \ldots, c_N$, not all zero, that make the sum

$$c\, U + c_1 V_1 + \cdots + c_N V_N = 0 \tag{1.102}$$

vanish. Now if c were 0, then the set $\{V_i\}$ would be linearly dependent. Thus $c \neq 0$, and so we may divide by c and express the arbitrary vector U as a linear combination of the vectors V_i

$$U = -\frac{1}{c}\left(c_1 V_1 + \cdots + c_N V_N\right) = u_1 V_1 + \cdots + u_N V_N \tag{1.103}$$

with $u_k = -c_k/c$. So a set of linearly independent vectors $\{V_i\}$ that is maximal in a space S can represent every vector U in S as a linear combination

$U = u_1 V_1 + \ldots + u_N V_N$ of its vectors. The set $\{V_i\}$ **spans** the space S; it is a **complete** set of vectors in the space S.

A set of vectors $\{V_i\}$ that spans a vector space S provides a **basis** for that space because the set lets us represent an arbitrary vector U in S as a linear combination of the basis vectors $\{V_i\}$. If the vectors of a basis are linearly dependent, then at least one of them is superfluous, and so it is convenient to have the vectors of a basis be linearly independent.

1.9 Dimension of a vector space

If $V_1, \ldots, V_N$ and $W_1, \ldots, W_M$ are two maximal sets of N and M linearly independent vectors in a vector space S, then $N = M$.

Suppose $M < N$. Since the Us are complete, they span S, and so we may express each of the N vectors V_i in terms of the M vectors W_j

$$V_i = \sum_{j=1}^{M} A_{ij} W_j. \tag{1.104}$$

Let A_j be the vector with components A_{ij}. There are $M < N$ such vectors, and each has $N > M$ components. So it is always possible to find a nonzero N-dimensional vector C with components c_i that is orthogonal to all M vectors A_j

$$\sum_{i=1}^{N} c_i A_{ij} = 0. \tag{1.105}$$

Thus the linear combination

$$\sum_{i=1}^{N} c_i V_i = \sum_{i=1}^{N} \sum_{j=1}^{M} c_i \, A_{ij} \, W_j = 0 \tag{1.106}$$

vanishes, which implies that the N vectors V_i are linearly dependent. Since these vectors are by assumption linearly independent, it follows that $N \le M$.

Similarly, one may show that $M \le N$. Thus $M = N$.

The number of vectors in a maximal set of linearly independent vectors in a vector space S is the **dimension** of the vector space. Any N linearly independent vectors in an N-dimensional space form a **basis** for it.

1.10 Orthonormal vectors

Suppose the vectors $V_1, V_2, \ldots, V_N$ are linearly independent. Then we can make out of them a set of N vectors U_i that are orthonormal

$$(U_i, U_j) = \delta_{ij}. \tag{1.107}$$

There are many ways to do this, because there are many such sets of orthonormal vectors. We will use the Gram–Schmidt method. We set

$$U_1 = \frac{V_1}{\sqrt{(V_1, V_1)}}, \tag{1.108}$$

so the first vector U_1 is normalized. Next we set $u_2 = V_2 + c_{12} U_1$ and require that u_2 be orthogonal to U_1

$$0 = (U_1, u_2) = (U_1, c_{12} U_1 + V_2) = c_{12} + (U_1, V_2). \tag{1.109}$$

Thus $c_{12} = -(U_1, V_2)$, and so

$$u_2 = V_2 - (U_1, V_2)\, U_1. \tag{1.110}$$

The normalized vector U_2 then is

$$U_2 = \frac{u_2}{\sqrt{(u_2, u_2)}}. \tag{1.111}$$

We next set $u_3 = V_3 + c_{13} U_1 + c_{23} U_2$ and ask that u_3 be orthogonal to U_1

$$0 = (U_1, u_3) = (U_1, c_{13} U_1 + c_{23} U_2 + V_3) = c_{13} + (U_1, V_3) \tag{1.112}$$

and also to U_2

$$0 = (U_2, u_3) = (U_2, c_{13} U_1 + c_{23} U_2 + V_3) = c_{23} + (U_2, V_3). \tag{1.113}$$

So $c_{13} = -(U_1, V_3)$ and $c_{23} = -(U_2, V_3)$, and we have

$$u_3 = V_3 - (U_1, V_3)\, U_1 - (U_2, V_3)\, U_2. \tag{1.114}$$

The normalized vector U_3 then is

$$U_3 = \frac{u_3}{\sqrt{(u_3, u_3)}}. \tag{1.115}$$

We may continue in this way until we reach the last of the N linearly independent vectors. We require the kth unnormalized vector u_k

$$u_k = V_k + \sum_{i=1}^{k-1} c_{ik}\, U_i \tag{1.116}$$

to be orthogonal to the $k - 1$ vectors U_i and find that $c_{ik} = -(U_i, V_k)$ so that

$$u_k = V_k - \sum_{i=1}^{k-1} (U_i, V_k)\, U_i. \tag{1.117}$$

The normalized vector then is

$$U_k = \frac{u_k}{\sqrt{(u_k, u_k)}}. \tag{1.118}$$

A basis is more convenient if its vectors are orthonormal.

1.11 Outer products

From any two vectors f and g, we may make an operator A that takes any vector h into the vector f with coefficient (g, h)

$$A\,h = f\,(g, h). \tag{1.119}$$

Since for any vectors e, h and numbers z, w

$$A\,(z\,h + w\,e) = f\,(g, z\,h + w\,e) = zf\,(g, h) + wf\,(g, e) = z\,A\,h + w\,A\,e \tag{1.120}$$

it follows that A is linear.

If in some basis f, g, and h are vectors with components f_i, g_i, and h_i, then the linear transformation is

$$(Ah)_i = \sum_{j=1}^{N} A_{ij}\,h_j = f_i \sum_{j=1}^{N} g_j^*\,h_j \tag{1.121}$$

and in that basis A is the matrix with entries

$$A_{ij} = f_i\,g_j^*. \tag{1.122}$$

It is the **outer product** of the vectors f and g.

Example 1.20 (Outer product) If in some basis the vectors f and g are

$$f = \begin{pmatrix} 2 \\ 3 \end{pmatrix} \quad \text{and} \quad g = \begin{pmatrix} i \\ 1 \\ 3i \end{pmatrix} \tag{1.123}$$

then their outer product is the matrix

$$A = \begin{pmatrix} 2 \\ 3 \end{pmatrix} \begin{pmatrix} -i & 1 & -3i \end{pmatrix} = \begin{pmatrix} -2i & 2 & -6i \\ -3i & 3 & -9i \end{pmatrix}. \tag{1.124}$$

Dirac developed a notation that handles outer products very easily. □

Example 1.21 (Outer products) If the vectors $f = |f\rangle$ and $g = |g\rangle$ are

$$|f\rangle = \begin{pmatrix} a \\ b \\ c \end{pmatrix} \quad \text{and} \quad |g\rangle = \begin{pmatrix} z \\ w \end{pmatrix} \tag{1.125}$$

then their outer products are

$$|f\rangle\langle g| = \begin{pmatrix} az^* & aw^* \\ bz^* & bw^* \\ cz^* & cw^* \end{pmatrix} \quad \text{and} \quad |g\rangle\langle f| = \begin{pmatrix} za^* & zb^* & zc^* \\ wa^* & wb^* & wc^* \end{pmatrix} \tag{1.126}$$

as well as

$$|f\rangle\langle f| = \begin{pmatrix} aa^* & ab^* & ac^* \\ ba^* & bb^* & bc^* \\ ca^* & cb^* & cc^* \end{pmatrix} \quad \text{and} \quad |g\rangle\langle g| = \begin{pmatrix} zz^* & zw^* \\ wz^* & ww^* \end{pmatrix}. \tag{1.127}$$

Students should feel free to write down their own examples. □

1.12 Dirac notation

Outer products are important in quantum mechanics, and so Dirac invented a notation for linear algebra that makes them easy to write. In his notation, a vector f is a **ket** $f = |f\rangle$. The new thing in his notation is the **bra** $\langle g|$. The inner product of two vectors (g,f) is the **bracket** $(g,f) = \langle g|f\rangle$. A matrix element (g, Af) is then $(g, Af) = \langle g|A|f\rangle$ in which the bra and ket bracket the operator. In Dirac notation, the outer product $A\,h = f\,(g,h)$ reads $A\,|h\rangle = |f\rangle\langle g|h\rangle$, so that the outer product A itself is $A = |f\rangle\langle g|$. Before Dirac, bras were implicit in the definition of the inner product, but they did not appear explicitly; there was no way to write the bra $\langle g|$ or the operator $|f\rangle\langle g|$.

If the kets $|n\rangle$ form an orthonormal basis in an N-dimensional vector space, then we can expand an arbitrary ket in the space as

$$|f\rangle = \sum_{n=1}^{N} c_n |n\rangle. \tag{1.128}$$

Since the basis vectors are orthonormal $\langle \ell|n\rangle = \delta_{\ell n}$, we can identify the coefficients c_n by forming the inner product

$$\langle \ell|f\rangle = \sum_{n=1}^{N} c_n\,\langle \ell|n\rangle = \sum_{n=1}^{N} c_n\,\delta_{\ell,n} = c_\ell. \tag{1.129}$$

The original expansion (1.128) then must be

$$|f\rangle = \sum_{n=1}^{N} c_n |n\rangle = \sum_{n=1}^{N} \langle n|f\rangle\,|n\rangle = \sum_{n=1}^{N} |n\rangle\,\langle n|f\rangle = \left(\sum_{n=1}^{N} |n\rangle\,\langle n|\right)|f\rangle. \tag{1.130}$$

Since this equation must hold for *every* vector $|f\rangle$ in the space, it follows that the sum of outer products within the parentheses is the identity operator for the space

$$I = \sum_{n=1}^{N} |n\rangle\,\langle n|. \tag{1.131}$$

Every set of kets $|\alpha_n\rangle$ that forms an orthonormal basis $\langle\alpha_n|\alpha_\ell\rangle = \delta_{n\ell}$ for the space gives us an equivalent representation of the identity operator

$$I = \sum_{n=1}^{N} |\alpha_n\rangle \langle\alpha_n| = \sum_{n=1}^{N} |n\rangle \langle n|. \tag{1.132}$$

Before Dirac, one could not write such equations. They provide for every vector $|f\rangle$ in the space the expansions

$$|f\rangle = \sum_{n=1}^{N} |\alpha_n\rangle \langle\alpha_n|f\rangle = \sum_{n=1}^{N} |n\rangle \langle n|f\rangle. \tag{1.133}$$

Example 1.22 (Inner-product rules) In Dirac's notation, the rules (1.73–1.76) of a positive-definite inner product are

$$\langle f|g\rangle = \langle g|f\rangle^* \tag{1.134}$$
$$\langle f|z_1 g_1 + z_2 g_2\rangle = z_1\langle f|g_1\rangle + z_2\langle f|g_2\rangle \tag{1.135}$$
$$\langle z_1 f_1 + z_2 f_2|g\rangle = z_1^*\langle f_1|g\rangle + z_2^*\langle f_2|g\rangle \tag{1.136}$$
$$\langle f|f\rangle \geq 0 \text{ and } \langle f|f\rangle = 0 \iff f = 0. \tag{1.137}$$

Usually states in Dirac notation are labeled $|\psi\rangle$ or by their quantum numbers $|n, l, m\rangle$, and one rarely sees plus signs or complex numbers or operators inside bras or kets. But one should. □

Example 1.23 (Gram–Schmidt) In Dirac notation, the formula (1.117) for the kth orthogonal linear combination of the vectors $|V_\ell\rangle$ is

$$|u_k\rangle = |V_k\rangle - \sum_{i=1}^{k-1} |U_i\rangle\langle U_i|V_k\rangle = \left(I - \sum_{i=1}^{k-1} |U_i\rangle\langle U_i|\right)|V_k\rangle \tag{1.138}$$

and the formula (1.118) for the kth orthonormal linear combination of the vectors $|V_\ell\rangle$ is

$$|U_k\rangle = \frac{|u_k\rangle}{\sqrt{\langle u_k|u_k\rangle}}. \tag{1.139}$$

The vectors $|U_k\rangle$ are not unique; they vary with the order of the $|V_k\rangle$. □

Vectors and linear operators are abstract. The numbers we compute with are inner products like $\langle g|f\rangle$ and $\langle g|A|f\rangle$. In terms of N orthonormal basis vectors $|n\rangle$ with $f_n = \langle n|f\rangle$ and $g_n^* = \langle g|n\rangle$, we can use the expansion (1.131) to write these inner products as

$$\langle g|f\rangle = \langle g|I|f\rangle = \sum_{n=1}^{N} \langle g|n\rangle\langle n|f\rangle = \sum_{n=1}^{N} g_n^* f_n,$$

$$\langle g|A|f\rangle = \langle g|IAI|f\rangle = \sum_{n,\ell=1}^{N} \langle g|n\rangle\langle n|A|\ell\rangle\langle \ell|f\rangle = \sum_{n,\ell=1}^{N} g_n^* A_{n\ell} f_\ell \quad (1.140)$$

in which $A_{n\ell} = \langle n|A|\ell\rangle$. We often gather the inner products $f_\ell = \langle \ell|f\rangle$ into a column vector f with components $f_\ell = \langle \ell|f\rangle$

$$f = \begin{pmatrix} \langle 1|f\rangle \\ \langle 2|f\rangle \\ \vdots \\ \langle N|f\rangle \end{pmatrix} = \begin{pmatrix} f_1 \\ f_2 \\ \vdots \\ f_3 \end{pmatrix} \quad (1.141)$$

and the $\langle n|A|\ell\rangle$ into a matrix A with matrix elements $A_{n\ell} = \langle n|A|\ell\rangle$. If we also line up the inner products $\langle g|n\rangle = \langle n|g\rangle^*$ in a row vector that is the transpose of the complex conjugate of the column vector g

$$g^\dagger = \left(\langle 1|g\rangle^*, \langle 2|g\rangle^*, \ldots, \langle N|g\rangle^*\right) = \left(g_1^*, g_2^*, \ldots, g_N^*\right) \quad (1.142)$$

then we can write inner products in matrix notation as $\langle g|f\rangle = g^\dagger f$ and as $\langle g|A|f\rangle = g^\dagger A f$.

If we switch to a different basis, say from $|n\rangle$s to $|\alpha_n\rangle$s, then the components of the column vectors change from $f_n = \langle n|f\rangle$ to $f'_n = \langle \alpha_n|f\rangle$, and similarly those of the row vectors $g^\dagger$ and of the matrix A change, but the bras, the kets, the linear operators, and the inner products $\langle g|f\rangle$ and $\langle g|A|f\rangle$ do not change because the identity operator is basis independent (1.132)

$$\langle g|f\rangle = \sum_{n=1}^{N} \langle g|n\rangle\langle n|f\rangle = \sum_{n=1}^{N} \langle g|\alpha_n\rangle\langle \alpha_n|f\rangle,$$

$$\langle g|A|f\rangle = \sum_{n,\ell=1}^{N} \langle g|n\rangle\langle n|A|\ell\rangle\langle \ell|f\rangle = \sum_{n,\ell=1}^{N} \langle g|\alpha_n\rangle\langle \alpha_n|A|\alpha_\ell\rangle\langle \alpha_\ell|f\rangle. \quad (1.143)$$

Dirac's outer products show how to change from one basis to another. The sum of outer products

$$U = \sum_{n=1}^{N} |\alpha_n\rangle\langle n| \quad (1.144)$$

maps the ket $|\ell\rangle$ of the orthonormal basis we started with into $|\alpha_\ell\rangle$

$$U|\ell\rangle = \sum_{n=1}^{N} |\alpha_n\rangle\langle n|\ell\rangle = \sum_{n=1}^{N} |\alpha_n\rangle\,\delta_{n\ell} = |\alpha_\ell\rangle. \quad (1.145)$$

Example 1.24 (A simple change of basis) If the ket $|\alpha_n\rangle$ of the new basis is simply $|\alpha_n\rangle = |n+1\rangle$ with $|\alpha_N\rangle = |N+1\rangle \equiv |1\rangle$ then the operator that maps the N kets $|n\rangle$ into the kets $|\alpha_n\rangle$ is

$$U = \sum_{n=1}^{N} |\alpha_n\rangle\langle n| = \sum_{n=1}^{N} |n+1\rangle\langle n|. \tag{1.146}$$

The square U^2 of U also changes the basis; it sends $|n\rangle$ to $|n+2\rangle$. The set of operators U^k for $k = 1, 2, \ldots, N$ forms a group known as Z_N. □

1.13 The adjoint of an operator

In Dirac's notation, the most general linear operator on an N-dimensional vector space is a sum of dyadics like $z\,|n\rangle\langle\ell|$ in which z is a complex number and the kets $|n\rangle$ and $|\ell\rangle$ are two of the N orthonormal kets that make up a basis for the space. The **adjoint** of this basic linear operator is

$$(z\,|n\rangle\langle\ell|)^\dagger = z^*\,|\ell\rangle\langle n|. \tag{1.147}$$

Thus with $z = \langle n|A|\ell\rangle$, the most general linear operator on the space is

$$A = IAI = \sum_{n,\ell=1}^{N} |n\rangle\langle n|A|\ell\rangle\langle\ell| \tag{1.148}$$

and its adjoint $A^\dagger$ is the operator $IA^\dagger I$

$$A^\dagger = \sum_{n,\ell=1}^{N} |n\rangle\langle n|A^\dagger|\ell\rangle\langle\ell| = \sum_{n,\ell=1}^{N} |\ell\rangle\langle n|A|\ell\rangle^*\langle n| = \sum_{n,\ell=1}^{N} |n\rangle\langle\ell|A|n\rangle^*\langle\ell|.$$

It follows that $\langle n|A^\dagger|\ell\rangle = \langle\ell|A|n\rangle^*$ so that the matrix $A^\dagger_{n\ell}$ that represents $A^\dagger$ in this basis is

$$A^\dagger_{n\ell} = \langle n|A^\dagger|\ell\rangle = \langle\ell|A|n\rangle^* = A^*_{\ell n} = A^{*\mathsf{T}}_{n\ell} \tag{1.149}$$

in agreement with our definition (1.28) of the adjoint of a matrix as the transpose of its complex conjugate, $A^\dagger = A^{*\mathsf{T}}$. We also have

$$\langle g|A^\dagger f\rangle = \langle g|A^\dagger|f\rangle = \langle f|A|g\rangle^* = \langle f|Ag\rangle^* = \langle Ag|f\rangle. \tag{1.150}$$

Taking the adjoint of the adjoint is by (1.147)

$$\left[(z\,|n\rangle\langle\ell|)^\dagger\right]^\dagger = \left[z^*\,|\ell\rangle\langle n|\right]^\dagger = z\,|n\rangle\langle\ell| \tag{1.151}$$

the same as doing nothing at all. This also follows from the matrix formula (1.149) because both $(A^*)^* = A$ and $(A^\mathsf{T})^\mathsf{T} = A$, and so

$$\left(A^\dagger\right)^\dagger = \left(A^{*\mathsf{T}}\right)^{*\,\mathsf{T}} = A, \tag{1.152}$$

the adjoint of the adjoint of a matrix is the original matrix.

Before Dirac, the adjoint $A^\dagger$ of a linear operator A was defined by

$$(g, A^\dagger f) = (A\,g, f) = (f, A\,g)^*. \tag{1.153}$$

This definition also implies that $A^{\dagger\dagger} = A$ since

$$(g, A^{\dagger\dagger} f) = (A^\dagger g, f) = (f, A^\dagger g)^* = (Af, g)^* = (g, Af). \tag{1.154}$$

We also have $(g, Af) = (g, A^{\dagger\dagger} f) = (A^\dagger g, f)$.

1.14 Self-adjoint or hermitian linear operators

An operator A that is equal to its adjoint, $A^\dagger = A$, is **self adjoint** or **hermitian**. In view of (1.149), the matrix elements of a self-adjoint linear operator A satisfy $\langle n|A^\dagger|\ell\rangle = \langle \ell|A|n\rangle^* = \langle n|A|\ell\rangle$ in any orthonormal basis. So a matrix that represents a hermitian operator is equal to the transpose of its complex conjugate

$$A_{n\ell} = \langle n|A|\ell\rangle = \langle n|A^\dagger|\ell\rangle = \langle \ell|A|n\rangle^* = A^{*T}_{n\ell} = A^\dagger_{n\ell}. \tag{1.155}$$

We also have

$$\langle g|\,A\,|f\rangle = \langle A\,g|f\rangle = \langle f|A\,g\rangle^* = \langle f|\,A\,|g\rangle^* \tag{1.156}$$

and in pre-Dirac notation

$$(g, Af) = (A\,g, f) = (f, A\,g)^*. \tag{1.157}$$

A matrix A_{ij} that is **real and symmetric** or **imaginary and antisymmetric** is hermitian. But a self-adjoint linear operator A that is represented by a matrix A_{ij} that is real and symmetric (or imaginary and antisymmetric) in one orthonormal basis will not in general be represented by a matrix that is real and symmetric (or imaginary and antisymmetric) in a different orthonormal basis, but it will be represented by a hermitian matrix in every orthonormal basis.

A ket $|a'\rangle$ is an **eigenvector** of a linear operator A with **eigenvalue** a' if $A|a'\rangle = a'|a'\rangle$. As we'll see in section 1.28, hermitian matrices have real eigenvalues and complete sets of orthonormal eigenvectors. Hermitian operators and matrices represent physical variables in quantum mechanics.

1.15 Real, symmetric linear operators

In quantum mechanics, we usually consider complex vector spaces, that is, spaces in which the vectors $|f\rangle$ are complex linear combinations

$$|f\rangle = \sum_{i=1}^{N} z_i\,|i\rangle \tag{1.158}$$

of complex orthonormal basis vectors $|i\rangle$.

But real vector spaces also are of interest. A real vector space is a vector space in which the vectors $|f\rangle$ are real linear combinations

$$|f\rangle = \sum_{n=1}^{N} x_n \, |n\rangle \tag{1.159}$$

of real orthonormal basis vectors, $x_n^* = x_n$ and $|n\rangle^* = |n\rangle$.

A real linear operator A on a real vector space

$$A = \sum_{n,m=1}^{N} |n\rangle\langle n|A|m\rangle\langle m| = \sum_{n,m=1}^{N} |n\rangle A_{nm} \langle m| \tag{1.160}$$

is represented by a real matrix $A_{nm}^* = A_{nm}$. A real linear operator A that is self adjoint on a real vector space satisfies the condition (1.157) of hermiticity but with the understanding that complex conjugation has no effect

$$(g, Af) = (A\,g, f) = (f, A\,g)^* = (f, A\,g). \tag{1.161}$$

Thus its matrix elements are symmetric, $\langle g|A|f\rangle = \langle f|A|g\rangle$. Since A is hermitian as well as real, the matrix A_{nm} that represents it (in a real basis) is real and hermitian, and so is symmetric $A_{nm} = A_{mn}^* = A_{mn}$.

1.16 Unitary operators

A **unitary operator** U is one whose adjoint is its inverse

$$U\,U^\dagger = U^\dagger\,U = I. \tag{1.162}$$

Any operator that changes from one orthonormal basis $|n\rangle$ to another $|\alpha_n\rangle$

$$U = \sum_{n=1}^{N} |\alpha_n\rangle\langle n| \tag{1.163}$$

is unitary since

$$\begin{aligned} UU^\dagger &= \sum_{n=1}^{N} |\alpha_n\rangle\langle n| \sum_{m=1}^{N} |m\rangle\langle\alpha_m| = \sum_{n,m=1}^{N} |\alpha_n\rangle\langle n|m\rangle\langle\alpha_m| \\ &= \sum_{n,m=1}^{N} |\alpha_n\rangle\delta_{n,m}\langle\alpha_m| = \sum_{n=1}^{N} |\alpha_n\rangle\langle\alpha_n| = I \end{aligned} \tag{1.164}$$

as well as

$$U^\dagger U = \sum_{m=1}^{N} |m\rangle\langle\alpha_m| \sum_{n=1}^{N} |\alpha_n\rangle\langle n| = \sum_{n=1}^{N} |n\rangle\langle n| = I. \tag{1.165}$$

A unitary operator maps any orthonormal basis $|n\rangle$ into another orthonormal basis $|\alpha_n\rangle$. For if $|\alpha_n\rangle = U|n\rangle$, then $\langle\alpha_n|\alpha_m\rangle = \delta_{n,m}$ (exercise 1.22). If we multiply the relation $|\alpha_n\rangle = U|n\rangle$ by the bra $\langle n|$ and then sum over the index n, we get

$$\sum_{n=1}^{N} |\alpha_n\rangle\langle n| = \sum_{n=1}^{N} U|n\rangle\langle n| = U\sum_{n=1}^{N} |n\rangle\langle n| = U. \tag{1.166}$$

Every unitary operator is a basis-changing operator, and *vice versa*.

Inner products do not change under unitary transformations because $\langle g|f\rangle = \langle g|U^\dagger\, U|f\rangle = \langle Ug|U|f\rangle = \langle Ug|Uf\rangle$, which in pre-Dirac notation is $(g,f) = (g, U^\dagger\, Uf) = (Ug, Uf)$.

Unitary matrices have unimodular determinants because the determinant of the product of two matrices is the product of their determinants (1.204) and because transposition doesn't change the value of a determinant (1.194)

$$1 = |I| = |UU^\dagger| = |U||U^\dagger| = |U||U^{\mathsf{T}}|^* = |U||U|^*. \tag{1.167}$$

A unitary matrix that is real is **orthogonal** and satisfies

$$OO^{\mathsf{T}} = O^{\mathsf{T}}O = I. \tag{1.168}$$

1.17 Hilbert space

We have mostly been talking about linear operators that act on finite-dimensional vector spaces and that can be represented by matrices. But infinite-dimensional vector spaces and the linear operators that act on them play central roles in electrodynamics and quantum mechanics. For instance, the Hilbert space $\mathcal{H}$ of all "wave" functions $\psi(\boldsymbol{x}, t)$ that are square integrable over three-dimensional space at all times t is of infinite dimension.

In one space dimension, the state $|x'\rangle$ represents a particle at position x' and is an eigenstate of the hermitian position operator x with eigenvalue x', that is, $x|x'\rangle = x'|x'\rangle$. These states form a basis that is orthogonal in the sense that $\langle x|x'\rangle = 0$ for $x \neq x'$ and normalized in the sense that $\langle x|x'\rangle = \delta(x - x')$ in which $\delta(x - x')$ is Dirac's delta function. The delta function $\delta(x - x')$ actually is a **functional** $\delta_{x'}$ that maps any suitably smooth function f into

$$\delta_{x'}[f] = \int \delta(x - x')f(x)\,dx = f(x'), \tag{1.169}$$

its value at x'.

Another basis for the Hilbert space of one-dimensional quantum mechanics is made of the states $|p\rangle$ of well-defined momentum. The state $|p'\rangle$ represents a particle or system with momentum p'. It is an eigenstate of the hermitian momentum operator p with eigenvalue p', that is, $p|p'\rangle = p'|p'\rangle$. The momentum states also are orthonormal in Dirac's sense, $\langle p|p'\rangle = \delta(p - p')$.

The operator that translates a system in space by a distance a is

$$U(a) = \int |x + a\rangle\langle x|\, dx. \tag{1.170}$$

It maps the state $|x'\rangle$ to the state $|x' + a\rangle$ and is unitary (exercise 1.23). Remarkably, this translation operator is an exponential of the momentum operator $U(a) = \exp(-i\,p\,a/\hbar)$ in which $\hbar = h/2\pi = 1.054 \times 10^{-34}$ Js is Planck's constant divided by 2π.

In two dimensions, with basis states $|x, y\rangle$ that are orthonormal in Dirac's sense, $\langle x, y|x', y'\rangle = \delta(x - x')\delta(y - y')$, the unitary operator

$$U(\theta) = \int |x\cos\theta - y\sin\theta, x\sin\theta + y\cos\theta\rangle\langle x, y|\, dxdy \tag{1.171}$$

rotates a system in space by the angle θ. This rotation operator is the exponential $U(\theta) = \exp(-i\theta\, L_z/\hbar)$ in which the z component of the angular momentum is $L_z = x\,p_y - y\,p_x$.

We may carry most of our intuition about matrices over to these unitary transformations that change from one infinite basis to another. But we must use common sense and keep in mind that infinite sums and integrals do not always converge.

1.18 Antiunitary, antilinear operators

Certain maps on states $|\psi\rangle \to |\psi'\rangle$, such as those involving time reversal, are implemented by operators K that are **antilinear**

$$K(z\psi + w\phi) = K(z|\psi\rangle + w|\phi\rangle) = z^* K|\psi\rangle + w^* K|\phi\rangle = z^* K\psi + w^* K\phi \tag{1.172}$$

and **antiunitary**

$$(K\phi, K\psi) = \langle K\phi|K\psi\rangle = (\phi, \psi)^* = \langle\phi|\psi\rangle^* = \langle\psi|\phi\rangle = (\psi, \phi). \tag{1.173}$$

In Dirac notation, these rules are $K(z|\psi\rangle) = z^*\langle\psi|$ and $K(w\langle\phi|) = w^*|\phi\rangle$.

1.19 Symmetry in quantum mechanics

In quantum mechanics, a symmetry is a map of states $|\psi\rangle \to |\psi'\rangle$ and $|\phi\rangle \to |\phi'\rangle$ that preserves probabilities

$$|\langle\phi'|\psi'\rangle|^2 = |\langle\phi|\psi\rangle|^2. \tag{1.174}$$

Eugene Wigner (1902–1995) showed that every symmetry in quantum mechanics can be represented either by an operator U that is linear and unitary or by an operator K that is antilinear and antiunitary. The antilinear, antiunitary case seems to occur only when the symmetry involves time reversal. Most symmetries are represented by operators that are linear and unitary. Unitary operators are of great importance in quantum mechanics. We use them to represent rotations, translations, Lorentz transformations, and internal-symmetry transformations.

1.20 Determinants

The **determinant** of a 2×2 matrix A is

$$\det A = |A| = A_{11}A_{22} - A_{21}A_{12}. \tag{1.175}$$

In terms of the 2×2 antisymmetric ($e_{ij} = -e_{ji}$) matrix $e_{12} = 1 = -e_{21}$ with $e_{11} = e_{22} = 0$, this determinant is

$$\det A = \sum_{i=1}^{2} \sum_{j=1}^{2} e_{ij} A_{i1} A_{j2}. \tag{1.176}$$

It's also true that

$$e_{k\ell} \det A = \sum_{i=1}^{2} \sum_{j=1}^{2} e_{ij} A_{ik} A_{j\ell}. \tag{1.177}$$

These definitions and results extend to any square matrix. If A is a 3×3 matrix, then its determinant is

$$\det A = \sum_{i,j,k=1}^{3} e_{ijk} A_{i1} A_{j2} A_{k3} \tag{1.178}$$

in which e_{ijk} is totally antisymmetric with $e_{123} = 1$, and the sums over i, j, and k run from 1 to 3. More explicitly, this determinant is

$$\begin{aligned}\det A &= \sum_{i,j,k=1}^{3} e_{ijk} A_{i1} A_{j2} A_{k3} \\ &= \sum_{i=1}^{3} A_{i1} \sum_{j,k=1}^{3} e_{ijk} A_{j2} A_{k3} \\ &= A_{11} \left(A_{22}A_{33} - A_{32}A_{23}\right) + A_{21} \left(A_{32}A_{13} - A_{12}A_{33}\right) \\ &\quad + A_{31} \left(A_{12}A_{23} - A_{22}A_{13}\right).\end{aligned} \tag{1.179}$$

The terms within parentheses are the 2×2 determinants (called **minors**) of the matrix A without column 1 and row i, multiplied by $(-1)^{1+i}$:

$$\begin{aligned}\det A &= A_{11}(-1)^2\,(A_{22}A_{33} - A_{32}A_{23}) + A_{21}(-1)^3\,(A_{12}A_{33} - A_{32}A_{13}) \\ &\quad + A_{31}(-1)^4\,(A_{12}A_{23} - A_{22}A_{13}) \\ &= A_{11}C_{11} + A_{21}C_{21} + A_{31}C_{31} \qquad (1.180)\end{aligned}$$

The minors multiplied by $(-1)^{1+i}$ are called **cofactors**:

$$\begin{aligned}C_{11} &= A_{22}A_{33} - A_{23}A_{32}, \\ C_{21} &= A_{32}A_{13} - A_{12}A_{33}, \\ C_{31} &= A_{12}A_{23} - A_{22}A_{13}. \qquad (1.181)\end{aligned}$$

Example 1.25 (Determinant of a 3×3 matrix) The determinant of a 3×3 matrix is the dot-product of the vector of its first row with the cross-product of the vectors of its second and third rows:

$$\begin{vmatrix} U_1 & U_2 & U_3 \\ V_1 & V_2 & V_3 \\ W_1 & W_2 & W_3 \end{vmatrix} = \sum_{i,j,k=1}^{3} e_{ijk} U_i V_j W_k = \sum_{i=1}^{3} U_i (\boldsymbol{V} \times \boldsymbol{W})_i = \boldsymbol{U} \cdot (\boldsymbol{V} \times \boldsymbol{W})$$

which is called the scalar triple product. □

Laplace used the totally antisymmetric symbol $e_{i_1 i_2 \dots i_N}$ with N indices and with $e_{123\dots N} = 1$ to define the determinant of an $N \times N$ matrix A as

$$\det A = \sum_{i_1, i_2, \dots, i_N = 1}^{N} e_{i_1 i_2 \dots i_N} A_{i_1 1} A_{i_2 2} \dots A_{i_N N} \qquad (1.182)$$

in which the sums over $i_1 \dots i_N$ run from 1 to N. In terms of cofactors, two forms of his expansion of this determinant are

$$\det A = \sum_{i=1}^{N} A_{ik} C_{ik} = \sum_{k=1}^{N} A_{ik} C_{ik} \qquad (1.183)$$

in which the first sum is over the row index i but not the (arbitrary) column index k, and the second sum is over the column index k but not the (arbitrary) row index i. The cofactor C_{ik} is $(-1)^{i+k} M_{ik}$ in which the minor M_{ik} is the determinant of the $(N-1) \times (N-1)$ matrix A without its ith row and kth column. It's also true that

$$e_{k_1 k_2 \dots k_N} \det A = \sum_{i_1, i_2, \dots, i_N = 1}^{N} e_{i_1 i_2 \dots i_N} A_{i_1 k_1} A_{i_2 k_2} \dots A_{i_N k_N}. \qquad (1.184)$$

The key feature of a determinant is that it is an *antisymmetric* combination of products of the elements A_{ik} of a matrix A. One implication of this antisymmetry is that the interchange of any two rows or any two columns changes the sign of the determinant. Another is that if one adds a multiple of one column to another column, for example a multiple xA_{i2} of column 2 to column 1, then the determinant

$$\det A' = \sum_{i_1,i_2,\ldots,i_n=1}^{N} e_{i_1 i_2 \ldots i_N} \left(A_{i_1 1} + xA_{i_1 2}\right) A_{i_2 2} \ldots A_{i_N N} \tag{1.185}$$

is unchanged. The reason is that the extra term $\delta \det A$ vanishes

$$\delta \det A = \sum_{i_1,i_2,\ldots,i_N=1}^{N} x\, e_{i_1 i_2 \ldots i_N} A_{i_1 2} A_{i_2 2} \ldots A_{i_N N} = 0 \tag{1.186}$$

because it is proportional to a sum of products of a factor $e_{i_1 i_2 \ldots i_N}$ that is antisymmetric in i_1 and i_2 and a factor $A_{i_1 2} A_{i_2 2}$ that is symmetric in these indices. For instance, when i_1 and i_2 are 5 & 7 and 7 & 5, the two terms cancel

$$e_{57\ldots i_N} A_{52} A_{72} \ldots A_{i_N N} + e_{75\ldots i_N} A_{72} A_{52} \ldots A_{i_N N} = 0 \tag{1.187}$$

because $e_{57\ldots i_N} = -e_{75\ldots i_N}$.

By repeated additions of $x_2 A_{i2}$, $x_3 A_{i3}$, etc. to A_{i1}, we can change the first column of the matrix A to a linear combination of all the columns

$$A_{i1} \longrightarrow A_{i1} + \sum_{k=2}^{N} x_k A_{ik} \tag{1.188}$$

without changing $\det A$. In this linear combination, the coefficients x_k are arbitrary. The analogous operation with arbitrary y_k

$$A_{i\ell} \longrightarrow A_{i\ell} + \sum_{k=1, k\neq \ell}^{N} y_k A_{ik} \tag{1.189}$$

replaces the ℓth column by a linear combination of all the columns without changing $\det A$.

Suppose that the columns of an $N \times N$ matrix A are linearly dependent (section 1.8), so that the linear combination of columns

$$\sum_{k=1}^{N} y_k A_{ik} = 0 \quad \text{for } i = 1, \ldots N \tag{1.190}$$

vanishes for some coefficients y_k not all zero. Suppose $y_1 \neq 0$. Then by adding suitable linear combinations of columns 2 through N to column 1, we could make all the modified elements A'_{i1} of column 1 vanish without changing $\det A$.

But then $\det A$ as given by (1.182) would vanish. **Thus the determinant of any matrix whose columns are linearly dependent must vanish.**

The converse also is true: if columns of a matrix are linearly independent, then the determinant of that matrix can not vanish. The reason is that any linearly independent set of vectors is complete (section 1.8). Thus if the columns of a matrix A are linearly independent and therefore complete, some linear combination of all columns 2 through N when added to column 1 will convert column 1 into a (nonzero) multiple of the N-dimensional column vector $(1,0,0,\ldots 0)$, say $(c_1,0,0,\ldots 0)$. Similar operations will convert column 2 into a (nonzero) multiple of the column vector $(0,1,0,\ldots 0)$, say $(0,c_2,0,\ldots 0)$. Continuing in this way, we may convert the matrix A to a matrix with nonzero entries along the main diagonal and zeros everywhere else. The determinant $\det A$ is then the product of the nonzero diagonal entries $c_1 c_2 \ldots c_N \neq 0$, and so $\det A$ can not vanish.

We may extend these arguments to the rows of a matrix. The addition to row k of a linear combination of the other rows

$$A_{ki} \longrightarrow A_{ki} + \sum_{\ell=1,\ell\neq k}^{N} z_\ell A_{\ell i} \tag{1.191}$$

does not change the value of the determinant. In this way, one may show that the determinant of a matrix vanishes if and only if its rows are linearly dependent. The reason why these results apply to the rows as well as to the columns is that the determinant of a matrix A may be defined either in terms of the columns as in definitions (1.182 & 1.184) or in terms of the rows:

$$\det A = \sum_{i_1,i_2,\ldots,i_N=1}^{N} e_{i_1 i_2 \ldots i_N} A_{1i_1} A_{2i_2} \ldots A_{Ni_N}, \tag{1.192}$$

$$e_{k_1 k_2 \ldots k_N} \det A = \sum_{i_1,i_2,\ldots,i_N=1}^{N} e_{i_1 i_2 \ldots i_N} A_{k_1 i_1} A_{k_2 i_2} \ldots A_{k_N i_N}. \tag{1.193}$$

These and other properties of determinants follow from a study of **permutations** (section 10.13). Detailed proofs are in Aitken (1959).

By comparing the row (1.182 & 1.184) and column (1.192 & 1.193) definitions of determinants, we see that the determinant of the transpose of a matrix is the same as the determinant of the matrix itself:

$$\det\left(A^{\mathsf{T}}\right) = \det A. \tag{1.194}$$

Let us return for a moment to Laplace's expansion (1.183) of the determinant $\det A$ of an $N \times N$ matrix A as a sum of $A_{ik}C_{ik}$ over the row index i with the column index k held fixed

$$\det A = \sum_{i=1}^{N} A_{ik} C_{ik} \tag{1.195}$$

in order to prove that

$$\delta_{k\ell} \det A = \sum_{i=1}^{N} A_{ik} C_{i\ell}. \tag{1.196}$$

For $k = \ell$, this formula just repeats Laplace's expansion (1.195). But for $k \neq \ell$, it is Laplace's expansion for the determinant of a matrix A' that is the same as A but with its ℓth column replaced by its kth one. Since the matrix A' has two identical columns, its determinant vanishes, which explains (1.196) for $k \neq \ell$.

This rule (1.196) provides a formula for the inverse of a matrix A whose determinant does not vanish. Such matrices are said to be **nonsingular**. The inverse A^{-1} of an $N \times N$ nonsingular matrix A is the transpose of the matrix of cofactors divided by $\det A$

$$\left(A^{-1}\right)_{\ell i} = \frac{C_{i\ell}}{\det A} \quad \text{or} \quad A^{-1} = \frac{C^{\mathsf{T}}}{\det A}. \tag{1.197}$$

To verify this formula, we use it for A^{-1} in the product $A^{-1}A$ and note that by (1.196) the ℓkth entry of the product $A^{-1}A$ is just $\delta_{\ell k}$

$$\left(A^{-1}A\right)_{\ell k} = \sum_{i=1}^{N} \left(A^{-1}\right)_{\ell i} A_{ik} = \sum_{i=1}^{N} \frac{C_{i\ell}}{\det A} A_{ik} = \delta_{\ell k}. \tag{1.198}$$

Example 1.26 (Inverting a 2×2 matrix) Let's apply our formula (1.197) to find the inverse of the general 2×2 matrix

$$A = \begin{pmatrix} a & b \\ c & d \end{pmatrix}. \tag{1.199}$$

We find then

$$A^{-1} = \frac{1}{ad - bc} \begin{pmatrix} d & -b \\ -c & a \end{pmatrix}, \tag{1.200}$$

which is the correct inverse as long as $ad \neq bc$. □

The simple example of matrix multiplication

$$\begin{pmatrix} a & b & c \\ d & e & f \\ g & h & i \end{pmatrix} \begin{pmatrix} 1 & x & y \\ 0 & 1 & z \\ 0 & 0 & 1 \end{pmatrix} = \begin{pmatrix} a & xa + b & ya + zb + c \\ d & xd + e & yd + ze + f \\ g & xg + h & yg + zh + i \end{pmatrix} \tag{1.201}$$

shows that the operations (1.189) on columns that don't change the value of the determinant can be written as matrix multiplication from the right by a matrix that has unity on its main diagonal and zeros below. Now consider the matrix product

$$\begin{pmatrix} A & 0 \\ -I & B \end{pmatrix} \begin{pmatrix} I & B \\ 0 & I \end{pmatrix} = \begin{pmatrix} A & AB \\ -I & 0 \end{pmatrix} \tag{1.202}$$

in which A and B are $N \times N$ matrices, I is the $N \times N$ identity matrix, and 0 is the $N \times N$ matrix of all zeros. The second matrix on the left-hand side has unity on its main diagonal and zeros below, and so it does not change the value of the determinant of the matrix to its left, which then must equal that of the matrix on the right-hand side:

$$\det \begin{pmatrix} A & 0 \\ -I & B \end{pmatrix} = \det \begin{pmatrix} A & AB \\ -I & 0 \end{pmatrix}. \tag{1.203}$$

By using Laplace's expansion (1.183) along the first column to evaluate the determinant on the left-hand side and his expansion along the last row to compute the determinant on the right-hand side, one finds that **the determinant of the product of two matrices is the product of the determinants**

$$\det A \, \det B = \det AB. \tag{1.204}$$

Example 1.27 (Two 2 × 2 matrices) When the matrices A and B are both 2×2, the two sides of (1.203) are

$$\det \begin{pmatrix} A & 0 \\ -I & B \end{pmatrix} = \det \begin{pmatrix} a_{11} & a_{12} & 0 & 0 \\ a_{21} & a_{22} & 0 & 0 \\ -1 & 0 & b_{11} & b_{12} \\ 0 & -1 & b_{21} & b_{22} \end{pmatrix}$$
$$= a_{11}a_{22} \det B - a_{21}a_{12} \det B = \det A \det B \tag{1.205}$$

and

$$\det \begin{pmatrix} A & AB \\ -I & 0 \end{pmatrix} = \det \begin{pmatrix} a_{11} & a_{12} & ab_{11} & ab_{12} \\ a_{21} & a_{22} & ab_{21} & ab_{22} \\ -1 & 0 & 0 & 0 \\ 0 & -1 & 0 & 0 \end{pmatrix}$$
$$= (-1)C_{42} = (-1)(-1) \det AB = \det AB \tag{1.206}$$

and so they give the product rule $\det A \det B = \det AB$. □

Often one uses the notation $|A| = \det A$ to denote a determinant. In this more compact notation, the obvious generalization of the product rule is

$$|ABC \ldots Z| = |A||B| \ldots |Z|. \tag{1.207}$$

The product rule (1.204) implies that $\det\left(A^{-1}\right)$ is $1/\det A$ since

$$1 = \det I = \det\left(AA^{-1}\right) = \det A \det\left(A^{-1}\right). \tag{1.208}$$

Incidentally, Gauss, Jordan, and modern mathematicians have developed much faster ways of computing determinants and matrix inverses than those (1.183 & 1.197) due to Laplace. Octave, Matlab, Maple, and Mathematica use these modern techniques, which also are freely available as programs in C and FORTRAN from www.netlib.org/lapack.

Example 1.28 (Numerical tricks) Adding multiples of rows to other rows does not change the value of a determinant, and interchanging two rows only changes a determinant by a minus sign. So we can use these operations, which leave determinants invariant, to make a matrix **upper triangular**, a form in which its determinant is just the product of the factors on its diagonal. For instance, to make the matrix

$$A = \begin{pmatrix} 1 & 2 & 1 \\ -2 & -6 & 3 \\ 4 & 2 & -5 \end{pmatrix} \tag{1.209}$$

upper triangular, we add twice the first row to the second row

$$\begin{pmatrix} 1 & 2 & 1 \\ 0 & -2 & 5 \\ 4 & 2 & -5 \end{pmatrix} \tag{1.210}$$

and then subtract four times the first row from the third

$$\begin{pmatrix} 1 & 2 & 1 \\ 0 & -2 & 5 \\ 0 & -6 & -9 \end{pmatrix}. \tag{1.211}$$

Next, we subtract three times the second row from the third

$$\begin{pmatrix} 1 & 2 & 1 \\ 0 & -2 & 5 \\ 0 & 0 & -24 \end{pmatrix}. \tag{1.212}$$

We now find as the determinant of A the product of its diagonal elements:

$$|A| = 1(-2)(-24) = 48. \tag{1.213}$$

The Matlab command is $d = \det(A)$. □

1.21 Systems of linear equations

Suppose we wish to solve the system of N linear equations

$$\sum_{k=1}^{N} A_{ik} x_k = y_i \tag{1.214}$$

for N unknowns x_k. In matrix notation, with A an $N \times N$ matrix and x and y N-vectors, this system of equations is $Ax = y$. If the matrix A is **nonsingular**, that is, if $\det(A) \neq 0$, then it has an inverse A^{-1} given by (1.197), and we may multiply both sides of $Ax = y$ by A^{-1} and so find x as $x = A^{-1}y$. When A is nonsingular, this is the unique solution to (1.214).

When A is singular, $\det(A) = 0$, and so its columns are linearly dependent (section 1.20). In this case, the linear dependence of the columns of A implies that $Az = 0$ for some nonzero vector z. Thus if x is a solution, so that $Ax = y$, then $A(x + cz) = Ax + cAz = y$ implies that $x + cz$ for all c also is a solution. So if $\det(A) = 0$, then there may be solutions, but there can be no unique solution. Whether equation (1.214) has any solutions when $\det(A) = 0$ depends on whether the vector y can be expressed as a linear combination of the columns of A. Since these columns are linearly dependent, they span a subspace of fewer than N dimensions, and so (1.214) has solutions only when the N-vector y lies in that subspace.

A system of $M < N$ equations

$$\sum_{k=1}^{N} A_{ik} x_k = y_i \quad \text{for} \quad i = 1, 2, \ldots, M \tag{1.215}$$

in N unknowns is **under-determined**. As long as at least M of the N columns A_{ik} of the matrix A are linearly independent, such a system always has solutions, but they will not be unique.

1.22 Linear least squares

Suppose we have a system of $M > N$ equations in N unknowns x_k

$$\sum_{k=1}^{N} A_{ik} x_k = y_i \quad \text{for} \quad i = 1, 2, \ldots, M. \tag{1.216}$$

This problem is **over-determined** and, in general, has no solution, but it does have an approximate solution due to Carl Gauss (1777–1855).

If the matrix A and the vector y are real, then Gauss's solution is the N values x_k that minimize the sum E of the squares of the errors

$$E = \sum_{i=1}^{M} \left(y_i - \sum_{k=1}^{N} A_{ik} x_k \right)^2 . \tag{1.217}$$

The minimizing values x_k make the N derivatives of E vanish

$$\frac{\partial E}{\partial x_\ell} = 0 = \sum_{i=1}^{M} 2 \left(y_i - \sum_{k=1}^{N} A_{ik} x_k \right) (-A_{i\ell}) \tag{1.218}$$

or in matrix notation $A^{\mathsf{T}} y = A^{\mathsf{T}} A x$. Since A is real, the matrix $A^{\mathsf{T}} A$ is nonnegative (1.38); if it also is positive (1.39), then it has an inverse, and our **least-squares solution** is

$$x = \left(A^{\mathsf{T}} A \right)^{-1} A^{\mathsf{T}} y. \tag{1.219}$$

If the matrix A and the vector y are complex, and if the matrix $A^\dagger A$ is positive, then one may show (exercise 1.25) that Gauss's solution is

$$x = \left(A^\dagger A \right)^{-1} A^\dagger y. \tag{1.220}$$

1.23 Lagrange multipliers

The maxima and minima of a function $f(x)$ of several variables $x_1, x_2, \ldots, x_n$ are among the points at which its gradient vanishes

$$\nabla f(x) = 0. \tag{1.221}$$

These are the stationary points of f.

Example 1.29 (Minimum) For instance, if $f(x) = x_1^2 + 2x_2^2 + 3x_3^2$, then its minimum is at

$$\nabla f(x) = (2x_1, 4x_2, 6x_3) = 0 \tag{1.222}$$

that is, at $x_1 = x_2 = x_3 = 0$. □

But how do we find the extrema of $f(x)$ if x must satisfy a constraint? We use a Lagrange multiplier (Joseph-Louis Lagrange, 1736–1813).

In the case of one constraint $c(x) = 0$, we no longer expect the gradient $\nabla f(x)$ to vanish, but its projection $dx \cdot \nabla f(x)$ must vanish in those directions dx that preserve the constraint. So $dx \cdot \nabla f(x) = 0$ for all dx that make the dot-product $dx \cdot \nabla c(x)$ vanish. This means that $\nabla f(x)$ and $\nabla c(x)$ must be parallel. So the extrema of $f(x)$ subject to the constraint $c(x) = 0$ satisfy two equations

$$\nabla f(x) = \lambda \, \nabla c(x) \quad \text{and} \quad c(x) = 0. \tag{1.223}$$

These equations define the extrema of the unconstrained function

$$L(x,\lambda) = f(x) - \lambda\, c(x) \tag{1.224}$$

of the $n+1$ variables $x, \ldots, x_n, \lambda$

$$\nabla L(x,\lambda) = \nabla f(x) - \lambda\, \nabla c(x) = 0 \quad \text{and} \quad \frac{\partial L(x,\lambda)}{\partial \lambda} = -\, c(x) = 0. \tag{1.225}$$

The variable λ is a **Lagrange multiplier**.

In the case of k constraints $c_1(x) = 0, \ldots, c_k(x) = 0$, the projection ∇f must vanish in those directions dx that preserve all the constraints. So $dx \cdot \nabla f(x) = 0$ for all dx that make all $dx \cdot \nabla c_j(x) = 0$ for $j = 1, \ldots, k$. The gradient ∇f will satisfy this requirement if it's a linear combination

$$\nabla f = \lambda_1 \nabla c_1 + \cdots + \lambda_k \nabla c_k \tag{1.226}$$

of the k gradients because then $dx \cdot \nabla f$ will vanish if $dx \cdot \nabla c_j = 0$ for $j = 1, \ldots, k$. The extrema also must satisfy the constraints

$$c_1(x) = 0, \ldots, c_k(x) = 0. \tag{1.227}$$

Equations (1.226 & 1.227) define the extrema of the unconstrained function

$$L(x,\lambda) = f(x) - \lambda_1\, c_1(x) + \cdots \lambda_k\, c_k(x) \tag{1.228}$$

of the $n+k$ variables x and λ

$$\nabla L(x,\lambda) = \nabla f(x) - \lambda\, \nabla c_1(x) - \cdots - \lambda\, \nabla c_k(x) = 0 \tag{1.229}$$

and

$$\frac{\partial L(x,\lambda)}{\partial \lambda_j} = -\, c_j(x) = 0 \quad \text{for} \quad j = 1, \ldots, k. \tag{1.230}$$

Example 1.30 (Constrained extrema and eigenvectors) Suppose we want to find the extrema of a real, symmetric quadratic form $f(x) = x^{\mathsf{T}} A\, x$ subject to the constraint $c(x) = x \cdot x - 1$, which says that the vector x is of unit length. We form the function

$$L(x,\lambda) = x^{\mathsf{T}} A\, x - \lambda\,(x \cdot x - 1) \tag{1.231}$$

and since the matrix A is real and symmetric, we find its unconstrained extrema as

$$\nabla L(x,\lambda) = 2A\, x - 2\lambda\, x = 0 \quad \text{and} \quad x \cdot x = 1. \tag{1.232}$$

The extrema of $f(x) = x^{\mathsf{T}} A\, x$ subject to the constraint $c(x) = x \cdot x - 1$ are the normalized **eigenvectors**

$$A\, x = \lambda\, x \quad \text{and} \quad x \cdot x = 1 \tag{1.233}$$

of the real, symmetric matrix A. □

1.24 Eigenvectors

If a linear operator A maps a nonzero vector $|u\rangle$ into a multiple of itself

$$A|u\rangle = \lambda|u\rangle \tag{1.234}$$

then the vector $|u\rangle$ is an **eigenvector** of A with **eigenvalue** λ. (The German adjective *eigen* means special or proper.)

If the vectors $\{|k\rangle\}$ for $k = 1, \ldots, N$ form a basis for the vector space in which A acts, then we can write the identity operator for the space as $I = |1\rangle\langle 1| + \cdots + |N\rangle\langle N|$. By inserting this formula for I twice into the eigenvector equation (1.234), we can write it as

$$\sum_{\ell=1}^{N} \langle k|A|\ell\rangle\langle\ell|u\rangle = \lambda\, \langle k|u\rangle. \tag{1.235}$$

In matrix notation, with $A_{k\ell} = \langle k|A|\ell\rangle$ and $u_\ell = \langle\ell|u\rangle$, this is $A\, u = \lambda\, u$.

Example 1.31 (Eigenvalues of an orthogonal matrix) The matrix equation

$$\begin{pmatrix} \cos\theta & \sin\theta \\ -\sin\theta & \cos\theta \end{pmatrix} \begin{pmatrix} 1 \\ \pm i \end{pmatrix} = e^{\pm i\theta} \begin{pmatrix} 1 \\ \pm i \end{pmatrix} \tag{1.236}$$

tells us that the eigenvectors of this 2×2 orthogonal matrix are the 2-tuples $(1, \pm i)$ with eigenvalues $e^{\pm i\theta}$. The eigenvalues λ of a unitary (and of an orthogonal) matrix are unimodular, $|\lambda| = 1$ (exercise 1.26). □

Example 1.32 (Eigenvalues of an antisymmetric matrix) Let us consider an eigenvector equation for a matrix A that is antisymmetric

$$\sum_{k=1}^{N} A_{ik}\, u_k = \lambda\, u_i. \tag{1.237}$$

The antisymmetry $A_{ik} = -A_{ki}$ of A implies that

$$\sum_{i,k=1}^{N} u_i\, A_{ik}\, u_k = 0. \tag{1.238}$$

Thus the last two relations imply that

$$0 = \sum_{i,k=1}^{N} u_i\, A_{ik}\, u_k = \lambda \sum_{i=1}^{N} u_i^2 = 0. \tag{1.239}$$

Thus either the eigenvalue λ or the dot-product of the eigenvector with itself vanishes.

A subspace $c_\ell|u_\ell\rangle + \cdots + c_r|u_r\rangle$ spanned by any set of eigenvectors of a matrix A is left invariant by its action, that is

$$A\left(c_\ell|u_\ell\rangle + \cdots + c_r|u_r\rangle\right) = c_\ell\lambda_\ell|u_\ell\rangle + \cdots + c_r\lambda_r|u_r\rangle. \tag{1.240}$$

Eigenvectors span **invariant subspaces**. □

1.25 Eigenvectors of a square matrix

Let A be an $N \times N$ matrix with complex entries A_{ik}. A vector V with N entries V_k (not all zero) is an **eigenvector** of A with **eigenvalue** λ if

$$AV = \lambda V \iff \sum_{k=1}^{N} A_{ik} V_k = \lambda V_i. \tag{1.241}$$

Every $N \times N$ matrix A has N eigenvectors $V^{(\ell)}$ and eigenvalues λ_ℓ

$$AV^{(\ell)} = \lambda_\ell V^{(\ell)} \tag{1.242}$$

for $\ell = 1 \ldots N$. To see why, we write the top equation (1.241) as

$$\sum_{k=1}^{N} \left(A_{ik} - \lambda\delta_{ik}\right) V_k = 0 \tag{1.243}$$

or in matrix notation as $(A - \lambda I)\, V = 0$ in which I is the $N \times N$ matrix with entries $I_{ik} = \delta_{ik}$. This equation and (1.243) say that the columns of the matrix $A - \lambda I$, considered as vectors, are linearly dependent, as defined in section 1.8. We saw in section 1.20 that the columns of a matrix $A - \lambda I$ are linearly dependent if and only if the determinant $|A - \lambda I|$ vanishes. Thus a nonzero solution of the eigenvalue equation (1.241) exists if and only if the determinant

$$\det(A - \lambda I) = |A - \lambda I| = 0 \tag{1.244}$$

vanishes. This requirement that the determinant of $A - \lambda I$ vanishes is called the **characteristic equation**. For an $N \times N$ matrix A, it is a polynomial equation of the Nth degree in the unknown eigenvalue λ

$$\begin{aligned} |A - \lambda I| \equiv P(\lambda, A) &= |A| + \cdots + (-1)^{N-1}\lambda^{N-1}\,\mathrm{Tr}A + (-1)^N\lambda^N \\ &= \sum_{k=0}^{N} p_k\,\lambda^k = 0 \end{aligned} \tag{1.245}$$

in which $p_0 = |A|$, $p_{N-1} = (-1)^{N-1}\mathrm{Tr}A$, and $p_N = (-1)^N$. (All the p_ks are basis independent.) By the fundamental theorem of algebra (section 5.9), the characteristic equation always has N roots or solutions λ_ℓ lying somewhere in the complex plane. Thus the characteristic polynomial has the factored form

$$P(\lambda, A) = (\lambda_1 - \lambda)(\lambda_2 - \lambda)\ldots(\lambda_N - \lambda). \tag{1.246}$$

For every root λ_ℓ, there is a nonzero eigenvector $V^{(\ell)}$ whose components $V_k^{(\ell)}$ are the coefficients that make the N vectors $A_{ik} - \lambda_\ell \, \delta_{ik}$ that are the columns of the matrix $A - \lambda_\ell I$ sum to zero in (1.243). Thus **every $N \times N$ matrix has N eigenvalues λ_ℓ and N eigenvectors $V^{(\ell)}$.**

The $N \times N$ diagonal matrix $D_{k\ell} = \delta_{k\ell} \, \lambda_\ell$ is the **canonical form** of the matrix A; the matrix $V_{k\ell} = V_k^{(\ell)}$ whose columns are the eigenvectors $V^{(\ell)}$ of A is the **modal matrix**; and $AV = VD$.

Example 1.33 (The canonical form of a 3 × 3 matrix) If in Matlab we set $A =$ [0 1 2; 3 4 5; 6 7 8] and enter $[V, D] = \text{eig}(A)$, then we get

$$V = \begin{pmatrix} 0.1648 & 0.7997 & 0.4082 \\ 0.5058 & 0.1042 & -0.8165 \\ 0.8468 & -0.5913 & 0.4082 \end{pmatrix} \quad \text{and} \quad D = \begin{pmatrix} 13.3485 & 0 & 0 \\ 0 & -1.3485 & 0 \\ 0 & 0 & 0 \end{pmatrix}$$

and one may check that $AV = VD$. □

Setting $\lambda = 0$ in the factored form (1.246) of $P(\lambda, A)$ and in the characteristic equation (1.245), we see that **the determinant of every $N \times N$ matrix is the product of its N eigenvalues**

$$P(0, A) = |A| = p_0 = \lambda_1 \lambda_2 \ldots \lambda_N. \tag{1.247}$$

These N roots usually are all different, and when they are, the eigenvectors $V^{(\ell)}$ are linearly independent. The first eigenvector is trivially linearly independent. Let's assume that the first $K < N$ eigenvectors are linearly independent; we'll show that the first $K+1$ eigenvectors are linearly independent. If they were linearly dependent, then there would be $K + 1$ numbers c_ℓ, not all zero, such that

$$\sum_{\ell=1}^{K+1} c_\ell V^{(\ell)} = 0. \tag{1.248}$$

First we multiply this equation from the left by the linear operator A and use the eigenvalue equation (1.242)

$$A \sum_{\ell=1}^{K+1} c_\ell \, V^{(\ell)} = \sum_{\ell=1}^{K+1} c_\ell \, AV^{(\ell)} = \sum_{\ell=1}^{K+1} c_\ell \, \lambda_\ell \, V^{(\ell)} = 0. \tag{1.249}$$

Now we multiply the same equation (1.248) by λ_{K+1}

$$\sum_{\ell=1}^{K+1} c_\ell \, \lambda_N \, V^{(\ell)} = 0 \tag{1.250}$$

and subtract the product (1.250) from (1.249). The terms with $\ell = K+1$ cancel leaving

$$\sum_{\ell=1}^{K} c_\ell \left(\lambda_\ell - \lambda_N\right) V^{(\ell)} = 0 \tag{1.251}$$

in which all the factors $(\lambda_\ell - \lambda_{K+1})$ are different from zero since by assumption all the eigenvalues are different. But this last equation says that the first K eigenvectors are linearly dependent, which contradicts our assumption that they were linearly independent. This contradiction tells us that **if all N eigenvectors of an $N \times N$ square matrix have different eigenvalues, then they are linearly independent**.

An eigenvalue λ that is a single root of the characteristic equation (1.245) is associated with a single eigenvector; it is called a **simple eigenvalue**. An eigenvalue λ that is an nth root of the characteristic equation is associated with n eigenvectors; it is said to be an ***n*-fold degenerate eigenvalue** or to have **algebraic multiplicity** n. Its **geometric multiplicity** is the number $n' \le n$ of linearly independent eigenvectors with eigenvalue λ. A matrix with $n' < n$ for any eigenvalue λ is **defective**. Thus an $N \times N$ matrix with fewer than N linearly independent eigenvectors is defective.

Example 1.34 (A defective 2×2 matrix) Each of the 2×2 matrices

$$\begin{pmatrix} 0 & 1 \\ 0 & 0 \end{pmatrix} \quad \text{and} \quad \begin{pmatrix} 0 & 0 \\ 1 & 0 \end{pmatrix} \tag{1.252}$$

has only one linearly independent eigenvector and so is defective. □

Suppose A is an $N \times N$ matrix that is not defective. We may use its N linearly independent eigenvectors $V^{(\ell)} = |\ell\rangle$ to define the columns of an $N \times N$ matrix S as $S_{k\ell} = V_k^{(\ell)}$. In terms of S, the eigenvalue equation (1.242) takes the form

$$\sum_{k=1}^{N} A_{ik} S_{k\ell} = \lambda_\ell S_{i\ell}. \tag{1.253}$$

Since the columns of S are linearly independent, the determinant of S does not vanish – the matrix S is **nonsingular** – and so its inverse S^{-1} is well defined by (1.197). So we may multiply this equation by S^{-1} and get

$$\sum_{i,k=1}^{N} \left(S^{-1}\right)_{ni} A_{ik} S_{k\ell} = \sum_{i=1}^{N} \lambda_\ell \left(S^{-1}\right)_{ni} S_{i\ell} = \lambda_n \delta_{n\ell} = \lambda_\ell \tag{1.254}$$

or in matrix notation

$$S^{-1} A S = A^{(\mathrm{d})} \tag{1.255}$$

in which $A^{(d)}$ is the diagonal form of the matrix A in which its eigenvalues λ_ℓ are arranged along its main diagonal with zeros elsewhere. This equation (1.255) is a **similarity transformation**. Thus **every nondefective square matrix can be diagonalized by a similarity transformation** $S^{-1}AS = A^{(d)}$ and can be generated from its diagonal form by the inverse $A = SA^{(d)}S^{-1}$ of that similarity transformation. By using the product rule (1.207), we see that the determinant of any nondefective square matrix is the product of its eigenvalues

$$|A| = |SA^{(d)}S^{-1}| = |S|\,|A^{(d)}|\,|S^{-1}| = |SS^{-1}|\,|A^{(d)}| = |A^{(d)}| = \prod_{\ell=1}^{N} \lambda_\ell, \tag{1.256}$$

which is a special case of (1.247).

1.26 A matrix obeys its characteristic equation

Every square matrix obeys its characteristic equation (1.245). That is, the characteristic equation

$$P(\lambda, A) = |A - \lambda I| = \sum_{k=0}^{N} p_k\, \lambda^k = 0 \tag{1.257}$$

remains true when the matrix A replaces the variable λ

$$P(A, A) = \sum_{k=0}^{N} p_k\, A^k = 0. \tag{1.258}$$

To see why, we use the formula (1.197) for the inverse of the matrix $A - \lambda I$

$$(A - \lambda I)^{-1} = \frac{C(\lambda, A)^{\mathsf{T}}}{|A - \lambda I|} \tag{1.259}$$

in which $C(\lambda, A)^{\mathsf{T}}$ is the transpose of the matrix of cofactors of the matrix $A - \lambda I$. Since $|A - \lambda I| = P(\lambda, A)$, we have, rearranging,

$$(A - \lambda I)\, C(\lambda, A)^{\mathsf{T}} = |A - \lambda I|\, I = P(\lambda, A)\, I. \tag{1.260}$$

The transpose of the matrix of cofactors of the matrix $A - \lambda I$ is a polynomial in λ with matrix coefficients

$$C(\lambda, A)^{\mathsf{T}} = C_0 + C_1\lambda + \cdots + C_{N-1}\lambda^{N-1}. \tag{1.261}$$

The left-hand side of equation (1.260) is then

$$\begin{aligned}(A - \lambda I)C(\lambda, A)^{\mathsf{T}} = {} & AC_0 + (AC_1 - C_0)\lambda + (AC_2 - C_1)\lambda^2 + \cdots \\ & + (AC_{N-1} - C_{N-2})\lambda^{N-1} - C_{N-1}\lambda^N. \end{aligned} \tag{1.262}$$

Equating equal powers of λ on both sides of (1.260), we have, using (1.257) and (1.262),

$$\begin{aligned} AC_0 &= p_0 I, \\ AC_1 - C_0 &= p_1 I, \\ AC_2 - C_1 &= p_2 I, \\ \cdots &= \cdots \\ AC_{N-1} - C_{N-2} &= p_{N-1} I, \\ -C_{N-1} &= p_N I. \end{aligned} \tag{1.263}$$

We now multiply from the left the first of these equations by I, the second by A, the third by A^2, ..., and the last by A^N and then add the resulting equations. All the terms on the left-hand sides cancel, while the sum of those on the right gives $P(A, A)$. Thus a square matrix A obeys its characteristic equation $0 = P(A, A)$ or

$$0 = \sum_{k=0}^{N} p_k \, A^k = |A| \, I + p_1 A + \cdots + (-1)^{N-1} (\text{Tr} A) \, A^{N-1} + (-1)^N \, A^N, \tag{1.264}$$

a result known as the **Cayley–Hamilton theorem** (Arthur Cayley, 1821–1895, and William Hamilton, 1805–1865). This derivation is due to Israel Gelfand (1913–2009) (Gelfand, 1961, pp. 89–90).

Because every $N \times N$ matrix A obeys its characteristic equation, its Nth power A^N can be expressed as a linear combination of its lesser powers

$$A^N = (-1)^{N-1} \left(|A| \, I + p_1 A + p_2 A^2 + \cdots + (-1)^{N-1} (\text{Tr} A) \, A^{N-1} \right). \tag{1.265}$$

For instance, the square A^2 of every 2×2 matrix is given by

$$A^2 = -|A| I + (\text{Tr} A) A. \tag{1.266}$$

Example 1.35 (Spin-one-half rotation matrix) If $\boldsymbol{\theta}$ is a real 3-vector and $\boldsymbol{\sigma}$ is the 3-vector of Pauli matrices (1.32), then the square of the traceless 2×2 matrix $A = \boldsymbol{\theta} \cdot \boldsymbol{\sigma}$ is

$$(\boldsymbol{\theta} \cdot \boldsymbol{\sigma})^2 = - \, |\boldsymbol{\theta} \cdot \boldsymbol{\sigma}| \; I = - \begin{vmatrix} \theta_3 & \theta_1 - i\theta_2 \\ \theta_1 + i\theta_2 & -\theta_3 \end{vmatrix} \; I = \theta^2 \, I \tag{1.267}$$

in which $\theta^2 = \boldsymbol{\theta} \cdot \boldsymbol{\theta}$. One may use this identity to show (exercise (1.28)) that

$$\exp\left(-i\boldsymbol{\theta} \cdot \boldsymbol{\sigma}/2\right) = \cos(\theta/2) \, I - i\hat{\boldsymbol{\theta}} \cdot \boldsymbol{\sigma} \; \sin(\theta/2) \tag{1.268}$$

in which $\hat{\boldsymbol{\theta}}$ is a unit 3-vector. For a spin-one-half object, this matrix represents a right-handed rotation of θ radians about the axis $\hat{\boldsymbol{\theta}}$. □

1.27 Functions of matrices

What sense can we make of a function f of an $N \times N$ matrix A and how would we compute it? One way is to use the characteristic equation (1.265) to express every power of A in terms of I, A, ..., A^{N-1} and the coefficients $p_0 = |A|, p_1, p_2, \ldots, p_{N-2}$, and $p_{N-1} = (-1)^{N-1}\mathrm{Tr}A$. Then if $f(x)$ is a polynomial or a function with a convergent power series

$$f(x) = \sum_{k=0}^{\infty} c_k\, x^k \tag{1.269}$$

in principle we may express $f(A)$ in terms of N functions $f_k(\boldsymbol{p})$ of the coefficients $\boldsymbol{p} \equiv (p_0, \ldots, p_{N-1})$ as

$$f(A) = \sum_{k=0}^{N-1} f_k(\boldsymbol{p})\, A^k. \tag{1.270}$$

The identity (1.268) for $\exp(-i\boldsymbol{\theta} \cdot \boldsymbol{\sigma}/2)$ is an $N = 2$ example of this technique, which can become challenging when $N > 3$.

Example 1.36 (The 3×3 rotation matrix) In exercise (1.29), one finds the characteristic equation (1.264) for the $3{\times}3$ matrix $-i\boldsymbol{\theta} \cdot \boldsymbol{J}$ in which $(J_k)_{ij} = i\epsilon_{ikj}$, and ϵ_{ijk} is totally antisymmetric with $\epsilon_{123} = 1$. The generators J_k satisfy the commutation relations $[J_i, J_j] = i\epsilon_{ijk}J_k$ in which sums over repeated indices from 1 to 3 are understood. In exercise (1.31), one uses this characteristic equation for $-i\boldsymbol{\theta} \cdot \boldsymbol{J}$ to show that the $3{\times}3$ real orthogonal matrix $\exp(-i\boldsymbol{\theta} \cdot \boldsymbol{J})$, which represents a right-handed rotation by θ radians about the axis $\hat{\boldsymbol{\theta}}$, is

$$\exp(-i\boldsymbol{\theta} \cdot \boldsymbol{J}) = \cos\theta\, I - i\hat{\boldsymbol{\theta}} \cdot \boldsymbol{J}\, \sin\theta + (1 - \cos\theta)\, \hat{\boldsymbol{\theta}} (\hat{\boldsymbol{\theta}})^{\mathsf{T}} \tag{1.271}$$

or

$$\exp(-i\boldsymbol{\theta} \cdot \boldsymbol{J})_{ij} = \delta_{ij}\cos\theta - \sin\theta\, \epsilon_{ijk}\hat{\theta}_k + (1 - \cos\theta)\, \hat{\theta}_i \hat{\theta}_j \tag{1.272}$$

in terms of indices. □

Direct use of the characteristic equation can become unwieldy for larger values of N. Fortunately, another trick is available if A is a nondefective square matrix, and if the power series (1.269) for $f(x)$ converges. For then A is related to its diagonal form $A^{(\mathrm{d})}$ by a similarity transformation (1.255), and we may define $f(A)$ as

$$f(A) = S f(A^{(\mathrm{d})}) S^{-1} \tag{1.273}$$

in which $f(A^{(\mathrm{d})})$ is the diagonal matrix with entries $f(a_\ell)$

$$f(A^{(d)}) = \begin{pmatrix} f(a_1) & 0 & 0 & \dots \\ 0 & f(a_2) & 0 & \dots \\ \vdots & \vdots & \vdots & \vdots \\ 0 & 0 & \dots & f(a_N) \end{pmatrix}, \tag{1.274}$$

in which $a_1, a_2, \dots, a_N$ are the eigenvalues of the matrix A. This definition makes sense if $f(A)$ is a series in powers of A because then

$$f(A) = \sum_{n=0}^{\infty} c_n A^n = \sum_{n=0}^{\infty} c_n \left(S A^{(d)} S^{-1} \right)^n. \tag{1.275}$$

So since $S^{-1} S = I$, we have $\left(S A^{(d)} S^{-1}\right)^n = S \left(A^{(d)}\right)^n S^{-1}$ and thus

$$f(A) = S \left[\sum_{n=0}^{\infty} c_n \left(A^{(d)} \right)^n \right] S^{-1} = S f(A^{(d)}) S^{-1}, \tag{1.276}$$

which is (1.273).

Example 1.37 (The time-evolution operator) In quantum mechanics, the time-evolution operator is the exponential $\exp(-iHt/\hbar)$ where $H = H^\dagger$ is a hermitian linear operator, the hamiltonian (William Rowan Hamilton, 1805–1865), and $\hbar = h/(2\pi) = 1.054 \times 10^{-34}$ Js where h is constant (Max Planck, 1858–1947). As we'll see in the next section, hermitian operators are never defective, so H can be diagonalized by a similarity transformation

$$H = S H^{(d)} S^{-1}. \tag{1.277}$$

The diagonal elements of the diagonal matrix $H^{(d)}$ are the **energies** E_ℓ of the states of the system described by the hamiltonian H. The time-evolution operator $U(t)$ then is

$$U(t) = S \exp(-iH^{(d)} t/\hbar) S^{-1}. \tag{1.278}$$

For a three-state system with angular frequencies $\omega_i = E_i/\hbar$, it is

$$U(t) = S \begin{pmatrix} e^{-i\omega_1 t} & 0 & 0 \\ 0 & e^{-i\omega_2 t} & \\ 0 & 0 & e^{-i\omega_3 t} \end{pmatrix} S^{-1} \tag{1.279}$$

in which the angular frequencies are $\omega_\ell = E_\ell/\hbar$. □

Example 1.38 (Entropy) The **entropy** S of a system described by a density operator ρ is the trace

$$S = -k \operatorname{Tr}(\rho \ln \rho) \tag{1.280}$$

in which $k = 1.38 \times 10^{-23}$ J/K is the constant named after Ludwig Boltzmann (1844–1906). The density operator ρ is hermitian, nonnegative, and of unit trace.

Since ρ is hermitian, the matrix that represents it is never defective (section 1.28), and so it can be diagonalized by a similarity transformation $\rho = S\,\rho^{(d)}\,S^{-1}$. By (1.24), $\mathrm{Tr}ABC = \mathrm{Tr}BCA$, so we can write S as

$$S = -k\mathrm{Tr}\left(S\,\rho^{(d)}\,S^{-1}\,S\,\ln(\rho^{(d)})\,S^{-1}\right) = -k\mathrm{Tr}\left(\rho^{(d)}\,\ln(\rho^{(d)})\right). \tag{1.281}$$

A vanishing eigenvalue $\rho_k^{(d)} = 0$ contributes nothing to this trace since $\lim_{x\to 0} x \ln x = 0$. If the system has three states, populated with probabilities ρ_i, the elements of $\rho^{(d)}$, then the sum

$$\begin{aligned} S &= -k\left(\rho_1 \ln \rho_1 + \rho_2 \ln \rho_2 + \rho_3 \ln \rho_3\right) \\ &= k\left[\rho_1 \ln\left(1/\rho_1\right) + \rho_2 \ln\left(1/\rho_2\right) + \rho_3 \ln\left(1/\rho_3\right)\right] \end{aligned} \tag{1.282}$$

is its entropy. □

1.28 Hermitian matrices

Hermitian matrices have very nice properties. By definition (1.30), a hermitian matrix A is square and unchanged by hermitian conjugation $A^\dagger = A$. Since it is square, the results of section 1.25 ensure that an $N \times N$ hermitian matrix A has N eigenvectors $|n\rangle$ with eigenvalues a_n

$$A|n\rangle = a_n|n\rangle. \tag{1.283}$$

In fact, all its eigenvalues are real. To see why, we take the adjoint

$$\langle n|A^\dagger = a_n^*\langle n| \tag{1.284}$$

and use the property $A^\dagger = A$ to find

$$\langle n|A^\dagger = \langle n|A = a_n^*\langle n|. \tag{1.285}$$

We now form the inner product of both sides of this equation with the ket $|n\rangle$ and use the eigenvalue equation (1.283) to get

$$\langle n|A|n\rangle = a_n\langle n|n\rangle = a_n^*\langle n|n\rangle, \tag{1.286}$$

which (since $\langle n|n\rangle > 0$) tells us that the eigenvalues are real

$$a_n^* = a_n. \tag{1.287}$$

Since $A^\dagger = A$, the matrix elements of A between two of its eigenvectors satisfy

$$a_m^*\langle m|n\rangle = \left(a_m\langle n|m\rangle\right)^* = \langle n|A|m\rangle^* = \langle m|A^\dagger|n\rangle = \langle m|A|n\rangle = a_n\langle m|n\rangle, \tag{1.288}$$

which implies that

$$\left(a_m^* - a_n\right)\langle m|n\rangle = 0. \tag{1.289}$$

But by (1.287), the eigenvalues a_m are real, and so we have

$$(a_m - a_n) \langle m|n\rangle = 0, \tag{1.290}$$

which tells us that when the eigenvalues are different, the eigenvectors are orthogonal. In the absence of a symmetry, all n eigenvalues usually are different, and so the eigenvectors usually are mutually orthogonal.

When two or more eigenvectors $|n_\alpha\rangle$ of a hermitian matrix have the same eigenvalue a_n, their eigenvalues are said to be **degenerate**. In this case, any linear combination of the degenerate eigenvectors also will be an eigenvector with the same eigenvalue a_n

$$A\left(\sum_{\alpha\in D} c_\alpha |n_\alpha\rangle\right) = a_n \left(\sum_{\alpha\in D} c_\alpha |n_\alpha\rangle\right) \tag{1.291}$$

where D is the set of labels α of the eigenvectors with the same eigenvalue. If the degenerate eigenvectors $|n_\alpha\rangle$ are linearly independent, then we may use the Gramm–Schmidt procedure (1.108–1.118) to choose the coefficients c_α so as to construct degenerate eigenvectors that are orthogonal to each other and to the nondegenerate eigenvectors. We then may normalize these mutually orthogonal eigenvectors.

But two related questions arise. Are the degenerate eigenvectors $|n_\alpha\rangle$ linearly independent? And if so, what orthonormal linear combinations of them should we choose for a given physical problem? Let's consider the second question first.

We know (section 1.16) that unitary transformations preserve the orthonormality of a basis. Any unitary transformation that commutes with the matrix A

$$[A, U] = 0 \tag{1.292}$$

maps each set of orthonormal degenerate eigenvectors of A into another set of orthonormal degenerate eigenvectors of A with the same eigenvalue because

$$AU|n_\alpha\rangle = UA|n_\alpha\rangle = a_n\, U|n_\alpha\rangle. \tag{1.293}$$

So there's a huge spectrum of choices for the orthonormal degenerate eigenvectors of A with the same eigenvalue. What is the right set for a given physical problem?

A sensible way to proceed is to add to the matrix A a second hermitian matrix B multiplied by a tiny, real scale factor ϵ

$$A(\epsilon) = A + \epsilon B. \tag{1.294}$$

The matrix B must completely break whatever symmetry led to the degeneracy in the eigenvalues of A. Ideally, the matrix B should be one that represents a modification of A that is physically plausible and relevant to the problem at hand. The hermitian matrix $A(\epsilon)$ then will have N different eigenvalues $a_n(\epsilon)$ and N orthonormal nondegenerate eigenvectors

$$A(\epsilon)|n_\beta, \epsilon\rangle = a_{n_\beta}(\epsilon)|n_\beta, \epsilon\rangle. \tag{1.295}$$

These eigenvectors $|n_\beta, \epsilon\rangle$ of $A(\epsilon)$ are orthogonal to each other

$$\langle n_\beta, \epsilon|n_{\beta'}, \epsilon\rangle = \delta_{\beta,\beta'} \tag{1.296}$$

and to the eigenvectors of $A(\epsilon)$ with other eigenvalues, and they remain so as we take the limit

$$|n_\beta\rangle = \lim_{\epsilon\to 0} |n_\beta, \epsilon\rangle. \tag{1.297}$$

We may choose them as the orthogonal degenerate eigenvectors of A. Since one always may find a crooked hermitian matrix B that breaks any particular symmetry, it follows that every $N \times N$ hermitian matrix A possesses N orthonormal eigenvectors, which are complete in the vector space in which A acts. (Any N linearly independent vectors span their N-dimensional vector space, as explained in section 1.9.)

Now let's return to the first question and again show that an $N \times N$ hermitian matrix has N orthogonal eigenvectors. To do this, we'll first show that the space of vectors orthogonal to an eigenvector $|n\rangle$ of a hermitian operator A

$$A|n\rangle = \lambda|n\rangle \tag{1.298}$$

is **invariant** under the action of A – that is, $\langle n|y\rangle = 0$ implies $\langle n|A|y\rangle = 0$. We use successively the definition of $A^\dagger$, the hermiticity of A, the eigenvector equation (1.298), the definition of the inner product, and the reality of the eigenvalues of a hermitian matrix:

$$\langle n|A|y\rangle = \langle A^\dagger n|y\rangle = \langle An|y\rangle = \langle \lambda n|y\rangle = \bar{\lambda}\langle n|y\rangle = \lambda\langle n|y\rangle = 0. \tag{1.299}$$

Thus the space of vectors orthogonal to an eigenvector of a hermitian operator A is invariant under the action of that operator.

Now a hermitian operator A acting on an N-dimensional vector space S is represented by an $N \times N$ hermitian matrix, and so it has at least one eigenvector $|1\rangle$. The subspace of S consisting of all vectors orthogonal to $|1\rangle$ is an $(N-1)$-dimensional vector space S_{N-1} that is invariant under the action of A. On this space S_{N-1}, the operator A is represented by an $(N-1) \times (N-1)$ hermitian matrix A_{N-1}. This matrix has at least one eigenvector $|2\rangle$. The subspace of S_{N-1} consisting of all vectors orthogonal to $|2\rangle$ is an $(N-2)$-dimensional vector space S_{N-2} that is invariant under the action of A. On S_{N-2}, the operator A is represented by an $(N-2) \times (N-2)$ hermitian matrix A_{N-2}, which has at least one eigenvector $|3\rangle$. By construction, the vectors $|1\rangle$, $|2\rangle$, and $|3\rangle$ are mutually orthogonal. Continuing in this way, we see that A **has** N **orthogonal eigenvectors** $|k\rangle$ for $k = 1, 2, \ldots, N$. Thus no hermitian matrix is defective.

The N orthogonal eigenvectors $|k\rangle$ of an $N \times N$ matrix A can be normalized and used to write the $N \times N$ identity operator I as

$$I = \sum_{k=1}^{N} |k\rangle\langle k|. \tag{1.300}$$

On multiplying from the left by the matrix A, we find

$$A = AI = A \sum_{k=1}^{N} |k\rangle\langle k| = \sum_{k=1}^{N} a_k |k\rangle\langle k|, \tag{1.301}$$

which is the diagonal form of the hermitian matrix A. This expansion of A as a sum over outer products of its eigenstates multiplied by their eigenvalues exhibits the possible values a_k of the physical quantity represented by the matrix A when selective, nondestructive measurements $|k\rangle\langle k|$ of the quantity A are done.

The hermitian matrix A is diagonal in the basis of its eigenstates $|k\rangle$

$$A_{kj} = \langle k|A|j\rangle = a_k \delta_{kj}. \tag{1.302}$$

But in any other basis $|\alpha_k\rangle$, the matrix A appears as

$$A_{k\ell} = \langle \alpha_k|A|\alpha_\ell\rangle = \sum_{n=1}^{N} \langle \alpha_k|n\rangle a_n \langle n|\alpha_\ell\rangle. \tag{1.303}$$

The unitary matrix $U_{kn} = \langle \alpha_k|n\rangle$ relates the matrix $A_{k\ell}$ in an arbitrary basis to its diagonal form $A = UA^{(d)}U^\dagger$ in which $A^{(d)}$ is the diagonal matrix $A^{(d)}_{nm} = a_n\,\delta_{nm}$. An arbitrary $N \times N$ hermitian matrix A can be diagonalized by a unitary transformation.

A matrix that is **real and symmetric** is hermitian; so is one that is **imaginary and antisymmetric**. A real, symmetric matrix R can be diagonalized by an **orthogonal transformation**

$$R = O\,R^{(d)}O^{\mathsf{T}} \tag{1.304}$$

in which the matrix O is a real unitary matrix, that is, an orthogonal matrix (1.168).

Example 1.39 (The seesaw mechanism) Suppose we wish to find the eigenvalues of the real, symmetric mass matrix

$$\mathcal{M} = \begin{pmatrix} 0 & m \\ m & M \end{pmatrix} \tag{1.305}$$

in which m is an ordinary mass and M is a huge mass. The eigenvalues μ of this hermitian mass matrix satisfy $\det(\mathcal{M} - \mu I) = \mu(\mu - M) - m^2 = 0$ with solutions $\mu_\pm = \left(M \pm \sqrt{M^2 + 4m^2}\right)/2$. The larger mass $\mu_+ \approx M + m^2/M$ is approximately the huge mass M and the smaller mass $\mu_- \approx -m^2/M$ is very tiny. The physical mass of a fermion is the absolute value of its mass parameter, here m^2/M.

The product of the two eigenvalues is the constant $\mu_+\mu_- = \det \mathcal{M} = -m^2$ so as μ_- goes down, μ_+ must go up. In 1975, Gell-Mann, Ramond, Slansky, and Jerry Stephenson invented this "**seesaw**" mechanism as an explanation of why neutrinos have such small masses, less than 1 eV/c^2. If $mc^2 = 10$ MeV, and $\mu_- c^2 \approx 0.01$ eV, which is a plausible light-neutrino mass, then the rest energy of the huge mass would be $Mc^2 = 10^7$ GeV. This huge mass would point at new physics, beyond the standard model. Yet the small masses of the neutrinos may be related to the weakness of their interactions. □

If we return to the orthogonal transformation (1.304) and multiply column ℓ of the matrix O and row ℓ of the matrix O^{T} by $\sqrt{|R^{(d)}_\ell|}$, then we arrive at the **congruency transformation** of Sylvester's theorem

$$R = C\,\hat{R}^{(d)}C^{\mathsf{T}} \tag{1.306}$$

in which the diagonal entries $\hat{R}^{(d)}_\ell$ are either ± 1 or 0 because the matrices C and C^{T} have absorbed the moduli $|R^{(d)}_\ell|$.

Example 1.40 (Equivalence principle) If G is a real, symmetric 4×4 matrix then there's a real 4×4 matrix $D = C^{\mathsf{T}-1}$ such that

$$G_d = D^{\mathsf{T}} G D = \begin{pmatrix} g_1 & 0 & 0 & 0 \\ 0 & g_2 & 0 & 0 \\ 0 & 0 & g_3 & 0 \\ 0 & 0 & 0 & g_4 \end{pmatrix} \tag{1.307}$$

in which the diagonal entries g_i are ± 1 or 0. Thus there's a real 4×4 matrix D that casts the real nonsingular symmetric metric g_{ik} of space-time at any given point into the diagonal metric $\eta_{j\ell}$ of flat space-time by the congruence

$$g_d = D^{\mathsf{T}} g D = \begin{pmatrix} -1 & 0 & 0 & 0 \\ 0 & 1 & 0 & 0 \\ 0 & 0 & 1 & 0 \\ 0 & 0 & 0 & 1 \end{pmatrix} = \eta. \tag{1.308}$$

Usually one needs different Ds at different points. Since one can implement the congruence by changing coordinates, it follows that in any gravitational field, one may choose free-fall coordinates in which all physical laws take the same form as in special relativity without acceleration or gravitation at least over suitably small volumes of space-time (section 11.39). □

1.29 Normal matrices

The largest set of matrices that can be diagonalized by a unitary transformation is the set of **normal** matrices. These are square matrices that commute with their adjoints

$$[A, A^\dagger] = AA^\dagger - A^\dagger A = 0. \tag{1.309}$$

This broad class of matrices includes not only hermitian matrices but also unitary matrices since

$$[U, U^\dagger] = UU^\dagger - U^\dagger U = I - I = 0. \tag{1.310}$$

To see why a normal matrix can be diagonalized by a unitary transformation, let us consider an $N \times N$ normal matrix V which (since it is square (section 1.25)) has N eigenvectors $|n\rangle$ with eigenvalues v_n

$$(V - v_n I)\,|n\rangle = 0. \tag{1.311}$$

The square of the norm (1.80) of this vector must vanish

$$\|\, (V - v_n I)\,|n\rangle \,\|^2 = \langle n|\,(V - v_n I)^\dagger\,(V - v_n I)\,|n\rangle = 0. \tag{1.312}$$

But since V is normal, we also have

$$\langle n|\,(V - v_n I)^\dagger\,(V - v_n I)\,|n\rangle = \langle n|\,(V - v_n I)\,(V - v_n I)^\dagger\,|n\rangle. \tag{1.313}$$

So the square of the norm of the vector $\left(V^\dagger - v_n^* I\right)|n\rangle = (V - v_n I)^\dagger\,|n\rangle$ also vanishes $\|\,\left(V^\dagger - v_n^* I\right)|n\rangle\,\|^2 = 0$, which tells us that $|n\rangle$ also is an eigenvector of $V^\dagger$ with eigenvalue v_n^*

$$V^\dagger|n\rangle = v_n^*|n\rangle \quad \text{and so} \quad \langle n|V = v_n\langle n|. \tag{1.314}$$

If now $|m\rangle$ is an eigenvector of V with eigenvalue v_m

$$V|m\rangle = v_m|m\rangle = 0 \tag{1.315}$$

then we have

$$\langle n|V|m\rangle = v_m\langle n|m\rangle \tag{1.316}$$

and from (1.314)

$$\langle n|V|m\rangle = v_n\langle n|m\rangle. \tag{1.317}$$

Subtracting (1.316) from (1.317), we get

$$(v_n - v_m)\,\langle m|n\rangle = 0, \tag{1.318}$$

which shows that **any two eigenvectors of a normal matrix V with different eigenvalues are orthogonal.**

Usually, all N eigenvalues of an $N \times N$ normal matrix are different. In this case, all the eigenvectors are orthogonal and may be individually normalized.

But even when a set D of eigenvectors has the same (degenerate) eigenvalue, one may use the argument (1.291–1.297) to find a suitable set of orthonormal eigenvectors with that eigenvalue. Thus **every $N \times N$ normal matrix has N orthonormal eigenvectors.** It follows then from the argument of equations (1.300–1.303) that every $N \times N$ normal matrix V can be diagonalized by an $N \times N$ unitary matrix U

$$V = UV^{(d)}U^{\dagger} \tag{1.319}$$

whose nth column $U_{kn} = \langle \alpha_k | n \rangle$ is the eigenvector $|n\rangle$ in the arbitrary basis $|\alpha_k\rangle$ of the matrix $V_{k\ell} = \langle \alpha_k | V | \alpha_\ell \rangle$ as in (1.303).

Since the eigenstates $|n\rangle$ of a normal matrix A

$$A|n\rangle = a_n|n\rangle \tag{1.320}$$

are complete and orthonormal, we can write the identity operator I as

$$I = \sum_{n=1}^{N} |n\rangle\langle n|. \tag{1.321}$$

The product AI is A itself, so

$$A = AI = A \sum_{n=1}^{N} |n\rangle\langle n| = \sum_{n=1}^{N} a_n \, |n\rangle\langle n|. \tag{1.322}$$

It follows therefore that if f is a function, then $f(A)$ is

$$f(A) = \sum_{n=1}^{N} f(a_n) \, |n\rangle\langle n|, \tag{1.323}$$

which is simpler than the expression (1.273) for an arbitrary nondefective matrix. This is a good way to think about functions of normal matrices.

Example 1.41 How do we handle the operator $\exp(-iHt/\hbar)$ that translates states in time by t? The hamiltonian H is hermitian and so is normal. Its orthonormal eigenstates $|n\rangle$ are the energy levels E_n

$$H|n\rangle = E_n|n\rangle. \tag{1.324}$$

So we apply (1.323) with $A \to H$ and get

$$e^{-iHt/\hbar} = \sum_{n=1}^{N} e^{-iE_n t/\hbar} \, |n\rangle\langle n|, \tag{1.325}$$

which lets us compute the time evolution of any state $|\psi\rangle$ as

$$e^{-iHt/\hbar}|\psi\rangle = \sum_{n=1}^{N} e^{-iE_n t/\hbar}\,|n\rangle\langle n|\psi\rangle \tag{1.326}$$

if we know the eigenstates $|n\rangle$ and eigenvalues E_n of the hamiltonian H. □

The determinant $|V|$ of a normal matrix V satisfies the identities

$$|V| = \exp\left[\mathrm{Tr}(\ln V)\right], \quad \ln|V| = \mathrm{Tr}(\ln V), \quad \text{and} \quad \delta \ln|V| = \mathrm{Tr}\left(V^{-1}\delta V\right). \tag{1.327}$$

1.30 Compatible normal matrices

Two normal matrices A and B that **commute**

$$[A, B] \equiv AB - BA = 0 \tag{1.328}$$

are said to be **compatible**. Since these operators are normal, they have complete sets of orthonormal eigenvectors. If $|u\rangle$ is an eigenvector of A with eigenvalue z, then so is $B|u\rangle$ since

$$AB|u\rangle = BA|u\rangle = Bz|u\rangle = z\,B|u\rangle. \tag{1.329}$$

We have seen that any normal matrix A can be written as a sum (1.322) of outer products

$$A = \sum_{n=1}^{N} |a_n\rangle a_n \langle a_n| \tag{1.330}$$

of its orthonormal eigenvectors $|a_n\rangle$, which are complete in the N-dimensional vector space $\mathcal{S}$ on which A acts. Suppose now that the eigenvalues a_n of A are nondegenerate, and that B is another normal matrix acting on $\mathcal{S}$ and that the matrices A and B are compatible. Then in the basis provided by the eigenvectors (or eigenstates) $|a_n\rangle$ of the matrix A, the matrix B must satisfy

$$0 = \langle a_n|AB - BA|a_k\rangle = (a_n - a_k)\,\langle a_n|B|a_k\rangle, \tag{1.331}$$

which says that $\langle a_n|B|a_k\rangle$ is zero unless $a_n = a_k$. Thus if the eigenvalues a_n of the operator A are nondegenerate, then the operator B is diagonal

$$B = IBI = \sum_{n=1}^{N} |a_n\rangle\langle a_n|B \sum_{k=1}^{N} |a_k\rangle\langle a_k| = \sum_{n=1}^{N} |a_n\rangle\langle a_n|B|a_n\rangle\langle a_n| \tag{1.332}$$

in the $|a_n\rangle$ basis. Moreover B maps each eigenket $|a_k\rangle$ of A into

$$B|a_k\rangle = \sum_{n=1}^{N} |a_n\rangle\langle a_n|B|a_n\rangle\langle a_n|a_k\rangle = \sum_{n=1}^{N} |a_n\rangle\langle a_n|B|a_n\rangle\delta_{nk} = \langle a_k|B|a_k\rangle|a_k\rangle, \tag{1.333}$$

which says that each eigenvector $|a_k\rangle$ of the matrix A also is an eigenvector of the matrix B with eigenvalue $\langle a_k|B|a_k\rangle$. Thus **two compatible normal matrices can be simultaneously diagonalized** if one of them has nondegenerate eigenvalues.

If A's eigenvalues a_n are degenerate, each eigenvalue a_n may have d_n orthonormal eigenvectors $|a_n, k\rangle$ for $k = 1, \ldots, d_n$. In this case, the matrix elements $\langle a_n, k|B|a_m, k'\rangle$ of B are zero unless the eigenvalues are the same, $a_n = a_m$. The matrix representing the operator B in this basis consists of square, $d_n \times d_n$, normal submatrices $\langle a_n, k|B|a_n, k'\rangle$ arranged along its main diagonal; it is said to be in **block-diagonal form**. Since each submatrix is a $d_n \times d_n$ normal matrix, we may find linear combinations $|a_n, b_k\rangle$ of the degenerate eigenvectors $|a_n, k\rangle$ that are orthonormal eigenvectors of both compatible operators

$$A|a_n, b_k\rangle = a_n|a_n, b_k\rangle \quad \text{and} \quad B|a_n, b_k\rangle = b_k|a_n, b_k\rangle. \tag{1.334}$$

Thus one can simultaneously diagonalize any two compatible operators.

The converse also is true: if the operators A and B can be simultaneously diagonalized as in (1.334), then they commute

$$AB|a_n, b_k\rangle = Ab_k|a_n, b_k\rangle = a_n b_k|a_n, b_k\rangle = a_n B|a_n, b_k\rangle = BA|a_n, b_k\rangle$$

and so are compatible. Normal matrices can be simultaneously diagonalized if and only if they are compatible, that is, if and only if they commute.

In quantum mechanics, compatible hermitian operators represent physical observables that can be measured simultaneously to arbitrary precision (in principle). A set of compatible hermitian operators $\{A, B, C, \ldots\}$ is said to be **complete** if to every set of eigenvalues $\{a_n, b_k, c_\ell, \ldots\}$ there is only a single eigenvector $|a_n, b_k, c_\ell, \ldots\rangle$.

Example 1.42 (Compatible photon observables) The state of a photon is completely characterized by its momentum and its angular momentum about its direction of motion. For a photon, the momentum operator $\boldsymbol{P}$ and the dot-product $\boldsymbol{J} \cdot \boldsymbol{P}$ of the angular momentum $\boldsymbol{J}$ with the momentum form a complete set of compatible hermitian observables. Incidentally, because its mass is zero, the angular momentum $\boldsymbol{J}$ of a photon about its direction of motion can have only two values $\pm\hbar$, which correspond to its two possible states of circular polarization. □

Example 1.43 (Thermal density operator) A **density operator** ρ is the most general description of a quantum-mechanical system. It is hermitian, positive, and of unit trace. Since it is hermitian, it can be diagonalized (section 1.28)

$$\rho = \sum_n |n\rangle\langle n|\rho|n\rangle\langle n| \tag{1.335}$$

and its eigenvalues $\rho_n = \langle n|\rho|n\rangle$ are real. Each ρ_n is the probability that the system is in the state $|n\rangle$ and so is nonnegative. The unit-trace rule

$$\sum_n \rho_n = 1 \tag{1.336}$$

ensures that these probabilities add up to one – the system is in some state.

The mean value of an operator F is the trace, $\langle F\rangle = \text{Tr}(\rho F)$. So the average energy E is the trace, $E = \langle H\rangle = \text{Tr}(\rho H)$. The **entropy operator** S is the negative logarithm of the density operator multiplied by Boltzmann's constant $S = -k\ln\rho$, and the mean entropy S is $S = \langle S\rangle = -k\text{Tr}(\rho\ln\rho)$.

A density operator that describes a system in thermal equilibrium at a constant temperature T is time independent and so commutes with the hamiltonian, $[\rho, H] = 0$. Since ρ and H commute, they are compatible operators (1.328), and so they can be simultaneously diagonalized. Each eigenstate $|n\rangle$ of ρ is an eigenstate of H; its energy E_n is its eigenvalue, $H|n\rangle = E_n|n\rangle$.

If we have no information about the state of the system other than its mean energy E, then we take ρ to be the density operator that maximizes the mean entropy S while respecting the constraints

$$c_1 = \sum_n \rho_n - 1 = 0 \quad \text{and} \quad c_2 = \text{Tr}(\rho H) - E = 0. \tag{1.337}$$

We introduce two Lagrange multipliers (section 1.23) and maximize the unconstrained function

$$\begin{aligned} L(\rho, \lambda_1, \lambda_2) &= S - \lambda_1\, c_1 - \lambda_2\, c_2 \\ &= -k\sum_n \rho_n \ln\rho_n - \lambda_1\left[\sum_n \rho_n - 1\right] - \lambda_2\left[\sum_n \rho_n E_n - E\right] \end{aligned} \tag{1.338}$$

by setting its derivatives with respect to ρ_n, λ_1, and λ_2 equal to zero

$$\frac{\partial L}{\partial \rho_n} = -k\,(\ln\rho_n + 1) - \lambda_1 - \lambda_2 E_n = 0, \tag{1.339}$$

$$\frac{\partial L}{\partial \lambda_1} = \sum_n \rho_n - 1 = 0, \tag{1.340}$$

$$\frac{\partial L}{\partial \lambda_2} = \sum_n \rho_n E_n - E = 0. \tag{1.341}$$

The first (1.339) of these conditions implies that

$$\rho_n = \exp\left[-(\lambda_1 + \lambda_2 E_n + k)/k\right]. \tag{1.342}$$

We satisfy the second condition (1.340) by choosing λ_1 so that

$$\rho_n = \frac{\exp(-\lambda_2 E_n/k)}{\sum_n \exp(-\lambda_2 E_n/k)}. \tag{1.343}$$

Setting $\lambda_2 = 1/T$, we define the temperature T so that ρ satisfies the third condition (1.341). Its eigenvalue ρ_n then is

$$\rho_n = \frac{\exp(-E_n/kT)}{\sum_n \exp(-E_n/kT)}. \tag{1.344}$$

In terms of the inverse temperature $\beta \equiv 1/(kT)$, the density operator is

$$\rho = \frac{e^{-\beta H}}{\mathrm{Tr}\left(e^{-\beta H}\right)}, \tag{1.345}$$

which is the **Boltzmann distribution**. □

1.31 The singular-value decomposition

Every complex $M \times N$ rectangular matrix A is the product of an $M \times M$ unitary matrix U, an $M \times N$ rectangular matrix Σ that is zero except on its main diagonal, which consists of its nonnegative singular values S_k, and an $N \times N$ unitary matrix $V^\dagger$

$$A = U\,\Sigma\,V^\dagger. \tag{1.346}$$

This singular-value decomposition (SVD) is a key theorem of matrix algebra.

Suppose A is a linear operator that maps vectors in an N-dimensional vector space V_N into vectors in an M-dimensional vector space V_M. The spaces V_N and V_M will have infinitely many orthonormal bases $\{|n,a\rangle \in V_N\}$ and $\{|m,b\rangle \in V_M\}$ labeled by continuous parameters a and b. Each pair of bases provides a resolution of the identity operator I_N for V_N and I_M for V_M

$$I_N = \sum_{n=1}^{N} |n,a\rangle\langle n,a| \quad \text{and} \quad I_M = \sum_{m=1}^{M} |m,b\rangle\langle m,b|. \tag{1.347}$$

These identity operators give us many ways of writing the linear operator A

$$A = I_M A I_N = \sum_{m=1}^{M}\sum_{n=1}^{N} |m,b\rangle\langle m,b|A|n,a\rangle\langle n,a|, \tag{1.348}$$

in which the $\langle m,b|A|n,a\rangle$ are the elements of a complex $M \times N$ matrix. The singular-value decomposition of the linear operator A is a choice among all these expressions for I_N and I_M that expresses A as

$$A = \sum_{k=1}^{\min(M,N)} |U_k\rangle S_k \langle V_k| \tag{1.349}$$

in which the $\min(M, N)$ singular values S_k are nonnegative

$$S_k \geq 0. \tag{1.350}$$

Let's use the notation $|An\rangle \equiv A|n\rangle$ for the image of a vector $|n\rangle$ in an orthonormal basis $\{|n\rangle\}$ of V_N under the map A. We seek a special orthonormal basis $\{|n\rangle\}$ of V_N that has the property that the vectors $|An\rangle$ are orthogonal. This special basis $\{|n\rangle\}$ of V_N is the set of N orthonormal eigenstates of the $N \times N$ (nonnegative) hermitian operator $A^\dagger A$

$$A^\dagger A|n\rangle = e_n|n\rangle. \tag{1.351}$$

For since $A|n'\rangle = |An'\rangle$ and $A^\dagger A|n\rangle = e_n|n\rangle$, it follows that

$$\langle An'|An\rangle = \langle n'|A^\dagger A|n\rangle = e_n\langle n'|n\rangle = e_n\delta_{n'n}, \tag{1.352}$$

which shows that the vectors $|An\rangle$ are orthogonal and that their eigenvalues $e_n = \langle An|An\rangle$ are nonnegative. This is the essence of the singular-value decomposition.

If $N = M$, so that matrices $\langle m, b|A|n, a\rangle$ representing the linear operator A are square, then the $N = M$ singular values S_n are the nonnegative square-roots of the eigenvalues e_n

$$S_n = \sqrt{e_n} = \sqrt{\langle An|An\rangle} \geq 0. \tag{1.353}$$

We therefore may normalize each vector $|An\rangle$ whose singular value S_n is positive as

$$|m_n\rangle = \frac{1}{S_n}|An\rangle \quad \text{for} \quad S_n > 0 \tag{1.354}$$

so that the vectors $\{|m_n\rangle\}$ with positive singular values are orthonormal

$$\langle m_{n'}|m_n\rangle = \delta_{n',n}. \tag{1.355}$$

If only $P < N$ of the singular values are positive, then we may augment this set of P vectors $\{|m_n\rangle\}$ with $N - P = M - P$ new normalized vectors $|m_{n'}\rangle$ that are orthogonal to each other and to the P vectors defined by (1.354) (with positive singular values $S_n > 0$) so that the set of $N = M$ vectors $\{|m_n\rangle, |m_{n'}\}\rangle$ are complete and orthonormal in the space $V_{M=N}$.

If $N > M$, then A maps the N-dimensional space V_N into the smaller M-dimensional space V_M, and so A must annihilate $N - M$ basis vectors

$$A|n'\rangle = 0 \quad \text{for} \quad M < n' \leq N. \tag{1.356}$$

In this case, there are only M singular values S_n of which Z may be zero. The Z vectors $|An\rangle = A|n\rangle$ with vanishing S_ns are vectors of length zero; for these values of n, the matrix A maps the vector $|n\rangle$ to the zero vector. If there are more than $N - M$ zero-length vectors $|An\rangle = A|n\rangle$, then we must replace the extra ones by new normalized vectors $|m_{n'}\rangle$ that are orthogonal to each other and to the vectors defined by (1.354) so that we have M orthonormal vectors in the augmented set $\{|m_n\rangle, |m_{n'}\rangle\}$. These vectors then form a basis for V_M.

When $N \le M$, there are only N singular values S_n of which Z may be zero. If Z of the S_ns vanish, then one must add $Q = Z + M - N$ new normalized vectors $|m_{n'}\rangle$ that are orthogonal to each other and to the vectors defined by (1.354)

$$\langle m_{n'}|m_n\rangle = \frac{1}{S_n}\langle m_{n'}|A|n\rangle = 0 \quad \text{for} \quad n' > N - Z \quad \text{and} \quad S_n > 0 \tag{1.357}$$

so that we have M orthonormal vectors in the augmented set $\{|m_n\rangle, |m_{n'}\rangle\}$. These vectors then form a basis for V_M.

In both cases, $N > M$ and $M \ge N$, there are $\min(M, N)$ singular values, Z of which may be zero. We may choose the new vectors $\{|m_{n'}\rangle\}$ arbitrarily – as long as the augmented set $\{|m_n\rangle, |m_{n'}\rangle\}$ includes all the vectors defined by (1.354) and forms an orthonormal basis for V_M.

We now have two special orthonormal bases: the N N-dimensional eigenvectors $|n\rangle \in V_N$ that satisfy (1.351) and the M M-dimensional vectors $|m_n\rangle \in V_M$. To make the singular-value decomposition of the linear operator A, we choose as the identity operators I_N for the N-dimensional space V_N and I_M for the M-dimensional space V_M the sums

$$I_N = \sum_{n=1}^{N} |n\rangle\langle n| \quad \text{and} \quad I_M = \sum_{n'=1}^{M} |m_{n'}\rangle\langle m_{n'}|. \tag{1.358}$$

The singular-value decomposition of A then is

$$A = I_M A I_N = \sum_{n'=1}^{M} |m_{n'}\rangle\langle m_{n'}|A \sum_{n=1}^{N} |n\rangle\langle n|. \tag{1.359}$$

There are $\min(M, N)$ singular values S_n, all nonnegative. For the positive singular values, equations (1.352 & 1.354) show that the matrix element $\langle m_{n'}|A|n\rangle$ vanishes unless $n' = n$

$$\langle m_{n'}|A|n\rangle = \frac{1}{S_{n'}}\langle An'|An\rangle = S_{n'}\,\delta_{n'n}. \tag{1.360}$$

For the Z vanishing singular values, equation (1.353) shows that $A|n\rangle = 0$ and so

$$\langle m_{n'}|A|n\rangle = 0. \tag{1.361}$$

Thus only the $\min(M, N) - Z$ singular values that are positive contribute to the singular-value decomposition (1.359). If $N > M$, then there can be at most M nonzero eigenvalues e_n. If $N \le M$, there can be at most N nonzero e_ns. The final form of the singular-value decomposition then is a sum of dyadics weighted by the positive singular values

$$A = \sum_{n=1}^{\min(M,N)} |m_n\rangle S_n \langle n| = \sum_{n=1}^{\min(M,N)-Z} |m_n\rangle S_n \langle n|. \tag{1.362}$$

The vectors $|m_n\rangle$ and $|n\rangle$ respectively are the left and right singular vectors. The nonnegative numbers S_n are the singular values.

The linear operator A maps the $\min(M, N)$ right singular vectors $|n\rangle$ into the $\min(M, N)$ left singular vectors $S_n|m_n\rangle$ scaled by their singular values

$$A|n\rangle = S_n|m_n\rangle \tag{1.363}$$

and its adjoint $A^\dagger$ maps the $\min(M, N)$ left singular vectors $|m_n\rangle$ into the $\min(M, N)$ right singular vectors $|n\rangle$ scaled by their singular values

$$A^\dagger|m_n\rangle = S_n|n\rangle. \tag{1.364}$$

The N-dimensional vector space V_N is the **domain** of the linear operator A. If $N > M$, then A annihilates $N - M + Z$ of the basis vectors $|n\rangle$. The **null space** or **kernel** of A is the space spanned by the basis vectors $|n\rangle$ that A annihilates. The vector space spanned by the left singular vectors $|m_n\rangle$ with nonzero singular values $S_n > 0$ is the **range** or **image** of A. It follows from the singular-value decomposition (1.362) that the dimension N of the domain is equal to the dimension of the kernel $N - M + Z$ plus that of the range $M - Z$, a result called the **rank-nullity theorem**.

Incidentally, the vectors $|m_n\rangle$ are the eigenstates of the hermitian matrix $A\,A^\dagger$ as one may see from the explicit product of the expansion (1.362) with its adjoint

$$\begin{aligned} A\,A^\dagger &= \sum_{n=1}^{\min(M,N)} |m_n\rangle S_n \langle n| \sum_{n'=1}^{\min(M,N)} |n'\rangle S_{n'} \langle m_{n'}| \\ &= \sum_{n=1}^{\min(M,N)} \sum_{n'=1}^{\min(M,N)} |m_n\rangle S_n \, \delta_{nn'} S_{n'} \langle m_{n'}| \\ &= \sum_{n=1}^{\min(M,N)} |m_n\rangle S_n^2 \langle m_n|, \end{aligned} \tag{1.365}$$

which shows that $|m_n\rangle$ is an eigenvector of $A\,A^\dagger$ with eigenvalue $e_n = S_n^2$,

$$A\,A^\dagger|m_n\rangle = S_n^2|m_n\rangle. \tag{1.366}$$

The SVD expansion (1.362) usually is written as a product of three explicit matrices, $A = U\Sigma V^\dagger$. The middle matrix Σ is an $M \times N$ matrix with the $\min(M, N)$ singular values $S_n = \sqrt{e_n}$ on its main diagonal and zeros elsewhere.

By convention, one writes the S_n in decreasing order with the biggest S_n as entry Σ_{11}. The first matrix U and the third matrix $V^\dagger$ depend upon the bases one uses to represent the linear operator A. If these basis vectors are $|\alpha_k\rangle$ and $|\beta_\ell\rangle$, then

$$A_{k\ell} = \langle \alpha_k | A | \beta_\ell \rangle = \sum_{n=1}^{\min(M,N)} \langle \alpha_k | m_n \rangle S_n \langle n | \beta_\ell \rangle \tag{1.367}$$

so that the k, nth entry in the matrix U is $U_{kn} = \langle \alpha_k | m_n \rangle$. The columns of the matrix U are the left singular vectors of the matrix A:

$$\begin{pmatrix} U_{1n} \\ U_{2n} \\ \vdots \\ U_{Mn} \end{pmatrix} = \begin{pmatrix} \langle \alpha_1 | m_n \rangle \\ \langle \alpha_2 | m_n \rangle \\ \vdots \\ \langle \alpha_M | m_n \rangle \end{pmatrix}. \tag{1.368}$$

Similarly, the n, ℓth entry of the matrix $V^\dagger$ is $\left(V^\dagger\right)_{n,\ell} = \langle n | \beta_\ell \rangle$. Thus $V_{\ell,n} = \left(V^\top\right)_{n,\ell} = \langle n | \beta_\ell \rangle^* = \langle \beta_\ell | n \rangle$. The columns of the matrix V are the right singular vectors of the matrix A

$$\begin{pmatrix} V_{1n} \\ V_{2n} \\ \vdots \\ V_{Nn} \end{pmatrix} = \begin{pmatrix} \langle \beta_1 | n \rangle \\ \langle \beta_2 | n \rangle \\ \vdots \\ \langle \beta_N | n \rangle \end{pmatrix}. \tag{1.369}$$

Since the columns of U and of V respectively are M and N orthonormal vectors, both of these matrices are unitary, that is $U^\dagger U = I_M$ and $V^\dagger V = I_N$ are the $M \times M$ and $N \times N$ identity matrices. The matrix form of the singular-value decomposition of A then is

$$A_{k\ell} = \sum_{m=1}^{M} \sum_{n=1}^{N} U_{km} \Sigma_{mn} V^\dagger_{n\ell} = \sum_{n=1}^{\min(M,N)} U_{kn} S_n V^\dagger_{n\ell} \tag{1.370}$$

or in matrix notation

$$A = U \Sigma V^\dagger. \tag{1.371}$$

The usual statement of the SVD theorem is: Every $M \times N$ complex matrix A can be written as the matrix product of an $M \times M$ unitary matrix U, an $M \times N$ matrix Σ that is zero except for its $\min(M, N)$ nonnegative diagonal elements, and an $N \times N$ unitary matrix $V^\dagger$

$$A = U \, \Sigma \, V^\dagger. \tag{1.372}$$

The first $\min(M, N)$ diagonal elements of S are the singular values S_k. They are real and nonnegative. The first $\min(M, N)$ columns of U and V are the left and right singular vectors of A. The last $\max(N - M, 0) + Z$ columns (1.369) of the matrix V span the null space or kernel of A, and the first $\min(M, N) - Z$ columns (1.368) of the matrix U span the range of A.

Example 1.44 (Singular-value decomposition of a 2×3 matrix) If A is

$$A = \begin{pmatrix} 0 & 1 & 0 \\ 1 & 0 & 1 \end{pmatrix} \tag{1.373}$$

then the positive hermitian matrix $A^\dagger A$ is

$$A^\dagger A = \begin{pmatrix} 1 & 0 & 1 \\ 0 & 1 & 0 \\ 1 & 0 & 1 \end{pmatrix}. \tag{1.374}$$

The normalized eigenvectors and eigenvalues of $A^\dagger A$ are

$$|1\rangle = \frac{1}{\sqrt{2}} \begin{pmatrix} 1 \\ 0 \\ 1 \end{pmatrix}, \; e_1 = 2; \quad |2\rangle = \begin{pmatrix} 0 \\ 1 \\ 0 \end{pmatrix}, \; e_2 = 1; \quad |3\rangle = \frac{1}{\sqrt{2}} \begin{pmatrix} -1 \\ 0 \\ 1 \end{pmatrix}, \; e_3 = 0. \tag{1.375}$$

The third eigenvalue e_3 had to vanish because A is a 3×2 matrix.

The vector $A|1\rangle$ is (as a row vector) $|A1\rangle = A|1\rangle = (0, \sqrt{2})$, and its norm is $\sqrt{\langle 1|A^\dagger A|1\rangle} = \sqrt{2}$, so the normalized vector $|m_1\rangle$ is $|m_1\rangle = |A1\rangle/\sqrt{2} = (0, 1)$. Similarly, the vector $|m_2\rangle$ is $|m_2\rangle = A|2\rangle/\sqrt{\langle 2|A^\dagger A|2\rangle} = (1, 0)$. The SVD of A then is

$$A = \sum_{n=1}^{2} |m_n\rangle S_n \langle n| = U \Sigma V^\dagger \tag{1.376}$$

where $S_n = \sqrt{e_n}$. The unitary matrices $U_{k,n} = \langle \alpha_k | m_n \rangle$ and $V_{k,n} = \langle \beta_k | n \rangle$ are

$$U = \begin{pmatrix} 0 & 1 \\ 1 & 0 \end{pmatrix} \quad \text{and} \quad V = \frac{1}{\sqrt{2}} \begin{pmatrix} 1 & 0 & -1 \\ 0 & \sqrt{2} & 0 \\ 1 & 0 & 1 \end{pmatrix} \tag{1.377}$$

and the diagonal matrix Σ is

$$\Sigma = \begin{pmatrix} \sqrt{2} & 0 & 0 \\ 0 & 1 & 0 \end{pmatrix}. \tag{1.378}$$

So finally the SVD of $A = U \Sigma V^\dagger$ is

$$A = \begin{pmatrix} 0 & 1 \\ 1 & 0 \end{pmatrix} \begin{pmatrix} \sqrt{2} & 0 & 0 \\ 0 & 1 & 0 \end{pmatrix} \frac{1}{\sqrt{2}} \begin{pmatrix} 1 & 0 & 1 \\ 0 & \sqrt{2} & 0 \\ -1 & 0 & 1 \end{pmatrix}. \tag{1.379}$$

The null space or kernel of A is the set of vectors that are real multiples c

$$N_A = \frac{c}{\sqrt{2}} \begin{pmatrix} -1 \\ 0 \\ 1 \end{pmatrix} \tag{1.380}$$

of the third column of the matrix V displayed in (1.377). □

Example 1.45 (Matlab's SVD) Matlab's command [U,S,V] = svd(X) performs the singular-value decomposition of the matrix X. For instance

```
>> X = rand(3,3) + i*rand(3,3)

      0.6551 + 0.2551i  0.4984 + 0.8909i   0.5853 + 0.1386i
  X = 0.1626 + 0.5060i  0.9597 + 0.9593i   0.2238 + 0.1493i
      0.1190 + 0.6991i  0.3404 + 0.5472i   0.7513 + 0.2575i
>> [U,S,V] = svd(X)

      -0.3689 - 0.4587i  0.4056 - 0.2075i  0.4362 - 0.5055i
  U = -0.3766 - 0.5002i -0.5792 - 0.2810i  0.0646 + 0.4351i
      -0.2178 - 0.4626i  0.1142 + 0.6041i -0.5938 - 0.0901i

      2.2335        0        0
  S =      0   0.7172        0
           0        0   0.3742

      -0.4577            0.5749            0.6783
  V = -0.7885 - 0.0255i -0.6118 - 0.0497i -0.0135 + 0.0249i
      -0.3229 - 0.2527i  0.3881 + 0.3769i -0.5469 - 0.4900i.
```

The singular values are 2.2335, 0.7172, and 0.3742. □

We may use the SVD to solve, when possible, the matrix equation

$$A\,|x\rangle = |y\rangle \tag{1.381}$$

for the N-dimensional vector $|x\rangle$ in terms of the M-dimensional vector $|y\rangle$ and the $M \times N$ matrix A. Using the SVD expansion (1.362), we have

$$\sum_{n=1}^{\min(M,N)} |m_n\rangle S_n\,\langle n|x\rangle = |y\rangle. \tag{1.382}$$

The orthonormality (1.355) of the vectors $|m_n\rangle$ then tells us that

$$S_n\,\langle n|x\rangle = \langle m_n|y\rangle. \tag{1.383}$$

If the singular value is positive $S_n > 0$ whenever $\langle m_n|y\rangle \neq 0$, then we may divide by the singular value to get $\langle n|x\rangle = \langle m_n|y\rangle / S_n$ and so find the solution

$$|x\rangle = \sum_{n=1}^{\min(M,N)} \frac{\langle m_n|y\rangle}{S_n}\,|n\rangle. \tag{1.384}$$

But this solution is not always available or unique.

For instance, if for some n' the inner product $\langle m_{n'}|y\rangle \neq 0$ while the singular value $S_{n'} = 0$, then there is no solution to equation (1.381). This problem often occurs when $M > N$.

Example 1.46 Suppose A is the 3×2 matrix

$$A = \begin{pmatrix} r_1 & p_1 \\ r_2 & p_2 \\ r_3 & p_3 \end{pmatrix} \tag{1.385}$$

and the vector $|y\rangle$ is the cross-product $|y\rangle = \boldsymbol{L} = \boldsymbol{r} \times \boldsymbol{p}$. Then no solution $|x\rangle$ exists to the equation $A|x\rangle = |y\rangle$ (unless $\boldsymbol{r}$ and $\boldsymbol{p}$ are parallel) because $A|x\rangle$ is a linear combination of the vectors $\boldsymbol{r}$ and $\boldsymbol{p}$ while $|y\rangle = \boldsymbol{L}$ is perpendicular to both $\boldsymbol{r}$ and $\boldsymbol{p}$. □

Even when the matrix A is square, the equation (1.381) sometimes has no solutions. For instance, if A is a square matrix that vanishes, $A = 0$, then (1.381) has no solutions whenever $|y\rangle \neq 0$. And when $N > M$, as in for instance

$$\begin{pmatrix} a & b & c \\ d & e & f \end{pmatrix} \begin{pmatrix} x_1 \\ x_2 \\ x_3 \end{pmatrix} = \begin{pmatrix} y_1 \\ y_2 \end{pmatrix} \tag{1.386}$$

the solution (1.384) is never unique, for we may add to it any linear combination of the vectors $|n\rangle$ that A annihilates for $M < n \leq N$

$$|x\rangle = \sum_{n=1}^{\min(M,N)} \frac{\langle m_n|y\rangle}{S_n} |n\rangle + \sum_{n=M+1}^{N} x_n |n\rangle. \tag{1.387}$$

These are the vectors $|n\rangle$ for $M < n \leq N$ which A maps to zero since they do not occur in the sum (1.362), which stops at $n = \min(M, N) < N$.

Example 1.47 (The CKM matrix) In the standard model, the mass matrix of the d, s, and b quarks is a 3×3 complex, symmetric matrix M. Since M is symmetric ($M = M^{\mathsf{T}}$), its adjoint is its complex conjugate, $M^{\dagger} = M^*$. So the right singular vectors $|n\rangle$ are the eigenstates of M^*M as in (1.351)

$$M^* M|n\rangle = S_n^2|n\rangle \tag{1.388}$$

and the left singular vectors $|m_n\rangle$ are the eigenstates of MM^* as in (1.366)

$$MM^*|m_n\rangle = \left(M^* M\right)^* |m_n\rangle = S_n^2|m_n\rangle. \tag{1.389}$$

Thus the left singular vectors are just the complex conjugates of the right singular vectors, $|m_n\rangle = |n\rangle^*$. But this means that the unitary matrix V is the complex conjugate of the unitary matrix U, so the SVD of M is (Autonne, 1915)

$$M = U \Sigma U^{\mathsf{T}}. \tag{1.390}$$

The masses of the quarks then are the nonnegative singular values S_n along the diagonal of the matrix Σ. By redefining the quark fields, one may make the (CKM) matrix U real – except for a single complex phase, which causes a violation of charge-conjugation-parity (CP) symmetry. A similar matrix determines the neutrino masses. □

1.32 The Moore–Penrose pseudoinverse

Although a matrix A has an inverse A^{-1} if and only if it is square and has a nonzero determinant, one may use the singular-value decomposition to make a pseudoinverse A^+ for an arbitrary $M \times N$ matrix A. If the singular-value decomposition of the matrix A is

$$A = U \, \Sigma \, V^\dagger \tag{1.391}$$

then the Moore–Penrose pseudoinverse (Eliakim H. Moore, 1862–1932, Roger Penrose, 1931–) is

$$A^+ = V \, \Sigma^+ \, U^\dagger \tag{1.392}$$

in which Σ^+ is the transpose of the matrix Σ with every nonzero entry replaced by its inverse (and the zeros left as they are). One may show that the pseudoinverse A^+ satisfies the four relations

$$\begin{aligned} A \, A^+ \, A = A \quad &\text{and} \quad A^+ \, A \, A^+ = A^+, \\ \left(A \, A^+\right)^\dagger = A \, A^+ \quad &\text{and} \quad \left(A^+ \, A\right)^\dagger = A^+ \, A \end{aligned} \tag{1.393}$$

and that it is the only matrix that does so.

Suppose that all the singular values of the $M \times N$ matrix A are positive. In this case, if A has more rows than columns, so that $M > N$, then the product AA^+ is the $N \times N$ identity matrix I_N

$$A^+ A = V^\dagger \Sigma^+ \Sigma V = V^\dagger I_N V = I_N \tag{1.394}$$

and AA^+ is an $M \times M$ matrix that is not the identity matrix I_M. If instead A has more columns than rows, so that $N > M$, then AA^+ is the $M \times M$ identity matrix I_M

$$AA^+ = U \Sigma \Sigma^+ U^\dagger = U I_M U^\dagger = I_M \tag{1.395}$$

but A^+A is an $N \times N$ matrix that is not the identity matrix I_N. If the matrix A is square with positive singular values, then it has a true inverse A^{-1} which is equal to its pseudoinverse

$$A^{-1} = A^+. \tag{1.396}$$

If the columns of A are linearly independent, then the matrix $A^\dagger A$ has an inverse, and the pseudoinverse is

$$A^+ = \left(A^\dagger A\right)^{-1} A^\dagger. \tag{1.397}$$

The solution (1.220) to the complex least-squares method used this pseudoinverse.

If the rows of A are linearly independent, then the matrix $AA^\dagger$ has an inverse, and the pseudoinverse is

$$A^+ = A^\dagger \left(AA^\dagger\right)^{-1}. \tag{1.398}$$

If both the rows and the columns of A are linearly independent, then the matrix A has an inverse A^{-1} which is its pseudoinverse

$$A^{-1} = A^+. \tag{1.399}$$

Example 1.48 (The pseudoinverse of a 2×3 matrix) The pseudoinverse A^+ of the matrix A

$$A = \begin{pmatrix} 0 & 1 & 0 \\ 1 & 0 & 1 \end{pmatrix} \tag{1.400}$$

with singular-value decomposition (1.379) is

$$\begin{aligned} A^+ &= V\,\Sigma^+\,U^\dagger \\ &= \frac{1}{\sqrt{2}} \begin{pmatrix} 1 & 0 & -1 \\ 0 & \sqrt{2} & 0 \\ 1 & 0 & 1 \end{pmatrix} \begin{pmatrix} 1/\sqrt{2} & 0 \\ 0 & 1 \\ 0 & 0 \end{pmatrix} \begin{pmatrix} 0 & 1 \\ 1 & 0 \end{pmatrix} \\ &= \begin{pmatrix} 0 & 1/2 \\ 1 & 0 \\ 0 & 1/2 \end{pmatrix}, \end{aligned} \tag{1.401}$$

which satisfies the four conditions (1.393). The product $A\,A^+$ gives the 2×2 identity matrix

$$A\,A^+ = \begin{pmatrix} 0 & 1 & 0 \\ 1 & 0 & 1 \end{pmatrix} \begin{pmatrix} 0 & 1/2 \\ 1 & 0 \\ 0 & 1/2 \end{pmatrix} = \begin{pmatrix} 1 & 0 \\ 0 & 1 \end{pmatrix}, \tag{1.402}$$

which is an instance of (1.395). Moreover, the rows of A are linearly independent, and so the simple rule (1.398) works:

$$A^+ = A^\dagger \left(AA^\dagger\right)^{-1}$$

$$= \begin{pmatrix} 1 & 0 \\ 0 & 1 \\ 1 & 0 \end{pmatrix} \left(\begin{pmatrix} 0 & 1 & 0 \\ 1 & 0 & 1 \end{pmatrix} \begin{pmatrix} 1 & 0 \\ 0 & 1 \\ 1 & 0 \end{pmatrix} \right)^{-1} = \begin{pmatrix} 1 & 0 \\ 0 & 1 \\ 1 & 0 \end{pmatrix} \begin{pmatrix} 0 & 1 \\ 2 & 0 \end{pmatrix}^{-1}$$

$$= \begin{pmatrix} 1 & 0 \\ 0 & 1 \\ 1 & 0 \end{pmatrix} \begin{pmatrix} 0 & 1/2 \\ 1 & 0 \end{pmatrix} = \begin{pmatrix} 0 & 1/2 \\ 1 & 0 \\ 0 & 1/2 \end{pmatrix}, \tag{1.403}$$

which is (1.401).

The columns of the matrix A are not linearly independent, however, and so the simple rule (1.397) fails. Thus the product A^+A

$$A^+A = \begin{pmatrix} 0 & 1/2 \\ 1 & 0 \\ 0 & 1/2 \end{pmatrix} \begin{pmatrix} 0 & 1 & 0 \\ 1 & 0 & 1 \end{pmatrix} = \frac{1}{2} \begin{pmatrix} 1 & 0 & 1 \\ 0 & 2 & 0 \\ 1 & 0 & 1 \end{pmatrix} \tag{1.404}$$

is not the 3×3 identity matrix which it would be if (1.397) held. □

1.33 The rank of a matrix

Four equivalent definitions of the **rank** $R(A)$ of an $M \times N$ matrix A are:

1 the number of its linearly independent rows,
2 the number of its linearly independent columns,
3 the number of its nonzero singular values, and
4 the number of rows in its biggest square nonsingular submatrix.

A matrix of rank zero has no nonzero singular values and so is zero.

Example 1.49 (Rank) The 3×4 matrix

$$A = \begin{pmatrix} 1 & 0 & 1 & -2 \\ 2 & 2 & 0 & 2 \\ 4 & 3 & 1 & 1 \end{pmatrix} \tag{1.405}$$

has three rows, so its rank can be at most 3. But twice the first row added to thrice the second row equals twice the third row or

$$2r_1 + 3r_2 - 2r_3 = 0 \tag{1.406}$$

so $R(A) \le 2$. The first two rows obviously are not parallel, so they are linearly independent. Thus the number of linearly independent rows of A is 2, and so A has rank 2. □

1.34 Software

Free, high-quality software for virtually all numerical problems in linear algebra are available in LAPACK – the Linear Algebra PACKage. The FORTRAN version is available at the web-site www.netlib.org/lapack/ and the C++ version at math.nist.gov/tnt/.

Matlab is a superb commercial program for numerical problems. A free GNU version of it is available at www.gnu.org/software/octave/. Maple and Mathematica are good commercial programs for symbolic problems.

1.35 The tensor/direct product

The **tensor product** (also called the **direct product**) is simple, but it can confuse students if they see it for the first time in a course on quantum mechanics. The tensor product is used to describe composite systems, such as an angular momentum composed of orbital and spin angular momenta.

If A is an $M \times N$ matrix with elements A_{ij} and Λ is a $K \times L$ matrix with elements $\Lambda_{\alpha\beta}$, then their **direct product** $C = A \otimes \Lambda$ is an $MK \times NL$ matrix with elements $C_{i\alpha,j\beta} = A_{ij}\,\Lambda_{\alpha\beta}$. This direct-product matrix $A \otimes \Lambda$ maps the vector $V_{j\beta}$ into the vector

$$W_{i\alpha} = \sum_{j=1}^{N} \sum_{\beta=1}^{L} C_{i\alpha,j\beta}\, V_{j\beta} = \sum_{j=1}^{N} \sum_{\beta=1}^{L} A_{ij}\,\Lambda_{\alpha\beta}\, V_{j\beta}. \tag{1.407}$$

In this sum, the second indices of A and Λ match those of the vector V. The most important case is when both A and Λ are square matrices, as will be their product $C = A \otimes \Lambda$. We'll focus on this case in the rest of this section.

The key idea here is that the direct product is a product of two operators that act on two different spaces. The operator A acts on the space S spanned by the N kets $|i\rangle$, and the operator Λ acts on the space Σ spanned by the K kets $|\alpha\rangle$. Let us assume that both operators map into these spaces, so that we may write them as

$$A = I_S A I_S = \sum_{i,j=1}^{N} |i\rangle\langle i|A|j\rangle\langle j| \tag{1.408}$$

and as

$$\Lambda = I_\Sigma \Lambda I_\Sigma = \sum_{\alpha,\beta=1}^{K} |\alpha\rangle\langle\alpha|\Lambda|\beta\rangle\langle\beta|. \tag{1.409}$$

Then the direct product $C = A \otimes \Lambda$

$$C = A \otimes \Lambda = \sum_{i,j=1}^{N} \sum_{\alpha,\beta=1}^{K} |i\rangle \otimes |\alpha\rangle\, \langle i|A|j\rangle\langle\alpha|\Lambda|\beta\rangle\, \langle j| \otimes \langle\beta| \tag{1.410}$$

acts on the direct product of the two vector spaces $S \otimes \Sigma$, which is spanned by the direct-product kets $|i, \alpha\rangle = |i\rangle\, |\alpha\rangle = |i\rangle \otimes |\alpha\rangle$.

In general, the direct-product space $S \otimes \Sigma$ is much bigger than the spaces S and Σ. For although $S \otimes \Sigma$ is spanned by the direct-product kets $|i\rangle \otimes |\alpha\rangle$, most vectors in the space $S \otimes \Sigma$ are of the form

$$|\psi\rangle = \sum_{i=1}^{N} \sum_{\alpha=1}^{K} \psi(i, \alpha)|i\rangle \otimes |\alpha\rangle \tag{1.411}$$

and not the direct product $|s\rangle \otimes |\sigma\rangle$ of a pair of vectors $|s\rangle \in S$ and $|\sigma\rangle \in \Sigma$

$$\begin{aligned} |s\rangle \otimes |\sigma\rangle &= \left(\sum_{i=1}^{N} s_i |i\rangle\right) \otimes \left(\sum_{\alpha=1}^{K} \sigma_\alpha |\alpha\rangle\right) \\ &= \sum_{i=1}^{N} \sum_{\alpha=1}^{K} s_i \sigma_\alpha |i\rangle \otimes |\alpha\rangle. \end{aligned} \tag{1.412}$$

Using the simpler notation $|i, \alpha\rangle$ for $|i\rangle \otimes |\alpha\rangle$, we may write the action of the direct-product operator $A \otimes \Lambda$ on the state

$$|\psi\rangle = \sum_{i=1}^{N} \sum_{\alpha=1}^{K} |i, \alpha\rangle\langle i, \alpha|\psi\rangle \tag{1.413}$$

as

$$(A \otimes \Lambda)|\psi\rangle = \sum_{i,j=1}^{N} \sum_{\alpha,\beta=1}^{K} |i, \alpha\rangle\, \langle i|A|j\rangle\langle\alpha|\Lambda|\beta\rangle\, \langle j, \beta|\psi\rangle. \tag{1.414}$$

Example 1.50 (States of the hydrogen atom) Suppose the states $|n, \ell, m\rangle$ are the eigenvectors of the hamiltonian H, the square $\boldsymbol{L}^2$ of the orbital angular momentum $\boldsymbol{L}$, and the third component of the orbital angular momentum L_3 for a hydrogen atom without spin:

$$\begin{aligned} H|n, \ell, m\rangle &= E_n|n, \ell, m\rangle, \\ \boldsymbol{L}^2|n, \ell, m\rangle &= \hbar^2 \ell(\ell + 1)|n, \ell, m\rangle, \\ L_3|n, \ell, m\rangle &= \hbar m|n, \ell, m\rangle. \end{aligned} \tag{1.415}$$

Suppose the states $|\sigma\rangle$ for $\sigma = \pm$ are the eigenstates of the third component S_3 of the operator $\boldsymbol{S}$ that represents the spin of the electron

$$S_3|\sigma\rangle = \sigma \frac{\hbar}{2}|\sigma\rangle. \tag{1.416}$$

Then the direct- or tensor-product states

$$|n, \ell, m, \sigma\rangle \equiv |n, \ell, m\rangle \otimes |\sigma\rangle \equiv |n, \ell, m\rangle|\sigma\rangle \tag{1.417}$$

represent a hydrogen atom including the spin of its electron. They are eigenvectors of all four operators H, $\boldsymbol{L}^2$, L_3, and S_3:

$$H|n,\ell,m,\sigma\rangle = E_n|n,\ell,m,\sigma\rangle, \qquad \boldsymbol{L}^2|n,\ell,m,\sigma\rangle = \hbar^2\ell(\ell+1)|n,\ell,m,\sigma\rangle,$$
$$L_3|n,\ell,m,\sigma\rangle = \hbar m|n,\ell,m,\sigma\rangle, \qquad S_3|n,\ell,m,\sigma\rangle = \sigma\hbar|n,\ell,m,\sigma\rangle. \tag{1.418}$$

Suitable linear combinations of these states are eigenvectors of the square $\boldsymbol{J}^2$ of the composite angular momentum $\boldsymbol{J} = \boldsymbol{L} + \boldsymbol{S}$ as well as of J_3, L_3, and S_3. □

Example 1.51 (Adding two spins) The smallest positive value of angular momentum is $\hbar/2$. The spin-one-half angular momentum operators $\boldsymbol{S}$ are represented by three 2×2 matrices

$$S_a = \frac{\hbar}{2}\sigma_a \tag{1.419}$$

in which the σ_a are the Pauli matrices

$$\sigma_1 = \begin{pmatrix} 0 & 1 \\ 1 & 0 \end{pmatrix}, \quad \sigma_2 = \begin{pmatrix} 0 & -i \\ i & 0 \end{pmatrix}, \quad \text{and} \quad \sigma_3 = \begin{pmatrix} 1 & 0 \\ 0 & -1 \end{pmatrix}. \tag{1.420}$$

Consider two spin operators $\boldsymbol{S}^{(1)}$ and $\boldsymbol{S}^{(2)}$ acting on two spin-one-half systems. The states $|\pm\rangle_1$ are eigenstates of $S_3^{(1)}$, and the states $|\pm\rangle_2$ are eigenstates of $S_3^{(2)}$

$$S_3^{(1)}|\pm\rangle_1 = \pm\frac{\hbar}{2}|\pm\rangle_1 \quad \text{and} \quad S_3^{(2)}|\pm\rangle_2 = \pm\frac{\hbar}{2}|\pm\rangle_2. \tag{1.421}$$

Then the direct-product states $|\pm,\pm\rangle = |\pm\rangle_1|\pm\rangle_2 = |\pm\rangle_1 \otimes |\pm\rangle_2$ are eigenstates of both $S_3^{(1)}$ and $S_3^{(2)}$

$$S_3^{(1)}|\pm,s_2\rangle = \pm\frac{\hbar}{2}\,|+,s_2\rangle \quad \text{and} \quad S_3^{(2)}|s_1,\pm\rangle = \pm\frac{\hbar}{2}\,|s_1,\pm\rangle. \tag{1.422}$$

These states also are eigenstates of the third component of the spin operator of the combined system

$$S_3 = S_3^{(1)} + S_3^{(2)}, \quad \text{that is} \quad S_3|s_1,s_2\rangle = \frac{\hbar}{2}\,(s_1+s_2)\,|s_1,s_2\rangle. \tag{1.423}$$

Thus $S_3|+,+\rangle = \hbar|+,+\rangle$, and $S_3|-,-\rangle = -\hbar|-,-\rangle$, while $S_3|+,-\rangle = 0$ and $S_3|-,+\rangle = 0$.

Now let's consider the effect of the operator S_1^2 on the state $|++\rangle$

$$\begin{aligned} S_1^2|++\rangle &= \left(S_1^{(1)} + S_1^{(2)}\right)^2|++\rangle = \frac{\hbar^2}{4}\left(\sigma_1^{(1)} + \sigma_1^{(2)}\right)^2|++\rangle \\ &= \frac{\hbar^2}{2}\left(1 + \sigma_1^{(1)}\sigma_1^{(2)}\right)|++\rangle = \frac{\hbar^2}{2}\left(|++\rangle + \sigma_1^{(1)}|+\rangle\sigma_1^{(2)}|+\rangle\right) \\ &= \frac{\hbar^2}{2}\left(|++\rangle + |--\rangle\right). \end{aligned} \tag{1.424}$$

The rest of this example will be left to exercise 1.36. □

1.36 Density operators

A general quantum-mechanical system is represented by a **density operator** ρ that is hermitian $\rho^\dagger = \rho$, of unit trace $\text{Tr}\rho = 1$, and positive $\langle\psi|\rho|\psi\rangle \geq 0$ for all kets $|\psi\rangle$.

If the state $|\psi\rangle$ is normalized, then $\langle\psi|\rho|\psi\rangle$ is the nonnegative probability that the system is in that state. This probability is real because the density matrix is hermitian. If $\{|n\rangle\}$ is any complete set of orthonormal states,

$$I = \sum_n |n\rangle\langle n|, \tag{1.425}$$

then the probability that the system is in the state $|n\rangle$ is

$$p_n = \langle n|\rho|n\rangle = \text{Tr}\,(\rho|n\rangle\langle n|)\,. \tag{1.426}$$

Since $\text{Tr}\rho = 1$, the sum of these probabilities is unity

$$\sum_n p_n = \sum_n \langle n|\rho|n\rangle = \text{Tr}\left(\rho \sum_n |n\rangle\langle n|\right) = \text{Tr}\,(\rho I) = \text{Tr}\rho = 1. \tag{1.427}$$

A system that is measured to be in a state $|n\rangle$ cannot simultaneously be measured to be in an orthogonal state $|m\rangle$. The probabilities sum to unity because the system must be in some state.

Since the density operator ρ, is hermitian, it has a complete, orthonormal set of eigenvectors $|k\rangle$, all of which have nonnegative eigenvalues ρ_k

$$\rho|k\rangle = \rho_k|k\rangle. \tag{1.428}$$

They afford for it the expansion

$$\rho = \sum_{k=1}^{N} \rho_k |k\rangle\langle k| \tag{1.429}$$

in which the eigenvalue ρ_k is the probability that the system is in the state $|k\rangle$.

1.37 Correlation functions

We can define two Schwarz inner products for a density matrix ρ. If $|f\rangle$ and $|g\rangle$ are two states, then the inner product

$$(f,g) \equiv \langle f|\rho|g\rangle \tag{1.430}$$

for $g = f$ is nonnegative, $(f,f) = \langle f|\rho|f\rangle \geq 0$, and satisfies the other conditions (1.73, 1.74, & 1.76) for a Schwarz inner product.

The second Schwarz inner product applies to operators A and B and is defined (Titulaer and Glauber, 1965) as

$$(A, B) = \mathrm{Tr}\left(\rho A^\dagger B\right) = \mathrm{Tr}\left(B\rho A^\dagger\right) = \mathrm{Tr}\left(A^\dagger B\rho\right). \tag{1.431}$$

This inner product is nonnegative when $A = B$ and obeys the other rules (1.73, 1.74, & 1.76) for a Schwarz inner product.

These two degenerate inner products are not inner products in the strict sense of (1.73–1.79), but they are Schwarz inner products, and so (1.92–1.93) they satisfy the Schwarz inequality (1.93)

$$(f,f)(g,g) \geq |(f,g)|^2. \tag{1.432}$$

Applied to the first, vector, Schwarz inner product (1.430), the Schwarz inequality gives

$$\langle f|\rho|f\rangle\langle g|\rho|g\rangle \geq |\langle f|\rho|g\rangle|^2, \tag{1.433}$$

which is a useful property of density matrices. Application of the Schwarz inequality to the second, operator, Schwarz inner product (1.431) gives (Titulaer and Glauber, 1965)

$$\mathrm{Tr}\left(\rho A^\dagger A\right)\mathrm{Tr}\left(\rho B^\dagger B\right) \geq \left|\mathrm{Tr}\left(\rho A^\dagger B\right)\right|^2. \tag{1.434}$$

The operator $E_i(x)$ that represents the ith component of the electric field at the point x is the hermitian sum of the "positive-frequency" part $E_i^{(+)}(x)$ and its adjoint $E_i^{(-)}(x) = (E_i^{(+)}(x))^\dagger$

$$E_i(x) = E_i^{(+)}(x) + E_i^{(-)}(x). \tag{1.435}$$

Glauber has defined the first-order correlation function $G_{ij}^{(1)}(x, y)$ as (Glauber, 1963b)

$$G_{ij}^{(1)}(x, y) = \mathrm{Tr}\left(\rho E_i^{(-)}(x)E_j^{(+)}(y)\right) \tag{1.436}$$

or in terms of the operator inner product (1.431) as

$$G_{ij}^{(1)}(x, y) = \left(E_i^{(+)}(x), E_j^{(+)}(y)\right). \tag{1.437}$$

By setting $A = E_i^{(+)}(x)$, etc., it follows then from the Schwarz inequality (1.434) that the correlation function $G_{ij}^{(1)}(x, y)$ is bounded by (Titulaer and Glauber, 1965)

$$|G_{ij}^{(1)}(x, y)|^2 \leq G_{ii}^{(1)}(x, x)G_{jj}^{(1)}(y, y). \tag{1.438}$$

Interference fringes are sharpest when this inequality is saturated:

$$|G_{ij}^{(1)}(x, y)|^2 = G_{ii}^{(1)}(x, x)G_{jj}^{(1)}(y, y), \tag{1.439}$$

which can occur only if the correlation function $G^{(1)}_{ij}(x,y)$ factorizes (Titulaer and Glauber, 1965)

$$G^{(1)}_{ij}(x,y) = \mathcal{E}^*_i(x)\mathcal{E}_j(y) \tag{1.440}$$

as it does when the density operator is an outer product of coherent states

$$\rho = |\{\alpha_k\}\rangle\langle\{\alpha_k\}|, \tag{1.441}$$

which are eigenstates of $E^{(+)}_i(x)$ with eigenvalue $\mathcal{E}_i(x)$ (Glauber, 1963b, a)

$$E^{(+)}_i(x)|\{\alpha_k\}\rangle = \mathcal{E}_i(x)|\{\alpha_k\}\rangle. \tag{1.442}$$

The higher-order correlation functions

$$G^{(n)}_{i_1\dots i_{2n}}(x_1\dots x_{2n}) = \text{Tr}\left(\rho E^{(-)}_{i_1}(x_1)\dots E^{(-)}_{i_n}(x_n)E^{(+)}_{i_{n+1}}(x_{n+1})\dots E^{(+)}_{i_{2n}}(x_n)\right) \tag{1.443}$$

satisfy similar inequalities (Glauber, 1963b), which also follow from the Schwarz inequality (1.434).

Exercises

1.1 Why is the most complicated function of two Grassmann numbers a polynomial with at most four terms as in (1.12)?
1.2 Derive the cyclicity (1.24) of the trace from (1.23).
1.3 Show that $(AB)^{\mathsf{T}} = B^{\mathsf{T}}A^{\mathsf{T}}$, which is (1.26).
1.4 Show that a real hermitian matrix is symmetric.
1.5 Show that $(AB)^\dagger = B^\dagger A^\dagger$, which is (1.29).
1.6 Show that the matrix (1.40) is positive on the space of all real 2-vectors but not on the space of all complex 2-vectors.
1.7 Show that the two 4×4 matrices (1.45) satisfy Grassmann's algebra (1.11) for $N = 2$.
1.8 Show that the operators $a_i = \theta_i$ defined in terms of the Grassmann matrices (1.45) and their adjoints $a_i^\dagger = \theta_i^\dagger$ satisfy the anticommutation relations (1.46) of the creation and annihilation operators for a system with two fermionic states.
1.9 Derive (1.64) from (1.61–1.63).
1.10 Fill in the steps leading to the formulas (1.69) for the vectors b'_1 and b'_2 and the formula (1.70) for the matrix a'.
1.11 Show that the antilinearity (1.76) of the inner product follows from its first two properties (1.73 & 1.74).
1.12 Show that the Minkowski product $(x,y) = x^0y^0 - \boldsymbol{x}\cdot\boldsymbol{y}$ of two 4-vectors x and y is an inner product that obeys the rules (1.73, 1.74, and 1.79).
1.13 Show that if $f = 0$, then the linearity (1.74) of the inner product implies that (f,f) and (g,f) vanish.

1.14 Show that the condition (1.75) of being positive definite implies nondegeneracy (1.79).

1.15 Show that the nonnegativity (1.77) of the Schwarz inner product implies the condition (1.78). Hint: the inequality $(f - \lambda g, f - \lambda g) \ge 0$ must hold for every complex λ and for all vectors f and g.

1.16 Show that the inequality (1.97) follows from the Schwarz inequality (1.96).

1.17 Show that the inequality (1.99) follows from the Schwarz inequality (1.98).

1.18 Use the Gram–Schmidt method to find orthonormal linear combinations of the three vectors

$$s_1 = \begin{pmatrix} 1 \\ 0 \\ 0 \end{pmatrix}, \quad s_2 = \begin{pmatrix} 1 \\ 1 \\ 0 \end{pmatrix}, \quad s_3 = \begin{pmatrix} 1 \\ 1 \\ 1 \end{pmatrix}. \tag{1.444}$$

1.19 Now use the Gram–Schmidt method to find orthonormal linear combinations of the same three vectors but in a different order

$$s_1' = \begin{pmatrix} 1 \\ 1 \\ 1 \end{pmatrix}, \quad s_2' = \begin{pmatrix} 1 \\ 1 \\ 0 \end{pmatrix}, \quad s_3' = \begin{pmatrix} 1 \\ 0 \\ 0 \end{pmatrix}. \tag{1.445}$$

Did you get the same orthonormal vectors as in the previous exercise?

1.20 Derive the linearity (1.120) of the outer product from its definition (1.119).

1.21 Show that a linear operator A that is represented by a hermitian matrix (1.155) in an orthonormal basis satisfies $(g, Af) = (A\,g, f)$.

1.22 Show that a unitary operator maps one orthonormal basis into another.

1.23 Show that the integral (1.170) defines a unitary operator that maps the state $|x'\rangle$ to the state $|x' + a\rangle$.

1.24 For the 2×2 matrices

$$A = \begin{pmatrix} 1 & 2 \\ 3 & -4 \end{pmatrix} \quad \text{and} \quad B = \begin{pmatrix} 2 & -1 \\ 4 & -3 \end{pmatrix} \tag{1.446}$$

verify equations (1.202–1.204).

1.25 Derive the least-squares solution (1.220) for complex A, x, and y when the matrix $A^\dagger A$ is positive.

1.26 Show that the eigenvalues λ of a unitary matrix are unimodular, that is, $|\lambda| = 1$.

1.27 What are the eigenvalues and eigenvectors of the two defective matrices (1.252)?

1.28 Use (1.267) to derive expression (1.268) for the 2×2 rotation matrix $\exp(-i\boldsymbol{\theta} \cdot \boldsymbol{\sigma}/2)$.

1.29 Compute the characteristic equation for the matrix $-i\boldsymbol{\theta} \cdot \boldsymbol{J}$ in which the generators are $(J_k)_{ij} = i\epsilon_{ikj}$ and ϵ_{ijk} is totally antisymmetric with $\epsilon_{123} = 1$.

1.30 Show that the sum of the eigenvalues of a normal antisymmetric matrix vanishes.

1.31 Use the characteristic equation of exercise 1.29 to derive identities (1.271) and (1.272) for the 3×3 real orthogonal matrix $\exp(-i\boldsymbol{\theta} \cdot \boldsymbol{J})$.

1.32 Consider the 2×3 matrix A

$$A = \begin{pmatrix} 1 & 2 & 3 \\ -3 & 0 & 1 \end{pmatrix}. \tag{1.447}$$

Perform the singular value decomposition $A = USV^{\mathrm{T}}$, where V^{T} is the transpose of V. Find the singular values and the real orthogonal matrices U and V. Students may use Lapack, Octave, Matlab, Maple or any other program to do this exercise.

1.33 Consider the 6×9 matrix A with elements

$$A_{j,k} = x + x^j + i(y - y^k) \tag{1.448}$$

in which $x = 1.1$ and $y = 1.02$. Find the singular values, and the first left and right singular vectors. Students may use Lapack, Octave, Matlab, Maple or any other program to do this exercise.

1.34 Show that the totally antisymmetric Levi-Civita symbol ϵ_{ijk} satisfies the useful relation

$$\sum_{i=1}^{3} \epsilon_{ijk}\, \epsilon_{inm} = \delta_{jn}\, \delta_{km} - \delta_{jm}\, \delta_{kn}. \tag{1.449}$$

1.35 Consider the hamiltonian

$$H = \tfrac{1}{2}\hbar\omega\sigma_3 \tag{1.450}$$

where σ_3 is defined in (1.420). The entropy S of this system at temperature T is

$$S = -k\mathrm{Tr}\,[\rho \ln(\rho)] \tag{1.451}$$

in which the density operator ρ is

$$\rho = \frac{e^{-H/(kT)}}{\mathrm{Tr}\left[e^{-H/(kT)}\right]}. \tag{1.452}$$

Find expressions for the density operator ρ and its entropy S.

1.36 Find the action of the operator $\boldsymbol{S}^2 = \left(\boldsymbol{S}^{(1)} + \boldsymbol{S}^{(2)}\right)^2$ defined by (1.419) on the four states $|\pm\pm\rangle$ and then find the eigenstates and eigenvalues of $\boldsymbol{S}^2$ in the space spanned by these four states.

1.37 A system that has three fermionic states has three creation operators $a_i^\dagger$ and three annihilation operators a_k which satisfy the anticommutation relations $\{a_i, a_k^\dagger\} = \delta_{ik}$ and $\{a_i, a_k\} = \{a_i^\dagger, a_k^\dagger\} = 0$ for $i, k = 1, 2, 3$. The eight states of the system are $|v, u, t\rangle \equiv (a_3^\dagger)^t (a_2^\dagger)^u (a_1^\dagger)^v |0,0,0\rangle$. We can represent them by eight 8-vectors, each of which has seven 0s with a 1 in position $5v+3u+t$. How big should the matrices that represent the creation and annihilation operators be? Write down the three matrices that represent the three creation operators.

1.38 Show that the Schwarz inner product (1.430) is degenerate because it can violate (1.79) for certain density operators and certain pairs of states.

1.39 Show that the Schwarz inner product (1.431) is degenerate because it can violate (1.79) for certain density operators and certain pairs of operators.

1.40 The coherent state $|\{\alpha_k\}\rangle$ is an eigenstate of the annihilation operator a_k with eigenvalue α_k for each mode k of the electromagnetic field, $a_k|\{\alpha_k\}\rangle = \alpha_k|\{\alpha_k\}\rangle$. The positive-frequency part $E_i^{(+)}(x)$ of the electric field is a linear combination of the annihilation operators

$$E_i^{(+)}(x) = \sum_k a_k \, \mathcal{E}_i^{(+)}(k) \, e^{i(kx-\omega t)}. \tag{1.453}$$

Show that $|\{\alpha_k\}\rangle$ is an eigenstate of $E_i^{(+)}(x)$ as in (1.442) and find its eigenvalue $\mathcal{E}_i(x)$.

2

Fourier series

2.1 Complex Fourier series

The phases $\exp(inx)/\sqrt{2\pi}$, one for each integer n, are orthonormal on an interval of length 2π

$$\int_0^{2\pi} \left(\frac{e^{imx}}{\sqrt{2\pi}}\right)^* \frac{e^{inx}}{\sqrt{2\pi}}\, dx = \int_0^{2\pi} \frac{e^{i(n-m)x}}{2\pi}\, dx = \delta_{m,n} \tag{2.1}$$

where $\delta_{n,m} = 1$ if $n = m$, and $\delta_{n,m} = 0$ if $n \neq m$. So if a function $f(x)$ is a sum of these phases

$$f(x) = \sum_{n=-\infty}^{\infty} f_n \frac{e^{inx}}{\sqrt{2\pi}} \tag{2.2}$$

then their orthonormality (2.1) gives the nth coefficient f_n as the integral

$$\int_0^{2\pi} \frac{e^{-inx}}{\sqrt{2\pi}} f(x)\, dx = \int_0^{2\pi} \frac{e^{-inx}}{\sqrt{2\pi}} \sum_{m=-\infty}^{\infty} f_m \frac{e^{imx}}{\sqrt{2\pi}}\, dx = \sum_{m=-\infty}^{\infty} \delta_{n,m} f_m = f_n \tag{2.3}$$

(Joseph Fourier, 1768–1830).

The **Fourier series** (2.2) is **periodic** with period 2π because the phases are periodic with period 2π, $\exp(in(x + 2\pi)) = \exp(inx)$. Thus even if the function $f(x)$ which we use in (2.3) to make the Fourier coefficients f_n is not periodic, its Fourier series (2.2) will nevertheless be strictly periodic, as illustrated by Figs. 2.2 & 2.4.

If the Fourier series (2.2) converges uniformly (section 2.7), then the term-by-term integration implicit in the formula (2.3) for f_n is permitted.

How is the Fourier series for the complex-conjugate function $f^*(x)$ related to the series for $f(x)$? The complex conjugate of the Fourier series (2.2) is

$$f^*(x) = \sum_{n=-\infty}^{\infty} f_n^* \frac{e^{-inx}}{\sqrt{2\pi}} = \sum_{n=-\infty}^{\infty} f_{-n}^* \frac{e^{inx}}{\sqrt{2\pi}} \tag{2.4}$$

so the coefficients $f_n(f^*)$ for $f^*(x)$ are related to those $f_n(f)$ for $f(x)$ by

$$f_n(f^*) = f_{-n}^*(f). \tag{2.5}$$

Thus if the function $f(x)$ is real, then

$$f_n(f) = f_n(f^*) = f_{-n}^*(f). \tag{2.6}$$

Dropping all reference to the functions, we see that the Fourier coefficients f_n for a real function $f(x)$ satisfy

$$f_n = f_{-n}^*. \tag{2.7}$$

Example 2.1 (Fourier series by inspection) The doubly exponential function $\exp(\exp(ix))$ has the Fourier series

$$\exp\left(e^{ix}\right) = \sum_{n=0}^{\infty} \frac{1}{n!} e^{inx} \tag{2.8}$$

in which $n! = n(n-1)\ldots 1$ is n-factorial with $0! \equiv 1$. □

Example 2.2 (Beats) The sum of two sines $f(x) = \sin \omega_1 x + \sin \omega_2 x$ of similar frequencies $\omega_1 \approx \omega_2$ is the product (exercise 2.1)

$$f(x) = 2\cos \tfrac{1}{2}(\omega_1 - \omega_2)x \; \sin \tfrac{1}{2}(\omega_1 + \omega_2)x \tag{2.9}$$

in which the first factor $\cos \frac{1}{2}(\omega_1 - \omega_2)x$ is the *beat* which modulates the second factor $\sin \frac{1}{2}(\omega_1 + \omega_2)x$, as illustrated by Fig. 2.1.

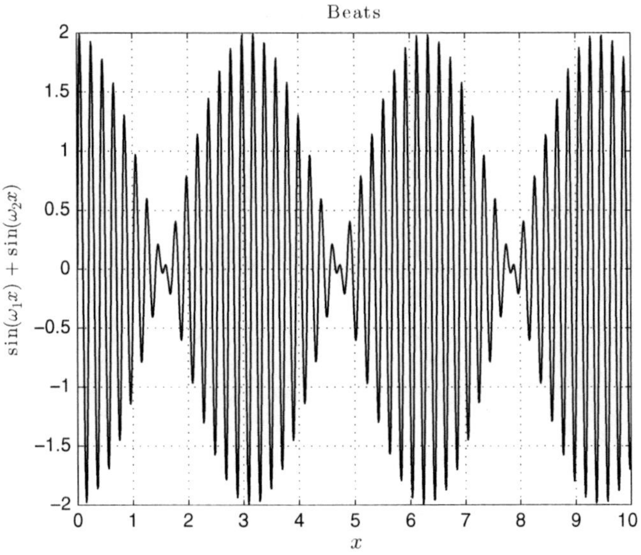

Figure 2.1 The curve $\sin \omega_1 x + \sin \omega_2 x$ for $\omega_1 = 30$ and $\omega_2 = 32$.

□

2.2 The interval

In equations (2.1–2.3), we singled out the interval $[0, 2\pi]$, but to represent a periodic function with period 2π, we could have used any interval of length 2π, such as the interval $[-\pi, \pi]$ or $[r, r + 2\pi]$

$$f_n = \int_r^{r+2\pi} e^{-inx} f(x) \, \frac{dx}{\sqrt{2\pi}}. \tag{2.10}$$

This integral is independent of its lower limit r as long as the function $f(x)$ is periodic with period 2π. The choice $r = -\pi$ often is convenient. With this choice of interval, the coefficient f_n is the integral (2.3) shifted by $-\pi$

$$f_n = \int_{-\pi}^{\pi} e^{-inx} f(x) \, \frac{dx}{\sqrt{2\pi}}. \tag{2.11}$$

But if the function $f(x)$ is not periodic with period 2π, then the Fourier coefficients (2.10) do depend upon r.

2.3 Where to put the 2πs

In equations (2.2 & 2.3), we used the orthonormal functions $\exp(inx)/\sqrt{2\pi}$, and so we had factors of $1/\sqrt{2\pi}$ in both equations. If one gets tired of having so many explicit square-roots, then one may set $d_n = f_n/\sqrt{2\pi}$ and write (2.2) and (2.3) as

$$f(x) = \sum_{n=-\infty}^{\infty} d_n\, e^{inx} \quad \text{and} \quad d_n = \frac{1}{2\pi} \int_0^{2\pi} dx\, e^{-inx} f(x). \tag{2.12}$$

One also may use the rules

$$f(x) = \frac{1}{2\pi} \sum_{n=-\infty}^{\infty} c_n e^{inx} \quad \text{and} \quad c_n = \int_{-\pi}^{\pi} f(x) e^{-inx}\, dx. \tag{2.13}$$

Example 2.3 (Fourier series for $\exp(-m|x|)$) Let's compute the Fourier series for the real function $f(x) = \exp(-m|x|)$ on the interval $(-\pi, \pi)$. Using (2.10) for the shifted interval and the 2π-placement convention (2.12), we find for the coefficient d_n

$$d_n = \int_{-\pi}^{\pi} \frac{dx}{2\pi}\, e^{-inx}\, e^{-m|x|}, \tag{2.14}$$

which we may split into the two pieces

$$d_n = \int_{-\pi}^{0} \frac{dx}{2\pi}\, e^{(m-in)x} + \int_0^{\pi} \frac{dx}{2\pi}\, e^{-(m+in)x}. \tag{2.15}$$

After doing the integrals, we find

$$d_n = \frac{1}{\pi} \frac{m}{m^2+n^2} \left[1 - (-1)^n e^{-\pi m}\right]. \tag{2.16}$$

Here, since m is real, $d_n = d_n^*$, but also $d_n = d_{-n}$. So the coefficients d_n satisfy the condition (2.7) that holds when the function $f(x)$ is real, $d_n = d_{-n}^*$. The Fourier series for $\exp(-m|x|)$ with d_n given by (2.16) is

$$\begin{aligned} e^{-m|x|} &= \sum_{n=-\infty}^{\infty} d_n e^{inx} = \sum_{n=-\infty}^{\infty} \frac{1}{\pi} \frac{m}{m^2+n^2} \left[1 - (-1)^n e^{-\pi m}\right] e^{inx} \\ &= \frac{1}{m\pi} \left(1 - e^{-\pi m}\right) \\ &\quad + 2 \sum_{n=1}^{\infty} \frac{1}{\pi} \frac{m}{m^2+n^2} \left[1 - (-1)^n e^{-\pi m}\right] \cos(nx). \end{aligned} \tag{2.17}$$

In Fig. 2.2, the 10-term (dashes) Fourier series for $m = 2$ is plotted from $x = -2\pi$ to $x = 2\pi$. The function $\exp(-2|x|)$ itself is represented by a solid line. Although it is not periodic, its Fourier series is periodic with period 2π. The 10-term Fourier series represents the function $\exp(-2|x|)$ so well that the 100-term series would have been hard to distinguish from the function.

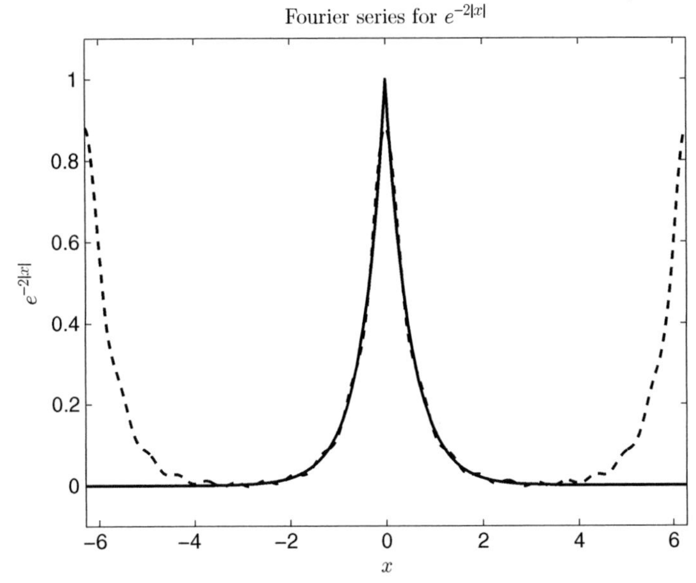

Figure 2.2 The 10-term (dashes) Fourier series (2.17) for the function $\exp(-2|x|)$ on the interval $(-\pi, \pi)$ are plotted from -2π to 2π. All Fourier series are periodic, but the function $\exp(-2|x|)$ (solid) is not.

In what follows, we usually won't bother to use different letters to distinguish between the symmetric (2.2 & 2.3) and asymmetric (2.12 & 2.13) conventions on the placement of the 2πs. □

2.4 Real Fourier series for real functions

The Fourier series outlined above are simple and apply to functions that are continuous and periodic – whether complex or real. If the function $f(x)$ is real, then by (2.7) $d_{-n} = d_n^*$, whence $d_0 = d_0^*$, so d_0 is real. Thus the Fourier series (2.12) for a real function $f(x)$ is

$$
\begin{aligned}
f(x) &= d_0 + \sum_{n=1}^{\infty} d_n\, e^{inx} + \sum_{n=-\infty}^{-1} d_n\, e^{inx} \\
&= d_0 + \sum_{n=1}^{\infty} \left[d_n\, e^{inx} + d_{-n}\, e^{-inx}\right] = d_0 + \sum_{n=1}^{\infty} \left[d_n\, e^{inx} + d_n^*\, e^{-inx}\right] \\
&= d_0 + \sum_{n=1}^{\infty} d_n\,(\cos nx + i \sin nx) + d_n^*\,(\cos nx - i \sin nx) \\
&= d_0 + \sum_{n=1}^{\infty} (d_n + d_n^*)\, \cos nx + i(d_n - d_n^*)\, \sin nx. \qquad (2.18)
\end{aligned}
$$

Let's write d_n as

$$d_n = \frac{1}{2}(a_n - ib_n), \quad \text{so that} \quad a_n = d_n + d_n^* \quad \text{and} \quad b_n = i(d_n - d_n^*). \tag{2.19}$$

Then the Fourier series (2.18) for a real function $f(x)$ is

$$f(x) = \frac{a_0}{2} + \sum_{n=1}^{\infty} a_n \cos nx + b_n \sin nx. \tag{2.20}$$

What are the formulas for a_n and b_n? By (2.19 & 2.12), the coefficient a_n is

$$a_n = \int_0^{2\pi} \left[e^{-inx} f(x) + e^{inx} f^*(x) \right] \frac{dx}{2\pi} = \int_0^{2\pi} \frac{\left(e^{-inx} + e^{inx} \right)}{2} f(x) \frac{dx}{2\pi} \tag{2.21}$$

since the function $f(x)$ is real. So the coefficient a_n of $\cos nx$ in (2.20) is the cosine integral of $f(x)$

$$a_n = \int_0^{2\pi} \cos nx\, f(x) \frac{dx}{\pi} = \int_{-\pi}^{\pi} \cos nx\, f(x) \frac{dx}{\pi}. \tag{2.22}$$

Similarly, (2.19 & 2.12) and the reality of $f(x)$ imply that the coefficient b_n is the sine integral of $f(x)$

$$b_n = \int_0^{2\pi} i \frac{\left(e^{-inx} - e^{inx} \right)}{2} f(x) \frac{dx}{\pi} = \int_0^{2\pi} \sin nx\, f(x) \frac{dx}{\pi} = \int_{-\pi}^{\pi} \sin nx\, f(x) \frac{dx}{\pi}. \tag{2.23}$$

The real Fourier series (2.20) and the cosine (2.22) and sine (2.23) integrals for the coefficients a_n and b_n also follow from the orthogonality relations

$$\int_0^{2\pi} \sin mx \sin nx \, dx = \begin{cases} \pi & \text{if } n = m \neq 0 \\ 0 & \text{otherwise,} \end{cases} \tag{2.24}$$

$$\int_0^{2\pi} \cos mx \cos nx \, dx = \begin{cases} \pi & \text{if } n = m \neq 0 \\ 2\pi & \text{if } n = m = 0 \\ 0 & \text{otherwise, and} \end{cases} \tag{2.25}$$

$$\int_0^{2\pi} \sin mx \cos nx \, dx = 0, \tag{2.26}$$

which hold for integer values of n and m.

What if a function $f(x)$ is not periodic? The Fourier series for a function that is not periodic is itself strictly periodic. In such cases, the Fourier series differs somewhat from the function near the ends of the interval and differs markedly from it outside the interval, where the series but not the function is periodic.

Example 2.4 (The Fourier series for x^2) The function x^2 is even and so the integrals (2.23) for its sine Fourier coefficients b_n all vanish. Its cosine coefficients a_n are given by (2.22)

$$a_n = \int_{-\pi}^{\pi} \cos nx f(x) \frac{dx}{\pi} = \int_{-\pi}^{\pi} \cos nx \, x^2 \frac{dx}{\pi}. \tag{2.27}$$

Integrating twice by parts, we find for $n \neq 0$

$$a_n = -\frac{2}{n} \int_{-\pi}^{\pi} x \sin nx \frac{dx}{\pi} = (-1)^n \frac{4}{n^2} \tag{2.28}$$

and

$$a_0 = \int_{-\pi}^{\pi} x^2 \frac{dx}{\pi} = \frac{2\pi^2}{3}. \tag{2.29}$$

Equation (2.20) now gives for x^2 the cosine Fourier series

$$x^2 = \frac{a_0}{2} + \sum_{n=1}^{\infty} a_n \cos nx = \frac{\pi^2}{3} + 4 \sum_{n=1}^{\infty} (-1)^n \frac{\cos nx}{n^2}. \tag{2.30}$$

This series rapidly converges within the interval $(-\pi, \pi)$ as shown in Fig. 2.3, but not near its endpoints $\pm\pi$.

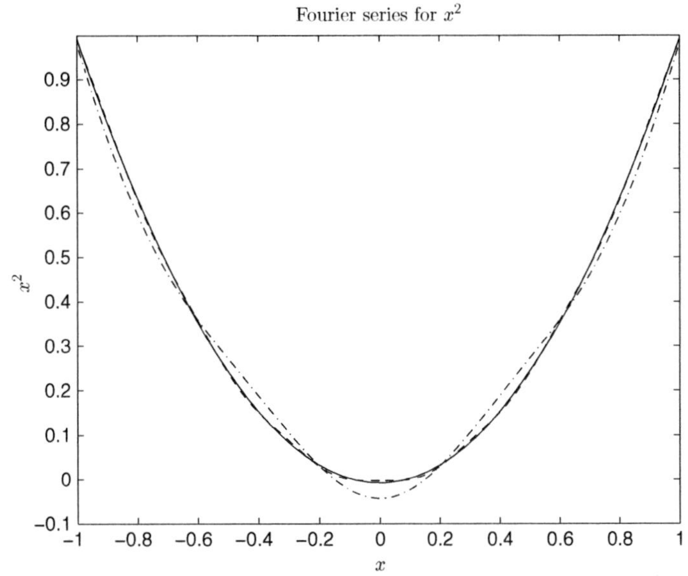

Figure 2.3 The function x^2 (solid) and its Fourier series of seven terms (dot dash) and 20 terms (dashes). The Fourier series (2.30) for x^2 quickly converges well inside the interval $(-\pi, \pi)$.

□

Example 2.5 (The Gibbs overshoot) The function $f(x) = x$ on the interval $(-\pi, \pi)$ is not periodic. So we expect trouble if we represent it as a Fourier series. Since x is an odd function, equation (2.22) tells us that the coefficients a_n all vanish. By (2.23), the b_ns are

$$b_n = \int_{-\pi}^{\pi} \frac{dx}{\pi}\, x \sin nx = 2\,(-1)^{n+1}\,\frac{1}{n}. \tag{2.31}$$

As shown in Fig. 2.4, the series

$$\sum_{n=1}^{\infty} 2\,(-1)^{n+1}\,\frac{1}{n}\,\sin nx \tag{2.32}$$

differs by about 2π from the function $f(x) = x$ for $-3\pi < x < -\pi$ and for $\pi < x < 3\pi$ because the series is periodic while the function x isn't.

Within the interval $(-\pi, \pi)$, the series with 100 terms is very accurate except for $x \gtrsim -\pi$ and $x \lesssim \pi$, where it overshoots by about 9% of the 2π discontinuity, a defect called the **Gibbs phenomenon** or the **Gibbs overshoot** (J. Willard Gibbs, 1839–1903; incidentally Gibbs's father successfully defended the Africans of the schooner *Amistad*). Any time we use a Fourier series to represent an aperiodic function, a Gibbs phenomenon will occur near the endpoints of the interval.

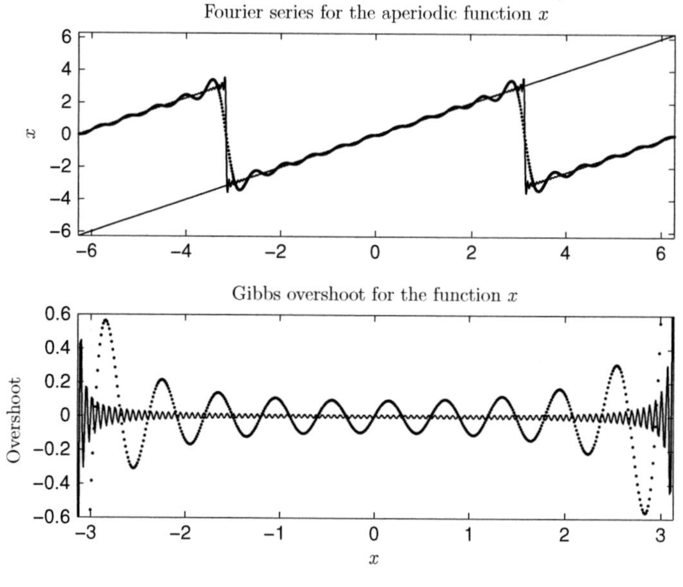

Figure 2.4 (top) The Fourier series (2.32) for the function x (solid line) with ten terms (dots) and 100 terms (solid curve) for $-2\pi < x < 2\pi$. The Fourier series is periodic, but the function x is not. (bottom) The differences between x and the ten-term (dots) and the 100-term (solid curve) on $(-\pi, \pi)$ exhibit a Gibbs overshoot of about 9% at $x \gtrsim -\pi$ and at $x \lesssim \pi$. □

2.5 Stretched intervals

If the interval of periodicity is of length L instead of 2π, then we may use the phases $\exp(i2\pi n/\sqrt{L})$, which are orthonormal on the interval $(0, L)$

$$\int_0^L dx \left(\frac{e^{i2\pi nx/L}}{\sqrt{L}}\right)^* \frac{e^{i2\pi mx/L}}{\sqrt{L}} = \delta_{nm}. \tag{2.33}$$

The Fourier series

$$f(x) = \sum_{n=-\infty}^{\infty} f_n \frac{e^{i2\pi nx/L}}{\sqrt{L}} \tag{2.34}$$

is periodic with period L. The coefficient f_n is the integral

$$f_n = \int_0^L \frac{e^{-i2\pi nx/L}}{\sqrt{L}} f(x)\, dx. \tag{2.35}$$

These relations (2.33–2.35) generalize to the interval $[0, L]$ our earlier formulas (2.1–2.3) for the interval $[0, 2\pi]$.

If the function $f(x)$ is periodic $f(x \pm L) = f(x)$ with period L, then we may shift the domain of integration by any real number r

$$f_n = \int_r^{L+r} \frac{e^{-i2\pi nx/L}}{\sqrt{L}} f(x)\, dx. \tag{2.36}$$

An obvious choice is $r = -L/2$ for which (2.34) and (2.35) give

$$f(x) = \sum_{n=-\infty}^{\infty} f_n \frac{e^{i2\pi nx/L}}{\sqrt{L}} \quad \text{and} \quad f_n = \int_{-L/2}^{L/2} \frac{e^{-i2\pi nx/L}}{\sqrt{L}} f(x)\, dx. \tag{2.37}$$

If the function $f(x)$ is real, then on the interval $[0, L]$ in place of equations (2.20, 2.22, & 2.23) one has

$$f(x) = \frac{a_0}{2} + \sum_{n=1}^{\infty} a_n \cos\left(\frac{2\pi nx}{L}\right) + b_n \sin\left(\frac{2\pi nx}{L}\right), \tag{2.38}$$

$$a_n = \frac{2}{L} \int_0^L dx \cos\left(\frac{2\pi nx}{L}\right) f(x), \tag{2.39}$$

and

$$b_n = \frac{2}{L} \int_0^L dx \sin\left(\frac{2\pi nx}{L}\right) f(x). \tag{2.40}$$

The corresponding orthogonality relations, which follow from equations (2.24, 2.25, & 2.26), are:

$$\int_0^L dx \sin\left(\frac{2\pi mx}{L}\right) \sin\left(\frac{2\pi nx}{L}\right) = \begin{cases} L/2 & \text{if } n = m \neq 0 \\ 0 & \text{otherwise} \end{cases} \tag{2.41}$$

$$\int_0^L dx \cos\left(\frac{2\pi mx}{L}\right) \cos\left(\frac{2\pi nx}{L}\right) = \begin{cases} L/2 & \text{if } n = m \neq 0 \\ L & \text{if } n = m = 0 \\ 0 & \text{otherwise} \end{cases} \tag{2.42}$$

$$\int_0^L dx \sin\left(\frac{2\pi mx}{L}\right) \cos\left(\frac{2\pi nx}{L}\right) = 0. \tag{2.43}$$

They hold for integer values of n and m, and they imply equations (2.38–2.40).

2.6 Fourier series in several variables

On the interval $[-L, L]$, the Fourier-series formulas (2.34 & 2.35) are

$$f(x) = \sum_{n=-\infty}^{\infty} f_n \frac{e^{i\pi nx/L}}{\sqrt{2L}} \tag{2.44}$$

$$f_n = \int_{-L}^{L} \frac{e^{-i\pi nx/L}}{\sqrt{2L}} f(x)\, dx. \tag{2.45}$$

We may generalize these equations from a single variable to N variables $x = (x_1, \ldots, x_N)$ with $n \cdot x = n_1 x_1 + \cdots + n_N x_N$

$$f(x) = \sum_{n_1=-\infty}^{\infty} \cdots \sum_{n_N=-\infty}^{\infty} f_n \frac{e^{i\pi n \cdot x/L}}{(2L)^{N/2}} \tag{2.46}$$

$$f_n = \int_{-L}^{L} dx_1 \ldots \int_{-L}^{L} dx_N \frac{e^{-i\pi n \cdot x/L}}{(2L)^{N/2}} f(x). \tag{2.47}$$

2.7 How Fourier series converge

A Fourier series represents a function $f(x)$ as the limit of a sequence of functions $f_N(x)$ given by

$$f_N(x) = \sum_{k=-N}^{N} f_k \frac{e^{i2\pi kx/L}}{\sqrt{L}} \quad \text{in which} \quad f_k = \int_0^L f(x)\, e^{-i2\pi kx/L} \frac{dx}{\sqrt{L}}. \tag{2.48}$$

Since the exponentials are periodic with period L, a Fourier series always is periodic. So if the function $f(x)$ is not periodic, then its Fourier series will represent the periodic extension f_p of f defined by $f_p(x + nL) = f(x)$ for $0 \leq x \leq L$.

A sequence of functions $f_N(x)$ **converges** to a function $f(x)$ on a (**closed**) interval $[a, b]$ if for every $\epsilon > 0$ and each point $a \le x \le b$ there exists an integer $N(\epsilon, x)$ such that

$$\left|f(x) - f_N(x)\right| < \epsilon \quad \text{for all} \quad N > N(\epsilon, x). \tag{2.49}$$

If this holds for an $N(\epsilon)$ that is independent of $x \in [a, b]$, then the sequence of functions $f_N(x)$ **converges uniformly** to $f(x)$ on the interval $[a, b]$.

A function $f(x)$ is **continuous** on an *open* interval (a, b) if for every point $a < x < b$ the two limits

$$f(x-0) \equiv \lim_{0<\epsilon\to 0} f(x-\epsilon) \quad \text{and} \quad f(x+0) \equiv \lim_{0<\epsilon\to 0} f(x+\epsilon) \tag{2.50}$$

agree; it also is continuous on the *closed* interval $[a, b]$ if $f(a+0) = f(a)$ and $f(b-0) = f(b)$. A function continuous on $[a, b]$ is bounded there.

If a sequence of continuous functions $f_N(x)$ converges uniformly to a function $f(x)$ on a closed interval $a \le x \le b$, then we know that $|f_N(x) - f(x)| < \epsilon$ for $N > N(\epsilon)$, and so

$$\left|\int_a^b f_N(x)\,dx - \int_a^b f(x)\,dx\right| \le \int_a^b \left|f_N(x) - f(x)\right|\,dx < (b-a)\epsilon. \tag{2.51}$$

Thus one may integrate a uniformly convergent sequence of continuous functions on a closed interval $[a, b]$ term by term

$$\lim_{N\to\infty} \int_a^b f_N(x)\,dx = \int_a^b f(x)\,dx. \tag{2.52}$$

A function is **piecewise continuous** on $[a, b]$ if it is continuous there except for finite jumps from $f(x-0)$ to $f(x+0)$ at a finite number of points x. At such jumps, we shall *define* the periodically extended function f_p to be the mean $f_p(x) = [f_p(x-0) + f_p(x+0)]/2$.

Fourier's convergence theorem (Courant, 1937, p. 439) The Fourier series of a function $f(x)$ that is piecewise continuous with a piecewise continuous first derivative converges to its periodic extension $f_p(x)$. This convergence is uniform on every closed interval on which the function $f(x)$ is continuous (and absolute if the function $f(x)$ has no discontinuities). Examples 2.11 and 2.12 illustrate this result.

A function whose kth derivative is continuous is in **class C^k**. On the interval $[-\pi, \pi]$, its Fourier coefficients (2.13) are

$$f_n = \int_{-\pi}^{\pi} f(x)\, e^{-inx}\,dx. \tag{2.53}$$

If f is both periodic and in C^k, then one integration by parts gives

$$f_n = \int_{-\pi}^{\pi} \left\{ \frac{d}{dx} \left[f(x) \frac{e^{-inx}}{-in} \right] - f'(x) \frac{e^{-inx}}{-in} \right\} dx = \int_{-\pi}^{\pi} f'(x) \frac{e^{-inx}}{in} dx$$

and k integrations by parts give

$$f_n = \int_{-\pi}^{\pi} f^{(k)}(x) \frac{e^{-inx}}{(in)^k} dx \tag{2.54}$$

since the derivatives $f^{(\ell)}(x)$ of a C^k periodic function also are periodic. Moreover, if $f^{(k+1)}$ is piecewise continuous, then

$$\begin{aligned} f_n &= \int_{-\pi}^{\pi} \left\{ \frac{d}{dx} \left[f^{(k)}(x) \frac{e^{-inx}}{-(in)^{k+1}} \right] - f^{(k+1)}(x) \frac{e^{-inx}}{-(in)^{k+1}} \right\} dx \\ &= \int_{-\pi}^{\pi} f^{(k+1)}(x) \frac{e^{-inx}}{(in)^{k+1}} dx. \end{aligned} \tag{2.55}$$

Since $f^{(k+1)}(x)$ is piecewise continuous on the closed interval $[-\pi, \pi]$, it is bounded there in absolute value by, let us say, M. So the Fourier coefficients of a C^k periodic function with $f^{(k+1)}$ piecewise continuous are bounded by

$$|f_n| \le \frac{1}{n^{k+1}} \int_{-\pi}^{\pi} |f^{(k+1)}(x)| \, dx \le \frac{2\pi M}{n^{k+1}}. \tag{2.56}$$

We often can carry this derivation one step further. In most simple examples, the piecewise continuous periodic function $f^{(k+1)}(x)$ actually is piecewise continuously differentiable between its successive jumps at x_j. In this case, the derivative $f^{(k+2)}(x)$ is a piecewise continuous function plus a sum of a finite number of delta functions with finite coefficients. Thus we can integrate once more by parts. If for instance the function $f^{(k+1)}(x)$ jumps J times between $-\pi$ and π by $\Delta f_j^{(k+1)}$, then the Fourier coefficients are

$$\begin{aligned} f_n &= \int_{-\pi}^{\pi} f^{(k+2)}(x) \frac{e^{-inx}}{(in)^{k+2}} dx \\ &= \sum_{j=1}^{J} \int_{x_j}^{x_{j+1}} f_{\mathrm{s}}^{(k+2)}(x) \frac{e^{-inx}}{(in)^{k+2}} dx + \sum_{j=1}^{J} \Delta f_j^{(k+2)} \frac{e^{-inx_j}}{(in)^{k+2}} \end{aligned} \tag{2.57}$$

in which the subscript s means that we've separated out the delta functions. The Fourier coefficients then would be bounded by

$$|f_n| \le \frac{2\pi M}{n^{k+2}} \tag{2.58}$$

in which M is related to the maximum absolute values of $f_{\mathrm{s}}^{(k+2)}(x)$ and of the $\Delta f_j^{(k+1)}$. The Fourier series of periodic C^k functions converge very rapidly if k is big.

Example 2.6 (Fourier series of a C^0 function) The function defined by

$$f(x) = \begin{cases} 0, & -\pi \le x < 0 \\ x, & 0 \le x < \pi/2 \\ \pi - x, & \pi/2 \le x \le \pi \end{cases} \tag{2.59}$$

is continuous on the interval $[-\pi, \pi]$ and its first derivative is piecewise continuous on that interval. By (2.56), its Fourier coefficients f_n should be bounded by M/n. In fact they are (exercise 2.8)

$$f_n = \int_{-\pi}^{\pi} f(x) e^{-inx} \frac{dx}{\sqrt{2\pi}} = \frac{(-1)^{n+1}}{\sqrt{2\pi}} \frac{(i^n - 1)^2}{n^2} \tag{2.60}$$

bounded by $2\sqrt{2/\pi}/n^2$ in agreement with the stronger inequality (2.58). □

Example 2.7 (Fourier series for a C^1 function) The function defined by $f(x) = 1 + \cos 2x$ for $|x| \le \pi/2$ and $f(x) = 0$ for $|x| \ge \pi/2$ has a periodic extension f_p that is continuous with a continuous first derivative and a piecewise continuous second derivative. Its Fourier coefficients (2.53)

$$f_n = \int_{-\pi/2}^{\pi/2} (1 + \cos 2x) e^{-inx} \frac{dx}{\sqrt{2\pi}} = \frac{8 \sin n\pi/2}{\sqrt{2\pi}(4n - n^3)}$$

satisfy the inequalities (2.56) and (2.58) for $k = 1$. □

Example 2.8 (The Fourier series for $\cos \mu x$) The Fourier series for the even function $f(x) = \cos \mu x$ has only cosines with coefficients (2.22)

$$\begin{aligned} a_n &= \int_{-\pi}^{\pi} \cos nx \cos \mu x \frac{dx}{\pi} = \int_0^{\pi} [\cos(\mu + n)x + \cos(\mu - n)x] \frac{dx}{\pi} \\ &= \frac{1}{\pi} \left[\frac{\sin(\mu + n)\pi}{\mu + n} + \frac{\sin(\mu - n)\pi}{\mu - n} \right] = \frac{2}{\pi} \frac{\mu(-1)^n}{\mu^2 - n^2} \sin \mu\pi. \end{aligned} \tag{2.61}$$

Thus, whether or not μ is an integer, the series (2.20) gives us

$$\cos \mu x = \frac{2\mu \sin \mu\pi}{\pi} \left(\frac{1}{2\mu^2} - \frac{\cos x}{\mu^2 - 1^2} + \frac{\cos 2x}{\mu^2 - 2^2} - \frac{\cos 3x}{\mu^2 - 3^2} + - \cdots \right), \tag{2.62}$$

which is continuous at $x = \pm\pi$ (Courant, 1937, chap. IX). □

Example 2.9 (The sine as an infinite product) In our series (2.62) for $\cos \mu x$, we set $x = \pi$, divide by $\sin \mu\pi$, replace μ with x, and so find for the cotangent the expansion

$$\cot \pi x = \frac{2x}{\pi} \left(\frac{1}{2x^2} + \frac{1}{x^2 - 1^2} + \frac{1}{x^2 - 2^2} + \frac{1}{x^2 - 3^2} + \cdots \right) \tag{2.63}$$

or equivalently

$$\cot \pi x - \frac{1}{\pi x} = -\frac{2x}{\pi}\left(\frac{1}{1^2 - x^2} + \frac{1}{2^2 - x^2} + \frac{1}{3^2 - x^2} + \cdots\right). \tag{2.64}$$

For $0 \le x \le q < 1$, the absolute value of the nth term on the right is less than $2q/(\pi(n^2 - q^2))$. Thus this series converges uniformly on $[0, x]$, and so we may integrate it term by term. We find (exercise 2.11)

$$\pi \int_0^x \left(\cot \pi t - \frac{1}{\pi t}\right) dt = \ln \frac{\sin \pi x}{\pi x} = \sum_{n=1}^{\infty} \int_0^x \frac{-2t\, dt}{n^2 - t^2} = \sum_{n=1}^{\infty} \ln \left[1 - \frac{x^2}{n^2}\right]. \tag{2.65}$$

Exponentiating, we get the infinite-product formula

$$\frac{\sin \pi x}{\pi x} = \exp\left[\sum_{n=1}^{\infty} \ln\left(1 - \frac{x^2}{n^2}\right)\right] = \prod_{n=1}^{\infty}\left(1 - \frac{x^2}{n^2}\right) \tag{2.66}$$

for the sine, from which one can derive the infinite product (exercise 2.12)

$$\cos \pi x = \prod_{n=1}^{\infty}\left(1 - \frac{x^2}{\left(n - \frac{1}{2}\right)^2}\right) \tag{2.67}$$

for the cosine (Courant, 1937, chap. IX). □

Fourier series can represent a much wider class of functions than those that are continuous. If a function $f(x)$ is square integrable on an interval $[a, b]$, then its N-term Fourier series $f_N(x)$ will converge to $f(x)$ **in the mean**, that is

$$\lim_{N\to\infty} \int_a^b dx |f(x) - f_N(x)|^2 = 0. \tag{2.68}$$

What happens to the convergence of a Fourier series if we integrate or differentiate term by term? If we integrate the series

$$f(x) = \sum_{n=-\infty}^{\infty} f_n \frac{e^{i2\pi n x/L}}{\sqrt{L}}, \tag{2.69}$$

then we get a series

$$F(x) = \int_a^x dx' f(x') = f_0 \frac{(x-a)}{\sqrt{L}} - i\frac{\sqrt{L}}{2\pi} \sum_{n=-\infty}^{\infty} \frac{f_n}{n}\left(e^{i2\pi n x/L} - e^{i2\pi n a/L}\right) \tag{2.70}$$

that converges **better** because of the extra factor of $1/n$. An integrated function $f(x)$ is smoother, and so its Fourier series converges better.

But if we differentiate the same series, then we get a series

$$f'(x) = i\frac{2\pi}{L^{3/2}} \sum_{n=-\infty}^{\infty} n f_n \, e^{i2\pi nx/L} \tag{2.71}$$

that converges **less well** because of the extra factor of n. A differentiated function is rougher, and so its Fourier series converges less well.

2.8 Quantum-mechanical examples

Suppose a particle of mass m is trapped in an infinitely deep one-dimensional square well of potential energy

$$V(x) = \begin{cases} 0 & \text{if } 0 < x < L \\ \infty & \text{otherwise.} \end{cases} \tag{2.72}$$

The hamiltonian operator is

$$H = -\frac{\hbar^2}{2m}\frac{d^2}{dx^2} + V(x), \tag{2.73}$$

in which $\hbar$ is Planck's constant divided by 2π. This tiny bit of action, $\hbar = 1.055 \times 10^{-34}$ J s, sets the scale at which quantum mechanics becomes important; quantum-mechanical corrections to classical predictions often are important in processes whose action is less than $\hbar$.

An eigenfunction $\psi(x)$ of the hamiltonian H with energy E satisfies the equation $H\psi(x) = E\psi(x)$, which breaks into two simple equations:

$$-\frac{\hbar^2}{2m}\frac{d^2\psi(x)}{dx^2} = E\psi(x) \quad \text{for} \quad 0 < x < L \tag{2.74}$$

and

$$-\frac{\hbar^2}{2m}\frac{d^2\psi(x)}{dx^2} + \infty\,\psi(x) = E\psi(x) \quad \text{for} \quad x < 0 \quad \text{and for} \quad x > L. \tag{2.75}$$

Every solution of these equations with finite energy E must vanish outside the interval $0 < x < L$. So we must find solutions of the first equation (2.74) that satisfy the boundary conditions

$$\psi(x) = 0 \quad \text{for} \quad x \le 0 \text{ and } x \ge L. \tag{2.76}$$

For any integer $n \neq 0$, the function

$$\psi_n(x) = \sqrt{\frac{2}{L}}\,\sin\left(\frac{\pi n x}{L}\right) \quad \text{for} \quad x \in [0, L] \tag{2.77}$$

and $\psi_n(x) = 0$ for $x \notin (0, L)$ satisfies the boundary conditions (2.76). When inserted into equation (2.74)

$$-\frac{\hbar^2}{2m}\frac{d^2}{dx^2}\psi_n(x) = \frac{\hbar^2}{2m}\left(\frac{n\pi}{L}\right)^2 \psi_n(x) = E_n\psi_n(x) \tag{2.78}$$

it reveals its energy to be $E_n = (n\pi\hbar/L)^2/2m$.

These eigenfunctions $\psi_n(x)$ are complete in the sense that they span the space of all functions $f(x)$ that are square-integrable on the interval $(0, L)$ and vanish at its endpoints. They provide for such functions the **sine Fourier series**

$$f(x) = \sum_{n=1}^{\infty} f_n \sqrt{\frac{2}{L}} \sin\left(\frac{\pi n x}{L}\right), \tag{2.79}$$

which also is the Fourier series for a function that is odd $f(-x) = -f(x)$ on the interval $(-L, L)$ and zero at both ends.

Example 2.10 (Time evolution of an initially piecewise continuous wave function) Suppose now that at time $t = 0$ the particle is confined to the middle half of the well with the square-wave wave function

$$\psi(x, 0) = \sqrt{\frac{2}{L}} \quad \text{for} \quad \frac{L}{4} < x < \frac{3L}{4} \tag{2.80}$$

and zero otherwise. This piecewise continuous wave function is discontinuous at $x = L/4$ and at $x = 3L/4$. Since the functions $\langle x|n\rangle = \psi_n(x)$ are orthonormal on $[0, L]$

$$\int_0^L dx \sqrt{\frac{2}{L}} \sin\left(\frac{\pi n x}{L}\right) \sqrt{\frac{2}{L}} \sin\left(\frac{\pi m x}{L}\right) = \delta_{nm}, \tag{2.81}$$

the coefficients f_n in the Fourier series

$$\psi(x, 0) = \sum_{n=1}^{\infty} f_n \sqrt{\frac{2}{L}} \sin\left(\frac{\pi n x}{L}\right) \tag{2.82}$$

are the inner products

$$f_n = \langle n|\psi, 0\rangle = \int_0^L dx \sqrt{\frac{2}{L}} \sin\left(\frac{\pi n x}{L}\right) \psi(x, 0). \tag{2.83}$$

They are proportional to $1/n$ in accord with (2.58)

$$f_n = \frac{2}{L}\int_{L/4}^{3L/4} dx \sin\left(\frac{\pi n x}{L}\right) = \frac{2}{\pi n}\left[\cos\left(\frac{\pi n}{4}\right) - \cos\left(\frac{3\pi n}{4}\right)\right]. \tag{2.84}$$

Figure 2.5 plots the wave function $\psi(x, 0)$ (2.80, straight solid lines) and its ten-term (solid curve) and 100-term (dashes) Fourier series (2.82) for an interval of length $L = 2$. Gibbs's overshoot reaches 1.093 at $x = 0.52$ for

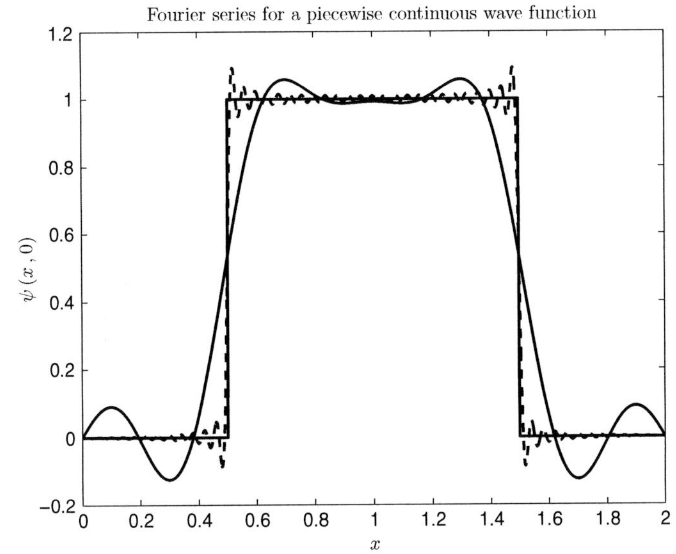

Figure 2.5 The piecewise continuous wave function $\psi(x,0)$ for $L = 2$ (2.80, straight solid lines) and its Fourier series (2.82) with ten terms (solid curve) and 100 terms (dashes). Gibbs overshoots occur near the discontinuities at $x = 1/2$ and $x = 3/2$.

100 terms and 1.0898 at $x = 0.502$ for 1000 terms (not shown), amounting to about 9% of the unit discontinuity at $x = 1/2$. A similar overshoot occurs at $x = 3/2$.

How does $\psi(x,0)$ evolve with time? Since $\psi_n(x)$, the Fourier component (2.77), is an eigenfunction of H with energy E_n, the time-evolution operator $U(t) = \exp(-iHt/\hbar)$ takes $\psi(x,0)$ into

$$\psi(x,t) = e^{-iHt/\hbar}\,\psi(x,0) = \sum_{n=1}^{\infty} f_n \sqrt{\frac{2}{L}}\,\sin\left(\frac{\pi n x}{L}\right)\,e^{-iE_n t/\hbar}. \tag{2.85}$$

Because $E_n = (n\pi\hbar)^2/2m$, the wave function at time t is

$$\psi(x,t) = \sum_{n=1}^{\infty} f_n \sqrt{\frac{2}{L}}\,\sin\left(\frac{\pi n x}{L}\right)\,e^{-i\hbar(n\pi)^2 t/(2mL^2)}. \tag{2.86}$$

It is awkward to plot complex functions, so Fig. 2.6 displays the probability distributions $P(x,t) = |\psi(x,t)|^2$ of the 1000-term Fourier series (2.86) for the wave function $\psi(x,t)$ at $t = 0$ (thick curve), $t = 10^{-3}\,\tau$ (medium curve), and $\tau = 2mL^2/\hbar$ (thin curve). The discontinuities in the initial wave function $\psi(x,0)$ cause both the Gibbs overshoots at $x = 1/2$ and $x = 3/2$ seen in the series for $\psi(x,0)$ plotted in Fig. 2.5 and the choppiness of the probability distribution $P(x,t)$ exhibited in Fig. (2.6).

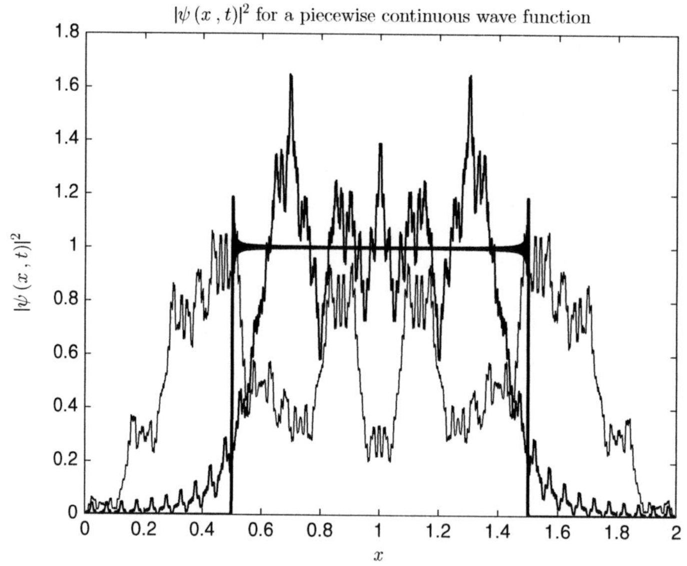

Figure 2.6 For an interval of length $L = 2$, the probability distributions $P(x,t) = |\psi(x,t)|^2$ of the 1000-term Fourier series (2.86) for the wave function $\psi(x,t)$ at $t = 0$ (thick curve), $t = 10^{-3}\,\tau$ (medium curve), and $\tau = 2mL^2/\hbar$ (thin curve). The jaggedness of $P(x,t)$ arises from the two discontinuities in the initial wave function $\psi(x,0)$ (2.87) at $x = L/4$ and $x = 3L/4$.

□

Example 2.11 (Time evolution of a continuous function) What does the Fourier series of a continuous function look like? How does it evolve with time? Let us take as the wave function at $t = 0$ the function

$$\psi(x,0) = \frac{2}{\sqrt{L}} \sin\left(\frac{2\pi(x - L/4)}{L}\right) \quad \text{for} \quad \frac{L}{4} < x < \frac{3L}{4} \tag{2.87}$$

and zero otherwise. This initial wave function is a continuous function with a piecewise continuous first derivative on the interval $[0, L]$, and it satisfies the periodic boundary condition $\psi(0,0) = \psi(L,0)$. It therefore satisfies the conditions of Fourier's convergence theorem (Courant, 1937, p. 439), and so its Fourier series converges uniformly (and absolutely) to $\psi(x,0)$ on $[0, L]$.

As in equation (2.83), the Fourier coefficients f_n are given by the integrals

$$f_n = \int_0^L dx \sqrt{\frac{2}{L}} \sin\left(\frac{\pi n x}{L}\right) \psi(x,0), \tag{2.88}$$

which now take the form

$$f_n = \frac{2\sqrt{2}}{L} \int_{L/4}^{3L/4} dx \sin\left(\frac{\pi n x}{L}\right) \sin\left(\frac{2\pi(x - L/4)}{L}\right). \tag{2.89}$$

Doing the integral, one finds for f_n that for $n \neq 2$

$$f_n = -\frac{\sqrt{2}}{\pi}\frac{4}{n^2-4}\left[\sin(3n\pi/4) + \sin(n\pi/4)\right] \tag{2.90}$$

while $c_2 = 0$. These Fourier coefficients satisfy the inequalities (2.56) and (2.58) for $k = 0$. The factor of $1/n^2$ in f_n guarantees the absolute convergence of the series

$$\psi(x,0) = \sum_{n=1}^{\infty} f_n \sqrt{\frac{2}{L}}\,\sin\left(\frac{\pi n x}{L}\right) \tag{2.91}$$

because asymptotically the coefficient f_n is bounded by $f_n \leq A/n^2$ where A is a constant ($A = 144/(5\pi\sqrt{L})$ will do) and the sum of $1/n^2$ converges to the Riemann zeta function (4.92)

$$\sum_{n=1}^{\infty} \frac{1}{n^2} = \zeta(2) = \frac{\pi^2}{6}. \tag{2.92}$$

Figure 2.7 plots the ten-term Fourier series (2.91) for $\psi(x,0)$ for $L = 2$. Because this series converges absolutely and uniformly on $[0, 2]$, the 100-term and 1000-term series were too close to $\psi(x,0)$ to be seen clearly in the figure and so were omitted.

As time goes by, the wave function $\psi(x,t)$ evolves from $\psi(x,0)$ to

$$\psi(x,t) = \sum_{n=1}^{\infty} f_n \sqrt{\frac{2}{L}}\,\sin\left(\frac{\pi n x}{L}\right)\, e^{-i\hbar(n\pi)^2 t/(2mL^2)} \tag{2.93}$$

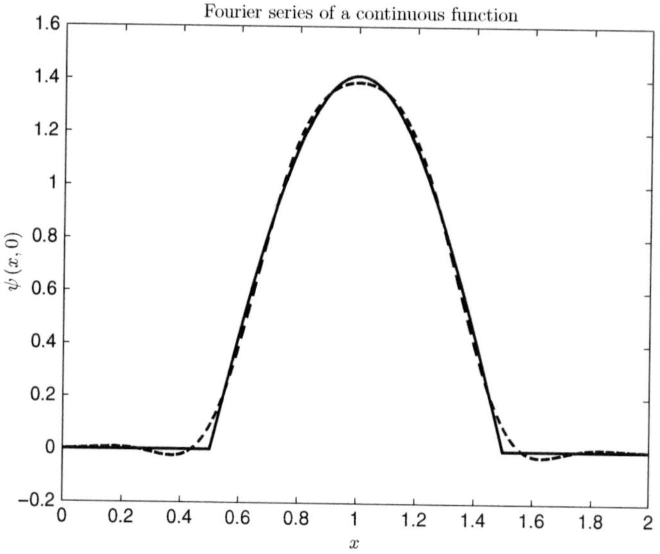

Figure 2.7 The continuous wave function $\psi(x,0)$ (2.87, solid) and its ten-term Fourier series (2.90–2.91, dashes) are plotted for the interval $[0, 2]$.

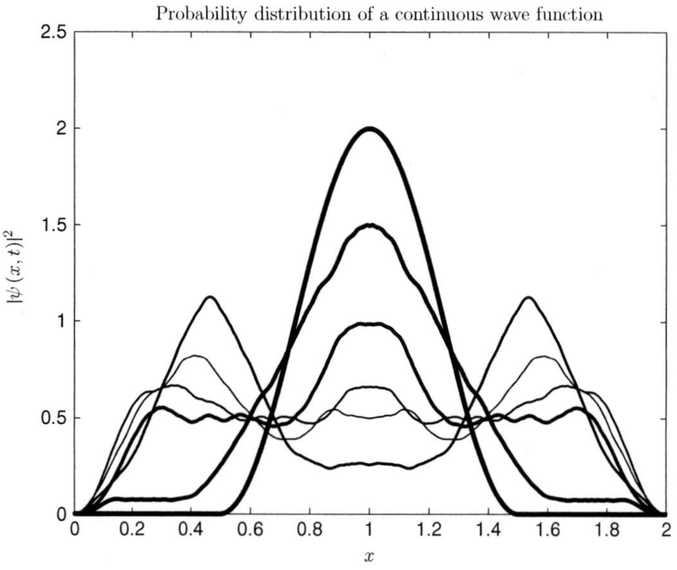

Figure 2.8 For the interval $[0, 2]$, the probability distributions $P(x, t) = |\psi(x, t)|^2$ of the 1000-term Fourier series (2.93) for the wave function $\psi(x, t)$ at $t = 0, 10^{-2}\,\tau$, $10^{-1}\,\tau$, $\tau = 2mL^2/\hbar$, 10τ, and 100τ are plotted as successively thinner curves.

in which the Fourier coefficients are given by (2.90). Because $\psi(x, 0)$ is continuous and periodic with a piecewise continuous first derivative, its evolution in time is much calmer than that of the piecewise continuous square wave (2.80). Figure 2.8 shows this evolution in successively thinner curves at times $t = 0$, $10^{-2}\,\tau$, $10^{-1}\,\tau$, $\tau = 2mL^2/\hbar$, 10τ, and 100τ. The curves at $t = 0$ and $t = 10^{-2}\,\tau$ are smooth, but some wobbles appear at $t = 10^{-1}\,\tau$ and at $t = \tau$ due to the discontinuities in the first derivative of $\psi(x, 0)$ at $x = 0.5$ and at $x = 1.5$. □

Example 2.12 (Time evolution of a smooth wave function) Finally, let's try a wave function $\psi(x, 0)$ that is periodic and infinitely differentiable on $[0, L]$. An infinitely differentiable function is said to be **smooth** or C^∞. The infinite square-well potential $V(x)$ of equation (2.72) imposes the periodic boundary conditions $\psi(0, 0) = \psi(L, 0) = 0$, so we try

$$\psi(x, 0) = \sqrt{\frac{2}{3L}}\left[1 - \cos\left(\frac{2\pi x}{L}\right)\right]. \tag{2.94}$$

Its Fourier series

$$\psi(x, 0) = \sqrt{\frac{1}{6L}}\left(2 - e^{2\pi i x/L} - e^{-2\pi i x/L}\right) \tag{2.95}$$

has coefficients that satisfy the upper bounds (2.56) by vanishing for $|n| > 1$.

The coefficients of the Fourier sine series for the wave function $\psi(x, 0)$ are given by the integrals (2.83)

$$
\begin{aligned}
f_n &= \int_0^L dx \sqrt{\frac{2}{L}} \sin\left(\frac{\pi n x}{L}\right) \psi(x,0) \\
&= \frac{2}{\sqrt{3}L} \int_0^L dx \sin\left(\frac{\pi n x}{L}\right)\left[1 - \cos\left(\frac{2\pi x}{L}\right)\right] \\
&= \frac{8\left[(-1)^n - 1\right]}{\pi\sqrt{3}\, n(n^2-4)}
\end{aligned} \tag{2.96}
$$

with all the even coefficients zero, $c_{2n} = 0$. The f_ns are proportional to $1/n^3$, which is more than enough to ensure the absolute and uniform convergence of its Fourier sine series

$$
\psi(x,0) = \sum_{n=1}^{\infty} f_n \sqrt{\frac{2}{L}} \sin\left(\frac{\pi n x}{L}\right). \tag{2.97}
$$

As time goes by, it evolves to

$$
\psi(x,t) = \sum_{n=1}^{\infty} f_n \sqrt{\frac{2}{L}} \sin\left(\frac{\pi n x}{L}\right) e^{-i\hbar(n\pi)^2 t/(2mL^2)} \tag{2.98}
$$

and remains absolutely convergent for all times t.

The effects of the absolute and uniform convergence with $f_n \propto 1/n^3$ are obvious in the graphs. Figure 2.9 shows (for $L = 2$) that only ten terms are required

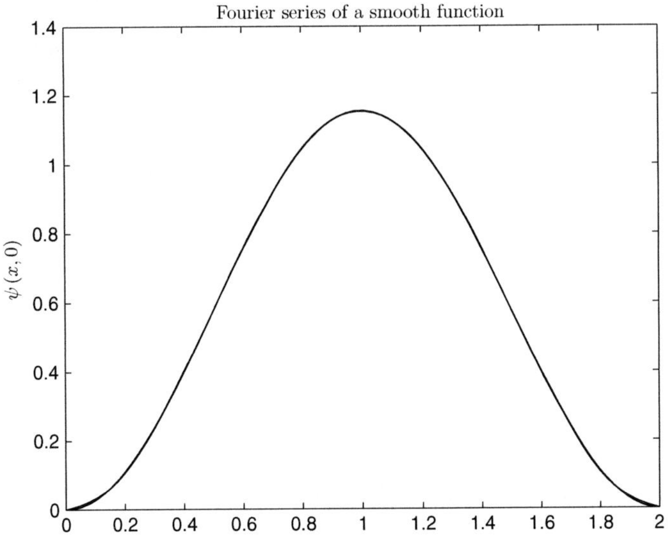

Figure 2.9 The wave function $\psi(x,0)$ (2.94) is infinitely differentiable, and so the first ten terms of its uniformly convergent Fourier series (2.97) offer a very good approximation to it.

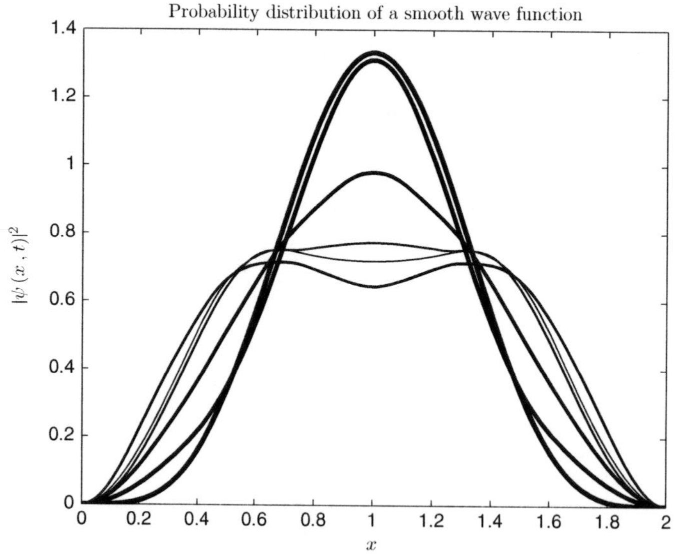

Figure 2.10 The probability distributions $P(x,t) = |\psi(x,t)|^2$ of the 1000-term Fourier series (2.98) for the wave function $\psi(x,t)$ at $t = 0,\ 10^{-2}\tau,\ 10^{-1}\tau,\ \tau = 2mL^2/\hbar,\ 10\tau$, and 100τ are plotted as successively thinner curves. The time evolution is calm because the wave function $\psi(x,0)$ is smooth.

to nearly overlap the initial wave function $\psi(x,0)$. Figure 2.10 shows that the evolution of the probability distribution $|\psi(x,t)|^2$ with time is smooth, with no sign of the jaggedness of Fig. 2.6 or the wobbles of Fig. 2.8. Because $\psi(x,0)$ is smooth and periodic, it evolves calmly as time passes. □

2.9 Dirac notation

Vectors $|j\rangle$ that are orthonormal, $\langle k|j\rangle = \delta_{k,j}$ span a vector space and express the identity operator I of the space as (1.132)

$$I = \sum_{j=1}^{N} |j\rangle\langle j|. \tag{2.99}$$

Multiplying from the right by any vector $|g\rangle$ in the space, we get

$$|g\rangle = I|g\rangle = \sum_{j=1}^{N} |j\rangle\langle j|g\rangle, \tag{2.100}$$

which says that every vector $|g\rangle$ in the space has an expansion (1.133) in terms of the N orthonormal basis vectors $|j\rangle$. The coefficients $\langle j|g\rangle$ of the expansion are inner products of the vector $|g\rangle$ with the basis vectors $|j\rangle$.

These properties of finite-dimensional vector spaces also are true of infinite-dimensional vector spaces of functions. We may use as basis vectors the phases $\exp(inx)/\sqrt{2\pi}$. They are orthonormal with inner product (2.1)

$$(m, n) = \int_0^{2\pi} \left(\frac{e^{imx}}{\sqrt{2\pi}} \right)^* \frac{e^{inx}}{\sqrt{2\pi}} \, dx = \int_0^{2\pi} \frac{e^{i(n-m)x}}{2\pi} \, dx = \delta_{m,n}, \tag{2.101}$$

which in Dirac notation with $\langle x|n\rangle = \exp(inx)/\sqrt{2\pi}$ and $\langle m|x\rangle = \langle x|m\rangle^*$ is

$$\langle m|n\rangle = \int_0^{2\pi} \langle m|x\rangle \langle x|n\rangle \, dx = \int_0^{2\pi} \frac{e^{i(n-m)x}}{2\pi} \, dx = \delta_{m,n}. \tag{2.102}$$

The identity operator for Fourier's space of functions is

$$I = \sum_{n=-\infty}^{\infty} |n\rangle\langle n|. \tag{2.103}$$

So we have

$$|f\rangle = I|f\rangle = \sum_{n=-\infty}^{\infty} |n\rangle\langle n|f\rangle \tag{2.104}$$

and

$$\langle x|f\rangle = \langle x|I|f\rangle = \sum_{n=-\infty}^{\infty} \langle x|n\rangle\langle n|f\rangle = \sum_{n=-\infty}^{\infty} \frac{e^{inx}}{\sqrt{2\pi}} \langle n|f\rangle, \tag{2.105}$$

which with $\langle n|f\rangle = f_n$ is the Fourier series (2.2). The coefficients $\langle n|f\rangle = f_n$ are the inner products (2.3)

$$\langle n|f\rangle = \int_0^{2\pi} \langle n|x\rangle\langle x|f\rangle \, dx = \int_0^{2\pi} \frac{e^{-inx}}{\sqrt{2\pi}} \langle x|f\rangle \, dx = \int_0^{2\pi} \frac{e^{-inx}}{\sqrt{2\pi}} f(x) \, dx. \tag{2.106}$$

2.10 Dirac's delta function

A Dirac delta function is a (continuous, linear) map from a space of functions into the real or complex numbers. It is a **functional** that associates a number with each function in the function space. Thus $\delta(x - y)$ associates the number $f(y)$ with the function $f(x)$. We may write this association as

$$f(y) = \int f(x) \, \delta(x - y) \, dx. \tag{2.107}$$

Delta functions pop up all over physics. The inner product of two of the kets $|x\rangle$ that appear in the Fourier-series formulas (2.105) and (2.106) is a delta function, $\langle x|y\rangle = \delta(x-y)$. The formula (2.106) for the coefficient $\langle n|f\rangle$ becomes obvious if we write the identity operator for functions defined on the interval $[0, 2\pi]$ as

$$I = \int_0^{2\pi} |x\rangle\langle x|\, dx \tag{2.108}$$

for then

$$\langle n|f\rangle = \langle n|I|f\rangle = \int_0^{2\pi} \langle n|x\rangle\langle x|f\rangle\, dx = \int_0^{2\pi} \frac{e^{-inx}}{\sqrt{2\pi}} \langle x|f\rangle\, dx. \tag{2.109}$$

The equation $|y\rangle = I|y\rangle$ with the identity operator (2.108) gives

$$|y\rangle = I|y\rangle = \int_0^{2\pi} |x\rangle\langle x|y\rangle\, dx. \tag{2.110}$$

Multiplying (2.108) from the right by $|f\rangle$ and from the left by $\langle y|$, we get

$$f(y) = \langle y|I|f\rangle = \int_0^{2\pi} \langle y|x\rangle\langle x|f\rangle\, dx = \int_0^{2\pi} \langle y|x\rangle f(x)\, dx. \tag{2.111}$$

These relations (2.110) and (2.111) say that the inner product $\langle y|x\rangle$ is a **delta function**, $\langle y|x\rangle = \langle x|y\rangle = \delta(x-y)$.

The Fourier-series formulas (2.105) and (2.106) lead to a statement about the completeness of the phases $\exp(inx)/\sqrt{2\pi}$

$$f(x) = \sum_{n=-\infty}^{\infty} f_n \frac{e^{inx}}{\sqrt{2\pi}} = \sum_{n=-\infty}^{\infty} \int_0^{2\pi} \frac{e^{-iny}}{\sqrt{2\pi}} f(y) \frac{e^{inx}}{\sqrt{2\pi}}\, dy. \tag{2.112}$$

Interchanging and rearranging, we have

$$f(x) = \int_0^{2\pi} \left(\sum_{n=-\infty}^{\infty} \frac{e^{in(x-y)}}{2\pi} \right) f(y)\, dy. \tag{2.113}$$

But the phases (2.112) are **periodic** with period 2π, so we also have

$$f(x + 2\pi i) = \int_0^{2\pi} \left(\sum_{n=-\infty}^{\infty} \frac{e^{in(x-y)}}{2\pi} \right) f(y)\, dy. \tag{2.114}$$

Thus we arrive at the **Dirac comb**

$$\sum_{n=-\infty}^{\infty} \frac{e^{in(x-y)}}{2\pi} = \sum_{\ell=-\infty}^{\infty} \delta(x - y - 2\pi\ell) \tag{2.115}$$

or more simply

$$\sum_{n=-\infty}^{\infty} \frac{e^{inx}}{2\pi} = \frac{1}{2\pi}\left[1 + 2\sum_{n=1}^{\infty} \cos(nx)\right] = \sum_{\ell=-\infty}^{\infty} \delta(x - 2\pi\ell). \tag{2.116}$$

Example 2.13 (Dirac's comb) The sum of the first 100,000 terms of this cosine series (2.116) for the Dirac comb is plotted for the interval $(-15, 15)$ in Fig. 2.11. Gibbs overshoots appear at the discontinuities. The integral of the first 100,000 terms from -15 to 15 is 5.0000.

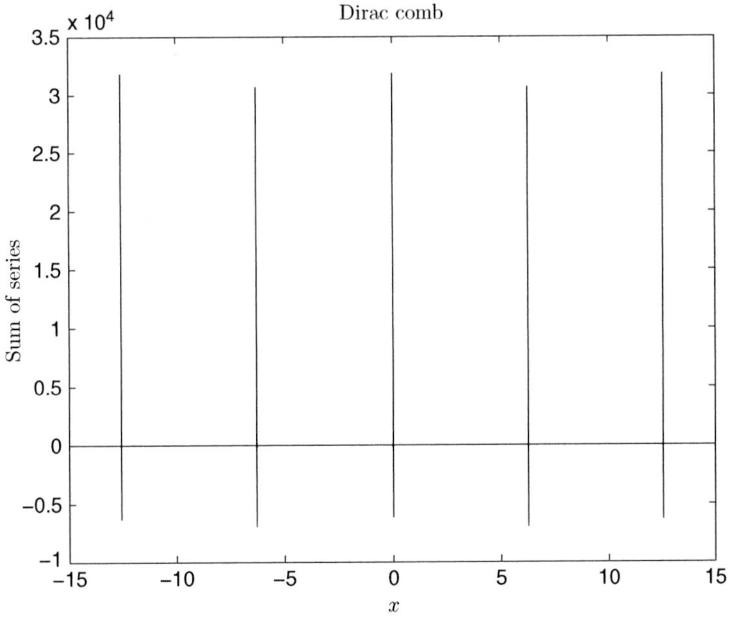

Figure 2.11 The sum of the first 100,000 terms of this cosine series (2.116) for the Dirac comb is plotted for $-15 \le x \le 15$.

□

The stretched Dirac comb is

$$\sum_{n=-\infty}^{\infty} \frac{e^{2\pi in(x-y)/L}}{L} = \sum_{\ell=-\infty}^{\infty} \delta(x - y - \ell L). \tag{2.117}$$

Example 2.14 (Parseval's identity) Using our formula (2.35) for the Fourier coefficients of a stretched interval, we can relate a sum of products $f_n^* \, g_n$ of the Fourier coefficients of the functions $f(x)$ and $g(x)$ to an integral of the product $f^*(x)\, g(x)$

$$\sum_{n=-\infty}^{\infty} f_n^* \, g_n = \sum_{n=-\infty}^{\infty} \int_0^L dx \, \frac{e^{i2\pi nx/L}}{\sqrt{L}} f^*(x) \int_0^L dy \, \frac{e^{-i2\pi ny/L}}{\sqrt{L}} \, g(y). \tag{2.118}$$

This sum contains Dirac's comb (2.117) and so

$$\begin{aligned} \sum_{n=-\infty}^{\infty} f_n^* \, g_n &= \int_0^L dx \int_0^L dy f^*(x)\, g(y) \frac{1}{L} \sum_{n=-\infty}^{\infty} e^{i2\pi n(x-y)/L} \\ &= \int_0^L dx \int_0^L dy f^*(x)\, g(y) \sum_{\ell=-\infty}^{\infty} \delta(x - y - 2\pi \ell L). \end{aligned} \tag{2.119}$$

But because only the $\ell = 0$ tooth of the comb lies in the interval $[0, L]$, we have more simply

$$\sum_{n=-\infty}^{\infty} f_n^* \, g_n = \int_0^L dx \int_0^L dy f^*(x)\, g(y)\, \delta(x-y) = \int_0^L dx f^*(x)\, g(x). \tag{2.120}$$

In particular, if the two functions are the same, then

$$\sum_{n=-\infty}^{\infty} |f_n|^2 = \int_0^L dx \, |f(x)|^2, \tag{2.121}$$

which is **Parseval's identity**. Thus if a function is **square integrable** on an interval, then the sum of the squares of the absolute values of its Fourier coefficients is the integral of the square of its absolute value. □

Example 2.15 (Derivatives of delta functions) Delta functions and other generalized functions or distributions map smooth functions that vanish at infinity into numbers in ways that are linear and continuous. Derivatives of delta functions are defined so as to allow integration by parts. Thus the nth derivative of the delta function $\delta^{(n)}(x-y)$ maps the function $f(x)$ to $(-1)^n$ times its nth derivative $f^{(n)}(y)$ at y

$$\int \delta^{(n)}(x-y) f(x)\, dx = \int \delta(x-y)\, (-1)^n f^{(n)}(x)\, dx = (-1)^n f^{(n)}(y) \tag{2.122}$$

with no surface term.

Example 2.16 (The equation $xf(x) = a$) Dirac's delta function sometimes appears unexpectedly. For instance, the general solution to the equation

$xf(x) = a$ is $f(x) = a/x + b\,\delta(x)$ where b is an arbitrary constant (Dirac, 1967, sec. 15, Waxman and Peck, 1998). Similarly, the general solution to the equation $x^2 f(x) = a$ is $f(x) = a/x^2 + b\,\delta(x)/x + c\,\delta(x) + d\,\delta'(x)$ in which $\delta'(x)$ is the derivative of the delta function, and b, c, and d are arbitrary constants. □

2.11 The harmonic oscillator

The hamiltonian for the harmonic oscillator is

$$H = \frac{p^2}{2m} + \frac{1}{2} m\omega^2 q^2. \tag{2.123}$$

The commutation relation $[q, p] \equiv qp - pq = i\hbar$ implies that the **lowering** and **raising** operators

$$a = \sqrt{\frac{m\omega}{2\hbar}} \left(q + \frac{ip}{m\omega} \right) \quad \text{and} \quad a^\dagger = \sqrt{\frac{m\omega}{2\hbar}} \left(q - \frac{ip}{m\omega} \right) \tag{2.124}$$

obey the commutation relation $[a, a^\dagger] = 1$. In terms of a and $a^\dagger$, which also are called the **annihilation** and **creation** operators, the hamiltonian H has the simple form

$$H = \hbar\omega \left(a^\dagger a + \tfrac{1}{2} \right). \tag{2.125}$$

There is a unique state $|0\rangle$ that is annihilated by the operator a, as may be seen by solving the differential equation

$$\langle q'|a|0\rangle = \sqrt{\frac{m\omega}{2\hbar}} \langle q'| \left(q + \frac{ip}{m\omega} \right) |0\rangle = 0. \tag{2.126}$$

Since $\langle q'|q = q'\langle q'|$ and

$$\langle q'|p|0\rangle = \frac{\hbar}{i} \frac{d\langle q'|0\rangle}{dq'} \tag{2.127}$$

the resulting differential equation is

$$\frac{d\langle q'|0\rangle}{dq'} = -\frac{m\omega}{\hbar} q' \langle q'|0\rangle. \tag{2.128}$$

Its suitably normalized solution is the wave function for the ground state of the harmonic oscillator

$$\langle q'|0\rangle = \left(\frac{m\omega}{\pi\hbar} \right)^{1/4} \exp\left(-\frac{m\omega q'^2}{2\hbar} \right). \tag{2.129}$$

For $n = 0, 1, 2, \ldots$, the nth eigenstate of the hamiltonian H is

$$|n\rangle = \frac{1}{\sqrt{n!}} \left(a^\dagger\right)^n |0\rangle \tag{2.130}$$

where $n! \equiv n(n-1)\ldots 1$ is n-**factorial** and $0! = 1$. Its energy is

$$H|n\rangle = \hbar\omega \left(n + \tfrac{1}{2}\right) |n\rangle. \tag{2.131}$$

The identity operator is

$$I = \sum_{n=0}^{\infty} |n\rangle\langle n|. \tag{2.132}$$

An arbitrary state $|\psi\rangle$ has an expansion in terms of the eigenstates $|n\rangle$

$$|\psi\rangle = I|\psi\rangle = \sum_{n=0}^{\infty} |n\rangle\langle n|\psi\rangle \tag{2.133}$$

and evolves in time like a Fourier series

$$|\psi, t\rangle = e^{-iHt/\hbar}|\psi\rangle = e^{-iHt/\hbar} \sum_{n=0}^{\infty} |n\rangle\langle n|\psi\rangle = e^{-i\omega t/2} \sum_{n=0}^{\infty} e^{-in\omega t}|n\rangle\langle n|\psi\rangle \tag{2.134}$$

with wave function

$$\psi(q, t) = \langle q|\psi, t\rangle = e^{-i\omega t/2} \sum_{n=0}^{\infty} e^{-in\omega t} \langle q|n\rangle\langle n|\psi\rangle. \tag{2.135}$$

The wave functions $\langle q|n\rangle$ of the energy eigenstates are related to the Hermite polynomials (example 8.6)

$$H_n(x) = (-1)^n e^{x^2} \frac{d^n}{dx^n} e^{-x^2} \tag{2.136}$$

by a change of variables $x = \sqrt{m\omega/\hbar}\, q \equiv sq$ and a normalization factor

$$\psi_n(q) = \langle q|n\rangle = \frac{\sqrt{s}\, e^{-(sq)^2/2}}{\sqrt{2^n n! \sqrt{\pi}}} H_n(sq) = \frac{m\omega}{\pi\hbar}^{1/4} \frac{e^{-m\omega q^2/2\hbar}}{\sqrt{2^n n!}} H_n\left(\left(\frac{m\omega}{\hbar}\right)^{1/2} q\right). \tag{2.137}$$

The **coherent state** $|\alpha\rangle$

$$|\alpha\rangle = e^{-|\alpha|^2/2} e^{\alpha a^\dagger} |0\rangle = e^{-|\alpha|^2/2} \sum_{n=0}^{\infty} \frac{\alpha^n}{\sqrt{n!}} |n\rangle \tag{2.138}$$

is an eigenstate $a|\alpha\rangle = \alpha|\alpha\rangle$ of the lowering (or annihilation) operator a with eigenvalue α. Its time evolution is simply

$$|\alpha, t\rangle = e^{-i\omega t/2} e^{-|\alpha|^2/2} \sum_{n=0}^{\infty} \frac{\left(\alpha e^{-i\omega t}\right)^n}{\sqrt{n!}} |n\rangle = e^{-i\omega t/2}\, |\alpha e^{-i\omega t}\rangle. \tag{2.139}$$

2.12 Nonrelativistic strings

If we clamp the ends of a nonrelativistic string at $x = 0$ and $x = L$, then the amplitude $y(x, t)$ will obey the boundary conditions

$$y(0, t) = y(L, t) = 0 \tag{2.140}$$

and the wave equation

$$v^2 \frac{\partial^2 y}{\partial x^2} = \frac{\partial^2 y}{\partial t^2} \tag{2.141}$$

as long as $y(x, t)$ remains small. The functions

$$y_n(x, t) = \sin \frac{n\pi x}{L} \left(f_n \sin \frac{n\pi vt}{L} + d_n \cos \frac{n\pi vt}{L} \right) \tag{2.142}$$

satisfy this wave equation (2.141) and the boundary conditions (2.140). They represent waves traveling along the x-axis with speed v.

The space S_L of functions $f(x)$ that satisfy the boundary condition (2.140) is spanned by the functions $\sin(n\pi x/L)$. One may use the integral formula

$$\int_0^L \sin \frac{n\pi x}{L} \sin \frac{m\pi x}{L} \, dx = \frac{L}{2} \, \delta_{nm} \tag{2.143}$$

to derive for any function $f \in S_L$ the Fourier series

$$f(x) = \sum_{n=1}^{\infty} f_n \sin \frac{n\pi x}{L} \tag{2.144}$$

with coefficients

$$f_n = \frac{2}{L} \int_0^L \sin \frac{n\pi x}{L} f(x) \, dx \tag{2.145}$$

and the representation

$$\sum_{m=-\infty}^{\infty} \delta(x - z - 2mL) = \frac{2}{L} \sum_{n=1}^{\infty} \sin \frac{n\pi x}{L} \sin \frac{n\pi z}{L} \tag{2.146}$$

for the Dirac comb on S_L.

2.13 Periodic boundary conditions

Periodic boundary conditions are often convenient. For instance, rather than studying an infinitely long one-dimensional system, we might study the same system, but of length L. The ends cause effects not present in the infinite system. To avoid them, we imagine that the system forms a circle and impose the periodic boundary condition

$$\psi(x \pm L, t) = \psi(x, t). \tag{2.147}$$

In three dimensions, the analogous conditions are

$$\begin{aligned} \psi(x \pm L, y, z, t) &= \psi(x, y, z, t), \\ \psi(x, y \pm L, z, t) &= \psi(x, y, z, t), \\ \psi(x, y, z \pm L, t) &= \psi(x, y, z, t). \end{aligned} \tag{2.148}$$

The eigenstates $|\boldsymbol{p}\rangle$ of the free hamiltonian $H = \boldsymbol{p}^2/2m$ have wave functions

$$\psi_{\boldsymbol{p}}(\boldsymbol{x}) = \langle \boldsymbol{x}|\boldsymbol{p}\rangle = e^{i\boldsymbol{x}\cdot\boldsymbol{p}/\hbar}/(2\pi\hbar)^{3/2}. \tag{2.149}$$

The periodic boundary conditions (2.148) require that each component p_i of the momentum satisfy $Lp_i/\hbar = 2\pi n_i$ or

$$\boldsymbol{p} = \frac{2\pi\hbar\boldsymbol{n}}{L} = \frac{h\boldsymbol{n}}{L} \tag{2.150}$$

where $\boldsymbol{n}$ is a vector of integers, which may be positive or negative or zero.

Periodic boundary conditions arise naturally in the study of solids. The atoms of a perfect crystal are at the vertices of a **Bravais** lattice

$$\boldsymbol{x}_i = \boldsymbol{x}_0 + \sum_{i=1}^{3} n_i \boldsymbol{a}_i \tag{2.151}$$

in which the three vectors $\boldsymbol{a}_i$ are the **primitive vectors** of the lattice and the n_i are three integers. The hamiltonian of such an infinite crystal is invariant under translations in space by

$$\sum_{i=1}^{3} n_i \boldsymbol{a}_i. \tag{2.152}$$

To keep the notation simple, let's restrict ourselves to a cubic lattice with lattice spacing a. Then since the momentum operator $\boldsymbol{p}$ generates translations in space, the invariance of H under translations by $a\,\boldsymbol{n}$

$$\exp(ia\boldsymbol{n}\cdot\boldsymbol{p})H\exp(-ia\boldsymbol{n}\cdot\boldsymbol{p}) = H \tag{2.153}$$

implies that $\exp(ia\boldsymbol{n}\cdot\boldsymbol{p})$ and H are compatible observables $[\exp(ia\boldsymbol{n}\cdot\boldsymbol{p}), H] = 0$. As explained in section 1.30, it follows that we may choose the eigenstates of H also to be eigenstates of $\boldsymbol{p}$

$$e^{ia\boldsymbol{p}\cdot\boldsymbol{n}/\hbar}|\psi\rangle = e^{ia\boldsymbol{k}\cdot\boldsymbol{n}}\,|\psi\rangle, \tag{2.154}$$

which implies that

$$\psi(\boldsymbol{x} + a\boldsymbol{n}, t) = e^{ia\boldsymbol{k}\cdot\boldsymbol{n}}\,\psi(\boldsymbol{x}, t). \tag{2.155}$$

Setting

$$\psi(\boldsymbol{x}) = e^{i\boldsymbol{k}\cdot\boldsymbol{x}}\, u(\boldsymbol{x}) \tag{2.156}$$

we see that condition (2.155) implies that $u(\boldsymbol{x})$ is periodic

$$u(\boldsymbol{x} + a\boldsymbol{n}) = u(\boldsymbol{x}). \tag{2.157}$$

For a general Bravais lattice, this **Born–von Karman** periodic boundary condition is

$$u\left(\boldsymbol{x} + \sum_{i=1}^{3} n_i \boldsymbol{a}_i, t\right) = u(\boldsymbol{x}, t). \tag{2.158}$$

Equations (2.155) and (2.157) are known as **Bloch's theorem**.

Exercises

2.1 Show that $\sin \omega_1 x + \sin \omega_2 x$ is the same as (2.9).

2.2 Find the Fourier series for the function $\exp(ax)$ on the interval $-\pi < x \le \pi$.

2.3 Find the Fourier series for the function $(x^2 - \pi^2)^2$ on the same interval $(-\pi, \pi]$.

2.4 Find the Fourier series for the function $(1 + \cos x) \sin ax$ on the interval $(-\pi, \pi]$.

2.5 Show that the Fourier series for the function $x \cos x$ on the interval $[-\pi, \pi]$ is

$$x \cos x = -\frac{1}{2} \sin x + 2 \sum_{n=2}^{\infty} \frac{(-1)^n\, n}{n^2 - 1} \sin nx. \tag{2.159}$$

2.6 (a) Show that the Fourier series for the function $|x|$ on the interval $[-\pi, \pi]$ is

$$|x| = \frac{\pi}{2} - \frac{4}{\pi} \sum_{n=0}^{\infty} \frac{\cos(2n+1)x}{(2n+1)^2}. \tag{2.160}$$

(b) Use this result to find a neat formula for $\pi^2/8$. Hint: set $x = 0$.

2.7 Show that the Fourier series for the function $|\sin x|$ on the interval $[-\pi, \pi]$ is

$$|\sin x| = \frac{2}{\pi} - \frac{4}{\pi} \sum_{n=1}^{\infty} \frac{\cos 2nx}{4n^2 - 1}. \tag{2.161}$$

2.8 Show that the Fourier coefficients of the C^0 function (2.59) on the interval $[-\pi, \pi]$ are given by (2.60).

2.9 Find by inspection the Fourier series for the function $\exp[\exp(-ix)]$.

2.10 Fill in the steps in the computation (2.28) of the Fourier series for x^2.

2.11 Do the first integral in equation (2.65). Hint: differentiate $\ln(\sin x/x)$.

2.12 Use the infinite-product formula (2.66) for the sine and the relation $\cos \pi x = \sin 2\pi x/(2 \sin \pi x)$ to derive the infinite-product formula (2.67) for the cosine. Hint:

$$\prod_{n=1}^{\infty}\left[1-\frac{x^2}{\frac{1}{4}n^2}\right] = \prod_{n=1}^{\infty}\left[1-\frac{x^2}{\frac{1}{4}(2n-1)^2}\right]\left[1-\frac{x^2}{\frac{1}{4}(2n)^2}\right]. \tag{2.162}$$

2.13 What's the general solution to the equation $x^3 f(x) = a$?

2.14 Suppose we wish to approximate the real square-integrable function $f(x)$ by the Fourier series with N terms

$$f_N(x) = \frac{a_0}{2} + \sum_{n=1}^{N} \left(a_n \cos nx + b_n \sin nx\right). \tag{2.163}$$

Then the error

$$E_N = \int_0^{2\pi} [f(x) - f_N(x)]^2 \, dx \tag{2.164}$$

will depend upon the $2N + 1$ coefficients a_n and b_n. The best coefficients minimize this error and satisfy the conditions

$$\frac{\partial E_N}{\partial a_n} = \frac{\partial E_N}{\partial b_n} = 0. \tag{2.165}$$

By using these conditions, find them.

2.15 Find the Fourier series for the function

$$f(x) = \theta(a^2 - x^2) \tag{2.166}$$

on the interval $[-\pi, \pi]$ for the case $a^2 < \pi^2$. The **Heaviside step function** $\theta(x)$ is zero for $x < 0$, one-half for $x = 0$, and unity for $x > 0$ (Oliver Heaviside, 1850–1925). The value assigned to $\theta(0)$ seldom matters, and you need not worry about it in this problem.

2.16 Derive or infer the formula (2.117) for the stretched Dirac comb.

2.17 Use the commutation relation $[q, p] = i\hbar$ to show that the annihilation and creation operators (2.124) satisfy the commutation relation $[a, a^\dagger] = 1$.

2.18 Show that the state $|n\rangle = (a^\dagger)^n|0\rangle/\sqrt{n!}$ is an eigenstate of the hamiltonian (2.125) with energy $\hbar\omega(n + 1/2)$.

2.19 Show that the coherent state $|\alpha\rangle$ (2.138) is an eigenstate of the annihilation operator a with eigenvalue α.

2.20 Derive equations (2.145 & 2.146) from the expansion (2.144) and the integral formula (2.143).

2.21 Consider a string like the one described in section 2.12, which satisfies the boundary conditions (2.140) and the wave equation (2.141). The string is at rest at time $t = 0$

$$y(x, 0) = 0 \tag{2.167}$$

and is struck precisely at $t = 0$ and $x = a$ so that

$$\left.\frac{\partial y(x,t)}{\partial t}\right|_{t=0} = Lv_0\delta(x-a). \tag{2.168}$$

Find $y(x,t)$ and $\dot{y}(x,t)$, where the dot means time derivative.

2.22 Same as exercise (2.21), but now the initial conditions are

$$u(x,0) = f(x) \quad \text{and} \quad \dot{u}(x,0) = g(x) \tag{2.169}$$

in which $f(0) = f(L) = 0$ and $g(0) = g(L) = 0$. Find the motion of the amplitude $u(x,t)$ of the string.

2.23 (a) Find the Fourier series for the function $f(x) = x^2$ on the interval $[-\pi, \pi]$. (b) Use your result at $x = \pi$ to show that

$$\sum_{n=1}^{\infty} \frac{1}{n^2} = \frac{\pi^2}{6} \tag{2.170}$$

which is the value of Riemann's zeta function (4.92) $\zeta(x)$ at $x = 2$.

3

Fourier and Laplace transforms

The complex exponentials $\exp(i2\pi nx/L)$ are orthonormal and easy to differentiate (and to integrate), but they are periodic with period L. If one wants to represent functions that are not periodic, then a better choice is the complex exponentials $\exp(ikx)$, where k is an arbitrary real number. These orthonormal functions are the basis of the Fourier transform. The choice of complex k leads to the transforms of Laplace, Mellin, and Bromwich.

3.1 The Fourier transform

The interval $[-L/2, L/2]$ is arbitrary in the Fourier series pair (2.37)

$$f(x) = \sum_{n=-\infty}^{\infty} f_n \frac{e^{i2\pi nx/L}}{\sqrt{L}} \quad \text{and} \quad f_n = \int_{-L/2}^{L/2} f(x) \frac{e^{-i2\pi nx/L}}{\sqrt{L}} \, dx. \tag{3.1}$$

What happens when we stretch this interval without limit, letting $L \to \infty$?

We may use the **nearest-integer function** $[y]$ to convert the coefficients f_n into a continuous function $\hat{f}(y) \equiv f_{[y]}$ such that $\hat{f}(y) = f_n$ when $|y - n| < 1/2$. In terms of this function $\hat{f}(y)$, the Fourier series (3.1) for the function $f(x)$ is

$$f(x) = \sum_{n=-\infty}^{\infty} \int_{n-1/2}^{n+1/2} \hat{f}(y) \frac{e^{i2\pi [y]x/L}}{\sqrt{L}} \, dy = \int_{-\infty}^{\infty} \hat{f}(y) \frac{e^{i2\pi [y]x/L}}{\sqrt{L}} \, dy. \tag{3.2}$$

Since $[y]$ and y differ by no more than $1/2$, the absolute value of the difference between $\exp(i\pi [y]x/L)$ and $\exp(i\pi yx/L)$ for fixed x is

$$\left| e^{i2\pi [y]x/L} - e^{i2\pi yx/L} \right| = \left| e^{i2\pi ([y]-y)x/L} - 1 \right| \approx \frac{\pi |x|}{L}, \tag{3.3}$$

which goes to zero as $L \to \infty$. So in this limit, we may replace $[y]$ by y and express $f(x)$ as

$$f(x) = \int_{-\infty}^{\infty} \hat{f}(y) \, \frac{e^{i2\pi yx/L}}{\sqrt{L}} \, dy. \tag{3.4}$$

We now change variables to $k = 2\pi y/L$ and find for $f(x)$ the integral

$$f(x) = \int_{-\infty}^{\infty} \hat{f}\left(\frac{Lk}{2\pi}\right) \frac{e^{ikx}}{\sqrt{L}} \frac{L}{2\pi} \, dk = \int_{-\infty}^{\infty} \sqrt{\frac{L}{2\pi}} \, \hat{f}\left(\frac{Lk}{2\pi}\right) e^{ikx} \, \frac{dk}{\sqrt{2\pi}}. \tag{3.5}$$

So in terms of the Fourier transform $\tilde{f}(k)$ defined as

$$\tilde{f}(k) = \sqrt{\frac{L}{2\pi}} \, \hat{f}\left(\frac{Lk}{2\pi}\right) \tag{3.6}$$

the integral (3.5) for $f(x)$ is the inverse Fourier transform

$$f(x) = \int_{-\infty}^{\infty} \tilde{f}(k) \, e^{ikx} \, \frac{dk}{\sqrt{2\pi}}. \tag{3.7}$$

To find $\tilde{f}(k)$, we use its definition (3.6), the definition (3.1) of f_n, our formula $\hat{f}(y) = f_{[y]}$, and the inequality $|2\pi \, [Lk/2\pi] \, /L - k| \le \pi/2L$ to write

$$\tilde{f}(k) = \sqrt{\frac{L}{2\pi}} \, f_{\left[\frac{Lk}{2\pi}\right]} = \sqrt{\frac{L}{2\pi}} \int_{-L/2}^{L/2} f(x) \, \frac{e^{-i2\pi\left[\frac{Lk}{2\pi}\right]\frac{x}{L}}}{\sqrt{L}} \, dx \approx \int_{-L/2}^{L/2} f(x) \, e^{-ikx} \, \frac{dx}{\sqrt{2\pi}}.$$

This formula becomes exact in the limit $L \to \infty$,

$$\tilde{f}(k) = \int_{-\infty}^{\infty} f(x) \, e^{-ikx} \, \frac{dx}{\sqrt{2\pi}} \tag{3.8}$$

and so we have the Fourier transformations

$$f(x) = \int_{-\infty}^{\infty} \tilde{f}(k) \, e^{ikx} \, \frac{dk}{\sqrt{2\pi}} \quad \text{and} \quad \tilde{f}(k) = \int_{-\infty}^{\infty} f(x) \, e^{-ikx} \, \frac{dx}{\sqrt{2\pi}}. \tag{3.9}$$

The function $\tilde{f}(k)$ is the **Fourier transform** of $f(x)$, and $f(x)$ is the **inverse Fourier transform** of $\tilde{f}(k)$.

In these symmetrical relations (3.9), the distinction between a Fourier transform and an inverse Fourier transform is entirely a matter of convention. There is no rule for which sign, ikx or $-ikx$, goes with which transform or for where to put the 2πs. Thus one often sees

$$f(x) = \int_{-\infty}^{\infty} \tilde{f}(k) \, e^{\pm ikx} \, dk \quad \text{and} \quad \tilde{f}(k) = \int_{-\infty}^{\infty} f(x) \, e^{\mp ikx} \, \frac{dx}{2\pi} \tag{3.10}$$

as well as

$$f(x) = \int_{-\infty}^{\infty} \tilde{f}(k)\, e^{\pm ikx} \frac{dk}{2\pi} \quad \text{and} \quad \tilde{f}(k) = \int_{-\infty}^{\infty} f(x)\, e^{\mp ikx}\, dx. \tag{3.11}$$

One often needs to relate a function's Fourier series to its Fourier transform. So let's compare the Fourier series (3.1) for the function $f(x)$ on the interval $[-L/2, L/2]$ with its Fourier transform (3.9) in the limit of large L

$$f(x) = \sum_{n=-\infty}^{\infty} f_n \frac{e^{i2\pi nx/L}}{\sqrt{L}} = \sum_{n=-\infty}^{\infty} f_n \frac{e^{ik_n x}}{\sqrt{L}} = \int_{-\infty}^{\infty} \tilde{f}(k) e^{ikx} \frac{dk}{\sqrt{2\pi}} \tag{3.12}$$

in which $k_n = 2\pi n/L = 2\pi y/L$. Now $f_n = \hat{f}(y)$, and so by the definition (3.6) of $\tilde{f}(k)$, we have $f_n = \hat{f}(Lk/2\pi) = \sqrt{2\pi/L}\,\tilde{f}(k)$. Thus, to get the Fourier series from the Fourier transform, we multiply the series by $2\pi/L$ and use the Fourier transform at k_n divided by $\sqrt{2\pi}$

$$f(x) = \frac{1}{\sqrt{L}} \sum_{n=-\infty}^{\infty} f_n\, e^{ik_n x} = \frac{2\pi}{L} \sum_{n=-\infty}^{\infty} \frac{\tilde{f}(kn)}{\sqrt{2\pi}} e^{ik_n x}. \tag{3.13}$$

Going the other way, we have

$$f(x) = \int_{-\infty}^{\infty} \tilde{f}(k) e^{ikx} \frac{dk}{\sqrt{2\pi}} = \frac{L}{2\pi} \int_{-\infty}^{\infty} \frac{f_{[Lk/2\pi]}}{\sqrt{L}} e^{ikx} dk. \tag{3.14}$$

Example 3.1 (The Fourier transform of a gaussian is a gaussian) The Fourier transform of the gaussian $f(x) = \exp(-m^2 x^2)$ is

$$\tilde{f}(k) = \int_{-\infty}^{\infty} \frac{dx}{\sqrt{2\pi}}\, e^{-ikx}\, e^{-m^2 x^2}. \tag{3.15}$$

We complete the square in the exponent:

$$\tilde{f}(k) = e^{-k^2/4m^2} \int_{-\infty}^{\infty} \frac{dx}{\sqrt{2\pi}}\, e^{-m^2 (x+ik/2m^2)^2}. \tag{3.16}$$

As we shall see in section 5.14 when we study analytic functions, we may shift x to $x - ik/2m^2$, so the term $ik/2m^2$ in the exponential has no effect on the value of the x-integral.

$$\tilde{f}(k) = e^{-k^2/4m^2} \int_{-\infty}^{\infty} \frac{dx}{\sqrt{2\pi}}\, e^{-m^2 x^2} = \frac{1}{\sqrt{2}\,m}\, e^{-k^2/4m^2}. \tag{3.17}$$

Thus, the Fourier transform of a gaussian is another gaussian

$$\tilde{f}(k) = \int_{-\infty}^{\infty} \frac{dx}{\sqrt{2\pi}}\, e^{-ikx}\, e^{-m^2 x^2} = \frac{1}{\sqrt{2}\,m}\, e^{-k^2/4m^2}. \tag{3.18}$$

But the two gaussians are very different: if the gaussian $f(x) = \exp(-m^2 x^2)$ decreases slowly as $x \to \infty$ because m is small (or quickly because m is big), then its gaussian Fourier transform $\tilde{f}(k) = \exp(-k^2/4m^2)/m\sqrt{2}$ decreases quickly as $k \to \infty$ because m is small (or slowly because m is big).

Can we invert $\tilde{f}(k)$ to get $f(x)$? The inverse Fourier transform (3.7) says

$$f(x) = \int_{-\infty}^{\infty} \frac{dk}{\sqrt{2\pi}} \tilde{f}(k)\, e^{ikx} = \int_{-\infty}^{\infty} \frac{dk}{\sqrt{2\pi}} \frac{1}{m\sqrt{2}} e^{ikx-k^2/4m^2}. \tag{3.19}$$

By again completing the square in the exponent

$$f(x) = e^{-m^2x^2} \int_{-\infty}^{\infty} \frac{dk}{\sqrt{2\pi}} \frac{1}{m\sqrt{2}} e^{-(k-i2m^2x)^2/4m^2} \tag{3.20}$$

and shifting the variable of integration k to $k + i2m^2x$, we find

$$f(x) = e^{-m^2x^2} \int_{-\infty}^{\infty} \frac{dk}{\sqrt{2\pi}} \frac{1}{m\sqrt{2}} e^{-k^2/(4m^2)} = e^{-m^2x^2}, \tag{3.21}$$

which is reassuring.

Using (3.17) for $\tilde{f}(k)$ and the connections (3.12–3.14) between Fourier series and transforms, we see that a Fourier series for this gaussian is in the limit of $L \gg x$

$$f(x) = e^{-m^2x^2} = \frac{2\pi}{L} \sum_{n=-\infty}^{\infty} \frac{1}{\sqrt{4\pi}\, m} e^{-k_n^2/(4m^2)} e^{ik_nx} \tag{3.22}$$

in which $k_n = 2\pi n/L$. □

3.2 The Fourier transform of a real function

If the function $f(x)$ is real, then its Fourier transform (3.8)

$$\tilde{f}(k) = \int_{-\infty}^{\infty} \frac{dx}{\sqrt{2\pi}} f(x)\, e^{-ikx} \tag{3.23}$$

obeys the relation

$$\tilde{f}^*(k) = \tilde{f}(-k). \tag{3.24}$$

For since $f^*(x) = f(x)$, the complex conjugate of (3.23) is

$$\tilde{f}^*(k) = \int_{-\infty}^{\infty} \frac{dx}{\sqrt{2\pi}} f(x)\, e^{ikx} = \tilde{f}(-k). \tag{3.25}$$

It follows that a real function $f(x)$ satisfies the relation

$$f(x) = \frac{1}{\pi} \int_0^{\infty} dk \int_{-\infty}^{\infty} f(y)\, \cos k(y-x)\, dy \tag{3.26}$$

(exercise 3.1) as well as

$$f(x) = \frac{2}{\pi} \int_0^{\infty} \cos kx\, dk \int_0^{\infty} f(y)\, \cos ky\, dy \tag{3.27}$$

if it is even, and

$$f(x) = \frac{2}{\pi} \int_0^\infty \sin kx \, dk \int_0^\infty f(y) \sin ky \, dy \tag{3.28}$$

if it is odd (exercise 3.2).

Example 3.2 (Dirichlet's discontinuous factor) Using (3.27), one may write the square wave

$$f(x) = \begin{cases} 1, & |x| < 1 \\ \frac{1}{2}, & |x| = 1 \\ 0, & |x| > 1 \end{cases} \tag{3.29}$$

as Dirichlet's discontinuous factor

$$f(x) = \frac{2}{\pi} \int_0^\infty \frac{\sin k \cos kx}{k} \, dk \tag{3.30}$$

(exercise 3.3). □

Example 3.3 (Even and odd exponentials) By using the Fourier-transform formulas (3.27 & 3.28), one may show that the Fourier transform of the even exponential $\exp(-\beta|x|)$ is

$$e^{-\beta|x|} = \frac{2}{\pi} \int_0^\infty \frac{\beta \cos kx}{\beta^2 + k^2} \, dk \tag{3.31}$$

while that of the odd exponential $x \exp(-\beta|x|)/|x|$ is

$$\frac{x}{|x|} e^{-\beta|x|} = \frac{2}{\pi} \int_0^\infty \frac{k \sin kx}{\beta^2 + k^2} \, dk \tag{3.32}$$

(exercise 3.4). □

3.3 Dirac, Parseval, and Poisson

Combining the basic equations (3.9) that define the Fourier transform, we may do something apparently useless: we may write the function $f(x)$ in terms of itself as

$$f(x) = \int_{-\infty}^{\infty} \frac{dk}{\sqrt{2\pi}} \tilde{f}(k) \, e^{ikx} = \int_{-\infty}^{\infty} \frac{dk}{\sqrt{2\pi}} \, e^{ikx} \int_{-\infty}^{\infty} \frac{dy}{\sqrt{2\pi}} \, e^{-iky} f(y). \tag{3.33}$$

Let's compare this equation

$$f(x) = \int_{-\infty}^{\infty} dy \left(\int_{-\infty}^{\infty} \frac{dk}{2\pi} \exp[ik(x-y)] \right) f(y) \tag{3.34}$$

with one (2.107) that describes Dirac's delta function

$$f(x) = \int_{-\infty}^{\infty} dy\, \delta(x-y) f(y). \tag{3.35}$$

Thus for functions with sensible Fourier transforms, the delta function is

$$\delta(x-y) = \int_{-\infty}^{\infty} \frac{dk}{2\pi} \exp[ik(x-y)]. \tag{3.36}$$

The inner product (f, g) of two functions, $f(x)$ with Fourier transform $\tilde{f}(k)$ and $g(x)$ with Fourier transform $\hat{g}(k)$, is

$$\langle f|g\rangle = (f, g) = \int_{-\infty}^{\infty} dx f^*(x)\, g(x). \tag{3.37}$$

Since $f(x)$ and $g(x)$ are related to $\tilde{f}(k)$ and $\tilde{g}(k)$ by the Fourier transform (3.8), their inner product (f, g) is

$$\begin{aligned}(f, g) &= \int_{-\infty}^{\infty} dx \int_{-\infty}^{\infty} \frac{dk}{\sqrt{2\pi}} \left(\tilde{f}(k)\, e^{ikx}\right)^* \int_{-\infty}^{\infty} \frac{dk'}{\sqrt{2\pi}} \tilde{g}(k')\, e^{ik'x} \\ &= \int_{-\infty}^{\infty} dk \int_{-\infty}^{\infty} dk' \int_{-\infty}^{\infty} \frac{dx}{2\pi}\, e^{ix(k'-k)} \tilde{f}^*(k)\, \tilde{g}(k') \\ &= \int_{-\infty}^{\infty} dk \int_{-\infty}^{\infty} dk'\, \delta(k'-k) \tilde{f}^*(k)\, \tilde{g}(k') = \int_{-\infty}^{\infty} dk \tilde{f}^*(k)\, \tilde{g}(k).\end{aligned} \tag{3.38}$$

Thus we arrive at **Parseval's relation**

$$(f, g) = \int_{-\infty}^{\infty} dx f^*(x)\, g(x) = \int_{-\infty}^{\infty} dk \tilde{f}^*(k)\, \tilde{g}(k) = (\tilde{f}, \tilde{g}), \tag{3.39}$$

which says that the inner product of two functions is the same as the inner product of their Fourier transforms. The Fourier transform is a unitary transform. In particular, if $f = g$, then

$$\langle f|f\rangle = (f, f) = \int_{-\infty}^{\infty} dx\, |f(x)|^2 = \int_{-\infty}^{\infty} dk\, |\tilde{f}(k)|^2 \tag{3.40}$$

(Marc-Antoine Parseval des Chênes, 1755–1836).

In fact, one may show that the Fourier transform maps the space of (Lebesgue) square-integrable functions onto itself in a one-to-one manner. Thus the natural space for the Fourier transform is the space of square-integrable functions, and so the representation (3.36) of Dirac's delta function is suitable for continuous square-integrable functions.

This may be a good place to say a few words about how to evaluate integrals involving delta functions of more complicated arguments, such as

$$J = \int \delta(g(x)) f(x)\, dx. \tag{3.41}$$

To see how this works, let's assume that $g(x)$ vanishes at a single point x_0 at which its derivative $g'(x_0) \neq 0$ isn't zero. Then the integral J involves f only as $f(x_0)$, which we can bring outside as a prefactor

$$J = f(x_0) \int \delta(g(x))\, dx. \tag{3.42}$$

Near x_0 the function $g(x)$ is approximately $g'(x_0)(x - x_0)$, and so the integral is

$$J = f(x_0) \int \delta(g'(x_0)(x - x_0))\, dx. \tag{3.43}$$

Since the delta function is nonnegative, we can write

$$\begin{aligned} J &= \frac{f(x_0)}{|g'(x_0)|} \int \delta(g'(x_0)(x - x_0))|g'(x_0)|\, dx \\ &= \frac{f(x_0)}{|g'(x_0)|} \int \delta(g - g_0)\, dg = \frac{f(x_0)}{|g'(x_0)|}. \end{aligned} \tag{3.44}$$

Thus for a function $g(x)$ that has a single zero, we have

$$\int \delta(g(x)) f(x)\, dx = \frac{f(x_0)}{|g'(x_0)|}. \tag{3.45}$$

If $g(x)$ has several zeros x_{0k}, then we must sum over them

$$\int \delta(g(x)) f(x)\, dx = \sum_k \frac{f(x_{0k})}{|g'(x_{0k})|}. \tag{3.46}$$

Replacing the dummy variable n by $-k$ in our Dirac-comb formula (2.116), we find

$$\sum_{k=-\infty}^{\infty} \frac{e^{-ikx}}{2\pi} = \sum_{\ell=-\infty}^{\infty} \delta(x - 2\pi\ell). \tag{3.47}$$

Multiplying both sides of this comb by a function $f(x)$ and integrating over the real line, we have

$$\sum_{k=-\infty}^{\infty} \int_{-\infty}^{\infty} \frac{e^{-ikx}}{2\pi} f(x)\, dx = \sum_{\ell=-\infty}^{\infty} \int_{-\infty}^{\infty} \delta(x - 2\pi\ell) f(x)\, dx. \tag{3.48}$$

Our delta function formula (2.107) or (3.34) and our Fourier-transform relations (3.9) now give us the **Poisson summation formula**

$$\frac{1}{\sqrt{2\pi}} \sum_{k=-\infty}^{\infty} \tilde{f}(k) = \sum_{\ell=-\infty}^{\infty} f(2\pi\ell), \tag{3.49}$$

in which k and ℓ are summed over all the integers. The stretched version of the Poisson summation formula is

$$\frac{\sqrt{2\pi}}{L} \sum_{k=-\infty}^{\infty} \tilde{f}\,(2\pi k/L) = \sum_{\ell=-\infty}^{\infty} f(\ell L). \tag{3.50}$$

Both sides of these formulas make sense for continuous functions that are square integrable on the real line.

Example 3.4 (Poisson and Gauss) In example 3.1, we saw that the gaussian $f(x) = \exp(-m^2x^2)$ has $\tilde{f}(k) = \exp(-k^2/4m^2)/\sqrt{2}\,m$ as its Fourier transform. So in this case, the Poisson summation formula (3.49) gives

$$\frac{1}{2\sqrt{\pi}\,m} \sum_{k=-\infty}^{\infty} e^{-k^2/4m^2} = \sum_{\ell=-\infty}^{\infty} e^{-(2\pi \ell m)^2}. \tag{3.51}$$

For $m \gg 1$, the left-hand sum converges slowly, while the right-hand sum converges quickly. For $m \ll 1$, the right-hand sum converges slowly, while the left-hand sum converges quickly. □

A sum that converges slowly in space often converges quickly in momentum space. **Ewald summation** is a technique for summing electrostatic energies, which fall off only with a power of the distance, by summing their Fourier transforms (Darden *et al.*, 1993).

3.4 Fourier derivatives and integrals

By differentiating the inverse Fourier-transform relation (3.7)

$$f(x) = \int_{-\infty}^{\infty} \frac{dk}{\sqrt{2\pi}} \tilde{f}(k)\, e^{ikx} \tag{3.52}$$

we see that the Fourier transform of the derivative $f'(x)$ is $ik\tilde{f}(k)$

$$f'(x) = \int_{-\infty}^{\infty} \frac{dk}{\sqrt{2\pi}}\, ik\tilde{f}(k)\, e^{ikx}. \tag{3.53}$$

Differentiation with respect to x corresponds to multiplication by ik.

We may repeat the process and express the second derivative as

$$f''(x) = \int_{-\infty}^{\infty} \frac{dk}{\sqrt{2\pi}} (-k^2)\tilde{f}(k)\, e^{ikx} \tag{3.54}$$

and the nth derivative as

$$f^{(n)}(x) = \int_{-\infty}^{\infty} \frac{dk}{\sqrt{2\pi}} (ik)^n\, \tilde{f}(k)\, e^{ikx}. \tag{3.55}$$

The indefinite integral of the inverse Fourier transform (3.52) is

$$ {}^{\prime}f(x) \equiv \int^{x} dx_1 f(x_1) = \int_{-\infty}^{\infty} \frac{dk}{\sqrt{2\pi}} \tilde{f}(k) \frac{e^{ikx}}{ik} \tag{3.56} $$

and the nth indefinite integral is

$$ {}^{(n)}f(x) \equiv \int^{x} dx_1 \ldots \int^{x_{n-1}} dx_n f(x_n) = \int_{-\infty}^{\infty} \frac{dk}{\sqrt{2\pi}} \tilde{f}(k) \frac{e^{ikx}}{(ik)^n}. \tag{3.57} $$

Whether these derivatives and integrals converge better or worse than $f(x)$ depends upon the behavior of $\hat{f}(k)$ near $k = 0$ and as $|k| \to \infty$.

Example 3.5 (Momentum and momentum space) Let's write the inverse Fourier transform (3.7) with ψ instead of f and with the wave number k replaced by $k = p/\hbar$

$$ \psi(x) = \int_{-\infty}^{\infty} \tilde{\psi}(k) \, e^{ikx} \frac{dk}{\sqrt{2\pi}} = \int_{-\infty}^{\infty} \frac{\tilde{\psi}(p/\hbar)}{\sqrt{\hbar}} e^{ipx/\hbar} \frac{dp}{\sqrt{2\pi\hbar}}. \tag{3.58} $$

For a normalized wave function $\psi(x)$, Parseval's relation (3.40) implies

$$ 1 = \int_{-\infty}^{\infty} |\psi(x)|^2 \, dx = \int_{-\infty}^{\infty} |\tilde{\psi}(k)|^2 \, dk = \int_{-\infty}^{\infty} |\frac{\tilde{\psi}(p/\hbar)}{\sqrt{\hbar}}|^2 \, dp, \tag{3.59} $$

or with $\psi(x) = \langle x|\psi\rangle$ and $\varphi(p) = \langle p|\psi\rangle = \tilde{\psi}(p/\hbar)/\sqrt{\hbar}$

$$ \begin{aligned} 1 = \langle\psi|\psi\rangle &= \int_{-\infty}^{\infty} |\psi(x)|^2 \, dx = \int_{-\infty}^{\infty} \langle\psi|x\rangle\langle x|\psi\rangle \, dx \\ &= \int_{-\infty}^{\infty} \langle\psi|p\rangle\langle p|\psi\rangle \, dp = \int_{-\infty}^{\infty} |\varphi(p)|^2 \, dp. \end{aligned} \tag{3.60} $$

The inner product of any two states $|\psi\rangle$ and $|\phi\rangle$ is

$$ \begin{aligned} \langle\psi|\phi\rangle &= \int_{-\infty}^{\infty} \psi^*(x)\phi(x) \, dx = \int_{-\infty}^{\infty} \langle\psi|x\rangle\langle x|\phi\rangle \, dx \\ &= \int_{-\infty}^{\infty} \psi^*(p)\phi(p) \, dp = \int_{-\infty}^{\infty} \langle\psi|p\rangle\langle p|\phi\rangle \, dp \end{aligned} \tag{3.61} $$

so the outer products $|x\rangle\langle x|$ and $|p\rangle\langle p|$ can represent the identity operator

$$ I = \int_{-\infty}^{\infty} dx \, |x\rangle\langle x| = \int_{-\infty}^{\infty} dp \, |p\rangle\langle p|. \tag{3.62} $$

The Fourier transform (3.58) relating the wave function in momentum space to that in position space is

$$ \psi(x) = \int_{-\infty}^{\infty} e^{ipx/\hbar} \, \varphi(p) \frac{dp}{\sqrt{2\pi\hbar}} \tag{3.63} $$

and the inverse Fourier transform is

$$\varphi(p) = \int_{-\infty}^{\infty} e^{-ipx/\hbar}\, \psi(x)\, \frac{dx}{\sqrt{2\pi\hbar}}. \tag{3.64}$$

In Dirac notation, the first equation (3.63) of this pair is

$$\psi(x) = \langle x|\psi\rangle = \int_{-\infty}^{\infty} \langle x|p\rangle\langle p|\psi\rangle\, dp = \int_{-\infty}^{\infty} \frac{e^{ipx/\hbar}}{\sqrt{2\pi\hbar}}\, \varphi(p)\, dp \tag{3.65}$$

so we identify $\langle x|p\rangle$ with

$$\langle x|p\rangle = \frac{e^{ipx/\hbar}}{\sqrt{2\pi\hbar}}, \tag{3.66}$$

which in turn is consistent with the delta function relation (3.36)

$$\begin{aligned}\delta(x-y) = \langle x|y\rangle &= \int_{-\infty}^{\infty} \langle x|p\rangle\langle p|y\rangle\, dp = \int_{-\infty}^{\infty} \frac{e^{ipx/\hbar}}{\sqrt{2\pi\hbar}}\, \frac{e^{-ipy/\hbar}}{\sqrt{2\pi\hbar}}\, dp \\ &= \int_{-\infty}^{\infty} \frac{e^{ip(x-y)/\hbar}}{2\pi\hbar}\, dp = \int_{-\infty}^{\infty} e^{ik(x-y)}\, \frac{dk}{2\pi}.\end{aligned} \tag{3.67}$$

If we differentiate $\psi(x)$ as given by (3.65), then we find as in (3.53)

$$\frac{\hbar}{i}\frac{d}{dx}\psi(x) = \int_{-\infty}^{\infty} p\, \varphi(p)\, e^{ipx/\hbar}\, \frac{dp}{\sqrt{2\pi\hbar}} \tag{3.68}$$

or

$$\frac{\hbar}{i}\frac{d}{dx}\psi(x) = \langle x|p|\psi\rangle = \int_{-\infty}^{\infty} \langle x|p|p'\rangle\, \langle p'|\psi\rangle\, dp' = \int_{-\infty}^{\infty} p'\, \varphi(p')\, e^{ip'x/\hbar}\, \frac{dp'}{\sqrt{2\pi\hbar}}$$

in Dirac notation. □

Example 3.6 (The uncertainty principle) Let's first normalize the gaussian $\psi(x) = N\, \exp(-(x/a)^2)$ to unity over the real axis

$$1 = N^2 \int_{-\infty}^{\infty} e^{-2(x/a)^2}\, dx = \sqrt{\frac{\pi}{2}}\, a\, N^2, \tag{3.69}$$

which gives $N^2 = \sqrt{2/\pi}/a$. So the normalized wave-function is

$$\psi(x) \equiv \langle x|\psi\rangle = \left(\frac{2}{\pi}\right)^{1/4} \frac{1}{\sqrt{a}}\, e^{-(x/a)^2}. \tag{3.70}$$

The **mean value** $\langle A\rangle$ of an operator A in a state $|\psi\rangle$ is

$$\langle A\rangle \equiv \langle\psi|A|\psi\rangle. \tag{3.71}$$

More generally, the mean value of an operator A for a system described by a density operator ρ is the trace

$$\langle A\rangle \equiv \mathrm{Tr}\,(\rho A)\,. \tag{3.72}$$

Since the gaussian (3.70) is an even function of x (that is, $\psi(-x) = \psi(x)$), the mean value of the position operator x in the state (3.70) vanishes

$$\langle x \rangle = \langle \psi | x | \psi \rangle = \int_{-\infty}^{\infty} x \, |\psi(x)|^2 \, dx = 0. \tag{3.73}$$

The **variance** of an operator A with mean value $\langle A \rangle$ in a state $|\psi\rangle$ is the mean value of the square of the difference $A - \langle A \rangle$

$$(\Delta A)^2 \equiv \langle \psi | \, (A - \langle A \rangle)^2 \, | \psi \rangle. \tag{3.74}$$

For a system with density operator ρ, the variance of A is

$$(\Delta A)^2 \equiv \mathrm{Tr}\left[\rho \, (A - \langle A \rangle)^2\right]. \tag{3.75}$$

Since $\langle x \rangle = 0$, the variance of the position operator x is

$$\begin{aligned}(\Delta x)^2 &= \langle \psi | (x - \langle x \rangle)^2 | \psi \rangle = \langle \psi | x^2 | \psi \rangle \\ &= \int_{-\infty}^{\infty} x^2 \, |\psi(x)|^2 \, dx = \frac{a^2}{4}.\end{aligned} \tag{3.76}$$

We can use the Fourier transform to find the variance of the momentum operator. By (3.64), the wave-function $\varphi(p)$ in momentum space is

$$\varphi(p) = \langle p | \psi \rangle = \int_{-\infty}^{\infty} \langle p | x \rangle \langle x | \psi \rangle \, dx. \tag{3.77}$$

By (3.66), the inner product $\langle p | x \rangle = \langle x | p \rangle^*$ is $\langle p | x \rangle = e^{-ipx/\hbar}/\sqrt{2\pi\hbar}$, so

$$\varphi(p) = \langle p | \psi \rangle = \int_{-\infty}^{\infty} \frac{dx}{\sqrt{2\pi\hbar}} \, e^{-ipx/\hbar} \langle x | \psi \rangle. \tag{3.78}$$

Thus by (3.69 & 3.70), $\varphi(p)$ is the Fourier transform

$$\varphi(p) = \int_{-\infty}^{\infty} \frac{dx}{\sqrt{2\pi\hbar}} \, e^{-ipx/\hbar} \left(\frac{2}{\pi}\right)^{1/4} \frac{1}{\sqrt{a}} \, e^{-(x/a)^2}. \tag{3.79}$$

Using our formula (3.18) for the Fourier transform of a gaussian, we get

$$\varphi(p) = \sqrt{\frac{a}{2\hbar}} \left(\frac{2}{\pi}\right)^{1/4} e^{-(ap)^2/(2\hbar)^2}. \tag{3.80}$$

Since the gaussian $\varphi(p)$ is an even function of p, the mean value $\langle p \rangle$ of the momentum operator vanishes, like that of the position operator. So the variance of the momentum operator is

$$\begin{aligned}(\Delta p)^2 &= \langle \psi | (p - \langle p \rangle)^2 | \psi \rangle = \langle \psi | p^2 \, | \psi \rangle = \int_{-\infty}^{\infty} p^2 \, |\varphi(p)|^2 \, dp \\ &= \sqrt{\frac{2}{\pi}} \int_{-\infty}^{\infty} p^2 \, \frac{a}{2\hbar} \, e^{-(ap)^2/2\hbar^2} \, dp = \frac{\hbar^2}{a^2}.\end{aligned} \tag{3.81}$$

Thus in this case, the product of the two variances is

$$(\Delta x)^2\,(\Delta p)^2 = \frac{a^2}{4}\frac{\hbar^2}{a^2} = \frac{\hbar^2}{4}. \tag{3.82}$$

This is an example of **Heisenberg's uncertainty principle**

$$\Delta x\,\Delta p \geq \frac{\hbar}{2}, \tag{3.83}$$

which follows from the Fourier-transform relations between the conjugate variables x and p.

The state $|\psi\rangle$ of a free particle at time $t = 0$

$$|\psi, 0\rangle = \int_{-\infty}^{\infty} |p\rangle\langle p|\psi\rangle\, dp = \int_{-\infty}^{\infty} |p\rangle\varphi(p)\, dp \tag{3.84}$$

evolves under the influence of the hamiltonian $H = p^2/(2m)$ to the state

$$e^{-iHt/\hbar}|\psi, 0\rangle = \int_{-\infty}^{\infty} e^{-iHt/\hbar}|p\rangle\,\varphi(p)\, dp = \int_{-\infty}^{\infty} e^{-ip^2t/(2\hbar m)}|p\rangle\,\varphi(p)\, dp \tag{3.85}$$

at time t. □

Example 3.7 (The characteristic function) If $P(x)$ is a **probability distribution** normalized to unity over the range of x

$$\int P(x)\, dx = 1 \tag{3.86}$$

then its Fourier transform is the **characteristic function**

$$\chi(k) = \hat{P}(k) = \int e^{ikx} P(x)\, dx. \tag{3.87}$$

The **expected value** of a function $f(x)$ is the integral

$$E[f(x)] = \int f(x)\, P(x)\, dx. \tag{3.88}$$

So the **characteristic function** $\chi(k) = E[\exp(ikx)]$ is the expected value of the exponential $\exp(ikx)$, and its derivatives at $k = 0$ are the **moments** $E[x^n] \equiv \mu_n$ of the probability distribution

$$E[x^n] = \int x^n\, P(x)\, dx = (-i)^n \left.\frac{d^n\chi(k)}{dk^n}\right|_{k=0}. \tag{3.89}$$

We shall pick up this thread in section 13.12. □

3.5 Fourier transforms in several dimensions

If $f(x_1, x_2)$ is a function of two variables, then its double Fourier transform $\tilde{f}(k_1, k_2)$ is

$$\tilde{f}(k_1,k_2) = \int_{-\infty}^{\infty} \frac{dx_1}{\sqrt{2\pi}} \int_{-\infty}^{\infty} \frac{dx_2}{\sqrt{2\pi}} e^{-ik_1x_1-ik_2x_2} f(x_1,x_2). \tag{3.90}$$

By twice using the Fourier representation (3.36) of Dirac's delta function, we may invert this double Fourier transformation

$$\begin{aligned}
&\int_{-\infty}^{\infty}\int_{-\infty}^{\infty} \frac{dk_1 dk_2}{2\pi} e^{i(k_1x_1+k_2x_2)} \tilde{f}(k_1,k_2) \\
&= \int_{-\infty}^{\infty}\int_{-\infty}^{\infty} \frac{dk_1 dk_2}{2\pi} \int_{-\infty}^{\infty}\int_{-\infty}^{\infty} \frac{dx_1' dx_2'}{2\pi} e^{ik_1(x_1-x_1')+ik_2(x_2-x_2')} f(x_1',x_2') \\
&= \int_{-\infty}^{\infty} \frac{dk_2}{2\pi} \int_{-\infty}^{\infty}\int_{-\infty}^{\infty} dx_1' dx_2'\, e^{ik_2(x_2-x_2')}\, \delta(x_1 - x_1') f(x_1',x_2') \\
&= \int_{-\infty}^{\infty}\int_{-\infty}^{\infty} dx_1' dx_2'\, \delta(x_1 - x_1')\delta(x_2 - x_2') f(x_1',x_2') = f(x_1,x_2).
\end{aligned} \tag{3.91}$$

That is

$$f(x_1,x_2) = \int_{-\infty}^{\infty}\int_{-\infty}^{\infty} \frac{dk_1 dk_2}{2\pi} e^{i(k_1x_1+k_2x_2)} \tilde{f}(k_1,k_2). \tag{3.92}$$

The Fourier transform of a function $f(x_1,\ldots,x_n)$ of n variables is

$$\tilde{f}(k_1,\ldots,k_n) = \int_{-\infty}^{\infty} \cdots \int_{-\infty}^{\infty} \frac{dx_1 \ldots dx_n}{(2\pi)^{n/2}} e^{-i(k_1x_1+\cdots+k_nx_n)} f(x_1,\ldots,x_n) \tag{3.93}$$

and its inverse is

$$f(x_1,\ldots,x_n) = \int_{-\infty}^{\infty} \cdots \int_{-\infty}^{\infty} \frac{dk_1 \ldots dk_n}{(2\pi)^{n/2}} e^{i(k_1x_1+\cdots+k_nx_n)} \tilde{f}(k_1,\ldots,k_n), \tag{3.94}$$

in which all the integrals run from $-\infty$ to ∞.

If we generalize the relations (3.12–3.14) between Fourier series and transforms from one to n dimensions, then we find that the Fourier series corresponding to the Fourier transform (3.94) is

$$f(x_1,\ldots,x_n) = \left(\frac{2\pi}{L}\right)^n \sum_{j_1=-\infty}^{\infty} \cdots \sum_{j_n=-\infty}^{\infty} e^{i(k_{j_1}x_1+\cdots+k_{j_n}x_n)} \frac{\tilde{f}(k_{j_1},\ldots,k_{j_n})}{(2\pi)^{n/2}} \tag{3.95}$$

in which $k_{j_\ell} = 2\pi j_\ell/L$. Thus, for $n = 3$ we have

$$f(\boldsymbol{x}) = \frac{(2\pi)^3}{V} \sum_{j_1=-\infty}^{\infty} \sum_{j_2=-\infty}^{\infty} \sum_{j_3=-\infty}^{\infty} e^{i\boldsymbol{k_j}\cdot\boldsymbol{x}} \frac{\tilde{f}(\boldsymbol{k_j})}{(2\pi)^{3/2}}, \tag{3.96}$$

in which $\boldsymbol{k_j} = (k_{j_1}, k_{j_2}, k_{j_3})$ and $V = L^3$ is the volume of the box.

Example 3.8 (The Feynman propagator) For a spinless quantum field of mass m, Feynman's propagator is the four-dimensional Fourier transform

$$\Delta_F(x) = \int \frac{\exp(ik \cdot x)}{k^2 + m^2 - i\epsilon} \frac{d^4k}{(2\pi)^4} \tag{3.97}$$

where $k \cdot x = \boldsymbol{k} \cdot \boldsymbol{x} - k^0 x^0$, all physical quantities are in **natural units** ($c = \hbar = 1$), and $x^0 = ct = t$. The tiny imaginary term $-i\epsilon$ makes $\Delta_F(x - y)$ proportional to the mean value in the vacuum state $|0\rangle$ of the **time-ordered product** of the fields $\phi(x)$ and $\phi(y)$ (section 5.34)

$$\begin{aligned} -i\,\Delta_F(x-y) &= \langle 0|\mathcal{T}\left[\phi(x)\phi(y)\right]|0\rangle \qquad (3.98)\\ &\equiv \theta(x^0 - y^0)\langle 0|\phi(x)\phi(y)|0\rangle + \theta(y^0 - x^0)\langle 0|\phi(y)\phi(x)|0\rangle \end{aligned}$$

in which $\theta(a) = (a + |a|)/2|a|$ is the Heaviside function (2.166). □

3.6 Convolutions

The convolution of $f(x)$ with $g(x)$ is the integral

$$f * g(x) = \int_{-\infty}^{\infty} \frac{dy}{\sqrt{2\pi}} f(x-y)\, g(y). \tag{3.99}$$

The convolution product is symmetric

$$f * g(x) = g * f(x) \tag{3.100}$$

because, setting $z = x - y$, we have

$$\begin{aligned} f * g(x) &= \int_{-\infty}^{\infty} \frac{dy}{\sqrt{2\pi}} f(x-y)\, g(y) = -\int_{\infty}^{-\infty} \frac{dz}{\sqrt{2\pi}} f(z)\, g(x-z) \\ &= \int_{-\infty}^{\infty} \frac{dz}{\sqrt{2\pi}} g(x-z) f(z) = g * f(x). \end{aligned} \tag{3.101}$$

Convolutions may look strange at first, but they often occur in physics in the three-dimensional form

$$F(\boldsymbol{x}) = \int G(\boldsymbol{x} - \boldsymbol{x}')\, S(\boldsymbol{x}')\, d^3\boldsymbol{x}, \tag{3.102}$$

in which G is a Green's function and S is a source.

Example 3.9 (Gauss's law) The divergence of the electric field $\boldsymbol{E}$ is the microscopic charge density ρ divided by the electric permittivity of the vacuum $\epsilon_0 = 8.854 \times 10^{-12}$ F/m, that is, $\nabla \cdot \boldsymbol{E} = \rho/\epsilon_0$. This constraint is known as Gauss's law. If the charges and fields are independent of time, then the electric field $\boldsymbol{E}$ is the gradient of a scalar potential $\boldsymbol{E} = -\nabla\phi$. These last two equations imply that ϕ obeys Poisson's equation

$$-\nabla^2\phi = \frac{\rho}{\epsilon_0}. \tag{3.103}$$

We may solve this equation by using Fourier transforms as described in section 3.13. If $\tilde{\phi}(\boldsymbol{k})$ and $\tilde{\rho}(\boldsymbol{k})$ respectively are the Fourier transforms of $\phi(\boldsymbol{x})$ and $\rho(\boldsymbol{x})$, then Poisson's differential equation (3.103) gives

$$\begin{aligned} -\nabla^2\phi(\boldsymbol{x}) &= -\nabla^2 \int e^{i\boldsymbol{k}\cdot\boldsymbol{x}}\,\tilde{\phi}(\boldsymbol{k})\,d^3\boldsymbol{k} = \int \boldsymbol{k}^2\, e^{i\boldsymbol{k}\cdot\boldsymbol{x}}\,\tilde{\phi}(\boldsymbol{k})\,d^3\boldsymbol{k} \\ &= \frac{\rho(\boldsymbol{x})}{\epsilon_0} = \int e^{i\boldsymbol{k}\cdot\boldsymbol{x}}\,\frac{\tilde{\rho}(\boldsymbol{k})}{\epsilon_0}\,d^3\boldsymbol{k}, \end{aligned} \tag{3.104}$$

which implies the algebraic equation $\tilde{\phi}(\boldsymbol{k}) = \tilde{\rho}(\boldsymbol{k})/\epsilon_0\boldsymbol{k}^2$, which is an instance of (3.163). Performing the inverse Fourier transformation, we find for the scalar potential

$$\begin{aligned} \phi(\boldsymbol{x}) &= \int e^{i\boldsymbol{k}\cdot\boldsymbol{x}}\,\tilde{\phi}(\boldsymbol{k})\,d^3\boldsymbol{k} = \int e^{i\boldsymbol{k}\cdot\boldsymbol{x}}\,\frac{\tilde{\rho}(\boldsymbol{k})}{\epsilon_0\,\boldsymbol{k}^2}\,d^3\boldsymbol{k} \qquad (3.105) \\ &= \int e^{i\boldsymbol{k}\cdot\boldsymbol{x}}\,\frac{1}{\boldsymbol{k}^2}\int e^{-i\boldsymbol{k}\cdot\boldsymbol{x}'}\,\frac{\rho(\boldsymbol{x}')}{\epsilon_0}\,\frac{d^3\boldsymbol{x}'d^3\boldsymbol{k}}{(2\pi)^3} = \int G(\boldsymbol{x}-\boldsymbol{x}')\,\frac{\rho(\boldsymbol{x}')}{\epsilon_0}\,d^3\boldsymbol{x}', \end{aligned}$$

in which

$$G(\boldsymbol{x}-\boldsymbol{x}') = \int \frac{d^3\boldsymbol{k}}{(2\pi)^3}\,\frac{1}{\boldsymbol{k}^2}\,e^{i\boldsymbol{k}\cdot(\boldsymbol{x}-\boldsymbol{x}')}. \tag{3.106}$$

This function $G(\boldsymbol{x}-\boldsymbol{x}')$ is the Green's function for the differential operator $-\nabla^2$ in the sense that

$$-\nabla^2 G(\boldsymbol{x}-\boldsymbol{x}') = \int \frac{d^3\boldsymbol{k}}{(2\pi)^3}\,e^{i\boldsymbol{k}\cdot(\boldsymbol{x}-\boldsymbol{x}')} = \delta^{(3)}(\boldsymbol{x}-\boldsymbol{x}'). \tag{3.107}$$

This Green's function ensures that expression (3.105) for $\phi(\boldsymbol{x})$ satisfies Poisson's equation (3.103). To integrate (3.106) and compute $G(\boldsymbol{x}-\boldsymbol{x}')$, we use spherical coordinates with the z-axis parallel to the vector $\boldsymbol{x}-\boldsymbol{x}'$

$$\begin{aligned} G(\boldsymbol{x}-\boldsymbol{x}') &= \int \frac{d^3\boldsymbol{k}}{(2\pi)^3}\,\frac{1}{\boldsymbol{k}^2}\,e^{i\boldsymbol{k}\cdot(\boldsymbol{x}-\boldsymbol{x}')} = \int_0^\infty \frac{dk}{(2\pi)^2}\int_{-1}^{1} d\cos\theta\; e^{ik|\boldsymbol{x}-\boldsymbol{x}'|\cos\theta} \\ &= \int_0^\infty \frac{dk}{(2\pi)^2}\,\frac{e^{ik|\boldsymbol{x}-\boldsymbol{x}'|}-e^{-ik|\boldsymbol{x}-\boldsymbol{x}'|}}{ik|\boldsymbol{x}-\boldsymbol{x}'|} \\ &= \frac{1}{2\pi^2|\boldsymbol{x}-\boldsymbol{x}'|}\int_0^\infty \frac{\sin k|\boldsymbol{x}-\boldsymbol{x}'|\,dk}{k} = \frac{1}{2\pi^2|\boldsymbol{x}-\boldsymbol{x}'|}\int_0^\infty \frac{\sin k\,dk}{k}. \end{aligned} \tag{3.108}$$

In example 5.35 of section 5.18 on Cauchy's principal value, we'll show that

$$\int_0^\infty \frac{\sin k}{k}\,dk = \frac{\pi}{2}. \tag{3.109}$$

Using this result, we have

$$\int \frac{d^3k}{(2\pi)^3} \frac{1}{k^2} e^{ik\cdot(x-x')} = G(x - x') = \frac{1}{4\pi |x - x'|}. \tag{3.110}$$

Finally, by substituting this formula for $G(\boldsymbol{x} - \boldsymbol{x}')$ into Equation (3.105), we find that the Fourier transform $\phi(x)$ of the product $\hat{\rho}(\boldsymbol{k})/\boldsymbol{k}^2$ of the functions $\hat{\rho}(\boldsymbol{k})$ and $1/\boldsymbol{k}^2$ is the convolution

$$\phi(\boldsymbol{x}) = \frac{1}{4\pi\epsilon_0} \int \frac{\rho(\boldsymbol{x}')}{|\boldsymbol{x} - \boldsymbol{x}'|} d^3x' \tag{3.111}$$

of their Fourier transforms $1/|\boldsymbol{x} - \boldsymbol{x}'|$ and $\rho(\boldsymbol{x}')$. The Fourier transform of the product of any two functions is the convolution of their Fourier transforms, as we'll see in the next section (George Green, 1793–1841). □

Example 3.10 (The magnetic vector potential) The magnetic induction $\boldsymbol{B}$ has zero divergence (as long as there are no magnetic monopoles) and so may be written as the curl $\boldsymbol{B} = \nabla \times \boldsymbol{A}$ of a vector potential $\boldsymbol{A}$. For time-independent currents, Ampère's law is $\nabla \times \boldsymbol{B} = \mu_0 \boldsymbol{J}$, in which $\mu_0 = 1/(\epsilon_0 c^2) = 4\pi \times 10^{-7}$ N A^{-2} is the permeability of the vacuum. It follows that in the Coulomb gauge $\nabla \cdot \boldsymbol{A} = 0$, the magnetostatic vector potential $\boldsymbol{A}$ satisfies the equation

$$\nabla \times \boldsymbol{B} = \nabla \times (\nabla \times \boldsymbol{A}) = \nabla(\nabla \cdot \boldsymbol{A}) - \nabla^2 \boldsymbol{A} = -\nabla^2 \boldsymbol{A} = \mu_0 \boldsymbol{J}. \tag{3.112}$$

Applying the Fourier-transform technique (3.103–3.111), we find that the Fourier transforms of $\boldsymbol{A}$ and $\boldsymbol{J}$ satisfy the algebraic equation

$$\hat{\boldsymbol{A}}(\boldsymbol{k}) = \mu_0 \frac{\hat{\boldsymbol{J}}(\boldsymbol{k})}{\boldsymbol{k}^2}, \tag{3.113}$$

which is an instance of (3.163). Performing the inverse Fourier transform, we see that $\boldsymbol{A}$ is the convolution

$$\boldsymbol{A}(\boldsymbol{x}) = \frac{\mu_0}{4\pi} \int d^3x' \frac{\boldsymbol{J}(\boldsymbol{x}')}{|\boldsymbol{x} - \boldsymbol{x}'|}. \tag{3.114}$$

If in the solution (3.111) of Poisson's equation, $\rho(\boldsymbol{x})$ is translated by $\boldsymbol{a}$, then so is $\phi(\boldsymbol{x})$. That is, if $\rho'(\boldsymbol{x}) = \rho(\boldsymbol{x} + \boldsymbol{a})$ then $\phi'(\boldsymbol{x}) = \phi(\boldsymbol{x} + \boldsymbol{a})$. Similarly, if the current $\boldsymbol{J}(\boldsymbol{x})$ in (3.114) is translated by $\boldsymbol{a}$, then so is the potential $\boldsymbol{A}(\boldsymbol{x})$. **Convolutions respect translational invariance.** That's one reason why they occur so often in the formulas of physics. □

3.7 The Fourier transform of a convolution

The Fourier transform of the convolution $f * g$ is the product of the Fourier transforms $\tilde{f}$ and $\tilde{g}$:

$$\widetilde{f * g}(k) = \tilde{f}(k)\,\tilde{g}(k). \tag{3.115}$$

To see why, we form the Fourier transform $\widetilde{f * g}(k)$ of the convolution $f * g(x)$

$$\widetilde{f * g}(k) = \int_{-\infty}^{\infty} \frac{dx}{\sqrt{2\pi}} e^{-ikx} f * g(x)$$
$$= \int_{-\infty}^{\infty} \frac{dx}{\sqrt{2\pi}} e^{-ikx} \int_{-\infty}^{\infty} \frac{dy}{\sqrt{2\pi}} f(x-y)\, g(y). \tag{3.116}$$

Now we write $f(x-y)$ and $g(y)$ in terms of their Fourier transforms $\tilde{f}(p)$ and $\tilde{g}(q)$

$$\widetilde{f * g}(k) = \int_{-\infty}^{\infty} \frac{dx}{\sqrt{2\pi}} e^{-ikx} \int_{-\infty}^{\infty} \frac{dy}{\sqrt{2\pi}} \int_{-\infty}^{\infty} \frac{dp}{\sqrt{2\pi}} \tilde{f}(p)\, e^{ip(x-y)} \int_{-\infty}^{\infty} \frac{dq}{\sqrt{2\pi}} \tilde{g}(q)\, e^{iqy} \tag{3.117}$$

and use the representation (3.36) of Dirac's delta function twice to get

$$\widetilde{f * g}(k) = \int_{-\infty}^{\infty} \frac{dy}{2\pi} \int_{-\infty}^{\infty} dp \int_{-\infty}^{\infty} dq\, \delta(p-k) \tilde{f}(p)\, \tilde{g}(q)\, e^{i(q-p)y}$$
$$= \int_{-\infty}^{\infty} dp \int_{-\infty}^{\infty} dq\, \delta(p-k)\, \delta(q-p) \tilde{f}(p)\, \tilde{g}(q)$$
$$= \int_{-\infty}^{\infty} dp\, \delta(p-k) \tilde{f}(p)\, \tilde{g}(p) = \tilde{f}(k)\, \tilde{g}(k), \tag{3.118}$$

which is (3.115). Examples 3.9 and 3.10 were illustrations of this result.

3.8 Fourier transforms and Green's functions

A Green's function $G(x)$ for a differential operator P turns into a delta function when acted upon by P, that is, $PG(x) = \delta(x)$. If the differential operator is a polynomial $P(\partial) \equiv P(\partial_1, \ldots, \partial_n)$ in the derivatives $\partial_1, \ldots, \partial_n$ with constant coefficients, then a suitable Green's function $G(x) \equiv G(x_1, \ldots, x_n)$ will satisfy

$$P(\partial) G(x) = \delta^{(n)}(x). \tag{3.119}$$

Expressing both $G(x)$ and $\delta^{(n)}(x)$ as Fourier transforms, we get

$$P(\partial) G(x) = \int d^n k\, P(ik)\, e^{ik\cdot x}\, \tilde{G}(k) = \delta^{(n)}(x) = \int \frac{d^n k}{(2\pi)^n} e^{ik\cdot x}, \tag{3.120}$$

which gives us the algebraic equation

$$\tilde{G}(k) = \frac{1}{(2\pi)^n\, P(ik)}. \tag{3.121}$$

Thus the Green's function G_P for the differential operator $P(\partial)$ is

$$G_P(x) = \int \frac{d^n k}{(2\pi)^n} \frac{e^{ik\cdot x}}{P(ik)}. \tag{3.122}$$

Example 3.11 (Green and Yukawa) In 1935, Hideki Yukawa (1907–1981) proposed the partial differential equation

$$P_Y(\partial)G_Y(\boldsymbol{x}) \equiv (-\Delta + m^2)G_Y(\boldsymbol{x}) = (-\nabla^2 + m^2)G_Y(\boldsymbol{x}) = \delta(\boldsymbol{x}). \tag{3.123}$$

Our (3.122) gives as the Green's function for $P_Y(\partial)$ the Yukawa potential

$$G_Y(x) = \int \frac{d^3k}{(2\pi)^3} \frac{e^{i\boldsymbol{k}\cdot\boldsymbol{x}}}{P_Y(ik)} = \int \frac{d^3k}{(2\pi)^3} \frac{e^{i\boldsymbol{k}\cdot\boldsymbol{x}}}{\boldsymbol{k}^2 + m^2} = \frac{e^{-mr}}{4\pi r}, \tag{3.124}$$

an integration done in example 5.21. □

3.9 Laplace transforms

The Laplace transform $f(s)$ of a function $F(t)$ is the integral

$$f(s) = \int_0^\infty dt\, e^{-st}\, F(t). \tag{3.125}$$

Because the integration is over positive values of t, the exponential $\exp(-st)$ falls off rapidly with the real part of s. As $\mathrm{Re}\, s$ increases, the Laplace transform $f(s)$ becomes smoother and smaller. For $\mathrm{Re}\, s > 0$, the exponential $\exp(-st)$ lets many functions $F(t)$ that are not integrable over the half line $[0, \infty)$ have well-behaved Laplace transforms.

For instance, the function $F(t) = 1$ is not integrable over the half line $[0, \infty)$, but its Laplace transform

$$f(s) = \int_0^\infty dt\, e^{-st}\, F(t) = \int_0^\infty dt\, e^{-st} = \frac{1}{s} \tag{3.126}$$

is well defined for $\mathrm{Re}\, s > 0$ and square integrable for $\mathrm{Re}\, s > \epsilon$.

The function $F(t) = \exp(kt)$ diverges exponentially for $\mathrm{Re}\, k > 0$, but its Laplace transform

$$f(s) = \int_0^\infty dt\, e^{-st}\, F(t) = \int_0^\infty dt\, e^{-(s-k)t} = \frac{1}{s-k} \tag{3.127}$$

is well defined for $\mathrm{Re}\, s > k$ with a simple pole at $s = k$ (section 5.10) and is square integrable for $\mathrm{Re}\, s > k + \epsilon$.

The Laplace transforms of $\cosh kt$ and $\sinh kt$ are

$$f(s) = \int_0^\infty dt\, e^{-st} \cosh kt = \frac{1}{2} \int_0^\infty dt\, e^{-st} \left(e^{kt} + e^{-kt}\right) = \frac{s}{s^2 - k^2} \tag{3.128}$$

and

$$f(s) = \int_0^\infty dt\, e^{-st} \sinh kt = \frac{1}{2} \int_0^\infty dt\, e^{-st} \left(e^{kt} - e^{-kt}\right) = \frac{k}{s^2 - k^2}. \tag{3.129}$$

The Laplace transform of $\cos \omega t$ is

$$f(s) = \int_0^\infty dt\, e^{-st} \cos \omega t = \frac{1}{2} \int_0^\infty dt\, e^{-st} \left(e^{i\omega t} + e^{-i\omega t} \right) = \frac{s}{s^2 + \omega^2} \tag{3.130}$$

and that of $\sin \omega t$ is

$$f(s) = \int_0^\infty dt\, e^{-st} \sin \omega t = \frac{1}{2i} \int_0^\infty dt\, e^{-st} \left(e^{i\omega t} - e^{-i\omega t} \right) = \frac{\omega}{s^2 + \omega^2}. \tag{3.131}$$

Example 3.12 (Lifetime of a fluorophore) Fluorophores are molecules that emit visible light when excited by photons. The probability $P(t, t')$ that a fluorophore with a lifetime τ will emit a photon at time t if excited by a photon at time t' is

$$P(t, t') = \tau\, e^{-(t-t')/\tau}\, \theta(t - t') \tag{3.132}$$

in which $\theta(t - t') = (t - t' + |t - t'|)/2|t - t'|$ is the Heaviside function. One way to measure the lifetime τ of a fluorophore is to modulate the exciting laser beam at a frequency $\nu = 2\pi\omega$ of the order of 60 MHz and to detect the phase-shift ϕ in the light $L(t)$ emitted by the fluorophore. That light is the integral of $P(t, t')$ times the modulated beam $\sin \omega t$ or equivalently the convolution of $e^{-t/\tau}\theta(t)$ with $\sin \omega t$

$$\begin{aligned} L(t) &= \int_{-\infty}^{\infty} P(t, t') \sin(\omega t')\, dt' = \int_{-\infty}^{\infty} \tau\, e^{-(t-t')/\tau} \theta(t - t') \sin(\omega t')\, dt' \\ &= \int_{-\infty}^{t} \tau\, e^{-(t-t')/\tau} \sin(\omega t')\, dt'. \end{aligned} \tag{3.133}$$

Letting $u = t - t'$ and using the trigonometric formula

$$\sin(a - b) = \sin a \cos b - \cos a \sin b \tag{3.134}$$

we may relate this integral to the Laplace transforms of a sine (3.131) and a cosine (3.130)

$$\begin{aligned} L(t) &= -\tau \int_0^\infty e^{-u/\tau} \sin \omega(u - t)\, du \\ &= -\tau \int_0^\infty e^{-u/\tau} \left(\sin \omega u \cos \omega t - \cos \omega u \sin \omega t \right) du \\ &= \tau \left(\frac{\sin(\omega t)/\tau}{1/\tau^2 + \omega^2} - \frac{\omega \cos \omega t}{1/\tau^2 + \omega^2} \right). \end{aligned} \tag{3.135}$$

Setting $\cos \phi = (1/\tau)/\sqrt{1/\tau^2 + \omega^2}$ and $\sin \phi = \omega/\sqrt{1/\tau^2 + \omega^2}$, we have

$$L(t) = \frac{\tau}{\sqrt{1/\tau^2 + \omega^2}} \left(\sin \omega t \cos \phi - \cos \omega t \sin \phi \right) = \frac{\tau}{\sqrt{1/\tau^2 + \omega^2}} \sin(\omega t - \phi). \tag{3.136}$$

The phase-shift ϕ then is given by

$$\phi = \arcsin \frac{\omega}{\sqrt{1/\tau^2 + \omega^2}} \leq \frac{\pi}{2}. \tag{3.137}$$

So by inverting this formula, we get the lifetime of the fluorophore

$$\tau = (1/\omega) \tan \phi \tag{3.138}$$

in terms of the phase-shift ϕ, which is much easier to measure. □

3.10 Derivatives and integrals of Laplace transforms

The derivatives of a Laplace transform $f(s)$ are by its definition (3.125)

$$\frac{d^n f(s)}{ds^n} = \int_0^\infty dt \, (-t)^n \, e^{-st} \, F(t). \tag{3.139}$$

They usually are well defined if $f(s)$ is well defined. For instance, if we differentiate the Laplace transform $f(s) = 1/s$ of the function $F(t) = 1$ as given by (3.126), then we find

$$(-1)^n \frac{d^n s^{-1}}{ds^n} = \frac{n!}{s^{n+1}} = \int_0^\infty dt \, e^{-st} \, t^n, \tag{3.140}$$

which tells us that the Laplace transform of t^n is $n!/s^{n+1}$.

The result of differentiating the function $F(t)$ also has a simple form. Integrating by parts, we find for the Laplace transform of $F'(t)$

$$\begin{aligned} \int_0^\infty dt \, e^{-st} \, F'(t) &= \int_0^\infty dt \left\{ \frac{d}{dt} \left[e^{-st} \, F(t) \right] - F(t) \frac{d}{dt} e^{-st} \right\} \\ &= - F(0) + \int_0^\infty dt \, F(t) \, s \, e^{-st} \\ &= - F(0) + s f(s). \end{aligned} \tag{3.141}$$

The indefinite integral of the Laplace transform (3.125) is

$${}^{(1)}f(s) \equiv \int ds_1 \, f(s_1) = \int_0^\infty dt \, \frac{e^{-st}}{(-t)} \, F(t) \tag{3.142}$$

and its nth indefinite integral is

$${}^{(n)}f(s) \equiv \int ds_n \ldots \int ds_1 \, f(s_1) = \int_0^\infty dt \, \frac{e^{-st}}{(-t)^n} \, F(t). \tag{3.143}$$

If $f(s)$ is a well-behaved function, then these indefinite integrals usually are well defined, except possibly at $s = 0$.

3.11 Laplace transforms and differential equations

Suppose we wish to solve the differential equation

$$P(d/ds) f(s) = j(s). \tag{3.144}$$

By writing $f(s)$ and $j(s)$ as Laplace transforms

$$f(s) = \int_0^\infty e^{-st}\, F(t)\, dt$$
$$j(s) = \int_0^\infty e^{-st}\, J(t)\, dt \tag{3.145}$$

and using the formula (3.139) for the nth derivative of a Laplace transform, we see that the differential equation (3.144) amounts to

$$\int_0^\infty e^{-st}\, P(-t)\, F(t)\, dt = \int_0^\infty e^{-st}\, J(t)\, dt, \tag{3.146}$$

which is equivalent to the algebraic equation

$$F(t) = \frac{J(t)}{P(-t)}. \tag{3.147}$$

A particular solution to the inhomogeneous equation (3.144) is then the Laplace transform of this ratio

$$f(s) = \int_0^\infty e^{-st}\, \frac{J(t)}{P(-t)}\, dt. \tag{3.148}$$

A fairly general solution of the associated homogeneous equation

$$P(d/ds) f(s) = 0 \tag{3.149}$$

is the Laplace transform

$$f(s) = \int_0^\infty e^{-st}\, \delta(P(-t))\, H(t)\, dt, \tag{3.150}$$

in which the function $H(t)$ is arbitrary. Thus our solution of the inhomogeneous equation (3.144) is the sum of the two

$$f(s) = \int_0^\infty e^{-st}\, \frac{J(t)}{P(-t)}\, dt + \int_0^\infty e^{-st}\, \delta(P(-t))\, H(t)\, dt. \tag{3.151}$$

One may generalize this method to differential equations in n variables. But to carry out this procedure, one must be able to find the inverse Laplace transform $J(t)$ of the source function $j(s)$ as outlined in the next section.

3.12 Inversion of Laplace transforms

How do we invert the Laplace transform

$$f(s) = \int_0^\infty dt\, e^{-st}\, F(t)? \tag{3.152}$$

First we extend the Laplace transform from real s to $s + iu$

$$f(s+iu) = \int_0^\infty dt\, e^{-(s+iu)t}\, F(t). \tag{3.153}$$

Next we choose s to be sufficiently positive that the integral

$$\int_{-\infty}^{\infty} \frac{du}{2\pi}\, e^{(s+iu)t} f(s+iu) = \int_{-\infty}^{\infty} \frac{du}{2\pi} \int_0^\infty dt'\, e^{(s+iu)t}\, e^{-(s+iu)t'}\, F(t') \tag{3.154}$$

converges, and then we apply to it the delta function formula (3.36)

$$\begin{aligned} \int_{-\infty}^{\infty} \frac{du}{2\pi}\, e^{(s+iu)t} f(s+iu) &= \int_0^\infty dt'\, e^{s(t-t')}\, F(t') \int_{-\infty}^{\infty} \frac{du}{2\pi}\, e^{iu(t-t')} \\ &= \int_0^\infty dt'\, e^{s(t-t')}\, F(t')\, \delta(t-t') = F(t). \end{aligned} \tag{3.155}$$

So our inversion formula is

$$F(t) = e^{st} \int_{-\infty}^{\infty} \frac{du}{2\pi}\, e^{iut} f(s+iu) \tag{3.156}$$

for sufficiently large s. Some call this inversion formula a Bromwich integral, others a Fourier–Mellin integral.

3.13 Application to differential equations

Let us consider a linear partial differential equation in n variables

$$P(\partial_1, \ldots, \partial_n) f(x_1, \ldots, x_n) = g(x_1, \ldots, x_n), \tag{3.157}$$

in which P is a polynomial in the derivatives

$$\partial_j \equiv \frac{\partial}{\partial x_j} \tag{3.158}$$

with constant coefficients. If $g = 0$, the equation is homogeneous; otherwise it is inhomogeneous. We expand solution and source as integral transforms

$$\begin{aligned} f(x_1, \ldots, x_n) &= \int \tilde{f}(k_1, \ldots, k_n)\, e^{i(k_1 x_1 + \cdots + k_n x_n)} d^n k, \\ g(x_1, \ldots, x_n) &= \int \tilde{g}(k_1, \ldots, k_n)\, e^{i(k_1 x_1 + \cdots + k_n x_n)} d^n k, \end{aligned} \tag{3.159}$$

in which the k integrals may run from $-\infty$ to ∞ as in a Fourier transform or up the imaginary axis from 0 to ∞ as in a Laplace transform.

The correspondence (3.55) between differentiation with respect to x_j and multiplication by ik_j tells us that ∂_j^m acting on f gives

$$\partial_j^m f(x_1, \ldots, x_n) = \int \tilde{f}(k_1, \ldots, k_n)\,(ik_j)^m\, e^{i(k_1 x_1 + \cdots + k_n x_n)}\, d^n k. \tag{3.160}$$

If we abbreviate $f(x_1, \ldots, x_n)$ by $f(x)$ and do the same for g, then we may write our partial differential equation (3.157) as

$$\begin{aligned} P(\partial_1, \ldots, \partial_n) f(x) &= \int \tilde{f}(k)\, P(ik_1, \ldots, ik_n)\, e^{i(k_1 x_1 + \cdots + k_n x_n)}\, d^n k \\ &= \int \tilde{g}(k)\, e^{i(k_1 x_1 + \cdots + k_n x_n)}\, d^n k. \end{aligned} \tag{3.161}$$

Thus the inhomogeneous partial differential equation

$$P(\partial_1, \ldots, \partial_n) f_i(x_1, \ldots, x_n) = g(x_1, \ldots, x_n) \tag{3.162}$$

becomes an algebraic equation in k-space

$$P(ik_1, \ldots, ik_n) \tilde{f}_i(k_1, \ldots, k_n) = \tilde{g}(k_1, \ldots, k_n) \tag{3.163}$$

where $\tilde{g}(k_1, \ldots, k_n)$ is the mixed Fourier–Laplace transform of $g(x_1, \ldots, x_n)$. So one solution of the inhomogeneous differential equation (3.157) is

$$f_i(x_1, \ldots, x_n) = \int e^{i(k_1 x_1 + \cdots + k_n x_n)}\, \frac{\tilde{g}(k_1, \ldots, k_n)}{P(ik_1, \ldots, ik_n)}\, d^n k. \tag{3.164}$$

The space of solutions to the **homogeneous** form of equation (3.157)

$$P(\partial_1, \ldots, \partial_n) f_h(x_1, \ldots, x_n) = 0 \tag{3.165}$$

is vast. We will focus on those that satisfy the algebraic equation

$$P(ik_1, \ldots, ik_n) \tilde{f}_h(k_1, \ldots, k_n) = 0 \tag{3.166}$$

and that we can write in terms of Dirac's delta function as

$$\tilde{f}_h(k_1, \ldots, k_n) = \delta(P(ik_1, \ldots, ik_n))\, h(k_1, \ldots, k_n), \tag{3.167}$$

in which the function $h(k)$ is arbitrary. That is

$$f_h(x) = \int e^{i(k_1 x_1 + \cdots + k_n x_n)} \delta(P(ik_1, \ldots, ik_n))\, h(k)\, d^n k. \tag{3.168}$$

Our solution to the differential equation (3.157) then is a sum of a particular solution (3.164) of the inhomogeneous equation (3.163) and our solution (3.168) of the associated homogeneous equation (3.165)

$$f(x_1,\ldots,x_n) = \int e^{i(k_1x_1+\cdots+k_nx_n)} \left[\frac{\tilde g(k_1,\ldots,k_n)}{P(ik_1,\ldots,ik_n)} + \delta(P(ik_1,\ldots,ik_n))\, h(k_1,\ldots,k_n) \right] d^nk \tag{3.169}$$

in which $h(k_1,\ldots,k_n)$ is an arbitrary function. The wave equation and the diffusion equation will provide examples of this formula

$$f(x) = \int e^{ik\cdot x} \left[\frac{\tilde g(k)}{P(ik)} + \delta(P(ik))h(k) \right] d^nk. \tag{3.170}$$

Example 3.13 (Wave equation for a scalar field) A free scalar field $\phi(x)$ of mass m in flat space-time obeys the wave equation

$$\left(\nabla^2 - \partial_t^2 - m^2\right)\phi(x) = 0 \tag{3.171}$$

in natural units ($\hbar = c = 1$). We may use a four-dimensional Fourier transform to represent the field $\phi(x)$ as

$$\phi(x) = \int e^{ik\cdot x}\, \tilde\phi(k)\, \frac{d^4k}{(2\pi)^2}, \tag{3.172}$$

in which $k\cdot x = \boldsymbol{k}\cdot\boldsymbol{x} - k^0 t$ is the Lorentz-invariant inner product.

The homogeneous wave equation (3.171) then says

$$\left(\nabla^2 - \partial_t^2 - m^2\right)\phi(x) = \int \left(-\boldsymbol{k}^2 + (k^0)^2 - m^2\right) e^{ik\cdot x}\, \tilde\phi(k)\, \frac{d^4k}{(2\pi)^2} = 0, \tag{3.173}$$

which implies the algebraic equation

$$\left(-\boldsymbol{k}^2 + (k^0)^2 - m^2\right)\tilde\phi(k) = 0 \tag{3.174}$$

an instance of (3.166). Our solution (3.168) is

$$\phi(x) = \int \delta\left(-\boldsymbol{k}^2 + (k^0)^2 - m^2\right) e^{ik\cdot x}\, h(k)\, \frac{d^4k}{(2\pi)^2}, \tag{3.175}$$

in which $h(k)$ is an arbitrary function. The argument of the delta function

$$g_k(k^0) \equiv (k^0)^2 - \boldsymbol{k}^2 - m^2 = \left(k^0 - \sqrt{\boldsymbol{k}^2+m^2}\right)\left(k^0 + \sqrt{\boldsymbol{k}^2+m^2}\right) \tag{3.176}$$

has zeros at $k^0 = \pm\sqrt{\boldsymbol{k}^2+m^2} \equiv \pm\omega_k$ with $|g_k'(\pm\omega_k)| = 2\omega_k$. So using our formula (3.46) for integrals involving delta functions of functions, we have

$$\phi(x) = \int \left[e^{i(\boldsymbol{k}\cdot\boldsymbol{x}-\omega_k t)}\, h_+(\boldsymbol{k}) + e^{i(\boldsymbol{k}\cdot\boldsymbol{x}+\omega_k t)}\, h_-(\boldsymbol{k}) \right] \frac{d^3k}{(2\pi)^2 2\omega_k} \tag{3.177}$$

where $h_\pm(\boldsymbol{k}) \equiv h(\pm\omega_k, \boldsymbol{k})$. Since ω_k is an even function of $\boldsymbol{k}$, we can write

$$\phi(x) = \int \left[e^{i(\boldsymbol{k}\cdot\boldsymbol{x}-\omega_k t)}\, h_+(\boldsymbol{k}) + e^{-i(\boldsymbol{k}\cdot\boldsymbol{x}-\omega_k t)}\, h_-(-\boldsymbol{k}) \right] \frac{d^3k}{(2\pi)^2 2\omega_k}. \tag{3.178}$$

If $\phi(x)$ is a real-valued classical field, then (3.24) tells us that $h_-(-\boldsymbol{k}) = h_+(\boldsymbol{k})^*$. If it is a hermitian quantum field, then $h_-(-\boldsymbol{k}) = h_+^\dagger(\boldsymbol{k})$. One sets $a(\boldsymbol{k}) \equiv h_+(\boldsymbol{k})/\sqrt{4\pi\omega_k}$ and writes $\phi(x)$ as an integral over $a(\boldsymbol{k})$, an **annihilation** operator,

$$\phi(x) = \int \left[e^{i(\boldsymbol{k}\cdot\boldsymbol{x}-\omega_k t)}\, a(\boldsymbol{k}) + e^{-i(\boldsymbol{k}\cdot\boldsymbol{x}-\omega_k t)}\, a^\dagger(\boldsymbol{k}) \right] \frac{d^3k}{\sqrt{(2\pi)^3 2\omega_k}}, \tag{3.179}$$

and its adjoint $a^\dagger(\boldsymbol{k})$, a **creation** operator.

The momentum π canonically conjugate to the field is its time derivative

$$\pi(x) = -i \int \left[e^{i(\boldsymbol{k}\cdot\boldsymbol{x}-\omega_k t)}\, a(\boldsymbol{k}) - e^{-i(\boldsymbol{k}\cdot\boldsymbol{x}-\omega_k t)}\, a^\dagger(\boldsymbol{k}) \right] \sqrt{\frac{\omega_k}{2(2\pi)^3}}\, d^3k. \tag{3.180}$$

If the operators a and $a^\dagger$ obey the commutation relations

$$[a(\boldsymbol{k}), a^\dagger(\boldsymbol{k}')] = \delta(\boldsymbol{k}-\boldsymbol{k}') \quad \text{and} \quad [a(\boldsymbol{k}), a(\boldsymbol{k}')] = [a^\dagger(\boldsymbol{k}), a^\dagger(\boldsymbol{k}')] = 0 \tag{3.181}$$

then the field $\phi(\boldsymbol{x}, t)$ and its conjugate momentum $\pi(\boldsymbol{y}, t)$ satisfy (exercise 3.11) the equal-time commutation relations

$$[\phi(\boldsymbol{x}, t), \pi(\boldsymbol{y}, t)] = i\delta(\boldsymbol{x}-\boldsymbol{y}) \quad \text{and} \quad [\phi(\boldsymbol{x}, t), \phi(\boldsymbol{y}, t)] = [\pi(\boldsymbol{x}, t), \pi(\boldsymbol{y}, t)] = 0, \tag{3.182}$$

which generalize the commutation relations of quantum mechanics

$$[q_j, p_\ell] = i\hbar\delta_{j,\ell} \quad \text{and} \quad [q_j, q_\ell] = [p_j, p_\ell] = 0 \tag{3.183}$$

for a set of coordinates q_j and conjugate momenta p_ℓ. □

Example 3.14 (Fourier series for a scalar field) For a field defined in a cube of volume $V = L^3$, one often imposes periodic boundary conditions (section 2.13) in which a displacement of any spatial coordinate by $\pm L$ does not change the value of the field. A Fourier series can represent a periodic field. Using the relationship (3.96) between Fourier-transform and Fourier-series representations in three dimensions, we expect the Fourier series representation for the field (3.179) to be

$$\begin{aligned} \phi(x) &= \frac{(2\pi)^3}{V} \sum_{\boldsymbol{k}} \frac{1}{\sqrt{(2\pi)^3 2\omega_k}} \left[a(\boldsymbol{k}) e^{i(\boldsymbol{k}\cdot\boldsymbol{x}-\omega_k t)} + a^\dagger(\boldsymbol{k}) e^{-i(\boldsymbol{k}\cdot\boldsymbol{x}-\omega_k t)} \right] \\ &= \sum_{\boldsymbol{k}} \frac{1}{\sqrt{2\omega_k V}} \sqrt{\frac{(2\pi)^3}{V}} \left[a(\boldsymbol{k}) e^{i(\boldsymbol{k}\cdot\boldsymbol{x}-\omega_k t)} + a^\dagger(\boldsymbol{k}) e^{-i(\boldsymbol{k}\cdot\boldsymbol{x}-\omega_k t)} \right], \end{aligned} \tag{3.184}$$

in which the sum over $\boldsymbol{k} = (2\pi/L)(\ell, n, m)$ is over all (positive and negative) integers ℓ, n, and m. One can set

$$a_{\boldsymbol{k}} \equiv \sqrt{\frac{(2\pi)^3}{V}}\, a(\boldsymbol{k}) \tag{3.185}$$

and write the field as

$$\phi(x) = \sum_{\boldsymbol{k}} \frac{1}{\sqrt{2\omega_k V}} \left[a_{\boldsymbol{k}}\, e^{i(\boldsymbol{k}\cdot\boldsymbol{x} - \omega_k t)} + a_{\boldsymbol{k}}^\dagger\, e^{-i(\boldsymbol{k}\cdot\boldsymbol{x} - \omega_k t)} \right]. \tag{3.186}$$

The commutator of Fourier-series annihilation and creation operators is by (3.36, 3.181, & 3.185)

$$\begin{aligned}
[a_{\boldsymbol{k}}, a_{\boldsymbol{k}'}^\dagger] &= \frac{(2\pi)^3}{V}\,[a(\boldsymbol{k}), a^\dagger(\boldsymbol{k}')] = \frac{(2\pi)^3}{V}\,\delta(\boldsymbol{k} - \boldsymbol{k}') \\
&= \frac{(2\pi)^3}{V} \int e^{i(\boldsymbol{k}-\boldsymbol{k}')\cdot\boldsymbol{x}} \frac{d^3x}{(2\pi)^3} = \frac{(2\pi)^3}{V}\,\frac{V}{(2\pi)^3}\,\delta_{\boldsymbol{k},\boldsymbol{k}'} = \delta_{\boldsymbol{k},\boldsymbol{k}'},
\end{aligned} \tag{3.187}$$

in which the Kronecker delta $\delta_{\boldsymbol{k},\boldsymbol{k}'}$ is $\delta_{\ell,\ell'}\delta_{n,n'}\delta_{m,m'}$. □

Example 3.15 (Diffusion) The flow rate $\boldsymbol{J}$ (per unit area, per unit time) of a fixed number of randomly moving particles, such as molecules of a gas or a liquid, is proportional to the negative gradient of their density $\rho(\boldsymbol{x}, t)$

$$\boldsymbol{J}(\boldsymbol{x}, t) = -D\nabla\rho(\boldsymbol{x}, t) \tag{3.188}$$

where D is the **diffusion constant**, an equation known as **Fick's law** (Adolf Fick, 1829–1901). Since the number of particles is conserved, the 4-vector $J = (\rho, \boldsymbol{J})$ obeys the conservation law

$$\frac{\partial}{\partial t} \int \rho(\boldsymbol{x}, t)\, d^3\boldsymbol{x} = -\oint \boldsymbol{J}(\boldsymbol{x}, t)\cdot d\boldsymbol{a} = -\int \nabla\cdot\boldsymbol{J}(\boldsymbol{x}, t) d^3\boldsymbol{x}, \tag{3.189}$$

which with Fick's law (3.188) gives the **diffusion equation**

$$\dot\rho(\boldsymbol{x}, t) = -\nabla\cdot\boldsymbol{J}(\boldsymbol{x}, t) = D\nabla^2\rho(\boldsymbol{x}, t) \quad \text{or} \quad \left(D\nabla^2 - \partial_t\right)\rho(\boldsymbol{x}, t) = 0. \tag{3.190}$$

Fourier had in mind such equations when he invented his transform.

If we write the density $\rho(\boldsymbol{x}, t)$ as the transform

$$\rho(\boldsymbol{x}, t) = \int e^{i\boldsymbol{k}\cdot\boldsymbol{x} + i\omega t}\,\tilde\rho(\boldsymbol{k}, \omega)\, d^3\boldsymbol{k} d\omega \tag{3.191}$$

then the diffusion equation becomes

$$\left(D\nabla^2 - \partial_t\right)\rho(\boldsymbol{x}, t) = \int e^{i\boldsymbol{k}\cdot\boldsymbol{x} + i\omega t}\left(-D\boldsymbol{k}^2 - i\omega\right)\tilde\rho(\boldsymbol{k}, \omega)\, d^3\boldsymbol{k} d\omega = 0, \tag{3.192}$$

which implies the algebraic equation

$$\left(D\boldsymbol{k}^2 + i\omega\right)\tilde\rho(\boldsymbol{k}, \omega) = 0. \tag{3.193}$$

Our solution (3.168) of this homogeneous equation is

$$\rho(\boldsymbol{x}, t) = \int e^{i\boldsymbol{k}\cdot\boldsymbol{x}+i\omega t}\delta\left(-D\boldsymbol{k}^2 - i\omega\right) h(\boldsymbol{k}, \omega)\, d^3\boldsymbol{k} d\omega, \tag{3.194}$$

in which $h(\boldsymbol{k}, \omega)$ is an arbitrary function. Dirac's delta function requires ω to be imaginary $\omega = iD\boldsymbol{k}^2$, with $D\boldsymbol{k}^2 > 0$. So the ω-integration is up the imaginary axis. It is a Laplace transform, and we have

$$\rho(\boldsymbol{x}, t) = \int_{-\infty}^{\infty} e^{i\boldsymbol{k}\cdot\boldsymbol{x} - D\boldsymbol{k}^2 t}\, \tilde{\rho}(\boldsymbol{k})\, d^3\boldsymbol{k}, \tag{3.195}$$

in which $\tilde{\rho}(\boldsymbol{k}) \equiv h(\boldsymbol{k}, iD\boldsymbol{k}^2)$. Thus the function $\tilde{\rho}(\boldsymbol{k})$ is the Fourier transform of the initial density $\rho(\boldsymbol{x}, 0)$

$$\rho(\boldsymbol{x}, 0) = \int_{-\infty}^{\infty} e^{i\boldsymbol{k}\cdot\boldsymbol{x}}\tilde{\rho}(\boldsymbol{k})\, d^3\boldsymbol{k}. \tag{3.196}$$

So if the initial density $\rho(\boldsymbol{x}, 0)$ is concentrated at $\boldsymbol{y}$

$$\rho(\boldsymbol{x}, 0) = \delta(\boldsymbol{x} - \boldsymbol{y}) = \int_{-\infty}^{\infty} e^{i\boldsymbol{k}\cdot(\boldsymbol{x}-\boldsymbol{y})}\, \frac{d^3\boldsymbol{k}}{(2\pi)^3} \tag{3.197}$$

then its Fourier transform $\tilde{\rho}(\boldsymbol{k})$ is

$$\tilde{\rho}(\boldsymbol{k}) = \frac{e^{-i\boldsymbol{k}\cdot\boldsymbol{y}}}{(2\pi)^3} \tag{3.198}$$

and at later times the density $\rho(\boldsymbol{x}, t)$ is given by (3.195) as

$$\rho(\boldsymbol{x}, t) = \int_{-\infty}^{\infty} e^{i\boldsymbol{k}\cdot(\boldsymbol{x}-\boldsymbol{y}) - D\boldsymbol{k}^2 t}\, \frac{d^3\boldsymbol{k}}{(2\pi)^3}. \tag{3.199}$$

Using our formula (3.18) for the Fourier transform of a gaussian, we find

$$\rho(\boldsymbol{x}, t) = \frac{1}{(4\pi Dt)^{3/2}}\, e^{-(\boldsymbol{x}-\boldsymbol{y})^2/(4Dt)}. \tag{3.200}$$

Since the diffusion equation is linear, it follows (exercise 3.12) that an arbitrary initial distribution $\rho(\boldsymbol{y}, 0)$ evolves to the distribution

$$\rho(\boldsymbol{x}, t) = \frac{1}{(4\pi Dt)^{3/2}} \int e^{-(\boldsymbol{x}-\boldsymbol{y})^2/(4Dt)}\, \rho(\boldsymbol{y}, 0)\, d^3\boldsymbol{y} \tag{3.201}$$

at time t. Such **convolutions** often occur in physics (section 3.6). □

Exercises

3.1 Show that the Fourier integral formula (3.26) for real functions follows from (3.9) and (3.25).

3.2 Show that the Fourier integral formula (3.26) for real functions implies (3.27) if f is even and (3.28) if it is odd.

3.3 Derive the formula (3.30) for the square wave (3.29).

3.4 By using the Fourier-transform formulas (3.27 & 3.28), derive the formulas (3.31) and (3.32) for the even and odd extensions of the exponential $\exp(-\beta|x|)$.

3.5 For the state $|\psi, t\rangle$ given by equations (3.85 & 3.80), find the wave-function $\psi(x,t) = \langle x|\psi, t\rangle$ at time t. Then find the variance of the position operator at that time. Does it grow as time goes by?

3.6 At time $t=0$, a particle of mass m is in a gaussian superposition of momentum eigenstates centered at $p = \hbar K$:

$$\psi(x,0) = N \int_{-\infty}^{\infty} e^{ikx} e^{-l^2(k-K)^2} \, dk. \tag{3.202}$$

(a) Shift k by K and do the integral. Where is the particle most likely to be found? (b) At time t, the wave function $\psi(x,t)$ is $\psi(x,0)$ but with ikx replaced by $ikx - i\hbar k^2 t/2m$. Shift k by K and do the integral. Where is the particle most likely to be found? (c) Does the wave packet spread out like t or like $\sqrt{t}$ as in classical diffusion?

3.7 Express a probability distribution $P(x)$ as the Fourier transform of its characteristic function (3.87).

3.8 Express the characteristic function (3.87) of a probability distribution as a power series in its moments (3.89).

3.9 Find the characteristic function (3.87) of the gaussian probability distribution

$$P_G(x, \mu, \sigma) = \frac{1}{\sigma\sqrt{2\pi}} \exp\left(-\frac{(x-\mu)^2}{2\sigma^2}\right). \tag{3.203}$$

3.10 Find the moments $\mu_n = E[x^n]$ for $n = 0, \ldots, 3$ of the gaussian probability distribution $P_G(x, \mu, \sigma)$.

3.11 Show that the commutation relations (3.181) of the annihilation and creation operators imply the equal-time commutation relations (3.182) for the field ϕ and its conjugate momentum π.

3.12 Use the linearity of the diffusion equation and equations (3.197–3.200) to derive the general solution (3.201) of the diffusion equation.

3.13 Derive (3.112) from $\boldsymbol{B} = \nabla \times \boldsymbol{A}$ and Ampère's law $\nabla \times \boldsymbol{B} = \mu_0 \boldsymbol{J}$.

3.14 Derive (3.113) from (3.112).

3.15 Derive (3.114) from (3.113).

3.16 Use the Green's function relations (3.107) and (3.108) to show that (3.114) satisfies (3.112).

3.17 Show that the Laplace transform of t^{z-1} is the gamma function (4.55) divided by s^z

$$f(s) = \int_0^{\infty} e^{-st} \, t^{z-1} \, dt = s^{-z} \, \Gamma(z). \tag{3.204}$$

3.18 Compute the Laplace transform of $1/\sqrt{t}$. Hint: let $t = u^2$.

4

Infinite series

4.1 Convergence

A sequence of partial sums

$$S_N = \sum_{n=0}^{N} c_n \tag{4.1}$$

converges to a number S if for every $\epsilon > 0$ there exists an integer $N(\epsilon)$ such that

$$|S - S_N| < \epsilon \quad \text{for all} \quad N > N(\epsilon). \tag{4.2}$$

The number S is then said to be the limit of the **convergent** infinite series

$$S = \sum_{n=0}^{\infty} c_n = \lim_{N\to\infty} S_N = \lim_{N\to\infty} \sum_{n=0}^{N} c_n. \tag{4.3}$$

Some series converge; others wander or **oscillate**; and others **diverge**.

A series whose absolute values converge

$$S = \sum_{n=0}^{\infty} |c_n| \tag{4.4}$$

is said to converge **absolutely**. A convergent series that is not absolutely convergent is said to converge **conditionally**.

Example 4.1 (Two infinite series) The series of inverse factorials converges to the number $e = 2.718281828\ldots$

$$\sum_{n=0}^{\infty} \frac{1}{n!} = e. \tag{4.5}$$

But the harmonic series of inverse integers diverges

$$\sum_{k=1}^{\infty} \frac{1}{k} \to \infty \tag{4.6}$$

as one may see by grouping its terms

$$1 + \frac{1}{2} + \left(\frac{1}{3} + \frac{1}{4}\right) + \left(\frac{1}{5} + \frac{1}{6} + \frac{1}{7} + \frac{1}{8}\right) + \cdots \geq 1 + \frac{1}{2} + \frac{1}{2} + \frac{1}{2} + \cdots \tag{4.7}$$

to form a series that obviously diverges. This series up to $1/n$ approaches the natural logarithm $\ln n$ to within a constant

$$\gamma = \lim_{n\to\infty} \left(\sum_{k=1}^{n} \frac{1}{k} - \ln n \right) = 0.5772156649\ldots \tag{4.8}$$

known as the Euler–Mascheroni constant (Leonhard Euler, 1707–1783; Lorenzo Mascheroni, 1750–1800). □

4.2 Tests of convergence

The **Cauchy criterion** for the convergence of a sequence S_N is that for every $\epsilon > 0$ there is an integer $N(\epsilon)$ such that for $N > N(\epsilon)$ and $M > N(\epsilon)$ one has

$$|S_N - S_M| < \epsilon. \tag{4.9}$$

Cauchy's criterion is equivalent to the defining condition (4.2).

Suppose the convergent series

$$\sum_{n=0}^{\infty} b_n \tag{4.10}$$

has only positive terms $b_n \geq 0$, and that $|c_n| \leq b_n$ for all n. Then the series

$$\sum_{n=0}^{\infty} c_n \tag{4.11}$$

also (absolutely) converges. This is the **comparison test**.

Similarly, if for all n, the inequality $0 \leq c_n \leq b_n$ holds and the series of numbers c_n diverges, then so does the series of numbers b_n.

If for some N, the terms c_n satisfy

$$|c_n|^{1/n} \leq x < 1 \tag{4.12}$$

for all $n > N$, then the series

$$\sum_{n=0}^{\infty} c_n \tag{4.13}$$

converges by the **Cauchy root test**.

In the **ratio test of d'Alembert**, the series $\sum_n c_n$ converges if

$$\lim_{n\to\infty}\left|\frac{c_{n+1}}{c_n}\right| = r < 1 \tag{4.14}$$

and diverges if $r > 1$.

Probably the most useful test is the **Intel test**, in which one writes a computer program to sum the first N terms of the series and then runs it for $N = 100$, $N = 10,000$, $N = 1,000,000, \ldots$, as seems appropriate.

4.3 Convergent series of functions

A sequence of partial sums

$$S_N(z) = \sum_{n=0}^{N} f_n(z) \tag{4.15}$$

of functions $f_n(z)$ **converges** to a function $S(z)$ on a set D if for every $\epsilon > 0$ and every $z \in D$, there exists an integer $N(\epsilon, z)$ such that

$$|S(z) - S_N(z)| < \epsilon \quad \text{for all} \quad N > N(\epsilon, z). \tag{4.16}$$

The numbers z may be real or complex. The function $S(z)$ is said to be the limit on D of the **convergent** infinite series of functions

$$S(z) = \sum_{n=0}^{\infty} f_n(z). \tag{4.17}$$

A sequence of partial sums $S_N(z)$ of functions converges **uniformly** on the set D if the integers $N(\epsilon, z)$ can be chosen independently of the point $z \in D$, that is, if for every $\epsilon > 0$ and every $z \in D$, there exists an integer $N(\epsilon)$ such that

$$|S(z) - S_N(z)| < \epsilon \quad \text{for all} \quad N > N(\epsilon). \tag{4.18}$$

The limit (2.52) of the integral over a closed interval $a \le x \le b$ of a uniformly convergent sequence of partial sums $S_N(x)$ of continuous functions is equal to the integral of the limit

$$\lim_{N\to\infty} \int_a^b S_N(x)\, dx = \int_a^b S(x)\, dx. \tag{4.19}$$

A real or complex-valued function $f(x)$ of a real variable x is **square integrable** on an interval $[a, b]$ if the integral

$$\int_a^b |f(x)|^2\, dx \tag{4.20}$$

exists and is finite. A sequence of partial sums

$$S_N(x) = \sum_{n=0}^{N} f_n(x) \tag{4.21}$$

of square-integrable functions $f_n(x)$ **converges in the mean** to a function $S(x)$ if

$$\lim_{N\to\infty} \int_a^b |S(x) - S_N(x)|^2 \, dx = 0. \tag{4.22}$$

Convergence in the mean sometimes is defined as

$$\lim_{N\to\infty} \int_a^b \rho(x)\, |S(x) - S_N(x)|^2 \, dx = 0, \tag{4.23}$$

in which $\rho(x) \geq 0$ is a weight function that is positive except at isolated points where it may vanish. If the functions f_n are real, then this definition of convergence in the mean takes the slightly simpler form

$$\lim_{N\to\infty} \int_a^b \rho(x)\, (S(x) - S_N(x))^2 \, dx = 0. \tag{4.24}$$

4.4 Power series

A power series is a series of functions with $f_n(z) = c_n\, z^n$

$$S(z) = \sum_{n=0}^{\infty} c_n\, z^n. \tag{4.25}$$

By the ratio test (4.14), this power series converges if

$$\lim_{n\to\infty} \left| \frac{c_{n+1} z^{n+1}}{c_n z^n} \right| = |z| \lim_{n\to\infty} \left| \frac{c_{n+1}}{c_n} \right| \equiv \frac{|z|}{R} < 1, \tag{4.26}$$

that is, if z lies within a circle of radius R

$$|z| < R \tag{4.27}$$

given by

$$R = \left(\lim_{n\to\infty} \frac{|c_{n+1}|}{|c_n|} \right)^{-1}. \tag{4.28}$$

Within this circle, the convergence is uniform and absolute.

Example 4.2 (geometric series) For any positive integer N, the simple identity

$$(1-z)(1+z+z^2+\cdots+z^N) = 1 - z^{N+1} \tag{4.29}$$

implies that

$$S_N(z) = \sum_{n=0}^{N} z^n = \frac{1-z^{N+1}}{1-z}. \tag{4.30}$$

For $|z| < 1$, the term $|z^{N+1}| \to 0$ as $N \to \infty$, and so the power series

$$S_\infty(z) = \sum_{n=0}^{\infty} z^n = \frac{1}{1-z} \tag{4.31}$$

converges to the function $1/(1-z)$ as long as the absolute value of z is less than unity. The radius of convergence R is unity in agreement with the estimate (4.28)

$$R = \left(\lim_{n\to\infty} \frac{|c_{n+1}|}{|c_n|}\right)^{-1} = \left(\lim_{n\to\infty} 1\right)^{-1} = 1. \tag{4.32}$$

For tiny z, the approximation

$$\frac{1}{1 \pm z} \approx 1 \mp z \tag{4.33}$$

is useful. □

Example 4.3 (Credit) If a man deposits \$100 in a bank, he has a **credit** of \$100. Suppose banks are required to retain as reserves 10% of their deposits and are free to lend the other 90%. Then the bank getting the \$100 deposit can lend out \$90 to a borrower. That borrower can deposit \$90 in another bank. That bank can then lend \$81 to another borrower. Now three people have credits of \$100 + \$90 + \$81 = \$271. This multiplication of money is the miracle of credit.

If P is the original deposit and r is the fraction of deposits that banks must retain as reserves, then the total credit due to P is

$$P + P(1-r) + P(1-r)^2 + \cdots = P\sum_{n=0}^{\infty}(1-r)^n = P\left(\frac{1}{1-(1-r)}\right) = \frac{P}{r}. \tag{4.34}$$

An initial deposit of $P = \$100$ with $r = 10\%$ can produce total credits of $P/r = \$1000$. A reserve requirement of 1% can lead to total credits of \$10,000. Since banks charge a higher rate of interest on money they lend than the rate they pay to their depositors, bank profits soar as $r \to 0$. This is why bankers love deregulation.

The funds all the banks hold in reserve due to a deposit P is

$$Pr + (1-r)Pr + (1-r)^2 Pr + \cdots = Pr\sum_{n=0}^{\infty}(1-r)^n = P, \tag{4.35}$$

P itself. □

4.5 Factorials and the gamma function

For any positive integer n, the product

$$n! \equiv n(n-1)(n-2)\dots 3\cdot 2\cdot 1 \tag{4.36}$$

is ***n*-factorial**, with zero-factorial defined as unity

$$0! \equiv 1. \tag{4.37}$$

To estimate $n!$, one can use **Stirling's approximation**

$$n! \approx \sqrt{2\pi\, n}\,(n/e)^n \tag{4.38}$$

or **Ramanujan's correction** to it

$$n! \approx \sqrt{2\pi\, n}\,(n/e)^n \left(1 + 1/2n + 1/8n^2\right)^{1/6} \tag{4.39}$$

or Mermin's first

$$n! \approx \sqrt{2\pi\, n}\,\left(\frac{n}{e}\right)^n \exp\left(\frac{1}{12\,n}\right) \tag{4.40}$$

or second approximation

$$n! \approx \sqrt{2\pi\, n}\,\left(\frac{n}{e}\right)^n \exp\left(\frac{1}{12\,n} - \frac{1}{360\,n^3} + \frac{1}{1260\,n^5}\right), \tag{4.41}$$

which follow from his exact infinite-product formula

$$n! = \sqrt{2\pi\, n}\,\left(\frac{n}{e}\right)^n \prod_{j=1}^{\infty}\left(\frac{(1+1/j)^{j+1/2}}{e}\right). \tag{4.42}$$

Figure 4.1 plots the relative error of these estimates $E(n!)$ of $n!$

$$10^8\left(\frac{E(n!) - n!}{n!}\right) \tag{4.43}$$

magnified by 10^8, except for Stirling's formula (4.38), whose relative error is off the chart. Mermin's second approximation (4.40) is the most accurate, followed by Ramanujan's correction (4.39), and by Mermin's first approximation (4.40) (James Stirling, 1692–1770; Srinivasa Ramanujan, 1887–1920; N. David Mermin, 1935–).

The **binomial coefficient** is a ratio of factorials

$$\binom{n}{k} \equiv \frac{n!}{k!\,(n-k)!}. \tag{4.44}$$

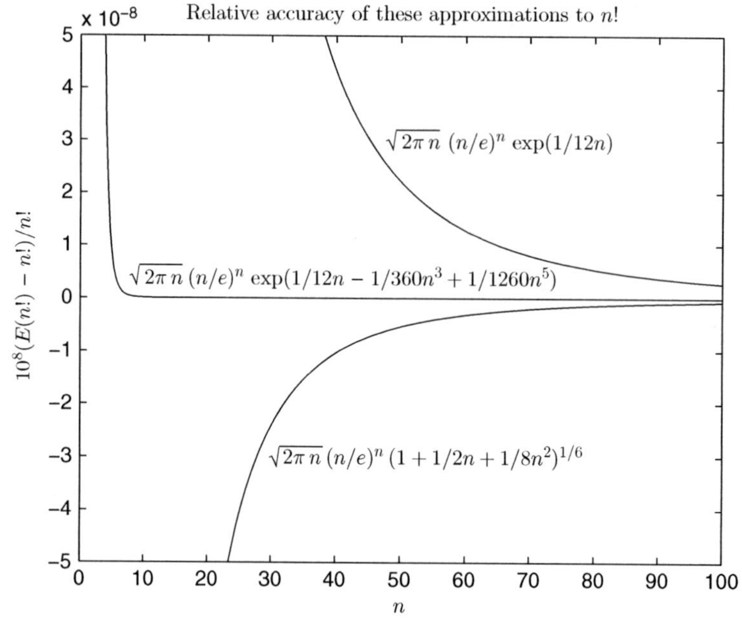

Figure 4.1 The magnified relative error $10^8[E(n!) - n!]/n!$ of Ramanujan's (4.39) and Mermin's (4.40 & 4.41) estimates $E(n!)$ of $n!$ are plotted for $n = 1, 2, \ldots, 100$.

Example 4.4 (The Leibniz rule) We can use the notation

$$f^{(n)}(x) \equiv \frac{d^n}{dx^n} f(x) \tag{4.45}$$

to state Leibniz's rule for the derivatives of the product of two functions

$$\frac{d^n}{dx^n}\,[f(x)\,g(x)] = \sum_{k=0}^{n} \binom{n}{k} f^{(k)}(x)\, g^{(n-k)}(x). \tag{4.46}$$

One may use mathematical induction to prove this rule, which is obviously true for $n = 0$ and $n = 1$ (exercise 4.4). □

Example 4.5 (The exponential function) The power series with coefficients $c_n = 1/n!$ defines the exponential function

$$e^z = \sum_{n=0}^{\infty} \frac{z^n}{n!}. \tag{4.47}$$

Formula (4.28) shows that the radius of convergence R of this power series is infinite

$$R = \left(\lim_{n\to\infty} \frac{|c_{n+1}|}{|c_n|} \right)^{-1} = \left(\lim_{n\to\infty} \frac{1}{n+1} \right)^{-1} = \infty. \tag{4.48}$$

The series converges uniformly and absolutely inside every circle. □

Example 4.6 (Bessel's series) For any integer n, the series

$$\begin{aligned} J_n(\rho) &= \frac{\rho^n}{2^n n!} \left[1 - \frac{\rho^2}{2(2n+2)} + \frac{\rho^4}{2 \cdot 4(2n+2)(2n+4)} - \cdots \right] \\ &= \left(\frac{\rho}{2}\right)^n \sum_{m=0}^{\infty} \frac{(-1)^m}{m!(m+n)!} \left(\frac{\rho}{2}\right)^{2m} \end{aligned} \tag{4.49}$$

defines the cylindrical Bessel function of the first kind, which is finite at the origin $\rho = 0$. This series converges even faster (exercise 4.5) than the one (4.47) for the exponential function. □

Double factorials also are useful and are defined as

$$(2n-1)!! \equiv (2n-1)(2n-3)(2n-5)\cdots 1, \tag{4.50}$$

$$(2n)!! \equiv 2n(2n-2)(2n-4)\cdots 2 \tag{4.51}$$

with 0!! and $(-1)!!$ both defined as unity

$$0!! = (-1)!! = 1. \tag{4.52}$$

Thus $5!! = 5 \cdot 3 \cdot 1 = 15$, and $6!! = 6 \cdot 4 \cdot 2 = 48$.

One may extend the definition (4.36) of n-factorial from positive integers to complex numbers by means of the integral formula

$$z! \equiv \int_0^\infty e^{-t}\, t^z \, dt \tag{4.53}$$

for $\mathrm{Re}\, z > -1$. In particular

$$0! = \int_0^\infty e^{-t} \, dt = 1, \tag{4.54}$$

which explains the definition (4.37). The factorial function $(z-1)!$ in turn defines the **gamma function** for $\mathrm{Re}\, z > 0$ as

$$\Gamma(z) = \int_0^\infty e^{-t}\, t^{z-1} \, dt = (z-1)! \tag{4.55}$$

as may be seen from (4.53). By differentiating this formula and integrating it by parts, we see that the gamma function satisfies the key identity

$$\Gamma(z+1) = \int_0^\infty \left(-\frac{d}{dt}e^{-t}\right) t^z\, dt = \int_0^\infty e^{-t}\left(\frac{d}{dt}t^z\right) dt = \int_0^\infty e^{-t}\, z\, t^{z-1}\, dt$$
$$= z\,\Gamma(z) \tag{4.56}$$

with $\Gamma(1) = 0! = 1$. We may use this identity (4.56) to extend the definition (5.102) of the gamma function in unit steps into the left half-plane

$$\Gamma(z) = \frac{1}{z}\,\Gamma(z+1) = \frac{1}{z}\,\frac{1}{z+1}\,\Gamma(z+2) = \frac{1}{z}\,\frac{1}{z+1}\,\frac{1}{z+2}\,\Gamma(z+3) = \cdots \tag{4.57}$$

as long as we avoid the negative integers and zero. This extension leads to Euler's definition

$$\Gamma(z) = \lim_{n\to\infty} \frac{1\cdot 2\cdot 3\cdots n}{z(z+1)(z+2)\cdots(z+n)}\, n^z \tag{4.58}$$

and to Weierstrass's (exercise 4.6)

$$\Gamma(z) = \frac{1}{z}\, e^{-\gamma z}\left[\prod_{n=1}^{\infty}\left(1+\frac{z}{n}\right)e^{-z/n}\right]^{-1} \tag{4.59}$$

(Karl Theodor Wilhelm Weierstrass, 1815–1897), and is an example of analytic continuation (section 5.12).

One may show (exercise 4.8) that another formula for $\Gamma(z)$ is

$$\Gamma(z) = 2\int_0^\infty e^{-t^2}\, t^{2z-1}\, dt \tag{4.60}$$

for $\operatorname{Re} z > 0$ and that

$$\Gamma(n+\tfrac{1}{2}) = \frac{(2n)!}{n!\, 2^{2n}}\sqrt{\pi}. \tag{4.61}$$

which implies (exercise 4.11) that

$$\Gamma\left(n+\frac{1}{2}\right) = \frac{(2n-1)!!}{2^n}\sqrt{\pi}. \tag{4.62}$$

Example 4.7 (Bessel function of nonintegral index) We can use the gamma-function formula (4.55) for $n!$ to extend the definition (4.49) of the Bessel function of the first kind $J_n(\rho)$ to nonintegral values ν of the index n. Replacing n by ν and $(m+n)!$ by $\Gamma(m+\nu+1)$, we get

$$J_\nu(\rho) = \left(\frac{\rho}{2}\right)^\nu \sum_{m=0}^{\infty} \frac{(-1)^m}{m!\,\Gamma(m+\nu+1)}\left(\frac{\rho}{2}\right)^{2m}, \tag{4.63}$$

which makes sense even for complex values of ν. □

Example 4.8 (Spherical Bessel function) The spherical Bessel function is defined as

$$j_\ell(\rho) \equiv \sqrt{\frac{\pi}{2\rho}}\, J_{\ell+1/2}(\rho). \tag{4.64}$$

For small values of its argument $|\rho| \ll 1$, the first term in the series (4.63) dominates and so (exercise 4.7)

$$j_\ell(\rho) \approx \frac{\sqrt{\pi}}{2}\left(\frac{\rho}{2}\right)^\ell \frac{1}{\Gamma(\ell+3/2)} = \frac{\ell\,!\,(2\rho)^\ell}{(2\ell+1)!} = \frac{\rho^\ell}{(2\ell+1)!!} \tag{4.65}$$

as one may show by repeatedly using the key identity $\Gamma(z+1) = z\,\Gamma(z)$. □

4.6 Taylor series

If the function $f(x)$ is a real-valued function of a real variable x with a continuous Nth derivative, then Taylor's expansion for it is

$$\begin{aligned} f(x+a) &= f(x) + af'(x) + \frac{a^2}{2}f''(x) + \cdots + \frac{a^{N-1}}{(N-1)!}f^{(N-1)} + E_N \\ &= \sum_{n=0}^{N-1} \frac{a^n}{n!} f^{(n)}(x) + E_N, \end{aligned} \tag{4.66}$$

in which the error E_N is

$$E_N = \frac{a^N}{N!} f^{(N)}(x+y) \tag{4.67}$$

for some $0 \le y \le a$.

For many functions $f(x)$ the errors go to zero, $E_N \to 0$, as $N \to \infty$; for these functions, the infinite Taylor series converges:

$$f(x+a) = \sum_{n=0}^{\infty} \frac{a^n}{n!} f^{(n)}(x) = \exp\left(a\frac{d}{dx}\right) f(x) = e^{iap/\hbar} f(x), \tag{4.68}$$

in which

$$p = \frac{\hbar}{i}\frac{d}{dx} \tag{4.69}$$

is the displacement operator or equivalently the momentum operator.

4.7 Fourier series as power series

The Fourier series (2.37)

$$f(x) = \sum_{n=-\infty}^{\infty} c_n \frac{e^{i2\pi nx/L}}{\sqrt{L}} \tag{4.70}$$

with coefficients (2.45)

$$c_n = \int_{-L/2}^{L/2} \frac{e^{-i2\pi nx/L}}{\sqrt{L}} f(x)\, dx \tag{4.71}$$

is a pair of power series

$$f(x) = \frac{1}{\sqrt{L}} \left(\sum_{n=0}^{\infty} c_n z^n + \sum_{n=1}^{\infty} c_{-n} (z^{-1})^n \right) \tag{4.72}$$

in the variables

$$z = e^{i2\pi x/L} \quad \text{and} \quad z^{-1} = e^{-i2\pi x/L}. \tag{4.73}$$

Formula (4.28) tells us that the radii of convergence of these two power series are given by

$$R_+^{-1} = \lim_{n\to\infty} \frac{|c_{n+1}|}{|c_n|} \quad \text{and} \quad R_-^{-1} = \lim_{n\to\infty} \frac{|c_{-n-1}|}{|c_{-n}|}. \tag{4.74}$$

Thus the pair of power series (4.72) will converge uniformly and absolutely as long as z satisfies the two inequalities

$$|z| < R_+ \quad \text{and} \quad \frac{1}{|z|} < R_-. \tag{4.75}$$

Since $|z| = 1$, the Fourier series (4.70) converges if $R_-^{-1} < |1| < R_+$.

Example 4.9 (A uniform and absolutely convergent Fourier series) The Fourier series

$$f(x) = \sum_{n=-\infty}^{\infty} \frac{1}{1+|n|^n} \frac{e^{i2\pi nx/L}}{\sqrt{L}} \tag{4.76}$$

converges uniformly and absolutely because $R_+ = R_- = \infty$.

4.8 The binomial series and theorem

The Taylor series for the function $f(x) = (1+x)^a$ is

$$
\begin{aligned}
(1+x)^a &= \sum_{n=0}^{\infty} \frac{x^n}{n!} \frac{d^n}{dx^n} (1+x)^a \Big|_{x=0} \\
&= 1 + ax + \frac{1}{2} a(a-1)x^2 + \cdots \\
&= 1 + \sum_{n=1}^{\infty} \frac{a(a-1)\cdots(a-n+1)}{n!} x^n. \qquad (4.77)
\end{aligned}
$$

If a is a positive integer $a = N$, then the nth power of x in this series is multiplied by a binomial coefficient (4.44)

$$
(1+x)^N = \sum_{n=0}^{N} \frac{N!}{n!(N-n)!} x^n = \sum_{n=0}^{N} \binom{N}{n} x^n. \qquad (4.78)
$$

The series (4.77) and (4.78) respectively imply (exercise 4.13)

$$
(x+y)^a = y^a + \sum_{n=1}^{\infty} \frac{a(a-1)\cdots(a-n+1)}{n!} x^n \, y^{a-n} \qquad (4.79)
$$

and

$$
(x+y)^N = \sum_{n=0}^{N} \binom{N}{n} x^n \, y^{N-n}. \qquad (4.80)
$$

We can use these versions of the **binomial theorem** to compute approximately or exactly.

Example 4.10 The phase difference $\Delta\phi$ between two highly relativistic neutrinos of momentum p going a distance L in a time $t \approx L$ varies with their masses m_1 and m_2 as

$$
\Delta\phi = t\,\Delta E = \frac{LE}{p} \Delta E = \frac{LE}{p} \left(\sqrt{p^2 + m_1^2} - \sqrt{p^2 + m_2^2} \right) \qquad (4.81)
$$

in natural units. We can approximate this phase by using the first two terms of the binomial expansion (4.79) with $y = 1$ and $x = m_i^2/p^2$

$$
\Delta\phi = LE \left(\sqrt{1 + m_1^2/p^2} - \sqrt{1 + m_2^2/p^2} \right) \approx \frac{LE\Delta m^2}{p^2} \approx \frac{L\Delta m^2}{E} \qquad (4.82)
$$

or in ordinary units $\Delta\phi \approx L\Delta m^2 c^3/(\hbar E)$. □

Example 4.11 We can use the binomial expansion (4.80) to compute

$$999^3 = \left(10^3 - 1\right)^3 = 10^9 - 3 \times 10^6 + 3 \times 10^3 - 1 = 997002999 \tag{4.83}$$

exactly. □

When a is not a positive integer, the series (4.77) does not terminate. For instance, the binomial series for $\sqrt{1+x}$ and $1/\sqrt{1+x}$ are (exercise 4.14)

$$\begin{aligned}(1+x)^{1/2} &= 1 + \sum_{n=1}^{\infty} \frac{\frac{1}{2}\left(\frac{1}{2}-1\right)\cdots\left(\frac{1}{2}-n+1\right)}{n!} x^n \\ &= 1 + \sum_{n=1}^{\infty} \frac{(-1)^{n-1}}{2^n} \frac{(2n-3)!!}{n!} x^n \end{aligned} \tag{4.84}$$

and

$$\begin{aligned}(1+x)^{-1/2} &= 1 + \sum_{n=1}^{\infty} \frac{-\frac{1}{2}\left(-\frac{3}{2}\right)\cdots\left(-\frac{1}{2}-n+1\right)}{n!} x^n \\ &= \sum_{n=0}^{\infty} \frac{(-1)^{n}}{2^n} \frac{(2n-1)!!}{n!} x^n. \end{aligned} \tag{4.85}$$

4.9 Logarithmic series

The Taylor series for the function $f(x) = \ln(1+x)$ is

$$\ln(1+x) = \sum_{n=0}^{\infty} \frac{x^n}{n!} \frac{d^n}{dx^n} \ln(1+x)|_{x=0} \tag{4.86}$$

in which

$$\begin{aligned} f^{(0)}(0) &= \ln(1+x)|_{x=0} = 0, \\ f^{(1)}(0) &= \left.\frac{1}{1+x}\right|_{x=0} = 1, \\ f^{(n)}(0) &= \left.\frac{(-1)^{n-1}\,(n-1)!}{(1+x)^n}\right|_{x=0} = (-1)^{n-1}\,(n-1)!. \end{aligned} \tag{4.87}$$

So the series for $\ln(1+x)$ is

$$\ln(1+x) = \sum_{n=1}^{\infty} \frac{(-1)^{n-1}\,x^n}{n} = x - \frac{1}{2}x^2 + \frac{1}{3}x^3 \pm \cdots, \tag{4.88}$$

which converges slowly for $-1 < x \le 1$. Letting $x \to -x$, we see that

$$\ln(1-x) = -\sum_{n=1}^{\infty} \frac{x^n}{n}. \tag{4.89}$$

So the series for the logarithm of the ratio $(1+x)/(1-x)$ is

$$\ln\left(\frac{1+x}{1-x}\right) = 2\sum_{n=0}^{\infty} \frac{x^{2n+1}}{2n+1}. \tag{4.90}$$

4.10 Dirichlet series and the zeta function

A **Dirichlet series** is one in which the nth term is proportional to $1/n^z$

$$f(z) = \sum_{n=1}^{\infty} \frac{c_n}{n^z}. \tag{4.91}$$

An important example is the **Riemann zeta function** $\zeta(z)$

$$\zeta(z) = \sum_{n=1}^{\infty} n^{-z}, \tag{4.92}$$

which converges for $\mathrm{Re}\, z > 1$.

Euler showed that for $\mathrm{Re}\, z > 1$, the Riemann zeta function is the infinite product

$$\zeta(z) = \prod_p \frac{1}{1-p^{-z}} \tag{4.93}$$

over all prime numbers $p = 2, 3, 5, 7, 11, \ldots$. Some specific values are $\zeta(2) = \pi^2/6 \approx 1.645$, $\zeta(4) = \pi^4/90 \approx 1.0823$, and $\zeta(6) = \pi^6/945 \approx 1.0173$.

Example 4.12 (Planck's distribution) Max Planck (1858–1947) showed that the electromagnetic energy in a closed cavity of volume V at a temperature T in the frequency interval $d\nu$ about ν is

$$dU(\beta, \nu, V) = \frac{8\pi h V}{c^3} \frac{\nu^3}{e^{\beta h\nu} - 1} d\nu, \tag{4.94}$$

in which $\beta = 1/(kT)$, $k = 1.3806503 \times 10^{-23}$ J/K is **Boltzmann's constant**, and $h = 6.626068 \times 10^{-34}$ Js is **Planck's constant**. The total energy then is the integral

$$U(\beta, V) = \frac{8\pi h V}{c^3} \int_0^{\infty} \frac{\nu^3}{e^{\beta h\nu} - 1} d\nu, \tag{4.95}$$

which we may do by letting $x = \beta h\nu$ and using the geometric series (4.31)

$$
\begin{aligned}
U(\beta, V) &= \frac{8\pi (kT)^4 V}{(hc)^3} \int_0^\infty \frac{x^3}{e^x - 1}\, dx \\
&= \frac{8\pi (kT)^4 V}{(hc)^3} \int_0^\infty \frac{x^3 e^{-x}}{1 - e^{-x}}\, dx \\
&= \frac{8\pi (kT)^4 V}{(hc)^3} \int_0^\infty x^3 e^{-x} \sum_{n=0}^\infty e^{-nx}\, dx. \qquad (4.96)
\end{aligned}
$$

The geometric series is absolutely and uniformly convergent for $x > 0$, and we may interchange the limits of summation and integration. After another change of variables, the gamma-function formula (5.102) gives

$$
\begin{aligned}
U(\beta, V) &= \frac{8\pi (kT)^4 V}{(hc)^3} \sum_{n=0}^\infty \int_0^\infty x^3 e^{-(n+1)x}\, dx \\
&= \frac{8\pi (kT)^4 V}{(hc)^3} \sum_{n=0}^\infty \frac{1}{(n+1)^4} \int_0^\infty y^3 e^{-y}\, dy \\
&= \frac{8\pi (kT)^4 V}{(hc)^3} 3!\, \zeta(4) = \frac{8\pi^5 (kT)^4 V}{15 (hc)^3}. \qquad (4.97)
\end{aligned}
$$

It follows that the power radiated by a "**black body**" is proportional to the fourth power of its temperature and to its area A

$$
P = \sigma\, A\, T^4, \qquad (4.98)
$$

in which

$$
\sigma = \frac{2\pi^5 k^4}{15\, h^3 c^2} = 5.670400(40) \times 10^{-8}\ \mathrm{W\, m^{-2}\, K^{-4}} \qquad (4.99)
$$

is **Stefan's constant**.

The number of photons in the black-body distribution (4.94) at inverse temperature β in the volume V is

$$
\begin{aligned}
N(\beta, V) &= \frac{8\pi V}{c^3} \int_0^\infty \frac{\nu^2}{e^{\beta h\nu} - 1}\, d\nu = \frac{8\pi V}{(c\beta h)^3} \int_0^\infty \frac{x^2}{e^x - 1}\, dx \\
&= \frac{8\pi V}{(c\beta h)^3} \int_0^\infty \frac{x^2 e^{-x}}{1 - e^{-x}}\, dx = \frac{8\pi V}{(c\beta h)^3} \int_0^\infty x^2 e^{-x} \sum_{n=0}^\infty e^{-nx}\, dx \\
&= \frac{8\pi V}{(c\beta h)^3} \sum_{n=0}^\infty \int_0^\infty x^2 e^{-(n+1)x}\, dx = \frac{8\pi V}{(c\beta h)^3} \sum_{n=0}^\infty \frac{1}{(n+1)^3} \int_0^\infty y^2 e^{-y}\, dy \\
&= \frac{8\pi V}{(c\beta h)^3} \zeta(3) 2! = \frac{8\pi (kT)^3 V}{(ch)^3} \zeta(3) 2!. \qquad (4.100)
\end{aligned}
$$

The mean energy $\langle E\rangle$ of a photon in the black-body distribution (4.94) is the energy $U(\beta, V)$ divided by the number of photons $N(\beta, V)$

$$\langle E\rangle = \langle h\nu\rangle = \frac{3!\,\zeta(4)}{2!\,\zeta(3)}\,kT = \frac{\pi^4}{30\,\zeta(3)}\,kT \tag{4.101}$$

or $\langle E\rangle \approx 2.70118\,kT$ since Apéry's constant $\zeta(3)$ is 1.2020569032 ... (Roger Apéry, 1916–1994). □

Example 4.13 (The Lerch transcendent) The **Lerch transcendent** is the series

$$\Phi(z, s, \alpha) = \sum_{n=0}^{\infty} \frac{z^n}{(n+\alpha)^s}. \tag{4.102}$$

It converges when $|z| < 1$ and $\mathrm{Re}\, s > 0$ and $\mathrm{Re}\, \alpha > 0$. □

4.11 Bernoulli numbers and polynomials

The **Bernoulli numbers** B_n are defined by the infinite series

$$\frac{x}{e^x - 1} = \sum_{n=0}^{\infty} \frac{x^n}{n!}\left[\frac{d^n}{dx^n}\frac{x}{e^x-1}\right]\bigg|_{x=0} = \sum_{n=0}^{\infty} B_n \frac{x^n}{n!} \tag{4.103}$$

for the **generating function** $x/(e^x - 1)$. They are the successive derivatives

$$B_n = \frac{d^n}{dx^n}\frac{x}{e^x-1}\bigg|_{x=0}. \tag{4.104}$$

So $B_0 = 1$ and $B_1 = -1/2$. The remaining odd Bernoulli numbers vanish

$$B_{2n+1} = 0 \quad \text{for } n > 0 \tag{4.105}$$

and the remaining even ones are given by Euler's zeta function (4.92) formula

$$B_{2n} = \frac{(-1)^{n-1}2(2n)!}{(2\pi)^{2n}}\,\zeta(2n) \quad \text{for } n > 0. \tag{4.106}$$

The Bernoulli numbers occur in the power series for many transcendental functions, for instance

$$\coth x = \frac{1}{x} + \sum_{k=1}^{\infty} \frac{2^{2k}B_{2k}}{(2k)!}\,x^{2k-1} \quad \text{for } x^2 < \pi^2. \tag{4.107}$$

Bernoulli's polynomials $B_n(y)$ are defined by the series

$$\frac{xe^{xy}}{e^x - 1} = \sum_{n=0}^{\infty} B_n(y)\,\frac{x^n}{n!} \tag{4.108}$$

for the **generating function** $xe^{xy}/(e^x - 1)$.

Some authors (Whittaker and Watson, 1927, p. 125–127) define Bernoulli's numbers instead by

$$B_n = \frac{2(2n)!}{(2\pi)^{2n}}\,\zeta(2n) = 4n\int_0^\infty \frac{t^{2n-1}\,dt}{e^{2\pi t}-1}, \tag{4.109}$$

a result due to Carda.

4.12 Asymptotic series

A series

$$s_n(x) = \sum_{k=0}^{n} \frac{a_k}{x^k} \tag{4.110}$$

is an **asymptotic** expansion for a real function $f(x)$ if the **remainder** R_n

$$R_n(x) = f(x) - s_n(x) \tag{4.111}$$

satisfies the condition

$$\lim_{x\to\infty} x^n R_n(x) = 0 \tag{4.112}$$

for fixed n. In this case, one writes

$$f(x) \approx \sum_{k=0}^{\infty} \frac{a_k}{x^k} \tag{4.113}$$

where the wavy equal sign indicates equality in the sense of (4.112). Some authors add the condition

$$\lim_{n\to\infty} x^n R_n(x) = \infty \tag{4.114}$$

for fixed x.

Example 4.14 (The asymptotic series for E_1) Let's develop an asymptotic expansion for the function

$$E_1(x) = \int_x^\infty e^{-y}\,\frac{dy}{y}, \tag{4.115}$$

which is related to the exponential-integral function

$$Ei(x) = \int_{-\infty}^{x} e^{y}\,\frac{dy}{y} \tag{4.116}$$

by the tricky formula $E_1(x) = -Ei(-x)$. Since

$$\frac{e^{-y}}{y} = -\frac{d}{dy}\left(\frac{e^{-y}}{y}\right) - \frac{e^{-y}}{y^2} \tag{4.117}$$

we may integrate by parts, getting

$$E_1(x) = \frac{e^{-x}}{x} - \int_x^\infty e^{-y}\,\frac{dy}{y^2}. \tag{4.118}$$

Integrating by parts again, we find

$$E_1(x) = \frac{e^{-x}}{x} - \frac{e^{-x}}{x^2} + 2\int_x^\infty e^{-y}\,\frac{dy}{y^3}. \tag{4.119}$$

Eventually, we develop the series

$$E_1(x) = e^{-x}\left(\frac{0!}{x} - \frac{1!}{x^2} + \frac{2!}{x^3} - \frac{3!}{x^4} + \frac{4!}{x^5} - \cdots\right) \tag{4.120}$$

with remainder

$$R_n(x) = (-1)^n\, n! \int_x^\infty e^{-y}\,\frac{dy}{y^{n+1}}. \tag{4.121}$$

Setting $y = u + x$, we have

$$R_n(x) = (-1)^n\,\frac{n!\,e^{-x}}{x^{n+1}} \int_0^\infty e^{-u}\,\frac{du}{\left(1+\frac{u}{x}\right)^{n+1}}, \tag{4.122}$$

which satisfies the condition (4.112) that defines an asymptotic series

$$\begin{aligned}
\lim_{x\to\infty} x^n R_n(x) &= \lim_{x\to\infty} (-1)^n\,\frac{n!\,e^{-x}}{x} \int_0^\infty e^{-u}\,\frac{du}{\left(1+\frac{u}{x}\right)^{n+1}} \\
&= \lim_{x\to\infty} (-1)^n\,\frac{n!\,e^{-x}}{x} \int_0^\infty e^{-u}\,du \\
&= \lim_{x\to\infty} (-1)^n\,\frac{n!\,e^{-x}}{x} = 0
\end{aligned} \tag{4.123}$$

for fixed n. □

Asymptotic series often occur in physics. In such physical problems, a small parameter λ usually plays the role of $1/x$. A perturbative series

$$S_n(\lambda) = \sum_{k=0}^{n} a_k\,\lambda^k \tag{4.124}$$

is an asymptotic expansion of the physical quantity $S(\lambda)$ if the remainder

$$R_n(\lambda) = S(\lambda) - S_n(\lambda) \tag{4.125}$$

satisfies for fixed n

$$\lim_{\lambda\to 0} \lambda^{-n}\,R_n(\lambda) = 0. \tag{4.126}$$

The WKB approximation and the Dyson series for quantum electrodynamics are asymptotic expansions in this sense.

4.13 Some electrostatic problems

Gauss's law $\nabla \cdot \boldsymbol{D} = \rho$ equates the divergence of the **electric displacement** $\boldsymbol{D}$ to the density ρ of **free charges** (charges that are free to move in or out of the dielectric medium – as opposed to those that are part of the medium and bound to it by molecular forces). In electrostatic problems, Maxwell's equations reduce to Gauss's law and the static form $\nabla \times \mathbf{E} = 0$ of Faraday's law, which implies that the electric field $\boldsymbol{E}$ is the gradient of an electrostatic potential $\boldsymbol{E} = -\nabla V$.

Across an interface with normal vector $\hat{\boldsymbol{n}}$ between two dielectrics, the tangential electric field is continuous while the normal electric displacement jumps by the surface charge density σ

$$\hat{\boldsymbol{n}} \times (\boldsymbol{E}_2 - \boldsymbol{E}_1) = 0 \quad \text{and} \quad \sigma = \hat{\boldsymbol{n}} \cdot (\boldsymbol{D}_2 - \boldsymbol{D}_1). \tag{4.127}$$

In a **linear dielectric**, the electric displacement $\boldsymbol{D}$ is proportional to the electric field $\boldsymbol{D} = \epsilon_m \boldsymbol{E}$, where the **permittivity** $\epsilon_m = \epsilon_0 + \chi_m = K_m \epsilon_0$ of the material differs from that of the vacuum ϵ_0 by the **electric susceptibility** χ_m and the **relative permittivity** K_m. The permittivity of the vacuum is the **electric constant** ϵ_0.

An electric field $\boldsymbol{E}$ exerts on a charge q a **force** $\boldsymbol{F} = q\boldsymbol{E}$ even in a dielectric medium. The electrostatic energy W of a system of linear dielectrics is the volume integral

$$W = \frac{1}{2} \int \boldsymbol{D} \cdot \boldsymbol{E} \, d^3 r. \tag{4.128}$$

Example 4.15 (Field of a charge near an interface) Consider two semi-infinite dielectrics of permittivities ϵ_1 and ϵ_2 separated by an infinite horizontal x-y-plane. What is the electrostatic potential due to a charge q in region 1 at a height h above the interface?

The easy way to solve this problem is to put an image charge q' at the same distance from the interface in region 2 so that the potential in region 1 is

$$V_1(\boldsymbol{r}) = \frac{1}{4\pi\epsilon_1} \left(\frac{q}{\sqrt{x^2 + y^2 + (z-h)^2}} + \frac{q'}{\sqrt{x^2 + y^2 + (z+h)^2}} \right). \tag{4.129}$$

This potential satisfies Gauss's law $\nabla \cdot \boldsymbol{D} = \rho$ in region 1. In region 2, the potential

$$V_2(\boldsymbol{r}) = \frac{1}{4\pi\epsilon_2} \frac{q''}{\sqrt{x^2 + y^2 + (z-h)^2}} \tag{4.130}$$

also satisfies Gauss's law. The continuity (4.127) of the tangential component of $\boldsymbol{E}$ tells us that the partial derivatives of V_1 and V_2 in the x (or y) direction must be the same at $z = 0$

$$\frac{\partial V_1(x,y,0)}{\partial x} = \frac{\partial V_2(x,y,0)}{\partial x}. \tag{4.131}$$

The discontinuity equation (4.127) for the electric displacement says that at the interface at $z = 0$ with no surface charge

$$\epsilon_1 \frac{\partial V_1(x,y,0)}{\partial z} = \epsilon_2 \frac{\partial V_2(x,y,0)}{\partial z}. \tag{4.132}$$

These two equations (4.131 & 4.132) allow one to solve for q' and q''

$$q' = \frac{\epsilon_1 - \epsilon_2}{\epsilon_1 + \epsilon_2} q \quad \text{and} \quad q'' = \frac{2\epsilon_2}{\epsilon_1 + \epsilon_2} q. \tag{4.133}$$

In the limit $h \to 0$, the potential in region 1 becomes

$$V_1(\boldsymbol{r}) = \frac{1}{4\pi\epsilon_1} \frac{q}{\sqrt{x^2 + y^2 + z^2}} \left(1 + \frac{\epsilon_1 - \epsilon_2}{\epsilon_1 + \epsilon_2}\right) = \frac{q}{4\pi\bar{\epsilon} r}, \tag{4.134}$$

in which $\bar{\epsilon} r$ is the mean permittivity $\bar{\epsilon} = (\epsilon_1 + \epsilon_2)/2$. Similarly, in region 2 the potential is

$$V_2(\boldsymbol{r}) = \frac{1}{4\pi\epsilon_2} \frac{q}{\sqrt{x^2 + y^2 + z^2}} \frac{2\epsilon_2}{\epsilon_1 + \epsilon_2} = \frac{q}{4\pi\bar{\epsilon} r} \tag{4.135}$$

in the limit $h \to 0$. □

Example 4.16 (A charge near a plasma membrane) A eukaryotic cell (the kind with a nucleus) is surrounded by a plasma membrane, which is a phospholipid bilayer about 5 nm thick. Both sides of the plasma membrane are in contact with salty water. The permittivity of the water is $\epsilon_w \approx 80\epsilon_0$ while that of the membrane considered as a simple lipid slab is $\epsilon_\ell \approx 2\epsilon_0$.

Let's think about the potential felt by an ion in the water outside a cell but near its membrane, and let us for simplicity imagine the membrane to be infinitely thick so that we can use the simple formulas we've derived. The potential due to the ion, if its charge is q, is then given by equation (4.129) with $\epsilon_1 = \epsilon_w$ and $\epsilon_2 = \epsilon_\ell$. The image-charge term in $V_1(\boldsymbol{r})$ is the potential due to the polarization of the membrane and the water by the ion. It is the potential felt by the ion. Since the image charge by (4.133) is $q' \approx q$, the potential the ion feels is $V_i(z) \approx q/8\pi e_w z$. The force on the ion then is

$$F = -qV_i'(z) = \frac{q^2}{8\pi e_w z}. \tag{4.136}$$

It always is positive no matter what the sign of the charge is. A lipid slab in water repels ions. Similarly, a charge in a lipid slab is attracted to the water outside the slab.

Now imagine an electric dipole in water near a lipid slab. Now there are two equal and opposite charges and two equal and opposite mirror charges. The net effect is that the slab repels the dipole. So lipids repel water molecules; they are said to be **hydrophobic**. This is one of the reasons why folding proteins move their hydrophobic amino acids inside and their polar or **hydrophilic** ones outside.

With some effort, one may use the method of images to compute the electric potential of a charge in or near a plasma membrane taken to be a lipid slab of finite thickness.

The electric potential in the lipid bilayer $V_\ell(\rho, z)$ of thickness t due to a charge q in the extracellular environment at a height h above the bilayer is

$$V_\ell(\rho, z) = \frac{q}{4\pi\epsilon_{w\ell}} \sum_{n=0}^{\infty} (pp')^n \left(\frac{1}{\sqrt{\rho^2 + (z - 2nt - h)^2}} - \frac{p'}{\sqrt{\rho^2 + (z + 2(n+1)t + h)^2}} \right), \tag{4.137}$$

in which $p = (\epsilon_w - \epsilon_\ell)/(\epsilon_w + \epsilon_\ell)$, $p' = (\epsilon_c - \epsilon_\ell)/(\epsilon_c + \epsilon_\ell)$, and $\epsilon_{w\ell} = (\epsilon_w + \epsilon_\ell)/2$. That in the extracellular environment is

$$V_w(\rho, z) = \frac{q}{4\pi\epsilon_w} \left(\frac{1}{r} + \frac{p}{\sqrt{\rho^2 + (z + h)^2}} - \frac{\epsilon_w \epsilon_\ell}{\epsilon_{w\ell}^2} \sum_{n=1}^{\infty} \frac{p^{n-1} p'^n}{\sqrt{\rho^2 + (z + 2nt + h)^2}} \right), \tag{4.138}$$

in which r is the distance from the charge q. Finally, the potential in the cytosol is

$$V_c(\rho, z) = \frac{q\epsilon_\ell}{4\pi\epsilon_{w\ell}\epsilon_{\ell c}} \sum_{n=0}^{\infty} \frac{(pp')^n}{\sqrt{\rho^2 + (z - 2nt - h)^2}} \tag{4.139}$$

where $\epsilon_{\ell c} = (\epsilon_\ell + \epsilon_c)/2$.

The first 1000 terms of these three series (4.137–4.139) are plotted in Fig. 4.2 for the case of a positive charge $q = |e|$ at $(\rho, z) = (0, 0)$ (top curve), $(0, 1)$ (middle curve), $(0, 2)$ (third curve), and (0, 6) nm (bottom curve). Although the potential $V(\rho, z)$ is continuous across the two interfaces, its normal derivative isn't due to the different dielectric constants in the three media. Because the potential is small and flat in the cytosol ($z < -5$ nm), charges in the extracellular environment ($z > 0$) are nearly decoupled from those in the cytosol.

Real plasma membranes are **phospholipid** bilayers. The lipids avoid the water and so are on the inside. The phosphate groups are dipoles (and phosphatidylserine is negatively charged). So a real membrane is a 4 nm thick lipid layer bounded on each side by dipole layers, each about 0.5 nm thick. The net effect is to weakly *attract* ions that are within 0.5 nm of the membrane.

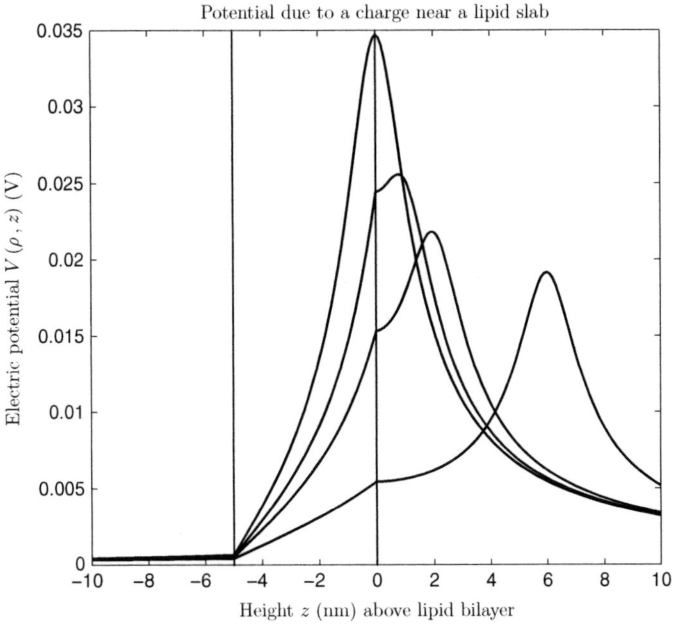

Figure 4.2 The electric potential $V(\rho, z)$ from (4.137–4.139) in volts for $\rho = 1$ nm as a function of the height z (nm) above (or below) a lipid slab for a unit charge $q = |e|$ at $(\rho, z) = (0, 0)$ (top curve), $(0, 1)$ (second curve), $(0, 2)$ (third curve), and $(0, 6)$ nm (bottom curve). The lipid slab extends from $z = 0$ to $z = -5$ nm, and the cytosol lies below $z = -5$ nm. The relative permittivities were taken to be $\epsilon_w/\epsilon_0 = \epsilon_c/\epsilon_0 = 80$ and $\epsilon_\ell/\epsilon_0 = 2$.

□

4.14 Infinite products

Weierstrass's definition (4.59) of the gamma function, Euler's formula (4.93) for the zeta function, and Mermin's formula (4.42) for $n!$ are useful infinite products. Other examples are the expansions of the trigonometric functions (2.66 & 2.67)

$$\sin z = z \prod_{n=1}^{\infty} \left[1 - \frac{z^2}{\pi^2 n^2} \right] \quad \text{and} \quad \cos z = \prod_{n=1}^{\infty} \left[1 - \frac{z^2}{\pi^2 (n - 1/2)^2} \right], \tag{4.140}$$

which imply those of the hyperbolic functions

$$\sinh z = z \prod_{n=1}^{\infty} \left[1 + \frac{z^2}{\pi^2 n^2} \right] \quad \text{and} \quad \cosh z = \prod_{n=1}^{\infty} \left[1 + \frac{z^2}{\pi^2 (n - 1/2)^2} \right]. \tag{4.141}$$

Exercises

4.1 Test the following series for convergence:

$$\text{(a)} \sum_{n=2}^{\infty} \frac{1}{(\ln n)^2}, \quad \text{(b)} \sum_{n=1}^{\infty} \frac{n!}{20^n}, \quad \text{(c)} \sum_{n=1}^{\infty} \frac{1}{n(n+2)}, \quad \text{(d)} \sum_{n=2}^{\infty} \frac{1}{n \ln n}.$$

In each case, say whether the series converges and how you found out.

4.2 Olber's paradox: assume a static universe with a uniform density of stars. With you at the origin, divide space into successive shells of thickness t, and assume that the stars in each shell subtend the same solid angle ω (as follows from the first assumption). Take into account the occulting of distant stars by nearer ones and show that the total solid angle subtended by all the stars would be 4π. The sky would be dazzlingly bright at night.

4.3 Use the geometric formula (4.30) to derive the trigonometric summation formula

$$\frac{1}{2} + \cos\alpha + \cos 2\alpha + \cdots + \cos n\alpha = \frac{\sin(n+\frac{1}{2})\alpha}{2\sin\frac{1}{2}\alpha}. \tag{4.142}$$

Hint: write $\cos n\alpha$ as $[\exp(in\alpha) + \exp(-in\alpha)]/2$.

4.4 Show that

$$\binom{n-1}{k} + \binom{n-1}{k-1} = \binom{n}{k} \tag{4.143}$$

and then use mathematical induction to prove Leibniz's rule (4.46).

4.5 (a) Find the radius of convergence of the series (4.49) for the Bessel function $J_n(\rho)$. (b) Show that this series converges even faster than the one (4.47) for the exponential function.

4.6 Use the formula (4.8) for the Euler–Mascheroni constant to show that Euler's definition (4.58) of the gamma function implies Weierstrass's (4.59).

4.7 Derive the approximation (4.65) for $j_\ell(\rho)$ for $|\rho| \ll 1$.

4.8 Derive formula (4.60) for the gamma function from its definition (4.55).

4.9 Use formula (4.60) to compute $\Gamma(1/2)$.

4.10 Show that $z! = \Gamma(z+1)$ diverges when z is a negative integer.

4.11 Derive formula (4.62) for $\Gamma((2n+1)/2)$.

4.12 Show that the area of the surface of the unit sphere in d dimensions is

$$A_d = 2\pi^{d/2}/\Gamma(d/2). \tag{4.144}$$

Hint: compute the integral of the gaussian $\exp(-x^2)$ in d dimensions using both rectangular and spherical coordinates. This formula (4.144) is used in dimensional regularization (Weinberg, 1995, p. 477).

4.13 Derive (4.80) from (4.78) and (4.79) from (4.77).

4.14 Derive the expansions (4.84 & 4.85) for $\sqrt{1+x}$ and $1/\sqrt{1+x}$.

4.15 Find the radii of convergence of the series (4.84) and (4.85).

4.16 Find the first three Bernoulli polynomials $B_n(y)$ by using their generating function (4.108).

4.17 How are the two definitions (4.106) and (4.109) of the Bernoulli numbers related?

4.18 Show that the Lerch transcendent $\Phi(z, s, \alpha)$ defined by the series (4.102) converges when $|z| < 1$ and $\mathrm{Re}\, s > 0$ and $\mathrm{Re}\, \alpha > 0$.

4.19 Langevin's classical formula for the electrical polarization of a gas or liquid of molecules of electric dipole moment p is

$$P(x) = Np\left(\frac{\cosh x}{\sinh x} - \frac{1}{x}\right) \tag{4.145}$$

where $x = pE/(kT)$, E is the applied electric field, and N is the number density of the molecules per unit volume. (a) Expand $P(x)$ for small x as an infinite power series involving the Bernoulli numbers. (b) What are the first three terms expressed in terms of familiar constants? (c) Find the saturation limit of $P(x)$ as $x \to \infty$.

4.20 Show that the energy of a charge q spread on the surface of a sphere of radius a in an infinite lipid of permittivity ϵ_ℓ is $W = q^2/8\pi\epsilon_\ell a$.

4.21 If the lipid of exercise 4.20 has finite thickness t and is surrounded on both sides by water of permittivity ϵ_w, then the image charges lower the energy W by (Parsegian, 1969)

$$\Delta W = \frac{q^2}{4\pi\epsilon_\ell\, t}\sum_{n=1}^{\infty}\frac{1}{n}\left(\frac{\epsilon_\ell - \epsilon_w}{\epsilon_\ell + \epsilon_w}\right)^n. \tag{4.146}$$

Sum this series. Hint: read section 4.9 carefully.

4.22 Consider a stack of three dielectrics of infinite extent in the x-y-plane separated by the two infinite x-y-planes $z = t/2$ and $z = -t/2$. Suppose the upper region $z > t/2$ is a uniform linear dielectric of permittivity ϵ_1, the central region $-t/2 < z < t/2$ is a uniform linear dielectric of permittivity ϵ_2, and the lower region $z < -t/2$ is a uniform linear dielectric of permittivity ϵ_3. Suppose the lower infinite x-y-plane $z = -t/2$ has a uniform surface charge density $-\sigma$, while the upper plane $z = t/2$ has a uniform surface charge density σ. What is the energy per unit area of this system? What is the pressure on the second dielectric? What is the capacitance of the stack?

5

Complex-variable theory

5.1 Analytic functions

A complex-valued function $f(z)$ of a complex variable z is **differentiable** at z with derivative $f'(z)$ if the limit

$$f'(z) = \lim_{z' \to z} \frac{f(z') - f(z)}{z' - z} \tag{5.1}$$

exists as z' approaches z from **any direction** in the complex plane. The limit must exist no matter how or from what direction z' approaches z.

If the function $f(z)$ is differentiable in a small disk around a point z_0, then $f(z)$ is said to be **analytic** at z_0 (and at all points inside the disk).

Example 5.1 (Polynomials) If $f(z) = z^n$ for some integer n, then for tiny dz and $z' = z + dz$, the difference $f(z') - f(z)$ is

$$f(z') - f(z) = (z + dz)^n - z^n \approx n z^{n-1}\, dz \tag{5.2}$$

and so the limit

$$\lim_{z' \to z} \frac{f(z') - f(z)}{z' - z} = \lim_{dz \to 0} \frac{n z^{n-1}\, dz}{dz} = n z^{n-1} \tag{5.3}$$

exists and is $n z^{n-1}$ *independently* of how z' approaches z. Thus the function z^n is analytic at z for all z with derivative

$$\frac{dz^n}{dz} = n z^{n-1}. \tag{5.4}$$

A function that is analytic everywhere is **entire**. All polynomials

$$P(z) = \sum_{n=0}^{N} c_n \, z^n \tag{5.5}$$

are entire. □

Example 5.2 (A function that's not analytic) To see what can go wrong when a function is not analytic, consider the function $f(x, y) = x^2 + y^2 = z\bar{z}$ for $z = x + iy$. If we compute its derivative at $(x, y) = (1, 0)$ by setting $x = 1 + \epsilon$ and $y = 0$, then the limit is

$$\lim_{\epsilon \to 0} \frac{f(1+\epsilon, 0) - f(1, 0)}{\epsilon} = \lim_{\epsilon \to 0} \frac{(1+\epsilon)^2 - 1}{\epsilon} = 2 \tag{5.6}$$

while if we instead set $x = 1$ and $y = \epsilon$, then the limit is

$$\lim_{\epsilon \to 0} \frac{f(1, \epsilon) - f(1, 0)}{i\epsilon} = \lim_{\epsilon \to 0} \frac{1 + \epsilon^2 - 1}{i\epsilon} = -i \lim_{\epsilon \to 0} \epsilon = 0. \tag{5.7}$$

So the derivative depends upon the direction through which $z \to 1$. □

5.2 Cauchy's integral theorem

If $f(z)$ is analytic at z_0, then near z_0 and to first order in $z - z_0$

$$f(z) \approx f(z_0) + f'(z_0)\,(z - z_0). \tag{5.8}$$

Let's compute the contour integral of $f(z)$ along a small circle of radius ϵ and center z_0. The points on the contour are

$$z = z_0 + \epsilon \, e^{i\theta} \tag{5.9}$$

for $\theta \in [0, 2\pi]$. So $dz = i\epsilon \, e^{i\theta} d\theta$, and the contour integral is

$$\oint f(z)\, dz = \int_0^{2\pi} \left[f(z_0) + f'(z_0)\,(z - z_0) \right] \, i\epsilon \, e^{i\theta} d\theta. \tag{5.10}$$

Since $z - z_0 = \epsilon \, e^{i\theta}$, the contour integral breaks into two pieces

$$\oint f(z)\, dz = f(z_0) \int_0^{2\pi} i\epsilon \, e^{i\theta} d\theta + f'(z_0) \int_0^{2\pi} \epsilon \, e^{i\theta} \, i\epsilon \, e^{i\theta} d\theta, \tag{5.11}$$

which vanish because the θ-integrals are zero. So the contour integral of the analytic function $f(z)$

$$\oint f(z)\, dz = 0 \tag{5.12}$$

is zero around the tiny circle – at least to order ϵ^2.

What about the contour integral of an analytic function $f(z)$ around a tiny square of size ϵ? Again we use the analyticity of $f(z)$ at $z = z_0$ to expand it as

$$f(z) \approx f(z_0) + f'(z_0)\,(z - z_0) \tag{5.13}$$

on the tiny square. The square contour consists of the four complex segments $dz_1 = \epsilon$, $dz_2 = i\,\epsilon$, $dz_3 = -\epsilon$, and $dz_4 = -i\,\epsilon$. The centers z_n of these segments are displaced by $z_1 - z_0 = -i\,\epsilon/2$, $z_2 - z_0 = \epsilon/2$, $z_3 - z_0 = i\,\epsilon/2$, and $z_4 - z_0 = -\epsilon/2$ from z_0. The integral of $f(z)$ around the square

$$\oint f(z)\,dz = \sum_{n=1}^{4} f(z_n)\,dz_n = \sum_{n=1}^{4} \left[f(z_0) + (z_n - z_0)f'(z_0)\right] dz_n \tag{5.14}$$

splits into two pieces

$$\oint f(z)\,dz = f(z_0)\,I_1 + f'(z_0)\,I_2. \tag{5.15}$$

The four segments dz_n form a path that goes around the square and ends where it started, so the first piece $f(z_0)I_1$ is zero

$$f(z_0)\,I_1 = f(z_0)\,[\epsilon + i\,\epsilon + (-\epsilon) + (-i\,\epsilon)] = 0. \tag{5.16}$$

And so is the second one $f'(z_0)\,I_2$

$$\begin{aligned} f'(z_0)I_2 &= f'(z_0)\,[(z_1 - z_0)dz_1 + (z_2 - z_0)dz_2 + (z_3 - z_0)dz_3 + (z_4 - z_0)dz_4] \\ &= f'(z_0)\,[(-i\,\epsilon/2)\,\epsilon + (\epsilon/2)\,i\,\epsilon + (i\,\epsilon/2)\,(-\epsilon) + (-\epsilon/2)\,(-i\,\epsilon)] \\ &= f'(z_0)\,(\epsilon^2/2)\,[-i + i - i + i] = 0. \end{aligned} \tag{5.17}$$

So the contour integral of an analytic function $f(z)$ around a tiny square of side ϵ is zero to order ϵ^2. Thus, the integral around such a square can be at most of order ϵ^3. This is very important. We'll use it to prove Cauchy's integral theorem.

Let's consider a function $f(z)$ that is analytic on a square of side L, as pictured in Fig. 5.1. The contour integral of $f(z)$ around the square can be expressed as the sum of L^2/ϵ^2 contour integrals around tiny squares of side ϵ. All interior integrals cancel, leaving the integral around the perimeter. Each contour integral around its tiny square is at most of order ϵ^3. So the sum of the L^2/ϵ^2 tiny contour integrals is at most $(L^2/\epsilon^2)\,\epsilon^3 = \epsilon\,L^2$, which vanishes as $\epsilon \to 0$. Thus the contour integral of a function $f(z)$ along the perimeter of a square of side L vanishes if $f(z)$ is analytic on the perimeter and inside the square. This is an example of Cauchy's integral theorem.

Suppose a function $f(z)$ is analytic in a region $\mathcal{R}$ and that I is a contour integral along a straight line within that region from z_1 to z_2. The contour integral of $f(z)$ around any square inside the region $\mathcal{R}$ of analyticity is zero. So by successively adding contour integrals around small squares to the straight-line

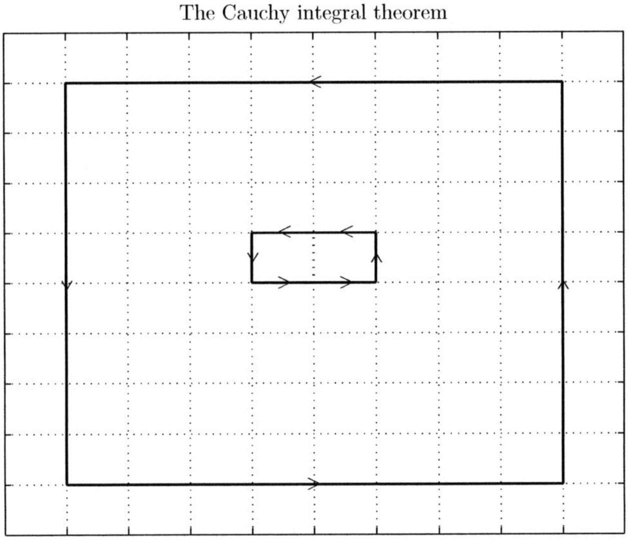
The Cauchy integral theorem

Figure 5.1 The sum of two contour integrals around two adjacent tiny squares is equal to the contour integral around the perimeter of the two tiny squares because the up integral along the right side of the left square cancels (dots) the down integral along the left side of the right square. A contour integral around a big $L \times L$ square is equal to the sum of the contour integrals around the L^2/ϵ^2 tiny $\epsilon \times \epsilon$ squares that tile the big square.

contour integral, one may deform the straight-line contour into an arbitrary contour from z_1 to z_2 without changing its value.

So a contour integral from z_1 to z_2 of a function $f(z)$ that is analytic in a region $\mathcal{R}$ remains invariant as we continuously deform the contour C to C' as long as these contours and all the intermediate contours lie entirely within the region $\mathcal{R}$ and have the same fixed endpoints z_1 and z_2 as in Fig. 5.2

$$I = \int_{z_1 C}^{z_2} f(z)\, dz = \int_{z_1 C'}^{z_2} f(z)\, dz. \tag{5.18}$$

Thus a contour integral depends upon its endpoints and upon the function $f(z)$ but not upon the actual contour as long as the deformations of the contour do not push it outside the region $\mathcal{R}$ in which $f(z)$ is analytic.

If the endpoints z_1 and z_2 are the same, then the contour C is closed, and we write the integral as

$$I = \oint_{z_1 C}^{z_1} f(z)\, dz \equiv \oint_C f(z)\, dz \tag{5.19}$$

with a little circle to denote that the contour is a closed loop. The value of that integral is independent of the contour as long as our deformations of the

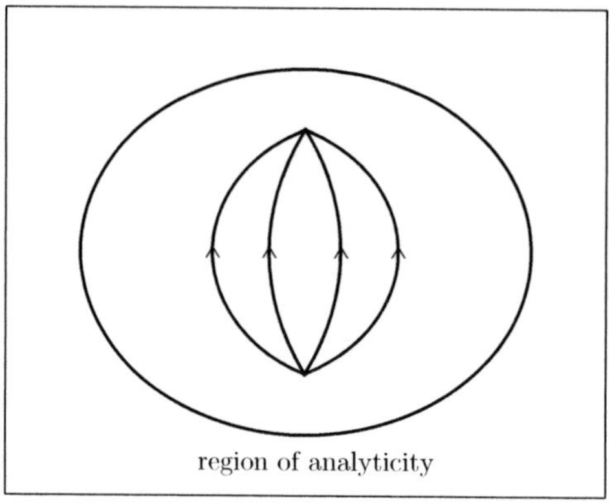

Figure 5.2 As long as the four contours are within the domain of analyticity of $f(z)$ and have the same endpoints, the four contour integrals of that function are all equal.

contour keep it within the domain of analyticity of the function and as long as the contour starts and ends at $z_1 = z_2$. Now suppose that the function $f(z)$ is analytic along the contour and at all points within it. Then we can shrink the contour, staying within the domain of analyticity of the function, until the area enclosed is zero and the contour is of zero length – all this without changing the value of the integral. But the value of the integral along such a null contour of zero length is zero. Thus the value of the original contour integral also must be zero

$$\oint_{z_1 C}^{z_1} f(z)\, dz = 0. \tag{5.20}$$

And so we arrive at **Cauchy's integral theorem**: The contour integral of a function $f(z)$ around a closed contour C lying entirely within the domain $\mathcal{R}$ of analyticity of the function vanishes

$$\oint_C f(z)\, dz = 0 \tag{5.21}$$

as long as the function $f(z)$ is analytic at all points within the contour.

A region in the complex plane is **simply connected** if we can shrink every loop in the region to a point while keeping the loop in the region. A slice of American cheese is simply connected, but a slice of Swiss cheese is not. A dime is simply connected, but a washer isn't. The surface of a sphere is simply connected, but the surface of a bagel isn't.

With this definition, we can restate the integral theorem of Cauchy: The contour integral of a function $f(z)$ around a closed contour C vanishes

$$\oint_C f(z)\, dz = 0 \tag{5.22}$$

if the contour lies within a simply connected domain of analyticity of the function $f(z)$ (Augustin-Louis Cauchy, 1789–1857).

If a region $\mathcal{R}$ is simply connected, then we may deform any contour C from z_1 to z_2 in $\mathcal{R}$ into any other contour C' from z_1 to z_2 in $\mathcal{R}$ while keeping the moving contour in the region $\mathcal{R}$. So another way of understanding the Cauchy integral theorem is to ask, what is the value of the contour integral

$$I_M = \int_{z_2 C'}^{z_1} f(z)\, dz\, ? \tag{5.23}$$

This integral is the same as the integral along C from z_1 to z_2, except for the sign of the dzs and the order in which the terms are added, and so

$$I_M = \int_{z_2 C'}^{z_1} f(z)\, dz = - \int_{z_1 C}^{z_2} f(z)\, dz. \tag{5.24}$$

Now consider a closed contour running along the contour C from z_1 to z_2 and backwards along C' from z_2 to z_1 all within a simply connected region $\mathcal{R}$ of analyticity. Since $I_M = -I$, the integral of $f(z)$ along this closed contour vanishes:

$$\oint f(z)\, dz = I + I_M = I - I = 0 \tag{5.25}$$

and we have again derived Cauchy's integral theorem.

Example 5.3 (Polynomials are entire functions) Every polynomial

$$P(z) = \sum_{n=0}^{N} c_n z^n \tag{5.26}$$

is entire (everywhere analytic), and so its integral along any closed contour

$$\oint P(z)\, dz = 0 \tag{5.27}$$

must vanish. □

5.3 Cauchy's integral formula

Let $f(z)$ be analytic in a simply connected region $\mathcal{R}$ and z_0 a point inside this region. We first will integrate the function $f(z)/(z-z_0)$ along a tiny closed counterclockwise contour around the point z_0. The contour is a circle of radius ϵ with center at z_0 with points $z = z_0 + \epsilon\, e^{i\theta}$ for $0 \le \theta \le 2\pi$, and $dz = i\epsilon\, e^{i\theta}\, d\theta$. Since $z - z_0 = \epsilon\, e^{i\theta}$, the contour integral in the limit $\epsilon \to 0$ is

$$\oint_{\epsilon} \frac{f(z)}{z-z_0}\, dz = \int_0^{2\pi} \frac{\left[f(z_0) + f'(z_0)\,(z-z_0)\right]}{z-z_0}\, i\epsilon\, e^{i\theta}\, d\theta$$
$$= \int_0^{2\pi} \frac{\left[f(z_0) + f'(z_0)\,\epsilon\, e^{i\theta}\right]}{\epsilon\, e^{i\theta}}\, i\epsilon\, e^{i\theta}\, d\theta$$
$$= \int_0^{2\pi} \left[f(z_0) + f'(z_0)\,\epsilon\, e^{i\theta}\right] i d\theta. \tag{5.28}$$

The θ-integral involving $f'(z_0)$ vanishes, and so we have

$$f(z_0) = \frac{1}{2\pi i} \oint_{\epsilon} \frac{f(z)}{z-z_0}\, dz, \tag{5.29}$$

which is a miniature version of Cauchy's integral formula.

Now consider the counterclockwise contour $\mathcal{C}'$ in Fig. 5.3, which is a big counterclockwise circle, a small clockwise circle, and two parallel straight lines, all within a simply connected region $\mathcal{R}$ in which $f(z)$ is analytic. The function $f(z)/(z-z_0)$ is analytic everywhere in $\mathcal{R}$ except at the point z_0. We can withdraw the contour $\mathcal{C}'$ to the left of the point z_0 and shrink it to a point without having the contour $\mathcal{C}'$ cross z_0. During this process, the integral of $f(z)/(z-z_0)$ does not change. Its final value is zero. So its initial value also is zero

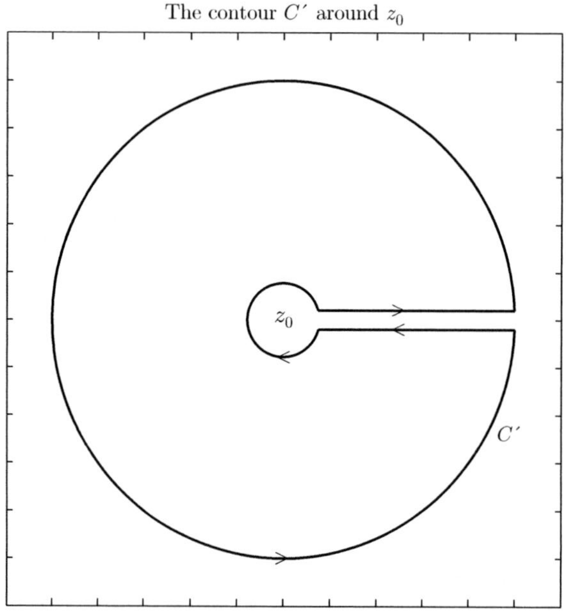

Figure 5.3 The full contour is the sum of a big counterclockwise contour and a small clockwise contour, both around z_0, and two straight lines that cancel.

$$0 = \frac{1}{2\pi i} \oint_{C'} \frac{f(z)}{z - z_0}\, dz. \tag{5.30}$$

We let the two straight-line segments approach each other so that they cancel. What remains of contour C' is a big counterclockwise contour C around z_0 and a tiny clockwise circle of radius ϵ around z_0. The tiny clockwise circle integral is the negative of the counterclockwise integral (5.29), so we have

$$0 = \frac{1}{2\pi i} \oint_{C'} \frac{f(z)}{z - z_0}\, dz = \frac{1}{2\pi i} \oint_{C} \frac{f(z)}{z - z_0}\, dz - \frac{1}{2\pi i} \oint_{\epsilon} \frac{f(z)}{z - z_0}\, dz. \tag{5.31}$$

Using the miniature result (5.29), we find

$$f(z_0) = \frac{1}{2\pi i} \oint_{C} \frac{f(z)}{z - z_0}\, dz, \tag{5.32}$$

which is **Cauchy's integral formula.**

We can use this formula to compute the first derivative $f'(z)$ of $f(z)$

$$\begin{aligned} f'(z) &= \frac{f(z + dz) - f(z)}{dz} \\ &= \frac{1}{2\pi i} \frac{1}{dz} \oint dz' f(z') \left(\frac{1}{z' - z - dz} - \frac{1}{z' - z} \right) \\ &= \frac{1}{2\pi i} \oint dz' \frac{f(z')}{(z' - z - dz)(z' - z)}. \end{aligned} \tag{5.33}$$

So in the limit $dz \to 0$, we get

$$f'(z) = \frac{1}{2\pi i} \oint dz' \frac{f(z')}{(z' - z)^2}. \tag{5.34}$$

The second derivative $f^{(2)}(z)$ of $f(z)$ then is

$$f^{(2)}(z) = \frac{2}{2\pi i} \oint dz' \frac{f(z')}{(z' - z)^3}. \tag{5.35}$$

And its nth derivative $f^{(n)}(z)$ is

$$f^{(n)}(z) = \frac{n!}{2\pi i} \oint dz' \frac{f(z')}{(z' - z)^{n+1}}. \tag{5.36}$$

In these formulas, the contour runs counterclockwise about the point z and lies within the simply connected domain $\mathcal{R}$ in which $f(z)$ is analytic.

Thus a function $f(z)$ that is analytic in a region $\mathcal{R}$ is infinitely differentiable there.

Example 5.4 (Schlaefli's formula for the Legendre polynomials) Rodrigues showed that the Legendre polynomial $P_n(x)$ is the nth derivative

$$P_n(x) = \frac{1}{2^n\, n!}\left(\frac{d}{dx}\right)^n (x^2-1)^n. \tag{5.37}$$

Schlaefli used this expression and Cauchy's integral formula (5.36) to represent $P_n(z)$ as the contour integral (exercise 5.8)

$$P_n(z) = \frac{1}{2^n\, 2\pi i}\oint \frac{(z'^2-1)^n}{(z'-z)^{n+1}}\, dz', \tag{5.38}$$

in which the contour encircles the complex point z counterclockwise. This formula tells us that at $z=1$ the Legendre polynomial is

$$P_n(1) = \frac{1}{2^n\, 2\pi i}\oint \frac{(z'^2-1)^n}{(z'-1)^{n+1}}\, dz' = \frac{1}{2^n\, 2\pi i}\oint \frac{(z'+1)^n}{(z'-1)}\, dz' = 1, \tag{5.39}$$

in which we applied Cauchy's integral formula (5.32) to $f(z) = (z+1)^n$. □

Example 5.5 (Bessel functions of the first kind) The counterclockwise integral around the unit circle $z = e^{i\theta}$ of the ratio z^m/z^n in which both m and n are integers is

$$\frac{1}{2\pi i}\oint dz\,\frac{z^m}{z^n} = \frac{1}{2\pi i}\int_0^{2\pi} ie^{i\theta}\,d\theta\, e^{i(m-n)\theta} = \frac{1}{2\pi}\int_0^{2\pi} d\theta\, e^{i(m+1-n)\theta}. \tag{5.40}$$

If $m+1-n \neq 0$, this integral vanishes because $\exp 2\pi i(m+1-n) = 1$

$$\frac{1}{2\pi}\int_0^{2\pi} d\theta\, e^{i(m+1-n)\theta} = \frac{1}{2\pi}\left[\frac{e^{i(m+1-n)\theta}}{i(m+1-n)}\right]_0^{2\pi} = 0. \tag{5.41}$$

If $m+1-n=0$, the exponential is unity, $\exp i(m+1-n)\theta = 1$, and the integral is $2\pi/2\pi = 1$. Thus the original integral is the Kronecker delta

$$\frac{1}{2\pi i}\oint dz\,\frac{z^m}{z^n} = \delta_{m+1,n}. \tag{5.42}$$

The generating function (9.5) for Bessel functions J_m of the first kind is

$$e^{t(z-1/z)/2} = \sum_{m=-\infty}^{\infty} z^m J_m(t). \tag{5.43}$$

Applying our integral formula (5.42) to it, we find

$$\frac{1}{2\pi i}\oint dz\, e^{t(z-1/z)/2}\,\frac{1}{z^{n+1}} = \frac{1}{2\pi i}\oint dz \sum_{m=-\infty}^{\infty}\frac{z^m}{z^{n+1}} J_m(t)$$
$$= \sum_{m=-\infty}^{\infty}\delta_{m+1,n+1}\, J_m(t) = J_n(t). \tag{5.44}$$

Thus, letting $z = e^{i\theta}$, we have

$$J_n(t) = \frac{1}{2\pi} \int_0^{2\pi} d\theta \, \exp\left[t \frac{\left(e^{i\theta} - e^{-i\theta}\right)}{2} - in\theta \right] \tag{5.45}$$

or more simply

$$J_n(t) = \frac{1}{2\pi} \int_0^{2\pi} d\theta \, e^{i(t \sin\theta - n\theta)} = \frac{1}{\pi} \int_0^{\pi} d\theta \, \cos(t \sin\theta - n\theta) \tag{5.46}$$

(exercise 5.3). □

5.4 The Cauchy–Riemann conditions

We can write any complex-valued function of two real variables x and y as $f = u + iv$ where $u(x, y)$ and $v(x, y)$ are real. If we use subscripts for partial differentiation $u_x = \partial u/\partial x$, $u_y = \partial u/\partial y$, and so forth, then the change in f due to small changes in x and y is $df = (u_x + iv_x)\, dx + (u_y + iv_y)\, dy$. But if f is a function of $z = x + iy$, rather than just of x and y, and if f is analytic at z, then the change in f due to small changes in x and y is

$$df = (u_x + iv_x)\, dx + (u_y + iv_y)\, dy = f'(z)\, dz = f'(z)(dx + idy). \tag{5.47}$$

Setting first dy and then dx equal to zero, we get first $u_x + iv_x = f'$ and then $-iu_y + v_y = f'$, which give us the **Cauchy–Riemann conditions**

$$u_x = v_y \quad \text{and} \quad u_y = -v_x. \tag{5.48}$$

These conditions (5.48) hold because if f is analytic at z, then its derivative f' is independent of the direction from which $dz \to 0$. Thus the derivatives in the x-direction and the iy-direction are the same

$$f_x = u_x + iv_x = f_y/i = \left(u_y + iv_y\right)/i = -iu_y + v_y, \tag{5.49}$$

which again gives us the Cauchy–Riemann conditions (5.48).

The directions in the x-y plane in which the real u and imaginary v parts of an analytic function change most rapidly are the vectors (u_x, u_y) and (v_x, v_y). The Cauchy–Riemann conditions (5.48) imply that these directions must be perpendicular

$$(u_x, u_y) \cdot (v_x, v_y) = u_x v_x + u_y v_y = v_y v_x - v_x v_y = 0. \tag{5.50}$$

The Cauchy–Riemann conditions (5.48) let us relate Cauchy's integral theorem (5.21) to Stokes's theorem in the x-y plane. The real and imaginary parts of a closed contour integral of a function $f = u + iv$

$$\oint_C f(z)\, dz = \oint_C (u + iv)(dx + idy) = \oint_C u\, dx - v\, dy + i \oint_C v\, dx + u\, dy \tag{5.51}$$

are loop integrals of the functions $\boldsymbol{a} = (u, -v, 0)$ and $\boldsymbol{b} = (v, u, 0)$. By Stokes's theorem, these loop integrals are surface integrals of $(\nabla \times \boldsymbol{a})_z$ and $(\nabla \times \boldsymbol{b})_z$ over the area enclosed by the contour C

$$\oint_C udx - vdy = \int_C \boldsymbol{a} \cdot (dx, dy, 0) = \int_S (\nabla \times \boldsymbol{a})_z \, dxdy = \int_S -v_x - u_y \, dxdy = 0,$$

$$\oint_C vdx + udy = \int_C \boldsymbol{b} \cdot (dx, dy, 0) = \int_S (\nabla \times \boldsymbol{b})_z \, dxdy = \int_S u_x - v_y \, dxdy = 0$$

which vanish by the Cauchy–Riemann conditions (5.48).

5.5 Harmonic functions

The Cauchy–Riemann conditions (5.48) tell us something about the Laplacian of the real part u of an analytic function $f = u + iv$. First, the second x-derivative u_{xx} is $u_{xx} = v_{yx} = v_{xy} = -u_{yy}$. So the real part u of an analytic function f is a **harmonic** function

$$u_{xx} + u_{yy} = 0 \tag{5.52}$$

that is, one with a vanishing Laplacian. Similarly $v_{xx} = -u_{yx} = -v_{yy}$, so the imaginary part of an analytic function f also is a harmonic function

$$v_{xx} + v_{yy} = 0. \tag{5.53}$$

A harmonic function $h(x, y)$ can have saddle points, but not local minima or maxima because at a local minimum both $h_{xx} > 0$ and $h_{yy} > 0$, while at a local maximum both $h_{xx} < 0$ and $h_{yy} < 0$. So in its domain of analyticity, the real and imaginary parts of an analytic function f have neither minima nor maxima.

For static fields, the electrostatic potential $\phi(x, y, z)$ is a harmonic function of the three spatial variables x, y, and z in regions that are free of charge because the electric field is $\mathbf{E} = -\nabla\phi$, and its divergence vanishes $\nabla \cdot \mathbf{E} = 0$ where the charge density is zero. Thus the Laplacian of the electrostatic potential $\phi(x, y, z)$ vanishes

$$\nabla \cdot \nabla\phi = \phi_{xx} + \phi_{yy} + \phi_{zz} = 0 \tag{5.54}$$

and $\phi(x, y, z)$ is harmonic where there is no charge. The location of each positive charge is a local maximum of the electrostatic potential $\phi(x, y, z)$ and the location of each negative charge is a local minimum of $\phi(x, y, z)$. But in the absence of charges, the electrostatic potential has neither local maxima nor local minima. Thus one can not trap charged particles with an electrostatic potential, a result known as Earnshaw's theorem.

We have seen (5.52 & 5.53) that the real and imaginary parts of an analytic function are harmonic functions with two-dimensional gradients that are mutually perpendicular (5.50). And we know that the electrostatic potential is a harmonic function. Thus the real part $u(x, y)$ (or the imaginary part $v(x, y)$) of any analytic function $f(z) = u(x, y) + iv(x, y)$ describes the electrostatic potential $\phi(x, y)$ for some electrostatic problem that does not involve the third spatial coordinate z. The surfaces of constant $u(x, y)$ are the equipotential surfaces, and since the two gradients are orthogonal, the surfaces of constant $v(x, y)$ are the electric field lines.

Example 5.6 (Two-dimensional potentials) The function

$$f(z) = u + iv = E\,z = E\,x + i\,E\,y \tag{5.55}$$

can represent a potential $V(x, y, z) = E\,x$ for which the electric-field lines $\boldsymbol{E} = -E\,\hat{\boldsymbol{x}}$ are lines of constant y. It also can represent a potential $V(x, y, z) = E\,y$ in which $\boldsymbol{E}$ points in the negative y-direction, which is to say along lines of constant x.

Another simple example is the function

$$f(z) = u + iv = z^2 = x^2 - y^2 + 2ixy \tag{5.56}$$

for which $u = x^2 - y^2$ and $v = 2xy$. This function gives us a potential $V(x, y, z)$ whose equipotentials are the hyperbolas $u = x^2 - y^2 = c^2$ and whose electric-field lines are the perpendicular hyperbolas $v = 2xy = d^2$. Equivalently, we may take these last hyperbolas $2xy = d^2$ to be the equipotentials and the other ones $x^2 - y^2 = c^2$ to be the lines of the electric field.

For a third example, we write the variable z as $z = re^{i\theta} = \exp(\ln r + i\theta)$ and use the function

$$f(z) = u(x, y) + iv(x, y) = -\frac{\lambda}{2\pi\epsilon_0} \ln z = -\frac{\lambda}{2\pi\epsilon_0} (\ln r + i\theta), \tag{5.57}$$

which describes the potential $V(x, y, z) = -(\lambda/2\pi\epsilon_0) \ln\sqrt{x^2 + y^2}$ due to a line of charge per unit length $\lambda = q/L$. The electric-field lines are the lines of constant v

$$\boldsymbol{E} = \frac{\lambda}{2\pi\epsilon_0} \frac{(x, y, 0)}{x^2 + y^2} \tag{5.58}$$

or equivalently of constant θ. □

5.6 Taylor series for analytic functions

Let's consider the contour integral of the function $f(z')/(z' - z)$ along a circle $\mathcal{C}$ inside a simply connected region $\mathcal{R}$ in which $f(z)$ is analytic. For any point z inside the circle, Cauchy's integral formula (5.32) tells us that

$$f(z) = \frac{1}{2\pi i} \oint_{\mathcal{C}} \frac{f(z')}{z' - z}\, dz'. \tag{5.59}$$

We add and subtract the center z_0 from the denominator $z' - z$

$$f(z) = \frac{1}{2\pi i} \oint_C \frac{f(z')}{z' - z_0 - (z - z_0)}\, dz' \tag{5.60}$$

and then factor the denominator

$$f(z) = \frac{1}{2\pi i} \oint_C \frac{f(z')}{(z' - z_0)\left(1 - \frac{z - z_0}{z' - z_0}\right)}\, dz'. \tag{5.61}$$

From Fig. 5.4, we see that the modulus of the ratio $(z - z_0)/(z' - z_0)$ is less than unity, and so the power series

$$\left(1 - \frac{z - z_0}{z' - z_0}\right)^{-1} = \sum_{n=0}^{\infty} \left(\frac{z - z_0}{z' - z_0}\right)^n \tag{5.62}$$

by (4.25–4.28) converges absolutely and uniformly on the circle. We therefore are allowed to integrate the series

$$f(z) = \frac{1}{2\pi i} \oint_C \frac{f(z')}{z' - z_0} \sum_{n=0}^{\infty} \left(\frac{z - z_0}{z' - z_0}\right)^n dz' \tag{5.63}$$

term by term

$$f(z) = \sum_{n=0}^{\infty} (z - z_0)^n \frac{1}{2\pi i} \oint_C \frac{f(z')\, dz'}{(z' - z_0)^{n+1}}. \tag{5.64}$$

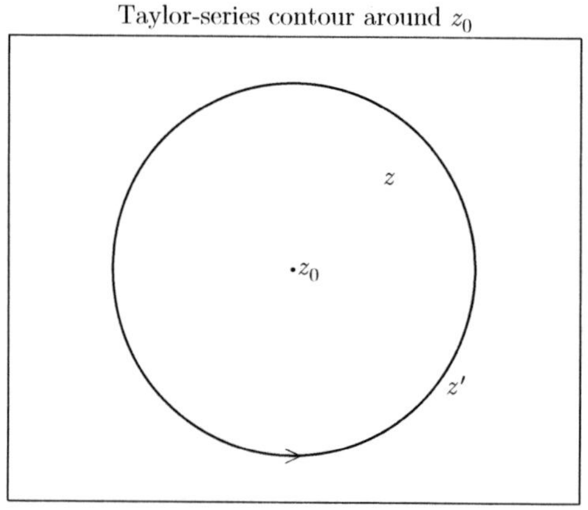

Figure 5.4 Contour of integral for the Taylor series (5.64).

By equation (5.36), the integral is just the nth derivative $f^{(n)}(z)$ divided by n-factorial. Thus the function $f(z)$ possesses the Taylor series

$$f(z) = \sum_{n=0}^{\infty} \frac{(z - z_0)^n}{n!} f^{(n)}(z_0), \tag{5.65}$$

which converges as long as the point z is inside a circle centered at z_0 that lies within a simply connected region $\mathcal{R}$ in which $f(z)$ is analytic.

5.7 Cauchy's inequality

Suppose a function $f(z)$ is analytic in a region that includes the disk $|z| \le R$ and that $f(z)$ is bounded by $|f(z)| \le M$ on the circle $z = R\,e^{i\theta}$ that is the perimeter of the disk. Then by using Cauchy's integral formula (5.36), we may bound the nth derivative $f^{(n)}(0)$ of $f(z)$ at $z = 0$ by

$$\begin{aligned} |f^{(n)}(0)| &\le \frac{n!}{2\pi} \oint \frac{|f(z)||dz|}{|z|^{n+1}} \\ &\le \frac{n!M}{2\pi} \int_0^{2\pi} \frac{R\,d\theta}{R^{n+1}} = \frac{n!M}{R^n} \end{aligned} \tag{5.66}$$

which is Cauchy's inequality.

5.8 Liouville's theorem

Suppose now that $f(z)$ is analytic everywhere (**entire**) and bounded by

$$|f(z)| \le M \quad \text{for all} \quad |z| \ge R_0. \tag{5.67}$$

Then by applying Cauchy's inequality (5.66) at successively larger values of R, we have

$$|f^{(n)}(0)| \le \lim_{R\to\infty} \frac{n!M}{R^n} = 0 \quad \text{for} \quad n \ge 1, \tag{5.68}$$

which shows that every derivative of $f(z)$ vanishes

$$f^{(n)}(0) = 0 \quad \text{for} \quad n \ge 1 \tag{5.69}$$

at $z = 0$. But then the Taylor series (4.66) about $z = 0$ for the function $f(z)$ consists of only a single term, and $f(z)$ is a constant

$$f(z) = \sum_{n=0}^{\infty} \frac{z^n}{n!} f^{(n)}(0) = f^{(0)}(0) = f(0). \tag{5.70}$$

So every bounded entire function is a constant, which is Liouville's theorem.

5.9 The fundamental theorem of algebra

Gauss applied Liouville's theorem to the function

$$f(z) = \frac{1}{P_N(z)} = \frac{1}{c_0 + c_1 z + c_2 z^2 + \cdots + c_N z^N} \tag{5.71}$$

which is the inverse of an arbitrary polynomial of order N. Suppose that the polynomial $P_N(z)$ had no zero, that is, no root anywhere in the complex plane. Then $f(z)$ would be analytic everywhere. Moreover, for sufficiently large $|z|$, the polynomial $P_N(z)$ is approximately $P_N(z) \approx c_N z^N$, and so $f(z)$ would be bounded by something like

$$|f(z)| \le \frac{1}{|c_N| R_0^N} \equiv M \quad \text{for all} \quad |z| \ge R_0. \tag{5.72}$$

So if $P_N(z)$ had no root, then the function $f(z)$ would be a bounded entire function and so would be a constant by Liouville's theorem (5.70). But of course, $f(z) = 1/P_N(z)$ is not a constant unless $N = 0$. Thus any polynomial $P_N(z)$ that is not a constant must have a root, a pole of $f(z)$, so that $f(z)$ is not entire. This is the only exit from the contradiction.

If the root of $P_N(z)$ is at $z = z_1$, then $P_N(z) = (z - z_1)\,P_{N-1}(z)$, in which $P_{N-1}(z)$ is a polynomial of order $N - 1$, and we may repeat the argument for its reciprocal $f_1(z) = 1/P_{N-1}(z)$. In this way, one arrives at the fundamental theorem of algebra: Every polynomial $P_N(z) = c_0 + c_1 z + \cdots + c_N z^N$ has N roots somewhere in the complex plane

$$P_N(z) = c_N\,(z - z_1)(z - z_2)\cdots(z - z_N). \tag{5.73}$$

5.10 Laurent series

Consider a function $f(z)$ that is analytic in a region that contains an outer circle C_1 of radius R_1, an inner circle C_2 of radius R_2, and the annulus between the two circles as in Fig. 5.5. We will integrate $f(z)$ along a contour C_{12} that encircles the point z in a counterclockwise fashion by following C_1 counterclockwise and C_2 clockwise and a line joining them in both directions. By Cauchy's integral formula (5.32), this contour integral yields $f(z)$

$$f(z) = \frac{1}{2\pi i} \oint_{C_{12}} \frac{f(z')}{z' - z}\, dz'. \tag{5.74}$$

The integrations in opposite directions along the line joining C_1 and C_2 cancel, and we are left with a counterclockwise integral around the outer circle C_1 and

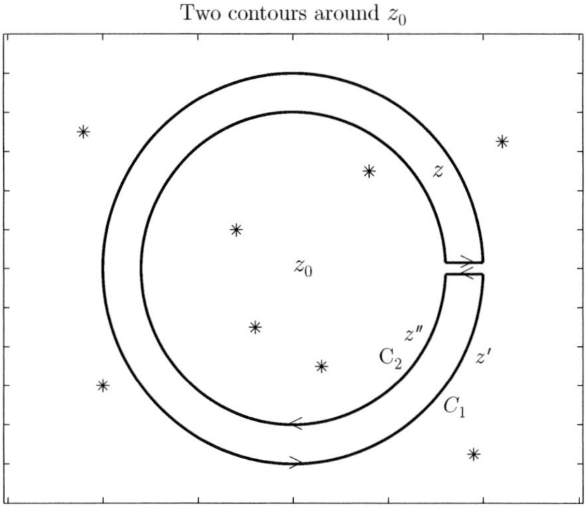

Figure 5.5 The contour consisting of two concentric circles with center at z_0 encircles the point z in a counterclockwise sense. The asterisks are poles or other singularities of the function $f(z)$.

a clockwise one around C_2 or *minus* a counterclockwise integral around C_2

$$f(z) = \frac{1}{2\pi i} \oint_{C_1} \frac{f(z')}{z' - z} dz' - \frac{1}{2\pi i} \oint_{C_2} \frac{f(z'')}{z'' - z} dz''. \tag{5.75}$$

Now from the figure (5.5), the center z_0 of the two concentric circles is closer to the points z'' on the inner circle C_2 than it is to z and also closer to z than to the points z' on C_1

$$\left| \frac{z'' - z_0}{z - z_0} \right| < 1 \quad \text{and} \quad \left| \frac{z - z_0}{z' - z_0} \right| < 1. \tag{5.76}$$

To use these inequalities, as we did in the series (5.62), we add and subtract z_0 from each of the denominators and absorb the minus sign before the second integral into its denominator

$$f(z) = \frac{1}{2\pi i} \oint_{C_1} \frac{f(z')}{z' - z_0 - (z - z_0)} dz' + \frac{1}{2\pi i} \oint_{C_2} \frac{f(z'')}{z - z_0 - (z'' - z_0)} dz''. \tag{5.77}$$

After factoring the two denominators

$$\begin{aligned} f(z) = {} & \frac{1}{2\pi i} \oint_{C_1} \frac{f(z')}{(z' - z_0)[1 - (z - z_0)/(z' - z_0)]} dz' \\ & + \frac{1}{2\pi i} \oint_{C_2} \frac{f(z'')}{(z - z_0)[1 - (z'' - z_0)/(z - z_0)]} dz'' \end{aligned} \tag{5.78}$$

we expand them, as in the series (5.62), in power series that converge absolutely and uniformly on the two contours

$$f(z) = \sum_{n=0}^{\infty} (z - z_0)^n \frac{1}{2\pi i} \oint_{C_1} \frac{f(z')}{(z' - z_0)^{n+1}} dz' + \sum_{m=0}^{\infty} \frac{1}{(z - z_0)^{m+1}} \frac{1}{2\pi i} \oint_{C_2} (z'' - z_0)^m f(z'') \, dz''. \tag{5.79}$$

Having removed the point z from the two integrals, we now apply cosmetics. Since the functions being integrated are analytic between the two circles, we may shift them to a common counterclockwise (ccw) contour C about any circle of radius $R_2 \le R \le R_1$ between the two circles C_1 and C_2. Then we set $m = -n-1$, or $n = -m-1$, so as to combine the two sums into one sum on n from $-\infty$ to ∞

$$f(z) = \sum_{n=-\infty}^{\infty} (z - z_0)^n \frac{1}{2\pi i} \oint_{C} \frac{f(z')}{(z' - z_0)^{n+1}} dz'. \tag{5.80}$$

This **Laurent series** often is written as

$$f(z) = \sum_{n=-\infty}^{\infty} a_n(z_0) \, (z - z_0)^n \tag{5.81}$$

with

$$a_n(z_0) = \frac{1}{2\pi i} \oint_C \frac{f(z) \, dz}{(z - z_0)^{n+1}} \tag{5.82}$$

(Pierre Laurent, 1813–1854). The coefficient $a_{-1}(z_0)$ is called the **residue** of the function $f(z)$ at z_0. Its significance will be discussed in section 5.13.

Most functions have Laurent series that start at some least integer L

$$f(z) = \sum_{n=L}^{\infty} a_n(z_0) \, (z - z_0)^n \tag{5.83}$$

rather than at $-\infty$. For such functions, we can pick off the coefficients a_n one by one without doing the integrals (5.82). The first one a_L is the limit

$$a_L(z_0) = \lim_{z \to z_0} (z - z_0)^{-L} f(z). \tag{5.84}$$

The second is given by

$$a_{L+1}(z_0) = \lim_{z \to z_0} (z - z_0)^{-L-1} \left[f(z) - (z - z_0)^L a_L(z_0) \right]. \tag{5.85}$$

The third requires two subtractions, and so forth.

5.11 Singularities

A function $f(z)$ that is analytic for all z is called **entire** or **holomorphic**. Entire functions have no singularities, except possibly as $|z| \to \infty$, which is called the **point at infinity**.

A function $f(z)$ has an **isolated singularity** at z_0 if it is analytic in a small disk about z_0 but not analytic that point.

A function $f(z)$ has a **pole** of order $n > 0$ at a point z_0 if $(z - z_0)^n f(z)$ is analytic at z_0 but $(z - z_0)^{n-1} f(z)$ has an isolated singularity at z_0. A pole of order $n = 1$ is called a **simple pole**. Poles are isolated singularities. A function is **meromorphic** if it is analytic for all z except for poles.

Example 5.7 (Poles) The function

$$f(z) = \prod_{j=1}^{n} \frac{1}{(z-j)^j} \tag{5.86}$$

has a pole of order j at $z = j$ for $j = 1, 2, \ldots, n$. It is meromorphic. □

An **essential singularity** is a pole of infinite order. If a function $f(z)$ has an essential singularity at z_0, then its Laurent series (5.80) really runs from $n = -\infty$ and not from $n = L$ as in (5.83). Essential singularities are spooky: if a function $f(z)$ has an essential singularity at w, then inside every disk around w, $f(z)$ takes on *every* complex number, with at most one exception, an infinite number of times – a result due to Picard (1856–1941).

Example 5.8 (An essential singularity) The function $f(z) = \exp(1/z)$ has an essential singularity at $z = 0$ because its Laurent series (5.80)

$$f(z) = e^{1/z} = \sum_{m=0}^{\infty} \frac{1}{m!} \frac{1}{z^m} = \sum_{n=-\infty}^{0} \frac{1}{|n|!} z^n \tag{5.87}$$

runs from $n = -\infty$. Near $z = 0$, $f(z) = \exp(1/z)$ takes on every complex number except 0 an infinite number of times. □

Example 5.9 (A meromorphic function with two poles) The function $f(z) = 1/z(z+1)$ has poles at $z = 0$ and at $z = -1$ but otherwise is analytic; it is meromorphic. We may expand it in a Laurent series (5.81–5.82)

$$f(z) = \frac{1}{z(z+1)} = \sum_{n=-\infty}^{\infty} a_n z^n \tag{5.88}$$

about $z = 0$ for $|z| < 1$. The coefficient a_n is the integral

$$a_n = \frac{1}{2\pi i} \oint_C \frac{dz}{z^{n+2}(z+1)}, \tag{5.89}$$

in which the contour C is a counterclockwise circle of radius $r < 1$. Since $|z| < 1$, we may expand $1/(1+z)$ as the series

$$\frac{1}{1+z} = \sum_{m=0}^{\infty} (-z)^m. \tag{5.90}$$

Doing the integrals, we find

$$a_n = \sum_{m=0}^{\infty} \frac{1}{2\pi i} \oint_C (-z)^m \frac{dz}{z^{n+2}} = \sum_{m=0}^{\infty} (-1)^m r^{m-n-1} \delta_{m,n+1} \tag{5.91}$$

for $n \ge -1$ and zero otherwise. So the Laurent series for $f(z)$ is

$$f(z) = \frac{1}{z(z+1)} = \sum_{n=-1}^{\infty} (-1)^{n+1} z^n. \tag{5.92}$$

The series starts at $n = -1$, not at $n = -\infty$, because $f(z)$ is meromorphic with only a simple pole at $z = 0$. □

Example 5.10 (The argument principle) Consider the counterclockwise integral

$$\frac{1}{2\pi i} \oint_C f(z) \frac{g'(z)}{g(z)}\, dz \tag{5.93}$$

along a contour C that lies inside a simply connected region R in which $f(z)$ is analytic and $g(z)$ meromorphic. If the function $g(z)$ has a zero or a pole of order n at $w \in R$ and no other singularity in R,

$$g(z) = a_n(w)(z-w)^n \tag{5.94}$$

then the ratio g'/g is

$$\frac{g'(z)}{g(z)} = \frac{n(z-w)^{n-1}}{(z-w)^n} = \frac{n}{z-w} \tag{5.95}$$

and the integral is

$$\frac{1}{2\pi i} \oint_C f(z) \frac{g'(z)}{g(z)}\, dz = \frac{1}{2\pi i} \oint_C f(z) \frac{n}{z-w}\, dz = n f(w). \tag{5.96}$$

Any function $g(z)$ meromorphic in R will possess a Laurent series

$$g(z) = \sum_{k=n}^{\infty} a_k(w)(z-w)^k \tag{5.97}$$

about each point $w \in R$. One may show (exercise 5.18) that as $z \to w$ the ratio g'/g again approaches (5.95). It follows that the integral (5.93) is a sum of $f(w_\ell)$ at the zeros of $g(z)$ minus a similar sum at the poles of $g(z)$

$$\frac{1}{2\pi i}\oint_C f(z)\,\frac{g'(z)}{g(z)}\,dz = \sum_\ell \frac{1}{2\pi i}\oint_C f(z)\,\frac{n_\ell}{z-w_\ell} = \sum_\ell n_\ell f(w_\ell) \tag{5.98}$$

in which $|n_\ell|$ is the multiplicity of the ℓth zero or pole. □

5.12 Analytic continuation

We saw in section 5.6 that a function $f(z)$ that is analytic within a circle of radius R about a point z_0 possesses a Taylor series (5.65)

$$f(z) = \sum_{n=0}^{\infty} \frac{(z-z_0)^n}{n!} f^{(n)}(z_0) \tag{5.99}$$

that converges for all z inside the disk $|z - z_0| < R$. Suppose z' is the singularity of $f(z)$ that is closest to z_0. Pick a point z_1 in the disk $|z - z_0| < R$ that is not on the line from z_0 to the nearest singularity z'. The function $f(z)$ is analytic at z_1 because z_1 is within the circle of radius R about the point z_0, and so $f(z)$ has a Taylor series expansion like (5.99) but about the point z_1. *Usually* the circle of convergence of this power series about z_1 will extend beyond the original disk $|z - z_0| < R$. If so, the two power series, one about z_0 and the other about z_1, define the function $f(z)$ and extend its domain of analyticity beyond the original disk $|z-z_0| < R$. Such an extension of the range of an analytic function is called **analytic continuation.**

We often can analytically continue a function more easily than by the successive construction of Taylor series.

Example 5.11 (The geometric series) The power series

$$f(z) = \sum_{n=0}^{\infty} z^n \tag{5.100}$$

converges and defines an analytic function for $|z| < 1$. But for such z, we may sum the series to

$$f(z) = \frac{1}{1-z}. \tag{5.101}$$

By summing the series (5.100), we have analytically continued the function $f(z)$ to the whole complex plane apart from its simple pole at $z = 1$. □

Example 5.12 (The gamma function) Euler's form of the gamma function is the integral

$$\Gamma(z) = \int_0^{\infty} e^{-t}\, t^{z-1}\, dt = (z-1)!, \tag{5.102}$$

which makes $\Gamma(z)$ analytic in the right half-plane $\text{Re}\, z > 0$. But by successively using the relation $\Gamma(z+1) = z\,\Gamma(z)$, we may extend $\Gamma(z)$ into the left half-plane

$$\Gamma(z) = \frac{1}{z}\,\Gamma(z+1) = \frac{1}{z}\,\frac{1}{z+1}\,\Gamma(z+2) = \frac{1}{z}\,\frac{1}{z+1}\,\frac{1}{z+2}\,\Gamma(z+3). \tag{5.103}$$

The last expression defines $\Gamma(z)$ as a function that is analytic for $\text{Re}\, z > -3$ apart from simple poles at $z = 0, -1$, and -2. Proceeding in this way, we may analytically continue the gamma function to the whole complex plane apart from the negative integers and zero. The analytically continued gamma function may be represented by the formula

$$\Gamma(z) = \frac{1}{z}\, e^{-\gamma z}\left[\prod_{n=1}^{\infty}\left(1+\frac{z}{n}\right)e^{-z/n}\right]^{-1} \tag{5.104}$$

due to Weierstrass. □

Example 5.13 (Dimensional regularization) The loop diagrams of quantum field theory involve badly divergent integrals like

$$I(4) = \int \frac{d^4q}{(2\pi)^4}\frac{\left(q^2\right)^a}{\left(q^2+\alpha^2\right)^b} \tag{5.105}$$

where often $a = 0$, $b = 2$, and $\alpha^2 > 0$. Gerardus 't Hooft (1946–) and Martinus J. G. Veltman (1931–) promoted the number of space-time dimensions from four to a complex number d. The resulting integral has the value (Srednicki, 2007, p. 102)

$$I(d) = \int \frac{d^dq}{(2\pi)^d}\frac{\left(q^2\right)^a}{\left(q^2+\alpha^2\right)^b} = \frac{\Gamma(b-a-d/2)\,\Gamma(a+d/2)}{(4\pi)^{d/2}\,\Gamma(b)\,\Gamma(d/2)}\frac{1}{(\alpha^2)^{b-a-d/2}} \tag{5.106}$$

and so defines a function of the complex variable d that is analytic everywhere except for simple poles at $d = 2(n - a + b)$ where $n = 0, 1, 2, \ldots, \infty$. At these poles, the formula

$$\Gamma(-n+z) = \frac{(-1)^n}{n!}\left(\frac{1}{z} - \gamma + \sum_{k=1}^{n}\frac{1}{k} + O(z)\right) \tag{5.107}$$

where $\gamma = 0.5772...$ is the Euler–Mascheroni constant (4.8) can be useful. □

5.13 The calculus of residues

A contour integral of an analytic function $f(z)$ does not change unless the endpoints move or the contour crosses a singularity or leaves the region of analyticity (section 5.2). Let us consider the integral of a function $f(z)$ along a counterclockwise contour C that encircles n poles at z_k for $k = 1, \ldots, n$ in a simply connected region $\mathcal{R}$ in which $f(z)$ is meromorphic. We may shrink the area within the contour C without changing the value of the integral until the area is infinitesimal and the contour is the sum of n tiny counterclockwise circles C_k around the n poles

$$\oint_C f(z)\,dz = \sum_{k=1}^{n} \oint_{C_k} f(z)\,dz. \tag{5.108}$$

These tiny counterclockwise integrals around the poles at z_i are the residues $a_{-1}(z_i)$ defined by (5.82) for $n = -1$ apart from the factor $2\pi i$. So the whole ccw integral is $2\pi i$ times the sum of the residues of the function $f(z)$ at the enclosed poles

$$\oint_C f(z)\,dz = 2\pi i \sum_{k=1}^{n} a_{-1}(z_k), \tag{5.109}$$

a result known as the **residue theorem**.

In general, one must do each tiny ccw integral about each pole z_i, but simple poles are an important special case. If w is a simple pole of the function $f(z)$, then near it $f(z)$ is given by its Laurent series (5.81) as

$$f(z) = \frac{a_{-1}(w)}{z-w} + \sum_{n=0}^{\infty} a_n(w)\,(z-w)^n. \tag{5.110}$$

In this case, its residue is by (5.84) with $L = -1$

$$a_{-1}(w) = \lim_{z\to w}\,(z-w)f(z), \tag{5.111}$$

which usually is easier to do than the integral (5.82)

$$a_{-1}(w) = \frac{1}{2\pi i}\oint_C f(z)dz. \tag{5.112}$$

Example 5.14 (Cauchy's integral formula) Suppose the function $f(z)$ is analytic within a region $\mathcal{R}$ and that C is a ccw contour that encircles a point $w \in \mathcal{R}$. Then the ccw contour C encircles the simple pole at w of the function $f(z)/(z-w)$, which is its only singularity in $\mathcal{R}$. By applying the residue theorem and formula (5.111) for the residue $a_{-1}(w)$ of the function $f(z)/(z-w)$, we find

$$\oint_C \frac{f(z)}{z-w}\,dz = 2\pi i\, a_{-1}(w) = 2\pi i \lim_{z\to w}\,(z-w)\,\frac{f(z)}{z-w} = 2\pi i f(w). \tag{5.113}$$

So Cauchy's integral formula (5.32) is an example of the calculus of residues. □

Example 5.15 (A meromorphic function) By the residue theorem (5.109), the integral of the function

$$f(z) = \frac{1}{z-1}\,\frac{1}{(z-2)^2} \tag{5.114}$$

along the circle $C = 4e^{i\theta}$ for $0 \le \theta \le 2\pi$ is the sum of the residues at $z = 1$ and $z = 2$

$$\oint_C f(z)\, dz = 2\pi i\,[a_{-1}(1) + a_{-1}(2)]\,. \tag{5.115}$$

The function $f(z)$ has a simple pole at $z = 1$, and so we may use the formula (5.111) to evaluate the residue $a_{-1}(1)$ as

$$a_{-1}(1) = \lim_{z\to 1} (z-1) f(z) = \lim_{z\to 1} \frac{1}{(z-2)^2} = 1 \tag{5.116}$$

instead of using Cauchy's integral formula (5.32) to do the integral of $f(z)$ along a tiny circle about $z = 1$, which gives the same result

$$a_{-1}(1) = \frac{1}{2\pi i} \oint \frac{dz}{z-1} \frac{1}{(z-2)^2} = \frac{1}{(1-2)^2} = 1. \tag{5.117}$$

The residue $a_{-1}(2)$ is the integral of $f(z)$ along a tiny circle about $z = 2$, which we do by using Cauchy's integral formula (5.34)

$$a_{-1}(2) = \frac{1}{2\pi i} \oint \frac{dz}{(z-2)^2} \frac{1}{z-1} = \frac{d}{dz} \frac{1}{z-1}\bigg|_{z=2} = -\frac{1}{(2-1)^2} = -1. \tag{5.118}$$

The sum of these two residues is zero, and so the integral (5.115) vanishes. Another way of evaluating this integral is to deform it, not into two tiny circles about the two poles, but rather into a huge circle $z = Re^{i\theta}$ and to notice that as $R \to \infty$ the modulus of this integral vanishes

$$\left|\oint f(z)\, dz\right| \approx \frac{2\pi}{R^2} \to 0. \tag{5.119}$$

This contour is an example of a ghost contour. □

5.14 Ghost contours

Often one needs to do an integral that is not a closed counterclockwise contour. Integrals along the real axis occur frequently. One sometimes can convert a line integral into a closed contour by adding a contour along which the integral vanishes, a **ghost contour**. We have just seen an example (5.119) of a ghost contour, and we shall see more of them in what follows.

Example 5.16 (Best case) Consider the integral

$$I = \int_{-\infty}^{\infty} \frac{1}{(x-i)(x-2i)(x-3i)}\, dx. \tag{5.120}$$

We could do the integral by adding a contour $Re^{i\theta}$ from $\theta = 0$ to $\theta = \pi$. In the limit $R \to \infty$, the integral of $1/[(z-i)(z-2i)(z-3i)]$ along this contour vanishes; it is a ghost contour. The original integral I and the ghost contour encircle the three poles, and so we could compute I by evaluating the residues at those poles. But we also could add a ghost contour around the lower half-plane. This contour and the real line encircle no poles. So we get $I = 0$ without doing any work at all. □

Example 5.17 (Fourier transform of a gaussian) During our computation of the Fourier transform of a gaussian (3.15–3.18), we promised to justify the shift in the variable of integration from x to $x + ik/2m^2$ in this chapter. So let us consider the contour integral of the entire function $f(z) = \exp(-m^2z^2)$ over the rectangular closed contour along the real axis from $-R$ to R and then from $z = R$ to $z = R + ic$ and then from there to $z = -R + ic$ and then to $z = -R$. Since the $f(z)$ is analytic within the contour, the integral is zero

$$\oint dz e^{-m^2z^2} = \int_{-R}^{R} e^{-m^2z^2} dz + \int_{R}^{R+ic} e^{-m^2z^2} dz + \int_{R+ic}^{-R+ic} e^{-m^2z^2} dz + \int_{-R+ic}^{-R} e^{-m^2z^2} dz = 0$$

for all finite positive values of R and so also in the limit $R \to \infty$. The two contours in the imaginary direction are of length c and are damped by the factor $\exp(-m^2R^2)$, and so they vanish in the limit $R \to \infty$. They are ghost contours. It follows then from this last equation in the limit $R \to \infty$ that

$$\int_{-\infty}^{\infty} dx\, e^{-m^2(x+ic)^2} = \int_{-\infty}^{\infty} dx\, e^{-m^2x^2} = \frac{\sqrt{\pi}}{m}, \tag{5.121}$$

which is the promised result.

It implies (exercise 5.20) that

$$\int_{-\infty}^{\infty} dx\, e^{-m^2(x+z)^2} = \int_{-\infty}^{\infty} dx\, e^{-m^2x^2} = \frac{\sqrt{\pi}}{m} \tag{5.122}$$

for $m > 0$ and arbitrary complex z. □

Example 5.18 (A cosine integral) To compute the integral

$$I = \int_0^{\infty} \frac{\cos x}{q^2 + x^2}\, dx \tag{5.123}$$

we use the evenness of the integrand to extend the integration

$$I = \frac{1}{2}\int_{-\infty}^{\infty} \frac{\cos x}{q^2 + x^2}\, dx, \tag{5.124}$$

write the cosine as $[\exp(ix) + \exp(-ix)]/2$, and factor the denominator

$$I = \frac{1}{4}\int_{-\infty}^{\infty} \frac{e^{ix}}{(x - iq)(x + iq)}\, dx + \frac{1}{4}\int_{-\infty}^{\infty} \frac{e^{-ix}}{(x - iq)(x + iq)}\, dx. \tag{5.125}$$

We promote x to a complex variable z and add the contours $z = Re^{i\theta}$ and $z = Re^{-i\theta}$ as θ goes from 0 to π respectively to the first and second integrals. The term $\exp(iz)dz/(q^2+z^2) = \exp(iR\cos\theta - R\sin\theta)iRe^{i\theta}\,d\theta/(q^2+R^2e^{2i\theta})$ vanishes in the limit $R \to \infty$, so the first contour is a ccw ghost contour. A similar argument applies to the second (clockwise) contour, and we have

$$I = \frac{1}{4} \oint \frac{e^{iz}}{(z - iq)(z + iq)}\, dz + \frac{1}{4} \oint \frac{e^{-iz}}{(z - iq)(z + iq)}\, dz. \tag{5.126}$$

The first integral picks up the pole at iq and the second the pole at $-iq$

$$I = \frac{i\pi}{2}\left(\frac{e^{-q}}{2iq} + \frac{e^{-q}}{2iq}\right) = \frac{\pi e^{-q}}{2q}. \tag{5.127}$$

So the value of the integral is $\pi e^{-q}/2q$. □

Example 5.19 (Third-harmonic microscopy) An ultra-short laser pulse intensely focused in a medium generates a third-harmonic electric field E_3 in the forward direction proportional to the integral (Boyd, 2000)

$$E_3 \propto \chi^{(3)} E_0^3 \int_{-\infty}^{\infty} e^{i\,\Delta k\, z} \frac{dz}{(1 + 2iz/b)^2} \tag{5.128}$$

along the axis of the beam as in Fig. 5.6. Here $b = 2\pi t_0^2 n/\lambda = kt_0^2$ in which $n = n(\omega)$ is the index of refraction of the medium, λ is the wave-length of the laser light in the medium, and t_0 is the transverse or waist radius of the gaussian beam, defined by $E(r) = E \exp(-r^2/t_0^2)$.

When the dispersion is normal, that is when $dn(\omega)/d\omega > 0$, the shift in the wave vector $\Delta k = 3\omega[n(\omega) - n(3\omega)]/c$ is negative. Since $\Delta k < 0$, the exponential is damped when $z = x + iy$ is in the lower half-plane (LHP)

$$e^{i\,\Delta k\, z} = e^{i\,\Delta k\,(x+iy)} = e^{i\,\Delta k\, x}\, e^{-\Delta k\, y}. \tag{5.129}$$

So as we did in example 5.18, we will add a contour around the lower half-plane ($z = Re^{i\theta}$, $\pi \le \theta \le 2\pi$, and $dz = iRe^{i\theta}d\theta$) because in the limit $R \to \infty$, the integral along it vanishes; it is a ghost contour.

The function $f(z) = \exp(i\,\Delta k\, z)/(1 + 2iz/b)^2$ has a double pole at $z = ib/2$, which is in the UHP since the length $b > 0$, but no singularity in the LHP $y < 0$. So the integral of $f(z)$ along the closed contour from $z = -R$ to $z = R$ and then along the ghost contour vanishes. But since the integral along the ghost contour vanishes, so does the integral from $-R$ to R. Thus when the dispersion is normal, the third-harmonic signal vanishes, $E_3 = 0$, as long as the medium with constant $\chi^{(3)}(z)$ effectively extends from $-\infty$ to ∞ so that its edges are in the unfocused region like the dotted lines of Fig. 5.6. But an edge with varying $\chi^{(3)}(z)$ in the focused region like the solid line of the figure does make a third-harmonic signal E_3. Third-harmonic microscopy lets us see edges or features instead of background. □

Example 5.20 (Green and Bessel) Let us evaluate the integral

$$I(x) = \int_{-\infty}^{\infty} dk\, \frac{e^{ikx}}{k^2 + m^2}, \tag{5.130}$$

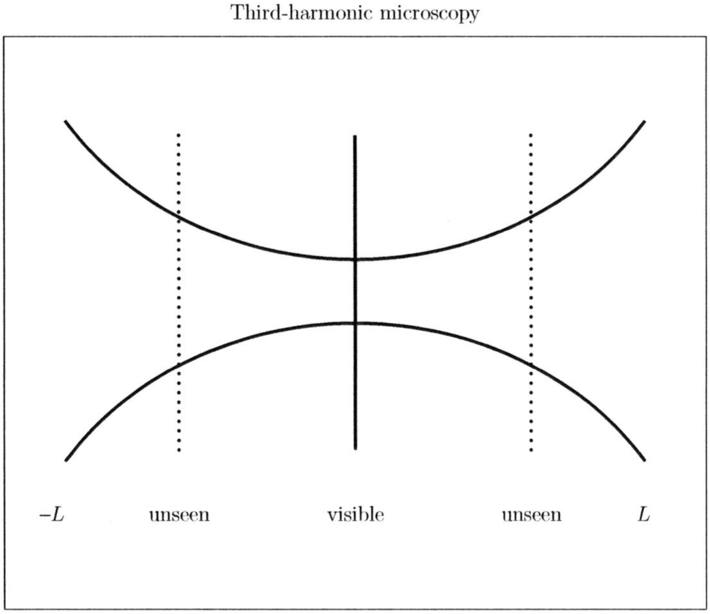

Figure 5.6 In the limit in which the distance L is much larger than the wave-length λ, the integral (5.128) is nonzero when an edge (solid line) lies where the beam is focused but not when a feature (dots) lies where the beam is not focused. Only features within the focused region are visible.

which is the Fourier transform of the function $1/(k^2 + m^2)$. If $x > 0$, then the exponential deceases with k in the upper half-plane. So as in example 5.18, the semicircular contour $k = R\,e^{i\theta}$, $0 \le \theta \le \pi$, and $dk = iRe^{i\theta}d\theta$ is a ghost contour. So if $x > 0$, then we can add this contour to the integral $I(x)$ without changing it. Thus $I(x)$ is equal to the closed contour integral along the real axis and the semicircular ghost contour

$$I(x) = \oint dk \frac{e^{ikx}}{k^2 + m^2} = \oint dk \frac{e^{ikx}}{(k + im)(k - im)}. \tag{5.131}$$

This closed contour encircles the simple pole at $k = im$ and no other singularity, and so we may shrink the contour into a tiny circle around the pole. Along that tiny circle, the function $e^{ikx}/(k + im)$ is simply $e^{-mx}/2im$, and so

$$I(x) = \frac{e^{-mx}}{2im} \oint \frac{dk}{k - im} = 2\pi i \frac{e^{-mx}}{2im} = \frac{\pi\, e^{-mx}}{m} \quad \text{for} \quad x > 0. \tag{5.132}$$

Similarly, if $x < 0$, we can add the semicircular ghost contour $k = R\,e^{i\theta}$, $\pi \le \theta \le 2\pi$, $dk = iRe^{i\theta}d\theta$ with k running around the perimeter of the lower half-plane. So if $x < 0$, then we can write the integral $I(x)$ as a shrunken closed contour that runs clockwise around the pole at $k = -im$

$$I(x) = \frac{e^{mx}}{-2im} \oint \frac{dk}{k+im} = -2\pi i \frac{e^{mx}}{-2im} = \frac{\pi e^{mx}}{m} \quad \text{for} \quad x < 0. \tag{5.133}$$

We combine the two cases (5.132) and (5.133) into the result

$$\int_{-\infty}^{\infty} dk \, \frac{e^{ikx}}{k^2 + m^2} = \frac{\pi}{m} \, e^{-m|x|}. \tag{5.134}$$

We can use this formula to develop an expression for the Green's function of the Laplacian in cylindrical coordinates. Setting $\boldsymbol{x}' = 0$ and $r = |\boldsymbol{x}| = \sqrt{\rho^2 + z^2}$ in the Coulomb Green's function (3.110), we have

$$G(r) = \frac{1}{4\pi r} = \frac{1}{4\pi\sqrt{\rho^2 + z^2}} = \int \frac{d^3\boldsymbol{k}}{(2\pi)^3} \frac{1}{\boldsymbol{k}^2} \, e^{i\boldsymbol{k}\cdot\boldsymbol{x}}. \tag{5.135}$$

The integral over the z-component of $\boldsymbol{k}$ is (5.134) with $m^2 = k_x^2 + k_y^2 \equiv k^2$

$$\int_{-\infty}^{\infty} dk_z \, \frac{e^{ik_z z}}{k_z^2 + k^2} = \frac{\pi}{k} \, e^{-k|z|}. \tag{5.136}$$

So with $k_x x + k_y y \equiv k\rho\cos\phi$, the Green's function is

$$\frac{1}{4\pi\sqrt{\rho^2 + z^2}} = \int_0^{\infty} \frac{\pi \, dk}{(2\pi)^3} \int_0^{2\pi} d\phi \, e^{ik\rho\cos\phi} \, e^{-k|z|}. \tag{5.137}$$

The ϕ integral is a representation (5.46, 9.7) of the Bessel function $J_0(k\rho)$

$$J_0(k\rho) = \int_0^{2\pi} \frac{d\phi}{2\pi} \, e^{ik\rho\cos\phi}. \tag{5.138}$$

Thus we arrive at Bessel's formula for the Coulomb Green's function

$$\frac{1}{4\pi\sqrt{\rho^2 + z^2}} = \int_0^{\infty} \frac{dk}{4\pi} \, J_0(k\rho) \, e^{-k|z|} \tag{5.139}$$

in cylindrical coordinates (Schwinger *et al.*, 1998, p. 166). □

Example 5.21 (Yukawa and Green) We saw in example 3.11 that the Green's function for Yukawa's differential operator (3.123) is

$$G_Y(\boldsymbol{x}) = \int \frac{d^3k}{(2\pi)^3} \frac{e^{i\boldsymbol{k}\cdot\boldsymbol{x}}}{\boldsymbol{k}^2 + m^2}. \tag{5.140}$$

Letting $\boldsymbol{k} \cdot \boldsymbol{x} = kr\cos\theta$ in which $r = |\boldsymbol{x}|$, we find

$$\begin{aligned} G_Y(r) &= \int_0^{\infty} \frac{k^2 dk}{(2\pi)^2} \int_{-1}^{1} \frac{e^{ikr\cos\theta}}{k^2 + m^2} \, d\cos\theta = \frac{1}{ir} \int_0^{\infty} \frac{dk}{(2\pi)^2} \frac{k}{k^2 + m^2} \left(e^{ikr} - e^{-ikr} \right) \\ &= \frac{1}{ir} \int_{-\infty}^{\infty} \frac{dk}{(2\pi)^2} \frac{k}{k^2 + m^2} \, e^{ikr} = \frac{1}{ir} \int_{-\infty}^{\infty} \frac{dk}{(2\pi)^2} \frac{k}{(k - im)(k + im)} \, e^{ikr}. \end{aligned}$$

We add a ghost contour that loops over the upper half-plane and get

$$G_Y(r) = \frac{2\pi i}{(2\pi)^2 ir}\frac{im}{2im}e^{-mr} = \frac{e^{-mr}}{4\pi r}, \tag{5.141}$$

which Yukawa proposed as the potential between two hadrons due to the exchange of a particle of mass m, the pion. Because the mass of the pion is 140 MeV, the range of the Yukawa potential is $\hbar/mc = 1.4 \times 10^{-15}$ m. □

Example 5.22 (The Green's function for the Laplacian in n dimensions) The Green's function for the Laplacian $-\triangle G(x) = \delta^{(n)}(x)$ is

$$G(x) = \int \frac{1}{k^2}\, e^{ik\cdot x}\, \frac{d^n k}{(2\pi)^n} \tag{5.142}$$

in n dimensions. We use the formula

$$\frac{1}{k^2} = \int_0^\infty e^{-\lambda k^2}\, d\lambda \tag{5.143}$$

to write it as a gaussian integral

$$G(x) = \int e^{-\lambda k^2 + ik\cdot x}\, d\lambda \frac{d^n k}{(2\pi)^n}. \tag{5.144}$$

We now complete the square in the exponent

$$-\lambda k^2 + ik\cdot x = -\lambda\,(k - ix/2\lambda)^2 - x^2/4\lambda \tag{5.145}$$

and use our gaussian formula (5.121) to write the Green's function as

$$\begin{aligned} G(x) &= \int_0^\infty d\lambda \int \frac{d^n k}{(2\pi)^n}\, e^{-x^2/4\lambda} e^{-\lambda(k-ix/2\lambda)^2} = \int_0^\infty d\lambda \int \frac{d^n k}{(2\pi)^n}\, e^{-x^2/4\lambda} e^{-\lambda k^2} \\ &= \int_0^\infty e^{-x^2/4\lambda} \frac{d\lambda}{(4\pi\lambda)^{n/2}} = \frac{(x^2)^{1-n/2}}{4\pi^{n/2}} \int_0^\infty e^{-\alpha} \alpha^{n/2-2} d\alpha \\ &= \frac{\Gamma(n/2-1)}{4\pi^{n/2}(x^2)^{(n/2-1)}}. \end{aligned} \tag{5.146}$$

Since $\Gamma(1/2) = \sqrt{\pi}$, this formula for $n = 3$ gives $G(x) = 1/4\pi|x|$, which is (3.110); since $\Gamma(1) = 1$, it gives

$$G(x) = \frac{1}{4\pi^2 x^2} \tag{5.147}$$

for $n = 4$. □

Example 5.23 (The Yukawa Green's function in n dimensions) The Yukawa Green's function which satisfies $(-\triangle + m^2)G(x) = \delta^{(n)}(x)$ in n dimensions is the integral (5.142) with k^2 replaced by $k^2 + m^2$

$$G(x) = \int \frac{1}{k^2 + m^2}\, e^{ik\cdot x}\, \frac{d^n k}{(2\pi)^n}. \tag{5.148}$$

Using the integral formula (5.143), we write it as a gaussian integral

$$G(x) = \int e^{-\lambda(k^2+m^2)+ik\cdot x} \frac{d\lambda d^n k}{(2\pi)^n}. \tag{5.149}$$

Completing the square as in (5.145), we have

$$\begin{aligned} G(x) &= \int e^{-x^2/4\lambda} e^{-\lambda(k-ix/2\lambda)^2-\lambda m^2} \frac{d\lambda d^n k}{(2\pi)^n} = \int e^{-x^2/4\lambda} e^{-\lambda(k^2+m^2)} \frac{d\lambda d^n k}{(2\pi)^n} \\ &= \int_0^\infty e^{-x^2/4\lambda-\lambda m^2} \frac{d\lambda}{(4\pi\lambda)^{n/2}}. \end{aligned} \tag{5.150}$$

We can relate this to a Bessel function by setting $\lambda = |x|/2m\exp(-\alpha)$

$$\begin{aligned} G(x) &= \frac{1}{(4\pi)^{n/2}} \left(\frac{2m}{x}\right)^{(n/2-1)} \int_{-\infty}^{\infty} e^{-mx\cosh\alpha+(n/2-1)\alpha} \, d\alpha \\ &= \frac{2}{(4\pi)^{n/2}} \left(\frac{2m}{x}\right)^{(n/2-1)} \int_0^{\infty} e^{-mx\cosh\alpha} \cosh(n/2-1)\alpha \, d\alpha \\ &= \frac{2}{(4\pi)^{n/2}} \left(\frac{2m}{x}\right)^{(n/2-1)} K_{n/2-1}(mx) \end{aligned} \tag{5.151}$$

where $x = |x| = \sqrt{x^2}$ and K is a modified Bessel function of the second kind (9.98). If $n = 3$, this is (exercise 5.27) the Yukawa potential (5.141). □

Example 5.24 (A Fourier transform) As another example, let's consider the integral

$$J(x) = \int_{-\infty}^{\infty} \frac{e^{ikx}}{(k^2+m^2)^2} \, dk. \tag{5.152}$$

We may add ghost contours as in the preceding example, but now the integrand has double poles at $k = \pm im$, and so we must use Cauchy's integral formula (5.36) for the case of $n = 1$, which is (5.34). For $x > 0$, we add a ghost contour in the UHP and find

$$\begin{aligned} J(x) &= \oint \frac{e^{ikx}}{(k+im)^2(k-im)^2} \, dk = 2\pi i \frac{d}{dk} \frac{e^{ikx}}{(k+im)^2}\bigg|_{k=im} \\ &= \frac{\pi}{2m^2}\left(x+\frac{1}{m}\right) e^{-mx}. \end{aligned} \tag{5.153}$$

If $x < 0$, then we add a ghost contour in the LHP and find

$$\begin{aligned} J(x) &= \oint \frac{e^{ikx}}{(k+im)^2(k-im)^2} \, dk = -2\pi i \frac{d}{dk} \frac{e^{ikx}}{(k-im)^2}\bigg|_{k=-im} \\ &= \frac{\pi}{2m^2}\left(-x+\frac{1}{m}\right) e^{mx}. \end{aligned} \tag{5.154}$$

Putting the two together, we get

$$J(x) = \int_{-\infty}^{\infty} \frac{e^{ikx}}{(k^2+m^2)^2} dk = \frac{\pi}{2m^2}\left(|x| + \frac{1}{m}\right) e^{-m|x|} \tag{5.155}$$

as the Fourier transform of $1/(k^2+m^2)^2$. □

Example 5.25 (Integral of a complex gaussian) As another example of the use of ghost contours, let us use one to do the integral

$$I = \int_{-\infty}^{\infty} e^{wx^2}\, dx, \tag{5.156}$$

in which the real part of the nonzero complex number $w = u + iv = \rho e^{i\phi}$ is negative or zero

$$u \le 0 \iff \frac{\pi}{2} \le \phi \le \frac{3\pi}{2}. \tag{5.157}$$

We first write the integral I as twice that along half the x-axis

$$I = 2\int_0^{\infty} e^{wx^2}\, dx. \tag{5.158}$$

If we promote x to a complex variable $z = re^{i\theta}$, then wz^2 will be negative if $\phi + 2\theta = \pi$, that is, if $\theta = (\pi - \phi)/2$ where in view of (5.157) θ lies in the interval $-\pi/4 \le \theta \le \pi/4$.

The closed pie-shaped contour of Fig. 5.7 (down the real axis from $z = 0$ to $z = R$, along the arc $z = R\exp(i\theta')$ as θ' goes from 0 to θ, and then down the line $z = r\exp(i\theta)$ from $z = R\exp(i\theta)$ to $z = 0$) encloses no singularities of the function $f(z) = \exp(wz^2)$. Hence the integral of $\exp(wz^2)$ along that contour vanishes.

To show that the arc is a ghost contour, we bound it by

$$\begin{aligned}\left|\int_0^{\theta} e^{(u+iv)R^2e^{2i\theta'}} R\, d\theta'\right| &\le \int_0^{\theta} \exp\left[uR^2\cos 2\theta' - vR^2 \sin 2\theta'\right] R\, d\theta' \\ &\le \int_0^{\theta} e^{-vR^2 \sin 2\theta'} R\, d\theta'. \end{aligned}\tag{5.159}$$

Here $v \sin 2\theta' \ge 0$, and so if v is positive, then so is θ'. Then $0 \le \theta' \le \pi/4$, and so $\sin(2\theta') \ge 4\theta'/\pi$. Thus since $u < 0$, we have the upper bound

$$\left|\int_0^{\theta} e^{(u+iv)R^2e^{2i\theta'}} R\, d\theta'\right| \le \int_0^{\theta} e^{-4vR^2\theta'/\pi} R\, d\theta' = \frac{\pi(e^{-4vR^2\theta'/\pi} - 1)}{4vR}, \tag{5.160}$$

which vanishes in the limit $R \to \infty$. (If v is negative, then so is θ', the pie-shaped contour is in the fourth quadrant, $\sin(2\theta') \le 4\theta'/\pi$, and the inequality (5.160) holds with absolute-value signs around the second integral.)

Since by Cauchy's integral theorem (5.22) the integral along the pie-shaped contour of Fig. 5.7 vanishes, it follows that

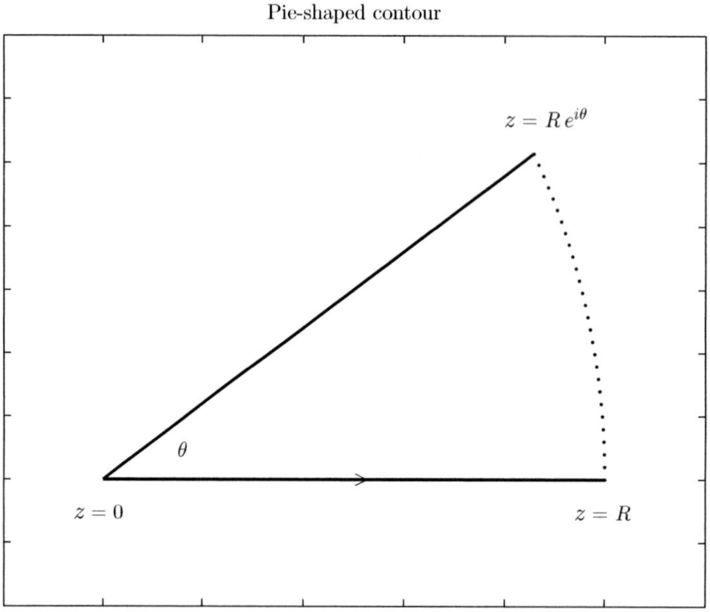

Figure 5.7 The integral of the entire function $\exp(wz^2)$ along the pie-shaped closed contour vanishes by Cauchy's theorem.

$$\frac{1}{2}I + \int_{Re^{i\theta}}^{0} e^{wz^2} dz = 0. \tag{5.161}$$

But the choice $\theta = (\pi - \phi)/2$ implies that on the line $z = r\exp(i\theta)$ the quantity wz^2 is negative, $wz^2 = -\rho r^2$. Thus with $dz = \exp(i\theta)dr$, we have

$$I = 2\int_{0}^{Re^{i\theta}} e^{wz^2} dz = 2e^{i\theta}\int_{0}^{R} e^{-\rho r^2} dr \tag{5.162}$$

so that as $R \to \infty$

$$I = 2e^{i\theta}\int_{0}^{\infty} e^{-\rho r^2} dr = e^{i\theta}\sqrt{\frac{\pi}{\rho}} = \sqrt{\frac{\pi}{\rho e^{-2i\theta}}}. \tag{5.163}$$

Finally, from $\theta = (\pi - \phi)/2$ and $w = \rho\exp(i\phi)$, we find that for $\mathrm{Re}\, w \le 0$

$$\int_{-\infty}^{\infty} e^{wx^2} dx = \sqrt{\frac{\pi}{-w}} \tag{5.164}$$

as long as $w \neq 0$. Shifting x by a complex number b, we still have

$$\int_{-\infty}^{\infty} e^{w(x-b)^2} dx = \sqrt{\frac{\pi}{-w}} \tag{5.165}$$

as long as Re $w < 0$. If $w = ia \neq 0$ and b is real, then

$$\int_{-\infty}^{\infty} e^{ia(x-b)^2}\, dx = \sqrt{\frac{i\pi}{a}}. \tag{5.166}$$

The simpler integral (5.122) applies when $m > 0$ and z is an arbitrary complex number

$$\int_{-\infty}^{\infty} e^{-m^2(x+z)^2}\, dx = \frac{\sqrt{\pi}}{m}. \tag{5.167}$$

These last two formulas are used in chapter 16 on path integrals. □

Let us try to express the line integral of a not necessarily analytic function $f(x,y) = u(x,y) + iv(x,y)$ along a closed ccw contour C as an integral over the surface enclosed by the contour. The contour integral is

$$\oint_C (u+iv)(dx+idy) = \oint_C (u\, dx - v\, dy) + i \oint_C (v\, dx + u\, dy). \tag{5.168}$$

Now since the contour C is counterclockwise, the differential dx is negative at the top of the curve with coordinates $(x, y_+(x))$ and positive at the bottom $(x, y_-(x))$. So the first line integral is the surface integral

$$\begin{aligned}\oint_C u\, dx &= \int [u(x, y_-(x)) - u(x, y_+(x))]\, dx \\ &= - \int \left[\int_{y_-(x)}^{y_+(x)} u_y(x,y) dy \right] dx \\ &= - \int u_y\, |dxdy| = - \int u_y\, da,\end{aligned} \tag{5.169}$$

in which $da = |dxdy|$ is a positive element of area. Similarly, we find

$$i \oint_C v\, dx = -i \int v_y\, |dxdy| = -i \int v_y\, da. \tag{5.170}$$

The dy integrals are then

$$- \oint_C v\, dy = - \int v_x\, |dxdy| = - \int v_x\, da, \tag{5.171}$$

$$i \oint_C u\, dy = i \int u_x\, |dxdy| = i \int u_x\, da. \tag{5.172}$$

Combining (5.168–5.172), we find

$$\oint_C (u+iv)(dx+idy) = - \int (u_y + v_x)\, da + i \int (-v_y + u_x)\, da. \tag{5.173}$$

This formula holds whether or not the function $f(x,y)$ is analytic. But if $f(x,y)$ is analytic on and within the contour C, then it satisfies the Cauchy–Riemann conditions (5.48) within the contour, and so both surface integrals vanish. The contour integral then is zero, which is Cauchy's integral theorem (5.21).

The contour integral of the function $f(x,y) = u(x,y) + iv(x,y)$ differs from zero (its value if $f(x,y)$ is analytic in $z = x + iy$) by the surface integrals of $u_y + v_x$ and $u_x - v_y$

$$\left|\oint_C f(z)dz\right|^2 = \left|\oint_C (u+iv)(dx+idy)\right|^2 = \left|\int (u_y + v_x)da\right|^2 + \left|\int (u_x - v_y)da\right|^2, \tag{5.174}$$

which vanish when $f = u + iv$ satisfies the Cauchy–Riemann conditions (5.48).

Example 5.26 (The integral of a nonanalytic function) The integral formula (5.173) can help us evaluate contour integrals of functions that are not analytic. The function

$$f(x,y) = \frac{1}{x + iy + i\epsilon}\,\frac{1}{1 + x^2 + y^2} \tag{5.175}$$

is the product of an analytic function $1/(z + i\epsilon)$, where ϵ is tiny and positive, and a nonanalytic real one $r(x,y) = 1/(1 + z^* z)$. The $i\epsilon$ pushes the pole in $u + iv = 1/(z + i\epsilon)$ into the lower half-plane. The real and imaginary parts of f are

$$U(x,y) = u(x,y)\,r(x,y) = \frac{x}{x^2 + (y+\epsilon)^2}\,\frac{1}{1 + x^2 + y^2} \tag{5.176}$$

and

$$V(x,y) = v(x,y)\,r(x,y) = \frac{-y-\epsilon}{x^2 + (y+\epsilon)^2}\,\frac{1}{1 + x^2 + y^2}. \tag{5.177}$$

We will use (5.173) to compute the contour integral I of f along the real axis from $-\infty$ to ∞ and then along the ghost contour $z = x + iy = Re^{i\theta}$ for $0 \le \theta \le \pi$ and $R \to \infty$ around the upper half-plane

$$I = \oint f(x,y)\,dz = \int_{-\infty}^{\infty} dx \int_0^{\infty} dy \left[-U_y - V_x + i\left(-V_y + U_x\right)\right]. \tag{5.178}$$

Since u and v satisfy the Cauchy–Riemann conditions (5.48), the terms in the area integral simplify to $-U_y - V_x = -ur_y - vr_x$ and $-V_y + U_x = -vr_y + ur_x$. So the integral I is

$$I = \int_{-\infty}^{\infty} dx \int_0^{\infty} dy \left[-ur_y - vr_x + i(-vr_y + ur_x)\right] \tag{5.179}$$

or explicitly

$$I = \int_{-\infty}^{\infty} dx \int_0^{\infty} dy \frac{-2\epsilon x - 2i(x^2 + y^2 + \epsilon y)}{\left[x^2 + (y+\epsilon)^2\right]\left(1 + x^2 + y^2\right)^2}. \tag{5.180}$$

We let $\epsilon \to 0$ and find

$$I = -2i \int_{-\infty}^{\infty} dx \int_{0}^{\infty} dy \frac{1}{\left(1 + x^2 + y^2\right)^2}. \tag{5.181}$$

Changing variables to $\rho^2 = x^2 + y^2$, we have

$$I = -4\pi i \int_{0}^{\infty} d\rho \, \frac{\rho}{(1+\rho^2)^2} = 2\pi i \int_{0}^{\infty} d\rho \, \frac{d}{d\rho} \frac{1}{1+\rho^2} = -2\pi i, \tag{5.182}$$

which is simpler than evaluating the integral (5.178) directly. □

5.15 Logarithms and cuts

By definition, a function f is single valued; it maps every number z in its domain into a unique image $f(z)$. A function that maps only one number z in its domain into each $f(z)$ in its range is said to be **one to one**. A one-to-one function $f(z)$ has a well-defined inverse function $f^{-1}(z)$.

The exponential function is one to one when restricted to the real numbers. It maps every real number x into a positive number $\exp(x)$. It has an inverse function $\ln(x)$ that maps every positive number $\exp(x)$ back into x. But the exponential function is not one to one on the complex numbers because $\exp(z + 2\pi ni) = \exp(z)$ for every integer n. The exponential function is **many to one**. Thus on the complex numbers, the exponential function has no inverse function. Its would-be inverse function $\ln(\exp(z))$ is $z + 2\pi ni$, which is not unique. It has in it an arbitrary integer n.

In other words, when exponentiated, the logarithm of a complex number z returns $\exp(\ln z) = z$. So if $z = r\exp(i\theta)$, then a suitable logarithm is $\ln z = \ln r + i\theta$. But what is θ? In the polar representation of z, the argument θ can just as well be $\theta + 2\pi n$ because both give $z = r\exp(i\theta) = r\exp(i\theta + i2\pi n)$. So $\ln r + i\theta + i2\pi n$ is a correct value for $\ln[r\exp(i\theta)]$ for every integer n.

People usually want *one* of the correct values of a logarithm, rather than all of them. Two conventions are common. In the first convention, the angle θ is zero along the positive real axis and increases continuously as the point z moves counterclockwise around the origin, until at points just below the positive real axis, $\theta = 2\pi - \epsilon$ is slightly less than 2π. In this convention, the value of θ drops by 2π as one crosses the positive real axis moving counterclockwise. This discontinuity on the positive real axis is called a **cut**.

The second common convention puts the cut on the negative real axis. Here the value of θ is the same as in the first convention when the point z is in the upper half-plane. But in the lower half-plane, θ decreases from 0 to $-\pi$ as the point z moves clockwise from the positive real axis to just below the negative real axis, where $\theta = -\pi + \epsilon$. As one crosses the negative real axis moving clockwise or up, θ jumps by 2π while crossing the cut. The two conventions agree in the upper half-plane but differ by 2π in the lower half-plane.

Sometimes it is convenient to place the cut on the positive or negative imaginary axis – or along a line that makes an arbitrary angle with the real axis. In any particular calculation, we are at liberty to define the polar angle θ by placing the cut anywhere we like, but we must not change from one convention to another in the same computation.

5.16 Powers and roots

The logarithm is the key to many other functions to which it passes its arbitrariness. For instance, any power a of $z = r\exp(i\theta)$ is defined as

$$z^a = \exp(a \ln z) = \exp[a(\ln r + i\theta + i2\pi n)] = r^a\, e^{ia\theta}\, e^{i2\pi na}. \tag{5.183}$$

So z^a is not unique unless a is an integer. The square-root, for example,

$$\sqrt{z} = \exp\left[\tfrac{1}{2}(\ln r + i\theta + i2\pi n)\right] = \sqrt{r}\, e^{i\theta/2}\, e^{in\pi} = (-1)^n \sqrt{r}\, e^{i\theta/2} \tag{5.184}$$

changes sign when we change θ by 2π as we cross a cut. The mth root

$$\sqrt[m]{z} = z^{1/m} = \exp\left(\frac{\ln z}{m}\right) \tag{5.185}$$

changes by $\exp(\pm 2\pi i/m)$ when we cross a cut and change θ by 2π. And when $a = u + iv$ is a complex number, z^a is

$$z^a = e^{a\ln z} = e^{(u+iv)(\ln r + i\theta + i2\pi n)} = r^{u+iv}\, e^{(-v+iu)(\theta+2\pi n)}, \tag{5.186}$$

which changes by $\exp[2\pi(-v + iu)]$ as we cross a cut.

Example 5.27 (i^i) The number $i = \exp(i\pi/2 + i2\pi n)$ for any integer n. So the general value of i^i is $i^i = \exp[i(i\pi/2 + i2\pi n)] = \exp(-\pi/2 - 2\pi n)$. □

One can define a sequence of mth-root functions

$$\left(z^{1/m}\right)_n = \exp\left(\frac{\ln r + i(\theta + 2\pi n)}{m}\right), \tag{5.187}$$

one for each integer n. These functions are the **branches** of the mth-root function. One can merge all the branches into one **multivalued** mth-root function. Using a convention for θ, one would extend the $n = 0$ branch to the $n = 1$ branch by winding counterclockwise around the point $z = 0$. One would encounter no discontinuity as one passed from one branch to another.

The point $z = 0$, where any cut starts, is called a **branch point** because, by winding around it, one passes smoothly from one branch to another. Such branches, introduced by Riemann, can be associated with any multivalued analytic function, not just with the mth root.

Example 5.28 (Explicit square-roots) If the cut in the square-root $\sqrt{z}$ is on the negative real axis, then an explicit formula for the square-root of $x + iy$ is

$$\sqrt{x+iy} = \sqrt{\frac{\sqrt{x^2+y^2}+x}{2}} + i\,\mathrm{sign}(y)\sqrt{\frac{\sqrt{x^2+y^2}-x}{2}}, \tag{5.188}$$

in which $\mathrm{sign}(y) = \mathrm{sgn}(y) = y/|y|$. On the other hand, if the cut in the square-root $\sqrt{z}$ is on the positive real axis, then an explicit formula for the square-root of $x + iy$ is

$$\sqrt{x+iy} = \mathrm{sign}(y)\sqrt{\frac{\sqrt{x^2+y^2}+x}{2}} + i\sqrt{\frac{\sqrt{x^2+y^2}-x}{2}} \tag{5.189}$$

(exercise 5.28). □

Example 5.29 (Cuts) Cuts are discontinuities, so people place them where they do the least harm. For the function

$$f(z) = \sqrt{z^2-1} = \sqrt{(z-1)(z+1)} \tag{5.190}$$

two principal conventions work well. We could put the cut in the definition of the angle θ along either the positive or the negative real axis. And we'd get a bonus: the sign discontinuity (a factor of -1) from $\sqrt{z-1}$ would cancel the one from $\sqrt{z+1}$ except for $-1 \le z \le 1$. So the function $f(z)$ would have a discontinuity or a cut only for $-1 \le z \le 1$.

But now suppose we had to work with the function

$$f(z) = \sqrt{z^2+1} = \sqrt{(z-i)(z+i)}. \tag{5.191}$$

If we used one of the usual conventions, we'd have two semi-infinite cuts. So we put the θ-cut on the positive or negative imaginary axis, and the function $f(z)$ now has a cut running along the imaginary axis only from $-i$ to i. □

Example 5.30 (Integral with a square-root) Consider the integral

$$I = \int_{-1}^{1} \frac{1}{(x-k)\sqrt{1-x^2}}\,dx, \tag{5.192}$$

in which the constant k lies anywhere in the complex plane but not on the interval $[-1,1]$. Let's promote x to a complex variable z and write the square-root as $\sqrt{1-x^2} = -i\sqrt{x^2-1} = -i\sqrt{(z-1)(z+1)}$. As in the last example (5.29), if in both of the square-roots we put the cut on the negative (or the positive) real axis, then the function $f(z) = 1/[(z-k)(-i)\sqrt{(z-1)(z+1)}]$ will be analytic everywhere except along a cut on the interval $[-1, 1]$ and at $z = k$. The circle $z = Re^{i\theta}$ for $0 \le \theta \le 2\pi$ is a ghost contour as $R \to \infty$. If we shrink-wrap this ccw contour around the pole at $z = k$ and the interval $[-1, 1]$, then we get $0 = -2I + 2\pi i/[(-i)\sqrt{k-1}\sqrt{k+1}]$ or

$$I = -\frac{\pi}{\sqrt{k-1}\sqrt{k+1}}. \tag{5.193}$$

So if $k = -2$, then $I = \pi/\sqrt{3}$, while if $k = 2$, then $I = -\pi/\sqrt{3}$. □

Example 5.31 (Contour integral with a cut) Let's compute the integral

$$I = \int_0^\infty \frac{x^a}{(x+1)^2}\, dx \tag{5.194}$$

for $-1 < a < 1$. We promote x to a complex variable z and put the cut on the positive real axis. Since

$$\lim_{|z|\to\infty} \frac{|z|^{a+1}}{|z+1|^2} = 0, \tag{5.195}$$

the integrand vanishes faster than $1/|z|$, and we may add two ghost contours, $\mathcal{G}_+$ counterclockwise around the upper half-plane and $\mathcal{G}_-$ counterclockwise around the lower half-plane, as shown in Fig. 5.8.

We add a contour C_- that runs from $-\infty$ to the double pole at $z = -1$, loops around that pole, and then runs back to $-\infty$; the two long contours along the negative real axis cancel because the cut in θ lies on the positive real axis. So the contour integral along C_- is just the clockwise integral around the double pole, which by Cauchy's integral formula (5.34) is

$$\oint_{C_-} \frac{z^a}{(z-(-1))^2}\, dz = -2\pi i \left.\frac{dz^a}{dz}\right|_{z=-1} = 2\pi i\, a\, e^{\pi a i}. \tag{5.196}$$

We also add the integral I_- from ∞ to 0 just below the real axis

$$I_- = \int_\infty^0 \frac{(x-i\epsilon)^a}{(x-i\epsilon+1)^2}\, dx = \int_\infty^0 \frac{\exp(a(\ln(x)+2\pi i))}{(x+1)^2}\, dx, \tag{5.197}$$

which is

$$I_- = -e^{2\pi a i} \int_0^\infty \frac{x^a}{(x+1)^2}\, dx = -e^{2\pi a i}\, I. \tag{5.198}$$

Now the sum of all these contour integrals is zero because it is a closed contour that encloses no singularity. So we have

$$0 = \left(1 - e^{2\pi a i}\right) I + 2\pi i\, a\, e^{\pi a i} \tag{5.199}$$

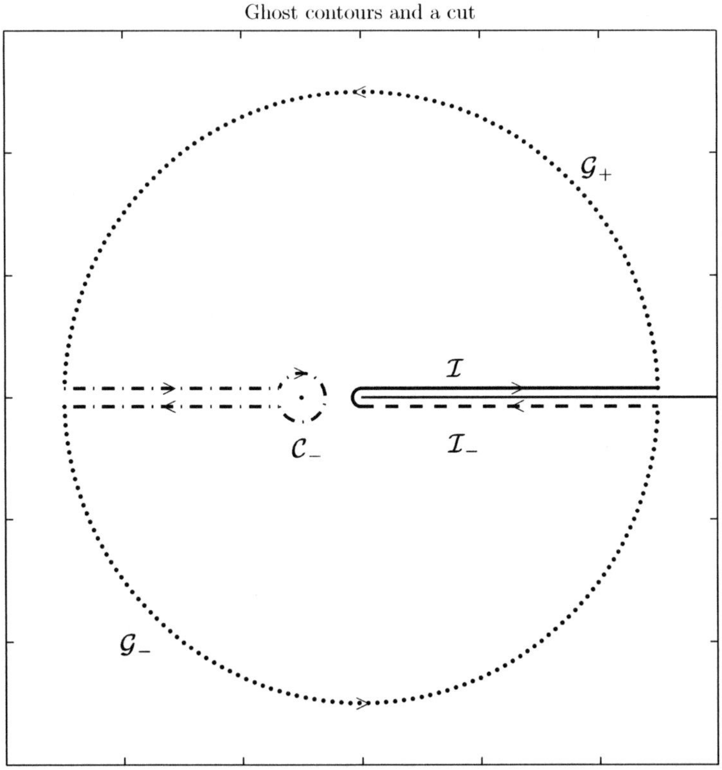

Figure 5.8 The integral of $f(z) = z^a/(z+1)$ along the ghost contours $\mathcal{G}_+$ and $\mathcal{G}_-$, the contour $\mathcal{C}_-$, the contour $\mathcal{I}_-$, and the contour $\mathcal{I}$ vanishes because the combined contour encircles no poles of $f(z)$. The cut (solid line) runs from the origin to infinity along the positive real axis.

or

$$I = \int_0^\infty \frac{x^a}{(x+1)^2}\, dx = \frac{\pi a}{\sin(\pi a)} \tag{5.200}$$

as the value of the integral (5.194). □

5.17 Conformal mapping

An analytic function $f(z)$ maps curves in the z plane into curves in the $f(z)$ plane. In general, this mapping preserves angles. To see why, we consider the angle $d\theta$ between two tiny complex lines $dz = \epsilon \exp(i\theta)$ and $dz' = \epsilon \exp(i\theta')$ that radiate from the same point z. This angle $d\theta = \theta' - \theta$ is the phase of the ratio

$$\frac{dz'}{dz} = \frac{\epsilon e^{i\theta'}}{\epsilon e^{i\theta}} = e^{i(\theta'-\theta)}. \tag{5.201}$$

Let's use $w = \rho e^{i\phi}$ for $f(z)$. Then the analytic function $f(z)$ maps dz into

$$dw = f(z + dz) - f(z) \approx f'(z)\, dz \tag{5.202}$$

and dz' into

$$dw' = f(z + dz') - f(z) \approx f'(z)\, dz'. \tag{5.203}$$

The angle $d\phi = \phi' - \phi$ between dw and dw' is the phase of the ratio

$$\frac{dw'}{dw} = \frac{e^{i\phi'}}{e^{i\phi}} = \frac{f'(z)\, dz'}{f'(z)\, dz} = \frac{dz'}{dz} = \frac{e^{i\theta'}}{e^{i\theta}} = e^{i(\theta'-\theta)}. \tag{5.204}$$

So as long as the derivative $f'(z)$ does not vanish, the angle in the w-plane is the same as the angle in the z-plane

$$d\phi = d\theta. \tag{5.205}$$

Analytic functions preserve angles. They are **conformal** maps.

What if $f'(z) = 0$? In this case, $dw \approx f''(z)\, dz^2/2$ and $dw' \approx f''(z)\, dz'^2/2$, and so the angle $d\phi = d\phi' - d\phi$ between these two tiny complex lines is the phase of the ratio

$$\frac{dw'}{dw} = \frac{e^{i\phi'}}{e^{i\phi}} = \frac{f''(z)\, dz'^2}{f''(z)\, dz^2} = \frac{dz'^2}{dz^2} = e^{2i(\theta'-\theta)}. \tag{5.206}$$

So angles are doubled, $d\phi = 2d\theta$.

In general, if the first nonzero derivative is $f^{(n)}(z)$, then

$$\frac{dw'}{dw} = \frac{e^{i\phi'}}{e^{i\phi}} = \frac{f^{(n)}(z)\, dz'^n}{f^{(n)}(z)\, dz^n} = \frac{dz'^n}{dz^n} = e^{ni(\theta'-\theta)} \tag{5.207}$$

and so $d\phi = nd\theta$. The angles increase by a factor of n.

Example 5.32 (z^n) The function $f(z) = cz^n$ has only one nonzero derivative at the origin $z = 0$

$$f^{(k)}(0) = c\, n!\, \delta_{nk} \tag{5.208}$$

so at $z = 0$ the conformal map $z \to cz^n$ scales angles by n, $d\phi = n\, d\theta$. □

For examples of conformal mappings see (Lin, 2011, section 3.5.7).

5.18 Cauchy's principal value

Suppose that $f(x)$ is differentiable or analytic at and near the point $x = 0$, and that we wish to evaluate the integral

$$K = \lim_{\epsilon \to 0} \int_{-a}^{b} dx\, \frac{f(x)}{x - i\epsilon} \tag{5.209}$$

for $a > 0$ and $b > 0$. First we regularize the pole at $x = 0$ by using a method devised by Cauchy

$$K = \lim_{\delta\to 0}\left[\lim_{\epsilon\to 0}\left(\int_{-a}^{-\delta} dx\,\frac{f(x)}{x-i\epsilon} + \int_{-\delta}^{\delta} dx\,\frac{f(x)}{x-i\epsilon} + \int_{\delta}^{b} dx\,\frac{f(x)}{x-i\epsilon}\right)\right]. \tag{5.210}$$

In the first and third integrals, since $|x| \ge \delta$, we may set $\epsilon = 0$

$$K = \lim_{\delta\to 0}\left(\int_{-a}^{-\delta} dx\frac{f(x)}{x} + \int_{\delta}^{b} dx\frac{f(x)}{x}\right) + \lim_{\delta\to 0}\lim_{\epsilon\to 0}\int_{-\delta}^{\delta} dx\,\frac{f(x)}{x-i\epsilon}. \tag{5.211}$$

We'll discuss the first two integrals before analyzing the last one.

The limit of the first two integrals is called **Cauchy's principal value**

$$P\int_{-a}^{b} dx\frac{f(x)}{x} \equiv \lim_{\delta\to 0}\left(\int_{-a}^{-\delta} dx\frac{f(x)}{x} + \int_{\delta}^{b} dx\frac{f(x)}{x}\right). \tag{5.212}$$

If the function $f(x)$ is nearly constant near $x = 0$, then the large negative values of $1/x$ for x slightly less than zero cancel the large positive values of $1/x$ for x slightly greater than zero. The point $x = 0$ is not special; Cauchy's principal value about $x = y$ is defined by the limit

$$P\int_{-a}^{b} dx\,\frac{f(x)}{x-y} \equiv \lim_{\delta\to 0}\left(\int_{-a}^{y-\delta} dx\,\frac{f(x)}{x-y} + \int_{y+\delta}^{b} dx\,\frac{f(x)}{x-y}\right). \tag{5.213}$$

Using Cauchy's principal value, we may write the quantity K as

$$K = P\int_{-a}^{b} dx\frac{f(x)}{x} + \lim_{\delta\to 0}\lim_{\epsilon\to 0}\int_{-\delta}^{\delta} dx\,\frac{f(x)}{x-i\epsilon}. \tag{5.214}$$

To evaluate the second integral, we use differentiability of $f(x)$ near $x = 0$ to write $f(x) = f(0) + xf'(0)$ and then extract the constants $f(0)$ and $f'(0)$

$$\begin{aligned}
\lim_{\delta\to 0}\lim_{\epsilon\to 0}\int_{-\delta}^{\delta} dx\,\frac{f(x)}{x-i\epsilon} &= \lim_{\delta\to 0}\lim_{\epsilon\to 0}\int_{-\delta}^{\delta} dx\frac{f(0)+xf'(0)}{x-i\epsilon}\\
&= f(0)\lim_{\delta\to 0}\lim_{\epsilon\to 0}\int_{-\delta}^{\delta}\frac{dx}{x-i\epsilon} + f'(0)\lim_{\delta\to 0}\lim_{\epsilon\to 0}\int_{-\delta}^{\delta}\frac{x\,dx}{x-i\epsilon}\\
&= f(0)\lim_{\delta\to 0}\lim_{\epsilon\to 0}\int_{-\delta}^{\delta}\frac{dx}{x-i\epsilon} + f'(0)\lim_{\delta\to 0} 2\delta\\
&= f(0)\lim_{\delta\to 0}\lim_{\epsilon\to 0}\int_{-\delta}^{\delta}\frac{dx}{x-i\epsilon}.
\end{aligned} \tag{5.215}$$

Now since $1/(z - i\epsilon)$ is analytic in the lower half-plane, we may deform the straight contour from $x = -\delta$ to $x = \delta$ into a tiny semicircle that avoids the point $x = 0$ by setting $z = \delta\, e^{i\theta}$ and letting θ run from π to 2π

$$K = P\int_{-a}^{b} dx \frac{f(x)}{x} + f(0) \lim_{\delta\to 0} \lim_{\epsilon\to 0} \int_{-\delta}^{\delta} dz \frac{1}{z - i\epsilon}. \tag{5.216}$$

We now can set $\epsilon = 0$ and so write K as

$$\begin{aligned} K &= P\int_{-a}^{b} dx \frac{f(x)}{x} + f(0) \lim_{\delta\to 0} \int_{\pi}^{2\pi} i\delta e^{i\theta} d\theta \frac{1}{\delta e^{i\theta}} \\ &= P\int_{-a}^{b} dx \frac{f(x)}{x} + i\pi f(0). \end{aligned} \tag{5.217}$$

Recalling the definition (5.209) of K, we have

$$\lim_{\epsilon\to 0} \int_{-a}^{b} dx \frac{f(x)}{x - i\epsilon} = P\int_{-a}^{b} dx \frac{f(x)}{x} + i\pi f(0) \tag{5.218}$$

for any function $f(x)$ that is differentiable at $x = 0$. Physicists write this as

$$\frac{1}{x - i\epsilon} = P\frac{1}{x} + i\pi\delta(x) \quad \text{and} \quad \frac{1}{x + i\epsilon} = P\frac{1}{x} - i\pi\delta(x) \tag{5.219}$$

or as

$$\frac{1}{x - y \pm i\epsilon} = P\frac{1}{x - y} \mp i\pi\delta(x - y). \tag{5.220}$$

Example 5.33 (Cauchy's trick) We use (5.219) to evaluate the integral

$$I = \int_{-\infty}^{\infty} dx \frac{1}{x + i\epsilon} \frac{1}{1 + x^2} \tag{5.221}$$

as

$$I = P\int_{-\infty}^{\infty} dx \frac{1}{x} \frac{1}{1 + x^2} - i\pi \int_{-\infty}^{\infty} dx \frac{\delta(x)}{1 + x^2}. \tag{5.222}$$

Because the function $1/x(1 + x^2)$ is odd, the principal part is zero. The integral over the delta function gives unity, so we have $I = -i\pi$. □

Example 5.34 (Cauchy's principal value) By explicit use of the formula

$$\int \frac{dx}{x^2 - a^2} = -\frac{1}{2a} \ln \frac{x + a}{x - a} \tag{5.223}$$

one may show (exercise 5.30) that

$$P\int_{0}^{\infty} \frac{dx}{x^2 - a^2} = \int_{0}^{a-\delta} \frac{dx}{x^2 - a^2} + \int_{a+\delta}^{\infty} \frac{dx}{x^2 - a^2} = 0, \tag{5.224}$$

a result we'll use in section 5.21. □

Example 5.35 ($\sin k/k$) To compute the integral

$$I = \int_0^\infty \frac{dk}{k} \sin k, \tag{5.225}$$

which we used to derive the formula (3.110) for the Green's function of the Laplacian in three dimensions, we first express I as an integral along the whole real axis

$$I = \int_0^\infty \frac{dk}{2ik} \left(e^{ik} - e^{-ik}\right) = \int_{-\infty}^\infty \frac{dk}{2ik} e^{ik}, \tag{5.226}$$

by which we actually mean the Cauchy principal part

$$I = \lim_{\delta\to 0} \left(\int_{-\infty}^{-\delta} dk \frac{e^{ik}}{2ik} + \int_\delta^\infty dk \frac{e^{ik}}{2ik} \right) = P \int_{-\infty}^\infty dk \frac{e^{ik}}{2ik}. \tag{5.227}$$

Using Cauchy's trick (5.219), we have

$$I = P \int_{-\infty}^\infty dk \frac{e^{ik}}{2ik} = \int_{-\infty}^\infty dk \frac{e^{ik}}{2i(k+i\epsilon)} + \int_{-\infty}^\infty dk \, i\pi \, \delta(k) \frac{e^{ik}}{2i}. \tag{5.228}$$

To the first integral, we add a ghost contour around the upper half-plane. For the contour from $k = L$ to $k = L + iH$ and then to $k = -L + iH$ and then down to $k = -L$, one may show (exercise 5.33) that the integral of $\exp(ik)/k$ vanishes in the double limit $L \to \infty$ and $H \to \infty$. With this ghost contour, the first integral therefore vanishes because the pole at $k = -i\epsilon$ is in the lower half-plane. The delta function in the second integral then gives $\pi/2$, so that

$$I = \oint dk \frac{e^{ik}}{2i(k+i\epsilon)} + \frac{\pi}{2} = \frac{\pi}{2} \tag{5.229}$$

as stated in (3.109). □

Example 5.36 (The Feynman propagator) Adding $\pm i\epsilon$ to the denominator of a pole term of an integral formula for a function $f(x)$ can slightly shift the pole into the upper or lower half-plane, causing the pole to contribute if a ghost contour goes around the upper half-plane or the lower half-plane. Such an $i\epsilon$ can impose a boundary condition on Green's function.

The Feynman propagator $\Delta_F(x)$ is a Green's function for the Klein–Gordon differential operator (Weinberg, 1995, pp. 274–280)

$$(m^2 - \Box)\Delta_F(x) = \delta^4(x) \tag{5.230}$$

in which $x = (x^0, \boldsymbol{x})$ and

$$\Box = \triangle - \frac{\partial^2}{\partial t^2} = \triangle - \frac{\partial^2}{\partial (x^0)^2} \tag{5.231}$$

is the four-dimensional version of the Laplacian $\triangle \equiv \nabla \cdot \nabla$. Here $\delta^4(x)$ is the four-dimensional Dirac delta function (3.36)

$$\delta^4(x) = \int \frac{d^4q}{(2\pi)^4} \exp[i(\boldsymbol{q}\cdot\boldsymbol{x} - q^0x^0)] = \int \frac{d^4q}{(2\pi)^4} e^{iqx}, \tag{5.232}$$

in which $qx = \boldsymbol{q}\cdot\boldsymbol{x} - q^0x^0$ is the Lorentz-invariant inner product of the 4-vectors q and x. There are many Green's functions that satisfy equation (5.230). Feynman's propagator $\Delta_F(x)$ is the one that satisfies boundary conditions that will become evident when we analyze the effect of its $i\epsilon$

$$\Delta_F(x) = \int \frac{d^4q}{(2\pi)^4} \frac{\exp(iqx)}{q^2+m^2-i\epsilon} = \int \frac{d^3q}{(2\pi)^3} \int_{-\infty}^{\infty} \frac{dq^0}{2\pi} \frac{e^{i\boldsymbol{q}\cdot\boldsymbol{x}-iq^0x^0}}{q^2+m^2-i\epsilon}. \tag{5.233}$$

The quantity $E_{\boldsymbol{q}} = \sqrt{\boldsymbol{q}^2+m^2}$ is the energy of a particle of mass m and momentum $\boldsymbol{q}$ in natural units with the speed of light $c = 1$. Using this abbreviation and setting $\epsilon' = \epsilon/2E_q$, we may write the denominator as

$$q^2+m^2-i\epsilon = \boldsymbol{q}\cdot\boldsymbol{q} - \left(q^0\right)^2 + m^2 - i\epsilon = \left(E_q - i\epsilon' - q^0\right)\left(E_q - i\epsilon' + q^0\right) + \epsilon'^2, \tag{5.234}$$

in which ϵ'^2 is negligible. Dropping the prime on ϵ, we do the q^0 integral

$$I(\boldsymbol{q}) = -\int_{-\infty}^{\infty} \frac{dq^0}{2\pi} e^{-iq^0x^0} \frac{1}{\left[q^0-(E_{\boldsymbol{q}}-i\epsilon)\right]\left[q^0-(-E_{\boldsymbol{q}}+i\epsilon)\right]}. \tag{5.235}$$

As shown in Fig. 5.9, the integrand

$$e^{-iq^0x^0} \frac{1}{\left[q^0-(E_{\boldsymbol{q}}-i\epsilon)\right]\left[q^0-(-E_{\boldsymbol{q}}+i\epsilon)\right]} \tag{5.236}$$

has poles at $E_{\boldsymbol{q}} - i\epsilon$ and at $-E_{\boldsymbol{q}} + i\epsilon$. When $x^0 > 0$, we can add a ghost contour that goes clockwise around the lower half-plane and get

$$I(\boldsymbol{q}) = ie^{-iE_{\boldsymbol{q}}x^0} \frac{1}{2E_{\boldsymbol{q}}}, \quad x^0 > 0. \tag{5.237}$$

When $x^0 < 0$, our ghost contour goes counterclockwise around the upper half-plane, and we get

$$I(\boldsymbol{q}) = ie^{iE_{\boldsymbol{q}}x^0} \frac{1}{2E_{\boldsymbol{q}}}, \quad x^0 < 0. \tag{5.238}$$

Using the step function $\theta(x) = (x+|x|)/2$, we combine (5.237) and (5.238)

$$-iI(\boldsymbol{q}) = \frac{1}{2E_{\boldsymbol{q}}}\left[\theta(x^0)\,e^{-iE_{\boldsymbol{q}}x^0} + \theta(-x^0)\,e^{iE_{\boldsymbol{q}}x^0}\right]. \tag{5.239}$$

In terms of the Lorentz-invariant function

$$\Delta_+(x) = \frac{1}{(2\pi)^3} \int \frac{d^3q}{2E_{\boldsymbol{q}}} \exp[i(\boldsymbol{q}\cdot\boldsymbol{x} - E_{\boldsymbol{q}}x^0)] \tag{5.240}$$

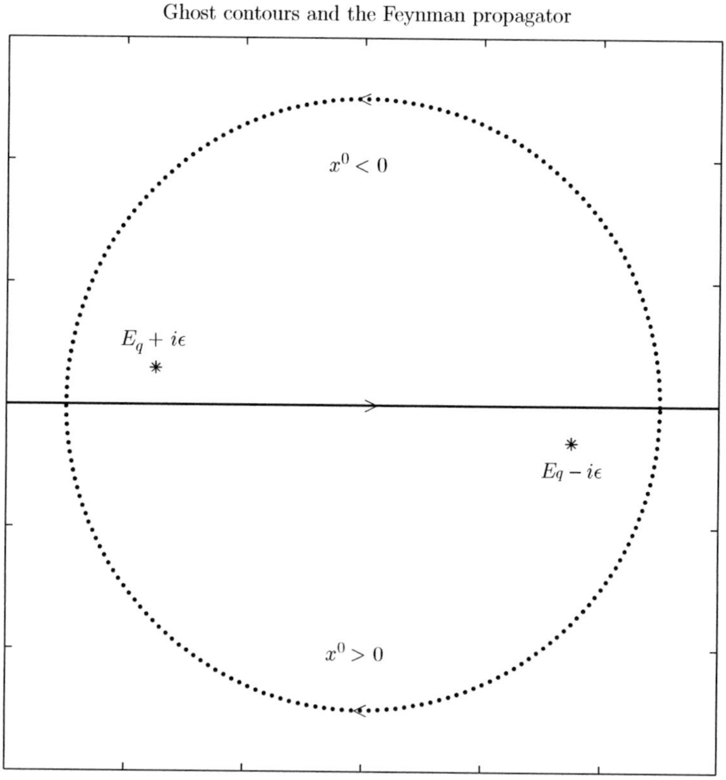

Figure 5.9 In equation (5.236), the function $f(q^0)$ has poles at $\pm(E_{\boldsymbol{q}} - i\epsilon)$, and the function $\exp(-iq^0x^0)$ is exponentially suppressed in the lower half-plane if $x^0 > 0$ and in the upper half-plane if $x^0 < 0$. So we can add a ghost contour (dots) in the LHP if $x^0 > 0$ and in the UHP if $x^0 < 0$.

and with a factor of $-i$, Feynman's propagator (5.233) is

$$-i\Delta_F(x) = \theta(x^0)\,\Delta_+(x) + \theta(-x^0)\,\Delta_+(\boldsymbol{x}, -x^0). \tag{5.241}$$

The integral (5.240) defining $\Delta_+(x)$ is insensitive to the sign of $\boldsymbol{q}$, and so

$$\begin{aligned}\Delta_+(-x) &= \frac{1}{(2\pi)^3}\int \frac{d^3q}{2E_{\boldsymbol{q}}}\exp[i(-\boldsymbol{q}\cdot\boldsymbol{x} + E_{\boldsymbol{q}}x^0)] \\ &= \frac{1}{(2\pi)^3}\int \frac{d^3q}{2E_{\boldsymbol{q}}}\exp[i(\boldsymbol{q}\cdot\boldsymbol{x} + E_{\boldsymbol{q}}x^0)] = \Delta_+(\boldsymbol{x}, -x^0).\end{aligned} \tag{5.242}$$

Thus we arrive at the standard form of the Feynman propagator

$$-i\Delta_F(x) = \theta(x^0)\,\Delta_+(x) + \theta(-x^0)\,\Delta_+(-x). \tag{5.243}$$

The annihilation operators $a(\boldsymbol{q})$ and the creation operators $a^\dagger(\boldsymbol{p})$ of a scalar field $\phi(x)$ satisfy the commutation relations

$$[a(\boldsymbol{q}), a^\dagger(\boldsymbol{p})] = \delta^3(\boldsymbol{q}-\boldsymbol{p}) \quad \text{and} \quad [a(\boldsymbol{q}), a(\boldsymbol{p})] = [a^\dagger(\boldsymbol{q}), a^\dagger(\boldsymbol{p})] = 0. \tag{5.244}$$

Thus the commutator of the positive-frequency part

$$\phi^+(x) = \int \frac{d^3p}{\sqrt{(2\pi)^3 2p^0}} \exp[i(\boldsymbol{p}\cdot\boldsymbol{x} - p^0x^0)]\, a(\boldsymbol{p}) \tag{5.245}$$

of a scalar field $\phi = \phi^+ + \phi^-$ with its negative-frequency part

$$\phi^-(y) = \int \frac{d^3q}{\sqrt{(2\pi)^3 2q^0}} \exp[-i(\boldsymbol{q}\cdot\boldsymbol{y} - q^0y^0)]\, a^\dagger(\boldsymbol{q}) \tag{5.246}$$

is the Lorentz-invariant function $\Delta_+(x-y)$

$$\begin{aligned} [\phi^+(x), \phi^-(y)] &= \int \frac{d^3p\, d^3q}{(2\pi)^3 2\sqrt{q^0p^0}}\, e^{ipx - iqy}\, [a(\boldsymbol{p}), a^\dagger(\boldsymbol{q})] \\ &= \int \frac{d^3p}{(2\pi)^3 2p^0}\, e^{ip(x-y)} = \Delta_+(x-y) \end{aligned} \tag{5.247}$$

in which $p(x-y) = \boldsymbol{p}\cdot(\boldsymbol{x}-\boldsymbol{y}) - p^0(x^0 - y^0)$.

At points x that are space-like, that is, for which $x^2 = \boldsymbol{x}^2 - (x^0)^2 \equiv r^2 > 0$, the Lorentz-invariant function $\Delta_+(x)$ depends only upon $r = +\sqrt{x^2}$ and has the value (Weinberg, 1995, p. 202)

$$\Delta_+(x) = \frac{m}{4\pi^2 r} K_1(mr), \tag{5.248}$$

in which the Hankel function K_1 is

$$K_1(z) = -\frac{\pi}{2}\,[J_1(iz) + iN_1(iz)] = \frac{1}{z} + \frac{z}{2j+2}\left[\ln\left(\frac{z}{2}\right) + \gamma - \frac{1}{2j+2}\right] + \cdots \tag{5.249}$$

where J_1 is the first Bessel function, N_1 is the first Neumann function, and $\gamma = 0.57721\ldots$ is the Euler–Mascheroni constant.

The Feynman propagator arises most simply as the mean value in the vacuum of the **time-ordered product** of the fields $\phi(x)$ and $\phi(y)$

$$\mathcal{T}\{\phi(x)\phi(y)\} \equiv \theta(x^0 - y^0)\phi(x)\phi(y) + \theta(y^0 - x^0)\phi(y)\phi(x). \tag{5.250}$$

The operators $a(\boldsymbol{p})$ and $a^\dagger(\boldsymbol{p})$ respectively annihilate the vacuum ket $a(\boldsymbol{p})|0\rangle = 0$ and bra $\langle 0|a^\dagger(\boldsymbol{p}) = 0$, and so by (5.245 & 5.246) do the positive- and negative-frequency parts of the field $\phi^+(z)|0\rangle = 0$ and $\langle 0|\phi^-(z) = 0$. Thus the mean value in the vacuum of the time-ordered product is

$$\begin{aligned} \langle 0|\mathcal{T}\{\phi(x)\phi(y)\}\,|0\rangle &= \langle 0|\theta(x^0 - y^0)\phi(x)\phi(y) + \theta(y^0 - x^0)\phi(y)\phi(x)|0\rangle \\ &= \langle 0|\theta(x^0 - y^0)\phi^+(x)\phi^-(y) + \theta(y^0 - x^0)\phi^+(y)\phi^-(x)|0\rangle \\ &= \langle 0|\theta(x^0 - y^0)[\phi^+(x), \phi^-(y)] \\ &\quad + \theta(y^0 - x^0)[\phi^+(y), \phi^-(x)]|0\rangle. \end{aligned} \tag{5.251}$$

But by (5.247), these commutators are $\Delta_+(x-y)$ and $\Delta_+(y-x)$. Thus the mean value in the vacuum of the time-ordered product

$$\langle 0|\mathcal{T}\{\phi(x)\phi(y)\}\,|0\rangle = \theta(x^0 - y^0)\Delta_+(x-y) + \theta(y^0 - x^0)\Delta_+(y-x)$$
$$= -i\Delta_F(x-y) \tag{5.252}$$

is the Feynman propagator (5.241) multiplied by $-i$. □

5.19 Dispersion relations

In many physical contexts, functions occur that are analytic in the upper half-plane (UHP). Suppose for instance that $\hat{f}(t)$ is a transfer function that determines an effect $e(t)$ due to a cause $c(t)$

$$e(t) = \int_{-\infty}^{\infty} dt'\,\hat{f}(t-t')\,c(t'). \tag{5.253}$$

If the system is **causal**, then the transfer function $\hat{f}(t-t')$ is zero for $t-t' < 0$, and so its Fourier transform

$$f(z) = \int_{-\infty}^{\infty} \frac{dt}{\sqrt{2\pi}}\hat{f}(t)\,e^{izt} = \int_{0}^{\infty} \frac{dt}{\sqrt{2\pi}}\hat{f}(t)\,e^{izt} \tag{5.254}$$

will be analytic in the upper half-plane and will shrink as the imaginary part of $z = x + iy$ increases.

So let us assume that the function $f(z)$ is analytic in the upper half-plane and on the real axis and further that

$$\lim_{r\to\infty} |f(re^{i\theta})| = 0 \quad \text{for} \quad 0 \le \theta \le \pi. \tag{5.255}$$

By Cauchy's integral formula (5.32), if z_0 lies in the upper half-plane, then $f(z_0)$ is given by the closed counterclockwise contour integral

$$f(z_0) = \frac{1}{2\pi i}\oint \frac{f(z)}{z-z_0}\,dz, \tag{5.256}$$

in which the contour runs along the real axis and then loops over the semicircle

$$\lim_{r\to\infty} re^{i\theta} \quad \text{for} \quad 0 \le \theta \le \pi. \tag{5.257}$$

Our assumption (5.255) about the behavior of $f(z)$ in the UHP implies that this contour (5.257) is a ghost contour because its modulus is bounded by

$$\lim_{r\to\infty} \frac{1}{2\pi}\int \frac{|f(re^{i\theta})|r}{r}\,d\theta = \lim_{r\to\infty} |f(re^{i\theta})| = 0. \tag{5.258}$$

So we may drop the ghost contour and write $f(z_0)$ as

$$f(z_0) = \frac{1}{2\pi i}\int_{-\infty}^{\infty} \frac{f(x)}{x-z_0}\,dx. \tag{5.259}$$

Letting the imaginary part y_0 of $z_0 = x_0 + iy_0$ shrink to ϵ

$$f(x_0) = \frac{1}{2\pi i} \int_{-\infty}^{\infty} \frac{f(x)}{x - x_0 - i\epsilon}\, dx \tag{5.260}$$

and using Cauchy's trick (5.220), we get

$$f(x_0) = \frac{1}{2\pi i}\, P \int_{-\infty}^{\infty} \frac{f(x)}{x - x_0}\, dx + \frac{i\pi}{2\pi i} \int_{-\infty}^{\infty} f(x)\, \delta(x - x_0)\, dx \tag{5.261}$$

or

$$f(x_0) = \frac{1}{2\pi i}\, P \int_{-\infty}^{\infty} \frac{f(x)}{x - x_0}\, dx + \frac{1}{2} f(x_0), \tag{5.262}$$

which is the **dispersion relation**

$$f(x_0) = \frac{1}{\pi i}\, P \int_{-\infty}^{\infty} \frac{f(x)}{x - x_0}\, dx. \tag{5.263}$$

If we break $f(z) = u(z) + iv(z)$ into its real $u(z)$ and imaginary $v(z)$ parts, then this dispersion relation (5.263)

$$\begin{aligned} u(x_0) + iv(x_0) &= \frac{1}{\pi i}\, P \int_{-\infty}^{\infty} \frac{u(x) + iv(x)}{x - x_0}\, dx \\ &= \frac{1}{\pi}\, P \int_{-\infty}^{\infty} \frac{v(x)}{x - x_0}\, dx - \frac{i}{\pi}\, P \int_{-\infty}^{\infty} \frac{u(x)}{x - x_0}\, dx \end{aligned} \tag{5.264}$$

breaks into its real and imaginary parts

$$u(x_0) = \frac{1}{\pi}\, P \int_{-\infty}^{\infty} \frac{v(x)}{x - x_0}\, dx \quad \text{and} \quad v(x_0) = -\frac{1}{\pi}\, P \int_{-\infty}^{\infty} \frac{u(x)}{x - x_0}\, dx, \tag{5.265}$$

which express u and v as **Hilbert transforms** of each other.

In applications of dispersion relations, the function $f(x)$ for $x < 0$ sometimes is either physically meaningless or experimentally inaccessible. In such cases, there may be a symmetry that relates $f(-x)$ to $f(x)$. For instance, if $f(x)$ is the Fourier transform of a real function $\hat{f}(k)$, then by equation (3.25) it obeys the symmetry relation

$$f^*(x) = u(x) - iv(x) = f(-x) = u(-x) + iv(-x), \tag{5.266}$$

which says that u is even, $u(-x) = u(x)$, and v odd, $v(-x) = -v(x)$. Using these symmetries, one may show (exercise 5.36) that the Hilbert transformations (5.265) become

$$u(x_0) = \frac{2}{\pi} P \int_0^{\infty} \frac{x\, v(x)}{x^2 - x_0^2}\, dx \quad \text{and} \quad v(x_0) = -\frac{2x_0}{\pi} P \int_0^{\infty} \frac{u(x)}{x^2 - x_0^2}\, dx, \tag{5.267}$$

which do not require input at negative values of x.

5.20 Kramers–Kronig relations

If we use $\sigma \boldsymbol{E}$ for the current density $\boldsymbol{J}$ and $\boldsymbol{E}(t) = e^{-i\omega t}\boldsymbol{E}$ for the electric field, then Maxwell's equation $\nabla \times \boldsymbol{B} = \mu \boldsymbol{J} + \epsilon\mu\dot{\boldsymbol{E}}$ becomes

$$\nabla \times \boldsymbol{B} = -i\omega\epsilon\mu\left(1 + i\frac{\sigma}{\epsilon\omega}\right)\boldsymbol{E} \equiv -i\omega n^2 \epsilon_0\mu_0 \boldsymbol{E} \tag{5.268}$$

and reveals the squared index of refraction as

$$n^2(\omega) = \frac{\epsilon\mu}{\epsilon_0\mu_0}\left(1 + i\frac{\sigma}{\epsilon\omega}\right). \tag{5.269}$$

The imaginary part of n^2 represents the scattering of light mainly by electrons. At high frequencies in nonmagnetic materials $n^2(\omega) \to 1$, and so Kramers and Kronig applied the Hilbert-transform relations (5.267) to the function $n^2(\omega) - 1$ in order to satisfy condition (5.255). Their relations are

$$\mathrm{Re}(n^2(\omega_0)) = 1 + \frac{2}{\pi} P \int_0^\infty \frac{\omega\,\mathrm{Im}(n^2(\omega))}{\omega^2 - \omega_0^2}\,d\omega \tag{5.270}$$

and

$$\mathrm{Im}(n^2(\omega_0)) = -\frac{2\omega_0}{\pi} P \int_0^\infty \frac{\mathrm{Re}(n^2(\omega)) - 1}{\omega^2 - \omega_0^2}\,d\omega. \tag{5.271}$$

What Kramers and Kronig actually wrote was slightly different from these dispersion relations (5.270 & 5.271). H. A. Lorentz had shown that the index of refraction $n(\omega)$ is related to the forward scattering amplitude $f(\omega)$ for the scattering of light by a density N of scatterers (Sakurai, 1982)

$$n(\omega) = 1 + \frac{2\pi c^2}{\omega^2} N f(\omega). \tag{5.272}$$

They used this formula to infer that the real part of the index of refraction approached unity in the limit of infinite frequency and applied the Hilbert transform (5.267)

$$\mathrm{Re}[n(\omega)] = 1 + \frac{2}{\pi} P \int_0^\infty \frac{\omega'\,\mathrm{Im}[n(\omega')]}{\omega'^2 - \omega^2}\,d\omega'. \tag{5.273}$$

The Lorentz relation (5.272) expresses the imaginary part $\mathrm{Im}[n(\omega)]$ of the index of refraction in terms of the imaginary part of the forward scattering amplitude $f(\omega)$

$$\mathrm{Im}[n(\omega)] = 2\pi(c/\omega)^2 N \mathrm{Im}[f(\omega)]. \tag{5.274}$$

And the **optical theorem**relates $\mathrm{Im}[f(\omega)]$ to the **total cross-section**

$$\sigma_{\text{tot}} = \frac{4\pi}{|\boldsymbol{k}|}\mathrm{Im}[f(\omega)] = \frac{4\pi c}{\omega}\mathrm{Im}[f(\omega)]. \tag{5.275}$$

Thus we have $\mathrm{Im}[n(\omega)] = cN\sigma_{\rm tot}/(2\omega)$, and by the Lorentz relation (5.272) $\mathrm{Re}[n(\omega)] = 1 + 2\pi(c/\omega)^2 N\mathrm{Re}[f(\omega)]$. Insertion of these formulas into the Kramers–Kronig integral (5.273) gives a dispersion relation for the real part of the forward scattering amplitude $f(\omega)$ in terms of the total cross-section

$$\mathrm{Re}[f(\omega)] = \frac{\omega^2}{2\pi^2 c} P \int_0^\infty \frac{\sigma_{\rm tot}(\omega')}{\omega'^2 - \omega^2} \, d\omega'. \tag{5.276}$$

5.21 Phase and group velocities

Suppose $A(\boldsymbol{x}, t)$ is the amplitude

$$A(\boldsymbol{x}, t) = \int e^{i(\boldsymbol{p}\cdot\boldsymbol{x} - Et)/\hbar} A(\boldsymbol{p}) \, d^3p = \int e^{i(\boldsymbol{k}\cdot\boldsymbol{x} - \omega t)} B(\boldsymbol{k}) \, d^3k \tag{5.277}$$

where $B(\boldsymbol{k}) = \hbar^3 A(\hbar \boldsymbol{k})$ varies slowly compared to the phase $\exp[i(\boldsymbol{k}\cdot\boldsymbol{x} - \omega t)]$. The **phase velocity** $\boldsymbol{v}_p$ is the linear relation $\boldsymbol{x} = \boldsymbol{v}_p t$ between $\boldsymbol{x}$ and t that keeps the phase $\phi = \boldsymbol{p}\cdot\boldsymbol{x} - Et$ constant as a function of the time

$$0 = \boldsymbol{p}\cdot d\boldsymbol{x} - E\,dt = (\boldsymbol{p}\cdot\boldsymbol{v}_p - E)\,dt \quad \Longleftrightarrow \quad \boldsymbol{v}_p = \frac{E}{p}\hat{\boldsymbol{p}} = \frac{\omega}{k}\hat{\boldsymbol{k}}, \tag{5.278}$$

in which $p = |\boldsymbol{p}|$, and $k = |\boldsymbol{k}|$. For light in the vacuum, $\boldsymbol{v}_p = c = (\omega/k)\,\hat{\boldsymbol{k}}$.

The **group velocity** $\boldsymbol{v}_g$ is the linear relation $\boldsymbol{x} = \boldsymbol{v}_g t$ between $\boldsymbol{x}$ and t that maximizes the amplitude $A(\boldsymbol{x}, t)$ by keeping the phase $\phi = \boldsymbol{p}\cdot\boldsymbol{x} - Et$ constant as a function of the momentum $\boldsymbol{p}$

$$\nabla_p(\boldsymbol{p}\boldsymbol{x} - Et) = \boldsymbol{x} - \nabla_p E(\boldsymbol{p})\, t = 0 \tag{5.279}$$

at the maximum of $A(\boldsymbol{p})$. This **condition of stationary phase** gives the group velocity as

$$\boldsymbol{v}_{\rm g} = \nabla_p E(\boldsymbol{p}) = \nabla_k \omega(\boldsymbol{k}). \tag{5.280}$$

If $E = \boldsymbol{p}^2/(2m)$, then $\boldsymbol{v}_{\rm g} = \boldsymbol{p}/m$.

When light traverses a medium with a complex index of refraction $n(\boldsymbol{k})$, the wave vector $\boldsymbol{k}$ becomes complex, and its (positive) imaginary part represents the scattering of photons in the forward direction, typically by the electrons of the medium. For simplicity, we'll consider the propagation of light through a medium in one dimension, that of the forward direction of the beam. Then the (real) frequency $\omega(k)$ and the (complex) wave-number k are related by $k = n(k)\,\omega(k)/c$, and the phase velocity of the light is

$$v_{\rm p} = \frac{\omega}{\mathrm{Re}(k)} = \frac{c}{\mathrm{Re}(n(k))}. \tag{5.281}$$

If we regard the index of refraction as a function of the frequency ω, instead of the wave-number k, then by differentiating the real part of the relation $\omega n(\omega) = ck$ with respect to ω, we find

$$n_r(\omega) + \omega \frac{dn_r(\omega)}{d\omega} = c\frac{dk_r}{d\omega}, \tag{5.282}$$

in which the subscript r means real part. Thus the group velocity (5.280) of the light is

$$v_g = \frac{d\omega}{dk_r} = \frac{c}{n_r(\omega) + \omega\, dn_r/d\omega}. \tag{5.283}$$

Optical physicists call the denominator the **group index of refraction**

$$n_g(\omega) = n_r(\omega) + \omega \frac{dn_r(\omega)}{d\omega} \tag{5.284}$$

so that as in the expression (5.281) for the phase velocity $v_p = c/n_r(\omega)$, the group velocity is $v_g = c/n_g(\omega)$.

In some media, the derivative $dn_r/d\omega$ is large and positive, and the group velocity v_g of light there can be much less than c (Steinberg *et al.*, 1993; Wang and Zhang, 1995) – as slow as 17 m/s (Hau *et al.*, 1999). This effect is called **slow light**. In certain other media, the derivative $dn/d\omega$ is so negative that the group index of refraction $n_g(\omega)$ is less than unity, and in them the group velocity v_g exceeds c! This effect is called **fast light**. In some media, the derivative $dn_r/d\omega$ is so negative that $dn_r/d\omega < -n_r(\omega)/\omega$, and then $n_g(\omega)$ is not only less than unity but also less than zero. In such a medium, the group velocity v_g of light is negative! This effect is called **backwards light**.

Sommerfeld and Brillouin (Brillouin, 1960, ch. II & III) anticipated fast light and concluded that it would not violate special relativity as long as the **signal velocity** – defined as the speed of the front of a square pulse – remained less than c. Fast light does not violate special relativity (Stenner *et al.*, 2003; Brunner *et al.*, 2004) (Léon Brillouin, 1889–1969; Arnold Sommerfeld, 1868–1951).

Slow, fast, and backwards light can occur when the frequency ω of the light is near a peak or **resonance** in the total cross-section σ_{tot} for the scattering of light by the atoms of the medium. To see why, recall that the index of refraction $n(\omega)$ is related to the forward scattering amplitude $f(\omega)$ and the density N of scatterers by the formula (5.272)

$$n(\omega) = 1 + \frac{2\pi c^2}{\omega^2} N f(\omega) \tag{5.285}$$

and that the real part of the forward scattering amplitude is given by the Kramers–Kronig integral (5.276) of the total cross-section

$$\mathrm{Re}(f(\omega)) = \frac{\omega^2}{2\pi^2 c} P \int_0^\infty \frac{\sigma_{tot}(\omega')\, d\omega'}{\omega'^2 - \omega^2}. \tag{5.286}$$

So the real part of the index of refraction is

$$n_r(\omega) = 1 + \frac{cN}{\pi} P \int_0^\infty \frac{\sigma_{tot}(\omega')\, d\omega'}{\omega'^2 - \omega^2}. \tag{5.287}$$

If the amplitude for forward scattering is of the Breit–Wigner form

$$f(\omega) = f_0 \frac{\Gamma/2}{\omega_0 - \omega - i\Gamma/2} \tag{5.288}$$

then by (5.285) the real part of the index of refraction is

$$n_r(\omega) = 1 + \frac{\pi c^2 N f_0 \Gamma(\omega_0 - \omega)}{\omega^2 \left[(\omega - \omega_0)^2 + \Gamma^2/4\right]} \tag{5.289}$$

and by (5.283) the group velocity is

$$v_g = c \left[1 + \frac{\pi c^2 N f_0 \Gamma\, \omega_0}{\omega^2} \frac{\left[(\omega - \omega_0)^2 - \Gamma^2/4\right]}{\left[(\omega - \omega_0)^2 + \Gamma^2/4\right]^2} \right]^{-1}. \tag{5.290}$$

This group velocity v_g is less than c whenever $(\omega - \omega_0)^2 > \Gamma^2/4$. But we get fast light $v_g > c$, if $(\omega - \omega_0)^2 < \Gamma^2/4$, and even backwards light, $v_g < 0$, if $\omega \approx \omega_0$ with $4\pi c^2 N f_0 / \Gamma \omega_0 \gg 1$. Robert W. Boyd's papers explain how to make slow and fast light (Bigelow *et al.*, 2003) and backwards light (Gehring *et al.*, 2006).

We can use the principal-part identity (5.224) to subtract

$$0 = \frac{cN}{\pi} P \int_0^\infty \frac{\sigma_{tot}(\omega)}{\omega'^2 - \omega^2}\, d\omega' \tag{5.291}$$

from the Kramers–Kronig integral (5.287) so as to write the index of refraction in the regularized form

$$n_r(\omega) = 1 + \frac{cN}{\pi} P \int_0^\infty \frac{\sigma_{tot}(\omega') - \sigma_{tot}(\omega)}{\omega'^2 - \omega^2}\, d\omega', \tag{5.292}$$

which we can differentiate and use in the group-velocity formula (5.283)

$$v_g(\omega) = c \left[1 + \frac{cN}{\pi} P \int_0^\infty \frac{\left[\sigma_{tot}(\omega') - \sigma_{tot}(\omega)\right](\omega'^2 + \omega^2)}{(\omega'^2 - \omega^2)^2}\, d\omega' \right]^{-1}. \tag{5.293}$$

5.22 The method of steepest descent

Suppose we want to approximate for big $x > 0$ the integral

$$I(x) = \int_a^b dz\, h(z) \exp(x f(z)), \tag{5.294}$$

in which the functions $h(z)$ and $f(z)$ are analytic in a simply connected region that includes the points a and b in its interior. The value of the integral $I(x)$ is independent of the contour between the endpoints a and b. In the limit $x \to \infty$, the integral $I(x)$ is dominated by the exponential. So the key factor is the real part u of $f = u + iv$. But since $f(z)$ is analytic, its real and imaginary parts $u(z)$ and $v(z)$ are harmonic functions which have no minima or maxima, only saddle points (5.52).

For simplicity, we'll assume that the real part $u(z)$ of $f(z)$ has only one saddle point between the points a and b. (If it has more than one, then we must repeat the computation that follows.) If w is the saddle point, then $u_x = u_y = 0$, which by the Cauchy–Riemann equations (5.48) implies that $v_x = v_y = 0$. Thus the derivative of the function f also vanishes at the saddle point $f'(w) = 0$, and so near w we may approximate $f(z)$ as

$$f(z) \approx f(w) + \frac{1}{2}(z - w)^2 f''(w). \tag{5.295}$$

Let's write the second derivative as $f''(w) = \rho\, e^{i\phi}$ and choose our contour through the saddle point w to be a straight line $z = w + y\, e^{i\theta}$ with θ fixed for z near w. As we vary y along this line, we want

$$(z - w)^2 f''(w) = y^2\, \rho\, e^{2i\theta}\, e^{i\phi} < 0 \tag{5.296}$$

so we keep $2\theta + \phi = \pi$ so that near $z = w$

$$f(z) \approx f(w) - \frac{1}{2}\rho\, y^2. \tag{5.297}$$

Since $z = w + y\, e^{i\theta}$, its differential is $dz = e^{i\theta}\, dy$, and the integral $I(x)$ is

$$\begin{aligned} I(x) &\approx \int_{-\infty}^{\infty} h(w) \exp\left\{x\left[f(w) + \frac{1}{2}(z - w)^2 f''(w)\right]\right\} dz \qquad (5.298)\\ &= h(w)\, e^{i\theta}\, e^{xf(w)} \int_{-\infty}^{\infty} \exp\left(-\tfrac{1}{2}x\rho y^2\right) dy = h(w)\, e^{i\theta}\, e^{xf(w)} \sqrt{\frac{2\pi}{x\rho}}. \end{aligned}$$

Moving the phase $e^{i\theta}$ inside the square-root

$$I(x) \approx h(w)\, e^{xf(w)} \sqrt{\frac{2\pi}{x\rho\, e^{-2i\theta}}} \tag{5.299}$$

and using $f''(w) = \rho\, e^{i\phi}$ and $2\theta + \phi = \pi$ to show that

$$\rho\, e^{-2i\theta} = \rho\, e^{i\phi - i\pi} = -\rho\, e^{i\phi} = -f''(w) \tag{5.300}$$

we get our formula for the saddle-point integral (5.294)

$$I(x) \approx \left(\frac{2\pi}{-xf''(w)}\right)^{1/2} h(w)\, e^{xf(w)}. \tag{5.301}$$

If there are n saddle points w_j for $j = 1, \ldots, n$, then the integral $I(x)$ is the sum

$$I(x) \approx \sum_{j=1}^{N} \left(\frac{2\pi}{-xf''(w_j)} \right)^{1/2} h(w_j)\, e^{xf(w_j)}. \tag{5.302}$$

5.23 The Abel–Plana formula and the Casimir effect

This section is optional on a first reading.

Suppose the function $f(z)$ is analytic and bounded for $n_1 \le \operatorname{Re} z \le n_2$. Let $\mathcal{C}_+$ and $\mathcal{C}_-$ be two contours that respectively run counterclockwise along the rectangles with vertices $n_1, n_2, n_2 + i\infty, n_1 + i\infty$ and $n_1, n_2, n_2 - i\infty, n_1 - i\infty$ indented with tiny semicircles and quarter-circles so as to avoid the integers $z = n_1, n_1 + 1, n_1 + 2, \ldots, n_2$ while keeping $\operatorname{Im} z > 0$ in the upper rectangle and $\operatorname{Im} z < 0$ in the lower one (and $n_1 < \operatorname{Re} z < n_2$). Then the contour integrals

$$\mathcal{I}_\pm = \int_{\mathcal{C}_\pm} \frac{f(z)}{e^{\mp 2\pi i z} - 1}\, dz = 0 \tag{5.303}$$

vanish by Cauchy's theorem (5.22) since the poles of the integrand lie outside the indented rectangles.

The absolute value of the exponential $\exp(-2\pi i z)$ is arbitrarily large on the top of the upper rectangle $\mathcal{C}_+$ where $\operatorname{Im} z = \infty$, and so that leg of the contour integral $\mathcal{I}_+$ vanishes. Similarly, the bottom leg of the contour integral $\mathcal{I}_-$ vanishes. Thus we can separate the difference $\mathcal{I}_+ - \mathcal{I}_-$ into a term T_x due to the integrals near the x-axis between n_1 and n_2, a term T_1 involving integrals between n_1 and $n_1 \pm i\infty$, and a term T_2 involving integrals between n_2 and $n_2 \pm i\infty$, that is, $0 = \mathcal{I}_+ - \mathcal{I}_- = T_x + T_1 + T_2$.

The term $T_x = I_x + S$ consists of the integrals I_x along the segments of the x-axis from n_1 to n_2 and a sum S over the tiny integrals along the semicircles and quarter-circles that avoid the integers from n_1 to n_2. Elementary algebra simplifies the integral I_x to

$$I_x = \int_{n_1}^{n_2} f(x) \left[\frac{1}{e^{-2\pi i x} - 1} + \frac{1}{e^{+2\pi i x} - 1} \right] dx = -\int_{n_1}^{n_2} f(x)\, dx. \tag{5.304}$$

The sum S is over the semicircles that avoid $n_1 + 1, \ldots, n_2 - 1$ and over the quarter-circles that avoid n_1 and n_2. For any integer $n_1 < n < n_2$, the integral along the semicircle of $\mathcal{C}_{n+}$ minus that along the semicircle of $\mathcal{C}_{n-}$, both around n, contributes to S the quantity

$$\begin{aligned} S_n &= \int_{S\mathcal{C}_{n+}} \frac{f(z)}{e^{-2\pi i z} - 1}\, dz - \int_{S\mathcal{C}_{n-}} \frac{f(z)}{e^{2\pi i z} - 1}\, dz \\ &= \int_{S\mathcal{C}_{n+}} \frac{f(z)}{e^{-2\pi i (z-n)} - 1}\, dz - \int_{S\mathcal{C}_{n-}} \frac{f(z)}{e^{2\pi i (z-n)} - 1}\, dz \end{aligned} \tag{5.305}$$

since $\exp(\pm 2\pi i n) = 1$. The first integral is clockwise in the upper half-plane, the second clockwise in the lower half-plane. So if we make both integrals counterclockwise, inserting minus signs, we find as the radii of these semicircles shrink to zero

$$S_n = \oint \frac{f(z)}{2\pi i(z-n)} dz = f(n). \tag{5.306}$$

One may show (exercise 5.39) that the quarter-circles around n_1 and n_2 contribute $(f(n_1) + f(n_2))/2$ to the sum S. Thus the term T_x is

$$T_x = \frac{1}{2} f(n_1) + \sum_{n=n_1+1}^{n_2-1} f(n) + \frac{1}{2} f(n_2) - \int_{n_1}^{n_2} f(x)\, dx. \tag{5.307}$$

Since $\exp(-2\pi i n_1) = 1$, the difference between the integrals along the imaginary axes above and below n_1 is (exercise 5.40)

$$T_1 = \int_{n_1+i\infty}^{n_1} \frac{f(z)}{e^{-2\pi i z} - 1}\, dz - \int_{n_1}^{n_1-i\infty} \frac{f(z)}{e^{2\pi i z} - 1}\, dz \tag{5.308}$$

$$= -i \int_0^{\infty} \frac{f(n_1 + iy) - f(n_1 - iy)}{e^{2\pi y} - 1}\, dy. \tag{5.309}$$

Similarly, the difference between the integrals along the imaginary axes above and below n_2 is (exercise 5.41)

$$T_2 = \int_{n_2}^{n_2+i\infty} \frac{f(z)}{e^{-2\pi i z} - 1}\, dz - \int_{n_2-i\infty}^{n_2} \frac{f(z)}{e^{2\pi i z} - 1}\, dz \tag{5.310}$$

$$= i \int_0^{\infty} \frac{f(n_2 + iy) - f(n_2 - iy)}{e^{2\pi y} - 1}\, dy. \tag{5.311}$$

Since $\mathcal{I}_+ - \mathcal{I}_- = T_x + T_1 + T_2 = 0$, we can use (5.307) and (5.309–5.311) to build the Abel–Plana formula (Whittaker and Watson, 1927, p. 145)

$$\frac{1}{2} f(n_1) + \sum_{n=n_1+1}^{n_2-1} f(n) + \frac{1}{2} f(n_2) - \int_{n_1}^{n_2} f(x)\, dx$$

$$= i \int_0^{\infty} \frac{f(n_1 + iy) - f(n_1 - iy) - f(n_2 + iy) + f(n_2 - iy)}{e^{2\pi y} - 1}\, dy \tag{5.312}$$

(Niels Abel, 1802–1829; Giovanni Plana, 1781–1864).

In particular, if $f(z) = z$, the integral over y vanishes, and the Abel–Plana formula (5.312) gives

$$\frac{1}{2} n_1 + \sum_{n=n_1+1}^{n_2-1} n + \frac{1}{2} n_2 = \int_{n_1}^{n_2} x\, dx, \tag{5.313}$$

which is an example of the trapezoidal rule.

Example 5.37 (The Casimir effect) The Abel–Plana formula provides one of the clearer formulations of the Casimir effect. We will assume that the hamiltonian for the electromagnetic field in empty space is a sum over two polarizations and an integral over all momenta of a symmetric product

$$H_0 = \sum_{s=1}^{2} \int \hbar\,\omega(\boldsymbol{k})\, \frac{1}{2} \left[a_s^\dagger(\boldsymbol{k}) a_s(\boldsymbol{k}) + a_s(\boldsymbol{k}) a_s^\dagger(\boldsymbol{k}) \right] d^3k \tag{5.314}$$

of the annihilation and creation operators $a_s(\boldsymbol{k})$ and $a_s^\dagger(\boldsymbol{k})$, which satisfy the commutation relations

$$[a_s(\boldsymbol{k}), a_{s'}^\dagger(\boldsymbol{k}')] = \delta_{ss'}\, \delta(\boldsymbol{k} - \boldsymbol{k}') \text{ and } [a_s(\boldsymbol{k}), a_{s'}(\boldsymbol{k}')] = 0 = [a_s^\dagger(\boldsymbol{k}), a_{s'}^\dagger(\boldsymbol{k}')]. \tag{5.315}$$

The vacuum state $|0\rangle$ has no photons, and so on it $a_s(\boldsymbol{k})|0\rangle = 0$ (and $\langle 0|a_s^\dagger(\boldsymbol{k}) = 0$). But because the operators in H_0 are symmetrically ordered, the energy E_0 of the vacuum as given by (5.314) is not zero; instead it is quarticly divergent

$$E_0 = \langle 0|H_0|0\rangle = \sum_{s=1}^{2} \int \hbar\,\omega(\boldsymbol{k})\, \frac{1}{2} \delta(\boldsymbol{0})\, d^3k = V \int \hbar\,\omega(\boldsymbol{k})\, \frac{d^3k}{(2\pi)^3}, \tag{5.316}$$

in which we used the delta function formula

$$\delta(\boldsymbol{k} - \boldsymbol{k}') = \int e^{\pm i(\boldsymbol{k}-\boldsymbol{k}')\cdot\boldsymbol{x}}\, \frac{d^3x}{(2\pi)^3} \tag{5.317}$$

to identify $\delta(\boldsymbol{0})$ as the volume V of empty space divided by $(2\pi)^3$. Since the photon has no mass, its (angular) frequency $\omega(\boldsymbol{k})$ is $c|\boldsymbol{k}|$, and so the energy density E_0/V is

$$\frac{E_0}{V} = \hbar c \int k^3 \frac{dk}{2\pi^2} = \hbar c \frac{K^4}{8\pi^2} = \frac{\hbar c}{8\pi^2} \frac{1}{d^4}, \tag{5.318}$$

in which we cut off the integral at some short distance $d = K^{-1}$ below which the hamiltonian (5.314) and the commutation relations (5.315) are no longer valid. But the energy density of empty space is

$$\Omega_\Lambda \rho_c = \Omega_\Lambda 3H_0^2/8\pi G \approx \frac{\hbar c}{8\pi^2} \frac{1}{(2.8 \times 10^{-5}\,\mathrm{m})^4}, \tag{5.319}$$

which corresponds to a distance scale d of 28 micrometers. Since quantum electrodynamics works well down to about 10^{-18} m, this distance scale is too big by thirteen orders of magnitude.

If the Universe were inside an enormous, perfectly conducting, metal cube of side L, then the tangential electric and normal magnetic fields would vanish on the surface of the cube $\boldsymbol{E}_t(\boldsymbol{r}, t) = 0 = \boldsymbol{B}_n(\boldsymbol{r}, t)$. The available wave-numbers of the electromagnetic field inside the cube then would be $\boldsymbol{k_n} = 2\pi(n_1, n_2, n_3)/L$, and the energy density would be

$$\frac{E_0}{V} = \frac{2\pi \hbar c}{L^4} \sum_{\boldsymbol{n}} \sqrt{\boldsymbol{n}^2}. \tag{5.320}$$

The Casimir effect exploits the difference between the continuous (5.318) and discrete (5.320) energy densities for the case of two metal plates of area A separated by a short distance $\ell \ll \sqrt{A}$.

If the plates are good conductors, then at low frequencies the boundary conditions $\boldsymbol{E}_t(\boldsymbol{r}, t) = 0 = \boldsymbol{B}_n(\boldsymbol{r}, t)$ hold, and the tangential electric and normal magnetic fields vanish on the surfaces of the metal plates. At high frequencies, above the plasma frequency $\omega_{\rm p}$ of the metal, these boundary conditions fail because the relative electric permittivity of the metal

$$\epsilon(\omega) \approx 1 - \frac{\omega_{\rm p}^2}{\omega^2}\left(1 - \frac{i}{\omega\tau}\right) \tag{5.321}$$

has a positive real part. Here τ is the mean time between electron collisions.

The modes that satisfy the low-frequency boundary conditions $\boldsymbol{E}_t(\boldsymbol{r}, t) = 0 = \boldsymbol{B}_n(\boldsymbol{r}, t)$ are (Bordag *et al.*, 2009, p. 30)

$$\omega(\boldsymbol{k}_\perp, n) \equiv c\sqrt{\boldsymbol{k}_\perp^2 + \left(\frac{\pi n}{\ell}\right)^2} \tag{5.322}$$

where $\boldsymbol{n} \cdot \boldsymbol{k}_\perp = 0$. The difference between the zero-point energies of these modes and those of the continuous modes in the absence of the two plates per unit area would be

$$\frac{E(\ell)}{A} = \frac{\pi \hbar c}{\ell} \int_0^\infty \frac{k_\perp dk_\perp}{2\pi} \left[\sum_{n=0}^{\infty} \sqrt{\frac{\ell^2 k_\perp^2}{\pi^2} + n^2} - \int_0^\infty \sqrt{\frac{\ell^2 k_\perp^2}{\pi^2} + x^2}\, dx - \frac{\ell k_\perp}{2\pi} \right] \tag{5.323}$$

if the boundary conditions held at all frequencies. With $p = \ell k_\perp/\pi$, we will represent the failure of these boundary conditions at the plasma frequency $\omega_{\rm p}$ by means of a cutoff function like $c(n) = (1 + n/n_{\rm p})^{-4}$ where $n_{\rm p} = \omega_{\rm p}\ell/\pi c$. In terms of such a cutoff function, the energy difference per unit area is

$$\frac{E(\ell)}{A} = \frac{\pi^2 \hbar c}{2\ell^3} \int_0^\infty p\, dp \left[\sum_{n=0}^{\infty} c(n)\sqrt{p^2 + n^2} - \int_0^\infty c(x)\sqrt{p^2 + x^2}\, dx - \frac{p}{2} \right]. \tag{5.324}$$

Since $c(n)$ falls off as $(n_{\rm p}/n)^4$ for $n \gg n_{\rm p}$, we may neglect terms in the sum and integral beyond some integer M that is much larger than $n_{\rm p}$

$$\frac{E(\ell)}{A} = \frac{\pi^2 \hbar c}{2\ell^3} \int_0^\infty p\, dp \left[\sum_{n=0}^{M} c(n)\sqrt{p^2 + n^2} - \int_0^M c(x)\sqrt{p^2 + x^2}\, dx - \frac{p}{2} \right]. \tag{5.325}$$

The function

$$f(z) = c(z)\sqrt{p^2 + z^2} = \frac{\sqrt{p^2 + z^2}}{(1 + z/n_p)^4} \tag{5.326}$$

is analytic in the right half-plane $\mathrm{Re}\, z = x > 0$ (exercise 5.42) and tends to zero $\lim_{x\to\infty} |f(x+iy)| \to 0$ as $\mathrm{Re}\, z = x \to \infty$. So we can apply the Abel–Plana formula (5.312) with $n_1 = 0$ and $n_2 = M$ to the term in the square brackets in (5.325) and get

$$\begin{aligned}\frac{E(\ell)}{A} = \frac{\pi^2 \hbar c}{2\ell^3} \int_0^\infty p\, dp \Bigg\{ \frac{c(M)}{2}\sqrt{p^2+M^2} \\ + i \int_0^\infty \Big[c(iy)\sqrt{p^2+(\epsilon+iy)^2} - c(-iy)\sqrt{p^2+(\epsilon-iy)^2} \\ - c(M+iy)\sqrt{p^2+(M+iy)^2} \\ + c(M-iy)\sqrt{p^2+(M-iy)^2}\Big] \frac{dy}{e^{2\pi y}-1} \Bigg\},\end{aligned} \tag{5.327}$$

in which the infinitesimal ϵ reminds us that the contour lies inside the right half-plane.

We now take advantage of the properties of the cutoff function $c(z)$. Since $M \gg n_p$, we can neglect the term $c(M)\sqrt{p^2+M^2}/2$. The denominator $\exp(2\pi y) - 1$ also allows us to neglect the terms $\mp c(M \mp iy)\sqrt{p^2+(M \mp iy)^2}$. We are left with

$$\begin{aligned}\frac{E(\ell)}{A} = \frac{\pi^2 \hbar c}{2\ell^3} \int_0^\infty p\, dp \\ \times i \int_0^\infty \Big[c(iy)\sqrt{p^2+(\epsilon+iy)^2} - c(-iy)\sqrt{p^2+(\epsilon-iy)^2}\Big] \frac{dy}{e^{2\pi y}-1}.\end{aligned} \tag{5.328}$$

Since the y integration involves the factor $1/(\exp(2\pi y) - 1)$, we can neglect the detailed behavior of the cutoff functions $c(iy)$ and $c(-iy)$ for $y > n_p$ where they differ appreciably from unity. The energy now is

$$\frac{E(\ell)}{A} = \frac{\pi^2 \hbar c}{2\ell^3} \int_0^\infty p\, dp \int_0^\infty i \frac{\sqrt{p^2+(\epsilon+iy)^2} - \sqrt{p^2+(\epsilon-iy)^2}}{e^{2\pi y}-1} dy. \tag{5.329}$$

When $y < p$, the square-roots with the ϵs cancel. But for $y > p$, they are

$$\sqrt{p^2 - y^2 \pm 2i\epsilon y} = \pm i\sqrt{y^2-p^2}. \tag{5.330}$$

Their difference is $2i\sqrt{y^2-p^2}$, and so $E(\ell)$ is

$$\frac{E(\ell)}{A} = \frac{\pi^2 \hbar c}{2\ell^3} \int_0^\infty p\, dp \int_0^\infty \frac{-2\sqrt{y^2-p^2}\,\theta(y-p)}{e^{2\pi y}-1} dy, \tag{5.331}$$

in which the Heaviside step function $\theta(x) \equiv (x + |x|)/(2|x|)$ keeps $y > p$

$$\frac{E(\ell)}{A} = -\frac{\pi^2 \hbar c}{\ell^3} \int_0^y p\, dp \int_0^\infty \frac{\sqrt{y^2-p^2}}{e^{2\pi y}-1} dy. \tag{5.332}$$

The p-integration is elementary, and so the energy difference is

$$\frac{E(\ell)}{A} = -\frac{\pi^2\hbar c}{3\ell^3}\int_0^\infty \frac{y^3\,dy}{e^{2\pi y}-1} = -\frac{\pi^2\hbar c}{3\ell^3}\frac{B_2}{8} = -\frac{\pi^2\hbar c}{720\,\ell^3}, \tag{5.333}$$

in which B_2 is the second Bernoulli number (4.109). The pressure pushing the plates together then is

$$p = -\frac{1}{A}\frac{\partial E(\ell)}{\partial \ell} = -\frac{\pi^2\hbar c}{240\,\ell^4}, \tag{5.334}$$

a result due to Casimir (Hendrik Brugt Gerhard Casimir, 1909–2000).

Although the Casimir effect is very attractive because of its direct connection with the symmetric ordering of the creation and annihilation operators in the hamiltonian (5.314), the reader should keep in mind that neutral atoms are mutually attractive, which is why most gases are diatomic, and that Lifshitz explained the effect in terms of the mutual attraction of the atoms in the metal plates (Lifshitz, 1956; Milonni and Shih, 1992) (Evgeny Mikhailovich Lifshitz, 1915–1985). □

5.24 Applications to string theory

This section is optional on a first reading.

String theory may or may not have anything to do with physics, but it does provide many amusing applications of complex-variable theory. The coordinates σ and τ of the world sheet of a string form a complex variable $z = e^{2(\tau - i\sigma)}$. The product of two operators $u(z)$ and $v(w)$ often has poles in $z - w$ as $z \to w$ but is well defined if z and w are radially ordered

$$\mathcal{R}\{u(z)v(w)\} \equiv u(z)\,v(w)\,\theta(|z| - |w|) + v(w)\,u(z)\,\theta(|w| - |z|), \tag{5.335}$$

in which $\theta(x) = (x + |x|)/2|x|$ is the step function. Since the modulus of $z = e^{2(\tau - i\sigma)}$ depends only upon τ, radial order is time order in τ_z and τ_w.

The modes L_n of the principal component of the energy–momentum tensor $T(z)$ are defined by its Laurent series

$$T(z) = \sum_{n=-\infty}^{\infty} \frac{L_n}{z^{n+2}} \tag{5.336}$$

and the inverse relation

$$L_n = \frac{1}{2\pi i}\oint z^{n+1}\,T(z)\,dz. \tag{5.337}$$

Thus the commutator of two modes involves two loop integrals

$$[L_m, L_n] = \left[\frac{1}{2\pi i}\oint z^{m+1}\,T(z)\,dz, \frac{1}{2\pi i}\oint w^{n+1}\,T(w)\,dw\right], \tag{5.338}$$

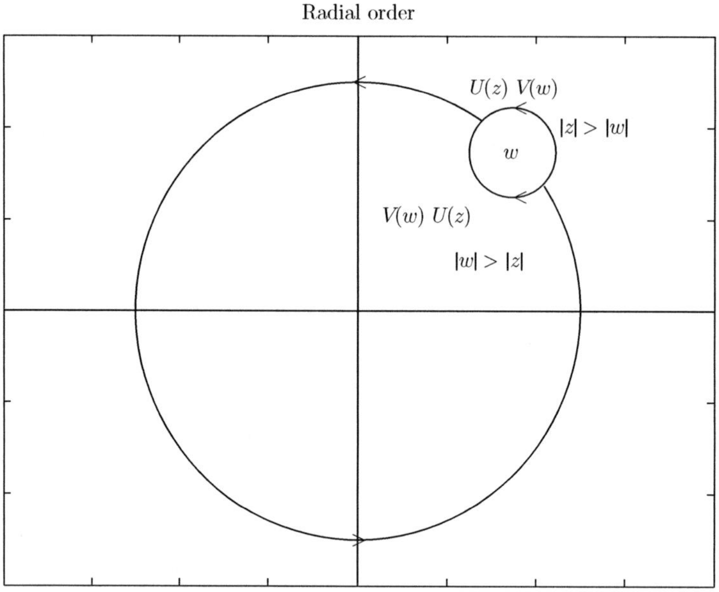

Figure 5.10 The two counterclockwise circles about the origin preserve radial order when z is near w by veering slightly to $|z| > |w|$ for the product $T(z)T(w)$ and to $|w| > |z|$ for the product $T(w)T(z)$.

which we may deform as long as we cross no poles. Let's hold w fixed and deform the z loop so as to keep the Ts radially ordered when z is near w as in Fig. 5.10. The operator-product expansion of the radially ordered product $\mathcal{R}\{T(z)T(w)\}$ is

$$\mathcal{R}\{T(z)T(w)\} = \frac{c/2}{(z-w)^4} + \frac{2}{(z-w)^2}T(w) + \frac{1}{z-w}T'(w) + \cdots, \tag{5.339}$$

in which the prime means derivative, c is a constant, and the dots denote terms that are analytic in z and w. The commutator introduces a minus sign that cancels most of the two contour integrals and converts what remains into an integral along a tiny circle C_w about the point w as in Fig. 5.10

$$[L_m, L_n] = \oint \frac{dw}{2\pi i} w^{n+1} \oint_{C_w} \frac{dz}{2\pi i} z^{m+1} \left[\frac{c/2}{(z-w)^4} + \frac{2T(w)}{(z-w)^2} + \frac{T'(w)}{z-w} \right]. \tag{5.340}$$

After doing the z-integral, which is left as a homework exercise (5.43), one may use the Laurent series (5.336) for $T(w)$ to do the w-integral, which one may choose to be along a tiny circle about $w = 0$, and so find the commutator

$$[L_m, L_n] = (m-n)\, L_{m+n} + \frac{c}{12}\, m(m^2-1)\, \delta_{m+n,0} \tag{5.341}$$

of the Virasoro algebra.

Exercises

5.1 Compute the two limits (5.6) and (5.7) of example 5.2 but for the function $f(x, y) = x^2 - y^2 + 2ixy$. Do the limits now agree? Explain.

5.2 Show that if $f(z)$ is analytic in a disk, then the integral of $f(z)$ around a tiny (isosceles) triangle of side $\epsilon \ll 1$ inside the disk is zero to order ϵ^2.

5.3 Derive the two integral representations (5.46) for Bessel's functions $J_n(t)$ of the first kind from the integral formula (5.45). Hint: think of the integral (5.45) as running from $-\pi$ to π.

5.4 Do the integral

$$\oint_C \frac{dz}{z^2-1}$$

in which the contour C is counterclockwise about the circle $|z| = 2$.

5.5 The function $f(z) = 1/z$ is analytic in the region $|z| > 0$. Compute the integral of $f(z)$ counterclockwise along the unit circle $z = e^{i\theta}$ for $0 \le \theta \le 2\pi$. The contour lies entirely within the domain of analyticity of the function $f(z)$. Did you get zero? Why? If not, why not?

5.6 Let $P(z)$ be the polynomial

$$P(z) = (z-a_1)(z-a_2)(z-a_3) \tag{5.342}$$

with roots a_1, a_2, and a_3. Let R be the maximum of the three moduli $|a_k|$. (a) If the three roots are all different, evaluate the integral

$$I = \oint_C \frac{dz}{P(z)} \tag{5.343}$$

along the counterclockwise contour $z = 2Re^{i\theta}$ for $0 \le \theta \le 2\pi$. (b) Same exercise, but for $a_1 = a_2 \ne a_3$.

5.7 Compute the integral of the function $f(z) = e^{az}/(z^2 - 3z + 2)$ along the counterclockwise contour $C_\square$ that follows the perimeter of a square of side 6 centered at the origin. That is, find

$$I = \oint_{C_\square} \frac{e^{az}}{z^2-3z+2}\, dz. \tag{5.344}$$

5.8 Use Cauchy's integral formula (5.36) and Rodrigues's expression (5.37) for Legendre's polynomial $P_n(x)$ to derive Schlaefli's formula (5.38).

5.9 Use Schlaefli's formula (5.38) for the Legendre polynomials and Cauchy's integral formula (5.32) to compute the value of $P_n(-1)$.

5.10 Evaluate the counterclockwise integral around the unit circle $|z| = 1$

$$\oint 3\sinh^2 2z - 4\cosh^3 z\, \frac{dz}{z}. \tag{5.345}$$

5.11 Evaluate the counterclockwise integral around the circle $|z| = 2$

$$\oint \frac{z^3}{z^4 - 1}\, dz. \tag{5.346}$$

5.12 Evaluate the contour integral of the function $f(z) = \sin wz/(z-5)^3$ along the curve $z = 6 + 4(\cos t + i \sin t)$ for $0 \le t \le 2\pi$.

5.13 Evaluate the contour integral of the function $f(z) = \sin wz/(z-5)^3$ along the curve $z = -6 + 4(\cos t + i \sin t)$ for $0 \le t \le 2\pi$.

5.14 Is the function $f(x, y) = x^2 + iy^2$ analytic?

5.15 Is the function $f(x, y) = x^3 - 3xy^2 + 3ix^2y - iy^3$ analytic? Is the function $x^3 - 3xy^2$ harmonic? Does it have a minimum or a maximum? If so, what are they?

5.16 Is the function $f(x, y) = x^2 + y^2 + i(x^2 + y^2)$ analytic? Is $x^2 + y^2$ a harmonic function? What is its minimum, if it has one?

5.17 Derive the first three nonzero terms of the Laurent series for $f(z) = 1/(e^z - 1)$ about $z = 0$.

5.18 Assume that a function $g(z)$ is meromorphic in R and has a Laurent series (5.97) about a point $w \in R$. Show that as $z \to w$, the ratio $g'(z)/g(z)$ becomes (5.95).

5.19 Find the poles and residues of the functions $1/\sin z$ and $1/\cos z$.

5.20 Derive the integral formula (5.122) from (5.121).

5.21 Show that if $\mathrm{Re}\, w < 0$, then for arbitrary complex z

$$\int_{-\infty}^{\infty} e^{w(x+z)^2}\, dx = \sqrt{\frac{\pi}{-w}}. \tag{5.347}$$

5.22 Use a ghost contour to evaluate the integral

$$\int_{-\infty}^{\infty} \frac{x \sin x}{x^2 + a^2}\, dx.$$

Show your work; do not just quote the result of a commercial math program.

5.23 For $a > 0$ and $b^2 - 4ac < 0$, use a ghost contour to do the integral

$$\int_{-\infty}^{\infty} \frac{dx}{ax^2 + bx + c}. \tag{5.348}$$

5.24 Show that

$$\int_0^{\infty} \cos ax\, e^{-x^2}\, dx = \frac{1}{2}\sqrt{\pi}\, e^{-a^2/4}. \tag{5.349}$$

5.25 Show that

$$\int_{-\infty}^{\infty} \frac{dx}{1 + x^4} = \frac{\pi}{\sqrt{2}}. \tag{5.350}$$

5.26 Evaluate the integral

$$\int_0^{\infty} \frac{\cos x}{1 + x^4}\, dx. \tag{5.351}$$

5.27 Show that the Yukawa Green's function (5.151) reproduces the Yukawa potential (5.141) when $n = 3$. Use $K_{1/2}(x) = \sqrt{\pi/2x}\, e^{-x}$ (9.99).

5.28 Derive the two explicit formulas (5.188) and (5.189) for the square-root of a complex number.

5.29 What is $(-i)^i$? What is the most general value of this expression?

5.30 Use the indefinite integral (5.223) to derive the principal-part formula (5.224).

5.31 The Bessel function $J_n(x)$ is given by the integral

$$J_n(x) = \frac{1}{2\pi i} \oint_C e^{(x/2)(z-1/z)} \frac{dz}{z^{n+1}} \tag{5.352}$$

along a counterclockwise contour about the origin. Find the generating function for these Bessel functions, that is, the function $G(x, z)$ whose Laurent series has the $J_n(x)$s as coefficients

$$G(x, z) = \sum_{n=-\infty}^{\infty} J_n(x)\, z^n. \tag{5.353}$$

5.32 Show that the Heaviside function $\theta(y) = (y + |y|)/(2|y|)$ is given by the integral

$$\theta(y) = \frac{1}{2\pi i} \int_{-\infty}^{\infty} e^{iyx} \frac{dx}{x - i\epsilon}, \tag{5.354}$$

in which ϵ is an infinitesimal positive number.

5.33 Show that the integral of $\exp(ik)/k$ along the contour from $k = L$ to $k = L + iH$ and then to $k = -L + iH$ and then down to $k = -L$ vanishes in the double limit $L \to \infty$ and $H \to \infty$.

5.34 Use a ghost contour and a cut to evaluate the integral

$$I = \int_{-1}^{1} \frac{dx}{(x^2 + 1)\sqrt{1 - x^2}} \tag{5.355}$$

by imitating example 5.30. Be careful when picking up the poles at $z = \pm i$. If necessary, use the explicit square-root formulas (5.188) and (5.189).

5.35 Redo the previous exercise (5.34) by defining the square-roots so that the cuts run from $-\infty$ to -1 and from 1 to ∞. Take advantage of the evenness of the integrand and integrate on a contour that is slightly above the whole real axis. Then add a ghost contour around the upper half-plane.

5.36 Show that if u is even and v is odd, then the Hilbert transforms (5.265) imply (5.267).

5.37 Show why the principal-part identity (5.224) lets one write the Kramers–Kronig integral (5.287) for the index of refraction in the regularized form (5.292).

5.38 Use the formula (5.283) for the group velocity and the regularized expression (5.292) for the real part of the index of refraction $n_r(\omega)$ to derive a formula (5.293) for the group velocity.

5.39 Show that the quarter-circles of the Abel–Plana contours $C_\pm$ contribute $\frac{1}{2}(f(n_1) + f(n_2))$ to the sum S in the formula $T_x = I_x + S$.

5.40 Derive the integral formula (5.309) from (5.308).

5.41 Derive the integral formula (5.311) from (5.310).

5.42 Show that the function (5.326) is analytic in the RHP $\mathrm{Re}\, z > 0$.

5.43 (a) Perform the z-integral in equation (5.340). (b) Use the result of part (a) to find the commutator $[L_m, L_n]$ of the **Virasoro algebra**. Hint: use the Laurent series (5.336).

5.44 Assume that $\epsilon(z)$ is analytic in a disk that contains a tiny circular contour C_w about the point w as in Fig 5.10. Do the contour integral

$$\oint_{C_w} \epsilon(z) \left[\frac{c/2}{(z-w)^4} + \frac{2T(w)}{(z-w)^2} + \frac{T'(w)}{z-w} \right] \frac{dz}{2\pi i} \tag{5.356}$$

and express your result in terms of $\epsilon(w)$, $T(w)$, and their derivatives.

6

Differential equations

6.1 Ordinary linear differential equations

There are many kinds of differential equation – linear and nonlinear, ordinary and partial, homogeneous and inhomogeneous. Any way of correctly solving any of them is fine. We start our overview with some definitions.

An operator of the form

$$L = \sum_{m=0}^{n} h_m(x) \frac{d^m}{dx^m} \tag{6.1}$$

is an ***n*****th-order, ordinary, linear differential operator**. It is *nth order* because the highest derivative is d^n/dx^n. It is *ordinary* because all the derivatives are with respect to the same independent variable x. It is *linear* because derivatives are linear operators

$$L\ [a_1 f_1(x) + a_2 f_2(x)] = a_1\, L f_1(x) + a_2\, L f_2(x). \tag{6.2}$$

If all the $h_m(x)$ in L are constants, independent of x, then L is an nth-order, ordinary, linear differential operator **with constant coefficients**.

Example 6.1 (Second-order linear differential operators) The operator

$$L = -\frac{d^2}{dx^2} + k^2 \tag{6.3}$$

is a second-order, linear differential operator with constant coefficients. The second-order linear differential operator

$$L = -\frac{d}{dx}\left(p(x)\frac{d}{dx}\right) + q(x) \tag{6.4}$$

is in **self-adjoint form** (section 6.27). □

The differential equation $Lf(x) = 0$ is **homogeneous** because each of its terms is linear in f or one of its derivatives $f^{(m)}$ – there is no term that is not proportional to f or one of its derivatives. The equation $Lf(x) = s(x)$ is **inhomogeneous** because of the source term $s(x)$.

If a differential equation is linear and homogeneous, then we can add solutions. If $f_1(x)$ and $f_2(x)$ are two solutions of the same linear homogeneous differential equation

$$Lf_1(x) = 0 \quad \text{and} \quad Lf_2(x) = 0 \tag{6.5}$$

then any linear combination of these solutions $f(x) = a_1 f_1(x) + a_2 f_2(x)$ with constant coefficients a_1 and a_2 also is a solution since

$$Lf(x) = L\,[a_1 f_1(x) + a_2 f_2(x)] = a_1\, Lf_1(x) + a_2\, Lf_2(x) = 0. \tag{6.6}$$

This additivity of solutions often makes it possible to find general solutions of linear homogeneous differential equations.

Example 6.2 (Sines and cosines) Two solutions of the second-order, linear, homogeneous, ordinary differential equation (ODE)

$$\left(\frac{d^2}{dx^2} + k^2\right) f(x) = 0 \tag{6.7}$$

are $\sin kx$ and $\cos kx$, and the most general solution is the linear combination $f(x) = a_1 \sin kx + a_2 \cos kx$. □

The functions $y_1(x), \ldots, y_n(x)$ are **linearly independent** if the only numbers $k_1, \ldots, k_n$ for which the linear combination vanishes for all x

$$k_1\, y_1(x) + k_2\, y_2(x) + \cdots + k_n\, y_n(x) = 0 \tag{6.8}$$

are $k_1 = \cdots = k_n = 0$. Otherwise they are **linearly dependent**.

Suppose that an nth-order homogeneous, linear, ordinary differential equation $Lf(x) = 0$ has n linearly independent solutions $f_j(x)$, and that all other solutions to this ODE are linear combinations of these n solutions. Then these n solutions are **complete** in the space of solutions of this equation and form a basis for this space. The **general solution** to $Lf(x) = 0$ is then a linear combination of the f_js with n arbitrary constant coefficients

$$f(x) = \sum_{j=1}^{n} a_j f_j(x). \tag{6.9}$$

With a source term $s(x)$, the differential equation $Lf(x) = 0$ becomes an **inhomogeneous** linear ordinary differential equation

$$Lf_i(x) = s(x). \tag{6.10}$$

If $f_{i1}(x)$ and $f_{i2}(x)$ are any two solutions of this inhomogeneous differential equation, then their difference $f_{i1}(x) - f_{i2}(x)$ is a solution of the associated homogeneous equation $Lf(x) = 0$

$$L\,[f_{i1}(x) - f_{i2}(x)] = Lf_{i1}(x) - Lf_{i2}(x) = s(x) - s(x) = 0. \tag{6.11}$$

Thus this difference must be given by the general solution (6.9) of the homogeneous equation for some constants a_j

$$f_{i1}(x) - f_{i2}(x) = \sum_{j=1}^{N} a_j f_j(x). \tag{6.12}$$

It follows therefore that every solution $f_{i1}(x)$ of the inhomogeneous differential equation (6.10) is the sum of a particular solution $f_{i2}(x)$ of that equation and some solution (6.9) of the associated homogeneous equation $Lf = 0$

$$f_{i1}(x) = f_{i2}(x) + \sum_{j=1}^{N} a_j f_j(x). \tag{6.13}$$

In other words, **the general solution of a linear inhomogeneous equation is a particular solution of that inhomogeneous equation plus the general solution of the associated homogeneous equation.**

A **nonlinear** differential equation is one in which a power $f^n(x)$ of the unknown function or of one of its derivatives $\left(f^{(k)}(x)\right)^n$ other than $n = 1$ or $n = 0$ appears or in which the unknown function f appears in some other nonlinear way. For instance, the equations

$$-f''(x) = f^3(x), \quad \left(f'(x)\right)^2 = f(x), \quad \text{and} \quad f'(x) = e^{-f(x)} \tag{6.14}$$

are nonlinear differential equations. We can't add two solutions of a nonlinear equation and expect to get a third solution. Nonlinear equations are much harder to solve.

6.2 Linear partial differential equations

An equation of the form

$$Lf(x) = \sum_{m_1,\ldots,m_k=0}^{n_1,\ldots,n_k} g_{m_1,\ldots,m_k}(x) \frac{\partial^{m_1+\cdots+m_k}}{\partial x_1^{m_1} \ldots \partial x_k^{m_k}} f(x) = 0 \tag{6.15}$$

in which x stands for $x_1, \ldots, x_k$ is a linear **partial** differential equation of order $n = n_1 + \cdots + n_k$ in the k variables $x_1, \ldots, x_k$. (A partial differential equation is a whole differential equation that has partial derivatives.)

Linear combinations of solutions of a linear homogeneous partial differential equation also are solutions of the equation. So if f_1 and f_2 are solutions of $Lf = 0$, and a_1 and a_2 are constants, then $f = a_1 f_1 + a_2 f_2$ is a solution since $Lf = a_1 \, Lf_1 + a_2 \, Lf_2 = 0$. Additivity of solutions is a property of all linear homogeneous differential equations, whether ordinary or partial.

The **general** solution $f(x) = f(x_1, \ldots, x_k)$ of a linear homogeneous partial differential equation (6.15) is a sum $f(x) = \sum_j a_j f_j(x)$ over a complete set of solutions $f_j(x)$ of the equation with arbitrary coefficients a_j. A linear partial differential equation $Lf_i(x) = s(x)$ with a source term $s(x) = s(x_1, \ldots, x_k)$ is an **inhomogeneous** linear partial differential equation because of the added source term.

Just as with ordinary differential equations, the difference $f_{i1} - f_{i2}$ of two solutions of the inhomogeneous linear partial differential equation $Lf_i = s$ is a solution of the associated homogeneous equation $Lf = 0$ (6.15)

$$L\,[f_{i1}(x) - f_{i2}(x)] = s(x) - s(x) = 0. \tag{6.16}$$

So we can expand this difference in terms of the complete set of solutions f_j of the inhomogeneous linear partial differential equation $Lf = 0$

$$f_{i1}(x) - f_{i2}(x) = \sum_j a_j f_j(x). \tag{6.17}$$

Thus the general solution of the inhomogeneous linear partial differential equation $Lf = s$ is the sum of a particular solution f_{i2} of $Lf = s$ and the general solution $\sum_j a_j f_j$ of the associated homogeneous equation (6.15)

$$f_{i1}(x) = f_{i2}(x) + \sum_j a_j f_j(x). \tag{6.18}$$

6.3 Notation for derivatives

One often uses primes or dots to denote derivatives as in

$$f' = \frac{df}{dx} \quad \text{or} \quad f'' = \frac{d^2f}{dx^2} \quad \text{and} \quad \dot{f} = \frac{df}{dt} \quad \text{or} \quad \ddot{f} = \frac{d^2f}{dt^2}.$$

For higher or partial derivatives, one sometimes uses superscripts

$$f^{(k)} = \frac{d^kf}{dx^k} \quad \text{and} \quad f^{(k,\ell)} = \frac{\partial^{k+\ell} f}{\partial x^k \partial y^\ell} \tag{6.19}$$

or subscripted letters, sometimes preceded by commas

$$f_x = f_{,x} = \frac{\partial f}{\partial x} \quad \text{and} \quad f_{xyy} = f_{,xyy} = \frac{\partial^3 f}{\partial x \partial y^2} \tag{6.20}$$

or subscripted indices, sometimes preceded by commas

$$f_{,k} = \partial_k f = \frac{\partial f}{\partial x_k} \quad \text{and} \quad f_{,k\ell} = \partial_k \partial_\ell f = \frac{\partial^2 f}{\partial x_k \partial x_\ell} \tag{6.21}$$

where the independent variables are $x = x_1, \ldots, x_n$.

In special relativity, one writes the time and space coordinates ct and $\boldsymbol{x}$ as x^0, x^1, x^2, and x^3 or as the 4-vector $(x^0, \boldsymbol{x})$. To form the invariant inner product $px \equiv \boldsymbol{x} \cdot \boldsymbol{x} - p^0 x^0 = \boldsymbol{x} \cdot \boldsymbol{x} - Et$ as $p_a x^a$ with a summed from 0 to 3, one attaches a minus sign to the time components of 4-vectors with lowered indices so that $p_0 = -p^0$ and $x_0 = -x^0$. The derivatives $\partial_a f$ and $\partial^a f$ are

$$\partial_a f = \frac{\partial f}{\partial x^a} \quad \text{and} \quad \partial^a f = \frac{\partial f}{\partial x_a}. \tag{6.22}$$

In rectangular coordinates, the gradient ∇ of a scalar f is

$$\nabla f = (f_x, f_y, f_z) = (f_{,x}, f_{,y}, f_{,z}) = (\partial_x f, \partial_y f, \partial_z f) = \left(\frac{\partial f}{\partial x}, \frac{\partial f}{\partial y}, \frac{\partial f}{\partial z} \right)$$

and the divergence of a vector $\boldsymbol{v} = (v_x, v_y, v_z)$ is the scalar

$$\nabla \cdot \boldsymbol{v} = v_{x,x} + v_{y,y} + v_{z,z} = \partial_x v_x + \partial_y v_y + \partial_z v_z = \frac{\partial v_x}{\partial x} + \frac{\partial v_y}{\partial y} + \frac{\partial v_z}{\partial z}. \tag{6.23}$$

Physicists sometimes write the Laplacian $\nabla \cdot \nabla f$ as $\nabla^2 f$ or as Δf.

Example 6.3 (Laplace's equation) The equation for the electrostatic potential in empty space is Laplace's equation

$$L\phi(x, y, z) = \nabla \cdot \nabla \phi(x, y, z) = \left(\frac{\partial^2}{\partial x^2} + \frac{\partial^2}{\partial y^2} + \frac{\partial^2}{\partial z^2} \right) \phi(x, y, z) = 0. \tag{6.24}$$

It is a second-order linear homogeneous partial differential equation. □

Example 6.4 (Poisson's equation) Poisson's equation for the electrostatic potential ϕ is

$$-\triangle \phi(x, y, z) \equiv -\left(\frac{\partial^2}{\partial x^2} + \frac{\partial^2}{\partial y^2} + \frac{\partial^2}{\partial z^2} \right) \phi(x, y, z) = \frac{\rho(x, y, z)}{\epsilon_0}, \tag{6.25}$$

in which ρ is the charge density and ϵ_0 is the electric constant. It is a second-order linear inhomogeneous partial differential equation. □

6.4 Gradient, divergence, and curl

In cylindrical coordinates, the change $d\boldsymbol{p}$ in a physical point $\boldsymbol{p}$ due to changes $d\rho$, $d\phi$, and dz in its coordinates is $d\boldsymbol{p} = \hat{\boldsymbol{\rho}}\, d\rho + \rho\, \hat{\boldsymbol{\phi}}\, d\phi + \hat{\boldsymbol{z}}\, dz$. In spherical coordinates, the change is $d\boldsymbol{p} = \hat{\boldsymbol{r}}\, dr + r\, \hat{\boldsymbol{\theta}}\, d\theta + r \sin\theta\, \hat{\boldsymbol{\phi}}\, d\phi$. In general **orthogonal** coordinates, the change $d\boldsymbol{p}$ in a physical point $\boldsymbol{p}$ due to changes du_i in its coordinates is $d\boldsymbol{p} = h_1\, \hat{\boldsymbol{e}}_1\, du_1 + h_2\, \hat{\boldsymbol{e}}_2\, du_2 + h_3\, \hat{\boldsymbol{e}}_3\, du_3$ where the basis vectors are orthonormal $\hat{\boldsymbol{e}}_k \cdot \hat{\boldsymbol{e}}_\ell = \delta_{k\ell}$. In cylindrical coordinates, the scale factors are $h_\rho = 1$, $h_\phi = \rho$, and $h_z = 1$, while in spherical coordinates they are $h_r = 1$, $h_\theta = r$, and $h_\phi = r \sin\theta$.

The dot-product of the **gradient** ∇f of a scalar function f with the change $d\boldsymbol{p}$ in the point is the change $df\ \nabla f \cdot d\boldsymbol{p}$ which is $df = \partial_1 f\, du_1 + \partial_2 f\, du_2 + \partial_3 f\, du_3$. So the gradient in orthogonal coordinates is

$$\nabla f = \frac{\hat{\boldsymbol{e}}_1}{h_1} \frac{\partial f}{\partial u_1} + \frac{\hat{\boldsymbol{e}}_2}{h_2} \frac{\partial f}{\partial u_2} + \frac{\hat{\boldsymbol{e}}_3}{h_3} \frac{\partial f}{\partial u_3}. \tag{6.26}$$

Thus the gradient in cylindrical coordinates is

$$\nabla f = \hat{\boldsymbol{\rho}} \frac{\partial f}{\partial \rho} + \frac{\hat{\boldsymbol{\phi}}}{\rho} \frac{\partial f}{\partial \phi} + \hat{\boldsymbol{z}} \frac{\partial f}{\partial z} \tag{6.27}$$

and in spherical coordinates it is

$$\nabla f = \hat{\boldsymbol{r}} \frac{\partial f}{\partial r} + \frac{\hat{\boldsymbol{\theta}}}{r} \frac{\partial f}{\partial \theta} + \frac{\hat{\boldsymbol{\phi}}}{r \sin\theta} \frac{\partial f}{\partial \phi}. \tag{6.28}$$

The **divergence** of a vector $\boldsymbol{v}$ at the center of a tiny cube is the surface integral dS of $\boldsymbol{v}$ over the boundary ∂dV of the cube divided by its tiny volume $dV = h_1 h_2 h_3\, du_1 du_2 du_3$. The surface integral dS is the sum of the differences of the integrals of v_1, v_2, and v_3 over the cube's three pairs of opposite faces $dS = [\partial(v_1 h_2 h_3)/\partial u_1 + \partial(v_2 h_1 h_3)/\partial u_2 + \partial(v_3 h_1 h_2)/\partial u_3]\, du_1 du_2 du_3$. So the divergence $\nabla \cdot \boldsymbol{v}$ is the ratio dS/dV, which is

$$\nabla \cdot \boldsymbol{v} = \frac{1}{h_1 h_2 h_3} \left[\frac{\partial(v_1 h_2 h_3)}{\partial u_1} + \frac{\partial(v_2 h_1 h_3)}{\partial u_2} + \frac{\partial(v_3 h_1 h_2)}{\partial u_3} \right]. \tag{6.29}$$

Thus the divergence in cylindrical coordinates is

$$\nabla \cdot \boldsymbol{v} = \frac{1}{\rho} \left[\frac{\partial(v_\rho \rho)}{\partial \rho} + \frac{\partial v_\phi}{\partial \phi} + \frac{\partial(v_z \rho)}{\partial z} \right] = \frac{1}{\rho} \frac{\partial(\rho v_\rho)}{\partial \rho} + \frac{1}{\rho} \frac{\partial v_\phi}{\partial \phi} + \frac{\partial v_z}{\partial z} \tag{6.30}$$

and in spherical coordinates it is

$$\nabla \cdot \boldsymbol{v} = \frac{1}{r^2} \frac{\partial(v_r r^2)}{\partial r} + \frac{1}{r \sin\theta} \frac{\partial(v_\theta \sin\theta)}{\partial \theta} + \frac{1}{r \sin\theta} \frac{\partial v_\phi}{\partial \phi}. \tag{6.31}$$

By assembling a large number of tiny cubes, one may create a finite volume V. The integral of the divergence $\nabla \cdot \boldsymbol{v}$ over the tiny volumes dV of the tiny cubes that make up the volume V is the sum of the surface integrals dS over the faces of these tiny cubes. The integrals over the interior faces cancel leaving just the surface integral over the boundary ∂V of the finite volume V. Thus we arrive at Stokes's theorem

$$\int_V \boldsymbol{\nabla} \cdot \boldsymbol{v} \, dV = \int_{\partial V} \boldsymbol{v} \cdot \boldsymbol{dS}. \tag{6.32}$$

The Laplacian is the divergence (6.29) of the gradient (6.26). So in general orthogonal coordinates it is

$$\Delta f = \boldsymbol{\nabla} \cdot \boldsymbol{\nabla f} = \frac{1}{h_1 h_2 h_3} \left[\sum_{k=1}^{3} \frac{\partial}{\partial u_k} \left(\frac{h_1 h_2 h_3}{h_k^2} \frac{\partial f}{\partial u_k} \right) \right]. \tag{6.33}$$

Thus in cylindrical coordinates, the Laplacian is

$$\Delta f = \frac{1}{\rho} \left[\frac{\partial}{\partial \rho} \left(\rho \frac{\partial f}{\partial \rho} \right) + \frac{1}{\rho} \frac{\partial^2 f}{\partial \phi^2} + \rho \frac{\partial^2 f}{\partial z^2} \right] = \frac{1}{\rho} \frac{\partial}{\partial \rho} \left(\rho \frac{\partial f}{\partial \rho} \right) + \frac{1}{\rho^2} \frac{\partial^2 f}{\partial \phi^2} + \frac{\partial^2 f}{\partial z^2} \tag{6.34}$$

and in spherical coordinates it is

$$\begin{aligned} \Delta f &= \frac{1}{r^2 \sin\theta} \left[\frac{\partial}{\partial r} \left(r^2 \sin\theta \frac{\partial f}{\partial r} \right) + \frac{\partial}{\partial \theta} \left(\sin\theta \frac{\partial f}{\partial \theta} \right) + \frac{\partial}{\partial \phi} \left(\frac{1}{\sin\theta} \frac{\partial f}{\partial \phi} \right) \right] \\ &= \frac{1}{r^2} \frac{\partial}{\partial r} \left(r^2 \frac{\partial f}{\partial r} \right) + \frac{1}{r^2 \sin\theta} \frac{\partial}{\partial \theta} \left(\sin\theta \frac{\partial f}{\partial \theta} \right) + \frac{1}{r^2 \sin^2\theta} \frac{\partial^2 f}{\partial \phi^2}. \end{aligned} \tag{6.35}$$

The area $\boldsymbol{dS}$ of a tiny square dS whose sides are the tiny perpendicular vectors $h_i \hat{\boldsymbol{e}}_i du_i$ and $h_j \hat{\boldsymbol{e}}_j du_j$ (no sum) is their cross-product

$$\boldsymbol{dS} = h_i \hat{\boldsymbol{e}}_i du_i \times h_j \hat{\boldsymbol{e}}_j du_j = \hat{\boldsymbol{e}}_k \, h_i h_j \, du_i du_j, \tag{6.36}$$

in which the perpendicular unit vectors $\hat{\boldsymbol{e}}_i$, $\hat{\boldsymbol{e}}_j$, and $\hat{\boldsymbol{e}}_k$ obey the right-hand rule. The dot-product of this area with the **curl** of a vector $\boldsymbol{v}$, which is $(\boldsymbol{\nabla} \times \boldsymbol{v}) \cdot dS = (\boldsymbol{\nabla} \times \boldsymbol{v})_k \, h_i h_j \, du_i du_j$, is the line integral dL of $\boldsymbol{v}$ along the boundary ∂dS of the square

$$(\boldsymbol{\nabla} \times \boldsymbol{v})_k \, h_i h_j du_i du_j = \left[\partial_i (h_j v_j) - \partial_j (h_i v_i) \right] du_i du_j. \tag{6.37}$$

Thus the kth component of the curl is

$$(\nabla \times v)_k = \frac{1}{h_i h_j} \left(\frac{\partial (h_j v_j)}{\partial u_i} - \frac{\partial (h_i v_i)}{\partial u_j} \right) \quad \text{(no sum)}. \tag{6.38}$$

In terms of the Levi-Civita symbol ϵ_{ijk}, which is totally antisymmetric with $\epsilon_{123} = 1$, the curl is

$$\boldsymbol{\nabla} \times \boldsymbol{v} = \frac{1}{2} \sum_{i,j,k=1}^{3} \epsilon_{ijk} \frac{\hat{\boldsymbol{e}}_k}{h_i h_j} \left[\frac{\partial (h_j v_j)}{\partial u_i} - \frac{\partial (h_i v_i)}{\partial u_j} \right] = \sum_{i,j,k=1}^{3} \epsilon_{ijk} \frac{\hat{\boldsymbol{e}}_k}{h_i h_j} \frac{\partial (h_j v_j)}{\partial u_i}, \tag{6.39}$$

in which the sums over i, j, and k run from 1 to 3. In rectangular coordinates, each scale factor $h_i = 1$, and the ith component of $\nabla \times v$ is

$$(\nabla \times v)_i = \sum_{j,k=1}^{3} \epsilon_{ijk} \frac{\partial v_k}{\partial x_j} = \sum_{j,k=1}^{3} \epsilon_{ijk} \partial_j v_k \tag{6.40}$$

or $(\nabla \times v)_i = \epsilon_{ijk}\partial_j v_k$ if we sum implicitly over j and k.

We can write the curl as a determinant

$$\nabla \times \boldsymbol{v} = \frac{1}{h_1 h_2 h_3} \begin{vmatrix} h_1 \hat{\boldsymbol{e}}_1 & h_2 \hat{\boldsymbol{e}}_2 & h_3 \hat{\boldsymbol{e}}_3 \\ \partial_1 & \partial_2 & \partial_3 \\ h_1 v_1 & h_2 v_2 & h_3 v_3 \end{vmatrix} . \tag{6.41}$$

Thus in cylindrical coordinates, where $h_1 = 1$, $h_2 = \rho$, and $h_3 = 1$, the curl is

$$\nabla \times \boldsymbol{v} = \frac{1}{\rho} \begin{vmatrix} \hat{\boldsymbol{\rho}} & \rho \hat{\boldsymbol{\phi}} & \hat{\boldsymbol{z}} \\ \partial_\rho & \partial_\phi & \partial_z \\ v_\rho & \rho v_\phi & v_z \end{vmatrix} \tag{6.42}$$

and in spherical coordinates, where $h_1 = 1$, $h_2 = r$, and $h_3 = r \sin\theta$, it is

$$\nabla \times \boldsymbol{v} = \frac{1}{r^2 \sin\theta} \begin{vmatrix} \hat{\boldsymbol{r}} & r\hat{\boldsymbol{\theta}} & r\sin\theta\, \hat{\boldsymbol{\phi}} \\ \partial_r & \partial_\theta & \partial_\phi \\ v_r & r\, v_\theta & r \sin\theta\, v_\phi \end{vmatrix} . \tag{6.43}$$

By assembling a large number of tiny squares, one may create an arbitrary finite surface S. The surface integral of the curl $\nabla \times \boldsymbol{v}$ over the tiny squares dS that make up the surface S is the sum of the line integrals dL over the sides of these tiny squares. The line integrals over the interior sides cancel leaving just the line integral along the boundary ∂S of the finite surface S. Thus we arrive at Stokes's theorem

$$\int_S \nabla \times \boldsymbol{v} \cdot dS = \int_{\partial S} \boldsymbol{v} \cdot d\boldsymbol{\ell}. \tag{6.44}$$

6.5 Separable partial differential equations

A linear partial differential equation (PDE) is **separable** if it can be decomposed into ordinary differential equations (ODEs). One then finds solutions to the ODEs and thus to the original PDE. The general solution to the PDE is then a sum over all of its linearly independent solutions with arbitrary coefficients. Sometimes the separability of a differential operator or of a differential equation depends upon the choice of coordinates.

Example 6.5 (The Helmholtz equation in two dimensions) In several coordinate systems, one can convert **Helmholtz's** linear homogeneous partial differential equation $-\nabla \cdot \nabla f(x) = -\Delta f(x) = k^2 f(x)$ into ordinary differential equations by writing the function $f(x)$ as a product of functions of a single variable.

In two dimensions and in **rectangular coordinates** (x, y), the function $f(x, y) = X(x)\, Y(y)$ is a solution of the Helmholtz equation as long as X and Y satisfy

$-X_a''(x) = a^2 X_a(x)$ and $-Y_b''(y) = b^2 Y_b(y)$ with $a^2 + b^2 = k^2$. One sets $X_a(x) = \alpha \sin ax + \beta \cos ax$ with a similar equation for $Y_b(y)$. Any linear combination of the functions $X_a(x)\, Y_b(y)$ with $a^2 + b^2 = k^2$ will be a solution of Helmholtz's equation $-\triangle f = k^2 f$.

The z-independent part of (6.34) is the Laplacian in **polar coordinates**

$$\nabla \cdot \nabla f = \triangle f = \frac{\partial^2 f}{\partial \rho^2} + \frac{1}{\rho}\frac{\partial f}{\partial \rho} + \frac{1}{\rho^2}\frac{\partial^2 f}{\partial \phi^2}, \tag{6.45}$$

in which Helmholtz's equation $-\triangle f = k^2 f$ also is separable. We let $f(\rho, \phi) = \mathrm{P}(\rho)\, \Phi(\phi)$ and get $\mathrm{P}''\Phi + \mathrm{P}'\Phi/\rho + \mathrm{P}\Phi''/\rho^2 = -k^2 \mathrm{P}\, \Phi$. Multiplying both sides by $\rho^2/\mathrm{P}\,\Phi$, we have

$$\rho^2 \frac{\mathrm{P}''}{\mathrm{P}} + \rho \frac{\mathrm{P}'}{\mathrm{P}} + \rho^2 k^2 = -\frac{\Phi''}{\Phi} = n^2, \tag{6.46}$$

in which the first three terms are functions of ρ, the fourth term $-\boldsymbol{\Phi}''/\Phi$ is a function of ϕ, and the last term n^2 is a constant. The constant n must be an integer if $\Phi_n(\phi) = a \sin(n\phi) + b\cos(n\phi)$ is to be single valued on the interval $[0, 2\pi]$. The function $\mathrm{P}_{k,n}(\rho) = J_n(k\rho)$ satisfies

$$\rho^2 \mathrm{P}_{k,n}'' + \rho \mathrm{P}_{k,n}' + \rho^2 k^2 \mathrm{P}_{k,n} = n^2 \mathrm{P}_{k,n}, \tag{6.47}$$

because the **Bessel function** of the first kind $J_n(x)$ satisfies

$$x^2 J_n'' + x J_n' + x^2 J_n = n^2 J_n, \tag{6.48}$$

which is Bessel's equation (9.4) (Friedrich Bessel, 1784–1846). So the product $f_{k,n}(\rho, \phi) = P_{k,n}(\rho)\, \Phi_n(\phi)$ is a solution to Helmholtz's equation $-\triangle f = k^2 f$, as is any linear combination of such products for different ns. □

Example 6.6 (The Helmholtz equation in three dimensions) In three dimensions and in **rectangular coordinates** $\boldsymbol{r} = (x, y, z)$, the function $f(x, y, z) = X(x)Y(y)Z(z)$ is a solution of the ODE $-\triangle f = k^2 f$ as long as X, Y, and Z satisfy $-X_a'' = a^2 X_a$, $-Y_b'' = b^2 Y_b$, and $-Z_c'' = c^2 Z_c$ with $a^2 + b^2 + c^2 = k^2$. We set $X_a(x) = \alpha \sin ax + \beta \cos ax$ and so forth. Arbitrary linear combinations of the products $X_a\, Y_b\, Z_c$ also are solutions of Helmholtz's equation $-\triangle f = k^2 f$ as long as $a^2 + b^2 + c^2 = k^2$.

In **cylindrical coordinates** (ρ, ϕ, z), the Laplacian (6.34) is

$$\nabla \cdot \nabla f = \triangle f = \frac{1}{\rho}\left[\left(\rho f_{,\rho}\right)_{,\rho} + \frac{1}{\rho} f_{,\phi\phi} + \rho f_{,zz}\right] \tag{6.49}$$

and so if we substitute $f(\rho, \phi, z) = P(\rho)\, \Phi(\phi)\, Z(z)$ into Helmholtz's equation $-\triangle f = \alpha^2 f$ and multiply both sides by $-\rho^2/P\,\Phi\, Z$, then we get

$$\frac{\rho^2}{f}\triangle f = \frac{\rho^2 P'' + \rho P'}{P} + \frac{\Phi''}{\Phi} + \rho^2 \frac{Z''}{Z} = -\alpha^2 \rho^2. \tag{6.50}$$

If we set $Z_k(z) = e^{kz}$, then this equation becomes (6.46) with k^2 replaced by $\alpha^2 + k^2$. Its solution then is

$$f(\rho, \phi, z) = J_n(\sqrt{\alpha^2 + k^2}\rho)\, e^{in\phi}\, e^{kz}, \tag{6.51}$$

in which n must be an integer if the solution is to apply to the full range of ϕ from 0 to 2π. The case in which $\alpha = 0$ corresponds to Laplace's equation with solution $f(\rho, \phi, z) = J_n(k\rho)e^{in\phi}e^{kz}$. We could have required Z to satisfy $Z'' = -k^2 Z$. The solution (6.51) then would be

$$f(\rho, \phi, z) = J_n(\sqrt{\alpha^2 - k^2}\,\rho)\, e^{in\phi}\, e^{ikz}. \tag{6.52}$$

But if $\alpha^2 - k^2 < 0$, we write this solution in terms of the **modified Bessel function** $I_n(x) = i^{-n} J_n(ix)$ (section 9.3) as

$$f(\rho, \phi, z) = I_n(\sqrt{k^2 - \alpha^2}\,\rho)\, e^{in\phi}\, e^{ikz}. \tag{6.53}$$

In **spherical coordinates**, the Laplacian (6.35) is

$$\triangle f = \frac{1}{r^2}\frac{\partial}{\partial r}\left(r^2 \frac{\partial f}{\partial r}\right) + \frac{1}{r^2 \sin\theta}\frac{\partial}{\partial \theta}\left(\sin\theta \frac{\partial f}{\partial \theta}\right) + \frac{1}{r^2 \sin^2\theta}\frac{\partial^2 f}{\partial \phi^2}. \tag{6.54}$$

If we set $f(r, \theta, \phi) = R(r)\, \Theta(\theta)\, \Phi_m(\phi)$ where $\Phi_m = e^{im\phi}$ and multiply both sides of the Helmholtz equation $-\triangle f = k^2 f$ by $-r^2/R\Theta\Phi$, then we get

$$\frac{\left(r^2 R'\right)'}{R} + \frac{\left(\sin\theta\, \Theta'\right)'}{\sin\theta\, \Theta} - \frac{m^2}{\sin^2\theta} = -k^2 r^2. \tag{6.55}$$

The first term is a function of r, the next two terms are functions of θ, and the last term is a constant. So we set the r-dependent terms equal to a constant $\ell(\ell+1) - k^2$ and the θ-dependent terms equal to $-\ell(\ell+1)$, and we require the **associated Legendre function** $\Theta_{\ell,m}(\theta)$ to satisfy (8.91)

$$\left(\sin\theta\, \Theta'_{\ell,m}\right)' / \sin\theta + \left[\ell(\ell+1) - m^2/\sin^2\theta\right] \Theta_{\ell,m} = 0. \tag{6.56}$$

If $\Phi(\phi) = e^{im\phi}$ is to be single valued for $0 \le \phi \le 2\pi$, then the parameter m must be an integer. As we'll see in chapter 8, the constant ℓ also must be an integer with $-\ell \le m \le \ell$ (example 6.29, section 8.12) if $\Theta_{\ell,m}(\theta)$ is to be single valued and finite for $0 \le \theta \le \pi$. The product $f = R\, \Theta\, \Phi$ then will obey Helmholtz's equation $-\triangle f = k^2 f$ if the radial function $R_{k,\ell}(r) = j_\ell(kr)$ satisfies

$$r^2 R''_{k,\ell} + 2r R'_{k,\ell} + \left[k^2 r^2 - \ell(\ell+1)\right] R_{k,\ell} = 0, \tag{6.57}$$

which it does because the **spherical Bessel function** $j_\ell(x)$ obeys Bessel's equation (9.63)

$$x^2 j''_\ell + 2x j'_\ell + [x^2 - \ell(\ell+1)] j_\ell = 0. \tag{6.58}$$

In three dimensions, Helmholtz's equation separates in 11 standard coordinate systems (Morse and Feshbach, 1953, pp. 655–664). □

6.6 Wave equations

You can easily solve some of the linear homogeneous partial differential equations of electrodynamics (Exercise 6.6) and quantum field theory.

Example 6.7 (The Klein–Gordon equation) In Minkowski space, the analog of the Laplacian in natural units ($\hbar = c = 1$) is (summing over a from 0 to 3)

$$\Box = \partial_a \partial^a = \Delta - \frac{\partial^2}{\partial x^{02}} = \Delta - \frac{\partial^2}{\partial t^2} \tag{6.59}$$

and the Klein–Gordon wave equation is

$$\left(\Box - m^2\right) A(x) = \left(\Delta - \frac{\partial^2}{\partial t^2} - m^2\right) A(x) = 0. \tag{6.60}$$

If we set $A(x) = B(px)$ where $px = p_a x^a = \boldsymbol{p} \cdot \boldsymbol{x} - p^0 x^0$, then the kth partial derivative of A is p_k times the first derivative of B

$$\frac{\partial}{\partial x^k} A(x) = \frac{\partial}{\partial x^k} B(px) = p_k B'(px) \tag{6.61}$$

and so the Klein–Gordon equation (6.60) becomes

$$\left(\Box - m^2\right) A = (\boldsymbol{p}^2 - (p^0)^2) B'' - m^2 B = p^2 B'' - m^2 B = 0, \tag{6.62}$$

in which $p^2 = \boldsymbol{p}^2 - (p^0)^2$. Thus if $B(p \cdot x) = \exp(ip \cdot x)$ so that $B'' = -B$, and if the energy–momentum 4-vector $(p^0, \boldsymbol{p})$ satisfies $p^2 + m^2 = 0$, then $A(x)$ will satisfy the Klein–Gordon equation. The condition $p^2 + m^2 = 0$ relates the energy $p^0 = \sqrt{\boldsymbol{p}^2 + m^2}$ to the momentum $\boldsymbol{p}$ for a particle of mass m. □

Example 6.8 (Field of a spinless boson) The quantum field

$$\phi(x) = \int \frac{d^3p}{\sqrt{2p^0(2\pi)^3}} \left[a(\boldsymbol{p}) e^{ipx} + a^\dagger(\boldsymbol{p}) e^{-ipx}\right] \tag{6.63}$$

describes spinless bosons of mass m. It satisfies the Klein–Gordon equation $\left(\Box - m^2\right) \phi(x) = 0$ because $p^0 = \sqrt{\boldsymbol{p}^2 + m^2}$. The operators $a(\boldsymbol{p})$ and $a^\dagger(\boldsymbol{p})$ respectively represent the annihilation and creation of the bosons and obey the commutation relations

$$[a(\boldsymbol{p}), a^\dagger(\boldsymbol{p}')] = \delta^3(\boldsymbol{p} - \boldsymbol{p}') \text{ and } [a(\boldsymbol{p}), a(\boldsymbol{p}')] = [a^\dagger(\boldsymbol{p}), a^\dagger(\boldsymbol{p}')] = 0 \tag{6.64}$$

in units with $\hbar = c = 1$. These relations make the field $\phi(x)$ and its time derivative $\dot{\phi}(y)$ satisfy **the canonical equal-time commutation relations**

$$[\phi(\boldsymbol{x}, t), \dot{\phi}(\boldsymbol{y}, t)] = i\delta^3(\boldsymbol{x} - \boldsymbol{y}) \text{ and } [\phi(\boldsymbol{x}, t), \phi(\boldsymbol{y}, t)] = [\dot{\phi}(\boldsymbol{x}, t), \dot{\phi}(\boldsymbol{y}, t)] = 0, \tag{6.65}$$

in which the dot means time derivative. □

Example 6.9 (Field of the photon) The electromagnetic field has four components, but in the Coulomb or radiation gauge $\nabla \cdot \boldsymbol{A}(x) = 0$, the component A_0 is a function of the charge density, and the vector potential $\boldsymbol{A}$ in the absence of charges and currents satisfies the wave equation $\Box \boldsymbol{A}(x) = 0$ for a spin-one massless particle. We write it as

$$\boldsymbol{A}(x) = \sum_{s=1}^{2} \int \frac{d^3p}{\sqrt{2p^0(2\pi)^3}} \left[\boldsymbol{e}(\boldsymbol{p}, s)\, a(\boldsymbol{p}, s)\, e^{ipx} + \boldsymbol{e}^*(\boldsymbol{p}, s)\, a^\dagger(\boldsymbol{p}, s)\, e^{-ipx} \right], \tag{6.66}$$

in which the sum is over the two possible polarizations s. The energy p^0 is equal to the modulus $|\boldsymbol{p}|$ of the momentum because the photon is massless, $p^2 = 0$. The dot-product of the polarization vectors $\boldsymbol{e}(\boldsymbol{p}, s)$ with the momentum vanishes $\boldsymbol{p} \cdot \boldsymbol{e}(\boldsymbol{p}, s) = 0$ so as to respect the gauge condition $\nabla \cdot \boldsymbol{A}(x) = 0$. The annihilation and creation operators obey the commutation relations

$$\begin{aligned} [a(\boldsymbol{p}, s), a^\dagger(\boldsymbol{p}', s')] &= \delta^3(\boldsymbol{p} - \boldsymbol{p}')\, \delta_{s,s'} \\ [a(\boldsymbol{p}, s), a(\boldsymbol{p}', s')] &= [a^\dagger(\boldsymbol{p}, s), a^\dagger(\boldsymbol{p}', s')] = 0 \end{aligned} \tag{6.67}$$

but the commutation relations of the vector potential $\boldsymbol{A}(x)$ involve the transverse delta function

$$\begin{aligned} \left[A_i(t, \boldsymbol{x}), \dot{A}_j(t, \boldsymbol{y}) \right] &= i\delta_{ij}\delta^{(3)}(\boldsymbol{x} - \boldsymbol{y}) + i \frac{\partial^2}{\partial x^i \partial x^j} \frac{1}{4\pi |\boldsymbol{x} - \boldsymbol{y}|} \\ &= i \int e^{i\boldsymbol{k}\cdot(\boldsymbol{x}-\boldsymbol{y})} \left(\delta_{ij} - \frac{k_i k_j}{\boldsymbol{k}^2} \right) \frac{d^3k}{(2\pi)^3} \end{aligned} \tag{6.68}$$

because of the Coulomb-gauge condition $\nabla \cdot \boldsymbol{A}(x) = 0$. □

Example 6.10 (Dirac's equation) Fields $\chi_b(x)$ that describe particles of spin one-half have four components, $b = 1, \ldots, 4$. In the absence of interactions, they satisfy the Dirac equation

$$\left(\gamma^a_{bc} \partial_a + m\delta_{bc} \right) \chi_c(x) = 0, \tag{6.69}$$

in which repeated indices are summed over – b, c from 1 to 4 and a from 0 to 3. In matrix notation, the **Dirac equation** is

$$\left(\gamma^a \partial_a + m \right) \chi(x) = 0. \tag{6.70}$$

The four Dirac gamma matrices are defined by the 16 rules

$$\{\gamma^a, \gamma^b\} \equiv \gamma^a\gamma^b + \gamma^b\gamma^a = 2\eta^{ab}, \tag{6.71}$$

in which η is the 4×4 diagonal matrix $\eta^{00} = \eta_{00} = -1$ and $\eta^{bc} = \eta_{bc} = \delta_{bc}$ for $b, c = 1, 2$, or 3.

If $\phi(x)$ is a 4-component field that satisfies the Klein–Gordon equation $(\Box - m^2)\phi = 0$, then the field $\chi(x) = (\gamma^b \partial_b - m)\phi(x)$ satisfies (exercise 6.7) the Dirac equation (6.70)

$$\begin{aligned}(\gamma^a \partial_a + m)\, \chi(x) &= (\gamma^a \partial_a + m)\,(\gamma^b \partial_b - m)\phi(x)\\ &= \left(\gamma^a \gamma^b \partial_a \partial_b - m^2\right) \phi(x)\\ &= \left[\tfrac{1}{2}\left(\{\gamma^a, \gamma^b\} + [\gamma^a, \gamma^b]\right) \partial_a \partial_b - m^2\right] \phi(x)\\ &= \left(\eta^{ab} \partial_a \partial_b - m^2\right) \phi(x) = (\Box - m^2)\phi(x) = 0.\end{aligned}$$

The simplest Dirac field is the Majorana field

$$\chi_b(x) = \int \frac{d^3p}{(2\pi)^{3/2}} \sum_s \left[u_b(\boldsymbol{p}, s)\, a(\boldsymbol{p}, s) e^{ipx} + v_b(\boldsymbol{p}, s)\, a^\dagger(\boldsymbol{p}, s) e^{-ipx}\right] \tag{6.72}$$

in which $p^0 = \sqrt{\boldsymbol{p}^2 + m^2}$, s labels the two spin states, and the operators a and $a^\dagger$ obey **the anticommutation relations**

$$\begin{aligned}\{a(\boldsymbol{p}, s), a^\dagger(\boldsymbol{p}', s')\} &\equiv a(\boldsymbol{p}, s)\, a^\dagger(\boldsymbol{p}', s') + a^\dagger(\boldsymbol{p}', s')\, a(\boldsymbol{p}, s) = \delta_{ss'}\, \delta(\boldsymbol{p} - \boldsymbol{p}'),\\ \{a(\boldsymbol{p}, s), a(\boldsymbol{p}', s')\} &= \{a^\dagger(\boldsymbol{p}, s), a^\dagger(\boldsymbol{p}', s')\} = 0.\end{aligned} \tag{6.73}$$

It describes a neutral particle of mass m.

If two Majorana fields χ_1 and χ_2 represent particles of the same mass, then one may combine them into one Dirac field

$$\psi(x) = \frac{1}{\sqrt{2}}\left[\chi_1(x) + i\chi_2(x)\right], \tag{6.74}$$

which describes a charged particle such as a quark or a lepton. □

6.7 First-order differential equations

The equation

$$\frac{dy}{dx} = f(x, y) = -\frac{P(x, y)}{Q(x, y)} \tag{6.75}$$

or *system*

$$P(x, y)\, dx + Q(x, y)\, dy = 0 \tag{6.76}$$

is a **first-order ordinary differential equation.**

6.8 Separable first-order differential equations

If in a first-order ordinary differential equation like (6.76) one can separate the dependent variable y from the independent variable x

$$F(x)\, dx + G(y)\, dy = 0 \tag{6.77}$$

then the equation (6.76) is **separable** and (6.77) is **separated.**

Once the variables are separated, one can integrate and so obtain an equation, called the **general integral**

$$0 = \int_{x_0}^{x} F(x')\, dx' + \int_{y_0}^{y} G(y')\, dy' \tag{6.78}$$

relating y to x and providing a solution $y(x)$ of the differential equation.

Example 6.11 (Zipf's law) In 1913, Auerbach noticed that many quantities are distributed as (Gell-Mann, 1994, pp. 92–100)

$$dn = -a \frac{dx}{x^{k+1}} \tag{6.79}$$

an ODE that is separable and separated. For $k \neq 0$, we may integrate this to $n + c = a/kx^k$ or

$$x = \left(\frac{a}{k(n+c)} \right)^{1/k} \tag{6.80}$$

in which c is a constant.

The case $k = 1$ occurs frequently $x = a/(n+c)$ and is called **Zipf's law**. With $c = 0$, it applies approximately to the populations of cities: if the largest city ($n = 1$) has population x, then the populations of the second, third, and fourth cities ($n = 2, 3, 4$) will be $x/2$, $x/3$, and $x/4$.

Again with $c = 0$, Zipf's law applies to the occurrence of numbers x in a table of some sort. Since $x = a/n$, the rank n of the number x is approximately $n = a/x$. So the number of numbers that occur with first digit d and, say, 4 trailing digits will be

$$\begin{aligned} n(d0000) - n(d9999) &= a \left(\frac{1}{d0000} - \frac{1}{d9999} \right) = a \left(\frac{9999}{d0000 \times d9999} \right) \\ &\approx a \left(\frac{10^4}{d(d+1)\, 10^8} \right) = \frac{a\, 10^{-4}}{d(d+1)}. \end{aligned} \tag{6.81}$$

The ratio of the number of numbers with first digit d to the number with first digit d' is then $d'(d'+1)/d(d+1)$. For example, the first digit is more likely to be 1 than 9 by a factor of 45. The German government uses such formulas to catch tax evaders. □

Example 6.12 (The logistic equation)

$$\frac{dy}{dt} = ay \left(1 - \frac{y}{Y} \right) \tag{6.82}$$

is separable and separated. It describes a wide range of phenomena whose evolution with time t is sigmoidal such as (Gell-Mann, 2008) the cumulative number of casualties in a war, the cumulative number of deaths in London's great plague, and the cumulative number of papers in an academic's career. It also describes the effect y on an animal of a given dose t of a drug.

With $f = y/Y$, the logistic equation (6.82) is $\dot{f} = af(1-f)$ or

$$a\,dt = \frac{df}{f(1-f)} = \frac{df}{f} + \frac{df}{1-f}, \tag{6.83}$$

which we may integrate to $a(t - t_h) = \ln[f/(1-f)]$. Taking the exponential of both sides, we find $\exp[a(t-t_h)] = f/(1-f)$, which we can solve for f

$$f(t) = \frac{e^{a(t-t_h)}}{1 + e^{a(t-t_h)}}. \tag{6.84}$$

The sigmoidal shape of $f(t)$ is like a smoothed Heaviside function. □

Example 6.13 (Lattice QCD) In lattice field theory, the beta function

$$\beta(g) \equiv -\frac{dg}{d\ln a} \tag{6.85}$$

tells us how we must adjust the coupling constant g in order to keep the physical predictions of the theory constant as we vary the lattice spacing a. In quantum chromodynamics $\beta(g) = -\beta_0 g^3 - \beta_1 g^5 + \cdots$ where

$$\beta_0 = \frac{1}{(4\pi)^2}\left(11 - \frac{2}{3}n_{\rm f}\right) \quad \text{and} \quad \beta_1 = \frac{1}{(4\pi)^4}\left(102 - 10\,n_{\rm f} - \frac{8}{3}n_{\rm f}\right), \tag{6.86}$$

in which $n_{\rm f}$ is the number of light quark flavors. Combining the definition (6.85) of the β-function with the first term of its expansion $\beta(g) = -\beta_0 g^3$ for small g, one arrives at the differential equation

$$\frac{dg}{d\ln a} = \beta_0\, g^3, \tag{6.87}$$

which one may integrate

$$\int d\ln a = \ln a + c = \int \frac{dg}{\beta_0 g^3} = -\frac{1}{2\beta_0 g^2} \tag{6.88}$$

to find

$$\Lambda\, a(g) = e^{-1/2\beta_0 g^2}, \tag{6.89}$$

in which Λ is a constant of integration. As g approaches 0, which is an essential singularity (section 5.11), the lattice spacing $a(g)$ goes to zero *very fast* (as long as $n_{\rm f} \le 16$). The inverse of this relation $g(a) \approx 1/\sqrt{\beta_0 \ln(1/a^2\Lambda^2)}$ shows that the coupling constant $g(a)$ slowly goes to zero as the lattice spacing (or shortest wave-length) a goes to zero. The strength of the interaction shrinks logarithmically as the energy $1/a$ increases in this lattice version of **asymptotic freedom**. □

6.9 Hidden separability

As long as each of the functions $P(x,y)$ and $Q(x,y)$ in the ODE

$$P(x,y)dx + Q(x,y)dy = U(x)V(y)dx + R(x)S(y)dy = 0 \tag{6.90}$$

can be factored $P(x,y) = U(x)V(y)$ and $Q(x,y) = R(x)S(y)$ into the product of a function of x times a function of y, then the ODE is separable. Following Ince (1956), we divide the ODE by $R(x)V(y)$, separate the variables

$$\frac{U(x)}{R(x)}\,dx + \frac{S(y)}{V(y)}\,dy = 0, \tag{6.91}$$

and integrate

$$\int \frac{U(x)}{R(x)}\,dx + \int \frac{S(y)}{V(y)}\,dy = C, \tag{6.92}$$

in which C is a constant of integration.

Example 6.14 (Hidden separability) We separate the variables in

$$x(y^2-1)\,dx - y(x^2-1)\,dy = 0 \tag{6.93}$$

by dividing by $(y^2-1)(x^2-1)$ so as to get

$$\frac{x}{x^2-1}\,dx - \frac{y}{y^2-1}\,dy = 0. \tag{6.94}$$

Integrating, we find $\ln(x^2-1) - \ln(y^2-1) = -\ln C$ or $C(x^2-1) = y^2-1$, which we solve for $y(x) = \sqrt{1 + C(x^2-1)}$. □

6.10 Exact first-order differential equations

The differential equation

$$P(x,y)\,dx + Q(x,y)\,dy = 0 \tag{6.95}$$

is **exact** if its left-hand side is the differential of some function $\phi(x,y)$

$$P\,dx + Q\,dy = d\phi = \phi_x\,dx + \phi_y\,dy. \tag{6.96}$$

We'll have more to say about the **exterior derivative** d in section 12.2.

The **criteria of exactness** are

$$P(x,y) = \frac{\partial\phi(x,y)}{\partial x} \equiv \phi_x(x,y) \quad \text{and} \quad Q(x,y) = \frac{\partial\phi(x,y)}{\partial y} \equiv \phi_y(x,y). \tag{6.97}$$

Thus, if the ODE (6.95) is exact, then

$$P_y(x, y) = \phi_{yx}(x, y) = \phi_{xy}(x, y) = Q_x(x, y), \tag{6.98}$$

which is called the **condition of integrability**. This condition implies that the ODE (6.95) is exact and integrable, as we'll see in section 6.11.

A first-order ODE that is separable and separated

$$P(x)dx + Q(y)dy = 0 \tag{6.99}$$

is exact because

$$P_y = 0 = Q_x. \tag{6.100}$$

But a first-order ODE may be exact without being separable.

Example 6.15 (Boyle's law) At a fixed temperature T, changes in the pressure P and volume V of an ideal gas are related by

$$PdV + VdP = 0. \tag{6.101}$$

This ODE is exact because $PdV + VdP = d(PV)$. Its integrated form is the **ideal-gas law**

$$PV = NkT, \tag{6.102}$$

in which N is the number of molecules in the gas and k is **Boltzmann's constant**, $k = 1.38066 \times 10^{-23}$ J/K $= 8.617385 \times 10^{-5}$ eV/K.

Incidentally, a more accurate formula, proposed by van der Waals (1837–1923) in his doctoral thesis in 1873, is

$$\left[P + \left(\frac{N}{V}\right)^2 a'\right]\left(V - Nb'\right) = NkT, \tag{6.103}$$

in which a' represents the mutual attraction of the molecules and has the dimensions of energy times volume and b' is the effective volume of a single molecule. This equation was one of many signs that molecules were real particles, independent of the imagination of chemists. Lamentably, most physicists refused to accept the reality of molecules until 1905 when Einstein related the viscous-friction coefficient ζ and the diffusion constant D to the energy kT of a thermal fluctuation by the equation $\zeta\, D = kT$, as explained in section 13.9 (Albert Einstein, 1879–1955). □

Example 6.16 (Human population growth) If the number of people rises as the square of the population, then $\dot{N} = N^2/b$. The separated and hence exact form of this differential equation is

$$\frac{dN}{N^2} = \frac{dt}{b}, \tag{6.104}$$

which we integrate to $N(t) = b/(T - t)$ where T is the time at which the population becomes infinite. With $T = 2025$ years and $b = 2 \times 10^{11}$ years, this formula is a fair model of the world's population between the years 1 and 1970. For a more accurate account, see von Foerster *et al.* (1960). □

6.11 The meaning of exactness

We can integrate the differentials of a first-order ODE

$$P(x,y)\,dx + Q(x,y)\,dy = 0 \tag{6.105}$$

along any contour $\mathcal{C}$ in the x-y plane, but in general we'd get a functional

$$\phi(x,y,\mathcal{C},x_0,y_0) = \int_{(x_0,y_0)\mathcal{C}}^{(x,y)} P(x',y')\,dx' + Q(x',y')\,dy' \tag{6.106}$$

that depends upon the contour $\mathcal{C}$ of integration as well as upon the endpoints (x_0, y_0) and (x, y).

But if the differential $Pdx + Qdy$ is exact, then it's the differential or exterior derivative $d\phi = P(x,y)\,dx + Q(x,y)\,dy$ of a function $\phi(x,y)$ that depends upon the variables x and y without any reference to a contour of integration. Thus if $Pdx + Qdy = d\phi$, then the contour integral (6.105) is

$$\int_{(x_0,y_0)\mathcal{C}}^{(x,y)} P(x',y')\,dx' + Q(x',y')\,dy' = \int_{(x_0,y_0)}^{(x,y)} d\phi = \phi(x,y) - \phi(x_0,y_0). \tag{6.107}$$

This integral defines a function $\phi(x,y,x_0,y_0) \equiv \phi(x,y) - \phi(x_0,y_0)$ whose differential vanishes $d\phi = Pdx + Qdy = 0$ according to the original differential equation (6.105). Thus the ODE and its exactness lead to an equation

$$\phi(x,y,x_0,y_0) = B \tag{6.108}$$

that we can solve for y, our solution of the ODE (6.105)

$$d\phi(x,y,x_0,y_0) = P(x,y)\,dx + Q(x,y)\,dy = 0. \tag{6.109}$$

Example 6.17 (Explicit use of exactness) We'll now explicitly use the criteria of exactness

$$P(x,y) = \frac{\partial\phi(x,y)}{\partial x} \equiv \phi_x(x,y) \quad\text{and}\quad Q(x,y) = \frac{\partial\phi(x,y)}{\partial y} \equiv \phi_y(x,y) \tag{6.110}$$

to integrate the general exact differential equation

$$P(x,y)\,dx + Q(x,y)\,dy = 0. \tag{6.111}$$

We use the first criterion $P = \phi_x$ to integrate the condition $\phi_x = P$ in the x-direction getting a known integral $R(x, y)$ and an unknown function $C(y)$

$$\phi(x, y) = \int P(x, y)\, dx + C(y) = R(x, y) + C(y). \tag{6.112}$$

The second criterion $Q = \phi_y$ tells us that

$$Q(x, y) = \phi_y(x, y) = R_y(x, y) + C_y(y). \tag{6.113}$$

We get $C(y)$ by integrating its known derivative $C_y = Q - R_y$

$$C(y) = \int Q(x, y) - R_y(x, y)\, dy + D. \tag{6.114}$$

We now put C into the formula $\phi = R + C$, which is (6.112). Setting $\phi = E$, a constant, we find an equation

$$\begin{aligned}\phi(x, y) &= R(x, y) + C(y) \\ &= R(x, y) + \int Q(x, y) - R_y(x, y)\, dy + D = E \end{aligned} \tag{6.115}$$

that we can solve for y. □

Example 6.18 (Using exactness) The functions P and Q in the differential equation

$$P(x, y)\, dx + Q(x, y)\, dy = \ln(y^2 + 1)\, dx + \frac{2y(x-1)}{y^2+1}\, dy = 0 \tag{6.116}$$

are factorized, so the ODE is separable. It's also exact since

$$P_y = \frac{2y}{y^2+1} = Q_x \tag{6.117}$$

and so we can apply the method just outlined. First, as in (6.112), we integrate $\phi_x = P$ in the x-direction

$$\phi(x, y) = \int \ln(y^2 + 1)\, dx + C(y) = x \ln(y^2 + 1) + C(y). \tag{6.118}$$

Then as in (6.113), we use $\phi_y = Q$

$$\phi(x, y)_y = \frac{2xy}{y^2+1} + C_y(y) = Q(x, y) = \frac{2y(x-1)}{y^2+1} \tag{6.119}$$

to find that $C_y = -2y/(y^2+1)$, which we integrate in the y-direction as in (6.114)

$$C(y) = -\ln(y^2 + 1) + D. \tag{6.120}$$

We now put $C(y)$ into our formula (6.118) for $\phi(x, y)$

$$\begin{aligned}\phi(x, y) &= x \ln(y^2 + 1) - \ln(y^2 + 1) + D \\ &= (x - 1) \ln(y^2 + 1) + D, \end{aligned} \tag{6.121}$$

which we set equal to a constant

$$\phi(x, y) = (x-1)\ln(y^2+1) + D = E \tag{6.122}$$

or more simply $(x-1)\ln(y^2+1) = F$. Unraveling this equation we find

$$y(x) = \left(e^{F/(x-1)} - 1\right)^{1/2} \tag{6.123}$$

as our solution to the differential equation (6.116). □

6.12 Integrating factors

With great luck, one might invent an **integrating factor** $\alpha(x, y)$ that makes an ordinary differential equation $P\,dx + Q\,dy = 0$ exact

$$\alpha\, P\, dx + \alpha\, Q\, dy = d\phi \tag{6.124}$$

and therefore integrable. Such an integrating factor α must satisfy both

$$\alpha\, P = \phi_x \quad \text{and} \quad \alpha\, Q = \phi_y \tag{6.125}$$

so that

$$(\alpha\, P)_y = \phi_{xy} = (\alpha\, Q)_x. \tag{6.126}$$

Example 6.19 (Two simple integrating factors) The ODE $ydx - xdy = 0$ is not exact, but $\alpha(x, y) = 1/x^2$ is an integrating factor. For after multiplying by α, we have

$$-\frac{y}{x^2}dx + \frac{1}{x}dy = 0 \tag{6.127}$$

so that $P = -y/x^2$, $Q = 1/x$, and

$$P_y = -\frac{1}{x^2} = Q_x, \tag{6.128}$$

which shows that (6.127) is exact.

Another integrating factor is $\alpha(x, y) = 1/xy$, which separates the variables

$$\frac{dx}{x} = \frac{dy}{y} \tag{6.129}$$

so that we can integrate and get $\ln(y/y_0) = \ln(x/x_0)$ or $\ln(yx_0/xy_0) = 0$, which implies that $y = (y_0/x_0)x$. □

6.13 Homogeneous functions

A function $f(x) = f(x_1, \ldots, x_k)$ of k variables x_i is **homogeneous** of degree n if

$$f(tx) = f(tx_1, \ldots, tx_k) = t^n f(x). \tag{6.130}$$

For instance, $z^2 \ln(x/y)$ is homogeneous of degree 2 because

$$(tz)^2 \ln(tx/ty) = t^2 \left(z^2 \ln(x/y)\right). \tag{6.131}$$

By differentiating (6.130) with respect to t, we find

$$\frac{d}{dt} f(tx) = \sum_{i=1}^{k} \frac{dtx_i}{dt} \frac{\partial f(tx)}{\partial tx_i} = \sum_{i=1}^{k} x_i \frac{\partial f(tx)}{\partial tx_i} = nt^{n-1} f(x). \tag{6.132}$$

Setting $t = 1$, we see that a function that is homogeneous of degree n satisfies

$$\sum_{i=1}^{k} x_i \frac{\partial f(x)}{\partial x_i} = n f(x), \tag{6.133}$$

which is one of Euler's many theorems.

6.14 The virial theorem

Consider N particles moving nonrelativistically in a potential $V(x)$ of $3N$ variables that is homogeneous of degree n. Their **virial** is the sum of the products of the coordinates x_i multiplied by the momenta p_i

$$G = \sum_{i=1}^{3N} x_i p_i. \tag{6.134}$$

In terms of the kinetic energy $T = (v_1 p_1 + \cdots + v_{3N} p_{3N})/2$, the time derivative of the virial is

$$\frac{dG}{dt} = \sum_{i=1}^{3N} (v_i p_i + x_i F_i) = 2T + \sum_{i=1}^{3N} x_i F_i, \tag{6.135}$$

in which the time derivative of a momentum $\dot{p}_i = F_i$ is a component of the force. We now form the infinite time average of both sides of this equation

$$\lim_{t \to \infty} \frac{G(t) - G(0)}{t} = \left\langle \frac{dG}{dt} \right\rangle = 2 \langle T \rangle + \left\langle \sum_{i=1}^{3N} x_i F_i \right\rangle. \tag{6.136}$$

If the particles are bound by a potential V, then it is reasonable to assume that the positions and momenta of the particles and their virial $G(t)$ are bounded for all times, and we will make this assumption. It follows that as $t \to \infty$, the time average of the time derivative $\dot{G}$ of the virial must vanish

$$0 = 2\langle T \rangle + \left\langle \sum_{i=1}^{3N} x_i F_i \right\rangle. \tag{6.137}$$

Newton's law

$$F_i = -\frac{\partial V(x)}{\partial x_i} \tag{6.138}$$

now implies that

$$2\langle T \rangle = \left\langle \sum_{i=1}^{3N} x_i \frac{\partial V(x)}{x_i} \right\rangle. \tag{6.139}$$

If, further, the potential $V(x)$ is a *homogeneous function of degree n*, then Euler's theorem (6.133) gives us $x_i \partial_i V = nV$ and the **virial theorem**

$$\langle T \rangle = \frac{n}{2} \langle V(x) \rangle. \tag{6.140}$$

The long-term time average of the kinetic energy of particles trapped in a homogeneous potential of degree n is $n/2$ times the long-term time average of their potential energy.

Example 6.20 (Coulomb forces) A $1/r$ gravitational or electrostatic potential is homogeneous of degree -1, and so the virial theorem asserts that particles bound in such wells must have long-term time averages that satisfy

$$\langle T \rangle = -\frac{1}{2} \langle V(x) \rangle. \tag{6.141}$$

In natural units ($\hbar = c = 1$), the energy of an electron of momentum p a distance r from a proton is $E = p^2/2m - e^2/r$ in which e is the charge of the electron. The uncertainty principle (example 3.6) gives us an approximate lower bound on the product $rp \gtrsim 1$, which we will use in the form $rp = 1$ to estimate the energy E of the ground state of the hydrogen atom. Using $1/r = p$, we have $E = p^2/2m - e^2 p$. Differentiating, we find the minimum of E is at $0 = p/m - e^2$. Thus the kinetic energy of the ground state is $T = p^2/2m = me^4/2$ while its potential energy is $V = -e^2 p = -me^4$. Since $T = -V/2$, these values satisfy the virial theorem. They give the ground-state energy as $E = -me^4/2 = -mc^2(e^2/\hbar c)^2 =$ 13.6 eV. □

Example 6.21 (Harmonic forces) Particles confined in a harmonic potential $V(r) = \sum_k m_k \omega_k^2 r_k^2$, which is homogeneous of degree 2, must have long-term time averages that satisfy $\langle T \rangle = \langle V(x) \rangle$. □

6.15 Homogeneous first-order ordinary differential equations

Suppose the functions $P(x, y)$ and $Q(x, y)$ in the first-order ODE

$$P(x, y)\,dx + Q(x, y)\,dy = 0 \tag{6.142}$$

are homogeneous of degree n (Ince, 1956). We change variables from x and y to x and $y(x) = xv(x)$ so that $dy = xdv + vdx$, and so

$$P(x, xv)dx + Q(x, xv)(xdv + vdx) = 0. \tag{6.143}$$

The homogeneity of $P(x, y)$ and $Q(x, y)$ implies that

$$x^n P(1, v)dx + x^n Q(1, v)(xdv + vdx) = 0. \tag{6.144}$$

Rearranging this equation, we are able to separate the variables

$$\frac{dx}{x} + \frac{Q(1, v)}{P(1, v) + vQ(1, v)}\,dv = 0. \tag{6.145}$$

We integrate this equation

$$\ln x + \int \frac{Q(1, v)}{P(1, v) + vQ(1, v)}\,dv = C \tag{6.146}$$

and find $v(x)$ and so too the solution $y(x) = xv(x)$.

Example 6.22 (Using homogeneity) In the differential equation

$$(x^2 - y^2)\,dx + 2xy\,dy = 0 \tag{6.147}$$

the coefficients of the differentials $P(x, y) = x^2 - y^2$ and $Q(x, y) = 2xy$ are homogeneous functions of degree $n = 2$, so the above method applies. With $y(x) = xv(x)$, we have

$$x^2(1 - v^2)dx + 2x^2v(vdx + xdv) = 0, \tag{6.148}$$

in which x^2 cancels out, leaving $(1 + v^2)dx + 2vxdv = 0$. Separating variables and integrating, we find

$$\int \frac{dx}{x} + \int \frac{2v\,dv}{1 + v^2} = \ln C \tag{6.149}$$

or $\ln(1 + v^2) + \ln x = \ln C$. So $(1 + v^2)x = C$, which leads to the general integral $x^2 + y^2 = Cx$ and so to $y(x) = \sqrt{Cx - x^2}$ as the solution of the ODE (6.147). □

6.16 Linear first-order ordinary differential equations

The general form of a linear first-order ODE is

$$\frac{dy}{dx} + r(x)y = s(x). \tag{6.150}$$

We always can find an integrating factor $\alpha(x)$ that makes

$$0 = \alpha(ry - s)dx + \alpha dy \tag{6.151}$$

exact. If $P \equiv \alpha(ry - s)$ and $Q \equiv \alpha$, then the condition (6.98) for this equation to be exact is $P_y = \alpha r = Q_x = \alpha_x$ or $\alpha_x/\alpha = r$. So

$$\frac{d\ln\alpha}{dx} = r, \tag{6.152}$$

which we integrate to

$$\alpha(x) = \alpha(x_0)\exp\left(\int_{x_0}^{x} r(x')dx'\right). \tag{6.153}$$

Now since $\alpha r = \alpha_x$, the original equation (6.150) multiplied by this integrating factor is

$$\alpha y_x + \alpha r y = \alpha y_x + \alpha_x y = (\alpha y)_x = \alpha s. \tag{6.154}$$

Integrating, we find

$$\alpha(x)y(x) = \alpha(x_0)y(x_0) + \int_{x_0}^{x} \alpha(x')s(x')dx' \tag{6.155}$$

so that

$$y(x) = \frac{\alpha(x_0)y(x_0)}{\alpha(x)} + \frac{1}{\alpha(x)}\int_{x_0}^{x} \alpha(x')s(x')dx', \tag{6.156}$$

in which $\alpha(x)$ is the exponential (6.153). More explicitly, $y(x)$ is

$$y(x) = \exp\left(-\int_{x_0}^{x} r(x')dx'\right)\left[y(x_0) + \int_{x_0}^{x} \exp\left(\int_{x_0}^{x'} r(x'')dx''\right)s(x')dx'\right]. \tag{6.157}$$

The first term in the square brackets multiplied by the prefactor $\alpha(x_0)/\alpha(x)$ is the general solution of the homogeneous equation $y_x + ry = 0$. The second term in the square brackets multiplied by the prefactor $\alpha(x_0)/\alpha(x)$ is a particular solution of the inhomogeneous equation $y_x + ry = s$. Thus equation (6.157) expresses the general solution of the inhomogeneous equation (6.150) as the sum of a particular solution of the inhomogeneous equation and the general solution of the associated homogeneous equation.

We were able to find an integrating factor α because the original equation (6.150) was *linear* in y. So we could set $P = \alpha(ry - s)$ and $Q = \alpha$. When P and Q are more complicated, integrating factors are harder to find or nonexistent.

Example 6.23 (Bodies falling in air) The downward speed v of a mass m in a gravitational field of constant acceleration g is described by the inhomogeneous first-order ODE $mv_t = mg - bv$, in which b represents air resistance. This equation is like (6.150) but with t instead of x as the independent variable, $r = b/m$, and $s = g$. Thus by (6.157), its solution is

$$v(t) = \frac{mg}{b} + \left(v(0) - \frac{mg}{b}\right) e^{-bt/m}. \tag{6.158}$$

The terminal speed mg/b is nearly 200 km/h for a falling man. A diving Peregrine falcon can exceed 320 km/h; so can a falling bullet. But mice can fall down mine shafts and run off unhurt, and insects and birds can fly.

If the falling bodies are microscopic, a statistical model is appropriate. The potential energy of a mass m at height h is $V = mgh$. The heights of particles at temperature T K follow Boltzmann's distribution (1.345)

$$P(h) = P(0)e^{-mgh/kT}, \tag{6.159}$$

in which $k = 1.380\,6504 \times 10^{-23}$ J/K $= 8.617\,343 \times 10^{-5}$eV/K is his constant. The probability depends exponentially upon the mass m and drops by a factor of e with the **scale height** $S = kT/mg$, which can be a few kilometers for a small molecule. □

Example 6.24 (R-C circuit) The **capacitance** C of a capacitor is the charge Q it holds (on each plate) divided by the applied voltage V, that is, $C = Q/V$. The current I through the capacitor is the time derivative of the charge $I = \dot{Q} = C\dot{V}$. The voltage across a **resistor** of R Ω (Ohms) through which a current I flows is $V = IR$ by Ohm's law. So if a time-dependent voltage $V(t)$ is applied to a capacitor in series with a resistor, then $V(t) = Q/C + IR$. The current I therefore obeys the first-order differential equation

$$\dot{I} + I/RC = \dot{V}/R \tag{6.160}$$

or (6.150) with $x \to t$, $y \to I$, $r \to 1/RC$, and $s \to \dot{V}/R$. Since r is a constant, the integrating factor $\alpha(x) \to \alpha(t)$ is

$$\alpha(t) = \alpha(t_0)\, e^{(t-t_0)/RC}. \tag{6.161}$$

Our general solution (6.157) of the linear first-order ODE gives us the expression

$$I(t) = e^{-(t-t_0)/(RC)} \left[I(t_0) + \int_{t_0}^{t} e^{(t'-t_0)/(RC)} \frac{\dot{V}(t')}{R}\, dt' \right] \tag{6.162}$$

for the current $I(t)$. □

Example 6.25 (Emission rate from fluorophores) A fluorophore is a molecule that emits light when illuminated. The frequency of the emitted photon usually is less than that of the incident one. Consider a population of N fluorophores of which N_+ are excited and can emit light and $N_- = N - N_+$ are unexcited. If the fluorophores are exposed to an illuminating photon flux I, and the cross-section for the excitation of an unexcited fluorophore is σ, then the rate at which unexcited fluorophores become excited is $I\sigma N_-$. The time derivative of the number of excited fluorophores is then

$$\dot{N}_+ = I\sigma N_- - \frac{1}{\tau} N_+ = -\frac{1}{\tau} N_+ + I\sigma\,(N - N_+), \tag{6.163}$$

in which $1/\tau$ is the decay rate (also the emission rate) of the excited fluorophores. Using the shorthand $a = I\sigma + 1/\tau$, we have $\dot{N}_+ = -aN_+ + I\sigma N$, which we solve using the general formula (6.157) with $r = a$ and $s = I\sigma N$

$$N_+(t) = e^{-at}\left[N_+(0) + \int_0^t e^{at'} I(t')\sigma N\,dt'\right]. \tag{6.164}$$

If the illumination $I(t)$ is constant, then by doing the integral we find

$$N_+(t) = \frac{I\sigma N}{a}\left(1 - e^{-at}\right) + N_+(0)e^{-at}. \tag{6.165}$$

The emission rate $E = N_+(t)/\tau$ of photons from the $N_+(t)$ excited fluorophores then is

$$E = \frac{I\sigma N}{a\tau}\left(1 - e^{-at}\right) + \frac{N_+(0)}{\tau}e^{-at}, \tag{6.166}$$

which with $a = I\sigma + 1/1\tau$ gives for the emission rate per fluorophore

$$\frac{E}{N} = \frac{I\sigma}{1 + I\sigma\tau}\left(1 - e^{-(I\sigma+1/\tau)t}\right) \tag{6.167}$$

if no fluorophores were excited at $t = 0$, so that $N_+(0) = 0$. □

6.17 Systems of differential equations

Actual physical problems often involve several differential equations. The motion of n particles in three dimensions is described by $3n$ equations, electrodynamics by the four Maxwell equations (11.82 & 11.83), and the concentrations of different molecular species in a cell by thousands of coupled differential equations.

This field is too vast to cover in these pages, but we may hint at some of its features by considering the motion of n particles in three dimensions as described by a lagrangian $L(q, \dot{q}, t)$, in which q stands for the $3n$ coordinates $q_1, q_2, \ldots, q_{3n}$ and $\dot{q}$ for their time derivatives. The action of a motion $q(t)$ is the time integral of the lagrangian

$$S = \int_{t_1}^{t_2} L(q, \dot{q}, t) \, dt. \tag{6.168}$$

If $q(t)$ changes by a little bit δq, then the first-order change in the action is

$$\delta S = \int_{t_1}^{t_2} \sum_{i=1}^{3n} \left[\frac{\partial L(q, \dot{q}, t)}{\partial q_i} \delta q_i(t) + \frac{\partial L(q, \dot{q}, t)}{\partial \dot{q}_i} \delta \dot{q}_i(t) \right] dt. \tag{6.169}$$

The change in $\dot{q}_i$ is

$$\delta \frac{dq_i}{dt} = \frac{d(q_i + \delta q_i)}{dt} - \frac{dq_i}{dt} = \frac{d\,\delta q_i}{dt}, \tag{6.170}$$

the time derivative of the change δq_i, so we have

$$\delta S = \int_{t_1}^{t_2} \sum_{i=1}^{3n} \left[\frac{\partial L(q, \dot{q}, t)}{\partial q_i} \delta q_i(t) + \frac{\partial L(q, \dot{q}, t)}{\partial \dot{q}_i} \frac{d\,\delta q_i(t)}{dt} \right] dt. \tag{6.171}$$

We can integrate this by parts

$$\delta S = \int_{t_1}^{t_2} \sum_{i=1}^{3n} \left[\left(\frac{\partial L}{\partial q_i} - \frac{d}{dt} \frac{\partial L}{\partial \dot{q}_i} \right) \delta q_i(t) \right] dt + \left[\sum_{i=1}^{3n} \frac{\partial L}{\partial \dot{q}_i} \delta q_i(t) \right]_{t_1}^{t_2}. \tag{6.172}$$

A classical process is one that makes the action **stationary** to first order in $\delta q(t)$ for changes that vanish at the endpoints $\delta q(t_1) = 0 = \delta q(t_2)$. Thus a classical process satisfies Lagrange's equations

$$\frac{d}{dt} \frac{\partial L}{\partial \dot{q}_i} - \frac{\partial L}{\partial q_i} = 0 \quad \text{for} \quad i = 1, \dots, 3n. \tag{6.173}$$

Moreover, if the lagrangian does not depend explicitly on the time t, as is usually the case, then the hamiltonian

$$H = \sum_{i=1}^{3n} \frac{\partial L}{\partial \dot{q}_i} \dot{q}_i - L \equiv \sum_{i=1}^{3n} p_i \dot{q}_i - L \tag{6.174}$$

does not change with time because its time derivative is the vanishing explicit time dependence of the lagrangian $-\partial L/\partial t = 0$. That is

$$\begin{aligned} \dot{H} &= \sum_{i=1}^{3n} \frac{d}{dt} \frac{\partial L}{\partial \dot{q}_i} \dot{q}_i + \frac{\partial L}{\partial \dot{q}_i} \ddot{q}_i - \dot{L} = \sum_{i=1}^{3n} \frac{\partial L}{\partial q_i} \dot{q}_i + \frac{\partial L}{\partial \dot{q}_i} \ddot{q}_i - \dot{L} \\ &= -\frac{\partial L}{\partial t} = 0. \end{aligned} \tag{6.175}$$

Example 6.26 (Small oscillations) The lagrangian

$$L = \sum_{i=1}^{3n} \frac{m_i}{2} \dot{x}_i^2 - V(x) \tag{6.176}$$

describes n particles of mass m_i interacting through a potential $U(q)$ that has no explicit time dependence. By letting $q_i = \sqrt{m_i/m}\, x_i$ we may scale the masses to the same value m and set $V(q) = U(x)$, so that we have

$$L = \frac{m}{2} \sum_{i=1}^{3n} \dot{q}_i^2 - V(q) = \frac{m}{2} \dot{q} \cdot \dot{q} - V(q), \tag{6.177}$$

which describes n particles of mass m interacting through a potential $V(q)$. The hamiltonian is conserved, and if it has a minimum energy H_0 at q_0, then its first derivatives there vanish. So near q_0 the potential V to lowest order is a quadratic form in the displacements $r_i \equiv q_i - q_{i0}$ from the minima, and the lagrangian, apart from the constant $V(q_0)$, is

$$L \approx \frac{m}{2} \sum_{i=1}^{3n} \dot{r}_i^2 - \frac{1}{2} \sum_{j,k=1}^{3n} r_j \, r_k \, \frac{\partial^2 V(q_0)}{\partial q_j \partial q_k}. \tag{6.178}$$

The matrix V'' of second derivatives is real and symmetric, and so we may diagonalize it $V'' = O^{\mathsf{T}}\, V_d''\, O$ by an orthogonal transformation O. The lagrangian is diagonal in the new coordinates $s = O\, r$

$$L \approx \frac{1}{2} \sum_{i=1}^{3n} \left(m \, \dot{s}_i^2 - V_{di}'' \, s_i^2 \right) \tag{6.179}$$

and Lagrange's equations are $m \, \ddot{s}_i = - V_{di}'' \, s_i$. These **normal modes** are uncoupled harmonic oscillators $s_i(t) = a_i \cos \sqrt{V_{di}''/m}\, t + b_i \sin \sqrt{V_{di}''/m}\, t$ with frequencies that are real because q_0 is the minimum of the potential. □

6.18 Singular points of second-order ordinary differential equations

If in the ODE $y'' = f(x, y, y')$, the acceleration $y'' = f(x_0, y, y')$ is finite for all finite y and y', then x_0 is **a regular point** of the ODE. If $y'' = f(x_0, y, y')$ is infinite for any finite y and y', then x_0 is **a singular point** of the ODE.

If a second-order ODE $y'' + P(x)y' + Q(x)y = 0$ is linear and homogeneous and both $P(x_0)$ and $Q(x_0)$ are finite, then x_0 is a regular point of the ODE. But if $P(x_0)$ or $Q(x_0)$ or both are infinite, then x_0 is a singular point.

Some singular points are regular. If $P(x)$ or $Q(x)$ diverges as $x \to x_0$, but both $(x - x_0)P(x)$ and $(x - x_0)^2 Q(x)$ remain finite as $x \to x_0$, then x_0 is **a**

regular singular point or equivalently a **nonessential singular point**. But if either $(x-x_0)P(x)$ or $(x-x_0)^2Q(x)$ diverges as $x \to x_0$, then x_0 is **an irregular singular point** or equivalently an **essential singularity**.

To treat the point at infinity, one sets $z = 1/x$. Then if $(2z - P(1/z))/z^2$ and $Q(1/z)/z^4$ remain finite as $z \to 0$, the point $x_0 = \infty$ is **a regular point** of the ODE. If they don't remain finite, but $(2z - P(1/z))/z$ and $Q(1/z)/z^2$ do remain finite as $z \to 0$, then $x_0 = \infty$ is **a regular singular point**. Otherwise the point at infinity is **an irregular singular point** or an **essential singularity**.

Example 6.27 (Legendre's equation) Its self-adjoint form is

$$\left[\left(1 - x^2\right) y'\right]' + \ell(\ell + 1)y = 0, \tag{6.180}$$

which is $(1 - x^2)y'' - 2xy' + \ell(\ell + 1)y = 0$ or

$$y'' - \frac{2x}{1 - x^2}y' + \frac{\ell(\ell + 1)}{1 - x^2}y = 0. \tag{6.181}$$

It has regular singular points at $x = \pm 1$ and $x = \infty$ (exercise 6.15). □

6.19 Frobenius's series solutions

Frobenius showed how to find a power-series solution of a second-order linear homogeneous ordinary differential equation $y'' + P(x)\,y' + Q(x)\,y = 0$ at any of its regular or regular singular points. Writing the equation in the form $x^2y'' + x\,p(x)\,y' + q(x)\,y = 0$, we will assume that p and q are polynomials or analytic functions, and that $x = 0$ is a regular or regular singular point of the ODE so that $p(0)$ and $q(0)$ are both finite.

We expand y as a power series in x about $x = 0$

$$y(x) = x^r \sum_{n=0}^{\infty} a_n\, x^n, \tag{6.182}$$

in which $a_0 \neq 0$ is the coefficient of the lowest power of x in $y(x)$. Differentiating, we have

$$y'(x) = \sum_{n=0}^{\infty} (r + n)\, a_n\, x^{r+n-1} \tag{6.183}$$

and

$$y''(x) = \sum_{n=0}^{\infty} (r + n)(r + n - 1)\, a_n\, x^{r+n-2}. \tag{6.184}$$

When we substitute the three series (6.182–6.184) into our differential equation $x^2y'' + xp(x)y' + q(x)y = 0$, we find

$$\sum_{n=0}^{\infty} [(n+r)(n+r-1) + (n+r)p(x) + q(x)]\, a_n x^{n+r} = 0. \tag{6.185}$$

If this equation is to be satisfied for all x, then the coefficient of every power of x must vanish. The lowest power of x is x^r, and it occurs when $n = 0$ with coefficient $[r(r-1+p(0)) + q(0)]\, a_0$. Thus since $a_0 \neq 0$, we have

$$r(r-1+p(0)) + q(0) = 0. \tag{6.186}$$

This quadratic **indicial equation** has two roots r_1 and r_2.

To analyze higher powers of x, we introduce the notation

$$p(x) = \sum_{j=0}^{\infty} p_j x^j \quad \text{and} \quad q(x) = \sum_{j=0}^{\infty} q_j x^j, \tag{6.187}$$

in which $p_0 = p(0)$ and $q_0 = q(0)$. The requirement (exercise 6.16) that the coefficient of x^{r+k} vanishes gives us a **recurrence relation**

$$a_k = -\left[\frac{1}{(r+k)(r+k-1+p_0) + q_0}\right] \sum_{j=0}^{k-1} \left[(j+r)p_{k-j} + q_{k-j}\right] a_j \tag{6.188}$$

that expresses a_k in terms of $a_0, a_1, \ldots, a_{k-1}$. When $p(x)$ and $q(x)$ are polynomials of low degree, these equations become much simpler.

Example 6.28 (Sines and cosines) To apply Frobenius's method to the ODE $y'' + \omega^2 y = 0$, we first write it in the form $x^2y'' + xp(x)y' + q(x)y = 0$, in which $p(x) = 0$ and $q(x) = \omega^2 x^2$. So both $p(0) = p_0 = 0$ and $q(0) = q_0 = 0$, and the indicial equation (6.186) is $r(r-1) = 0$ with roots $r_1 = 0$ and $r_2 = 1$.

We first set $r = r_1 = 0$. Since the ps and qs vanish except for $q_2 = \omega^2$, the recurrence relation (6.188) is $a_k = -q_2\, a_{k-2}/k(k-1) = -\omega^2 a_{k-2}/k(k-1)$. Thus $a_2 = -\omega^2 a_0/2$, and $a_{2n} = (-1)^n \omega^{2n} a_0/(2n)!$. The recurrence relation (6.188) gives no information about a_1, so to find the simplest solution, we set $a_1 = 0$. The recurrence relation $a_k = -\omega^2 a_{k-2}/k(k-1)$ then makes all the terms a_{2n+1} of odd index vanish. Our solution for the first root $r_1 = 0$ then is

$$y(x) = \sum_{n=0}^{\infty} a_n\, x^n = a_0 \sum_{n=0}^{\infty} (-1)^n \frac{(\omega x)^{2n}}{(2n)!} = a_0 \cos \omega x. \tag{6.189}$$

Similarly, the recurrence relation (6.188) for the second root $r_2 = 1$ is $a_k = -\omega^2 a_{k-2}/k(k+1)$, so that $a_{2n} = (-1)^n \omega^{2n} a_0/(2n+1)!$, and we again set all the terms of odd index equal to zero. Thus we have

$$y(x) = x \sum_{n=0}^{\infty} a_n x^n = \frac{a_0}{\omega} \sum_{n=0}^{\infty} (-1)^n \frac{(\omega x)^{2n+1}}{(2n+1)!} = \frac{a_0}{\omega} \sin \omega x \tag{6.190}$$

as our solution for the second root $r_2 = 1$. □

Frobenius's method sometimes shows that solutions exist only when a parameter in the ODE assumes a special value called an **eigenvalue**.

Example 6.29 (Legendre's equation) If one rewrites Legendre's equation $(1 - x^2)y'' - 2xy' + \lambda y = 0$ as $x^2 y'' + xpy' + qy = 0$, then one finds $p(x) = -2x^2/(1 - x^2)$ and $q(x) = x^2 \lambda/(1 - x^2)$, which are analytic but not polynomials. In this case, it is simpler to substitute the expansions (6.182–6.184) directly into Legendre's equation $(1 - x^2)y'' - 2xy' + \lambda y = 0$. We then find

$$\sum_{n=0}^{\infty} \left[(n+r)(n+r-1)(1-x^2)x^{n+r-2} - 2(n+r)x^{n+r} + \lambda x^{n+r} \right] a_n = 0.$$

The coefficient of the lowest power of x is $r(r-1)a_0$, and so the indicial equation is $r(r-1) = 0$. For $r = 0$, we shift the index n on the term $n(n-1)x^{n-2}a_n$ to $n = j+2$ and replace n by j in the other terms:

$$\sum_{j=0}^{\infty} \left\{ (j+2)(j+1)\, a_{j+2} - [j(j-1) + 2j - \lambda]\, a_j \right\} x^j = 0. \tag{6.191}$$

Since the coefficient of x^j must vanish, we get the recursion relation

$$a_{j+2} = \frac{j(j+1) - \lambda}{(j+2)(j+1)} a_j, \tag{6.192}$$

which for big j says that $a_{j+2} \approx a_j$. Thus the series (6.182) does not converge for $|x| \geq 1$ unless $\lambda = j(j+1)$ for some integer j in which case the series (6.182) is a Legendre polynomial (chapter 8). □

Frobenius's method also allows one to expand solutions about $x_0 \neq 0$

$$y(x) = (x - x_0)^k \sum_{n=0}^{\infty} a_n (x - x_0)^n. \tag{6.193}$$

6.20 Fuch's theorem

The method of Frobenius can run amok, especially if one expands about a singular point x_0. One can get only one solution or none at all. But Fuch has shown that if one applies Frobenius's method to a linear homogeneous second-order ODE and expands about a regular point or a regular singular point, then one always gets at least one power-series solution:

1 if the two roots of the indicial equation are equal, one gets only one solution;
2 if the two roots differ by a noninteger, one gets two solutions;
3 if the two roots differ by an integer, then the bigger root yields a solution.

Example 6.30 (Roots that differ by an integer) If one applies the method of Frobenius to Legendre's equation as in example 6.29, then one finds (exercise 6.18) that the $k = 0$ and $k = 1$ roots lead to the same solution. □

6.21 Even and odd differential operators

Under the parity transformation $x \to -x$, a typical term transforms as

$$x^n \left(\frac{d}{dx}\right)^p x^k = \frac{k!}{(k-p)!}\, x^{n+k-p} \to (-1)^{n+k-p} \frac{k!}{(k-p)!}\, x^{n+k-p} \tag{6.194}$$

and so the corresponding differential operator transforms as

$$x^n \left(\frac{d}{dx}\right)^p \to (-1)^{n-p}\, x^n \left(\frac{d}{dx}\right)^p. \tag{6.195}$$

The reflected form of the second-order linear differential operator

$$L(x) = h_0(x) + h_1(x)\frac{d}{dx} + h_2(x)\frac{d^2}{dx^2} \tag{6.196}$$

therefore is

$$L(-x) = h_0(-x) - h_1(-x)\frac{d}{dx} + h_2(-x)\frac{d^2}{dx^2}. \tag{6.197}$$

The operator $L(x)$ is **even** if it is unchanged by reflection, that is, if $h_0(-x) = h_0(x)$, $h_1(-x) = -h_1(x)$, and $h_2(-x) = h_2(x)$, so that

$$L(-x) = L(x). \tag{6.198}$$

It is **odd** if it changes sign under reflection, that is, if $h_0(-x) = -h_0(x)$, $h_1(-x) = h_1(x)$, and $h_2(-x) = -h_2(x)$, so that

$$L(-x) = -L(x). \tag{6.199}$$

Not every differential operator $L(x)$ is even or odd. But just as we can write every function $f(x)$ whose reflected form $f(-x)$ is well defined as the sum of $[f(x) + f(-x)]/2$, which is even, and $[f(x) - f(-x)]/2$, which is odd,

$$f(x) = \tfrac{1}{2}[f(x) + f(-x)] + \tfrac{1}{2}[f(x) - f(-x)] \tag{6.200}$$

so too we can write every differential operator $L(x)$ whose reflected form $L(-x)$ is well defined as the sum of one that is even and one that is odd

$$L(x) = \tfrac{1}{2}[L(x) + L(-x)] + \tfrac{1}{2}[L(x) - L(-x)]. \tag{6.201}$$

Many of the standard differential operators have $h_0 = 1$ and are even.

If $y(x)$ is a solution of the ODE $L(x)\,y(x) = 0$ and $L(-x)$ is well defined, then we have $L(-x)\,y(-x) = 0$. If further $L(-x) = \pm L(x)$, then $y(-x)$ also is a solution $L(x)\,y(-x) = 0$. Thus if a differential operator $L(x)$ has **a definite parity**, that is, if $L(x)$ is either even or odd, then $y(-x)$ is a solution if $y(x)$ is, and solutions come in pairs $y(x) \pm y(-x)$, one even, one odd.

6.22 Wronski's determinant

If the N functions $y_1(x), \ldots, y_N(x)$ are linearly dependent, then by (6.8) there is a set of coefficients $k_1, \ldots, k_N$, **not all zero**, such that the sum

$$0 = k_1\, y_1(x) + \cdots + k_N\, y_N(x) \tag{6.202}$$

vanishes for all x. Differentiating i times, we get

$$0 = k_1\, y_1^{(i)}(x) + \cdots + k_N\, y_N^{(i)}(x) \tag{6.203}$$

for all x. So if we use the y_j and their derivatives to define the matrix

$$Y_{ij}(x) \equiv y_j^{(i-1)}(x) \tag{6.204}$$

then we may express the linear dependence (6.202) and (6.203) of the functions $y_1, \ldots, y_N$ in matrix notation as $0 = Y(x)\,k$ for some nonzero vector $k = (k_1, k_2, \ldots, k_N)$. Since the matrix $Y(x)$ maps the nonzero vector k to zero, its determinant must vanish: $\det(Y(x)) \equiv |\,Y(x)| = 0$. This determinant

$$W(x) = |\,Y(x)| = \left|y_j^{(i-1)}(x)\right| \tag{6.205}$$

is called **Wronski's determinant** or **the wronskian**. It vanishes on an interval if and only if the functions $y_j(x)$ or their derivatives are linearly dependent on the interval.

6.23 A second solution

If we have one solution to a second-order linear homogeneous ODE, then we may use the wronskian to find a second solution. Here's how: if y_1 and y_2 are two linearly independent solutions of the second-order linear homogeneous ordinary differential equation

$$y''(x) + P(x)\,y'(x) + Q(x)\,y(x) = 0 \tag{6.206}$$

then their wronskian does not vanish

$$W(x) = \begin{vmatrix} y_1(x) & y_2(x) \\ y_1'(x) & y_2'(x) \end{vmatrix} = y_1(x)\,y_2'(x) - y_2(x)\,y_1'(x) \neq 0 \tag{6.207}$$

except perhaps at isolated points. Its derivative

$$\begin{aligned} W' &= y_1'\,y_2' + y_1\,y_2'' - y_2'\,y_1' - y_2\,y_1'' \\ &= y_1\,y_2'' - y_2\,y_1'' \end{aligned} \tag{6.208}$$

must obey

$$\begin{aligned} W' &= -y_1\,\left(P\,y_2' + Q\,y_2\right) + y_2\,\left(P\,y_1' + Q\,y_1\right) \\ &= -P\left(y_1\,y_2' - y_2\,y_1'\right) \end{aligned} \tag{6.209}$$

or $W'(x) = -P(x)\,W(x)$, which we integrate to

$$W(x) = W(x_0)\,\exp\left[-\int_{x_0}^{x} P(x')dx'\right]. \tag{6.210}$$

This is Abel's formula for the wronskian (Niels Abel, 1802–1829).

Having expressed the wronskian in terms of the known function $P(x)$, we now use it to find $y_2(x)$ from $y_1(x)$. We note that

$$W = y_1\,y_2' - y_2\,y_1' = y_1^2\,\frac{d}{dx}\left(\frac{y_2}{y_1}\right). \tag{6.211}$$

So

$$\frac{d}{dx}\left(\frac{y_2}{y_1}\right) = \frac{W}{y_1^2}, \tag{6.212}$$

which we integrate to

$$y_2(x) = y_1(x)\left[\int^x \frac{W(x')}{y_1^2(x')}dx' + c\right]. \tag{6.213}$$

Using our formula (6.210) for the wronskian, we find as the second solution

$$y_2(x) = y_1(x)\int^x \frac{1}{y_1^2(x')}\,\exp\left[-\int^{x'} P(x'')dx''\right]dx' \tag{6.214}$$

apart from additive and multiplicative constants.

In the important special case in which $P(x) = 0$ the wronskian is a constant, $W'(x) = 0$, and the second solution is simply

$$y_2(x) = y_1(x)\int^x \frac{dx'}{y_1^2(x')}. \tag{6.215}$$

By Fuchs's theorem, Frobenius's expansion about a regular point or a regular singular point yields at least one solution. From this solution, we can use Wronski's trick to find a second (linearly independent) solution. So we always get two linearly independent solutions if we expand a second-order linear homogeneous ODE about a regular point or a regular singular point.

6.24 Why not three solutions?

We have seen that a second-order linear homogeneous ODE has two linearly independent solutions. Why not three?

If y_1, y_2, and y_3 were three linearly independent solutions of the second-order linear homogeneous ODE

$$0 = y_j'' + P\,y_j' + Q\,y_j, \tag{6.216}$$

then their third-order wronskian

$$W = \begin{vmatrix} y_1 & y_2 & y_3 \\ y_1' & y_2' & y_3' \\ y_1'' & y_2'' & y_3'' \end{vmatrix} \tag{6.217}$$

would not vanish except at isolated points.

But the ODE (6.216) relates the second derivatives $y_j'' = -(P\,y_j' + Q\,y_j)$ to the y_j' and the y_j, and so the third row of this third-order wronskian is a linear combination of the first two rows. Thus it vanishes identically

$$W = \begin{vmatrix} y_1 & y_2 & y_3 \\ y_1' & y_2' & y_3' \\ -Py_1' - Qy_1 & -Py_2' - Qy_2 & -Py_3' - Qy_3 \end{vmatrix} = 0 \tag{6.218}$$

and so any three solutions of a second-order ODE (6.216) are linearly dependent.

One may extend this argument to show that an nth-order linear homogeneous ODE can have at most n linearly independent solutions. To do so, we'll use superscript notation (6.19) in which $y^{(n)}$ denotes the nth derivative of $y(x)$ with respect to x

$$y^{(n)} \equiv \frac{d^n y}{dx^n}. \tag{6.219}$$

Suppose there were $n+1$ linearly independent solutions y_j of the ODE

$$y^{(n)} + P_1\,y^{(n-1)} + P_2\,y^{(n-2)} + \cdots + P_{n-1}\,y' + P_n\,y = 0, \tag{6.220}$$

in which the P_ks are functions of x. Then we could form a wronskian of order $(n+1)$ in which row 1 would be $y_1, \ldots, y_{n+1}$, row 2 would be the first derivatives $y_1', \ldots, y_{n+1}'$, and row $n+1$ would be the nth derivatives $y_1^{(n)}, \ldots, y_{n+1}^{(n)}$. We could then replace each term $y_k^{(n)}$ in the last row by

$$y_k^{(n)} = -P_1\,y_k^{(n-1)} - P_2\,y_k^{(n-2)} - \cdots - P_{n-1}\,y_k' - P_n\,y_k. \tag{6.221}$$

But then the last row would be a linear combination of the first n rows, the determinant would vanish, and the $n+1$ solutions would be linearly dependent. This is why an nth-order linear homogeneous ODE can have at most n linearly independent solutions.

6.25 Boundary conditions

Since an nth-order linear homogeneous ordinary differential equation can have at most n linearly independent solutions, it follows that we can make a solution unique by requiring it to satisfy n boundary conditions. We'll see that the n arbitrary coefficients c_k of the general solution

$$y(x) = \sum_{k=1}^{n} c_k \, y_k(x) \tag{6.222}$$

of the differential equation (6.220) are fixed by the n boundary conditions

$$y(x_1) = b_1, \quad y(x_2) = b_2, \quad \ldots \quad y(x_n) = b_n \tag{6.223}$$

as long as the functions $y_k(x)$ are linearly independent, and as long as the matrix Y with entries $Y_{jk} = y_k(x_j)$ is nonsingular, that is, det $Y \neq 0$. In matrix notation, with B a vector with components b_j and C a vector with components c_k, the n boundary conditions (6.223) are

$$y(x_j) = \sum_{k=1}^{n} c_k \, y_k(x_j) = b_j \quad \text{or} \quad Y\,C = B. \tag{6.224}$$

Thus since det $Y \neq 0$, the coefficients are uniquely given by $C = Y^{-1}\,B$.

The boundary conditions can involve the derivatives $y_k^{(\ell_j)}(x_j)$. One may show (exercise 6.20) that in this case as long as the matrix $Y_{jk} = y_k^{(\ell_j)}(x_j)$ is nonsingular, the n boundary conditions

$$y^{(\ell_j)}(x_j) = \sum_{k=1}^{n} c_k \, y_k^{(\ell_j)}(x_j) = b_j \tag{6.225}$$

are $Y\,C = B$, and so the n coefficients are uniquely $C = Y^{-1}\,B$.

But what if all the b_j are zero? If all the boundary conditions are homogeneous $Y\,C = 0$, and det $Y \neq 0$, then $Y^{-1}\,Y\,C = C = 0$, and the only solution is $y_k(x) \equiv 0$. So there is no solution if $B = 0$ and the matrix Y is nonsingular. But if the $n \times n$ matrix Y has rank $n-1$, then (section 1.33) it maps a unique vector C to zero (apart from an overall factor). So if all the boundary conditions are homogeneous, and the matrix Y has rank $n-1$, then the solution $y = c_k y_k$ is unique. But if the rank of Y is less than $n-1$, the solution is not unique.

Since a matrix of rank zero vanishes identically, any nonzero 2×2 matrix Y must be of rank 1 or 2. Thus a second-order ODE with two homogeneous boundary conditions has either a unique solution or none at all.

Example 6.31 (Boundary conditions and eigenvalues) The solutions y_k of the differential equation $-y'' = k^2 y$ are $y_1(x) = \sin kx$ and $y_2(x) = \cos kx$. If we impose the boundary conditions $y(-a) = 0$ and $y(a) = 0$, then the matrix $Y_{jk} = y_k(x_j)$ is

$$Y = \begin{pmatrix} -\sin ka & \cos ka \\ \sin ka & \cos ka \end{pmatrix} \tag{6.226}$$

with determinant $\det Y = -2 \sin ka \cos ka = -\sin 2ka$. This determinant vanishes only if $ka = n\pi/2$ for some integer n, so if $ka \neq n\pi/2$, then no solution y of the differential equation $-y'' = k^2 y$ satisfies the boundary conditions $y(-a) = 0 = y(a)$. But if $ka = n\pi/2$, then there is a solution, and it is unique because for even (odd) n, the first (second) column of Y vanishes, but not the second (first), which implies that Y has rank 1. One may regard the condition $ka = n\pi/2$ either as determining the eigenvalue k^2 or as telling us what interval to use. □

6.26 A variational problem

For what functions $u(x)$ is the "energy" functional

$$E[u] \equiv \int_a^b \left[p(x) u'^2(x) + q(x) u^2(x) \right] dx \tag{6.227}$$

stationary? That is, for what functions u is $E[u + \delta u]$ unchanged to first order in δu when $u(x)$ is changed by an arbitrary but tiny function $\delta u(x)$ to $u(x) + \delta u(x)$? Our equations will be less cluttered if we drop explicit mention of the x-dependence of p, q, and u, which we assume to be real functions of x.

The first-order change in E is

$$\delta E[u] \equiv \int_a^b \left(p\, 2u'\, \delta u' + q\, 2u\, \delta u \right) dx, \tag{6.228}$$

in which the change in the derivative of u is $\delta u' = u' + (\delta u)' - u' = (\delta u)'$. Setting $\delta E = 0$ and integrating by parts, we have

$$\begin{aligned} 0 = \delta E &= \int_a^b \left[p\, u' (\delta u)' + q\, u\, \delta u \right] dx \\ &= \int_a^b \left[\left(p\, u' \delta u \right)' - \left(p\, u' \right)' \delta u + q\, u\, \delta u \right] dx \\ &= \int_a^b \left[- \left(p\, u' \right)' + q\, u \right] \delta u\, dx + \left[p\, u' \delta u \right]_a^b. \end{aligned} \tag{6.229}$$

So if E is to be stationary with respect to all tiny changes δu that vanish at the endpoints a and b, then u must satisfy the differential equation

$$L\,u = -\left(p\,u'\right)' + q\,u = 0. \tag{6.230}$$

If instead E is to be stationary with respect to ***all*** tiny changes δu, then u must satisfy the differential equation (6.230) as well as the **natural** boundary conditions

$$0 = p(b)\,u'(b) \quad \text{and} \quad 0 = p(a)\,u'(a). \tag{6.231}$$

If $p(a) \neq 0 \neq p(b)$, then these natural boundary conditions imply **Neumann's** boundary conditions

$$u'(a) = 0 \quad \text{and} \quad u'(b) = 0 \tag{6.232}$$

(Carl Neumann, 1832–1925).

6.27 Self-adjoint differential operators

If $p(x)$ and $q(x)$ are real, then the differential operator

$$L = -\frac{d}{dx}\left(p(x)\frac{d}{dx}\right) + q(x) \tag{6.233}$$

is formally **self adjoint**. Such operators are interesting because if we take any two functions u and v that are twice differentiable on an interval $[a, b]$ and integrate $v\,L\,u$ twice by parts over the interval, we get

$$\begin{aligned}
(v, L\,u) &= \int_a^b v\,L\,u\,dx = \int_a^b v\left[-\left(pu'\right)' + qu\right]dx \\
&= \int_a^b \left[pu'v' + uqv\right]dx - \left[vpu'\right]_a^b \\
&= \int_a^b \left[-(pv')' + qv\right]u\,dx + \left[puv' - vpu'\right]_a^b \\
&= \int_a^b (L\,v)\,u\,dx + \left[p(uv' - vu')\right]_a^b.
\end{aligned} \tag{6.234}$$

which is **Green's formula**

$$\int_a^b (vL\,u - u\,L\,v)\,dx = \left[p(uv' - vu')\right]_a^b = \left[p\,W(u, v)\right]_a^b \tag{6.235}$$

(George Green, 1793–1841).Its differential form is **Lagrange's identity**

$$vL\,u - u\,L\,v = \left[p\,W(u, v)\right]' \tag{6.236}$$

(Joseph-Louis Lagrange, 1736–1813). Thus if the twice-differentiable functions u and v satisfy boundary conditions at $x = a$ and $x = b$ that make the boundary term (6.235) vanish

$$\left[p(uv' - vu')\right]_a^b = [p\,W(u,v)]_a^b = 0 \tag{6.237}$$

then the real differential operator L is **symmetric**

$$(v, L\,u) = \int_a^b v\,L\,u\,dx = \int_a^b u\,L\,v\,dx = (u, L\,v). \tag{6.238}$$

A real linear operator A that acts in a real vector space and satisfies the analogous relation (1.161)

$$(g, Af) = (f, A\,g) \tag{6.239}$$

for all vectors in the space is said to be symmetric and **self adjoint**. In this sense, the differential operator (6.233) is self adjoint on the space of functions that satisfy the boundary condition (6.237).

In quantum mechanics, we often deal with wave functions that are complex. So keeping L real, let's replace u and v by twice-differentiable, complex-valued functions $\psi = u_1 + iu_2$ and $\chi = v_1 + iv_2$. If u_1, u_2, v_1, and v_2 satisfy boundary conditions at $x = a$ and $x = b$ that make the boundary terms (6.237) vanish

$$\left[p(u_i v_j' - v_j u_i')\right]_a^b = [p\,W(u_i, v_j)]_a^b = 0 \quad \text{for} \quad i,j = 1,2 \tag{6.240}$$

then (6.238) implies that

$$\int_a^b v_j\,L\,u_i\,dx = \int_a^b \left(L\,v_j\right)\,u_i\,dx \quad \text{for} \quad i,j = 1,2. \tag{6.241}$$

Under these assumptions, one may show (exercise 6.21) that the boundary condition (6.240) makes the complex boundary term vanish

$$[p\,W(\psi, \chi^*)]_a^b = [p\,(\psi\,\chi^{*\prime} - \psi'\,\chi^*)]_a^b = 0 \tag{6.242}$$

and (exercise 6.22) that since L is real, the identity (6.241) holds for complex functions

$$(\chi, L\,\psi) = \int_a^b \chi^* L\,\psi\,dx = \int_a^b (L\,\chi)^*\,\psi\,dx = (L\,\chi, \psi). \tag{6.243}$$

A linear operator A that satisfies the analogous relation (1.157)

$$(g, Af) = (A\,g, f) \tag{6.244}$$

is said to be self adjoint or **hermitian**. In this sense, the differential operator (6.233) is self adjoint on the space of functions that satisfy the boundary condition (6.242).

The formally self-adjoint differential operator (6.233) will satisfy the inner-product integral equations (6.238 or 6.243) only when the function p and the twice-differentiable functions u and v or ψ and χ conspire to make the boundary terms (6.237 or 6.242) vanish. This requirement leads us to define a self-adjoint differential system.

6.28 Self-adjoint differential systems

A self-adjoint **differential system** consists of a real formally self-adjoint differential operator, a differential equation on an interval, boundary conditions, and a set of twice-differentiable functions that obey them.

A second-order differential equation needs two boundary conditions to make a solution unique (section 6.25). In a **self-adjoint differential system**, the two boundary conditions are **linear and homogeneous** so that the set of all twice-differentiable functions u that satisfy them is a vector space. This space D is the **domain** of the system. For an interval $[a, b]$, **Dirichlet's boundary conditions** (Johann Dirichlet, 1805–1859) are

$$u(a) = 0 \quad \text{and} \quad u(b) = 0 \tag{6.245}$$

and **Neumann's** (6.232) are

$$u'(a) = 0 \quad \text{and} \quad u'(b) = 0. \tag{6.246}$$

We will require that the functions in the domain D all obey either Dirichlet or Neumann boundary conditions.

The **adjoint domain** D^* of a differential system is the set of all twice-differentiable functions v that make the boundary term (6.237) vanish

$$\left[p(uv' - vu')\right]_a^b = [pW(u, v)]_a^b = 0 \tag{6.247}$$

for all functions u that are in the domain D, that is, that satisfy either Dirichlet or Neumann boundary conditions.

A differential system is **regular** and **self adjoint** if the differential operator $Lu = -(pu')' + qu$ is formally self-adjoint, if the interval $[a, b]$ is finite, if p, p', and q are continuous real functions of x on the interval, if $p(x) > 0$ on $[a, b]$, and if the two domains D and D^* coincide, $D = D^*$.

One may show (exercises 6.23 and 6.24) that if D is the set of all twice-differentiable functions $u(x)$ on $[a, b]$ that satisfy either Dirichlet's boundary conditions (6.245) or Neumann's boundary conditions (6.246), and if the function $p(x)$ is continuous and positive on $[a, b]$, then the adjoint set D^* is the same as D. A real formally self-adjoint differential operator $Lu = -(pu')' + qu$ therefore forms together with Dirichlet (6.245) or Neumann (6.246) boundary conditions forms a regular and self-adjoint system if p, p', and q are real and continuous on a finite interval $[a, b]$, and p is positive on $[a, b]$.

Since any two functions u and v in the domain D of a regular and self-adjoint differential system make the boundary term (6.247) vanish, a real formally self-adjoint differential operator L is symmetric and self adjoint (6.238) on all functions in its domain

$$(v, L\,u) = \int_a^b v\,L\,u\,dx = \int_a^b u\,L\,v\,dx = (u, L\,v). \tag{6.248}$$

If functions in the domain are complex, then by (6.242 & 6.243) the operator L is self adjoint or hermitian

$$(\chi, L\,\psi) = \int_a^b \chi^* L\,\psi\,dx = \int_a^b (L\,\chi)^*\,\psi\,dx = (L\,\chi, \psi) \tag{6.249}$$

on all complex functions ψ and χ in its domain.

Example 6.32 (Sines and cosines) The differential system with the formally self-adjoint differential operator

$$L = -\frac{d^2}{dx^2} \tag{6.250}$$

on an interval $[a, b]$ and the differential equation $L\,u = -u'' = \lambda\,u$ has the function $p(x) = 1$. If we choose the interval to be $[-\pi, \pi]$ and the domain D to be the set of all functions that are twice differentiable on this interval and satisfy Dirichlet boundary conditions (6.245), then we get a self-adjoint differential system in which the domain includes linear combinations of $u_n(x) = \sin nx$. If instead we impose Neumann boundary conditions (6.246), then the domain D contains linear combinations of $u_n(x) = \cos nx$. In both cases, the system is regular and self adjoint. □

Some important differential systems are self adjoint but **singular** because the function $p(x)$ vanishes at one or both of the endpoints of the interval $[a, b]$ or because the interval is infinite, for instance $[0, \infty)$ or $(-\infty, \infty)$. In these singular, self-adjoint differential systems, the boundary term (6.247) vanishes if u and v are in the domain $D = D^*$.

Example 6.33 (Legendre's system) Legendre's formally self-adjoint differential operator is

$$L = -\frac{d}{dx}\left[(1 - x^2)\frac{d}{dx}\right] \tag{6.251}$$

and his differential equation is

$$L\,u = -\left[(1 - x^2)u'\right]' = \ell(\ell + 1)\,u \tag{6.252}$$

on the interval $[-1, 1]$. The function $p(x) = 1-x^2$ vanishes at both endpoints $x = \pm 1$, and so this self-adjoint system is singular. Because $p(\pm 1) = 0$, the boundary term (6.247) is zero as long as the functions u and v are differentiable on the interval. The domain D is the set of all functions that are twice differentiable on the interval $[-1, 1]$. □

Example 6.34 (Hermite's system) Hermite's formally self-adjoint differential operator is

$$L = -\frac{d^2}{dx^2} + x^2 \tag{6.253}$$

and his differential equation is

$$L\,u = -u'' + x^2\,u = (2n+1)\,u \tag{6.254}$$

on the interval $(-\infty, \infty)$. This system has $p(x) = 1$ and $q(x) = x^2$. It is self adjoint but singular because the interval is infinite. The domain D consists of all functions that are twice differentiable and that go to zero as $x \to \pm\infty$ faster than $1/x^{3/2}$, which ensures that the relevant integrals converge and that the boundary term (6.247) vanishes. □

6.29 Making operators formally self adjoint

We can make a generic real second-order linear homogeneous differential operator

$$L_0 = h_2 \frac{d^2}{dx^2} + h_1 \frac{d}{dx} + h_0 \tag{6.255}$$

formally self adjoint

$$L = -\frac{d}{dx}\left[p(x)\frac{d}{dx}\right] + q(x) = -p(x)\frac{d^2}{dx^2} - p'(x)\frac{d}{dx} + q(x) \tag{6.256}$$

by first dividing through by $-h_2(x)$

$$L_1 = -\frac{1}{h_2}L_0 = -\frac{d^2}{dx^2} - \frac{h_1}{h_2}\frac{d}{dx} - \frac{h_0}{h_2} \tag{6.257}$$

and then by multiplying L_1 by the positive prefactor

$$p(x) = \exp\left(\int^x \frac{h_1(y)}{h_2(y)}\,dy\right) > 0. \tag{6.258}$$

The product $p\,L_1$ then is formally self adjoint

$$L = p(x)\,L_1 = -\exp\left(\int^x \frac{h_1(y)}{h_2(y)}\,dy\right)\left[\frac{d^2}{dx^2} + \frac{h_1(x)}{h_2(x)}\frac{d}{dx} + \frac{h_0(x)}{h_2(x)}\right]$$

$$= -\frac{d}{dx}\left[\exp\left(\int^x \frac{h_1(y)}{h_2(y)}dy\right)\frac{d}{dx}\right] - \exp\left(\int^x \frac{h_1(y)}{h_2(y)}dy\right)\frac{h_0(x)}{h_2(x)}$$

$$= -\frac{d}{dx}\left(p\frac{d}{dx}\right) + q \tag{6.259}$$

with $q(x) = -p(x)\,h_0(x)/h_2(x)$. So we may turn any second-order linear homogeneous differential operator L into a formally self-adjoint operator L by multiplying it by

$$\rho(x) = -\frac{\exp\left(\int^x h_1(y)/h_2(y)dy\right)}{h_2(x)} = -\frac{p(x)}{h_2(x)}. \tag{6.260}$$

The two differential equations $L_0 u = 0$ and $Lu = \rho L_0 u = 0$ have the same solutions, and so we can restrict our attention to formally self-adjoint differential equations. But under the transformation (6.260), an eigenvalue equation $L_0 u = \lambda\, u$ becomes $Lu = \rho L_0 u = \rho\lambda u$, which is an eigenvalue equation

$$Lu = -(pu')' + qu = \lambda\,\rho\, u \tag{6.261}$$

with a **weight function** $\rho(x)$. Such an eigenvalue problem is known as a **Sturm–Liouville** problem (Jacques Sturm, 1803–1855; Joseph Liouville, 1809–1882). If $h_2(x)$ is negative (as for many positive operators), then the weight function $\rho(x) = -p(x)/h_2(x)$ is positive.

6.30 Wronskians of self-adjoint operators

We saw in (6.206–6.210) that if $y_1(x)$ and $y_2(x)$ are two linearly independent solutions of the ODE

$$y''(x) + P(x)\,y'(x) + Q(x)\,y(x) = 0 \tag{6.262}$$

then their wronskian $W(x) = y_1(x)\,y_2'(x) - y_2(x)\,y_1'(x)$ is

$$W(x) = W(x_0)\,\exp\left[-\int_{x_0}^x P(x')dx'\right]. \tag{6.263}$$

Thus if we convert the ODE (6.262) to its formally self-adjoint form

$$-\left[p(x)y'(x)\right]' + q(x)y(x) = -p(x)\frac{d^2y(x)}{dx^2} - p'(x)\frac{dy(x)}{dx} + q(x)y(x) = 0 \tag{6.264}$$

then $P(x) = p'(x)/p(x)$, and so the wronskian (6.263) is

$$W(x) = W(x_0)\,\exp\left[-\int_{x_0}^x p'(x')/p(x')dx'\right], \tag{6.265}$$

which we may integrate directly to

$$W(x) = W(x_0)\,\exp\left[-\ln\left[p(x)/p(x_0)\right]\right] = W(x_0)\,\frac{p(x_0)}{p(x)}. \tag{6.266}$$

We learned in (6.206–6.214) that if we had one solution $y_1(x)$ of the ODE (6.262 or 6.264), then we could find another solution $y_2(x)$ that is linearly independent of $y_1(x)$ as

$$y_2(x) = y_1(x) \int^x \frac{W(x')}{y_1^2(x')} dx'. \tag{6.267}$$

In view of (6.263), this is an iterated integral. But if the ODE is formally self adjoint, then the formula (6.266) reduces it to

$$y_2(x) = y_1(x) \int^x \frac{1}{p(x')\, y_1^2(x')}\, dx' \tag{6.268}$$

apart from a constant factor.

Example 6.35 (Legendre functions of the second kind) Legendre's self-adjoint differential equation (6.252) is

$$-\left[(1-x^2)\,y'\right]' = \ell(\ell+1)\,y \tag{6.269}$$

and an obvious solution for $\ell = 0$ is $y(x) \equiv P_0(x) = 1$. Since $p(x) = 1 - x^2$, the integral formula (6.268) gives us as a second solution

$$Q_0(x) = P_0(x) \int^x \frac{1}{p(x')\, P_0^2(x')}\, dx' = \int^x \frac{1}{(1-x^2)}\, dx' = \frac{1}{2} \ln\left(\frac{1+x}{1-x}\right). \tag{6.270}$$

This second solution $Q_0(x)$ is singular at both ends of the interval $[-1, 1]$ and so does not satisfy the Dirichlet (6.245) or Neumann (6.246) boundary conditions that make the system self adjoint or hermitian. □

6.31 First-order self-adjoint differential operators

The first-order differential operator

$$L = u\,\frac{d}{dx} + v \tag{6.271}$$

will be self adjoint if

$$\int_a^b \chi^* L\psi\, dx = \int_a^b \left(L^\dagger \chi\right)^* \psi\, dx = \int_a^b (L\chi)^*\, \psi\, dx. \tag{6.272}$$

Starting from the first term, we find

$$\begin{aligned}\int_a^b \chi^* L\psi\, dx &= \int_a^b \chi^* \left(u\,\psi' + v\psi\right) dx \\ &= \int_a^b \left[(-\chi^* u)' + \chi^* v\right] \psi\, dx + \left[\chi^* u \psi\right]_a^b\end{aligned}$$

$$= \int_a^b \left[(-\chi u^*)' + \chi v^*\right]^* \psi \, dx + \left[\chi^* u \psi\right]_a^b$$

$$= \int_a^b \left[-u^* \chi' + (v^* - u^{*\prime})\chi\right]^* \psi \, dx + \left[\chi^* u \psi\right]_a^b. \tag{6.273}$$

So if the boundary terms vanish

$$\left[\chi^* u \psi\right]_a^b = 0 \tag{6.274}$$

and if both $u^* = -u$ and $v^* - u^{*\prime} = v$, then

$$\int_a^b \chi^* L \psi \, dx = \int_a^b \left[u\chi' + v\chi\right]^* \psi \, dx = \int_a^b (L\chi)^* \psi \, dx \tag{6.275}$$

and so L will be self adjoint or hermitian, $L^\dagger = L$. The general form of a first-order self-adjoint linear operator is then

$$L = ir(x) \frac{d}{dx} + s(x) + \frac{i}{2} r'(x) \tag{6.276}$$

in which r and s are arbitrary real functions of x.

Example 6.36 (Momentum and angular momentum) The momentum operator

$$p = \frac{\hbar}{i} \frac{d}{dx} \tag{6.277}$$

has $r = -\hbar$, which is real, and $s = 0$ and so is formally self adjoint. The boundary terms (6.274) are zero if the functions ψ and χ vanish at a and b, which often are $\pm\infty$.

The angular-momentum operators $L_i = \epsilon_{ijk} \, x_j \, p_k$, where $p_k = -i\hbar \, \partial_k$, also are formally self adjoint because the total antisymmetry of ϵ_{ijk} ensures that j and k are different as they are summed from 1 to 3. □

Example 6.37 (Momentum in a magnetic field) In a magnetic field $\boldsymbol{B} = \nabla \times \boldsymbol{A}$, the differential operator

$$\frac{\hbar}{i} \nabla - e\boldsymbol{A} \tag{6.278}$$

that (in mks units) represents the kinetic momentum $m\boldsymbol{v}$ is formally self adjoint as is its Yang–Mills analog (11.471) when divided by i. □

6.32 A constrained variational problem

In quantum mechanics, we usually deal with normalizable wave-functions. So let's find the function $u(x)$ that minimizes the energy functional

$$E[u] = \int_a^b \left[p(x) \, u'^2(x) + q(x) \, u^2(x)\right] dx \tag{6.279}$$

subject to the constraint that $u(x)$ be normalized on $[a, b]$ with respect to a positive weight function $\rho(x)$

$$N[u] = \|u\|^2 = \int_a^b \rho(x)\, u^2(x)\, dx = 1. \tag{6.280}$$

Introducing λ as a Lagrange multiplier (section 1.23) and suppressing explicit mention of the x-dependence of the real functions p, q, ρ, and u, we minimize the unconstrained functional

$$\mathcal{E}[u, \lambda] = \int_a^b \left(p\, u'^2 + q\, u^2\right) dx - \lambda \left(\int_a^b \rho\, u^2\, dx - 1\right), \tag{6.281}$$

which will be stationary at the function u that minimizes it. The first-order change in $\mathcal{E}[u, \lambda]$ is

$$\delta\mathcal{E}[u, \lambda] = \int_a^b \left(p\, 2u'\, \delta u' + q\, 2u\, \delta u - \lambda\, \rho\, 2u\, \delta u\right) dx, \tag{6.282}$$

in which the change in the derivative of u is $\delta u' = u' + (\delta u)' - u' = (\delta u)'$. Setting $\delta\mathcal{E} = 0$ and integrating by parts, we have

$$\begin{aligned} 0 = \tfrac{1}{2}\,\delta\mathcal{E} &= \int_a^b \left[p\, u'(\delta u)' + (q - \lambda\, \rho)\, u\, \delta u\right] dx \\ &= \int_a^b \left[\left(p\, u' \delta u\right)' - \left(p\, u'\right)' \delta u + (q - \lambda\, \rho)\, u\, \delta u\right] dx \\ &= \int_a^b \left[-\left(p\, u'\right)' + (q - \lambda\, \rho)\, u\right] \delta u\, dx + \left[p\, u' \delta u\right]_a^b. \end{aligned} \tag{6.283}$$

So if $\mathcal{E}$ is to be stationary with respect to all tiny changes δu, then u must satisfy both the self-adjoint differential equation

$$0 = -\left(p\, u'\right)' + (q - \lambda\, \rho)\, u \tag{6.284}$$

and the **natural** boundary conditions

$$0 = p(b)\, u'(b) \quad \text{and} \quad 0 = p(a)\, u'(a). \tag{6.285}$$

If instead we require $\mathcal{E}[u, \lambda]$ to be stationary with respect to all variations δu that vanish at the endpoints, $\delta u(a) = \delta u(b) = 0$, then u must satisfy the differential equation (6.284) but need not satisfy the natural boundary conditions (6.285).

In both cases, the function $u(x)$ that minimizes the energy $E[u]$ subject to the normalization condition $N[u] = 1$ is an **eigenfunction** of the formally self-adjoint differential operator

$$L = -\frac{d}{dx}\left(p(x)\frac{d}{dx}\right) + q(x) \tag{6.286}$$

with **eigenvalue** λ

$$Lu = -\left(p\,u'\right)' + q\,u = \lambda\,\rho\,u. \tag{6.287}$$

The Lagrange multiplier λ has become an eigenvalue of a Sturm–Liouville equation (6.261).

Is the eigenvalue λ related to $E[u]$ and $N[u]$? To keep things simple, we restrict ourselves to a regular and self-adjoint differential system (section 6.28) consisting of the self-adjoint differential operator (6.286), the differential equation (6.287), and a domain $D = D^*$ of functions $u(x)$ that are twice differentiable on $[a, b]$ and that satisfy two homogeneous Dirichlet (6.245) or Neumann (6.246) boundary conditions on $[a, b]$. All functions u in the domain D therefore satisfy

$$\left[upu'\right]_a^b = 0. \tag{6.288}$$

We now multiply the Sturm–Liouville equation (6.287) from the left by u and integrate from a to b. After integrating by parts and noting the vanishing of the boundary terms (6.288), we find

$$\begin{aligned}\lambda \int_a^b \rho\,u^2\,dx &= \int_a^b u\,Lu\,dx = \int_a^b u\left[-\left(p\,u'\right)' + q\,u\right] dx\\ &= \int_a^b \left[p\,u'^2 + q\,u^2\right] dx - \left[upu'\right]_a^b\\ &= \int_a^b \left[p\,u'^2 + q\,u^2\right] dx = E[u].\end{aligned} \tag{6.289}$$

Thus in view of the normalization constraint (6.280), we see that the eigenvalue λ is the ratio of the energy $E[u]$ to the norm $N[u]$

$$\lambda = \frac{\displaystyle\int_a^b \left[p\,u'^2 + q\,u^2\right] dx}{\displaystyle\int_a^b \rho\,u^2\,dx} = \frac{E[u]}{N[u]}. \tag{6.290}$$

But is the function that minimizes the ratio

$$R[u] \equiv \frac{E[u]}{N[u]} \tag{6.291}$$

the eigenfunction u of the Sturm–Liouville equation (6.287)? And is the minimum of $R[u]$ the least eigenvalue λ of the Sturm–Liouville equation (6.287)? To see that the answers are *yes* and *yes*, we require $\delta R[u]$ to vanish

$$\delta R[u] = \frac{\delta E[u]}{N[u]} - \frac{E[u]\,\delta N[u]}{N^2[u]} = 0 \tag{6.292}$$

to first order in tiny changes $\delta u(x)$ that are zero at the endpoints of the interval, $\delta u(a) = \delta u(b) = 0$. Multiplying both sides by $N[u]$, we have

$$\delta E[u] = R[u]\, \delta N[u]. \tag{6.293}$$

Referring back to our derivation (6.281–6.283) of the Sturm–Liouville equation, we see that since $\delta u(a) = \delta u(b) = 0$, the change δE is

$$\begin{aligned} \delta E[u] &= 2 \int_a^b \left[-\,(p\,u')' + q\,u \right] \delta u\, dx + 2 \left[p\,u'\delta u \right]_a^b \\ &= 2 \int_a^b \left[-\,(p\,u')' + q\,u \right] \delta u\, dx \end{aligned} \tag{6.294}$$

while δN is

$$\delta N[u] = 2 \int_a^b \rho\, u\, \delta u\, dx. \tag{6.295}$$

Substituting these changes (6.294) and (6.295) into the condition (6.293) that $R[u]$ be stationary, we find that the integral

$$\int_a^b \left[-\,(p\,u')' + (q - R[u]\,\rho\,)\, u \right] \delta u\, dx = 0 \tag{6.296}$$

must vanish for all tiny changes $\delta u(x)$ that are zero at the endpoints of the interval. Thus on $[a, b]$, the function u that minimizes the ratio $R[u]$ must satisfy the Sturm–Liouville equation (6.287)

$$-\,(p\,u')' + q\,u = R[u]\,\rho\,u \tag{6.297}$$

with an eigenvalue $\lambda \equiv R[u]$ that is the minimum value of the ratio $R[u]$.

So the eigenfunction u_1 with the smallest eigenvalue λ_1 is the one that minimizes the ratio $R[u]$, and $\lambda_1 = R[u_1]$. What about other eigenfunctions with larger eigenvalues? How do we find the eigenfunction u_2 with the next smallest eigenvalue λ_2? Simple: we minimize $R[u]$ with respect to all functions u that are in the domain D and that are orthogonal to u_1.

Example 6.38 (Infinite square well) Let us consider a particle of mass m trapped in an interval $[a, b]$ by a potential that is V for $a < x < b$ but infinite for $x < a$ and for $x > b$. Because the potential is infinite outside the interval, the wave-function $u(x)$ will satisfy the boundary conditions

$$u(a) = u(b) = 0. \tag{6.298}$$

The mean value of the hamiltonian is then the energy functional

$$\langle u|H|u\rangle = E[u] = \int_a^b \left[p(x)\, u'^2(x) + q(x)\, u^2(x) \right] dx, \tag{6.299}$$

in which $p(x) = \hbar^2/2m$ and $q(x) = V$, a constant independent of x. Wave-functions in quantum mechanics are normalized when possible. So we need to minimize the functional

$$E[u] = \int_a^b \left[\frac{\hbar^2}{2m} u'^2(x) + V\, u^2(x) \right] dx \tag{6.300}$$

subject to the constraint

$$c = \int_a^b u^2(x)\, dx - 1 = 0 \tag{6.301}$$

for all tiny variations δu that vanish at the endpoints of the interval. The weight function $\rho(x) = 1$, and the eigenvalue equation (6.287) is

$$-\frac{\hbar^2}{2m} u'' + V\, u = \lambda\, u. \tag{6.302}$$

For any positive integer n, the normalized function

$$u_n(x) = \left(\frac{2}{b-a} \right)^{1/2} \sin\left(n\pi \frac{x-a}{b-a} \right) \tag{6.303}$$

satisfies the boundary conditions (6.298) and the eigenvalue equation (6.302) with energy eigenvalue

$$\lambda_n = E[u_n] = \frac{1}{2m} \left(\frac{n\pi\hbar}{b-a} \right)^2 + V. \tag{6.304}$$

The second eigenfunction u_2 minimizes the energy functional $E[u]$ over the space of normalized functions that satisfy the boundary conditions (6.298) and are orthogonal to the first eigenfunction u_1. The eigenvalue λ_2 is higher than λ_1 (four times higher). As the quantum number n increases, the energy $\lambda_n = E[u_n]$ goes to infinity as n^2. That $\lambda_n \to \infty$ as $n \to \infty$ is related (section 6.35) to the completeness of the eigenfunctions u_n. □

Example 6.39 (Bessel's system) Bessel's energy functional is

$$E[u] = \int_0^1 \left[x\, u'^2(x) + \frac{n^2}{x} u^2(x) \right] dx, \tag{6.305}$$

in which $n \geq 0$ is an integer. We seek the minimum of this functional over the set of twice-differentiable functions $u(x)$ on $[0, 1]$ that are normalized

$$N[u] = \|u\|^2 = \int_0^1 x\, u^2(x)\, dx = 1 \tag{6.306}$$

and that satisfy the boundary conditions $u(0) = 0$ for $n > 0$ and $u(1) = 0$. We'll use a Lagrange multiplier λ (section 1.23) and minimize the unconstrained functional $E[u] - \lambda\,(N[u] - 1)$. Proceeding as in (6.279–6.287), we find that u must obey the formally self-adjoint differential equation

$$L\,u = -(x\,u')' + \frac{n^2}{x}\,u = \lambda\,x\,u. \tag{6.307}$$

The ratio formula (6.290) and the positivity of Bessel's energy functional (6.305) tell us that the eigenvalues $\lambda = E[u]/N[u]$ are positive (exercise 6.25). As we'll see in a moment, the boundary conditions largely determine these eigenvalues $\lambda_{n,m} \equiv k_{n,m}^2$. By changing variables to $\rho = k_{n,m}x$ and letting $u(x) = J_n(\rho)$, we arrive (exercise 6.26) at

$$\frac{d^2 J_n}{d\rho^2} + \frac{1}{\rho}\frac{dJ_n}{d\rho} + \left(1 - \frac{n^2}{\rho^2}\right) J_n = 0, \tag{6.308}$$

which is Bessel's equation. The eigenvalues are determined by the condition $u(1) = J_n(k_{n,m}) = 0$; they are the squares of the zeros of $J_n(\rho)$. The eigenfunction of the self-adjoint differential equation (6.307) with eigenvalue $\lambda_{n,m} = k_{n,m}^2$ is $u_m(x) = J_n(k_{n,m}x)$. The parameter n labels the differential system; it is not an eigenvalue. Asymptotically as $m \to \infty$, one has (Courant and Hilbert, 1955, p. 416)

$$\lim_{m\to\infty} \frac{\lambda_{n,m}}{m^2\pi^2} = 1, \tag{6.309}$$

which shows that the eigenvalues $\lambda_{n,m}$ rise like m^2 as $m \to \infty$. □

Example 6.40 (Harmonic oscillator) We'll minimize the energy

$$E[u] = \int_{-\infty}^{\infty} \left[\frac{\hbar^2}{2m}\, u'^2(x) + \frac{1}{2} m\,\omega^2\, x^2\, u^2(x) \right] dx \tag{6.310}$$

subject to the normalization condition

$$N[u] = \|u\|^2 = \int_{-\infty}^{\infty} u^2(x)\, dx = 1. \tag{6.311}$$

We introduce λ as a Lagrange multiplier and find the minimum of the unconstrained function $E[u] - \lambda\,(N[u] - 1)$. Following equations (6.279–6.287), we find that u must satisfy Schrödinger's equation

$$-\frac{\hbar^2}{2m} u'' + \frac{1}{2} m\,\omega^2\, x^2\, u = \lambda u, \tag{6.312}$$

which we write as

$$\hbar\omega \left[\frac{m\omega}{2\hbar} \left(x - \frac{\hbar}{m\omega}\frac{d}{dx} \right) \left(x + \frac{\hbar}{m\omega}\frac{d}{dx} \right) + \frac{1}{2} \right] u = \lambda u. \tag{6.313}$$

The lowest eigenfunction u_0 is mapped to zero by the second factor

$$\left(x + \frac{\hbar}{m\omega}\frac{d}{dx} \right) u_0(x) = 0 \tag{6.314}$$

so its eigenvalue λ_0 is $\hbar\omega/2$. Integrating this differential equation, we get

$$u_0(x) = \left(\frac{m\omega}{\pi\hbar}\right)^{1/4} \exp\left(-\frac{m\omega x^2}{2\hbar}\right), \tag{6.315}$$

in which the prefactor is a normalization constant. As in section 2.11, one may get the higher eigenfunctions by acting on u_0 with powers of the first factor inside the square brackets (6.313)

$$u_n(x) = \frac{1}{\sqrt{n!}}\left(\frac{m\omega}{2\hbar}\right)^{n/2}\left(x - \frac{\hbar}{m\omega}\frac{d}{dx}\right)^n u_0(x). \tag{6.316}$$

The eigenvalue of u_n is $\lambda_n = \hbar\omega(n + 1/2)$. Again, $\lambda_n \to \infty$ as $n \to \infty$. □

6.33 Eigenfunctions and eigenvalues of self-adjoint systems

A **regular Sturm–Liouville system** is a set of regular and self-adjoint differential systems (section 6.28) that have the same differential operator, interval $[a, b]$, boundary conditions, and domain, and whose differential equations are of Sturm–Liouville (6.287) type

$$L\,\psi = -(p\,\psi')' + q\,\psi = \lambda\,\rho\,\psi, \tag{6.317}$$

each distinguished by an **eigenvalue** λ. The functions p, q, and ρ are real and continuous, p and ρ are positive on $[a, b]$, but the weight function ρ may vanish at isolated points of the interval.

Since the differential systems are self adjoint, the real or complex functions in the common domain D are twice differentiable on the interval $[a, b]$ and satisfy two homogeneous boundary conditions that make the boundary terms (6.247) vanish

$$p\,W(\psi', \psi^*)\big|_a^b = 0 \tag{6.318}$$

and so the differential operator L obeys the condition (6.249)

$$(\chi, L\,\psi) = \int_a^b \chi^* L\,\psi\,dx = \int_a^b (L\,\chi)^*\,\psi\,dx = (L\,\chi, \psi) \tag{6.319}$$

of being self adjoint or hermitian.

Let ψ_i and ψ_j be eigenfunctions of L with eigenvalues λ_i and λ_j

$$L\,\psi_i = \lambda_i\,\rho\,\psi_i \quad \text{and} \quad L\,\psi_j = \lambda_j\,\rho\,\psi_j \tag{6.320}$$

in a regular Sturm–Liouville system. Multiplying the first of these eigenvalue equations by ψ_j^* and the complex conjugate of the second by ψ_i, we get

$$\psi_j^* L\,\psi_i = \psi_j^*\lambda_i\,\rho\,\psi_i \quad \text{and} \quad \psi_i(L\,\psi_j)^* = \psi_i\lambda_j^*\,\rho\,\psi_j^*. \tag{6.321}$$

Integrating the difference of these equations over the interval $[a, b]$ and using (6.319) in the form $\int_a^b \psi_j^* L\, \psi_i \, dx = \int_a^b (L\, \psi_j)^* \, \psi_i \, dx$, we have

$$0 = \int_a^b \left[\psi_j^* L\, \psi_i - (L\, \psi_j)^* \, \psi_i \right] dx = \left(\lambda_i - \lambda_j^* \right) \int_a^b \psi_j^* \, \psi_i \, \rho \, dx. \tag{6.322}$$

Setting $i = j$, we find

$$0 = \left(\lambda_i^* - \lambda_i \right) \int_a^b \rho \, |\psi_i|^2 \, dx, \tag{6.323}$$

which, since the integral is positive, shows that the eigenvalue λ_i must be **real**. All the eigenvalues of a regular Sturm–Liouville system are real. Using $\lambda_j^* = \lambda_j$ in (6.322), we see that eigenfunctions that have different eigenvalues are **orthogonal** on the interval $[a, b]$ with weight function $\rho(x)$

$$0 = \left(\lambda_i - \lambda_j \right) \int_a^b \psi_j^* \, \rho \, \psi_i \, dx. \tag{6.324}$$

Since the differential operator L, the eigenvalues λ_i, and the weight function ρ are all real, we may write the first of the eigenvalue equations in (6.320) both as $L\, \psi_i = \lambda_i \, \rho \, \psi_i$ and as $L\, \psi_i^* = \lambda_i \, \rho \, \psi_i^*$. By adding these two equations, we see that the real part of ψ_i satisfies them, and by subtracting them, we see that the imaginary part of ψ_i also satisfies them. So it *might* seem that $\psi_i = u_i + i v_i$ is made of two real eigenfunctions with the same eigenvalue.

But each eigenfunction u_i in the domain D satisfies two homogeneous boundary conditions as well as its second-order differential equation

$$-\left(p\, u_i' \right)' + q\, u_i = \lambda_i \, \rho \, u_i \tag{6.325}$$

and so u_i is the unique solution in D to this equation. There can be no other eigenfunction in D with the same eigenvalue. In a regular Sturm–Liouville system, **there is no degeneracy**. All the eigenfunctions u_i are orthogonal and can be normalized on the interval $[a, b]$ with weight function $\rho(x)$

$$\int_a^b u_j^* \, \rho \, u_i \, dx = \delta_{ij}. \tag{6.326}$$

They may be taken to be **real**.

It is true that the eigenfunctions of a second-order differential equation come in pairs because one can use Wronski's formula (6.268)

$$y_2(x) = y_1(x) \int^x \frac{dx'}{p(x')\, y_1^2(x')} \tag{6.327}$$

to find a linearly independent second solution with the same eigenvalue. But the second solutions don't obey the boundary conditions of the domain. Bessel functions of the second kind, for example, are infinite at the origin.

A set of eigenfunctions u_i is **complete in the mean** in a space S of functions if every function $f \in S$ can be represented as a series

$$f(x) = \sum_{i=1}^{\infty} a_i u_i(x) \tag{6.328}$$

(called a Fourier series) that converges **in the mean**, that is

$$\lim_{N\to\infty} \int_a^b \left| f(x) - \sum_{i=1}^{N} a_i u_i(x) \right|^2 \rho(x)\, dx = 0. \tag{6.329}$$

The natural space S is the space $\mathcal{L}_2(a,b)$ of all functions f that are square-integrable on the interval $[a,b]$

$$\int_a^b |f(x)|^2 \, \rho(x)\, dx < \infty. \tag{6.330}$$

The orthonormal eigenfunctions of every regular Sturm–Liouville system on an interval $[a,b]$ are complete in the mean in $\mathcal{L}_2(a,b)$. The completeness of these eigenfunctions follows (section 6.35) from the fact that the eigenvalues λ_n of a regular Sturm–Liouville system are **unbounded**: when arranged in ascending order $\lambda_n < \lambda_{n+1}$ they go to infinity with the index n

$$\lim_{n\to\infty} \lambda_n = \infty \tag{6.331}$$

as we'll see in the next section.

6.34 Unboundedness of eigenvalues

We have seen (section 6.32) that the function $u(x)$ that minimizes the ratio

$$R[u] = \frac{E[u]}{N[u]} = \frac{\int_a^b \left[p\, u'^2 + q\, u^2 \right] dx}{\int_a^b \rho\, u^2 \, dx} \tag{6.332}$$

is a solution of the Sturm–Liouville equation

$$Lu = -\left(p\, u'\right)' + q\, u = \lambda\, \rho\, u \tag{6.333}$$

with eigenvalue

$$\lambda = \frac{E[u]}{N[u]}. \tag{6.334}$$

Let us call this least value of the ratio (6.332) λ_1; it also is the smallest eigenvalue of the differential equation (6.333). The second smallest eigenvalue λ_2 is the minimum of the same ratio (6.332) but for functions that are orthogonal to u_1

$$\int_a^b \rho\, u_1\, u_2\, dx = 0. \tag{6.335}$$

And λ_3 is the minimum of the ratio $R[u]$ but for functions that are orthogonal to both u_1 and u_2. Continuing in this way, we make a sequence of orthogonal eigenfunctions $u_n(x)$ (which we can normalize, $N[u_n] = 1$) with eigenvalues $\lambda_1 \leq \lambda_2 \leq \lambda_3 \leq \cdots \lambda_n$. How do the eigenvalues λ_n behave as $n \to \infty$?

Since the function $p(x)$ is positive for $a < x < b$, it is clear that the energy functional (6.279)

$$E[u] = \int_a^b \left[p\, u'^2 + q\, u^2 \right] dx \tag{6.336}$$

gets bigger as u'^2 increases. In fact, if we let the function $u(x)$ zigzag up and down about a given curve $\bar{u}$, then the kinetic energy $\int pu'^2 dx$ will rise but the potential energy $\int qu^2 dx$ will remain approximately constant. Thus by increasing the frequency of the zigzags, we can drive the energy $E[u]$ to infinity. For instance, if $u(x) = \sin x$, then its zigzag version $u_\omega(x) = u(x)(1 + 0.2 \sin \omega x)$ will have higher energy. The case of $\omega = 100$ is illustrated in Fig. 6.1. As $\omega \to \infty$, its energy $E[u_\omega] \to \infty$.

It is therefore intuitively clear (or at least plausible) that if the real functions $p(x)$, $q(x)$, and $\rho(x)$ are continuous on $[a, b]$ and if $p(x) > 0$ and $\rho(x) > 0$ on (a, b), then there are infinitely many energy eigenvalues λ_n, and that they increase without limit as $n \to \infty$

$$\lim_{n \to \infty} \lambda_n = \infty. \tag{6.337}$$

Courant and Hilbert (Richard Courant, 1888–1972, and David Hilbert, 1862–1943) provide several proofs of this result (Courant and Hilbert, 1955, pp. 397–429). One of their proofs involves the change of variables $f = (p\rho)^{1/4}$ and $v = fu$, after which the eigenvalue equation

$$L\, u = -\left(p\, u'\right)' + q\, u = \lambda \rho u \tag{6.338}$$

becomes $L_f\, v = -v'' + rv = \lambda_v v$ with $r = f''/f + q/\rho$. Were this $r(x)$ a constant, the eigenfunctions of L_f would be $v_n(x) = \sin(n\pi/(b - a))$ with eigenvalues

$$\lambda_{v_n} = \left(\frac{n\,\pi}{b - a} \right)^2 + r \tag{6.339}$$

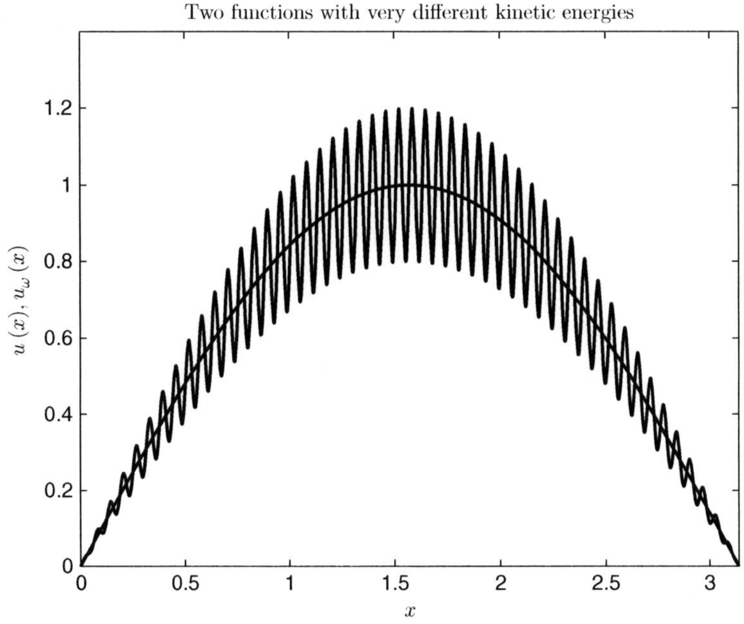

Figure 6.1 The energy functional $E[u]$ of equation (6.279) assigns a much higher energy to the function $u_\omega(x) = u(x)(1 + 0.2\sin(\omega x))$ (zigzag curve with $\omega = 100$) than to the function $u(x) = \sin(x)$ (smooth curve). As the frequency $\omega \to \infty$, the energy $E[u_2] \to \infty$.

rising as n^2. Courant and Hilbert show that as long as $r(x)$ is bounded for $a \le x \le b$, the actual eigenvalues of L_f are $\lambda_{v,n} = c\,n^2 + d_n$ in which d_n is bounded and that the eigenvalues λ_n of L differ from the $\lambda_{v,n}$ by a scale factor, so that they too diverge as $n \to \infty$

$$\lim_{n\to\infty} \frac{n^2}{\lambda_n} = g \tag{6.340}$$

where g is a constant.

6.35 Completeness of eigenfunctions

We have seen in section 6.34 that the eigenvalues of every regular Sturm–Liouville system when arranged in ascending order tend to infinity with the index n

$$\lim_{n\to\infty} \lambda_n = \infty. \tag{6.341}$$

We'll now use this property to show that the corresponding eigenfunctions $u_n(x)$ are complete in the mean (6.329) in the domain D of the system.

To do so, we follow Courant and Hilbert (Courant and Hilbert, 1955, pp. 397–428) and extend the energy E and norm N functionals to inner products on the domain of the system

$$E[f,g] \equiv \int_a^b \left[p(x)f'(x)\,g'(x) + q(x)f(x)\,g(x)\right]\,dx, \tag{6.342}$$

$$N[f,g] \equiv \int_a^b \rho(x)f(x)\,g(x)\,dx \tag{6.343}$$

for any f and g in D. Integrating $E[f,g]$ by parts, we have

$$\begin{aligned} E[f,g] &= \int_a^b \left[(pf\,g')' - f(pg')' + qf\,g\right]\,dx \\ &= \int_a^b \left[-f(pg')' + f\,q\,g\right]\,dx + \left[pf\,g'\right]_a^b \end{aligned} \tag{6.344}$$

or in terms of the self-adjoint differential operator L of the system

$$E[f,g] = \int_a^b f\,L\,g\,dx + \left[pf\,g'\right]_a^b. \tag{6.345}$$

Since the boundary term vanishes (6.288) when the functions f and g are in the domain D of the system, it follows that for f and g in D

$$E[f,g] = \int_a^b f\,L\,g\,dx. \tag{6.346}$$

We can use the first n orthonormal eigenfunctions u_k of the system

$$L\,u_k = \lambda_k\,\rho\,u_k \tag{6.347}$$

to approximate an arbitrary function in $f \in D$ as the linear combination

$$f(x) \sim \sum_{k=1}^{n} c_k\,u_k(x) \tag{6.348}$$

with coefficients c_k given by

$$c_k = N[f,u_k] = \int_a^b \rho\,f\,u_k\,dx. \tag{6.349}$$

We'll show that this series converges in the mean to the function f.

By construction (6.349), the remainder or error of the nth sum

$$r_n(x) = f(x) - \sum_{k=1}^{n} c_k\,u_k(x) \tag{6.350}$$

is orthogonal to the first n eigenfunctions

$$N[r_n, u_k] = 0 \quad \text{for} \quad k = 1, \ldots, n. \tag{6.351}$$

The next eigenfunction u_{n+1} minimizes the ratio

$$R[\phi] = \frac{E[\phi, \phi]}{N[\phi, \phi]} \tag{6.352}$$

over all ϕ that are orthogonal to the first n eigenfunctions u_k in the sense that $N[\phi, u_k] = 0$ for $k = 1, \ldots, n$. That minimum is the eigenvalue λ_{n+1}

$$R[u_{n+1}] = \lambda_{n+1}, \tag{6.353}$$

which therefore must be less than the ratio $R[r_n]$

$$\lambda_{n+1} \leq R[r_n] = \frac{E[r_n, r_n]}{N[r_n, r_n]}. \tag{6.354}$$

Thus the square of the norm of the remainder is bounded by the ratio

$$\|r_n\|^2 \equiv N[r_n, r_n] \leq \frac{E[r_n, r_n]}{\lambda_{n+1}}. \tag{6.355}$$

So since $\lambda_{n+1} \to \infty$ as $n \to \infty$, we're done if we can show that the energy $E[r_n, r_n]$ of the remainder is bounded.

This energy is

$$\begin{aligned} E[r_n, r_n] &= E\left[f - \sum_{k=1}^{n} c_k u_k, f - \sum_{k=1}^{n} c_k u_k\right] \\ &= E[f, f] - \sum_{k=1}^{n} c_k \left(E[f, u_k] + E[u_k, f]\right) + \sum_{k=1}^{n}\sum_{\ell=1}^{n} c_k c_\ell \, E[u_k, u_\ell] \\ &= E[f, f] - 2\sum_{k=1}^{n} c_k \, E[f, u_k] + \sum_{k=1}^{n}\sum_{\ell=1}^{n} c_k c_\ell \, E[u_k, u_\ell]. \end{aligned} \tag{6.356}$$

Since f and all the u_k are in the domain of the system, they satisfy the boundary condition (6.247 or 6.318), and so (6.345, 6.347, & 6.326) imply that

$$E[f, u_k] = \int_a^b f \, L u_k \, dx = \lambda_k \int_a^b \rho f \, u_k \, dx = \lambda_k \, c_k \tag{6.357}$$

and that

$$E[u_k, u_\ell] = \int_a^b u_k \, L u_\ell \, dx = \lambda_\ell \int_a^b \rho \, u_k \, u_\ell \, dx = \lambda_k \, \delta_{k,\ell}. \tag{6.358}$$

Using these relations to simplify our formula (6.356) for $E[r_n, r_n]$ we find

$$E[r_n, r_n] = E[f, f] - \sum_{k=1}^{n} \lambda_k c_k^2. \tag{6.359}$$

Since $\lambda_n \to \infty$ as $n \to \infty$, we can be sure that for high enough n, the sum

$$\sum_{k=1}^{n} \lambda_k c_k^2 > 0 \quad \text{for} \quad n > N \tag{6.360}$$

is positive. It follows from (6.359) that the energy of the remainder r_n is bounded by that of the function f

$$E[r_n, r_n] = E[f, f] - \sum_{k=1}^{n} \lambda_k c_k^2 \leq E[f, f]. \tag{6.361}$$

By substituting this upper bound $E[f, f]$ on $E[r_n, r_n]$ into our upper bound (6.355) on the squared norm $\|r_n\|^2$ of the remainder, we find

$$\|r_n\|^2 \leq \frac{E[f, f]}{\lambda_{n+1}}. \tag{6.362}$$

Thus since $\lambda_n \to \infty$ as $n \to \infty$, we see that the series (6.348) converges in the mean (section 4.3) to f

$$\lim_{n\to\infty} \|r_n\|^2 = \lim_{n\to\infty} \|f - \sum_{k=1}^{n} c_k u_k\|^2 \leq \lim_{n\to\infty} \frac{E[f, f]}{\lambda_{n+1}} = 0. \tag{6.363}$$

The eigenfunctions u_k of a regular Sturm–Liouville system are therefore complete in the mean in the domain D of the system. They span D.

It is a short step from spanning D to spanning the space $\mathcal{L}_2(a, b)$ of functions that are square integrable on the interval $[a, b]$ of the system. To take this step, we assume that the domain D is dense in $\mathcal{L}_2(a, b)$, that is, that for every function $g \in \mathcal{L}_2(a, b)$ there is a sequence of functions $f_n \in D$ that converges to it in the mean so that for any $\epsilon > 0$ there is an integer N_1 such that

$$\|g - f_n\|^2 \equiv \int_a^b |g(x) - f_n(x)|^2 \, \rho(x) \, dx < \epsilon \quad \text{for} \quad n > N_1. \tag{6.364}$$

Since $f_n \in D$, we can find a series of eigenfunctions u_k of the system that converges in the mean to f_n so that for any $\epsilon > 0$ there is an integer N_2 such that

$$\|f_n - \sum_{k=1}^{N} c_{n,k} \, u_k\|^2 \equiv \int_a^b \left| f_n(x) - \sum_{k=1}^{N} c_{n,k} \, u_k(x) \right|^2 \rho(x) \, dx < \epsilon \quad \text{for} \quad N > N_2. \tag{6.365}$$

The Schwarz inequality (1.99) applies to these inner products, and so

$$\|g - \sum_{k=1}^{N} c_{n,k} \, u_k\| \leq \|g - f_n\| + \|f_n(x) - \sum_{k=1}^{N} c_{n,k} \, u_k\|. \tag{6.366}$$

Combining the last three inequalities, we have for $n > N_1$ and $N > N_2$

$$\|g - \sum_{k=1}^{N} c_{n,k}\, u_k\| < 2\sqrt{\epsilon}. \tag{6.367}$$

So the eigenfunctions u_k of a regular Sturm–Liouville system span the space of functions that are square integrable on its interval $\mathcal{L}_2(a,b)$.

One may further show (Courant and Hilbert, 1955, p. 360; Stakgold, 1967, p. 220) that the eigenfunctions $u_k(x)$ of any regular Sturm–Liouville system form a complete orthonormal set in the sense that every function $f(x)$ that satisfies Dirichlet (6.245) or Neumann (6.246) boundary conditions and has a continuous first and a piecewise continuous second derivative may be expanded in a series

$$f(x) = \sum_{k=1}^{\infty} a_k\, u_k(x) \tag{6.368}$$

that converges absolutely and uniformly on the interval $[a,b]$ of the system.

Our discussion (6.341–6.363) of the completeness of the eigenfunctions of a regular Sturm–Liouville system was insensitive to the finite length of the interval $[a,b]$ and to the positivity of $p(x)$ on $[a,b]$. What was essential was the vanishing of the boundary terms (6.247), which can happen if p vanishes at the endpoints of a finite interval or if the functions u and v tend to zero as $|x| \to \infty$ on an infinite one. This is why the results of this section have been extended to singular Sturm–Liouville systems made of self-adjoint differential systems that are singular because the interval is infinite or has p vanishing at one or both of its ends.

If the eigenfunctions u_k are orthonormal with weight function $\rho(x)$

$$\delta_{k\ell} = \int_a^b u_k(x)\,\rho(x)\,u_\ell(x)\,dx \tag{6.369}$$

then the coefficients a_k of the expansion (6.348) are given by the integrals (6.349)

$$a_k = \int_a^b u_k(x)\,\rho(x) f(x)\,dx. \tag{6.370}$$

By combining equations (6.328) and (6.370), we have

$$f(x) = \sum_{k=1}^{\infty} \int_a^b u_k(y)\,\rho(y) f(y)\,dy\, u_k(x) \tag{6.371}$$

or rearranging

$$f(x) = \int_a^b f(y) \left[\sum_{k=1}^{\infty} u_k(y)\,u_k(x)\,\rho(y) \right] dy, \tag{6.372}$$

which implies the representation

$$\delta(x-y) = \rho(y) \sum_{k=1}^{\infty} u_k(x)\, u_k(y) \tag{6.373}$$

of Dirac's delta function. But since this series is nonzero only for $x = y$, the weight function $\rho(y)$ is just a scale factor, and we can write for $0 \le \alpha \le 1$

$$\delta(x-y) = \rho^{\alpha}(x)\, \rho^{1-\alpha}(y) \sum_{k=1}^{\infty} u_k(x)\, u_k(y). \tag{6.374}$$

These representations of the delta functional are suitable for functions f in the domain D of the regular Sturm–Liouville system.

Example 6.41 (A Bessel representation of the delta function) Bessel's nth system $L u = -(x u')' + n^2 u/x = \lambda\, x u$ has eigenvalues $\lambda = z_{n,k}^2$ that are the squares of the zeros of the Bessel function $J_n(x)$. The eigenfunctions

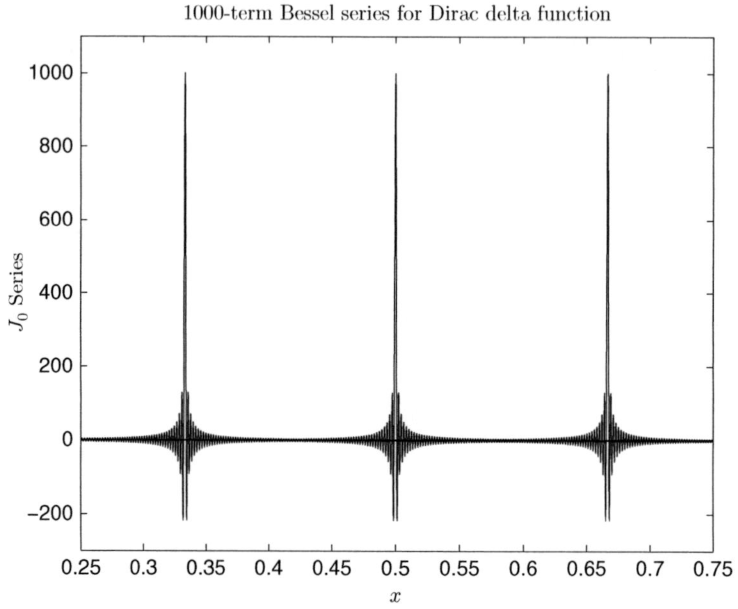

Figure 6.2 The sum of the first 1000 terms of the Bessel representation (6.376) for the Dirac delta function $\delta(x-y)$ is plotted for $y = 1/3$ and $\alpha = 0$, for $y = 1/2$ and $\alpha = 1/2$, and for $y = 2/3$ and $\alpha = 1$.

(section 9.1) that are orthonormal with weight function $\rho(x) = x$ are $u_k^{(n)}(x) = \sqrt{2}\, J_n(z_{n,k}x)/J_{n+1}(z_{n,k})$. Thus, by (6.374), we can represent Dirac's delta functional for functions in the domain D of Bessel's system as

$$\delta(x-y) = x^\alpha\, y^{1-\alpha} \sum_{k=1}^{\infty} u_k^{(n)}(x)\, u_k^{(n)}(y). \tag{6.375}$$

For $n = 0$, this Bessel representation is

$$\delta(x-y) = 2\, x^\alpha\, y^{1-\alpha} \sum_{k=1}^{\infty} \frac{J_0(z_{0,k}x) J_0(z_{0,k}y)}{J_1^2(z_{0,k})}. \tag{6.376}$$

Figure 6.2 plots the first 1000 terms of this sum (6.376) for $\alpha = 0$ and $y = 1/3$, for $\alpha = 1/2$ and $y = 1/2$, and for $\alpha = 1$ and $y = 2/3$. Figure 6.3 plots the first 10,000 terms of the same series but for $\alpha = 0$ and $y = 0.47$, for $\alpha = 1/2$ and $y = 1/2$, and for $\alpha = 1$ and $y = 0.53$. The integrals of these 10,000-term sums from 0 to 1 respectively are 0.9966, 0.9999, and 0.9999. These plots illustrate the Sturm–Liouville representation (6.374) of the delta function.

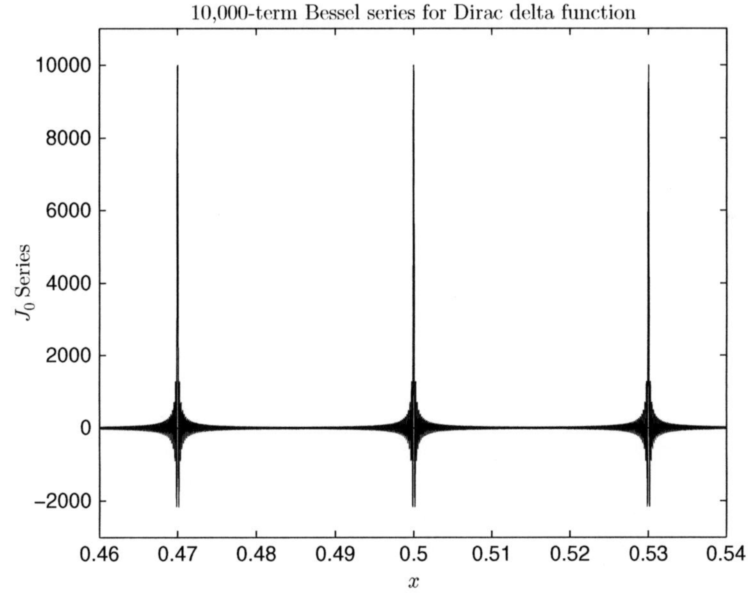

Figure 6.3 The sum of the first 10,000 terms of the Bessel representation (6.376) for the Dirac delta function $\delta(x-y)$ is plotted for $y = 0.47$ and $\alpha = 0$, for $y = 1/2$ and $\alpha = 1/2$, and for $y = 0.53$ and $\alpha = 1$.

□

6.36 The inequalities of Bessel and Schwarz

The inequality

$$\int_a^b \rho(x) \left| f(x) - \sum_{k=1}^{N} a_k u_k(x) \right|^2 dx \ge 0 \tag{6.377}$$

and the formula (6.370) for a_k lead (exercise 6.27) to Bessel's inequality

$$\int_a^b \rho(x)\, |f(x)|^2 \, dx \ge \sum_{k=1}^{\infty} |a_k|^2. \tag{6.378}$$

The argument we used to derive the Schwarz inequality (1.94) for vectors applies also to functions and leads to the Schwarz inequality

$$\int_a^b \rho(x)|f(x)|^2 \, dx \int_a^b \rho(x)|g(x)|^2 \, dx \ge \left| \int_a^b \rho(x) g^*(x) f(x) \, dx \right|^2. \tag{6.379}$$

6.37 Green's functions

Physics is full of equations of the form

$$L\, G(x) = \delta^{(n)}(x), \tag{6.380}$$

in which L is a differential operator in n variables. The solution $G(x)$ is a Green's function (section 3.8) for the operator L.

Example 6.42 (Poisson's Green's function) Probably the most important Green's function arises when the interaction is of long range as in gravity and electrodynamics. The divergence of the electric field is related to the charge density ρ by Gauss's law $\nabla \cdot \boldsymbol{E} = \rho/\epsilon_0$ where $\epsilon_0 = 8.854 \times 10^{-12}$ F/m is the electric constant. The electric field is $\boldsymbol{E} = -\nabla\phi - \dot{\boldsymbol{A}}$ in which ϕ is the scalar potential. In the Coulomb or radiation gauge, the divergence of $\boldsymbol{A}$ vanishes, $\nabla \cdot \boldsymbol{A} = 0$, and so $-\Delta\phi = -\nabla \cdot \nabla\phi = \rho/\epsilon_0$. The needed Green's function satisfies

$$-\Delta G(\boldsymbol{x}) = -\nabla \cdot \nabla G(\boldsymbol{x}) = \delta^{(3)}(\boldsymbol{x}) \tag{6.381}$$

and expresses the scalar potential ϕ as the integral

$$\phi(t, \boldsymbol{x}) = \int G(\boldsymbol{x} - \boldsymbol{x}') \, \frac{\rho(t, \boldsymbol{x}')}{\epsilon_0} \, d^3x'. \tag{6.382}$$

For when we apply (minus) the Laplacian to it, we get

$$\begin{aligned} -\triangle\phi(t,\boldsymbol{x}) &= -\int \triangle\, G(\boldsymbol{x}-\boldsymbol{x}')\,\frac{\rho(t,\boldsymbol{x}')}{\epsilon_0}\,d^3x' \\ &= \int \delta^{(3)}(\boldsymbol{x}-\boldsymbol{x}')\,\frac{\rho(t,\boldsymbol{x}')}{\epsilon_0}\,d^3x' = \frac{\rho(t,\boldsymbol{x})}{\epsilon_0}, \end{aligned} \tag{6.383}$$

which is **Poisson's equation**.

The reader might wonder how the potential $\phi(t,\boldsymbol{x})$ can depend upon the charge density $\rho(t,\boldsymbol{x}')$ at different points *at the same time*. The scalar potential is instantaneous because of the Coulomb gauge condition $\nabla\cdot\boldsymbol{A}=0$, which is not Lorentz invariant. The gauge-invariant physical fields $\boldsymbol{E}$ and $\boldsymbol{B}$ are not instantaneous and do describe Lorentz-invariant electrodynamics.

It is easy to find the Green's function $G(\boldsymbol{x})$ by expressing it as a Fourier transform

$$G(\boldsymbol{x}) = \int e^{i\boldsymbol{k}\cdot\boldsymbol{x}}\,g(\boldsymbol{k})\,d^3k \tag{6.384}$$

and by using the three-dimensional version

$$\delta^{(3)}(\boldsymbol{x}) = \int \frac{d^3k}{(2\pi)^3}\,e^{i\boldsymbol{k}\cdot\boldsymbol{x}} \tag{6.385}$$

of Dirac's delta function (3.36). If we insert these Fourier transforms into the equation (6.381) that defines the Green's function $G(\boldsymbol{x})$, then we find

$$\begin{aligned} -\triangle G(\boldsymbol{x}) &= -\triangle\int e^{i\boldsymbol{k}\cdot\boldsymbol{x}}\,g(\boldsymbol{k})\,d^3k \\ &= \int e^{i\boldsymbol{k}\cdot\boldsymbol{x}}\,\boldsymbol{k}^2\,g(\boldsymbol{k})\,d^3k = \delta^{(3)}(\boldsymbol{x}) = \int e^{i\boldsymbol{k}\cdot\boldsymbol{x}}\,\frac{d^3k}{(2\pi)^3}. \end{aligned} \tag{6.386}$$

Thus the Green's function $G(\boldsymbol{x})$ is the Fourier transform

$$G(\boldsymbol{x}) = \int \frac{e^{i\boldsymbol{k}\cdot\boldsymbol{x}}}{\boldsymbol{k}^2}\,\frac{d^3k}{(2\pi)^3}, \tag{6.387}$$

which we may integrate to

$$G(\boldsymbol{x}) = \frac{1}{4\pi|\boldsymbol{x}|} = \frac{1}{4\pi r} \tag{6.388}$$

where $r=|\boldsymbol{x}|$ is the length of the vector $\boldsymbol{x}$. This formula is generalized to n dimensions in example 5.22. □

Example 6.43 (Helmholtz's Green's functions) The Green's function for the Helmholtz equation $(-\triangle - m^2)V(\boldsymbol{x}) = \rho(\boldsymbol{x})$ must satisfy

$$(-\triangle - m^2)\,G_{\rm H}(\boldsymbol{x}) = \delta^{(3)}(\boldsymbol{x}). \tag{6.389}$$

By using the Fourier-transform method of the previous example, one may show that $G_{\rm H}$ is

$$G_{\rm H}(\boldsymbol{x}) = \frac{e^{imr}}{4\pi r}, \tag{6.390}$$

in which $r = |\boldsymbol{x}|$ and m has units of inverse length.

Similarly, the Green's function $G_{\rm mH}$ for the modified Helmholtz equation

$$(-\triangle + m^2)\, G_{\rm mH}(\boldsymbol{x}) = \delta^{(3)}(\boldsymbol{x}) \tag{6.391}$$

is (example 5.21)

$$G_{\rm mH}(\boldsymbol{x}) = \frac{e^{-mr}}{4\pi r}, \tag{6.392}$$

which is a Yukawa potential. □

Of these Green's functions, probably the most important is $G(\boldsymbol{x}) = 1/4\pi r$, which has the expansion

$$G(\boldsymbol{x} - \boldsymbol{x}') = \frac{1}{4\pi\,|\boldsymbol{x} - \boldsymbol{x}'|} = \sum_{\ell=0}^{\infty} \sum_{m=-\ell}^{\ell} \frac{1}{2\ell + 1} \frac{r_<^\ell}{r_>^{\ell+1}}\, Y_{\ell,m}(\theta, \phi) Y^*_{\ell,m}(\theta', \phi') \tag{6.393}$$

in terms of the spherical harmonics $Y_{\ell,m}(\theta, \phi)$. Here r, θ, and ϕ are the spherical coordinates of the point $\boldsymbol{x}$, and r', θ', and ϕ' are those of the point $\boldsymbol{x}'$; $r_>$ is the larger of r and r', and $r_<$ is the smaller of r and r'. If we substitute this expansion (6.393) into the formula (6.382) for the potential ϕ, then we arrive at the multipole expansion

$$\begin{aligned}\phi(t, \boldsymbol{x}) &= \int G(\boldsymbol{x} - \boldsymbol{x}')\, \frac{\rho(t, \boldsymbol{x}')}{\epsilon_0}\, d^3x' \qquad (6.394)\\ &= \sum_{\ell=0}^{\infty} \sum_{m=-\ell}^{\ell} \frac{1}{2\ell + 1} \int \frac{r_<^\ell}{r_>^{\ell+1}}\, Y_{\ell,m}(\theta, \phi) Y^*_{\ell,m}(\theta', \phi')\, \frac{\rho(t, \boldsymbol{x}')}{\epsilon_0}\, d^3x'.\end{aligned}$$

Physicists often use this expansion to compute the potential at $\boldsymbol{x}$ due to a localized, remote distribution of charge $\rho(t, \boldsymbol{x}')$. In this case, the integration is only over the restricted region where $\rho(t, \boldsymbol{x}') \neq 0$, and so $r_< = r'$ and $r_> = r$, and the multipole expansion is

$$\phi(t, \boldsymbol{x}) = \sum_{\ell=0}^{\infty} \frac{1}{2\ell + 1} \sum_{m=-\ell}^{\ell} \frac{Y_{\ell,m}(\theta, \phi)}{r^{\ell+1}} \int r'^{\ell}\, Y^*_{\ell,m}(\theta', \phi')\, \frac{\rho(t, \boldsymbol{x}')}{\epsilon_0}\, d^3x'. \tag{6.395}$$

In terms of the multipoles

$$Q_\ell^m = \int r'^{\ell}\, Y^*_{\ell,m}(\theta', \phi')\, \frac{\rho(t, \boldsymbol{x}')}{\epsilon_0}\, d^3x' \tag{6.396}$$

the potential is

$$\phi(t, \boldsymbol{x}) = \sum_{\ell=0}^{\infty} \frac{1}{2\ell + 1} \frac{1}{r^{\ell+1}} \sum_{m=-\ell}^{\ell} Q_\ell^m \, Y_{\ell,m}(\theta, \phi). \tag{6.397}$$

The spherical harmonics provide for the Legendre polynomial the expansion

$$P_\ell(\hat{\boldsymbol{x}} \cdot \hat{\boldsymbol{x}}') = \frac{4\pi}{2\ell + 1} \sum_{m=-\ell}^{\ell} Y_{\ell,m}(\theta, \phi) Y_{\ell,m}^*(\theta', \phi'), \tag{6.398}$$

which abbreviates the Green's function formula (6.393) to

$$G(\boldsymbol{x} - \boldsymbol{x}') = \frac{1}{4\pi |\boldsymbol{x} - \boldsymbol{x}'|} = \frac{1}{4\pi} \sum_{\ell=0}^{\infty} \frac{r_<^\ell}{r_>^{\ell+1}} P_\ell(\hat{\boldsymbol{x}} \cdot \hat{\boldsymbol{x}}'). \tag{6.399}$$

Example 6.44 (Feynman's propagator) The Feynman propagator

$$\Delta_F(x) = \int \frac{d^4 q}{(2\pi)^4} \frac{\exp(iqx)}{q^2 + m^2 - i\epsilon} \tag{6.400}$$

is a Green's function (5.230) for the operator $L = m^2 - \Box$

$$(m^2 - \Box)\Delta_F(x) = \delta^4(x). \tag{6.401}$$

By integrating over q^0 while respecting the $i\epsilon$ (example 5.36), one may write the propagator in terms of the Lorentz-invariant function

$$\Delta_+(x) = \frac{1}{(2\pi)^3} \int \frac{d^3 q}{2E_{\boldsymbol{q}}} \exp[i(\boldsymbol{q} \cdot \boldsymbol{x} - E_{\boldsymbol{q}} x^0)] \tag{6.402}$$

as (5.241)

$$\Delta_F(x) = i\theta(x^0)\,\Delta_+(x) + i\theta(-x^0)\,\Delta_+(\boldsymbol{x}, -x^0), \tag{6.403}$$

which for space-like x, that is, for $x^2 = \boldsymbol{x}^2 - (x^0)^2 \equiv r^2 > 0$, depends only upon $r = +\sqrt{x^2}$ and has the value (Weinberg, 1995, p. 202)

$$\Delta_+(x) = \frac{m}{4\pi^2 r} K_1(mr), \tag{6.404}$$

in which K_1 is the Hankel function (5.249). □

6.38 Eigenfunctions and Green's functions

The Green's function (6.387)

$$G(x - y) = \int \frac{dk}{2\pi} \frac{1}{k^2 + m^2} e^{ik(x-y)} \tag{6.405}$$

is based on the resolution (6.385) of the delta function

$$\delta(x-y) = \int \frac{dk}{2\pi}\, e^{ik\,(x-y)} \tag{6.406}$$

in terms of the eigenfunctions $\exp(ik\,x)$ of the differential operator $-\partial^2 + m^2$ with eigenvalues $k^2 + m^2$.

We may generalize this way of making Green's functions to a regular Sturm–Liouville system (section 6.33) with a differential operator L, eigenvalues λ_n

$$L\,u_n(x) = \lambda_n\,\rho(x)\,u_n(x), \tag{6.407}$$

and eigenfunctions $u_n(x)$ that are orthonormal with respect to a positive weight function $\rho(x)$

$$\delta_{n\ell} = (u_n, u_k) = \int \rho(x)\,u_n(x) u_k(x)\,dx \tag{6.408}$$

and that span in the mean the domain D of the system.

To make a Green's function $G(x-y)$ that satisfies

$$L\,G(x-y) = \delta(x-y) \tag{6.409}$$

we write $G(x-y)$ in terms of the complete set of eigenfunctions u_k as

$$G(x-y) = \sum_{k=1}^{\infty} \frac{u_k(x) u_k(y)}{\lambda_k} \tag{6.410}$$

so that the action $Lu_k = \lambda_k \rho u_k$ turns G into

$$L\,G(x-y) = \sum_{k=1}^{\infty} \frac{L\,u_k(x) u_k(y)}{\lambda_k} = \sum_{k=1}^{\infty} \rho(x)\,u_k(x)\,u_k(y) = \delta(x-y), \tag{6.411}$$

our $\alpha = 1$ series expansion (6.374) of the delta function.

6.39 Green's functions in one dimension

In one dimension, we can explicitly solve the inhomogeneous ordinary differential equation $Lf(x) = g(x)$ in which

$$L = -\frac{d}{dx}\left(p(x)\frac{d}{dx}\right) + q(x) \tag{6.412}$$

is formally self adjoint. We'll build a Green's function from two solutions u and v of the homogeneous equation $L\,u(x) = L\,v(x) = 0$ as

$$G(x,y) = \frac{1}{A}\left[\theta(x-y) u(y) v(x) + \theta(y-x) u(x) v(y)\right], \tag{6.413}$$

in which $\theta(x) = (x + |x|)/(2|x|)$ is **the Heaviside step function** (Oliver Heaviside, 1850–1925), and A is a constant which we'll presently identify. We'll show that the expression

$$f(x) = \int_a^b G(x, y)\, g(y)\, dy = \frac{v(x)}{A} \int_a^x u(y)\, g(y)\, dy + \frac{u(x)}{A} \int_x^b v(y)\, g(y)\, dy$$

solves our inhomogeneous equation. Differentiating, we find after a cancellation

$$f'(x) = \frac{v'(x)}{A} \int_a^x u(y)\, g(y)\, dy + \frac{u'(x)}{A} \int_x^b v(y)\, g(y)\, dy. \tag{6.414}$$

Differentiating again, we have

$$\begin{aligned} f''(x) &= \frac{v''(x)}{A} \int_a^x u(y)\, g(y)\, dy + \frac{u''(x)}{A} \int_x^b v(y)\, g(y)\, dy \\ &\quad + \frac{v'(x)u(x)g(x)}{A} - \frac{u'(x)v(x)g(x)}{A} \\ &= \frac{v''(x)}{A} \int_a^x u(y)\, g(y)\, dy + \frac{u''(x)}{A} \int_x^b v(y)\, g(y)\, dy \\ &\quad + \frac{W(x)}{A}\, g(x), \end{aligned} \tag{6.415}$$

in which $W(x)$ is the wronskian $W(x) = u(x)v'(x) - u'(x)v(x)$. The result (6.266) for the wronskian of two linearly independent solutions of a self-adjoint homogeneous ODE gives us $W(x) = W(x_0)\, p(x_0)/p(x)$. We set the constant $A = -W(x_0)p(x_0)$ so that the last term in (6.415) is $-g(x)/p(x)$. It follows that

$$Lf(x) = \frac{[Lv(x)]}{A} \int_a^x u(y)\, g(y)\, dy + \frac{[Lu(x)]}{A} \int_x^b v(y)\, g(y)\, dy + g(x) = g(x). \tag{6.416}$$

But $Lu(x) = Lv(x) = 0$, so we see that f satisfies our inhomogeneous equation $Lf(x) = g(x)$.

6.40 Nonlinear differential equations

The field of nonlinear differential equations is too vast to cover here, but we may hint at some of its features by considering some examples from cosmology and particle physics.

The Friedmann equations of general relativity (11.410 & 11.412) for the scale factor $a(t)$ of a homogeneous, isotropic universe are

$$\frac{\ddot{a}}{a} = -\frac{4\pi G}{3}(\rho + 3p) \quad \text{and} \quad \left(\frac{\dot{a}}{a}\right)^2 = \frac{8\pi G}{3}\rho - \frac{k}{a^2}, \tag{6.417}$$

in which k respectively is 1, 0, and -1 for closed, flat, and open geometries. (The scale factor $a(t)$ tells how much space has expanded or contracted by the time t.) These equations become more tractable when the energy density ρ is due to a single constituent whose pressure p is related to it by an equation of state $p = w\rho$. Conservation of energy $\dot{\rho} = -3(\rho + p)/a$ (11.426–11.431) then ensures (exercise 6.30) that the product $\rho\, a^{3(1+w)}$ is independent of time. The constant w respectively is $1/3$, 0, and -1 for radiation, matter, and the vacuum. The Friedmann equations then are

$$a^{2+3w}\,\ddot{a} = -\frac{4\pi G}{3}(1 + 3w)\,\rho\, a^{3(1+w)} \equiv -f, \tag{6.418}$$

where f is a constant that is positive when $w > -1/3$, and

$$a^{1+3w}\left(\dot{a}^2 + k\right) = \frac{2f}{1 + 3w}. \tag{6.419}$$

Example 6.45 (An open universe of radiation) Here $k = -1$ and the parameter $w = 1/3$, so the first-order Friedmann equation (6.419) becomes

$$\dot{a}^2 = \frac{f}{a^2} + 1. \tag{6.420}$$

The universe is expanding, so we take the positive square-root and get

$$dt = \frac{a\, da}{\sqrt{a^2 + f}}, \tag{6.421}$$

which leads to the general integral $t = \sqrt{a^2 + f} + C$. If we choose the constant of integration $C = -\sqrt{f}$, then we find

$$a(t) = \sqrt{\left(t + \sqrt{f}\right)^2 - f}, \tag{6.422}$$

a scale factor that vanishes at time zero and approaches t as $t \to \infty$. □

Example 6.46 (A closed universe of matter) Here $w = 0$ and $k = 1$, and so the first-order Friedmann equation (6.419) is

$$\dot{a}^2 = \frac{2f}{a} - 1. \tag{6.423}$$

Since the universe is expanding, we take the positive square-root

$$\dot{a} = \sqrt{\frac{2f}{a} - 1}, \tag{6.424}$$

which leads to the general integral

$$t = \int \frac{\sqrt{a}\,da}{\sqrt{2f - a}} = -\sqrt{a(2f - a)} - f \arcsin\left(1 - a/f\right) + C, \tag{6.425}$$

in which C is a constant of integration. □

Example 6.47 (An open universe of matter) Here $w = 0$ and $k = -1$, and so the first-order Friedmann equation (6.419) is $\dot{a}^2 = 2f/a + 1$, which leads to the general integral

$$t = \int \frac{\sqrt{a}\,da}{\sqrt{2f + a}} = \sqrt{a(2f + a)} - f \ln\left[\sqrt{a(2f + a)} + a + f\right] + C, \tag{6.426}$$

in which C is a constant of integration. □

The equations of particle physics are nonlinear. Physicists usually use perturbation theory to cope with the nonlinearities. But occasionally they focus on the nonlinearities and treat the quantum aspects classically or semi-classically. To keep things relatively simple, we'll work in a space-time of only two dimensions and consider a model field theory described by the action density

$$\mathcal{L} = \frac{1}{2}\left(\dot{\phi}^2 - \phi'^2\right) - V(\phi), \tag{6.427}$$

in which V is a simple function of the field ϕ. Lagrange's equation for this theory is

$$\ddot{\phi} - \phi'' = -\frac{dV}{d\phi}. \tag{6.428}$$

We can convert this partial differential equation to an ordinary one by making the field ϕ depend only upon the combination $u = x - vt$ rather than upon both x and t. We then have $\dot{\phi} = -v\phi_u$. With this restriction to traveling-wave solutions, Lagrange's equation reduces to

$$(1 - v^2)\phi_{uu} = \frac{dV}{d\phi}. \tag{6.429}$$

We multiply both sides of this equation by ϕ_u

$$(1 - v^2)\,\phi_u\,\phi_{uu} = \frac{dV}{d\phi}\,\phi_u. \tag{6.430}$$

and integrate both sides to get $(1 - v^2)\,\frac{1}{2}\,\phi_u^2 = V + E$, in which E is a constant of integration and a kind of energy

$$E = \tfrac{1}{2}\,(1 - v^2)\,\phi_u^2 - V(\phi). \tag{6.431}$$

We can convert (exercise 6.37) this equation into a problem of integration

$$u - u_0 = \int \frac{\sqrt{1 - v^2}}{\sqrt{2(E + V(\phi))}}\, d\phi. \tag{6.432}$$

By inverting the resulting equation relating u to ϕ, we may find the **soliton** solution $\phi(u - u_0)$, which is a lump of energy traveling with speed v.

Example 6.48 (Soliton of the ϕ^4 theory) To simplify the integration (6.432), we take as the action density

$$\mathcal{L} = \frac{1}{2}\left(\dot{\phi}^2 - \phi'^2\right) - \left[E - \frac{\lambda^2}{2}\left(\phi^2 - \phi_0^2\right)^2 - E\right]. \tag{6.433}$$

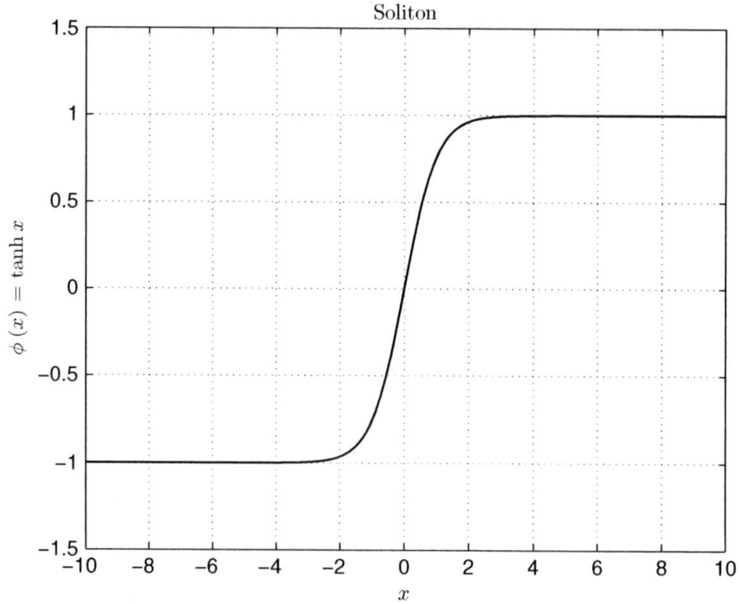

Figure 6.4 The field $\phi(x)$ of the soliton (6.435) at rest ($v = 0$) at position $x_0 = 0$ for $\lambda = 1 = \phi_0$. The energy density of the field vanishes when $\phi = \pm\phi_0 = \pm 1$. The energy of this soliton is concentrated at $x_0 = 0$.

Our formal solution (6.432) gives

$$u - u_0 = \pm \int \frac{\sqrt{1-v^2}}{\lambda\left(\phi^2 - \phi_0^2\right)}\, d\phi = \mp \frac{\sqrt{1-v^2}}{\lambda\phi_0} \tanh^{-1}(\phi/\phi_0) \tag{6.434}$$

or

$$\phi(x - vt) = \mp\phi_0 \tanh\left[\lambda\phi_0 \frac{x - x_0 - v(t - t_0)}{\sqrt{1-v^2}}\right], \tag{6.435}$$

which is a soliton (or an antisoliton) at $x_0 + v(t - t_0)$. A unit soliton at rest is plotted in Fig. 6.4. Its energy is concentrated at $x = 0$ where $|\phi^2 - \phi_0^2|$ is maximal. □

Exercises

6.1 In rectangular coordinates, the curl of a curl is by definition (6.40)

$$(\nabla \times (\nabla \times \boldsymbol{E}))_i = \sum_{j,k=1}^{3} \epsilon_{ijk}\partial_j(\nabla \times \boldsymbol{E})_k = \sum_{j,k,\ell,m=1}^{3} \epsilon_{ijk}\partial_j\epsilon_{k\ell m}\partial_\ell E_m. \tag{6.436}$$

Use Levi-Civita's identity (1.449) to show that

$$\nabla \times (\nabla \times \boldsymbol{E}) = \nabla(\nabla \cdot \boldsymbol{E}) - \triangle \boldsymbol{E}. \tag{6.437}$$

This formula defines $\triangle \boldsymbol{E}$ in any system of orthogonal coordinates.

6.2 Show that since the Bessel function $J_n(x)$ satisfies Bessel's equation (6.48), the function $P_n(\rho) = J_n(k\rho)$ satisfies (6.47).

6.3 Show that (6.58) implies that $R_{k,\ell}(r) = j_\ell(kr)$ satisfies (6.57).

6.4 Use (6.56, 6.57), and $\Phi''_m = -m^2\Phi_m$ to show in detail that the product $f(r,\theta,\phi) = R_{k,\ell}(r)\,\Theta_{\ell,m}(\theta)\,\Phi_m(\phi)$ satisfies $-\triangle f = k^2 f$.

6.5 Replacing Helmholtz's k^2 by $2m(E - V(r))/\hbar^2$, we get Schrödinger's equation

$$-(\hbar^2/2m)\triangle\psi(r,\theta,\phi) + V(r)\psi(r,\theta,\phi) = E\psi(r,\theta,\phi). \tag{6.438}$$

Let $\psi(r,\theta,\phi) = R_{n,\ell}(r)\Theta_{\ell,m}(\theta)e^{im\phi}$ in which $\Theta_{\ell,m}$ satisfies (6.56) and show that the radial function $R_{n,\ell}$ must obey

$$-\left(r^2 R'_{n,\ell}\right)'/r^2 + \left[\ell(\ell+1)/r^2 + 2mV/\hbar^2\right] R_{n,\ell} = 2mE_{n,\ell}R_{n,\ell}/\hbar^2. \tag{6.439}$$

6.6 Use the empty-space Maxwell's equations $\nabla \cdot \boldsymbol{B} = 0$, $\nabla \times \boldsymbol{E} + \dot{\boldsymbol{B}} = 0$, $\nabla \cdot \boldsymbol{E} = 0$, and $\nabla \times \boldsymbol{B} - \dot{\boldsymbol{E}}/c^2 = 0$ and the formula (6.437) to show that in vacuum $\triangle \boldsymbol{E} = \ddot{\boldsymbol{E}}/c^2$ and $\triangle \boldsymbol{B} = \ddot{\boldsymbol{B}}/c^2$.

6.7 Argue from symmetry and antisymmetry that $[\gamma^a, \gamma^{a'}]\partial_a\partial_{a'} = 0$, in which the sums over a and b run from 0 to 3.

6.8 Suppose a voltage $V(t) = V\sin(\omega t)$ is applied to a resistor of R (Ω) in series with a capacitor of capacitance C (F). If the current through the circuit at time $t = 0$ is zero, what is the current at time t?

6.9 (a) Is the ODE

$$\frac{(1+y^2)y\,dx + (1+x^2)x\,dy}{\left(1+x^2+y^2\right)^{3/2}} = 0$$

exact? (b) Find its general integral and solution $y(x)$. Use section 6.11.

6.10 (a) Separate the variables of the ODE $(1+y^2)y\,dx+(1+x^2)x\,dy = 0$. (b) Find its general integral and solution $y(x)$.

6.11 Find the general solution to the differential equation $y' + y/x = c/x$.

6.12 Find the general solution to the differential equation $y' + xy = ce^{-x^2/2}$.

6.13 James Bernoulli studied ODEs of the form $y' + py = qy^n$, in which p and q are functions of x. Division by y^n and the substitution $v = y^{1-n}$ gives $v' + (1-n)p = (1-n)q$, which is soluble as shown in section 6.16. Use this method to solve the ODE $y' - y/2x = 5x^2y^5$.

6.14 Integrate the ODE $(xy+1)\,dx + 2x^2(2xy-1)\,dy = 0$. Hint: use the variable $v(x) = xy(x)$ instead of $y(x)$.

6.15 Show that the points $x = \pm 1$ and ∞ are regular singular points of Legendre's equation (6.181).

6.16 Use the vanishing of the coefficient of every power of x in (6.185) and the notation (6.187) to derive the recurrence relation (6.188).

6.17 In example 6.29, derive the recursion relation for $r = 1$ and discuss the resulting eigenvalue equation.

6.18 In example 6.29, show that the solutions associated with the roots $r = 0$ and $r = 1$ are the same.

6.19 For a hydrogen atom, we set $V(r) = -e^2/4\pi\epsilon_0 r \equiv -q^2/r$ in (6.439) and get $(r^2R'_{n,\ell})' + \left[(2m/\hbar^2)\left(E_{n,\ell} + Zq^2/r\right)r^2 - \ell(\ell+1)\right]R_{n,\ell} = 0$. So at big r, $R''_{n,\ell} \approx -2mE_{n,\ell}R_{n,\ell}/\hbar^2$ and $R_{n,\ell} \sim \exp(-\sqrt{-2mE_{n,\ell}}r/\hbar)$. At tiny r, $(r^2R'_{n,\ell})' \approx \ell(\ell+1)R_{n,\ell}$ and $R_{n,\ell}(r) \sim r^\ell$. Set $R_{n,\ell}(r) = r^\ell \exp(-\sqrt{-2mE_{n,\ell}}r/\hbar)P_{n,\ell}(r)$ and apply the method of Frobenius to find the values of $E_{n,\ell}$ for which $R_{n,\ell}$ is suitably normalizable.

6.20 Show that as long as the matrix $\mathcal{Y}_{kj} = y_k^{(\ell_j)}(x_j)$ is nonsingular, the n boundary conditions

$$b_j = y^{(\ell_j)}(x_j) = \sum_{k=1}^{n} c_k\, y_k^{(\ell_j)}(x_j) \tag{6.440}$$

determine the n coefficients c_k of the expansion (6.222) to be

$$C^{\mathsf{T}} = B\,\mathcal{Y}^{-1} \quad \text{or} \quad C_k = \sum_{j=1}^{n} b_j \mathcal{Y}^{-1}_{jk}. \tag{6.441}$$

6.21 Show that if the real and imaginary parts u_1, u_2, v_1, and v_2 of ψ and χ satisfy boundary conditions at $x = a$ and $x = b$ that make the boundary term (6.235) vanish, then its complex analog (6.242) also vanishes.

6.22 Show that if the real and imaginary parts u_1, u_2, v_1, and v_2 of ψ and χ satisfy boundary conditions at $x = a$ and $x = b$ that make the boundary term (6.235) vanish, and if the differential operator L is real and self adjoint, then (6.238) implies (6.243).

6.23 Show that if D is the set of all twice-differentiable functions $u(x)$ on $[a, b]$ that satisfy Dirichlet's boundary conditions (6.245) and if the function $p(x)$ is continuous and positive on $[a, b]$, then the adjoint set D^* defined as the set of all twice-differentiable functions $v(x)$ that make the boundary term (6.247) vanish for all functions $u \in D$ is D itself.

6.24 Same as exercise (6.23) but for Neumann boundary conditions (6.246).

6.25 Use Bessel's equation (6.307) and the boundary conditions $u(0) = 0$ for $n > 0$ and $u(1) = 0$ to show that the eigenvalues λ are all positive.

6.26 Show that after the change of variables $u(x) = J_n(kx) = J_n(\rho)$ the self-adjoint differential equation (6.307) becomes Bessel's equation (6.308).

6.27 Derive Bessel's inequality (6.378) from the inequality (6.377).

6.28 Repeat example 6.41 using J_1s instead of J_0s. Hint: the *Mathematica* command Do[Print[N[BesselJZero[1, k], 10]], {k, 1, 100, 1}] gives the first 100 zeros $z_{1,k}$ of the Bessel function $J_1(x)$ to ten significant figures.

6.29 Derive the Yukawa potential (6.392) as the Green's function for the modified Helmholtz equation (6.391).

6.30 Derive the relation $\rho = \overline{\rho}(\overline{a}/a)^{3(1+w)}$ between the energy density ρ and the Robertson–Walker scale factor $a(t)$ from the conservation law $d\rho/da = -3(\rho + p)/a$ and the equation of state $p = w\rho$.

6.31 For a closed universe ($k = 1$) of radiation ($w = 1/3$), use Friedmann's equations (6.418 & 6.419) to derive the solution (11.448) subject to the boundary condition $a(0) = 0$. When does the universe collapse in a big crunch?

6.32 For a flat universe ($k = 0$) of matter ($w = 0$), use Friedmann's equations (6.418 & 6.419) to derive the solution (11.454) subject to the boundary condition $a(0) = 0$.

6.33 Derive the time evolution of $a(t)$ for a flat ($k = 0$) universe dominated by radiation ($w = 1/3$) subject to the boundary condition $a(0) = 0$. Use (6.419).

6.34 Derive the time evolution of $a(t)$ for an open ($k = -1$) universe with only dark energy ($w = -1$) subject to the boundary condition $a(0) = 0$. Use (6.419).

6.35 Use Friedmann's equations (6.418 & 6.419) to derive the evolution of the scale factor for a closed universe dominated by dark energy subject to the boundary condition $a(0) = \sqrt{3/8\pi G\rho}$ in which ρ is a constant density of dark energy.

6.36 Use Friedmann's equations (6.418 & 6.419) to derive the evolution of $a(t)$ for a flat ($k = 0$) expanding universe dominated by dark energy ($w = -1$) subject to the boundary condition $a(0) = \alpha$ in which ρ is a constant density of dark energy.

6.37 Derive the soliton solution (6.432) from the energy equation (6.431).

7

Integral equations

Differential equations when integrated become integral equations with built-in boundary conditions. Thus if we integrate the first-order ODE

$$\frac{du(x)}{dx} \equiv u_x(x) = p(x)\,u(x) + q(x) \tag{7.1}$$

then we get the integral equation

$$u(x) = \int_a^x p(y)\,u(y)\,dy + \int_a^x q(y)\,dy + u_0 \tag{7.2}$$

and the boundary condition $u(a) = f(a) = u_0$.

With a little more effort, we may integrate the second-order ODE

$$u'' = pu' + qu + r \tag{7.3}$$

(exercises 7.1 & 7.2) to

$$u(x) = f(x) + \int_a^x k(x,y)\,u(y)\,dy \tag{7.4}$$

with

$$k(x,y) = p(y) + (x-y)\left[q(y) - p'(y)\right] \tag{7.5}$$

and

$$f(x) = u(a) + (x-a)\left[u'(a) - p(a)\,u(a)\right] + \int_a^x (x-y)r(y)\,dy. \tag{7.6}$$

In some physical problems, integral equations arise independently of differential equations. Whatever their origin, integral equations tend to have properties more suitable to mathematical analysis because derivatives are unbounded operators.

7.1 Fredholm integral equations

An equation of the form

$$\int_a^b k(x,y)\,u(y)\,dy = \lambda\,u(x) + f(x) \tag{7.7}$$

for $a \le x \le b$ with a given **kernel** $k(x,y)$ and a specified function $f(x)$ is an **inhomogeneous Fredholm equation of the second kind** for the function $u(x)$ and the parameter λ (Erik Ivar Fredholm, 1866–1927).

If $f(x) = 0$, then it is a **homogeneous Fredholm equation of the second kind**

$$\int_a^b k(x,y)\,u(y)\,dy = \lambda\,u(x), \qquad a \le x \le b. \tag{7.8}$$

Such an equation typically has nontrivial solutions only for certain **eigenvalues** λ. Each solution $u(x)$ is an **eigenfunction.**

If $\lambda = 0$ but $f(x) \neq 0$, then equation (7.7) is an **inhomogeneous Fredholm equation of the first kind**

$$\int_a^b k(x,y)\,u(y)\,dy = f(x), \qquad a \le x \le b. \tag{7.9}$$

Finally, if both $\lambda = 0$ and $f(x) = 0$, then (7.7) is a **homogeneous Fredholm equation of the first kind**

$$\int_a^b k(x,y)\,u(y)\,dy = 0, \qquad a \le x \le b. \tag{7.10}$$

These Fredholm equations are **linear** because they involve only the first (and zeroth) power of the unknown function $u(x)$.

7.2 Volterra integral equations

If the kernel $k(x,y)$ in the equations (7.7–7.10) that define the Fredholm integral equations is **causal**, that is, if

$$k(x,y) = k(x,y)\,\theta(x-y), \tag{7.11}$$

in which $\theta(x) = (x + |x|)/2|x|$ is the Heaviside function, then the corresponding equations bear the name **Volterra** (Vito Volterra, 1860–1941). Thus, an equation of the form

$$\int_a^x k(x,y)\,u(y)\,dy = \lambda\,u(x) + f(x), \tag{7.12}$$

in which the **kernel** $k(x,y)$ and the function $f(x)$ are given, is an **inhomogeneous Volterra equation of the second kind** for the function $u(x)$ and the parameter λ.

If $f(x) = 0$, then it is a **homogeneous Volterra equation of the second kind**

$$\int_a^x k(x, y)\, u(y)\, dy = \lambda\, u(x). \tag{7.13}$$

Such an equation typically has nontrivial solutions only for certain **eigenvalues** λ. The solutions $u(x)$ are the **eigenfunctions**.

If $\lambda = 0$ but $f(x) \neq 0$, then equation (7.12) is an **inhomogeneous Volterra equation of the first kind**

$$\int_a^x k(x, y)\, u(y)\, dy = f(x). \tag{7.14}$$

Finally, if both $\lambda = 0$ and $f(x) = 0$, then it is a **homogeneous Volterra equation of the first kind**

$$\int_a^x k(x, y)\, u(y)\, dy = 0. \tag{7.15}$$

These Volterra equations are **linear** because they involve only the first (and zeroth) power of the unknown function $u(x)$.

In what follows, we'll mainly discuss Fredholm integral equations, since those of the Volterra type are a special case of the Fredholm type.

7.3 Implications of linearity

Because the Fredholm and Volterra integral equations are linear, one may add solutions of the homogeneous equations (7.8, 7.10, 7.13, & 7.15) and get new solutions. Thus if $u_1, u_2, \ldots$ are eigenfunctions

$$\int_a^b k(x, y)\, u_j(y)\, dy = \lambda\, u_j(x), \qquad a \le x \le b \tag{7.16}$$

with the same eigenvalue λ, then the sum $\sum_j a_j u_j(x)$ also is an eigenfunction with the same eigenvalue

$$\begin{aligned}\int_a^b k(x, y) \left(\sum_j a_j u_j(y) \right) dy &= \sum_j a_j \int_a^b k(x, y)\, u_j(y)\, dy \\ &= \sum_j a_j \lambda\, u_j(x) = \lambda \left(\sum_j a_j u_j(x) \right). \end{aligned} \tag{7.17}$$

It also is true that the difference between any two solutions $u_1^i(x)$ and $u_2^i(x)$ of one of the inhomogeneous Fredholm (7.7, 7.9) or Volterra (7.12, 7.14) equations is a solution of the associated homogeneous equation (7.8, 7.10, 7.13, or

7.15). Thus if $u_1^i(x)$ and $u_2^i(x)$ satisfy the inhomogeneous Fredholm equation of the second kind

$$\int_a^b k(x,y)\, u_j^i(y)\, dy = \lambda\, u_j^i(x) + f(x), \qquad j = 1,2 \tag{7.18}$$

then their difference $u_1^i(x) - u_2^i(x)$ satisfies the homogeneous Fredholm equation of the second kind

$$\int_a^b k(x,y)\, \left[u_1^i(y) - u_2^i(y)\right] dy = \lambda\, \left[u_1^i(x) - u_2^i(x)\right]. \tag{7.19}$$

Thus, the most general solution $u^i(x)$ of the inhomogeneous Fredholm equation of the second kind (7.18) is a particular solution $u_p^i(x)$ of that equation plus the general solution of the homogeneous Fredholm equation of the second kind (7.16)

$$u^i(x) = u_p^i(x) + \sum_j a_j u_j(x). \tag{7.20}$$

Linear integral equations are much easier to solve than nonlinear ones.

7.4 Numerical solutions

Let us break the real interval $[a, b]$ into N segments $[y_k, y_{k+1}]$ of equal length $\Delta y = (b-a)/N$ with $y_0 = a$, $y_k = a + k\,\Delta y$, and $y_N = b$. We'll also set $x_k = y_k$ and define U as the vector with entries $U_k = u(y_k)$ and K as the $(N+1) \times (N+1)$ square matrix with elements $K_{k\ell} = k(x_k, y_\ell)\,\Delta y$. Then we may approximate the homogeneous Fredholm equation of the second kind (7.8)

$$\int_a^b k(x,y)\, u(y)\, dy = \lambda\, u(x), \qquad a \le x \le b \tag{7.21}$$

as the algebraic equation

$$\sum_{\ell=0}^{N} K_{k,\ell}\, U_\ell = \lambda\, U_k \tag{7.22}$$

or, in matrix notation,

$$K\,U = \lambda\, U. \tag{7.23}$$

We saw in section 1.25 that every such equation has $N+1$ eigenvectors $U^{(\alpha)}$ and eigenvalues $\lambda^{(\alpha)}$, and that the eigenvalues $\lambda^{(\alpha)}$ are the solutions of the characteristic equation (1.244)

$$\det(K - \lambda^{(\alpha)} I) = \left| K - \lambda^{(\alpha)} I \right| = 0. \tag{7.24}$$

In general, as $N \to \infty$ and $\Delta y \to 0$, the number $N+1$ of eigenvalues $\lambda^{(\alpha)}$ and eigenvectors $U^{(\alpha)}$ becomes infinite.

We may apply the same technique to the inhomogeneous Fredholm equation of the first kind

$$\int_a^b k(x,y)\,u(y)\,dy = f(x) \quad \text{for} \quad a \le x \le b. \tag{7.25}$$

The resulting matrix equation is

$$K\,U = F, \tag{7.26}$$

in which the kth entry in the vector F is $F_k = f(x_k)$. This equation has the solution

$$U = K^{-1}\,F \tag{7.27}$$

as long as the matrix K is nonsingular, that is, as long as

$$\det K \neq 0. \tag{7.28}$$

This technique applied to the inhomogeneous Fredholm equation of the second kind

$$\int_a^b k(x,y)\,u(y)\,dy = \lambda\,u(x) + f(x) \tag{7.29}$$

leads to the matrix equation

$$K\,U = \lambda\,U + F. \tag{7.30}$$

The associated homogeneous matrix equation

$$K\,U = \lambda\,U \tag{7.31}$$

has $N+1$ eigenvalues $\lambda^{(\alpha)}$ and eigenvectors $U^{(\alpha)} \equiv |\alpha\rangle$. For any value of λ that is *not* one of the eigenvalues $\lambda^{(\alpha)}$, the matrix $K - \lambda I$ has a nonzero determinant and hence an inverse, and so the vector

$$U^i = (K - \lambda\,I)^{-1}\,F \tag{7.32}$$

is a solution of the inhomogeneous matrix equation (7.30).

If $\lambda = \lambda^{(\beta)}$ is one of the eigenvalues $\lambda^{(\alpha)}$ of the homogeneous matrix equation (7.31), then the matrix $K - \lambda^{(\beta)} I$ will not have an inverse, but it will have a pseudoinverse (section 1.32). If its singular-value decomposition (1.362) is

$$K - \lambda^{(\beta)} I = \sum_{n=1}^{N+1} |m_n\rangle S_n \langle n| \tag{7.33}$$

then its pseudoinverse (1.392) is

$$\left(K - \lambda^{(\beta)} I\right)^+ = \sum_{\substack{n=1 \\ S_n \neq 0}}^{N+1} |n\rangle S_n^{-1} \langle m_n|, \tag{7.34}$$

in which the sum is over the positive singular values. So if the vector F is a linear combination of the left singular vectors $|m_n\rangle$ whose singular values are positive

$$F = \sum_{\substack{n=1 \\ S_n \neq 0}}^{N+1} f_n |m_n\rangle \tag{7.35}$$

then the vector

$$U^i = \left(K - \lambda^{(\beta)} I\right)^+ F \tag{7.36}$$

will be a solution of the inhomogeneous matrix Fredholm equation (7.30). For in this case

$$\begin{aligned}\left(K - \lambda^{(\beta)} I\right) U^i &= \left(K - \lambda^{(\beta)} I\right) \left(K - \lambda^{(\beta)} I\right)^+ F \\ &= \sum_{n''=1}^{N+1} |m_{n''}\rangle S_{n''} \langle n''| \sum_{\substack{n'=1 \\ S_{n'} \neq 0}}^{N+1} |n'\rangle S_{n'}^{-1} \langle m_{n'}| \sum_{\substack{n=1 \\ S_n \neq 0}}^{N+1} f_n |m_n\rangle \\ &= \sum_{\substack{n=1 \\ S_n \neq 0}}^{N+1} f_n |m_n\rangle = F.\end{aligned} \tag{7.37}$$

The most general solution will be the sum of this particular solution of the inhomogeneous equation (7.30) and the most general solution of the homogeneous equation (7.31)

$$U = U^i + \sum_k f_{\beta,k}\, U^{(\beta,k)} = \left(K - \lambda^{(\beta)} I\right)^+ F + \sum_k f_{\beta,k}\, U^{(\beta,k)}. \tag{7.38}$$

Open-source programs are available in C++ (math.nist.gov/tnt/) and in FORTRAN (www.netlib.org/lapack/) that can solve such equations for the $N + 1$ eigenvalues $\lambda^{(\alpha)}$ and eigenvectors $U^{(\alpha)}$ and for the inverse K^{-1} for $N = 100$, 1000, 10,000, and so forth in milliseconds on a PC.

7.5 Integral transformations

Integral transformations (Courant and Hilbert, 1955, chap. VII) help us solve linear homogeneous differential equations like

$$L\,u + c\,u = 0, \tag{7.39}$$

in which L is a linear operator involving derivatives of $u(z)$ with respect to its complex argument $z = x + iy$ and c is a constant. We choose a **kernel** $K(z, w)$ analytic in both variables and write $u(z)$ as an integral along a contour in the complex w-plane weighted by an unknown function $v(w)$

$$u(z) = \int_C K(z, w)\, v(w)\, dw. \tag{7.40}$$

If the differential operator L commutes with the contour integration as it usually would, then our differential equation (7.39) is

$$\int_C [L\, K(z, w) + c\, K(z, w)]\; v(w)\, dw = 0. \tag{7.41}$$

The next step is to find a linear operator M that acting on $K(z, w)$ with w-derivatives (but no z-derivatives) gives L acting on $K(z, w)$

$$M\, K(z, w) = L\, K(z, w). \tag{7.42}$$

We then get an integral equation

$$\int_C [M\, K(z, w) + c\, K(z, w)]\; v(w)\, dw = 0 \tag{7.43}$$

involving w-derivatives that we can integrate by parts. We choose the contour C so that the resulting boundary terms vanish. By using our freedom to pick the kernel and the contour, we often can make the resulting differential equation for v simpler than the one (7.39) we started with.

Example 7.1 (Fourier, Laplace, and Euler kernels) We already are familiar with the most important integral transforms. In chapter 3, we learned that the kernel $K(z, w) = \exp(izw)$ leads to the Fourier transform

$$u(z) = \int_{-\infty}^{\infty} e^{izw}\; v(w)\, dw \tag{7.44}$$

and the kernel $K(z, w) = \exp(-zw)$ to the Laplace transform

$$u(z) = \int_0^{\infty} e^{-zw}\; v(w)\, dw \tag{7.45}$$

of section 3.9. Euler's kernel $K(z, w) = (z - w)^a$ occurs in many applications of Cauchy's integral theorem (5.21) and integral formula (5.36). These kernels help us solve differential equations. □

Example 7.2 (Bessel functions) The differential operator L for Bessel's equation (6.308)

$$z^2\, u'' + z\, u' + z^2\, u - \lambda^2\, u = 0 \tag{7.46}$$

is

$$L = z^2 \frac{d^2}{dz^2} + z \frac{d}{dz} + z^2 \tag{7.47}$$

and the constant c is $-\lambda^2$. If we choose $M = -d^2/dw^2$, then the kernel should satisfy (7.42)

$$LK - MK = z^2 K_{zz} + z K_z + z^2 K + K_{ww} = 0, \tag{7.48}$$

in which subscripts indicate differentiation as in (6.20). The kernel

$$K(z, w) = e^{\pm iz \sin w} \tag{7.49}$$

is a solution of (7.48) that is entire in both variables (exercise 7.3). In terms of it, our integral equation (7.43) is

$$\int_C \left[K_{ww}(z, w) + \lambda^2 K(z, w) \right] v(w)\, dw = 0. \tag{7.50}$$

We now integrate by parts once

$$\int_C \left[-K_w v' + \lambda^2 K v + \frac{dK_w v}{dw} \right] dw \tag{7.51}$$

and then again

$$\int_C \left[K \left(v'' + \lambda^2 v \right) + \frac{d(K_w v - K v')}{dw} \right] dw. \tag{7.52}$$

If we choose the contour so that $K_w v - K v' = 0$ at its ends, then the unknown function v must satisfy the differential equation

$$v'' + \lambda^2 v = 0, \tag{7.53}$$

which is vastly simpler than Bessel's; the solution $v(w) = \exp(i\lambda w)$ is an entire function of w for every complex λ. Our solution $u(z)$ then is

$$u(z) = \int_C K(z, w)\, v(w)\, dw = \int_C e^{\pm iz \sin w}\, e^{i\lambda w}\, dw. \tag{7.54}$$

For $\mathrm{Re}(z) > 0$ and any complex λ, the contour C_1 that runs from $-i\infty$ to the origin $w = 0$, then to $w = -\pi$, and finally up to $-\pi + i\infty$ has $K_w v - K v' = 0$ at its ends (exercise 7.4) provided we use the minus sign in the exponential. The function defined by this choice

$$H_\lambda^{(1)}(z) = -\frac{1}{\pi} \int_{C_1} e^{-iz \sin w + i\lambda w}\, dw \tag{7.55}$$

is the **first Hankel function** (Hermann Hankel, 1839–1873). The **second Hankel function** is defined for $\mathrm{Re}(z) > 0$ and any complex λ by a contour C_2 that runs from $\pi + i\infty$ to $w = \pi$, then to $w = 0$, and lastly to $-i\infty$

$$H_\lambda^{(2)}(z) = -\frac{1}{\pi} \int_{C_2} e^{-iz \sin w + i\lambda w}\, dw. \tag{7.56}$$

Because the integrand $\exp(-iz\sin w + i\lambda w)$ is an entire function of z and w, one may deform the contours C_1 and C_2 and analytically continue the Hankel functions beyond the right half-plane (Courant and Hilbert, 1955, chap. VII). One may verify (exercise 7.5) that the Hankel functions are related by complex conjugation

$$H_\lambda^{(1)}(z) = H_\lambda^{(2)*}(z) \tag{7.57}$$

when both $z > 0$ and λ are real. □

Exercises

7.1 Show that

$$\int_a^x dz \int_a^z dy\, f(y) = \int_a^x (x - y) f(y)\, dy. \tag{7.58}$$

Hint: differentiate both sides with respect to x.

7.2 Use this identity (7.58) to integrate (7.3) and derive equations (7.4, 7.5, & 7.6).

7.3 Show that the kernel $K(z, w) = \exp(\pm iz\sin w)$ satisfies the differential equation (7.48).

7.4 Show that for $\mathrm{Re}\, z > 0$ and arbitrary complex λ, the boundary terms in the integral (7.52) vanish for the two contours C_1 and C_2 that define the two Hankel functions.

7.5 Show that the Hankel functions are related by complex conjugation (7.57) when both $z > 0$ and λ are real.

8

Legendre functions

8.1 The Legendre polynomials

The monomials x^n span the space of functions $f(x)$ that have power-series expansions on an interval about the origin

$$f(x) = \sum_{n=0}^{\infty} c_n \, x^n = \sum_{n=0}^{\infty} \frac{f^{(n)}(0)}{n!} \, x^n. \tag{8.1}$$

They are complete but not orthogonal or normalized. We can make them into real, **orthogonal** polynomials $P_n(x)$ of degree n on the interval $[-1, 1]$

$$(P_n, P_m) = \int_{-1}^{1} P_n(x) \, P_m(x) \, dx = 0, \qquad n \neq m \tag{8.2}$$

by requiring that each $P_n(x)$ be orthogonal to all monomials x^m for $m < n$

$$\int_{-1}^{1} P_n(x) \, x^m \, dx = 0, \qquad m < n. \tag{8.3}$$

If we impose the **normalization** condition

$$P_n(1) = 1 \tag{8.4}$$

then they are unique and are the **Legendre polynomials** as in Fig. 8.1.

The coefficients a_k of the nth Legendre polynomial

$$P_n(x) = a_0 + a_1 x + \cdots + a_n x^n \tag{8.5}$$

must satisfy (exercise 8.3) the n conditions (8.3) of orthogonality

$$\int_{-1}^{1} P_n(x) \, x^m \, dx = \sum_{k=0}^{n} \frac{1 - (-1)^{m+k+1}}{m+k+1} \, a_k = 0 \quad \text{for} \quad 0 \leq m < n \tag{8.6}$$

and the normalization condition (8.4)

$$P_n(1) = a_0 + a_1 + \cdots + a_n = 1. \tag{8.7}$$

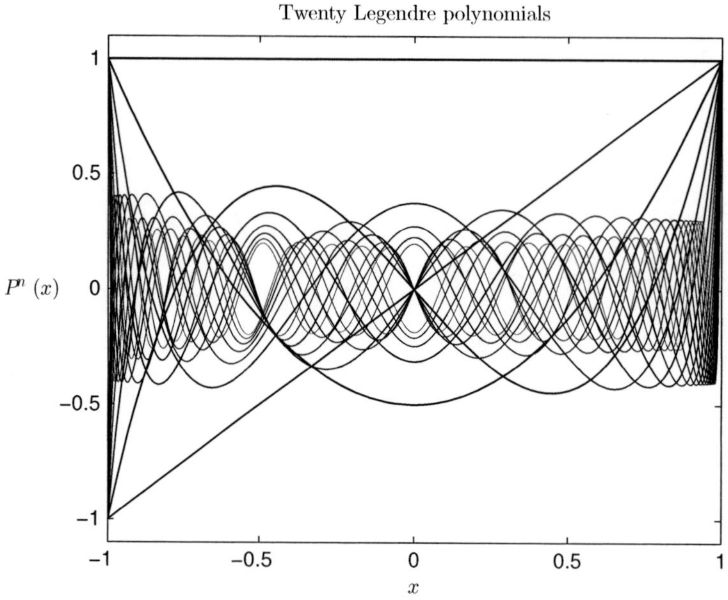

Figure 8.1 The first 20 Legendre polynomials in successively finer linewidths. The straight lines are $P_0(x) = 1$ and $P_1(x) = x$.

Example 8.1 (Building the Legendre polynomials) Conditions (8.6) and (8.7) give $P_0(x) = 1$ and $P_1(x) = x$. To make $P_2(x)$, we set $n = 2$ in the orthogonality condition (8.6) and find $2a_0 + 2a_2/3 = 0$ for $m = 0$, and $2a_1/3 = 0$ for $m = 1$. The normalization condition (8.7) then says that $a_0 + a_1 + a_2 = 1$. These three equations give $P_2(x) = (3x^2 - 1)/2$. Similarly, one finds $P_3(x) = (5x^3 - 3x)/2$ and $P_4(x) = (35x^4 - 30x^2 + 3)/8$. □

8.2 The Rodrigues formula

Perhaps the easiest way to compute the Legendre polynomials is to apply Leibniz's rule (4.46) to the **Rodrigues formula**

$$P_n(x) = \frac{1}{2^n n!} \frac{d^n (x^2 - 1)^n}{dx^n}, \tag{8.8}$$

which leads to (exercise 8.5)

$$P_n(x) = \frac{1}{2^n} \sum_{k=0}^{n} \binom{n}{k}^2 (x - 1)^{n-k} (x + 1)^k. \tag{8.9}$$

This formula at $x = 1$ is

$$P_n(1) = \frac{1}{2^n} \sum_{k=0}^{n} \binom{n}{k}^2 0^{n-k} 2^k = \frac{1}{2^n} \binom{n}{n}^2 2^n = 1, \tag{8.10}$$

which shows that Rodrigues got the normalization right (Benjamin Rodrigues, 1795–1851).

Example 8.2 (Using Rodrigues's formula) By (8.8) or (8.9) and with more effort, one finds

$$P_5(x) = \frac{1}{2^5 5!} \frac{d^5 (x^2-1)^5}{dx^5} = \frac{1}{8}\left(63x^5 - 70x^3 + 15x\right) \tag{8.11}$$

$$P_6(x) = \frac{1}{2^6} \sum_{k=0}^{6} \binom{6}{k}^2 (x-1)^{6-k}(x+1)^k = \frac{1}{16}\left(231x^6 - 315x^4 + 105x^2 - 5\right)$$

$$P_7(x) = \frac{(x-1)^7}{2^7} \sum_{k=0}^{7} \binom{7}{k}^2 \left(\frac{x+1}{x-1}\right)^k = \frac{1}{16}\left(429x^7 - 693x^5 + 315x^3 - 35x\right).$$

In MATLAB, mfun('P',n,x) returns the numerical value of $P_n(x)$. □

To check that the polynomial $P_n(x)$ generated by Rodrigues's formula (8.8) is orthogonal to x^m for $m < n$, we integrate $x^m P_n(x)$ by parts n times and drop all the surface terms (which vanish because $x^2 - 1$ is zero at $x = \pm 1$)

$$\int_{-1}^{1} x^m P_n(x)\, dx = \frac{1}{2^n n!} \int_{-1}^{1} x^m \frac{d^n}{dx^n}(x^2-1)^n\, dx$$
$$= \frac{(-1)^n}{2^n n!} \int_{-1}^{1} (x^2-1)^n \frac{d^n x^m}{dx^n}\, dx = 0 \quad \text{for} \quad n > m. \tag{8.12}$$

Thus the polynomial $P_n(x)$ generated by Rodrigues's formula (8.8) satisfies the orthogonality condition (8.3). It also satisfies the normalization condition (8.4) as shown by (8.10). The Rodrigues formula does generate Legendre's polynomials.

One may show (exercises 8.9, 8.10, & 8.11) that the inner product of two Legendre polynomials is

$$\int_{-1}^{1} P_n(x)\, P_m(x)\, dx = \frac{2}{2n+1}\, \delta_{nm}. \tag{8.13}$$

8.3 The generating function

In the expansion

$$g(t,x) = \left(1 - 2xt + t^2\right)^{-1/2} = \sum_{n=0}^{\infty} p_n(x)\, t^n \tag{8.14}$$

the coefficient $p_n(x)$ is the nth partial derivative of $g(t,x)$

$$p_n(x) = \frac{\partial^n}{\partial t^n} \left(1 - 2xt + t^2\right)^{-1/2} \Bigg|_{t=0} \tag{8.15}$$

and is a function of x alone. Explicit calculation shows that it is a polynomial of degree n.

To identify these polynomials $p_n(x)$, we use the integral formula

$$\int_{-1}^{1} g(t,x)\, g(v,x)\, dx = \int_{-1}^{1} \frac{dx}{\sqrt{1 - 2xt + t^2}\sqrt{1 - 2xv + v^2}} = \frac{1}{\sqrt{tv}} \ln \frac{1 + \sqrt{tv}}{1 - \sqrt{tv}} \tag{8.16}$$

and the logarithmic series (4.90)

$$\frac{1}{\sqrt{tv}} \ln \frac{1 + \sqrt{tv}}{1 - \sqrt{tv}} = \sum_{k=0}^{\infty} \frac{2}{2k+1} (tv)^k \tag{8.17}$$

to express the integral of $g(t,x)\, g(v,x)$ over the interval $-1 \le x \le 1$ as

$$\int_{-1}^{1} g(t,x)\, g(v,x)\, dx = \int_{-1}^{1} \sum_{n,m=0}^{\infty} p_n(x)\, p_m(x)\, t^n\, v^m\, dx = \sum_{k=0}^{\infty} \frac{2}{2k+1} (tv)^k. \tag{8.18}$$

Equating the coefficients of $t^n v^m$ in the second and third terms of this equation, we see that the polynomials $p_n(x)$ satisfy

$$\int_{-1}^{1} p_n(x)\, p_m(x)\, dx = \frac{2}{2n+1}\, \delta_{n,m}, \tag{8.19}$$

which is the inner product rule (8.13) obeyed by the Legendre polynomials. Next, setting $x = 1$ in the definition (8.14) of $g(t,x)$, we get from (4.31)

$$\frac{1}{1-t} = \sum_{n=0}^{\infty} t^n = \sum_{n=0}^{\infty} p_n(1)\, t^n, \tag{8.20}$$

which says that $p_n(1) = 1$ for all nonnegative integers $n = 0, 1, 2,$ and so forth. The polynomials $p_n(x)$ are therefore the Legendre polynomials $P_n(x)$, and the function $g(t,x)$ is their **generating function**

$$\frac{1}{\sqrt{1 - 2xt + t^2}} = \sum_{n=0}^{\infty} t^n\, P_n(x). \tag{8.21}$$

Example 8.3 (The Green's function for Poisson's equation) The Green's function (3.110) for the Laplacian is

$$G(\boldsymbol{R}-\boldsymbol{r}) = \frac{1}{4\pi|\boldsymbol{R}-\boldsymbol{r}|} = \frac{1}{4\pi\sqrt{R^2 - 2\boldsymbol{R}\cdot\boldsymbol{r} + r^2}}, \tag{8.22}$$

in which $R = |\boldsymbol{R}|$ and $r = |\boldsymbol{r}|$. It occurs throughout physics and satisfies

$$-\nabla^2 G(\boldsymbol{R}-\boldsymbol{r}) = \delta^{(3)}(\boldsymbol{R}-\boldsymbol{r}) \tag{8.23}$$

where the derivatives can act on $\boldsymbol{R}$ or on $\boldsymbol{r}$.

We set $x = \cos\theta = \boldsymbol{R}\cdot\boldsymbol{r}/rR$ and $t = r/R$, and then factor out $1/R$

$$\begin{aligned}\frac{1}{|\boldsymbol{R}-\boldsymbol{r}|} &= \frac{1}{\sqrt{R^2 - 2Rr\cos\theta + r^2}} \\ &= \frac{1}{R}\frac{1}{\sqrt{1 - 2(r/R)x + (r/R)^2}} \\ &= \frac{1}{R}\frac{1}{\sqrt{1 - 2xt + t^2}} = \frac{1}{R}g(t,x).\end{aligned} \tag{8.24}$$

With $t = r/R$ and so forth, this series is the well-known expansion

$$\frac{1}{|\boldsymbol{R}-\boldsymbol{r}|} = \frac{1}{R}\sum_{n=0}^{\infty}\left(\frac{r}{R}\right)^n P_n(\cos\theta) \tag{8.25}$$

of the Green's function $G(\boldsymbol{R}-\boldsymbol{r}) = 1/4\pi|\boldsymbol{R}-\boldsymbol{r}| = g(t,x)/4\pi R$. □

8.4 Legendre's differential equation

Apart from the prefactor $1/(2^n n!)$, the Legendre polynomial $P_n(x)$ is the nth derivative $u^{(n)}$ of $u = (x^2-1)^n$. Since $u' = 2nx(x^2-1)^{n-1}$, the function u satisfies $(x^2-1)u' = 2nxu$. Using Leibniz's rule (4.46) to differentiate $(n+1)$ times both sides of this equation $2nxu = (x^2-1)u'$, we find

$$(2nxu)^{(n+1)} = 2n\sum_{k=0}^{n+1}\binom{n+1}{k}x^{(k)}\,u^{(n+1-k)} = 2n\left(x\,u^{(n+1)} + (n+1)\,u^{(n)}\right) \tag{8.26}$$

and

$$\begin{aligned}\left((x^2-1)u'\right)^{(n+1)} &= \sum_{k=0}^{n+1}\binom{n+1}{k}(x^2-1)^{(k)}\,u^{(n+2-k)} \\ &= (x^2-1)u^{(n+2)} + 2(n+1)xu^{(n+1)} + n(n+1)u^{(n)}.\end{aligned} \tag{8.27}$$

Equating the two and setting $u^{(n)} = 2^n n! P_n$, we get

$$-\left[(1 - x^2)\, P_n'\right]' = n(n+1)\, P_n, \tag{8.28}$$

which is Legendre's equation in self-adjoint form.

The differential operator

$$L = -\frac{d}{dx}\, p(x)\, \frac{d}{dx} = -\frac{d}{dx}\,(1 - x^2)\, \frac{d}{dx} \tag{8.29}$$

is formally self adjoint and the real function $p(x) = 1 - x^2$ is positive on the open interval $(-1, 1)$ and vanishes at $x = \pm 1$, so Legendre's differential operator L, his differential equation (8.28), and the domain D of functions that are twice differentiable on the interval $[-1, 1]$ form a singular self-adjoint system (example 6.33). The Legendre polynomial $P_n(x)$ is an eigenfunction of L with eigenvalue $n(n+1)$ and weight function $w(x) = 1$. The orthogonality relation (6.324) tells us that eigenfunctions of a self-adjoint differential operator that have different eigenvalues are orthogonal on the interval $[-1, 1]$ with respect to the weight function $w(x)$. Thus $P_n(x)$ and $P_m(x)$ are orthogonal for $n \neq m$

$$\int_{-1}^{1} P_n(x)\, P_m(x)\, dx = \frac{2}{2n+1}\, \delta_{nm} \tag{8.30}$$

as we saw (8.13) directly from the Rodrigues formula.

The eigenvalues $n(n+1)$ increase without limit, and so the argument of section 6.35 shows that the eigenfunctions $P_n(x)$ are complete. Since the weight function of the Legendre polynomials is unity $w(x) = 1$, the expansion (6.374) of Dirac's delta function here is

$$\delta(x - x') = \sum_{n=0}^{\infty} \frac{2n+1}{2}\, P_n(x')\, P_n(x), \tag{8.31}$$

which leads to the Fourier–Legendre expansion

$$f(x) = \sum_{n=0}^{\infty} \frac{2n+1}{2}\, P_n(x) \int_{-1}^{1} P_n(x') f(x')\, dx' \tag{8.32}$$

at least for functions f that are twice differentiable on $[-1, 1]$.

Changing variables to $\cos\theta = x$, we have $(1 - x^2) = \sin^2\theta$ and

$$\frac{d}{d\theta} = \frac{d\cos\theta}{d\theta}\, \frac{d}{dx} = -\sin\theta\, \frac{d}{dx} \tag{8.33}$$

so that

$$\frac{d}{dx} = -\frac{1}{\sin\theta}\, \frac{d}{d\theta}. \tag{8.34}$$

Thus in spherical coordinates, Legendre's equation (8.28) appears as

$$\frac{1}{\sin\theta}\frac{d}{d\theta}\left[\sin\theta\,\frac{d}{d\theta}P_n(\cos\theta)\right] + n(n+1)\,P_n(\cos\theta) = 0. \tag{8.35}$$

8.5 Recurrence relations

The t-derivative of the generating function $g(t,x) = 1/\sqrt{1-2xt+t^2}$ is

$$\frac{\partial g(t,x)}{\partial t} = \frac{x-t}{(1-2xt+t^2)^{3/2}} = \sum_{n=1}^{\infty} n\,P_n(x)\,t^{n-1}, \tag{8.36}$$

which we can rewrite as

$$(1-2xt+t^2)\sum_{n=1}^{\infty} n\,P_n(x)\,t^{n-1} = (x-t)\,g(t,x) = (x-t)\sum_{n=0}^{\infty} P_n(x)\,t^n. \tag{8.37}$$

By equating the coefficients of t^n in the first and last of these expressions, we arrive at the **recurrence relation**

$$P_{n+1}(x) = \frac{1}{n+1}\left[(2n+1)\,x\,P_n(x) - n\,P_{n-1}(x)\right]. \tag{8.38}$$

Example 8.4 (Building the Legendre polynomials) Since $P_1(x) = x$ and $P_0(x) = 1$, this recurrence relation for $n = 1$ gives

$$P_2(x) = \tfrac{1}{2}\left[3\,x\,P_1(x) - P_0(x)\right] = \tfrac{1}{2}\left(3x^2 - 1\right). \tag{8.39}$$

Similarly for $n = 2$ it gives

$$P_3(x) = \tfrac{1}{3}\left[5\,x\,P_2(x) - 2\,P_1(x)\right] = \tfrac{1}{2}(5x^2 - 3x). \tag{8.40}$$

It builds Legendre polynomials faster than Rodrigues's formula (8.8). □

The x-derivative of the generating function is

$$\frac{\partial g(t,x)}{\partial x} = \frac{t}{(1-2xt+t^2)^{3/2}} = \sum_{n=1}^{\infty} P_n'(x)\,t^n, \tag{8.41}$$

which we can rewrite as

$$(1-2xt+t^2)\sum_{n=1}^{\infty} P_n'(x)\,t^n = t\,g(t,x) = \sum_{n=0}^{\infty} P_n(x)\,t^{n+1}. \tag{8.42}$$

Equating coefficients of t^n, we have

$$P_{n+1}'(x) + P_{n-1}'(x) = 2x\,P_n'(x) + P_n(x). \tag{8.43}$$

By differentiating the recurrence relation (8.38) and combining it with this last equation, we get

$$P'_{n+1}(x) - P'_{n-1}(x) = (2n+1)\,P_n(x). \tag{8.44}$$

The last two recurrence relations (8.43 & 8.44) lead to several more:

$$P'_{n+1}(x) = (n+1)\,P_n(x) + x\,P'_n(x), \tag{8.45}$$
$$P'_{n-1}(x) = -n\,P_n(x) + x\,P'_n(x), \tag{8.46}$$
$$(1-x^2)\,P'_n(x) = n\,P_{n-1}(x) - nx\,P_n(x), \tag{8.47}$$
$$(1-x^2)\,P'_n(x) = (n+1)x\,P_n(x) - (n+1)\,P_{n+1}(x). \tag{8.48}$$

By differentiating (8.48) and using (8.45) for P'_{n+1}, we recover Legendre's equation $-[(1-x^2)P'_n]' = n(n+1)P_n$.

8.6 Special values of Legendre's polynomials

At $x=-1$, the generating function is

$$g(t,-1) = \left(1+t^2+2t\right)^{-1/2} = \frac{1}{1+t} = \sum_{n=0}^{\infty}(-t)^n = \sum_{n=0}^{\infty} P_n(-1)\,t^n, \tag{8.49}$$

which implies that

$$P_n(-1) = (-1)^n \tag{8.50}$$

and reminds us of the normalization condition (8.4), $P_n(1)=1$.

The generating function $g(t,x)$ is even under the reflection of both independent variables, so

$$g(t,x) = \sum_{n=0}^{\infty} t^n\,P_n(x) = \sum_{n=0}^{\infty}(-t)^n\,P_n(-x) = g(-t,-x), \tag{8.51}$$

which implies that

$$P_n(-x) = (-1)^n\,P_n(x) \quad \text{whence} \quad P_{2n+1}(0) = 0. \tag{8.52}$$

With more effort, one can show that

$$P_{2n}(0) = (-1)^n\,\frac{(2n-1)!!}{(2n)!!} \quad \text{and that} \quad |P_n(x)| \le 1. \tag{8.53}$$

8.7 Schlaefli's integral

Schlaefli used Cauchy's integral formula (5.36) and Rodrigues's formula

$$P_n(x) = \frac{1}{2^n\, n!} \left(\frac{d}{dx}\right)^n (x^2 - 1)^n \tag{8.54}$$

to express $P_n(z)$ as a counterclockwise contour integral around the point z

$$P_n(z) = \frac{1}{2^n\, 2\pi i} \oint \frac{(z'^2 - 1)^n}{(z' - z)^{n+1}}\, dz'. \tag{8.55}$$

8.8 Orthogonal polynomials

Rodrigues's formula (8.8) generates other families of orthogonal polynomials. The n-th order polynomials R_n

$$R_n(x) = \frac{1}{e_n w(x)} \frac{d^n}{dx^n} \left[w(x)\, Q^n(x)\right] \tag{8.56}$$

are orthogonal on the interval from a to b with weight function $w(x)$

$$\int_a^b R_n(x)\, R_k(x)\, w(x)\, dx = N_n\, \delta_{nk} \tag{8.57}$$

as long as $Q(x)$ vanishes at a and b (exercise 8.8)

$$Q(a) = Q(b) = 0. \tag{8.58}$$

Example 8.5 (Jacobi's polynomials) The choice $Q(x) = (x^2 - 1)$ with weight function $w(x) = (1 - x)^\alpha (1 + x)^\beta$ and normalization $e_n = 2^n n!$ leads for $\alpha > -1$ and $\beta > -1$ to the Jacobi polynomials

$$P_n^{(\alpha,\beta)}(x) = \frac{1}{2^n n!}(1 - x)^{-\alpha}(1 + x)^{-\beta} \frac{d^n}{dx^n}\left[(1 - x)^\alpha (1 + x)^\beta (x^2 - 1)^n\right], \tag{8.59}$$

which are orthogonal on $[-1, 1]$

$$\int_{-1}^1 P_n^{(\alpha,\beta)}(x)\, P_m^{(\alpha,\beta)}(x)\, w(x)\, dx = \frac{2^{\alpha+\beta+1}\Gamma(n + \alpha + 1)\,\Gamma(n + \beta + 1)}{(2n + \alpha + \beta + 1)\,\Gamma(n + \alpha + \beta + 1)}\, \delta_{nm} \tag{8.60}$$

and satisfy the normalization condition

$$P_n^{(\alpha,\beta)}(1) = \binom{n + \alpha}{n} \tag{8.61}$$

and the differential equation

$$(1 - x^2)\, y'' + (\beta - \alpha - (\alpha + \beta + 2)x)\, y' + n(n + \alpha + \beta + 1)\, y = 0. \tag{8.62}$$

In terms of $R(x, y) = \sqrt{1 - 2xy + y^2}$, their generating function is

$$2^{\alpha+\beta}(1-y+R(x,y))^{-\alpha}(1+w+R(x,y))^{-\beta}/R(x,y)=\sum_{n=0}^{\infty}P_n^{(\alpha,\beta)}(x)y^n. \quad (8.63)$$

When $\alpha = \beta$, they are the Gegenbauer polynomials, which for $\alpha = \beta = \pm 1/2$ are the Chebyshev polynomials (of the second and first kind, respectively). For $\alpha = \beta = 0$, they are Legendre's polynomials. □

Example 8.6 (Hermite's polynomials) The choice $Q(x) = 1$ with weight function $w(x) = \exp(-x^2)$ leads to the Hermite polynomials

$$H_n(x)=(-1)^n e^{x^2}\frac{d^n}{dx^n}e^{-x^2}=e^{x^2/2}\left(x-\frac{d}{dx}\right)^n e^{-x^2/2}=2^n\,e^{-D^2/4}\,x^n \quad (8.64)$$

where $D = d/dx$ is the x-derivative. They are orthogonal on the real line

$$\int_{-\infty}^{\infty} H_n(x)\,H_m(x)\,e^{-x^2}\,dx=\sqrt{\pi}\;2^n\,n!\,\delta_{nm} \quad (8.65)$$

and satisfy the differential equation

$$y''-2\,x\,y'+2\,n\,y=0. \quad (8.66)$$

Their generating function is

$$e^{2xy-y^2}=\sum_{n=0}^{\infty}H_n(x)\frac{y^n}{n!}. \quad (8.67)$$

The nth excited state of the harmonic oscillator of mass m and angular frequency ω is proportional to $H_n(x)$ in which $x = \sqrt{m\omega/\hbar}\,q$ is the dimensionless position of the oscillator. □

Example 8.7 (Laguerre's polynomials) The choices $Q(x) = x$ and weight function $w(x) = x^\alpha e^{-x}$ lead to the generalized Laguerre polynomials

$$L_n^{(\alpha)}(x)=\frac{e^x}{n!\,x^\alpha}\frac{d^n}{dx^n}\left(e^{-x}x^{n+\alpha}\right). \quad (8.68)$$

They are orthogonal on the interval $[0,\infty)$

$$\int_0^{\infty} L_n^{(\alpha)}(x)\,L_m^{(\alpha)}(x)\,x^\alpha e^{-x}\,dx=\frac{\Gamma(n+\alpha+1)}{n!}\,\delta_{n,m} \quad (8.69)$$

and satisfy the differential equation

$$x\,y''+(\alpha+1-x)\,y'+n\,y=0. \quad (8.70)$$

Their generating function is

$$(1-y)^{-\alpha-1}\exp\left(\frac{x\,y}{y-1}\right)=\sum_{n=0}^{\infty}L_n^{(\alpha)}(x)\,y^n. \quad (8.71)$$

The radial wave-function for the state of the nonrelativistic hydrogen atom with quantum numbers n and ℓ is $\rho^\ell \, L_{n-\ell-1}^{2\ell+1}(\rho) \, e^{-\rho/2}$ in which $\rho = 2r/na_0$ and a_0 is the Bohr radius $a_0 = 4\pi\epsilon_0\hbar^2/m_e e^2$. □

8.9 The azimuthally symmetric Laplacian

We saw in section 6.5 that the Laplacian $\triangle = \nabla \cdot \nabla$ separates in spherical coordinates r, θ, ϕ. A system with no dependence on the angle ϕ is said to have **azimuthal symmetry**. An azimuthally symmetric function

$$f(r, \theta, \phi) = R_{k,\ell}(r) \, \Theta_\ell(\theta) \tag{8.72}$$

will be a solution of Helmholtz's equation

$$-\triangle f = k^2 f \tag{8.73}$$

if the functions $R_{k,\ell}(r)$ and $\Theta_\ell(\theta)$ satisfy

$$\frac{1}{r^2}\frac{d}{dr}\left(r^2 \frac{dR_{k,\ell}}{dr}\right) + \left[k^2 - \frac{\ell(\ell+1)}{r^2}\right] R_{k,\ell} = 0 \tag{8.74}$$

for a nonnegative integer ℓ and Legendre's equation (8.35)

$$\frac{1}{\sin\theta}\frac{d}{d\theta}\left(\sin\theta \frac{d\Theta_\ell}{d\theta}\right) + \ell(\ell+1)\Theta_\ell = 0 \tag{8.75}$$

so that we may set $\Theta_\ell(\theta) = P_\ell(\cos\theta)$. For $k > 0$, the solutions of the radial equation (8.74) that are regular at $r = 0$ are the spherical Bessel functions

$$R_{k,\ell}(r) = j_\ell(kr), \tag{8.76}$$

which are given by Rayleigh's formula (9.68)

$$j_\ell(x) = (-1)^\ell \, x^\ell \left(\frac{d}{xdx}\right)^\ell \left(\frac{\sin x}{x}\right). \tag{8.77}$$

So the general azimuthally symmetric solution of the Helmholtz equation (8.73) that is regular at $r = 0$ is

$$f(r, \theta) = \sum_{\ell=0}^{\infty} a_{k,\ell} \, j_\ell(kr) \, P_\ell(\cos\theta), \tag{8.78}$$

in which the $a_{k,\ell}$ are constants. If the solution need not be regular at the origin, then the Neumann functions

$$n_\ell(x) = -(-1)^\ell \, x^\ell \left(\frac{d}{xdx}\right)^\ell \left(\frac{\cos x}{x}\right) \tag{8.79}$$

must be included, and the general solution then is

$$f(r,\theta) = \sum_{\ell=0}^{\infty} \left[a_{k,\ell}\, j_\ell(kr) + b_{k,\ell}\, n_\ell(kr)\right] P_\ell(\cos\theta), \tag{8.80}$$

in which the $a_{k,\ell}$ and $b_{k,\ell}$ are constants.

When $k = 0$, Helmholtz's equation reduces to Laplace's

$$\triangle f = 0, \tag{8.81}$$

which describes the Coulomb-gauge electric potential in the absence of charges and the Newtonian gravitational potential in the absence of masses. Now the radial equation is simply

$$\frac{d}{dr}\left(r^2 \frac{dR_\ell}{dr}\right) = \ell(\ell+1)R_\ell \tag{8.82}$$

since $k = 0$. We try setting

$$R_\ell(r) = r^n, \tag{8.83}$$

which works if $n(n+1) = \ell(\ell+1)$, that is, if $n = \ell$ or $n = -(\ell+1)$. So the general solution to (8.81) is

$$f(r,\theta) = \sum_{\ell=0}^{\infty} \left[a_\ell\, r^\ell + b_\ell\, r^{-\ell-1}\right] P_\ell(\cos\theta). \tag{8.84}$$

If the solution must be regular at $r = 0$, then all the b_ℓs must vanish.

8.10 Laplacian in two dimensions

In section 6.5, we saw that Helmholtz's equation separates in cylindrical coordinates, and that the equation for $\mathrm{P}(\rho)$ is Bessel's equation (6.47). But if $\alpha = 0$, Helmholtz's equation reduces to Laplace's equation $\triangle f = 0$, and if the potential f also is independent of z, then simpler solutions exist. For now $\alpha = 0 = k$, and so if $\Phi_m'' = -m^2\Phi_m$, then equation (6.47) becomes

$$\rho \frac{d}{d\rho}\left(\rho \frac{d\mathrm{P}_m}{d\rho}\right) = m^2\, \mathrm{P}_m. \tag{8.85}$$

The function $\Phi(\phi)$ may be taken to be $\Phi(\phi) = \exp(im\phi)$ or a linear combination of $\cos(m\phi)$ and $\sin(m\phi)$. If the whole range of ϕ from 0 to 2π is physically relevant, then $\Phi(\phi)$ must be periodic, and so m must be an integer. To solve this equation (8.85) for P_m, we set $\mathrm{P}_m = \rho^n$ and get

$$n^2\, \rho^n = m^2\, \rho^n, \tag{8.86}$$

which says that $n = \pm m$. The general z-independent solution of Laplace's equation in cylindrical coordinates then is

$$f(\rho, \phi) = \sum_{m=0}^{\infty} (a_m \cos(m\phi) + b_m \sin(m\phi)) \left(c_m \rho^m + d_m \rho^{-m}\right). \tag{8.87}$$

8.11 The Laplacian in spherical coordinates

The Laplacian $\triangle$ separates in spherical coordinates, as we saw in section 6.5. Thus a function

$$f(r, \theta) = R_{k,\ell}(r)\, \Theta_{\ell,m}(\theta)\, \Phi_m(\phi) \tag{8.88}$$

will be a solution of the Helmholtz equation $-\triangle f = k^2 f$ if $R_{k,\ell}$ is a linear combination of the spherical Bessel functions j_ℓ (8.77) and n_ℓ (8.79)

$$R_{k,\ell}(r) = a_{k,\ell} j_\ell(kr) + b_{k,\ell} n_\ell(kr) \tag{8.89}$$

if $\Phi_m = e^{im\phi}$, and if $\Theta_{\ell,m}$ satisfies the associated Legendre equation

$$\frac{1}{\sin\theta} \frac{d}{d\theta}\left(\sin\theta \frac{d\Theta_{\ell,m}}{d\theta}\right) + \left[\ell(\ell+1) - \frac{m^2}{\sin^2\theta}\right] \Theta_{\ell,m} = 0. \tag{8.90}$$

8.12 The associated Legendre functions/polynomials

The associated Legendre functions $P_\ell^m(x) \equiv P_{\ell,m}(x)$ are polynomials in $\sin\theta$ and $\cos\theta$. They arise as solutions of the separated θ equation (8.90)

$$\frac{1}{\sin\theta} \frac{d}{d\theta}\left(\sin\theta \frac{dP_{\ell,m}}{d\theta}\right) + \left[\ell(\ell+1) - \frac{m^2}{\sin^2\theta}\right] P_{\ell,m} = 0 \tag{8.91}$$

of the Laplacian in spherical coordinates. In terms of $x = \cos\theta$, this self-adjoint ordinary differential equation (ODE) is

$$\left[(1-x^2)P'_{\ell,m}(x)\right]' + \left[\ell(\ell+1) - \frac{m^2}{1-x^2}\right] P_{\ell,m}(x) = 0. \tag{8.92}$$

To find the $P_{\ell,m}$s, we use Leibniz's rule (4.46) to differentiate Legendre's equation (8.28)

$$\left[(1-x^2)\, P'_\ell\right]' + \ell(\ell+1)\, P_\ell = 0 \tag{8.93}$$

m times, obtaining

$$(1-x^2)P_\ell^{(m+2)} - 2x(m+1)P_\ell^{(m+1)} + (\ell-m)(\ell+m+1)P_\ell^{(m)} = 0. \tag{8.94}$$

We may make this equation self adjoint by using the prefactor formula (6.258)

$$F = \frac{1}{1-x^2} \exp\left[(m+1)\int^x \frac{-2x'}{1-x'^2}\,dx'\right]$$
$$= \frac{1}{1-x^2} \exp\left[(m+1)\ln(1-x^2)\right] = (1-x^2)^m. \tag{8.95}$$

The resulting ordinary differential equation

$$\left[(1-x^2)^{m+1} P_\ell'^{(m)}\right]' + (1-x^2)^m(\ell-m)(\ell+m+1)P_\ell^{(m)} = 0 \tag{8.96}$$

is self adjoint, but it is not (8.92).

Instead, we define $P_{\ell,m}$ in terms of the mth derivative $P_\ell^{(m)}$ as

$$P_{\ell,m}(x) \equiv (1-x^2)^{m/2}\, P_\ell^{(m)}(x) \tag{8.97}$$

and compute the derivatives

$$P_\ell^{(m+1)} = \left(P'_{\ell,m} + \frac{mxP_{\ell,m}}{1-x^2}\right)(1-x^2)^{-m/2} \tag{8.98}$$
$$P_\ell^{(m+2)} = \left[P''_{\ell,m} + \frac{2mxP'_{\ell,m}}{1-x^2} + \frac{mP_{\ell,m}}{1-x^2} + \frac{m(m+2)x^2P_{\ell,m}}{(1-x^2)^2}\right](1-x^2)^{-m/2}.$$

When we put these three expressions in equation (8.94), we get the desired ODE (8.92). Thus the associated Legendre functions are

$$P_{\ell,m}(x) = (1-x^2)^{m/2}\, P_\ell^{(m)}(x) = (1-x^2)^{m/2}\, \frac{d^m}{dx^m} P_\ell(x). \tag{8.99}$$

They are simple polynomials in $x = \cos\theta$ and $\sqrt{1-x^2} = \sin\theta$

$$P_{\ell,m}(\cos\theta) = \sin^m\theta\, \frac{d^m}{d\cos^m\theta}\, P_\ell(\cos\theta). \tag{8.100}$$

It follows from Rodrigues's formula (8.8) for the Legendre polynomial $P_\ell(x)$ that $P_{\ell,m}(x)$ is given by the similar formula

$$P_{\ell,m}(x) = \frac{(1-x^2)^{m/2}}{2^\ell \ell!} \frac{d^{\ell+m}}{dx^{\ell+m}} (x^2-1)^\ell, \tag{8.101}$$

which tells us that under parity $P_\ell^m(x)$ changes by $(-1)^{\ell+m}$

$$P_{\ell,m}(-x) = (-1)^{\ell+m}\, P_{\ell,m}(x). \tag{8.102}$$

Rodrigues's formula (8.101) for the associated Legendre function makes sense as long as $\ell + m \geq 0$. This last condition is the requirement in quantum mechanics that m not be less than $-\ell$. And if m exceeds ℓ, then $P_{\ell,m}(x)$ is given by more than 2ℓ derivatives of a polynomial of degree 2ℓ; so $P_{\ell,m}(x) = 0$ if $m > \ell$. This last condition is the requirement in quantum mechanics that m not be greater than ℓ. So we have

$$-\ell \le m \le \ell. \tag{8.103}$$

One may show that

$$P_{\ell,-m}(x) = (-1)^m \frac{(\ell - m)!}{(\ell + m)!} P_{\ell,m}(x). \tag{8.104}$$

In fact, since m occurs only as m^2 in the ordinary differential equation (8.92), $P_{\ell,-m}(x)$ must be proportional to $P_{\ell,m}(x)$.

Under reflections, the parity of $P_{\ell,m}$ is $(-1)^{\ell+m}$, that is

$$P_{\ell,m}(-x) = (-1)^{\ell+m} P_{\ell,m}(x). \tag{8.105}$$

If $m \neq 0$, then $P_{\ell,m}(x)$ has a power of $\sqrt{1-x^2}$ in it, so

$$P_{\ell,m}(\pm 1) = 0. \tag{8.106}$$

We may consider either $\ell(\ell+1)$ or m^2 as the eigenvalue in the ODE (8.92)

$$\left[(1-x^2)P'_{\ell,m}(x)\right]' + \left[\ell(\ell+1) - \frac{m^2}{1-x^2}\right] P_{\ell,m}(x) = 0. \tag{8.107}$$

If $\ell(\ell+1)$ is the eigenvalue, then the weight function is unity, and since this ODE is self adjoint on the interval $[-1, 1]$ (at the ends of which $p(x) = (1-x^2) = 0$), the eigenfunctions $P_{\ell,m}(x)$ and $P_{\ell',m}(x)$ must be orthogonal on that interval when $\ell \neq \ell'$. The full integral formula is

$$\int_{-1}^{1} P_{\ell,m}(x)\, P_{\ell',m}(x)\, dx = \frac{2}{2\ell+1} \frac{(\ell+m)!}{(\ell-m)!} \delta_{\ell,\ell'}. \tag{8.108}$$

If m^2 for fixed ℓ is the eigenvalue, then the weight function is $1/(1-x^2)$, and the eigenfunctions $P_{\ell,m}(x)$ and $P_{\ell',m}(x)$ must be orthogonal on $[-1, 1]$ when $m \neq m'$. The full formula is

$$\int_{-1}^{1} P_{\ell,m}(x)\, P_{\ell,m'}(x) \frac{dx}{1-x^2} = \frac{(\ell+m)!}{m(\ell-m)!} \delta_{m,m'}. \tag{8.109}$$

8.13 Spherical harmonics

The spherical harmonic $Y_\ell^m(\theta, \phi) \equiv Y_{\ell,m}(\theta, \phi)$ is the product

$$Y_{\ell,m}(\theta, \phi) = \Theta_{\ell,m}(\theta)\, \Phi_m(\phi) \tag{8.110}$$

in which $\Theta_{\ell,m}(\theta)$ is proportional to the associated Legendre function $P_{\ell,m}$

$$\Theta_{\ell,m}(\theta) = (-1)^m \sqrt{\frac{2\ell+1}{2} \frac{(\ell-m)!}{(\ell+m)!}}\, P_{\ell,m}(\cos\theta) \tag{8.111}$$

and

$$\Phi_m(\phi) = \frac{e^{im\phi}}{\sqrt{2\pi}}. \tag{8.112}$$

The big square-root in the definition (8.111) ensures that

$$\int_0^{2\pi} d\phi \int_0^{\pi} \sin\theta \, d\theta \;\; Y^*_{\ell,m}(\theta,\phi)\, Y_{\ell',m'}(\theta,\phi) = \delta_{\ell\ell'}\, \delta_{mm'}. \tag{8.113}$$

In spherical coordinates, the parity transformation

$$\boldsymbol{x}' = -\boldsymbol{x} \tag{8.114}$$

is $r' = r$, $\theta' = \pi - \theta$, and $\phi' = \phi \pm \pi$. So under parity, $\cos\theta' = -\cos\theta$ and $\exp(im\phi') = (-1)^m \exp(im\phi)$. This factor of $(-1)^m$ cancels the m-dependence (8.102) of $P_{\ell,m}(\theta)$ under parity, so that under parity

$$Y_{\ell,m}(\theta',\phi') = Y_{\ell,m}(\pi-\theta,\phi\pm\pi) = (-1)^\ell\, Y_{\ell,m}(\theta,\phi). \tag{8.115}$$

Thus the parity of the state $|n,\ell,m\rangle$ is $(-1)^\ell$.

The spherical harmonics are complete on the unit sphere. They may be used to expand any smooth function $f(\theta,\phi)$ as

$$f(\theta,\phi) = \sum_{\ell=0}^{\infty} \sum_{m=-\ell}^{\ell} a_{\ell m}\, Y_{\ell,m}(\theta,\phi). \tag{8.116}$$

The orthonormality relation (8.113) says that the coefficients $a_{\ell m}$ are

$$a_{\ell m} = \int_0^{2\pi} d\phi \int_0^{\pi} \sin\theta \, d\theta \;\; Y^*_{\ell,m}(\theta,\phi) f(\theta,\phi). \tag{8.117}$$

Putting the last two equations together, we find

$$f(\theta,\phi) = \int_0^{2\pi} d\phi' \int_0^{\pi} \sin\theta' \, d\theta' \left[\sum_{\ell=0}^{\infty} \sum_{m=-\ell}^{\ell} Y^*_{\ell,m}(\theta',\phi')\, Y_{\ell,m}(\theta,\phi) \right] f(\theta',\phi') \tag{8.118}$$

and so we may identify the sum within the brackets as an angular delta function

$$\sum_{\ell=0}^{\infty} \sum_{m=-\ell}^{\ell} Y^*_{\ell,m}(\theta',\phi')\, Y_{\ell,m}(\theta,\phi) = \frac{1}{\sin\theta}\, \delta(\theta-\theta')\, \delta(\phi-\phi'), \tag{8.119}$$

which sometimes is abbreviated as

$$\sum_{\ell=0}^{\infty} \sum_{m=-\ell}^{\ell} Y^*_{\ell,m}(\Omega')\, Y_{\ell,m}(\Omega) = \delta^{(2)}(\Omega-\Omega'). \tag{8.120}$$

The spherical-harmonic expansion (8.116) of the Legendre polynomial $P_\ell(\hat{\boldsymbol{n}} \cdot \hat{\boldsymbol{n}}')$ of the cosine $\hat{\boldsymbol{n}} \cdot \hat{\boldsymbol{n}}'$ in which the polar angles of the unit vectors respectively are θ, ϕ and θ', ϕ' is the **addition theorem**

$$\begin{aligned} P_\ell(\hat{\boldsymbol{n}} \cdot \hat{\boldsymbol{n}}') &= \frac{4\pi}{2\ell+1} \sum_{m=-\ell}^{\ell} Y_{\ell,m}(\theta,\phi) Y^*_{\ell,m}(\theta',\phi') \\ &= \frac{4\pi}{2\ell+1} \sum_{m=-\ell}^{\ell} Y^*_{\ell,m}(\theta,\phi) Y_{\ell,m}(\theta',\phi'). \end{aligned} \tag{8.121}$$

Example 8.8 (CMB radiation) Instruments on the Wilkinson Microwave Anisotropy Probe (WMAP) and Planck satellites in orbit at the Lagrange point L_2 (in the Earth's shadow, 1.5×10^6 km farther from the Sun) have measured the temperature $T(\theta, \phi)$ of the cosmic microwave background (CMB) radiation as a function of the polar angles θ and ϕ in the sky as shown in Fig. 8.2. This radiation is photons last scattered when the visible Universe became transparent

Figure 8.2 The CMB temperature fluctuations over the celestial sphere as measured by the Planck satellite. The average temperature is 2.7255 K. White regions are warmer, and black ones colder by about 0.0005 degrees. © ESA and the Planck Collaboration.

at an age of 380,000 years and a temperature (3,000 K) cool enough for hydrogen atoms to be stable. This **initial transparency** is usually (and inexplicably) called **recombination**.

Since the spherical harmonics $Y_{\ell,m}(\theta,\phi)$ are complete on the sphere, we can expand the temperature as

$$T(\theta,\phi) = \sum_{\ell=0}^{\infty} \sum_{m=-\ell}^{\ell} a_{\ell,m}\, Y_{\ell,m}(\theta,\phi), \tag{8.122}$$

in which the coefficients are by (8.117)

$$a_{\ell,m} = \int_0^{2\pi} d\phi \int_0^{\pi} \sin\theta\, d\theta \quad Y^*_{\ell,m}(\theta,\phi)\, T(\theta,\phi). \tag{8.123}$$

The average temperature $\overline{T}$ contributes only to $a_{0,0} = \overline{T} = 2.7255$ K. The other coefficients describe the difference $\Delta T(\theta,\phi) = T(\theta,\phi) - \overline{T}$. The angular power spectrum is

$$C_\ell = \frac{1}{2\ell+1} \sum_{m=-\ell}^{\ell} |a_{\ell,m}|^2. \tag{8.124}$$

If we let the unit vector $\hat{\boldsymbol{n}}$ point in the direction θ, ϕ and use the addition theorem (8.121), then we can write the angular power spectrum as

$$C_\ell = \frac{1}{4\pi} \int d^2\hat{\boldsymbol{n}} \int d^2\hat{\boldsymbol{n}}' \; P_\ell(\hat{\boldsymbol{n}} \cdot \hat{\boldsymbol{n}}')\, T(\hat{\boldsymbol{n}})\, T(\hat{\boldsymbol{n}}'). \tag{8.125}$$

In Fig. 8.3, the measured values (arXiv:1303.5062) of the power spectrum $\mathcal{D}_\ell = \ell(\ell+1)\, C_\ell/2\pi$ are plotted against ℓ for $1 < \ell < 1300$ with the angles and distances *decreasing* with ℓ. The power spectrum is a snapshot at the moment of transparency of the temperature distribution of the plasma of photons, electrons, and nuclei undergoing acoustic oscillations. In these oscillations, gravity opposes radiation pressure, and $|\Delta T(\theta,\phi)|$ is maximal both when the oscillations are most compressed and when they are most rarefied. Regions that gravity has squeezed to maximum compression at transparency form the first and highest peak. Regions that have bounced off their first maximal compression and that radiation pressure has expanded to minimum density at transparency form the second peak. Those at their second maximum compression at transparency form the third peak, and so forth.

The solid curve is the prediction of an inflationary cosmological model with **cold dark matter** and a **cosmological constant** Λ. In this **ΛCDM** cosmology, the age of the visible Universe is 13.817 Gyr; the **Hubble constant** is $H_0 = 67.3$ km/sMpc; the total energy density of the Universe is enough to make the Universe flat as required by inflation; and the fractions of the energy density respectively due to baryons, **dark matter**, and **dark energy** are 4.9%, 26.6%, and 68.5% (Edwin Hubble, 1889–1953). □

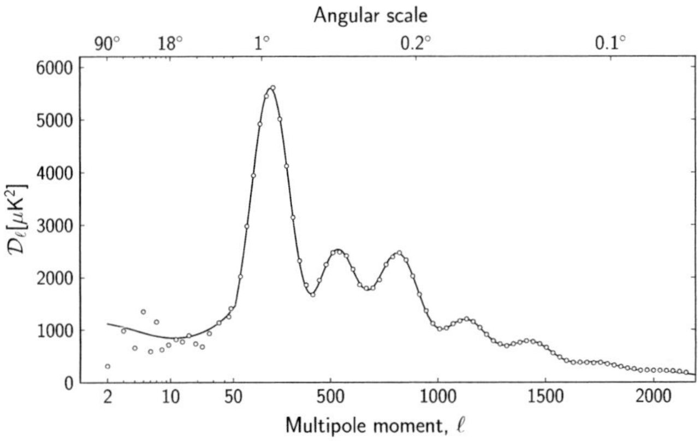

Figure 8.3 The power spectrum $\mathcal{D}_\ell = \ell(\ell+1)C_\ell/2\pi$ of the CMB temperature fluctuations in μK^2 as measured by the Planck Collaboration (arXiv:1303.5062) is plotted against the angular size and the multipole moment ℓ. The solid curve is the ΛCDM prediction.

Much is known about Legendre functions. The books *A Course of Modern Analysis* (Whittaker and Watson, 1927, chap. XV) and *Methods of Mathematical Physics* (Courant and Hilbert, 1955) are outstanding.

Exercises

8.1 Use conditions (8.6) and (8.7) to find $P_0(x)$ and $P_1(x)$.

8.2 Using the Gram–Schmidt method (section 1.10) to turn the functions x^n into a set of functions $L_n(x)$ that are orthonormal on the interval $[-1, 1]$ with inner product (8.2), find $L_n(x)$ for $n = 0, 1, 2,$ and 3. Isn't Rodrigues's formula (8.8) easier to use?

8.3 Derive the conditions (8.6–8.7) on the coefficients a_k of the Legendre polynomial $P_n(x) = a_0 + a_1 x + \cdots + a_n x^n$. Hint: first show that the orthogonality of the P_ns implies (8.12).

8.4 Use equations (8.6–8.7) to find $P_3(x)$ and $P_4(x)$.

8.5 In superscript notation (6.19), Leibniz's rule (4.46) for derivatives of products $u\,v$ of functions is

$$(uv)^{(n)} = \sum_{k=0}^{n} \binom{n}{k} u^{(n-k)}\, v^{(k)}. \tag{8.126}$$

Use it and Rodrigues's formula (8.8) to derive the explicit formula (8.9).

8.6 The product rule for derivatives in superscript notation (6.19) is

$$(uv)^{(n)} = \sum_{k=0}^{n} \binom{n}{k} u^{(n-k)}\, v^{(k)}. \tag{8.127}$$

Apply it to Rodrigues's formula (8.8) with $x^2 - 1 = (x-1)(x+1)$ and show that the Legendre polynomials satisfy $P_n(1) = 1$.

8.7 Use Cauchy's integral formula (5.36) and Rodrigues's formula (8.54) to derive Schlaefli's integral formula (8.55).

8.8 Show that the polynomials (8.56) are orthogonal (8.57) as long as they satisfy the endpoint condition (8.58).

8.9 Derive the orthogonality relation (8.2) from Rodrigues's formula (8.8).

8.10 (a) Use the fact that the quantities $w = x^2 - 1$ and $w_n = w^n$ vanish at the endpoints ± 1 to show by repeated integrations by parts that in superscript notation (6.19)

$$\int_{-1}^{1} w_n^{(n)} w_n^{(n)} dx = -\int_{-1}^{1} w_n^{(n-1)} w_n^{(n+1)} dx = (-1)^n \int_{-1}^{1} w_n w_n^{(2n)} dx. \tag{8.128}$$

(b) Show that the final integral is equal to

$$I_n = (2n)! \int_{-1}^{1} (1-x)^n\, (1+x)^n\, dx. \tag{8.129}$$

8.11 (a) Show by integrating by parts that $I_n = (n!)^2\, 2^{2n+1}/(2n+1)$.
(b) Prove (8.13).

8.12 Suppose that $P_n(x)$ and $Q_n(x)$ are two solutions of (8.28). Find an expression for their wronskian, apart from an over-all constant.

8.13 Use the method of sections 6.23 and 6.30 and the solution $f(r) = r^\ell$ to find a second solution of the ODE (8.82).

8.14 For a uniformly charged circle of radius a, find the resulting scalar potential $\phi(r,\theta)$ for $r < a$.

8.15 (a) Find the electrostatic potential $V(r,\theta)$ outside an uncharged perfectly conducting sphere of radius R in a vertical uniform static electric field that tends to $\boldsymbol{E} = E\hat{z}$ as $r \to \infty$. (b) Find the potential if the free charge on the sphere is q_f.

8.16 Derive (8.125) from (8.123) and (8.124).

9

Bessel functions

9.1 Bessel functions of the first kind

Friedrich Bessel (1784–1846) invented functions for problems with circular symmetry. The most useful ones are defined for any integer n by the series

$$J_n(z) = \frac{z^n}{2^n n!}\left[1 - \frac{z^2}{2(2n+2)} + \frac{z^4}{2\cdot 4(2n+2)(2n+4)} - \dots\right]$$
$$= \left(\frac{z}{2}\right)^n \sum_{m=0}^{\infty} \frac{(-1)^m}{m!\,(m+n)!} \left(\frac{z}{2}\right)^{2m}. \tag{9.1}$$

The first term of this series tells us that for small $|z| \ll 1$

$$J_n(z) \approx \frac{z^n}{2^n n!}. \tag{9.2}$$

The alternating signs in (9.1) make the waves plotted in Fig. 9.1, and we have for big $|z| \gg 1$ the approximation (Courant and Hilbert, 1955, chap. VII)

$$J_n(z) \approx \sqrt{\frac{2}{\pi z}}\, \cos\left(z - \frac{n\pi}{2} - \frac{\pi}{4}\right) + O(|z|^{-3/2}). \tag{9.3}$$

The $J_n(z)$ are entire transcendental functions. They obey Bessel's equation

$$\frac{d^2 J_n}{dz^2} + \frac{1}{z}\frac{dJ_n}{dz} + \left(1 - \frac{n^2}{z^2}\right) J_n = 0 \tag{9.4}$$

(6.308) as one may show (exercise 9.1) by substituting the series (9.1) into the differential equation (9.4). Their generating function is

$$\exp\left[\frac{z}{2}(u - 1/u)\right] = \sum_{n=-\infty}^{\infty} u^n J_n(z), \tag{9.5}$$

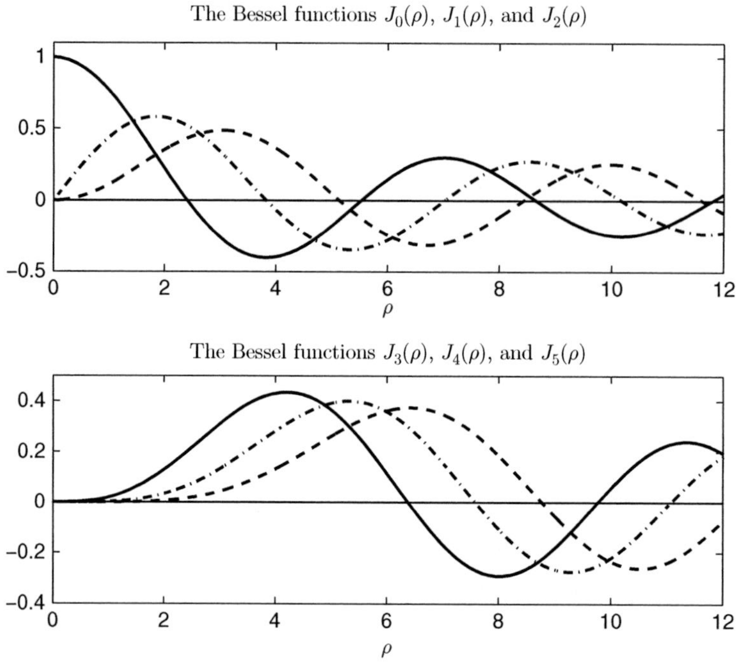

Figure 9.1 Top: plots of $J_0(\rho)$ (solid curve), $J_1(\rho)$ (dot-dash), and $J_2(\rho)$ (dashed) for real ρ. Bottom: plots of $J_3(\rho)$ (solid curve), $J_4(\rho)$ (dot-dash), and $J_5(\rho)$ (dashed). The points at which Bessel functions cross the ρ-axis are called **zeros** or **roots**; we use them to satisfy boundary conditions.

from which one may derive (exercise 9.5) the series expansion (9.1) and (exercise 9.6) the integral representation (5.46)

$$J_n(z) = \frac{1}{\pi}\int_0^{\pi} \cos(z\sin\theta - n\theta)\,d\theta = J_{-n}(-z) = (-1)^n J_{-n}(z) \tag{9.6}$$

for all complex z. For $n = 0$, this integral is (exercise 9.7) more simply

$$J_0(z) = \frac{1}{2\pi}\int_0^{2\pi} e^{iz\cos\theta}\,d\theta = \frac{1}{2\pi}\int_0^{2\pi} e^{iz\sin\theta}\,d\theta. \tag{9.7}$$

These integrals (exercise 9.8) give $J_n(0) = 0$ for $n \neq 0$, and $J_0(0) = 1$.

By differentiating the generating function (9.5) with respect to u and identifying the coefficients of powers of u, one finds the recursion relation

$$J_{n-1}(z) + J_{n+1}(z) = \frac{2n}{z}\,J_n(z). \tag{9.8}$$

Similar reasoning after taking the z derivative gives (exercise 9.10)

$$J_{n-1}(z) - J_{n+1}(z) = 2\,J_n'(z). \tag{9.9}$$

By using the gamma function (section 5.12), one may extend Bessel's equation (9.4) and its solutions $J_n(z)$ to nonintegral values of n

$$J_\upsilon(z) = \left(\frac{z}{2}\right)^\upsilon \sum_{m=0}^{\infty} \frac{(-1)^m}{m!\,\Gamma(m+\upsilon+1)} \left(\frac{z}{2}\right)^{2m}. \tag{9.10}$$

Letting $z = ax$, we arrive (exercise 9.11) at the self-adjoint form (6.307) of Bessel's equation

$$-\frac{d}{dx}\left(x\frac{d}{dx}J_n(ax)\right) + \frac{n^2}{x}J_n(ax) = a^2 x J_n(ax). \tag{9.11}$$

In the notation of equation (6.287), $p(x) = x$, a^2 is an eigenvalue, and $\rho(x) = x$ is a weight function. To have a self-adjoint system (section 6.28) on an interval $[0, b]$, we need the boundary condition (6.247)

$$0 = \left[p(J_n v' - J_n' v)\right]_0^b = \left[x(J_n v' - J_n' v)\right]_0^b \tag{9.12}$$

for all functions $v(x)$ in the domain D of the system. Since $p(x) = x$, $J_0(0) = 1$, and $J_n(0) = 0$ for integers $n > 0$, the terms in this boundary condition vanish at $x = 0$ as long as the domain consists of functions $v(x)$ that are continuous on the interval $[0, b]$. To make these terms vanish at $x = b$, we require that $J_n(ab) = 0$ and that $v(b) = 0$. So ab must be a zero $z_{n,m}$ of $J_n(z)$, that is $J_n(ab) = J_n(z_{n,m}) = 0$. With $a = z_{n,m}/b$, Bessel's equation (9.11) is

$$-\frac{d}{dx}\left(x\frac{d}{dx}J_n\left(z_{n,m}x/b\right)\right) + \frac{n^2}{x}J_n\left(z_{n,m}x/b\right) = \frac{z_{n,m}^2}{b^2}\, x\, J_n\left(z_{n,m}x/b\right). \tag{9.13}$$

For fixed n, the eigenvalue $a^2 = z_{n,m}^2/b^2$ is different for each positive integer m. Moreover as $m \to \infty$, the zeros $z_{n,m}$ of $J_n(x)$ rise as $m\pi$ as one might expect since the leading term of the asymptotic form (9.3) of $J_n(x)$ is proportional to $\cos(x - n\pi/2 - \pi/4)$, which has zeros at $m\pi + (n+1)\pi/2 + \pi/4$. It follows that the eigenvalues $a^2 \approx (m\pi)^2/b^2$ increase without limit as $m \to \infty$ in accordance with the general result of section 6.34. It follows then from the argument of section 6.35 and from the orthogonality relation (6.326) that for every fixed n, the eigenfunctions $J_n(z_{n,m}x/b)$, one for each zero, are complete in the mean, orthogonal, and normalizable on the interval $[0, b]$ with weight function $\rho(x) = x$

$$\int_0^b x\, J_n\left(\frac{z_{n,m}x}{b}\right) J_n\left(\frac{z_{n,m'}x}{b}\right) dx = \delta_{m,m'}\frac{b^2}{2}J_n'^2(z_{n,m}) = \delta_{m,m'}\frac{b^2}{2}J_{n+1}^2(z_{n,m}) \tag{9.14}$$

and a normalization constant (exercise 9.12) that depends upon the first derivative of the Bessel function or the square of the next Bessel function at the zero.

The analogous relation on an infinite interval is

$$\int_0^\infty x\, J_n(kx)\, J_n(k'x)\, dx = \frac{1}{k}\, \delta(k - k'). \tag{9.15}$$

One may generalize these relations (9.11–9.15) from integral n to real nonnegative ν (and to $\nu > -1/2$).

Example 9.1 (Bessel's drum) The top of a drum is a circular membrane with a fixed circumference $2\pi r_d$. The membrane's potential energy is approximately proportional to the extra area it has when it's not flat. Let $h(x, y)$ be the displacement of the membrane in the z direction normal to the x–y plane of the flat membrane, and let h_x and h_y denote its partial derivatives (6.20). The extra length of a line segment dx on the stretched membrane is $\sqrt{1 + h_x^2}\, dx$, and so the extra area of an element $dx\, dy$ is

$$dA \approx \left(\sqrt{1 + h_x^2 + h_y^2} - 1\right) dx\, dy \approx \frac{1}{2}\left(h_x^2 + h_y^2\right) dx\, dy. \tag{9.16}$$

The (nonrelativistic) kinetic energy of the area element is proportional to its speed squared. So if σ is the surface tension and μ the mass density of the membrane, then to lowest order in derivatives the action functional is

$$S[h] = \int \left[\frac{\mu}{2}\, h_t^2 - \frac{\sigma}{2}\left(h_x^2 + h_y^2\right)\right] dx\, dy\, dt. \tag{9.17}$$

We minimize this action for hs that vanish on the boundary $x^2 + y^2 = r_d^2$

$$0 = \delta S[h] = \int \left[\mu\, h_t\, \delta h_t - \sigma \left(h_x\, \delta h_x + h_y\, \delta h_y\right)\right] dx\, dy\, dt. \tag{9.18}$$

Since (6.170) $\delta h_t = (\delta h)_t$, $\delta h_x = (\delta h)_x$, and $\delta h_y = (\delta h)_y$, we can integrate by parts and get

$$0 = \delta S[h] = \int \left[-\, \mu\, h_{tt} + \sigma\left(h_{xx} + h_{yy}\right)\right] \delta h\, dx\, dy\, dt \tag{9.19}$$

apart from a surface term proportional to δh, which vanishes because $\delta h = 0$ on the circumference of the membrane. The membrane therefore obeys the wave equation

$$\mu\, h_{tt} = \sigma\left(h_{xx} + h_{yy}\right) \equiv \sigma\, \Delta h. \tag{9.20}$$

This equation is separable, and so letting $h(x, y, t) = s(t)\, v(x, y)$, we have

$$\frac{s_{tt}}{s} = \frac{\sigma}{\mu}\frac{\Delta v}{v} = -\,\omega^2. \tag{9.21}$$

The eigenvalues of the Helmholtz equation $-\Delta v = \lambda v$ give the angular frequencies as $\omega = \sqrt{\sigma\lambda/\mu}$. The time dependence then is

$$s(t) = a \sin\left(\sqrt{\sigma\lambda/\mu}\,(t - t_0)\right), \tag{9.22}$$

in which a and t_0 are constants.

In polar coordinates, Helmholtz's equation is separable (6.45–6.48)

$$-\Delta v = -v_{rr} - r^{-1} v_r - r^{-2} v_{\theta\theta} = \lambda v. \tag{9.23}$$

We set $v(r, \theta) = u(r)h(\theta)$ and find $-u''h - r^{-1}u'h - r^{-2}uh'' = \lambda uh$. After multiplying both sides by r^2/uh, we get

$$r^2 \frac{u''}{u} + r\frac{u'}{u} + \lambda r^2 = -\frac{h''}{h} = n^2. \tag{9.24}$$

The general solution for h then is $h(\theta) = b \sin(n(\theta - \theta_0))$ in which b and θ_0 are constants and n must be an integer so that h is single valued on the circumference of the membrane.

The function u thus is an eigenfunction of the self-adjoint differential equation (6.307) $-\left(r\,u'\right)' + n^2\,u/r = \lambda\,r\,u$ whose eigenvalues $\lambda \equiv z^2/r_d^2$ are all positive. By changing variables to $\rho = zr/r_d$ and letting $u(r) = J_n(\rho)$, we arrive (exercise 6.26) at

$$\frac{d^2 J_n}{d\rho^2} + \frac{1}{\rho}\frac{dJ_n}{d\rho} + \left(1 - \frac{n^2}{\rho^2}\right) J_n = 0, \tag{9.25}$$

which is Bessel's equation (6.308).

The eigenvalues $\lambda = z^2/r_d^2$ are determined by the boundary condition $u(r_d) = J_n(z) = 0$. For each integer $n \geq 0$, there are an infinite number of zeros $z_{n,m}$ at which the Bessel function vanishes, $J_n(z_{n,m}) = 0$. Thus $\lambda = \lambda_{n,m} = z_{n,m}^2/r_d^2$ and so the frequency is $\omega = (z_{n,m}/r_d)\sqrt{\sigma/\mu}$. The general solution to the wave equation (9.20) of the membrane then is

$$h(r, \theta, t) = \sum_{n=0}^{\infty} \sum_{m=1}^{\infty} c_{n,m} \sin\left[\frac{z_{n,m}}{r_d}\sqrt{\frac{\sigma}{\mu}}\,(t - t_0)\right] \sin\left[n(\theta - \theta_0)\right] J_n\left(z_{n,m}\frac{r}{r_d}\right). \tag{9.26}$$

For any n, the zeros $z_{n,m}$ are the square-roots of the dimensionless eigenvalues (6.309) and rise like $m\pi$ as $m \to \infty$. □

We learned in section 6.5 that in three dimensions Helmholtz's equation $-\Delta V = \alpha^2\,V$ separates in cylindrical coordinates (and in several other coordinate systems). That is, the function $V(\rho, \phi, z) = B(\rho)\Phi(\phi)Z(z)$ satisfies the equation

$$-\Delta V = -\frac{1}{\rho}\left[\left(\rho\,V_{,\rho}\right)_{,\rho} + \frac{1}{\rho}\,V_{,\phi\phi} + \rho\,V_{,zz}\right] = \alpha^2\,V \tag{9.27}$$

if $B(\rho)$ obeys Bessel's equation

$$\rho \frac{d}{d\rho}\left(\rho \frac{dB}{d\rho}\right) + \left((\alpha^2 + k^2)\rho^2 - n^2\right) B = 0 \tag{9.28}$$

and Φ and Z respectively satisfy

$$-\frac{d^2\Phi}{d\phi^2} = n^2\Phi(\phi) \quad \text{and} \quad \frac{d^2 Z}{dz^2} = k^2 Z(z) \tag{9.29}$$

or if $B(\rho)$ obeys the Bessel equation

$$\rho \frac{d}{d\rho}\left(\rho \frac{dB}{d\rho}\right) + \left((\alpha^2 - k^2)\rho^2 - n^2\right) B = 0 \tag{9.30}$$

and Φ and Z satisfy

$$-\frac{d^2\Phi}{d\phi^2} = n^2\Phi(\phi) \quad \text{and} \quad \frac{d^2 Z}{dz^2} = -k^2 Z(z). \tag{9.31}$$

In the first case (9.28 & 9.29), the solution V is

$$V_{k,n}(\rho,\phi,z) = J_n\left(\sqrt{\alpha^2 + k^2}\,\rho\right) e^{\pm in\phi} e^{\pm kz} \tag{9.32}$$

while in the second case (9.30 & 9.31), it is

$$V_{k,n}(\rho,\phi,z) = J_n\left(\sqrt{\alpha^2 - k^2}\,\rho\right) e^{\pm in\phi} e^{\pm ikz}. \tag{9.33}$$

In both cases, n must be an integer if the solution is to be single valued on the full range of ϕ from 0 to 2π.

When $\alpha = 0$, the Helmholtz equation reduces to Laplace's equation $\triangle V = 0$ of electrostatics, which the simpler functions

$$V_{k,n}(\rho,\phi,z) = J_n(k\rho)e^{\pm in\phi}e^{\pm kz} \quad \text{and} \quad V_{k,n}(\rho,\phi,z) = J_n(ik\rho)e^{\pm in\phi}e^{\pm ikz} \tag{9.34}$$

satisfy.

The product $i^{-\nu} J_\nu(ik\rho)$ is real and is known as the **modified Bessel function**

$$I_\nu(k\rho) \equiv i^{-\nu} J_\nu(ik\rho). \tag{9.35}$$

It occurs in various solutions of the **diffusion equation** $\triangle V = \alpha^2 V$. The function $V(\rho,\phi,z) = B(\rho)\Phi(\phi)Z(z)$ satisfies

$$\triangle V = \frac{1}{\rho}\left[\left(\rho\, V_{,\rho}\right)_{,\rho} + \frac{1}{\rho}\, V_{,\phi\phi} + \rho\, V_{,zz}\right] = \alpha^2\, V \tag{9.36}$$

if $B(\rho)$ obeys Bessel's equation

$$\rho \frac{d}{d\rho}\left(\rho \frac{dB}{d\rho}\right) - \left((\alpha^2 - k^2)\rho^2 + n^2\right) B = 0 \tag{9.37}$$

and Φ and Z respectively satisfy

$$-\frac{d^2\Phi}{d\phi^2} = n^2\Phi(\phi) \quad \text{and} \quad \frac{d^2Z}{dz^2} = k^2 Z(z) \tag{9.38}$$

or if $B(\rho)$ obeys the Bessel equation

$$\rho \frac{d}{d\rho}\left(\rho \frac{dB}{d\rho}\right) - \left((\alpha^2 + k^2)\rho^2 + n^2\right) B = 0 \tag{9.39}$$

and Φ and Z satisfy

$$-\frac{d^2\Phi}{d\phi^2} = n^2\Phi(\phi) \quad \text{and} \quad \frac{d^2Z}{dz^2} = -k^2 Z(z). \tag{9.40}$$

In the first case (9.37 & 9.38), the solution V is

$$V_{k,n}(\rho, \phi, z) = I_n\left(\sqrt{\alpha^2 - k^2}\, \rho\right) e^{\pm in\phi} e^{\pm kz} \tag{9.41}$$

while in the second case (9.39 & 9.40), it is

$$V_{k,n}(\rho, \phi, z) = I_n\left(\sqrt{\alpha^2 + k^2}\, \rho\right) e^{\pm in\phi} e^{\pm ikz}. \tag{9.42}$$

In both cases, n must be an integer if the solution is to be single valued on the full range of ϕ from 0 to 2π.

Example 9.2 (Charge near a membrane) We will use ρ to denote the density of **free charges** – those that are free to move into or out of a dielectric medium, as opposed to those that are part of the medium, bound in it by molecular forces. The time-independent Maxwell equations are Gauss's law $\nabla \cdot \boldsymbol{D} = \rho$ for the divergence of the electric displacement $\boldsymbol{D}$, and the static form $\nabla \times \mathbf{E} = 0$ of Faraday's law, which implies that the electric field $\boldsymbol{E}$ is the gradient of an electrostatic potential $\boldsymbol{E} = -\nabla V$.

Across an interface between two dielectrics with normal vector $\hat{\boldsymbol{n}}$, the tangential electric field is continuous, $\hat{\boldsymbol{n}} \times \boldsymbol{E}_2 = \hat{\boldsymbol{n}} \times \boldsymbol{E}_1$, while the normal component of the electric displacement jumps by the surface density σ of free charge, $\hat{\boldsymbol{n}} \cdot (\boldsymbol{D}_2 - \boldsymbol{D}_1) = \sigma$. In a linear dielectric, the electric displacement $\boldsymbol{D}$ is the electric field multiplied by the permittivity ϵ of the material, $\boldsymbol{D} = \epsilon\, \boldsymbol{E}$.

The membrane of a eukaryotic cell is a phospholipid bilayer whose area is some 3×10^8 nm^2, and whose thickness t is about 5 nm. On a scale of nanometers, the membrane is flat. We will take it to be a slab extending to infinity in the x and y directions. If the interface between the lipid bilayer and the extracellular salty water is at $z = 0$, then the cytosol extends thousands of nm down from

$z = -t = -5$ nm. We will ignore the phosphate head groups and set the permittivity ϵ_ℓ of the lipid bilayer to twice that of the vacuum $\epsilon_\ell \approx 2\epsilon_0$; the permittivity of the extracellular water and that of the cytosol are $\epsilon_w \approx \epsilon_c \approx 80\epsilon_0$.

We will compute the electrostatic potential V due to a charge q at a point $(0, 0, h)$ on the z-axis above the membrane. This potential is cylindrically symmetric about the z-axis, so $V = V(\rho, z)$. The functions $J_n(k\rho)\, e^{in\phi}\, e^{\pm kz}$ form a complete set of solutions of Laplace's equation but, due to the symmetry, we only need the $n = 0$ functions $J_0(k\rho)\, e^{\pm kz}$. Since there are no free charges in the lipid bilayer or in the cytosol, we may express the potential in the lipid bilayer V_ℓ and in the cytosol V_c as

$$
\begin{aligned}
V_\ell(\rho, z) &= \int_0^\infty dk\, J_0(k\rho) \left[m(k)\, e^{kz} + f(k)\, e^{-kz} \right], \\
V_c(\rho, z) &= \int_0^\infty dk\, J_0(k\rho)\, d(k)\, e^{kz}.
\end{aligned} \tag{9.43}
$$

The Green's function (3.110) for Poisson's equation $-\triangle G(\boldsymbol{x}) = \delta^{(3)}(\boldsymbol{x})$ in cylindrical coordinates is (5.139)

$$
G(\boldsymbol{x}) = \frac{1}{4\pi |\boldsymbol{x}|} = \frac{1}{4\pi \sqrt{\rho^2 + z^2}} = \int_0^\infty \frac{dk}{4\pi} J_0(k\rho)\, e^{-k|z|}. \tag{9.44}
$$

Thus we may expand the potential in the salty water as

$$
V_w(\rho, z) = \int_0^\infty dk\, J_0(k\rho) \left[\frac{q}{4\pi \epsilon_w} e^{-k|z-h|} + u(k)\, e^{-kz} \right]. \tag{9.45}
$$

Using $\hat{\boldsymbol{n}} \times \boldsymbol{E}_2 = \hat{\boldsymbol{n}} \times \boldsymbol{E}_1$ and $\hat{\boldsymbol{n}} \cdot (\boldsymbol{D}_2 - \boldsymbol{D}_1) = \sigma$, suppressing k, and setting $\beta \equiv q e^{-kh}/4\pi\epsilon_w$ and $y = e^{2kt}$, we get four equations

$$
\begin{aligned}
m + f - u = \beta \quad &\text{and} \quad \epsilon_\ell m - \epsilon_\ell f + \epsilon_w u = \epsilon_w \beta, \\
\epsilon_\ell m - \epsilon_\ell y f - \epsilon_c d = 0 \quad &\text{and} \quad m + y f - d = 0.
\end{aligned} \tag{9.46}
$$

In terms of the abbreviations $\epsilon_{w\ell} = (\epsilon_w + \epsilon_\ell)/2$ and $\epsilon_{c\ell} = (\epsilon_c + \epsilon_\ell)/2$ as well as $p = (\epsilon_w - \epsilon_\ell)/(\epsilon_w + \epsilon_\ell)$ and $p' = (\epsilon_c - \epsilon_\ell)/(\epsilon_c + \epsilon_\ell)$, the solutions are

$$
\begin{aligned}
u(k) = \beta\, \frac{p - p'/y}{1 - pp'/y} \quad &\text{and} \quad m(k) = \beta\, \frac{\epsilon_w}{\epsilon_{w\ell}} \frac{1}{1 - pp'/y}, \\
f(k) = -\beta\, \frac{\epsilon_w}{\epsilon_{w\ell}} \frac{p'/y}{1 - pp'/y} \quad &\text{and} \quad d(k) = \beta\, \frac{\epsilon_w \epsilon_\ell}{\epsilon_{w\ell}\epsilon_{c\ell}} \frac{1}{1 - pp'/y}.
\end{aligned} \tag{9.47}
$$

Inserting these solutions into the Bessel expansions (9.43) for the potentials, expanding their denominators

$$
\frac{1}{1 - pp'/y} = \sum_0^\infty (pp')^n\, e^{-2nkt}, \tag{9.48}
$$

and using the integral (9.44), we find that the potential V_w in the extracellular water of a charge q at $(0, 0, h)$ in the water is

$$V_{\mathrm{w}}(\rho, z)=\frac{q}{4\pi\epsilon_{\mathrm{w}}}\left[\frac{1}{r}+\frac{p}{\sqrt{\rho^2+(z+h)^2}}-\sum_{n=1}^{\infty}\frac{p'\left(1-p^2\right)(pp')^{n-1}}{\sqrt{\rho^2+(z+2nt+h)^2}}\right], \tag{9.49}$$

in which $r=\sqrt{\rho^2+(z-h)^2}$ is the distance to the charge q. The principal image charge pq is at $(0,0,-h)$. Similarly, the potential V_ℓ in the lipid bilayer is

$$V_\ell(\rho, z)=\frac{q}{4\pi\epsilon_{\mathrm{w}\ell}}\sum_{n=0}^{\infty}\left[\frac{(pp')^n}{\sqrt{\rho^2+(z-2nt-h)^2}}-\frac{p^n p'^{n+1}}{\sqrt{\rho^2+(z+2(n+1)t+h)^2}}\right] \tag{9.50}$$

and that in the cytosol is

$$V_{\mathrm{c}}(\rho, z)=\frac{q\,\epsilon_\ell}{4\pi\epsilon_{\mathrm{w}\ell}\epsilon_{\mathrm{c}\ell}}\sum_{n=0}^{\infty}\frac{(pp')^n}{\sqrt{\rho^2+(z-2nt-h)^2}}. \tag{9.51}$$

These potentials are the same as those of example 4.16, but this derivation is much simpler and less error prone than the method of images.

Since $p=(\epsilon_{\mathrm{w}}-\epsilon_\ell)/(\epsilon_{\mathrm{w}}+\epsilon_\ell)>0$, the principal image charge pq at $(0,0,-h)$ has the same sign as the charge q and so contributes a positive term proportional to pq^2 to the energy. So a lipid membrane repels a nearby charge in water no matter what the sign of the charge. A cell membrane is a phospholipid bilayer. The lipids avoid water and form a 4-nm-thick layer that lies between two 0.5-nm layers of phosphate groups which are electric dipoles. These electric dipoles cause the cell membrane to weakly *attract* ions that are within 0.5 nm of the membrane. □

Example 9.3 (Cylindrical wave-guides) An electromagnetic wave traveling in the z-direction down a cylindrical wave-guide looks like

$$\boldsymbol{E}\,e^{i(kz-\omega t)} \quad \text{and} \quad \boldsymbol{B}\,e^{i(kz-\omega t)}, \tag{9.52}$$

in which $\boldsymbol{E}$ and $\boldsymbol{B}$ depend upon ρ and ϕ

$$\boldsymbol{E}=E_\rho\hat{\boldsymbol{\rho}}+E_\phi\hat{\boldsymbol{\phi}}+E_z\hat{\boldsymbol{z}} \quad \text{and} \quad \boldsymbol{B}=B_\rho\hat{\boldsymbol{\rho}}+B_\phi\hat{\boldsymbol{\phi}}+B_z\hat{\boldsymbol{z}} \tag{9.53}$$

in cylindrical coordinates (11.164–11.169 & 11.241). If the wave-guide is an evacuated, perfectly conducting cylinder of radius r, then on the surface of the wave-guide the parallel components of $\boldsymbol{E}$ and the normal component of $\boldsymbol{B}$ must vanish, which leads to the boundary conditions

$$E_z(r,\phi)=0, \quad E_\phi(r,\phi)=0, \quad \text{and} \quad B_\rho(r,\phi)=0. \tag{9.54}$$

Since the E and B fields have subscripts, we will use commas to denote derivatives as in $\partial E_z/\partial\phi \equiv E_{z,\phi}$ and $\partial(\rho E_\phi)/\partial\rho \equiv (\rho E_\phi)_{,\rho}$ and so forth. In this notation, the vacuum forms $\nabla\times\boldsymbol{E}=-\dot{\boldsymbol{B}}$ and $\nabla\times\boldsymbol{B}=\dot{\boldsymbol{E}}/c^2$ of the Faraday and Maxwell–Ampère laws give us (exercise 9.14) the field equations

$$
\begin{aligned}
E_{z,\phi}/\rho - ikE_\phi &= i\omega B_\rho, & B_{z,\phi}/\rho - ikB_\phi &= -i\omega E_\rho/c^2, \\
ikE_\rho - E_{z,\rho} &= i\omega B_\phi, & ikB_\rho - B_{z,\rho} &= -i\omega E_\phi/c^2, \\
\left[(\rho E_\phi)_{,\rho} - E_{\rho,\phi}\right]/\rho &= i\omega B_z, & \left[(\rho B_\phi)_{,\rho} - B_{\rho,\phi}\right]/\rho &= -i\omega E_z/c^2. \qquad (9.55)
\end{aligned}
$$

Solving them for the ρ and ϕ components of $\boldsymbol{E}$ and $\boldsymbol{B}$ in terms of their z components (exercise 9.15), we find

$$
\begin{aligned}
E_\rho &= -i\frac{kE_{z,\rho} + \omega B_{z,\phi}/\rho}{k^2 - \omega^2/c^2}, & E_\phi &= -i\frac{kE_{z,\phi}/\rho - \omega B_{z,\rho}}{k^2 - \omega^2/c^2}, \\
B_\rho &= -i\frac{kB_{z,\rho} - \omega E_{z,\phi}/c^2\rho}{k^2 - \omega^2/c^2}, & B_\phi &= -i\frac{kB_{z,\phi}/\rho + \omega E_{z,\rho}/c^2}{k^2 - \omega^2/c^2}. \qquad (9.56)
\end{aligned}
$$

The fields E_z and B_z obey the separable wave equations (11.91), exercise 6.6,

$$
-\triangle E_z = -\ddot{E}_z/c^2 = \omega^2 E_z/c^2 \quad \text{and} \quad -\triangle B_z = -\ddot{B}_z/c^2 = \omega^2 B_z/c^2. \qquad (9.57)
$$

Because their z-dependence (9.52) is periodic, they are (exercise 9.16) linear combinations of $J_n(\sqrt{\omega^2/c^2 - k^2}\,\rho)e^{in\phi}e^{i(kz-\omega t)}$.

Modes with $B_z = 0$ are **transverse magnetic** or **TM** modes. For them the boundary conditions (9.54) will be satisfied if $\sqrt{\omega^2/c^2 - k^2}\,r$ is a zero $z_{n,m}$ of J_n. So the frequency $\omega_{n,m}(k)$ of the n, m TM mode is

$$
\omega_{n,m}(k) = c\sqrt{k^2 + z_{n,m}^2/r^2}. \qquad (9.58)
$$

Since the first zero of a Bessel function is $z_{0,1} \approx 2.4048$, the minimum frequency $\omega_{0,1}(0) = c\,z_{0,1}/r \approx 2.4048\,c/r$ occurs for $n = 0$ and $k = 0$. If the radius of the wave-guide is $r = 1$ cm, then $\omega_{0,1}(0)/2\pi$ is about 11 GHz, which is a microwave frequency with a wave-length of 2.6 cm. In terms of the frequencies (9.58), the field of a pulse moving in the $+z$-direction is

$$
E_z(\rho, \phi, z, t) = \sum_{n=0}^{\infty}\sum_{m=1}^{\infty}\int_0^\infty c_{n,m}(k)\,J_n\left(\frac{z_{n,m}\,\rho}{r}\right)e^{in\phi}\exp i\left[kz - \omega_{n,m}(k)t\right]dk. \qquad (9.59)
$$

Modes with $E_z = 0$ are **transverse electric** or **TE** modes. For them the boundary conditions (9.54) will be satisfied (exercise 9.18) if $\sqrt{\omega^2/c^2 - k^2}\,r$ is a zero $z'_{n,m}$ of J'_n. Their frequencies are $\omega_{n,m}(k) = c\sqrt{k^2 + z'^2_{n,m}/r^2}$. Since the first zero of a first derivative of a Bessel function is $z'_{1,1} \approx 1.8412$, the minimum frequency $\omega_{1,1}(0) = c\,z'_{1,1}/r \approx 1.8412\,c/r$ occurs for $n = 1$ and $k = 0$. If the radius of the wave-guide is $r = 1$ cm, then $\omega_{1,1}(0)/2\pi$ is about 8.8 GHz, which is a microwave frequency with a wave-length of 3.4 cm. □

Example 9.4 (Cylindrical cavity) The modes of an evacuated, perfectly conducting cylindrical cavity of radius r and height h are like those of a cylindrical wave-guide (example 9.3) but with extra boundary conditions

$$B_z(\rho,\phi,0,t) = 0 \quad \text{and} \quad B_z(\rho,\phi,h,t) = 0,$$
$$E_\rho(\rho,\phi,0,t) = 0 \quad \text{and} \quad E_\rho(\rho,\phi,h,t) = 0,$$
$$E_\phi(\rho,\phi,0,t) = 0 \quad \text{and} \quad E_\phi(\rho,\phi,h,t) = 0 \tag{9.60}$$

at the two ends of the cylinder. If ℓ is an integer and if $\sqrt{\omega^2/c^2 - \pi^2\ell^2/h^2}\,r$ is a zero $z'_{n,m}$ of J'_n, then the TE fields $E_z = 0$ and

$$B_z = J_n(z'_{n,m}\,\rho/r)\,e^{in\phi}\,\sin(\pi\,\ell z/h)\,e^{-i\omega t} \tag{9.61}$$

satisfy both these (9.60) boundary conditions at $z = 0$ and h and those (9.54) at $\rho = r$ as well as the separable wave equations (9.57). The frequencies of the resonant TE modes then are $\omega_{n,m,\ell} = c\sqrt{z'^2_{n,m}/r^2 + \pi^2\ell^2/h^2}$.

The TM modes are $B_z = 0$ and

$$E_z = J_n(z_{n,m}\,\rho/r)\,e^{in\phi}\,\sin(\pi\,\ell z/h)\,e^{-i\omega t} \tag{9.62}$$

with resonant frequencies $\omega_{n,m,\ell} = c\sqrt{z^2_{n,m}/r^2 + \pi^2\ell^2/h^2}$. □

9.2 Spherical Bessel functions of the first kind

If in Bessel's equation (9.4), one sets $n = \ell + 1/2$ and $j_\ell = \sqrt{\pi/2x}\,J_{\ell+1/2}$, then one may show (exercise 9.21) that

$$x^2 j''_\ell(x) + 2x j'_\ell(x) + [x^2 - \ell(\ell+1)]j_\ell(x) = 0, \tag{9.63}$$

which is the equation for the **spherical Bessel function** j_ℓ.

We saw in example 6.6 that by setting $V(r,\theta,\phi) = R_{k,\ell}(r)\,\Theta_{\ell,m}(\theta)\,\Phi_m(\phi)$ we could separate the variables of Helmholtz's equation $-\triangle V = k^2 V$ in spherical coordinates

$$\frac{r^2\triangle V}{V} = \frac{(r^2 R'_{k,\ell})'}{R_{k,\ell}} + \frac{(\sin\theta\,\Theta'_{\ell,m})'}{\sin\theta\,\Theta_{\ell,m}} + \frac{\Phi''}{\sin^2\theta\,\Phi} = -k^2r^2. \tag{9.64}$$

Thus if $\Phi_m(\phi) = e^{im\phi}$ so that $\Phi''_m = -m^2\Phi_m$, and if $\Theta_{\ell,m}$ satisfies the **associated Legendre equation** (8.91)

$$\sin\theta\left(\sin\theta\,\Theta'_{\ell,m}\right)' + [\ell(\ell+1)\sin^2\theta - m^2]\,\Theta_{\ell,m} = 0 \tag{9.65}$$

then the product $V(r,\theta,\phi) = R_{k,\ell}(r)\,\Theta_{\ell,m}(\theta)\,\Phi_m(\phi)$ will obey (9.64) because in view of (9.63) the radial function $R_{k,\ell}(r) = j_\ell(kr)$ satisfies

$$(r^2 R'_{k,\ell})' + [k^2r^2 - \ell(\ell+1)]R_{k,\ell} = 0. \tag{9.66}$$

In terms of the spherical harmonic $Y_{\ell,m}(\theta,\phi) = \Theta_{\ell,m}(\theta)\,\Phi_m(\phi)$, the solution is $V(r,\theta,\phi) = j_\ell(kr)\,Y_{\ell,m}(\theta,\phi)$.

Rayleigh's formula gives the spherical Bessel function

$$j_\ell(x) \equiv \sqrt{\frac{\pi}{2x}}\, J_{\ell+1/2}(x) \tag{9.67}$$

as the ℓth derivative of $\sin x/x$

$$j_\ell(x) = (-1)^\ell\, x^\ell \left(\frac{1}{x}\frac{d}{dx}\right)^\ell \left(\frac{\sin x}{x}\right) \tag{9.68}$$

(Lord Rayleigh (John William Strutt), 1842–1919). In particular, $j_0(x) = \sin x/x$ and $j_1(x) = \sin x/x^2 - \cos x/x$. Rayleigh's formula leads to the recursion relation (exercise 9.22)

$$j_{\ell+1}(x) = \frac{\ell}{x} j_\ell(x) - j_\ell'(x), \tag{9.69}$$

with which one can show (exercise 9.23) that the spherical Bessel functions as defined by Rayleigh's formula do satisfy their differential equation (9.66) with $x = kr$.

The spherical Bessel functions $j_\ell(kr)$ satisfy the self-adjoint Sturm–Liouville (6.333) equation (9.66)

$$-\, r^2 j_\ell'' - 2r j_\ell' + \ell(\ell+1) j_\ell = k^2 r^2 j_\ell \tag{9.70}$$

with eigenvalue k^2 and weight function $\rho = r^2$. If $j_\ell(z_{\ell,n}) = 0$, then the functions $j_\ell(kr) = j_\ell(z_{\ell,n} r/a)$ vanish at $r = a$ and form an orthogonal basis

$$\int_0^a j_\ell(z_{\ell,n} r/a)\, j_\ell(z_{\ell,m} r/a)\, r^2\, dr = \frac{a^3}{2}\, j_{\ell+1}^2(z_{\ell,n})\, \delta_{n,m} \tag{9.71}$$

for a self-adjoint system on the interval $[0, a]$. Moreover, since the eigenvalues $k_{\ell,n}^2 = z_{\ell,n}^2/a^2 \approx (n\pi)^2/a^2 \to \infty$ as $n \to \infty$, the eigenfunctions $j_\ell(z_{\ell,n} r/a)$ also are complete in the mean (section 6.35).

On an infinite interval, the analogous relation is

$$\int_0^\infty j_\ell(kr)\, j_\ell(k'r)\, r^2\, dr = \frac{\pi}{2k^2}\, \delta(k - k'). \tag{9.72}$$

If we write the spherical Bessel function $j_0(x)$ as the integral

$$j_0(z) = \frac{\sin z}{z} = \frac{1}{2}\int_{-1}^{1} e^{izx}\, dx \tag{9.73}$$

and use Rayleigh's formula (9.68), we may find an integral for $j_\ell(z)$

$$
\begin{aligned}
j_\ell(z) &= (-1)^\ell z^\ell \left(\frac{1}{z}\frac{d}{dz}\right)^\ell \left(\frac{\sin z}{z}\right) = (-1)^\ell z^\ell \left(\frac{1}{z}\frac{d}{dz}\right)^\ell \frac{1}{2}\int_{-1}^{1} e^{izx}\,dx \\
&= \frac{z^\ell}{2}\int_{-1}^{1} \frac{(1-x^2)^\ell}{2^\ell \ell!}\, e^{izx}\,dx = \frac{(-i)^\ell}{2}\int_{-1}^{1} \frac{(1-x^2)^\ell}{2^\ell \ell!}\,\frac{d^\ell}{dx^\ell} e^{izx}\,dx \\
&= \frac{(-i)^\ell}{2}\int_{-1}^{1} e^{izx}\,\frac{d^\ell}{dx^\ell}\frac{(x^2-1)^\ell}{2^\ell \ell!}\,dx = \frac{(-i)^\ell}{2}\int_{-1}^{1} P_\ell(x)\, e^{izx}\,dx
\end{aligned} \tag{9.74}
$$

(exercise 9.24) that contains Rodrigues's formula (8.8) for the Legendre polynomial $P_\ell(x)$. With $z = kr$ and $x = \cos\theta$, this formula

$$
i^\ell j_\ell(kr) = \frac{1}{2}\int_{-1}^{1} P_\ell(\cos\theta) e^{ikr\cos\theta}\, d\cos\theta \tag{9.75}
$$

and the Fourier–Legendre expansion (8.32) give

$$
\begin{aligned}
e^{ikr\cos\theta} &= \sum_{\ell=0}^{\infty} \frac{2\ell+1}{2} P_\ell(\cos\theta) \int_{-1}^{1} P_\ell(\cos\theta')\, e^{ikr\cos\theta'}\, d\cos\theta' \\
&= \sum_{\ell=0}^{\infty} (2\ell+1)\, P_\ell(\cos\theta)\, i^\ell j_\ell(kr).
\end{aligned} \tag{9.76}
$$

If θ, ϕ and θ', ϕ' are the polar angles of the vectors $\boldsymbol{r}$ and $\boldsymbol{k}$, then by using the addition theorem (8.121) we get

$$
e^{i\boldsymbol{k}\cdot\boldsymbol{r}} = \sum_{\ell=0}^{\infty} 4\pi\, i^\ell j_\ell(kr)\, Y_{\ell,m}(\theta,\phi)\, Y^*_{\ell,m}(\theta',\phi'). \tag{9.77}
$$

The series expansion (9.1) for J_n and the definition (9.67) of j_ℓ give us for small $|\rho| \ll 1$ the approximation

$$
j_\ell(\rho) \approx \frac{\ell!\,(2\rho)^\ell}{(2\ell+1)!} = \frac{\rho^\ell}{(2\ell+1)!!}. \tag{9.78}
$$

To see how $j_\ell(\rho)$ behaves for large $|\rho| \gg 1$, we use Rayleigh's formula (9.68) to compute $j_1(\rho)$ and notice that the derivative $d/d\rho$

$$
j_1(\rho) = -\frac{d}{d\rho}\left(\frac{\sin\rho}{\rho}\right) = -\frac{\cos\rho}{\rho} + \frac{\sin\rho}{\rho^2} \tag{9.79}
$$

adds a factor of $1/\rho$ when it acts on $1/\rho$ but not when it acts on $\sin\rho$. Thus the dominant term is the one in which all the derivatives act on the sine, and so for large $|\rho| \gg 1$, we have approximately

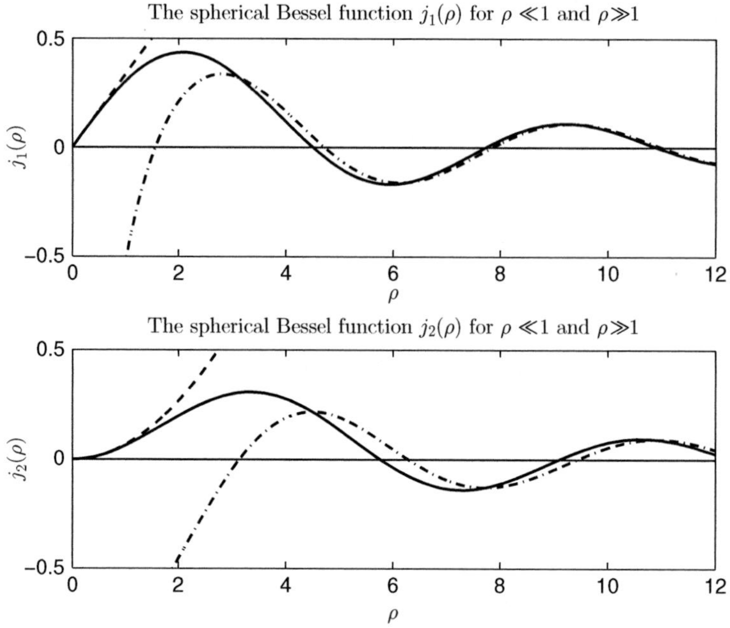

Figure 9.2 Top: plot of $j_1(\rho)$ (solid curve) and its approximations $\rho/3$ for small ρ (9.78, dashes) and $\sin(\rho - \pi/2)/\rho$ for big ρ (9.80, dot-dash). Bottom: plot of $j_2(\rho)$ (solid curve) and its approximations $\rho^2/15$ for small ρ (9.78, dashed) and $\sin(\rho-\pi)/\rho$ for big ρ (9.80, dot-dash). The values of ρ at which $j_\ell(\rho) = 0$ are the **zeros** or **roots** of j_ℓ; we use them to fit boundary conditions.

$$j_\ell(\rho) \approx (-1)^\ell \frac{1}{\rho} \frac{d^\ell \sin\rho}{d\rho^\ell} = \frac{1}{\rho} \sin\left(\rho - \ell\frac{\pi}{2}\right) \tag{9.80}$$

with an error that falls off as $1/\rho^2$. The quality of the approximation, which is exact for $\ell = 0$, is illustrated for $\ell = 1$ and 2 in Fig. 9.2.

Example 9.5 (Partial waves) Spherical Bessel functions occur in the wave-functions of free particles with well-defined angular momentum.

The hamiltonian $H_0 = \boldsymbol{p}^2/2m$ for a free particle of mass m and the square $\boldsymbol{L}^2$ of the orbital angular momentum operator are both invariant under rotations; thus they commute with the orbital angular-momentum operator $\boldsymbol{L}$. Since the operators H_0, $\boldsymbol{L}^2$, and L_z commute with each other, simultaneous eigenstates $|k,\ell,m\rangle$ of these **compatible** operators (section 1.30) exist

$$H_0\,|k,\ell,m\rangle = \frac{\boldsymbol{p}^2}{2m}\,|k,\ell,m\rangle = \frac{(\hbar k)^2}{2m}\,|k,\ell,m\rangle,$$

$$\boldsymbol{L}^2\,|k,\ell,m\rangle = \hbar^2\,\ell(\ell+1)\,|k,\ell,m\rangle, \quad \text{and} \quad L_z\,|k,\ell,m\rangle = \hbar m\,|k,\ell,m\rangle. \tag{9.81}$$

Their wave-functions are products of spherical Bessel functions and spherical harmonics (8.110)

$$\langle \boldsymbol{r}|k,\ell,m\rangle = \langle r,\theta,\phi|k,\ell,m\rangle = \sqrt{\frac{2}{\pi}}\, k\, j_\ell(kr)\; Y_{\ell,m}(\theta,\phi). \tag{9.82}$$

They satisfy the normalization condition

$$\begin{aligned}\langle k,\ell,m|k',\ell',m'\rangle &= \frac{2kk'}{\pi}\int_0^\infty j_\ell(kr) j_\ell(k'r)\, r^2\, dr \int Y_\ell^{m*}(\theta,\phi) Y_{\ell'}^{m'}(\theta,\phi)\, d\Omega \\ &= \delta(k-k')\,\delta_{\ell,\ell'}\,\delta_{m,m'}\end{aligned} \tag{9.83}$$

and the completeness relation

$$1 = \int_0^\infty dk \sum_{\ell=0}^{\infty}\sum_{m=-\ell}^{\ell} |k,\ell,m\rangle\langle k,\ell,m|. \tag{9.84}$$

Their inner products with an eigenstate $|\boldsymbol{k}'\rangle$ of a free particle of momentum $\boldsymbol{p}' = \hbar\boldsymbol{k}'$ are

$$\langle k,\ell,m|\boldsymbol{k}'\rangle = \frac{i^\ell}{k}\,\delta(k-k')\; Y_\ell^{m*}(\theta',\phi') \tag{9.85}$$

in which the polar coordinates of $\boldsymbol{k}'$ are θ', ϕ'.

Using the resolution (9.84) of the identity operator and the inner-product formulas (9.82 & 9.85), we recover the expansion (9.77)

$$\begin{aligned}\frac{e^{i\boldsymbol{k}'\cdot\boldsymbol{r}}}{(2\pi)^{3/2}} &= \langle\boldsymbol{r}|\boldsymbol{k}'\rangle = \int_0^\infty dk \sum_{\ell=0}^{\infty}\sum_{m=-\ell}^{\ell}\langle\boldsymbol{r}|k,\ell,m\rangle\langle k,\ell,m|\boldsymbol{k}'\rangle \\ &= \sum_{\ell=0}^{\infty}\sqrt{\frac{2}{\pi}}\; i^\ell\; j_\ell(kr)\; Y_\ell^m(\theta,\phi)\; Y_\ell^{m*}(\theta',\phi').\end{aligned} \tag{9.86}$$

The small kr approximation (9.78) and the definition (9.82) tell us that the probability that a particle with angular momentum $\hbar\ell$ about the origin has $r = |\boldsymbol{r}| \ll 1/k$ is

$$P(r) = \frac{2k^2}{\pi}\int_0^r j_\ell^2(kr) r^2 dr \approx \frac{2}{\pi((2\ell+1)!!)^2}\int_0^r (kr)^{2\ell+2} dr = \frac{(4\ell+6)(kr)^{2\ell+3}}{\pi((2\ell+3)!!)^2 k}, \tag{9.87}$$

which is very small for big ℓ and tiny k. So a short-range potential can only affect partial waves of low angular momentum. When physicists found that nuclei scattered low-energy hadrons into s-waves, they knew that the range of the nuclear force was short, about 10^{-15}m.

If the potential $V(r)$ that scatters a particle is of short range, then at big r the radial wave-function $u_\ell(r)$ of the scattered wave should look like that of a free particle (9.86), which by the big kr approximation (9.80) is

$$u_\ell^{(0)}(r) = j_\ell(kr) \approx \frac{\sin(kr - \ell\pi/2)}{kr} = \frac{1}{2ikr}\left[e^{i(kr-\ell\pi/2)} - e^{-i(kr-\ell\pi/2)}\right]. \tag{9.88}$$

Thus at big r the radial wave-function $u_\ell(r)$ differs from $u_\ell^{(0)}(r)$ only by a **phase shift** δ_ℓ

$$u_\ell(r) \approx \frac{\sin(kr - \ell\pi/2 + \delta_\ell)}{kr} = \frac{1}{2ikr}\left[e^{i(kr-\ell\pi/2+\delta_\ell)} - e^{-i(kr-\ell\pi/2+\delta_\ell)}\right]. \tag{9.89}$$

The phase shifts determine the **cross-section** σ to be (Cohen-Tannoudji *et al.*, 1977, chap. VIII)

$$\sigma = \frac{4\pi}{k^2}\sum_{\ell=0}^{\infty}(2\ell+1)\sin^2\delta_\ell. \tag{9.90}$$

If the potential $V(r)$ is negligible for $r > r_0$, then for momenta $k \ll 1/r_0$ the cross-section is $\sigma \approx 4\pi \sin^2\delta_0/k^2$. □

Example 9.6 (Quantum dots) The active region of some quantum dots is a CdSe sphere whose radius a is less than 2 nm. Photons from a laser excite electron–hole pairs, which fluoresce in nanoseconds.

I will model a quantum dot simply as an electron trapped in a sphere of radius a. Its wave-function $\psi(r,\theta,\phi)$ satisfies Schrödinger's equation

$$-\frac{\hbar^2}{2m}\triangle\psi = E\psi \tag{9.91}$$

with the boundary condition $\psi(a,\theta,\phi) = 0$. With $k^2 = 2mE/\hbar^2 = z_{\ell,n}^2/a^2$, the unnormalized eigenfunctions are

$$\psi_{n,\ell,m}(r,\theta,\phi) = j_\ell(z_{\ell,n}r/a)\, Y_{\ell,m}(\theta,\phi)\,\theta(a-r), \tag{9.92}$$

in which the Heaviside function $\theta(a-r)$ makes ψ vanish for $r > a$, and ℓ and m are integers with $-\ell \le m \le \ell$ because ψ must be single valued for all angles θ and ϕ.

The zeros $z_{\ell,n}$ of $j_\ell(x)$ fix the energy levels as $E_{n,\ell,m} = (\hbar z_{\ell,n}/a)^2/2m$. Since $z_{0,n} = n\pi$, the $\ell = 0$ levels are $E_{n,0,0} = (\hbar n\pi/a)^2/2m$. If the coupling to a photon is via a term like $\boldsymbol{p}\cdot\boldsymbol{A}$, then one expects $\Delta\ell = 1$. The energy gap from the $n, \ell = 1$ state to the $n = 1, \ell = 0$ ground state thus is

$$\Delta E_n = E_{n,1,0} - E_{1,0,0} = (z_{1,n}^2 - \pi^2)\frac{\hbar^2}{2ma^2}. \tag{9.93}$$

Inserting factors of c^2 and using $\hbar c = 197$ eV nm, and $mc^2 = 0.511$ MeV, we find from the zero $z_{1,2} = 7.72525$ that $\Delta E_2 = 1.89\ (\text{nm}/a)^2$ eV, which is red light if $a = 1$ nm. The next zero $z_{1,3} = 10.90412$ gives $\Delta E_3 = 4.14\ (\text{nm}/a)^2$ eV, which is in the visible if $1.2 < a < 1.5$ nm. The *Mathematica* command Do[Print[N[BesselJZero[1.5, k]]], {k, 1, 5, 1}] gives the first five zeros of $j_1(x)$ to six significant figures. □

9.3 Bessel functions of the second kind

In section 7.5 we derived integral representations (7.55 & 7.56) for the Hankel functions $H_\lambda^{(1)}(z)$ and $H_\lambda^{(2)}(z)$ for $\mathrm{Re}\, z > 0$. One may analytically continue them (Courant and Hilbert, 1955, chap. VII) to the upper

$$H_\lambda^{(1)}(z) = \frac{1}{\pi i} e^{-i\lambda/2} \int_{-\infty}^{\infty} e^{iz\cosh x - \lambda x}\, dx, \qquad \mathrm{Im} z \geq 0 \tag{9.94}$$

and lower

$$H_\lambda^{(2)}(z) = -\frac{1}{\pi i} e^{+i\lambda/2} \int_{-\infty}^{\infty} e^{-iz\cosh x - \lambda x}\, dx, \qquad \mathrm{Im} z \leq 0 \tag{9.95}$$

half z-planes. When both $z = \rho$ and $\lambda = \nu$ are real, the two Hankel functions are complex conjugates of each other

$$H_\nu^{(1)}(\rho) = H_\nu^{(2)*}(\rho). \tag{9.96}$$

Hankel functions, called **Bessel functions of the third kind**, are linear combinations of Bessel functions of the first $J_\lambda(z)$ and second $Y_\lambda(z)$ kind

$$\begin{aligned} H_\lambda^{(1)}(z) &= J_\lambda(z) + iY_\lambda(z), \\ H_\lambda^{(2)}(z) &= J_\lambda(z) - iY_\lambda(z). \end{aligned} \tag{9.97}$$

Bessel functions of the second kind are also called **Neumann functions**; the symbols $Y_\lambda(z) = N_\lambda(z)$ refer to the same function. They are infinite at $z = 0$ as illustrated in Fig. 9.3.

When $z = ix$ is imaginary, we get the **modified Bessel functions**

$$\begin{aligned} I_\alpha(x) &= i^{-\alpha} J_\alpha(ix) = \sum_{m=0}^{\infty} \frac{1}{m!\,\Gamma(m+\alpha+1)} \left(\frac{x}{2}\right)^{2m+\alpha}, \\ K_\alpha(x) &= \frac{\pi}{2} i^{\alpha+1} H_\alpha^{(1)}(ix) = \int_0^\infty e^{-x\cosh t} \cosh \alpha t\, dt. \end{aligned} \tag{9.98}$$

Some simple cases are

$$I_{-1/2}(z) = \sqrt{\frac{2}{\pi z}} \cosh z, \quad I_{1/2}(z) = \sqrt{\frac{2}{\pi z}} \sinh z, \quad \text{and} \quad K_{1/2}(z) = \sqrt{\frac{\pi}{2z}} e^{-z}. \tag{9.99}$$

When do we need to use these functions? If we are representing functions that are finite at the origin $\rho = 0$, then we don't need them. But if the point

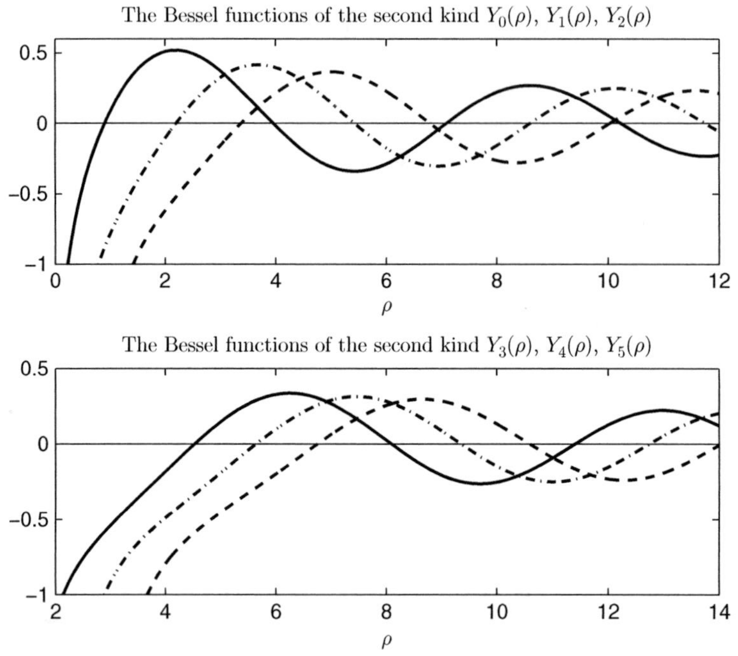

Figure 9.3 Top: $Y_0(\rho)$ (solid curve), $Y_1(\rho)$ (dot-dash), and $Y_2(\rho)$ (dashed) for $0 < \rho < 12$. Bottom: $Y_3(\rho)$ (solid curve), $Y_4(\rho)$ (dot-dash), and $Y_5(\rho)$ (dashed) for $2 < \rho < 14$. The points at which Bessel functions cross the ρ-axis are called **zeros** or **roots**; we use them to satisfy boundary conditions.

$\rho = 0$ lies outside the region of interest or if the function we are representing is infinite at that point, then we do need the $Y_\nu(\rho)$s.

Example 9.7 (Coaxial wave-guides) An ideal coaxial wave-guide is perfectly conducting for $\rho < r_0$ and $\rho > r$, and the waves occupy the region $r_0 < \rho < r$. Since points with $\rho = 0$ are not in the physical domain of the problem, the electric field $\boldsymbol{E}(\rho, \phi) \exp(i(kz - \omega t))$ is a linear combination of Bessel functions of the first and second kinds with

$$E_z(\rho, \phi) = \left[a\, J_n(\sqrt{\omega^2/c^2 - k^2}\, \rho) + b\, Y_n(\sqrt{\omega^2/c^2 - k^2}\, \rho) \right] \tag{9.100}$$

in the notation of example 9.3. A similar equation represents the magnetic field B_z. The fields $\boldsymbol{E}$ and $\boldsymbol{B}$ obey the equations and boundary conditions of example 9.3 as well as

$$E_z(r_0, \phi) = 0, \quad E_\phi(r_0, \phi) = 0, \quad \text{and} \quad B_\rho(r_0, \phi) = 0 \tag{9.101}$$

at $\rho = r_0$. In TM modes with $B_z = 0$, one may show (exercise 9.27) that the boundary conditions $E_z(r_0, \phi) = 0$ and $E_z(r, \phi) = 0$ can be satisfied if

$$J_n(x)\, Y_n(vx) - J_n(vx)\, Y_n(x) = 0 \tag{9.102}$$

in which $v = r/r_0$ and $x = \sqrt{\omega^2/c^2 - k^2}\, r_0$. One can use the Matlab code

```
n = 0.; v = 10.;
f=@(x)besselj(n,x).*bessely(n,v*x)-besselj(n,v*x).
     *bessely(n,x)
x=linspace(0,5,1000);
figure
plot(x,f(x)) % we use the figure to guess at the roots
grid on
options=optimset('tolx',1e-9);
fzero(f,0.3) % we tell fzero to look near 0.3
fzero(f,0.7)
fzero(f,1)
```

to find that for $n = 0$ and $v = 10$, the first three solutions are $x_{0,1} = 0.3314$, $x_{0,2} = 0.6858$, and $x_{0,3} = 1.0377$. Setting $n = 1$ and adjusting the guesses in the code, one finds $x_{1,1} = 0.3941$, $x_{1,2} = 0.7331$, and $x_{1,3} = 1.0748$. The corresponding dispersion relations are $\omega_{n,i}(k) = c\sqrt{k^2 + x_{n,i}^2/r_0^2}$. □

9.4 Spherical Bessel functions of the second kind

Spherical Bessel functions of the second kind are defined as

$$y_\ell(\rho) = \sqrt{\frac{\pi}{2\rho}}\, Y_{\ell+1/2}(\rho) \tag{9.103}$$

and Rayleigh formulas express them as

$$y_\ell(\rho) = (-1)^{\ell+1} \rho^\ell \left(\frac{d}{\rho\, d\rho}\right)^\ell \left(\frac{\cos\rho}{\rho}\right). \tag{9.104}$$

The term in which all the derivatives act on the cosine dominates at big ρ

$$y_\ell(\rho) \approx (-1)^{\ell+1}\, \frac{1}{\rho}\, \frac{d^\ell \cos\rho}{d\rho^\ell} = -\cos\left(\rho - \ell\pi/2\right)/\rho. \tag{9.105}$$

The second kind of spherical Bessel functions at small ρ are approximately

$$y_\ell(\rho) \approx -(2\ell - 1)!!/\rho^{\ell+1}. \tag{9.106}$$

They all are infinite at $x = 0$ as illustrated in Fig. 9.4.

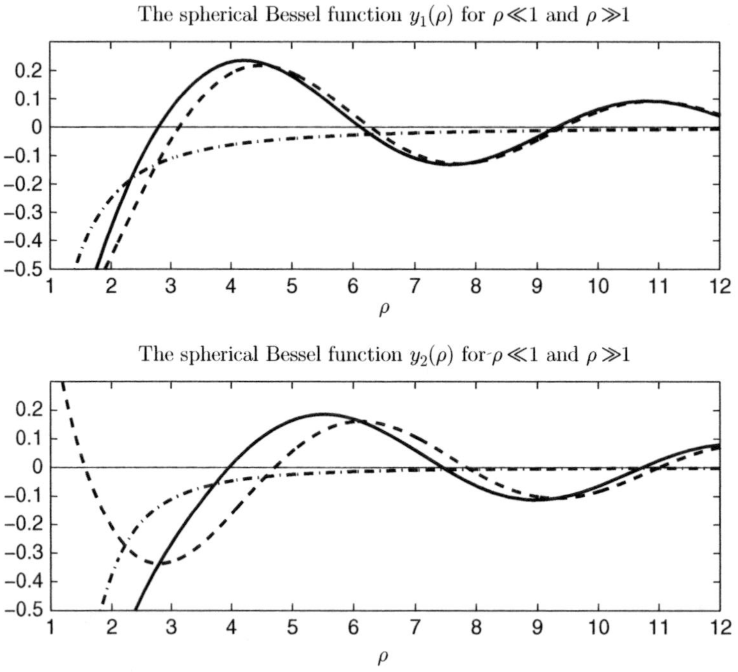

Figure 9.4 Top: plot of $y_1(\rho)$ (solid curve) and its approximations $-1/\rho^2$ for small ρ (9.106, dot-dash) and $-\cos(\rho - \pi/2)/\rho$ for big ρ (9.105, dashed). Bottom: plot of $y_2(\rho)$ (solid curve) and its approximations $-3/\rho^3$ for small ρ (9.106, dot-dash) and $-\cos(\rho - \pi)/\rho$ for big ρ (9.105, dashed). The values of ρ at which $y_\ell(\rho) = 0$ are the **zeros** or **roots** of y_ℓ; we use them to fit boundary conditions. All six plots run from $\rho = 1$ to $\rho = 12$.

Example 9.8 (Scattering off a hard sphere) In the notation of example 9.5, the potential of a hard sphere of radius r_0 is $V(r) = \infty\,\theta(r_0 - r)$ in which $\theta(x) = (x + |x|)/2|x|$ is Heaviside's function. Since the point $r = 0$ is not in the physical region, the scattered wave-function is a linear combination of spherical Bessel functions of the first and second kinds

$$u_\ell(r) = c_\ell\, j_\ell(kr) + d_\ell\, y_\ell(kr). \tag{9.107}$$

The boundary condition $u_\ell(kr_0) = 0$ fixes the ratio $v_\ell = d_\ell/c_\ell$ of the constants c_ℓ and d_ℓ. Thus for $\ell = 0$, Rayleigh's formulas (9.68 & 9.104) and the boundary condition say that $kr_0\, u_0(r_0) = c_0 \sin(kr_0) - d_0 \cos(kr_0) = 0$ or $d_0/c_0 = \tan kr_0$. The s-wave then is $u_0(kr) = c_0 \sin(kr - kr_0)/(kr \cos kr_0)$, which tells us that the tangent of the phase shift is $\tan \delta_0(k) = -kr_0$. By (9.90), the cross-section at low energy is $\sigma \approx 4\pi r_0^2$ or four times the classical value.

Similarly, one finds (exercise 9.28) that the tangent of the p-wave phase shift is

$$\tan \delta_1(k) = \frac{kr_0 \cos kr_0 - \sin kr_0}{\cos kr_0 - kr_0 \sin kr_0}. \tag{9.108}$$

For $kr_0 \ll 1$, we have $\delta_1(k) \approx -(kr_0)^3/3$; more generally the ℓth phase shift is $\delta_\ell(k) \approx -(kr_0)^{2\ell+1}/\{(2\ell+1)[(2\ell-1)!!]^2\}$ for a potential of range r_0 at low energy $k \ll 1/r_0$. □

Further reading

A great deal is known about Bessel functions. Students may find *Mathematical Methods for Physics and Engineering* (Riley *et al.*, 2006) as well as the classics *A Treatise on the Theory of Bessel Functions* (Watson, 1995), *A Course of Modern Analysis* (Whittaker and Watson, 1927, chap. XVII), and *Methods of Mathematical Physics* (Courant and Hilbert, 1955) of special interest.

Exercises

9.1 Show that the series (9.1) for $J_n(\rho)$ satisfies Bessel's equation (9.4).

9.2 Show that the generating function $\exp(z(u - 1/u)/2)$ for the Bessel functions is invariant under the substitution $u \to -1/u$.

9.3 Use the invariance of $\exp(z(u-1/u)/2)$ under $u \to -1/u$ to show that $J_{-n}(z) = (-1)^n J_n(z)$.

9.4 By writing the generating function (9.5) as the product of the exponentials $\exp(zu/2)$ and $\exp(-z/2u)$, derive the expansion

$$\exp\left[\frac{z}{2}\left(u - u^{-1}\right)\right] = \sum_{n=0}^{\infty} \sum_{m=-n}^{\infty} \left(\frac{z}{2}\right)^{m+n} \frac{u^{m+n}}{(m+n)!} \left(-\frac{z}{2}\right)^n \frac{u^{-n}}{n!}. \tag{9.109}$$

9.5 From this expansion (9.109) of the generating function (9.5), derive the power-series expansion (9.1) for $J_n(z)$.

9.6 In the formula (9.5) for the generating function $\exp(z(u - 1/u)/2)$, replace u by $\exp i\theta$ and then derive the integral representation (9.6) for $J_n(z)$. Start with the interval $[-\pi, \pi]$.

9.7 From the general integral representation (9.6) for $J_n(z)$, derive the two integral formulas (9.7) for $J_0(z)$.

9.8 Show that the integral representations (9.6 & 9.7) imply that for any integer $n \neq 0$, $J_n(0) = 0$, while $J_0(0) = 1$.

9.9 By differentiating the generating function (9.5) with respect to u and identifying the coefficients of powers of u, derive the recursion relation

$$J_{n-1}(z) + J_{n+1}(z) = \frac{2n}{z} J_n(z). \tag{9.110}$$

9.10 By differentiating the generating function (9.5) with respect to z and identifying the coefficients of powers of u, derive the recursion relation

$$J_{n-1}(z) - J_{n+1}(z) = 2\,J_n'(z). \tag{9.111}$$

9.11 Change variables to $z = ax$ and turn Bessel's equation (9.4) into the self-adjoint form (9.11).

9.12 If $y = J_n(ax)$, then equation (9.11) is $(xy')' + (xa^2 - n^2/x)y = 0$. Multiply this equation by xy', integrate from 0 to b, and so show that if $ab = z_{n,m}$ and $J_n(z_{n,m}) = 0$, then

$$2\int_0^b x\,J_n^2(ax)\,dx = b^2 J_n'^2(z_{n,m}), \tag{9.112}$$

which is the normalization condition (9.14).

9.13 Show that with $\lambda \equiv z^2/r_d^2$, the change of variables $\rho = zr/r_d$ and $u(r) = J_n(\rho)$ turns $-\left(r\,u'\right)' + n^2\,u/r = \lambda\,r\,u$ into (9.25).

9.14 Use the formula (6.42) for the curl in cylindrical coordinates and the vacuum forms $\nabla \times \boldsymbol{E} = -\dot{\boldsymbol{B}}$ and $\nabla \times \boldsymbol{B} = \dot{\boldsymbol{E}}/c^2$ of the laws of Faraday and Maxwell–Ampère to derive the field equations (9.55).

9.15 Derive equations (9.56) from (9.55).

9.16 Show that $J_n(\sqrt{\omega^2/c^2 - k^2}\,\rho)e^{in\phi}e^{i(kz-\omega t)}$ is a traveling-wave solution (9.52) of the wave equations (9.57).

9.17 Find expressions for the nonzero TM fields in terms of the formula (9.59) for E_z.

9.18 Show that the TE field $E_z = 0$ and $B_z = J_n(\sqrt{\omega^2/c^2 - k^2}\,\rho)e^{in\phi}e^{i(kz-\omega t)}$ will satisfy the boundary conditions (9.54) if $\sqrt{\omega^2/c^2 - k^2}\,r$ is a zero $z'_{n,m}$ of J'_n.

9.19 Show that if ℓ is an integer and if $\sqrt{\omega^2/c^2 - \pi^2\ell^2/h^2}\,r$ is a zero $z'_{n,m}$ of J'_n, then the fields $E_z = 0$ and $B_z = J_n(z'_{n,m}\rho/r)\,e^{in\phi}\,\sin(\ell\pi z/h)\,e^{-i\omega t}$ satisfy both the boundary conditions (9.54) at $\rho = r$ and those (9.60) at $z = 0$ and h as well as the wave equations (9.57). Hint: use Maxwell's equations $\nabla \times \boldsymbol{E} = -\dot{\boldsymbol{B}}$ and $\nabla \times \boldsymbol{B} = \dot{\boldsymbol{E}}/c^2$ as in (9.55).

9.20 Show that the resonant frequencies of the TM modes of the cavity of example 9.4 are $\omega_{n,m,\ell} = c\sqrt{z_{n,m}^2/r^2 + \pi^2\ell^2/h^2}$.

9.21 By setting $n = \ell + 1/2$ and $j_\ell = \sqrt{\pi/2x}\,J_{\ell+1/2}$, show that Bessel's equation (9.4) implies that the spherical Bessel function j_ℓ satisfies (9.63).

9.22 Show that Rayleigh's formula (9.68) implies the recursion relation (9.69).

9.23 Use the recursion relation (9.69) to show by induction that the spherical Bessel functions $j_\ell(x)$ as given by Rayleigh's formula (9.68) satisfy their differential equation (9.66) which with $x = kr$ is

$$-x^2 j_\ell'' - 2x j_\ell' + \ell(\ell+1) j_\ell = x^2 j_\ell. \tag{9.113}$$

Hint: start by showing that $j_0(x) = \sin(x)/x$ satisfies this equation. This problem involves some tedium.

9.24 Iterate the trick

$$\frac{d}{zdz}\int_{-1}^{1} e^{izx}\,dx = \frac{i}{z}\int_{-1}^{1} xe^{izx}\,dx = \frac{i}{2z}\int_{-1}^{1} e^{izx}\,d(x^2-1)$$
$$= -\frac{i}{2z}\int_{-1}^{1}(x^2-1)de^{izx} = \frac{1}{2}\int_{-1}^{1}(x^2-1)e^{izx}dx \quad (9.114)$$

to show that (Schwinger *et al.*, 1998, p. 227)

$$\left(\frac{d}{zdz}\right)^{\ell}\int_{-1}^{1} e^{izx}\,dx = \int_{-1}^{1}\frac{(x^2-1)^{\ell}}{2^{\ell}\ell!}\,e^{izx}\,dx. \quad (9.115)$$

9.25 Use the expansions (9.76 and 9.77) to show that the inner product of the ket $|\boldsymbol{r}\rangle$ that represents a particle at $\boldsymbol{r}$ with polar angles θ and ϕ and the one $|\boldsymbol{k}\rangle$ that represents a particle with momentum $\boldsymbol{p} = \hbar\boldsymbol{k}$ with polar angles θ' and ϕ' is, with $\boldsymbol{k}\cdot\boldsymbol{r} = kr\cos\theta$,

$$\langle\boldsymbol{r}|\boldsymbol{k}\rangle = \frac{1}{(2\pi)^{3/2}}e^{ikr\cos\theta} = \frac{1}{(2\pi)^{3/2}}\sum_{\ell=0}^{\infty}(2\ell+1)\,P_\ell(\cos\theta)\,i^\ell\,j_\ell(kr)$$
$$= \frac{1}{(2\pi)^{3/2}}e^{i\boldsymbol{k}\cdot\boldsymbol{r}} = \sqrt{\frac{2}{\pi}}\sum_{\ell=0}^{\infty} i^\ell\,j_\ell(kr)\,Y_{\ell,m}(\theta,\phi)\,Y^*_{\ell,m}(\theta',\phi'). \quad (9.116)$$

9.26 Show that $(-1)^\ell d^\ell \sin\rho/d\rho^\ell = \sin(\rho - \pi\ell/2)$ and so complete the derivation of the approximation (9.80) for $j_\ell(\rho)$ for big ρ.

9.27 In the context of examples 9.3 and 9.7, show that the boundary conditions $E_z(r_0,\phi) = 0$ and $E_z(r,\phi) = 0$ imply (9.102).

9.28 Show that for scattering off a hard sphere of radius r_0 as in example 9.8, the p-wave phase shift is given by (9.108).

10

Group theory

10.1 What is a group?

A group G is a set of objects f, g, h, ... and an operation called multiplication such that:

1 if $f \in G$ and $g \in G$, the product $fg \in G$ (**closure**);
2 if f, g, and h are in G, then $f(gh) = (fg)h$ (**associativity**);
3 there is an **identity** $e \in G$ such that if $g \in G$, then $ge = eg = g$;
4 every $g \in G$ has an **inverse** $g^{-1} \in G$ such that $gg^{-1} = g^{-1}g = e$.

Physical transformations naturally form groups. The product $T'\,T$ represents the transformation T followed by the transformation T'. And both $T''\,(T'\,T)$ and $(T''\,T')\,T$ represent the transformation T followed by the transformation T' and then by T''. So transformations are associative. The identity element e is the null transformation, the one that does nothing. The inverse T^{-1} is the transformation that reverses the effect of T. Such a set $\{T\}$ of transformations will form a group if any two successive transformations is a transformation in the set (closure). Closure occurs naturally when the criterion for membership in the group is that a transformation not change something. For if both T and T' leave that thing unchanged, then so will their product $T'\,T$.

Example 10.1 (Groups of coordinate transformations) The set of all transformations that leave invariant the distance from the origin of every point in n-dimensional space is the group $O(n)$ of **rotations** and **reflections**. The rotations in $\boldsymbol{R}^n$ form the group $SO(n)$.

The set of all transformations that leave invariant the spatial difference $\boldsymbol{x} - \boldsymbol{y}$ between every two points $\boldsymbol{x}$ and $\boldsymbol{y}$ in n-dimensional space is the group of **translations**. In this case, group multiplication is vector addition.

The set of all linear transformations that leave invariant the square of the Minkowski distance $x_1^2 + x_2^2 + x_3^2 - x_0^2$ between any 4-vector x and the origin is the **Lorentz group** (Hermann Minkowski, 1864–1909; Hendrik Lorentz, 1853–1928).

The set of all linear transformations that leave invariant the square of the Minkowski distance $(x_1-y_1)^2+(x_2-y_2)^2+(x_3-y_3)^2-(x^0-y^0)^2$ between any two 4-vectors x and y is the **Poincaré group**, which includes Lorentz transformations and translations (Henri Poincaré, 1854–1912). □

Except for the group of translations, the order of the physical transformations in these examples matters: the transformation $T'\,T$ is not in general the same as $T\,T'$. Such groups are called **nonabelian**. A group whose elements all **commute**

$$[T', T] \equiv T'\,T - T\,T' = 0 \tag{10.1}$$

is said to be **abelian** (Niels Abel, 1802–1829).

Matrices naturally form groups. Since matrix multiplication is associative, any set $\{D\}$ of $n \times n$ nonsingular matrices that includes the inverse D^{-1} of every matrix in the set as well as the identity matrix I automatically satisfies properties 2–4 with group multiplication defined as matrix multiplication. Only property 1, closure under multiplication, is uncertain. A set $\{D\}$ of matrices will form a group as long as the product of any two matrices is in the set. As with physical transformations, one way to ensure closure is to have every matrix leave something unchanged.

Example 10.2 (Orthogonal groups) The set of all $n \times n$ real matrices that leave the quadratic form $x_1^2 + x_2^2 + \cdots + x_n^2$ unchanged forms the orthogonal group $O(n)$ of all $n \times n$ orthogonal (1.36) matrices (exercises 10.1 & 10.2). The $n \times n$ orthogonal matrices that have unit determinant form the special orthogonal group $SO(n)$. The group $SO(3)$ describes rotations. □

Example 10.3 (Unitary groups) The set of all $n \times n$ complex matrices that leave invariant the quadratic form $x_1^* x_1 + x_2^* x_2 + \cdots + x_n^* x_n$ forms the unitary group $U(n)$ of all $n \times n$ unitary (1.35) matrices (exercises 10.3 & 10.4). Those of unit determinant form the special unitary group $SU(n)$ (exercise 10.5).

Like $SO(3)$, the group $SU(2)$ represents rotations. The group $SU(3)$ is the symmetry group of the strong interactions, quantum chromodynamics. Physicists have used the groups $SU(5)$ and $SO(10)$ to unify the electro weak and strong interactions; whether Nature also does so is unclear. □

The number of elements in a group is the **order** of the group. A **finite group** is a group with a finite number of elements, or equivalently a group of finite order.

Example 10.4 (Z_2 and Z_n) The **parity** group whose elements are 1 and -1 under ordinary multiplication is the finite group $\boldsymbol{Z_2}$. It is abelian and of order 2. The group Z_n for any positive integer n is made of the phases $\exp(i2k\pi/n)$ for $k = 1, 2, \ldots, n$. It is abelian and of order n. □

A group whose elements $g = g(\{\alpha\})$ depend continuously upon a set of parameters α_k is a **continuous group** or a **Lie group**. Continuous groups are of infinite order.

A group G of matrices D is **compact** if the (squared) norm as given by the trace

$$\mathrm{Tr}\left(D^\dagger D\right) \leq M \tag{10.2}$$

is bounded for all the $D \in G$.

Example 10.5 ($SO(n)$, $O(n)$, $SU(n)$, and $U(n)$) The groups $SO(n)$, $O(n)$, $SU(n)$, and $U(n)$ are continuous Lie groups of infinite order. Since for any matrix D in one of these groups

$$\mathrm{Tr}\left(D^\dagger D\right) = \mathrm{Tr}I = n \leq M \tag{10.3}$$

these groups also are compact. □

Example 10.6 (Noncompact groups) The set of all real $n \times n$ matrices forms the general linear group $GL(n, \mathbb{R})$; those of unit determinant form the special linear group $SL(n, \mathbb{R})$. The corresponding groups of matrices with complex entries are $GL(n, \mathbb{C})$ and $SL(n, \mathbb{C})$. These four groups have matrix elements that are unbounded; they are noncompact. They are continuous Lie groups of infinite order like the orthogonal and unitary groups. The group $SL(2, \mathbb{C})$ represents Lorentz transformations. □

10.2 Representations of groups

If one can associate with every element g of a group G a square matrix $D(g)$ and have matrix multiplication imitate group multiplication

$$D(f)\, D(g) = D(fg) \tag{10.4}$$

for all elements f and g of the group G, then the set of matrices $D(g)$ is said to form a **representation** of the group G. If the matrices of the representation are $n \times n$, then n is the **dimension** of the representation. The dimension of a representation also is the dimension of the vector space on which the matrices act. If the matrices $D(g)$ are unitary $D^\dagger(g) = D^{-1}(g)$, then they form a **unitary representation** of the group.

Compact groups possess finite-dimensional unitary representations; noncompact groups do not. A group of bounded (10.2) matrices is compact. An abstract group of elements $g(\{\alpha\})$ is compact if its space of parameters $\{\alpha\}$ is **closed and bounded.** (A set is closed if the limit of every convergent sequence of its points lies in the set. A set is **open** if each of its elements lies in a neighborhood that lies in the set. For example, the interval $[a, b] \equiv \{x|a \le x \le b\}$ is closed, and $(a, b) \equiv \{x|a < x < b\}$ is open.) The group of rotations is compact, but the group of translations and the Lorentz group are noncompact.

Every $n \times n$ matrix S that is nonsingular ($\det S \neq 0$) maps any $n \times n$ representation $D(g)$ of a group G into an **equivalent representation** $D'(g)$ through the **similarity transformation**

$$D'(g) = S^{-1} D(g) S, \tag{10.5}$$

which preserves the law of multiplication

$$\begin{aligned} D'(f)\, D'(g) &= S^{-1} D(f) S\, S^{-1} D(g) S \\ &= S^{-1} D(f)\, D(g) S = S^{-1} D(fg) S = D'(fg). \end{aligned} \tag{10.6}$$

A **proper subspace** W of a vector space V is a subspace of lower (but not zero) dimension. A proper subspace W is **invariant** under the action of a representation $D(g)$ if $D(g)$ maps every vector $v \in W$ to a vector $D(g)\, v = v' \in W$. A representation that has a **proper invariant subspace** is **reducible**. A representation that is not reducible is **irreducible**.

There is no need to keep track of several equivalent irreducible representations D, D', D'' of any group. So in what follows, we shall choose one of these equivalent irreducible representations and use it exclusively.

A representation is **completely reducible** if it is equivalent to a representation whose matrices are in **block-diagonal form**

$$\begin{pmatrix} D_1(g) & 0 & \dots \\ 0 & D_2(g) & \dots \\ \vdots & \vdots & \vdots \end{pmatrix} \tag{10.7}$$

in which each representation $D_i(g)$ irreducible. A representation in block-diagonal form is said to be a **direct sum** of the irreducible representations D_i

$$D_1 \oplus D_2 \oplus \cdots . \tag{10.8}$$

10.3 Representations acting in Hilbert space

A symmetry transformation g is a map (1.174) of states $\psi \to \psi'$ that preserves their inner products

$$|\langle \phi' | \psi' \rangle|^2 = |\langle \phi | \psi \rangle|^2 \tag{10.9}$$

and so their predicted probabilities. The action of a group G of symmetry transformations g on the Hilbert space of a quantum theory can be represented either by operators $U(g)$ that are linear and unitary (the usual case) or by ones $K(g)$ that are antilinear (1.172) and antiunitary (1.173), as in the case of time reversal. Wigner proved this theorem in the 1930s, and Weinberg improved it in his 1995 classic (Weinberg, 1995, p. 51) (Eugene Wigner, 1902–1995; Steven Weinberg, 1933–).

Two operators F_1 and F_2 that commute $F_1 F_2 = F_2 F_1$ are **compatible** (1.328). A set of compatible operators $F_1, F_2, \ldots$ is **complete** if to every set of eigenvalues there belongs only a single eigenvector (section 1.30).

Example 10.7 (Rotation operators) Suppose that the hamiltonian H, the square of the angular momentum $\boldsymbol{J}^2$, and its z-component J_z form a complete set of compatible observables, so that the identity operator can be expressed as a sum over the eigenvectors of these operators

$$I = \sum_{E,j,m} |E,j,m\rangle\langle E,j,m|. \tag{10.10}$$

Then the matrix element of a unitary operator $U(g)$ between two states $|\psi\rangle$ and $|\phi\rangle$ is

$$\langle\phi|U(g)|\psi\rangle = \langle\phi|\sum_{E',j',m'} |E',j',m'\rangle\langle E',j',m'|\, U(g) \sum_{E,j,m}|E,j,m\rangle\langle E,j,m|\psi\rangle. \tag{10.11}$$

Let H and $\boldsymbol{J}^2$ be invariant under the action of $U(g)$ so that $U^\dagger(g)HU(g) = H$ and $U^\dagger(g)\boldsymbol{J}^2 U(g) = \boldsymbol{J}^2$. Then $HU(g) = U(g)H$ and $\boldsymbol{J}^2 U(g) = U(g)\boldsymbol{J}^2$, and so if $H|E,j,m\rangle = E|E,j,m\rangle$ and $\boldsymbol{J}^2|E,j,m\rangle = j(j+1)|E,j,m\rangle$, we have

$$\begin{aligned} HU(g)|E,j,m\rangle &= U(g)H|E,j,m\rangle = EU(g)|E,j,m\rangle, \\ \boldsymbol{J}^2 U(g)|E,j,m\rangle &= U(g)\boldsymbol{J}^2|E,j,m\rangle = j(j+1)U(g)|E,j,m\rangle. \end{aligned} \tag{10.12}$$

Thus $U(g)$ can not change E or j, and so

$$\langle E',j',m'|U(g)|E,j,m\rangle = \delta_{E'E}\delta_{j'j}\langle j,m'|U(g)|j,m\rangle = \delta_{E'E}\delta_{j'j}D^{(j)}_{m'm}(g). \tag{10.13}$$

The matrix element (10.11) then is a single sum over E and j in which the irreducible representations $D^{(j)}_{m'm}(g)$ of the rotation group $SU(2)$ appear

$$\langle\phi|U(g)|\psi\rangle = \sum_{E,j,m',m} \langle\phi|E,j,m'\rangle D^{(j)}_{m'm}(g)\langle E,j,m|\psi\rangle. \tag{10.14}$$

This is how the block-diagonal form (10.7) usually appears in calculations. The matrices $D^{(j)}_{m'm}(g)$ inherit the unitarity of the operator $U(g)$. □

10.4 Subgroups

If all the elements of a group S also are elements of a group G, then S is a **subgroup** of G. Every group G has two **trivial subgroups** – the identity element e and the whole group G itself. Many groups have more interesting subgroups. For example, the rotations about a fixed axis is an abelian subgroup of the group of all rotations in three-dimensional space.

A subgroup $S \subset G$ is an **invariant** subgroup if every element s of the subgroup S is left inside the subgroup under the **action** of every element g of the whole group G, that is, if

$$g^{-1} s g = s' \in S \quad \text{for all} \quad g \in G. \tag{10.15}$$

This condition often is written as $g^{-1} S g = S$ for all $g \in G$ or as

$$S g = g S \qquad \text{for all } g \in G. \tag{10.16}$$

Invariant subgroups also are called **normal subgroups**.

A set $C \subset G$ is called a **conjugacy class** if it's invariant under the action of the whole group G, that is, if $Cg = g\,C$ or

$$g^{-1} C g = C \quad \text{for all } g \in G. \tag{10.17}$$

A subgroup that is the union of a set of conjugacy classes is invariant.

The **center** C of a group G is the set of all elements $c \in G$ that **commute** with every element g of the group, that is, their **commutators**

$$[c, g] \equiv cg - gc = 0 \tag{10.18}$$

vanish for all $g \in G$.

Example 10.8 (Centers are abelian subgroups) Does the center C always form an abelian subgroup of its group G? The product $c_1 c_2$ of any two elements c_1 and c_2 of the center must commute with every element g of G since $c_1 c_2 g = c_1 g c_2 = g c_1 c_2$. So the center is closed under multiplication. The identity element e commutes with every $g \in G$, so $e \in C$. If $c' \in C$, then $c'g = gc'$ for all $g \in G$, and so multiplication of this equation from the left and the right by c'^{-1} gives $gc'^{-1} = c'^{-1}g$, which shows that $c'^{-1} \in C$. The subgroup C is abelian because each of its elements commutes with all the elements of G including those of C itself. □

So the center of any group always is one of its abelian invariant subgroups. The center may be trivial, however, consisting either of the identity or of the whole group. But a group with a nontrivial center can not be simple or semisimple (section 10.23).

10.5 Cosets

If H is a subgroup of a group G, then for every element $g \in G$ the set of elements $Hg \equiv \{hg|h \in H, g \in G\}$ is a **right coset of the subgroup $H \subset G$**. (Here $\subset$ means *is a subset of* or equivalently *is contained in*.)

If H is a subgroup of a group G, then for every element $g \in G$ the set of elements gH is a **left coset of the subgroup $H \subset G$**.

The number of elements in a coset is the same as the number of elements of H, which is the order of H.

An element g of a group G is in one and only one right coset (and in one and only one left coset) of the subgroup $H \subset G$. For suppose instead that g were in two right cosets $g \in Hg_1$ and $g \in Hg_2$, so that $g = h_1 g_1 = h_2 g_2$ for suitable $h_1, h_2 \in H$ and $g_1, g_2 \in G$. Then since H is a (sub)group, we have $g_2 = h_2^{-1} h_1 g_1 = h_3 g_1$, which says that $g_2 \in Hg_1$. But this means that every element $hg_2 \in Hg_2$ is of the form $hg_2 = hh_3 g_1 = h_4 g_1 \in Hg_1$. So every element $hg_2 \in Hg_2$ is in Hg_1: the two right cosets are identical, $Hg_1 = Hg_2$.

The right (or left) cosets are the points of the **quotient coset space** G/H.

If H is an invariant subgroup of G, then by definition (10.16) $Hg = gH$ for all $g \in G$, and so the left cosets are the same sets as the right cosets. In this case, the coset space G/H is itself a group with multiplication defined by

$$\begin{aligned}(Hg_1)(Hg_2) &= \left\{h_i g_1 h_j g_2 | h_i, h_j \in H\right\} \\ &= \left\{h_i g_1 h_j g_1^{-1} g_1 g_2 | h_i, h_j \in H\right\} \\ &= \{h_i h_k g_1 g_2 | h_i, h_k \in H\} \\ &= \{h_\ell g_1 g_2 | h_\ell \in H\} = Hg_1 g_2,\end{aligned} \tag{10.19}$$

which is the multiplication rule of the group G. This group G/H is called the **factor group of G by H**.

10.6 Morphisms

An **isomorphism** is a one-to-one map between groups that respects their multiplication laws. For example, the relation between two equivalent representations

$$D'(g) = S^{-1} D(g) S \tag{10.20}$$

is an isomorphism (exercise 10.8). An **automorphism** is an isomorphism between a group and itself. The map $g_i \to g g_i g^{-1}$ is one to one because $g g_1 g^{-1} = g g_2 g^{-1}$ implies that $g g_1 = g g_2$, and so that $g_1 = g_2$. This map also preserves the law of multiplication since $g g_1 g^{-1} g g_2 g^{-1} = g g_1 g_2 g^{-1}$. So the map

$$G \to gGg^{-1} \tag{10.21}$$

is an automorphism. It is called an **inner automorphism** because g is an element of G. An automorphism not of this form (10.21) is called an **outer automorphism**.

10.7 Schur's lemma

Part 1 If D_1 and D_2 are inequivalent, irreducible representations of a group G, and if $D_1(g)A = AD_2(g)$ for some matrix A and for all $g \in G$, then the matrix A must vanish, $A = 0$.

Proof First suppose that A annihilates some vector $|x\rangle$, that is, $A|x\rangle = 0$. Let P be the projection operator into the subspace that A annihilates, which is of at least one dimension. This subspace, incidentally, is called the **null space** $\mathcal{N}(A)$ or the **kernel** of the matrix A. The representation D_2 must leave this null space $\mathcal{N}(A)$ invariant since

$$AD_2(g)P = D_1(g)AP = 0. \tag{10.22}$$

If $\mathcal{N}(A)$ were a proper subspace, then it would be a proper invariant subspace of the representation D_2, and so D_2 would be reducible, which is contrary to our assumption that D_1 and D_2 are irreducible. So the null space $\mathcal{N}(A)$ must be the whole space upon which A acts, that is, $A = 0$.

A similar argument shows that if $\langle y|A = 0$ for some bra $\langle y|$, then $A = 0$.

So either A is zero or it annihilates no ket and no bra. In the latter case, A must be square and invertible, which would imply that $D_2(g) = A^{-1}D_1(g)A$, that is, that D_1 and D_2 are equivalent representations, which is contrary to our assumption that they are inequivalent. The only way out is that A vanishes.

Part 2 If for a finite-dimensional, irreducible representation $D(g)$ of a group G, we have $D(g)A = AD(g)$ for some matrix A and for all $g \in G$, then $A = cI$. That is, any matrix that commutes with every element of a finite-dimensional, irreducible representation must be a multiple of the identity matrix.

Proof Every square matrix A has at least one eigenvector $|x\rangle$ and eigenvalue c so that $A|x\rangle = c|x\rangle$ because its characteristic equation $\det(A - cI) = 0$ always has at least one root by the fundamental theorem of algebra (5.73). So the null space $\mathcal{N}(A - cI)$ has dimension greater than zero. The assumption $D(g)A = AD(g)$ for all $g \in G$ implies that $D(g)(A - cI) = (A - cI)D(g)$ for all $g \in G$. Let P be the projection operator onto the null space $\mathcal{N}(A - cI)$. Then we have $(A - cI)D(g)P = D(g)(A - cI)P = 0$ for all $g \in G$, which implies that $D(g)P$ maps vectors into the null space $\mathcal{N}(A - cI)$. This null space therefore is invariant under $D(g)$, which means that D is reducible unless the null space $\mathcal{N}(A - cI)$ is the whole space. Since by assumption D is irreducible, it follows that $\mathcal{N}(A - cI)$ is the whole space, that is, that $A = cI$ (Issai Schur, 1875–1941).

Example 10.9 (Schur, Wigner, and Eckart) Suppose an arbitrary observable O is invariant under the action of the rotation group $SU(2)$ represented by unitary operators $U(g)$ for $g \in SU(2)$

$$U^{\dagger}(g) O U(g) = O \quad \text{or} \quad [O, U(g)] = 0. \tag{10.23}$$

These unitary rotation operators commute with the square $\boldsymbol{J}^2$ of the angular momentum $[\boldsymbol{J}^2, U] = 0$. Suppose that they also leave the hamiltonian H unchanged $[H, U] = 0$. Then as shown in example 10.7, the state $U|E, j, m\rangle$ is a sum of states all with the same values of j and E. It follows that

$$\sum_{m'} \langle E, j, m|O|E', j', m'\rangle \langle E', j', m'|U(g)|E', j', m''\rangle$$
$$= \sum_{m'} \langle E, j, m|U(g)|E, j, m'\rangle \langle E, j, m'|O|E', j', m''\rangle \tag{10.24}$$

or more simply in view of (10.13)

$$\sum_{m'} \langle E, j, m|O|E', j', m'\rangle D^{j'}(g)_{m'm''} = \sum_{m'} D^{(j)}(g)_{mm'} \langle E, j, m'|O|E', j', m''\rangle. \tag{10.25}$$

Now Part 1 of Schur's lemma tells us that the matrix $\langle E, j, m|O|E', j', m'\rangle$ must vanish unless the representations are equivalent, which is to say unless $j = j'$. So we have

$$\sum_{m'} \langle E, j, m|O|E', j, m'\rangle D^{j}(g)_{m'm''}$$
$$= \sum_{m'} D^{(j)}(g)_{mm'} \langle E, j, m'|O|E', j, m''\rangle. \tag{10.26}$$

Now Part 2 of Schur's lemma tells us that the matrix $\langle E, j, m|O|E', j, m'\rangle$ must be a multiple of the identity. Thus the symmetry of O under rotations simplifies the matrix element to

$$\langle E, j, m|O|E', j', m'\rangle = \delta_{jj'} \delta_{mm'} O_j(E, E'). \tag{10.27}$$

This result is a special case of the **Wigner–Eckart theorem** (Eugene Wigner, 1902–1995; Carl Eckart, 1902–1973). □

10.8 Characters

Suppose the $n \times n$ matrices $D_{ij}(g)$ form a representation of a group $G \ni g$. The **character** $\chi_D(g)$ of the matrix $D(g)$ is the trace

$$\chi_D(g) = \mathrm{Tr} D(g) = \sum_{i=1}^{n} D_{ii}(g). \tag{10.28}$$

Traces are cyclic, that is, $\mathrm{Tr}ABC = \mathrm{Tr}BCA = \mathrm{Tr}CAB$. So if two representations D and D' are equivalent, so that $D'(g) = S^{-1}D(g)S$, then they have the same characters because

$$\chi_{D'}(g) = \mathrm{Tr}D'(g) = \mathrm{Tr}\left(S^{-1}D(g)S\right) = \mathrm{Tr}\left(D(g)SS^{-1}\right) = \mathrm{Tr}D(g) = \chi_D(g). \tag{10.29}$$

If two group elements g_1 and g_2 are in the same conjugacy class, that is, if $g_2 = gg_1g^{-1}$ for all $g \in G$, then they have the same character in a given representation $D(g)$ because

$$\begin{aligned}\chi_D(g_2) &= \mathrm{Tr}D(g_2) = \mathrm{Tr}D(gg_1g^{-1}) = \mathrm{Tr}\left(D(g)D(g_1)D(g^{-1})\right)\\ &= \mathrm{Tr}\left(D(g_1)D^{-1}(g)D(g)\right) = \mathrm{Tr}D(g_1) = \chi_D(g_1).\end{aligned} \tag{10.30}$$

10.9 Tensor products

Suppose $D_1(g)$ is a k-dimensional representation of a group G, and $D_2(g)$ is an n-dimensional representation of the same group. Suppose the vectors $|\ell\rangle$ for $\ell = 1, \dots, k$ are the basis vectors of the k-dimensional space V_k on which $D_1(g)$ acts, and that the vectors $|m\rangle$ for $m = 1, \dots, n$ are the basis vectors of the n-dimensional space V_n on which $D_2(g)$ acts. The $k \times n$ vectors $|\ell, m\rangle$ are basis vectors for the kn-dimensional tensor-product space V_{kn}. The matrices $D_{D_1 \otimes D_2}(g)$ defined as

$$\langle \ell', m'|D_{D_1 \otimes D_2}(g)|\ell, m\rangle = \langle \ell'|D_1(g)|\ell\rangle\langle m'|D_2(g)|m\rangle \tag{10.31}$$

act in this kn-dimensional space V_{kn} and form a representation of the group G; this tensor-product representation usually is reducible. Many tricks help one to decompose reducible tensor-product representations into direct sums of irreducible representations (Georgi, 1999, p. 309).

Example 10.10 (Adding angular momenta) The addition of angular momenta illustrates both the tensor product and its reduction to a direct sum of irreducible representations. Let $D_{j_1}(g)$ and $D_{j_2}(g)$ respectively be the $(2j_1 + 1) \times (2j_1 + 1)$ and the $(2j_2 + 1) \times (2j_2 + 1)$ representations of the rotation group $SU(2)$. The tensor-product representation $D_{D_{j_1} \otimes D_{j_2}}$

$$\langle m_1', m_2'|D_{D_{j_1} \otimes D_{j_2}}|m_1, m_2\rangle = \langle m_1'|D_{j_1}(g)|m_1\rangle\langle m_2'|D_{j_2}(g)|m_2\rangle \tag{10.32}$$

is reducible into a direct sum of all the irreducible representations of $SU(2)$ from $D_{j_1+j_2}(g)$ down to $D_{|j_1-j_2|}(g)$ in integer steps:

$$D_{D_{j_1} \otimes D_{j_2}} = D_{j_1+j_2} \oplus D_{j_1+j_2-1} \oplus \cdots \oplus D_{|j_1-j_2|+1} \oplus D_{|j_1-j_2|}, \tag{10.33}$$

each irreducible representation occurring once in the direct sum. □

Example 10.11 (Adding two spins) When one adds $j_1 = 1/2$ to $j_2 = 1/2$, one finds that the tensor-product matrix $D_{D_{1/2} \otimes D_{1/2}}$ is equivalent to the direct sum $D_1 \oplus D_0$

$$D_{D_{1/2} \otimes D_{1/2}}(\boldsymbol{\theta}) = S^{-1} \begin{pmatrix} D_1(\boldsymbol{\theta}) & 0 \\ 0 & D_0(\boldsymbol{\theta}) \end{pmatrix} S \tag{10.34}$$

where the matrices S, D_1, and D_0 respectively are 4×4, 3×3, and 1×1. □

10.10 Finite groups

A **finite group** is one that has a finite number of elements. The number of elements in a group is the **order** of the group.

Example 10.12 (Z_2) The group $\boldsymbol{Z_2}$ consists of two elements e and p with multiplication rules

$$ee = e, \quad ep = p, \quad pe = p, \quad pp = e. \tag{10.35}$$

Clearly, Z_2 is abelian, and its order is 2. The identification $e \to 1$ and $p \to -1$ gives a 1-dimensional representation of the group Z_2 in terms of 1×1 matrices, which are just numbers. □

It is tedious to write the multiplication rules as individual equations. Normally people compress them into a multiplication table like this:

×	e	p
e	e	p
p	p	e

, (10.36)

A simple generalization of Z_2 is the group $\boldsymbol{Z_n}$ whose elements may be represented as $\exp(i2\pi m/n)$ for $m = 1, \ldots, n$. This group is also abelian, and its order is n.

Example 10.13 (Z_3) The multiplication table for $\boldsymbol{Z_3}$ is

×	e	a	b
e	e	a	b
a	a	b	e
b	b	e	a

, (10.37)

whichsays that $a^2 = b$, $b^2 = a$, and $ab = ba = e$. □

10.11 The regular representation

For any finite group G we can associate an orthonormal vector $|g_i\rangle$ with each element g_i of the group. So $\langle g_i|g_j\rangle = \delta_{ij}$. These orthonormal vectors $|g_i\rangle$ form a basis for a vector space whose dimension is the order of the group. The matrix $D(g_k)$ of the regular representation of G is defined to map any vector $|g_i\rangle$ into the vector $|g_k g_i\rangle$ associated with the product $g_k g_i$

$$D(g_k)|g_i\rangle = |g_k g_i\rangle. \tag{10.38}$$

Since group multiplication is associative, we have

$$D(g_j)D(g_k)|g_i\rangle = D(g_j)|g_k g_i\rangle = |g_j(g_k g_i)\rangle = |(g_j g_k)g_i)\rangle = D(g_j g_k)|g_i\rangle. \tag{10.39}$$

Because the vector $|g_i\rangle$ was an arbitrary basis vector, it follows that

$$D(g_j)D(g_k) = D(g_j g_k), \tag{10.40}$$

which means that the matrices $D(g)$ satisfy the closure criterion (10.4) for their being a representation of the group G. The matrix $D(g)$ has entries

$$[D(g)]_{ij} = \langle g_i|D(g)|g_j\rangle. \tag{10.41}$$

The sum of dyadics $|g_\ell\rangle\langle g_\ell|$ over all the elements g_ℓ of a finite group G is the unit matrix

$$\sum_{g_\ell \in G} |g_\ell\rangle\langle g_\ell| = I_n, \tag{10.42}$$

in which n is the order of G, that is, the number of elements in G. So by taking the m, n matrix element of the multiplication law (10.40), we find

$$\begin{aligned}
[D(g_j g_k)]_{m,n} &= \langle g_m|D(g_j g_k)|g_n\rangle = \langle g_m|D(g_j)D(g_k)|g_n\rangle \\
&= \sum_{g_\ell \in G} \langle g_m|D(g_j)|g_\ell\rangle\langle g_\ell|D(g_k)|g_n\rangle \\
&= \sum_{g_\ell \in G} [D(g_j)]_{m,\ell}[D(g_k)]_{\ell,n}.
\end{aligned} \tag{10.43}$$

Example 10.14 (Z_3's regular representation) The regular representation of Z_3 is

$$D(e) = \begin{pmatrix} 1 & 0 & 0 \\ 0 & 1 & 0 \\ 0 & 0 & 1 \end{pmatrix}, \quad D(a) = \begin{pmatrix} 0 & 0 & 1 \\ 1 & 0 & 0 \\ 0 & 1 & 0 \end{pmatrix}, \quad D(b) = \begin{pmatrix} 0 & 1 & 0 \\ 0 & 0 & 1 \\ 1 & 0 & 0 \end{pmatrix} \tag{10.44}$$

so $D(a)^2 = D(b)$, $D(b)^2 = D(a)$, and $D(a)D(b) = D(b)D(a) = D(e)$. □

10.12 Properties of finite groups

In his book (Georgi, 1999, ch. 1), Georgi proves the following theorems.

1 Every representation of a finite group is equivalent to a unitary representation.
2 Every representation of a finite group is completely reducible.
3 The irreducible representations of a finite abelian group are one dimensional.
4 If $D^{(a)}(g)$ and $D^{(b)}(g)$ are two unitary irreducible representations of dimensions n_a and n_b of a group G of N elements $g_1, \ldots, g_N$, then the functions

$$\sqrt{\frac{n_a}{N}}\, D^{(a)}_{jk}(g) \tag{10.45}$$

are orthonormal and complete in the sense that

$$\frac{n_a}{N} \sum_{j=1}^{N} D^{(a)*}_{ik}(g_j) D^{(b)}_{\ell m}(g_j) = \delta_{ab}\delta_{i\ell}\delta_{km}. \tag{10.46}$$

5 The order N of a finite group is the sum of the squares of the dimensions of its inequivalent irreducible representations

$$N = \sum_a n_a^2. \tag{10.47}$$

Example 10.15 (Z_N) The abelian cyclic group Z_N with elements

$$g_j = e^{2\pi ij/N} \tag{10.48}$$

has N one-dimensional irreducible representations

$$D^{(a)}(g_j) = e^{2\pi iaj/N} \tag{10.49}$$

for $a = 1, 2, \ldots, N$. Their orthonormality relation (10.46) is the Fourier formula

$$\frac{1}{N} \sum_{j=1}^{N} e^{-2\pi iaj/N} e^{2\pi ibj/N} = \delta_{ab}. \tag{10.50}$$

The n_a are all unity, there are N of them, and the sum of the n_a^2 is N as required by the sum rule (10.47). □

10.13 Permutations

The permutation group on n objects is called S_n. Permutations are made of **cycles** that change the order of some of the n objects. For instance, the permutation $(1\,2) = (2\,1)$ is a 2-cycle that means $x_1 \to x_2 \to x_1$; the unitary operator

$U((1\,2))$ that represents it interchanges states like this:

$$U((1\,2))|+,-\rangle = U((1\,2))|+,1\rangle\,|-,2\rangle = |-,1\rangle,|+,2\rangle = |-,+\rangle. \tag{10.51}$$

The 2-cycle (3 4) means $x_3 \to x_4 \to x_3$, it changes (a,b,c,d) into (a,b,d,c). The 3-cycle $(1\,2\,3) = (2\,3\,1) = (3\,1\,2)$ means $x_1 \to x_2 \to x_3 \to x_1$, it changes (a,b,c,d) into (b,c,a,d). The 4-cycle $(1\,3\,2\,4)$ means $x_1 \to x_3 \to x_2 \to x_4 \to x_1$ and changes (a,b,c,d) into (c,d,b,a). The 1-cycle (2) means $x_2 \to x_2$ and leaves everything unchanged.

The identity element of S_n is the product of 1-cycles $e = (1)(2)\ldots(n)$. The inverse of the cycle $(1\,3\,2\,4)$ must invert $x_1 \to x_3 \to x_2 \to x_4 \to x_1$, so it must be $(1\,4\,2\,3)$, which means $x_1 \to x_4 \to x_2 \to x_3 \to x_1$ so that it changes (c,d,b,a) back into (a,b,c,d). Every element of S_n has each integer from 1 to n in one and only one cycle. So an arbitrary element of S_n with ℓ_k k-cycles must satisfy

$$\sum_{k=1}^{n} k\,\ell_k = n. \tag{10.52}$$

10.14 Compact and noncompact Lie groups

Imagine rotating an object repeatedly. Notice that the biggest rotation is by an angle of $\pm\pi$ about some axis. The possible angles form a circle; the space of parameters is a circle. The parameter space of a compact group is compact – closed and bounded. The rotations form a **compact group**.

Now consider the translations. Imagine moving a pebble to the Sun, then moving it to the next-nearest star, then moving it to the nearest galaxy. If space is flat, then there is no limit to how far one can move a pebble. The parameter space of a noncompact group is not compact. The translations form a **noncompact group**.

We'll see that compact Lie groups possess unitary representations, with $N \times N$ unitary matrices $D(\alpha)$, while noncompact ones don't.

10.15 Lie algebra

Continuous groups can be very complicated. So one uses not only algebra but also calculus, and one studies the simplest part of the group – the elements $g(d\alpha)$ that are near the identity $e = g(0)$ for which all $\alpha_a = 0$.

If $D(g(\{\alpha_a\}))$ is a representation of a Lie group with parameters $\{\alpha_a\}$, it gets tedious to write $D(g(\{\alpha_a\}))$ over and over. So instead one writes $g(\alpha) = g(\{\alpha_a\})$ and

$$D(\alpha) = D(g(\alpha)) = D(g(\{\alpha_a\})) \tag{10.53}$$

leaving out the explicit mentions both of g and of $\{\alpha_a\}$.

Any matrix $D(d\alpha)$ representing a group element $g(d\alpha)$ that is near the identity is approximately

$$D(d\alpha) = I + i\sum_a t_a \, d\alpha_a \tag{10.54}$$

where the **generators** t_a of the group are the partial derivatives

$$t_a = -i \left. \frac{\partial}{\partial \alpha_a} D(\alpha) \right|_{\alpha=0} . \tag{10.55}$$

The i is inserted so that if the matrices $D(\alpha)$ are unitary, then the generators are hermitian matrices

$$t_a^\dagger = t_a. \tag{10.56}$$

Compact groups have finite-dimensional, unitary representations and hermitian generators.

Our formulas will look nicer if we adopt the convention that we sum over all indices that occur twice in a monomial. That is, we drop the summation symbol when summing over a repeated index so that (10.54) looks like this

$$D(d\alpha) = I + i \, t_a \, d\alpha_a. \tag{10.57}$$

Unless the parameters α_a are redundant, the $N(G)$ generators are linearly independent. They span a vector space over the real numbers and any linear combination $t = \alpha_a t_a$ may be called a generator.

By using the Gram–Schmidt procedure, we may make the $N(G)$ generators t_a orthogonal with respect to the inner product (1.86)

$$(t_a, t_b) = \mathrm{Tr}\left(t_a^\dagger t_b\right) = k \, \delta_{ab}, \tag{10.58}$$

in which k is a nonnegative normalization constant that in general depends upon the representation. The reason why we don't normalize the generators and so make k unity will become apparent shortly.

Since group multiplication is closed, any power $g^n(d\alpha) \in G$, and so we may take the limit

$$D(\alpha) = \lim_{n\to\infty} D^n(\alpha/n) = \left(I + \frac{i\alpha_a t_a}{n}\right)^n = e^{i\alpha_a t_a}. \tag{10.59}$$

This parametrization of a representation of a group is called the **exponential parametrization**.

Now for tiny ϵ the product

$$\begin{aligned} e^{i\epsilon t_b} e^{i\epsilon t_a} e^{-i\epsilon t_b} e^{-i\epsilon t_a} \approx & \left(1 + i\epsilon \, t_b - \frac{\epsilon^2}{2} t_b^2\right)\left(1 + i\epsilon \, t_a - \frac{\epsilon^2}{2} t_a^2\right) \\ & \times \left(1 - i\epsilon \, t_b - \frac{\epsilon^2}{2} t_b^2\right)\left(1 - i\epsilon \, t_a - \frac{\epsilon^2}{2} t_a^2\right) \end{aligned} \tag{10.60}$$

to order ϵ^2 is

$$e^{i\epsilon t_b}\, e^{i\epsilon t_a}\, e^{-i\epsilon t_b}\, e^{-i\epsilon t_a} \approx 1 + \epsilon^2(t_a\, t_b - t_b\, t_a) = 1 + \epsilon^2[t_a, t_b]. \tag{10.61}$$

Since this product represents a group element near the identity, the commutator must be a linear combination of generators of order ϵ^2

$$e^{i\epsilon t_b}\, e^{i\epsilon t_a}\, e^{-i\epsilon t_b}\, e^{-i\epsilon t_a} = e^{i\epsilon^2 f^c_{ab} t_c} \approx 1 + i\epsilon^2 f^c_{ab}\, t_c. \tag{10.62}$$

By matching (10.61) with (10.62) we have

$$[t_a, t_b] = i f^c_{ab}\, t_c. \tag{10.63}$$

The numbers f^c_{ab} are the **structure constants** of the group G.

By taking the trace of equation (10.63) multiplied by $t^\dagger_d$ and by using the orthogonality relation (10.58), we find

$$\mathrm{Tr}\left([t_a, t_b]\, t^\dagger_d\right) = i f^c_{ab}\, \mathrm{Tr}\left(t_c\, t^\dagger_d\right) = i f^c_{ab}\, k\, \delta_{cd} = ik f^d_{ab}, \tag{10.64}$$

which implies that the structure constant f^c_{ab} is the trace

$$f^c_{ab} = \left(-\frac{i}{k}\right) \mathrm{Tr}\left([t_a, t_b]\, t^\dagger_c\right). \tag{10.65}$$

Because of the antisymmetry of the commutator $[t_a, t_b]$, the structure constant f^c_{ab} is **antisymmetric in its lower indices**

$$f^c_{ab} = -f^c_{ba}. \tag{10.66}$$

From any $n \times n$ matrix A, one may make a hermitian matrix $A + A^\dagger$ and an antihermitian one $A - A^\dagger$. Thus, one may separate the $N(G)$ generators into a set that are hermitian $t^{(h)}_a$ and a set that are antihermitian $t^{(ah)}_a$. The exponential of any imaginary linear combination of $n \times n$ hermitian generators $D(\alpha) = \exp\left(i\alpha_a\, t^{(h)}_a\right)$ is an $n \times n$ unitary matrix since

$$D^\dagger(\alpha) = \exp\left(-i\alpha_a\, t^{\dagger(h)}_a\right) = \exp\left(-i\alpha_a\, t^{(h)}_a\right) = D^{-1}(\alpha). \tag{10.67}$$

A group with only hermitian generators is **compact** and has finite-dimensional unitary representations.

On the other hand, the exponential of any imaginary linear combination of antihermitian generators $D(\alpha) = \exp\left(i\alpha_a\, t^{(ah)}_a\right)$ is a real exponential of their hermitian counterparts $i\, t^{(ah)}_a$ whose squared norm

$$\|D(\alpha)\|^2 = \mathrm{Tr}\left[D(\alpha)^\dagger D(\alpha)\right] = \mathrm{Tr}\left[\exp\left(2\alpha_a\, i t^{(ah)}_a\right)\right] \tag{10.68}$$

grows exponentially and without limit as the parameters $\alpha_a \to \pm\infty$. A group with some antihermitian generators is **noncompact** and does not have finite-dimensional unitary representations. (The unitary representations of the translations and of the Lorentz and Poincaré groups are infinite dimensional.)

Compact Lie groups have hermitian generators, and so the structure-constant formula (10.65) reduces in this case to

$$f_{ab}^{c} = (-i/k)\mathrm{Tr}\left([t_a, t_b]\, t_c^{\dagger}\right) = (-i/k)\mathrm{Tr}\left([t_a, t_b]\, t_c\right). \tag{10.69}$$

Now, since the trace is cyclic, we have

$$\begin{aligned} f_{ac}^{b} &= (-i/k)\mathrm{Tr}\left([t_a, t_c]\, t_b\right) = (-i/k)\mathrm{Tr}\left(t_a t_c t_b - t_c t_a t_b\right) \\ &= (-i/k)\mathrm{Tr}\left(t_b t_a t_c - t_a t_b t_c\right) \\ &= (-i/k)\mathrm{Tr}\left([t_b, t_a]\, t_c\right) = f_{ba}^{c} = -f_{ab}^{c}. \end{aligned} \tag{10.70}$$

Interchanging a and b, we get

$$f_{bc}^{a} = f_{ab}^{c} = -f_{ba}^{c}. \tag{10.71}$$

Finally, interchanging b and c gives

$$f_{ab}^{c} = f_{ca}^{b} = -f_{ac}^{b}. \tag{10.72}$$

Combining (10.70, 10.71, & 10.72), we see that **the structure constants of a compact Lie group are totally antisymmetric**

$$f_{ac}^{b} = -f_{ca}^{b} = f_{ba}^{c} = -f_{ab}^{c} = -f_{bc}^{a} = f_{cb}^{a}. \tag{10.73}$$

Because of this antisymmetry, it is usual to lower the upper index

$$f_{abc} \equiv f_{ab}^{c} \tag{10.74}$$

and write the antisymmetry of the structure constants of compact Lie groups more simply as

$$f_{acb} = -f_{cab} = f_{bac} = -f_{abc} = -f_{bca} = f_{cba}. \tag{10.75}$$

For compact Lie groups, the generators are hermitian, and so the **structure constants f_{abc} are real**, as we may see by taking the complex conjugate of equation (10.69)

$$f_{abc}^{*} = (i/k)\mathrm{Tr}\left(t_c\, [t_b, t_a]\right) = (-i/k)\mathrm{Tr}\left([t_a, t_b]\, t_c\right) = f_{abc}. \tag{10.76}$$

All the representations of a given group must obey the same multiplication law, that of the group. Thus in exponential parametrization, if the representation D_1 satisfies (10.62)

$$e^{i\epsilon t_b}\, e^{i\epsilon t_a}\, e^{-i\epsilon t_b}\, e^{-i\epsilon t_a} \approx e^{i\epsilon^2 f_{ab}^{c} t_c}, \tag{10.77}$$

that is, if with ϵ_a being the vector with kth component $\epsilon\delta_{ak}$ and ϵ_b being the vector with kth component $\epsilon\delta_{bk}$, we have

$$D_1(\epsilon_b)\, D_1(\epsilon_a)\, D_1(-\epsilon_b)\, D_1(-\epsilon_a) \approx D_1(\epsilon^2 f^c_{ab}) \tag{10.78}$$

then any other representation D_2 must satisfy the same relation with 2 replacing 1:

$$D_2(\epsilon_b) D_2(\epsilon_a) D_2(-\epsilon_b) D_2(-\epsilon_a) \approx D_2(\epsilon^2 f^c_{ab}). \tag{10.79}$$

Such uniformity will occur if the **structure constants** (10.65) **are the same for all representations of a compact or a noncompact Lie group**. To ensure that this is so, we must allow each representation $D_r(\alpha)$ to have its own normalization parameter k_r in the trace relation (10.65). The structure constants f_{abc} then are a property of the group G and are independent of the particular representation $D_r(\alpha)$. This is why we didn't make the generators t_a orthonormal.

It follows from (10.63 & 10.74–10.76) that **the commutator of any two generators of a Lie group is a linear combination**

$$[t_a, t_b] = i f^c_{ab}\, t_c \tag{10.80}$$

of its generators t_c, and that the structure constants $f_{abc} = f^c_{ab}$ are real and totally antisymmetric if the group is compact.

Example 10.16 (Gauge transformation) The action density of a Yang–Mills theory is unchanged when a space-time dependent unitary matrix $U(x)$ changes a vector $\psi(x)$ of matter fields to $\psi'(x) = U(x)\psi(x)$. Terms like $\psi^\dagger\psi$ are invariant because $\psi^\dagger(x)U^\dagger(x)U(x)\psi(x) = \psi^\dagger(x)\psi(x)$, but how can kinetic terms like $\partial_i\psi^\dagger\, \partial^i\psi$ be made invariant? Yang and Mills introduced matrices A_i of gauge fields, replaced ordinary derivatives ∂_i by **covariant derivatives** $D_i \equiv \partial_i + A_i$, and required that $D'_i\psi' = UD_i\psi$ or that

$$\left(\partial_i + A'_i\right) U = \partial_i U + U\partial_i + A'_i U = U\left(\partial_i + A_i\right). \tag{10.81}$$

Their nonabelian gauge transformation is

$$\begin{aligned} \psi'(x) &= U(x)\psi(x) \\ A'_i(x) &= U(x)A_i(x)U^\dagger(x) - \left(\partial_i U(x)\right) U^\dagger(x). \end{aligned} \tag{10.82}$$

One often writes the unitary matrix as $U(x) = \exp(-ig\,\theta_a(x)\, t_a)$ in which g is a coupling constant, the functions $\theta_a(x)$ parametrize the gauge transformation, and the generators t_a belong to the representation that acts on the vector $\psi(x)$ of matter fields. □

10.16 The rotation group

The rotations and reflections in three-dimensional space form a compact group $O(3)$ whose elements R are real 3×3 matrices that leave invariant the dot-product of any two 3-vectors

$$(Rx) \cdot (Ry) = x^{\mathsf{T}} R^{\mathsf{T}} R y = x^{\mathsf{T}} I y = x \cdot y. \tag{10.83}$$

These matrices therefore are orthogonal (1.168)

$$R^{\mathsf{T}} R = I. \tag{10.84}$$

Taking the determinant of both sides and using the transpose (1.194) and product (1.207) rules, we have

$$(\det R)^2 = 1 \tag{10.85}$$

whence $\det R = \pm 1$. The subgroup with $\det R = 1$ is the group $SO(3)$. An $SO(3)$ element near the identity $R = I + \omega$ must satisfy

$$(I + \omega)^{\mathsf{T}} (I + \omega) = I. \tag{10.86}$$

Neglecting the tiny quadratic term, we find that the infinitesimal matrix ω is antisymmetric

$$\omega^{\mathsf{T}} = - \omega. \tag{10.87}$$

One complete set of real 3×3 antisymmetric matrices is

$$\omega_1 = \begin{pmatrix} 0 & 0 & 0 \\ 0 & 0 & -1 \\ 0 & 1 & 0 \end{pmatrix}, \quad \omega_2 = \begin{pmatrix} 0 & 0 & 1 \\ 0 & 0 & 0 \\ -1 & 0 & 0 \end{pmatrix}, \quad \omega_3 = \begin{pmatrix} 0 & -1 & 0 \\ 1 & 0 & 0 \\ 0 & 0 & 0 \end{pmatrix}, \tag{10.88}$$

which we may write as

$$[\omega_b]_{ac} = \epsilon_{abc}, \tag{10.89}$$

in which ϵ_{abc} is the **Levi-Civita symbol**, which is totally antisymmetric with $\epsilon_{123} = 1$ (Tullio Levi-Civita, 1873–1941). The ω_b are antihermitian, but we make them hermitian by multiplying by i

$$t_b = i\, \omega_b \tag{10.90}$$

so that $R = I - i\theta_b\, t_b$.

The three hermitian generators t_a satisfy (exercise 10.15) the commutation relations

$$[t_a, t_b] = i\, f_{abc}\, t_c \tag{10.91}$$

in which the structure constants are given by the Levi-Civita symbol ϵ_{abc}

$$f_{abc} = \epsilon_{abc} \tag{10.92}$$

so that

$$[t_a, t_b] = i\epsilon_{abc}\, t_c. \tag{10.93}$$

They are the generators of $SO(3)$ in the **adjoint representation** (section 10.20).

Physicists usually scale the generators by $\hbar$ and define the angular-momentum generator L_a as

$$L_a = \hbar t_a \tag{10.94}$$

so that the eigenvalues of the angular-momentum operators are the physical values of the angular momenta. With $\hbar$, the commutation relations are

$$[L_a, L_b] = i\hbar\, \epsilon_{abc}\, L_c. \tag{10.95}$$

The matrix that represents a right-handed rotation (of an object) of angle $\theta = |\boldsymbol{\theta}|$ about an axis $\boldsymbol{\theta}$ is

$$D(\boldsymbol{\theta}) = e^{-i\boldsymbol{\theta}\cdot \boldsymbol{t}} = e^{-i\boldsymbol{\theta}\cdot \boldsymbol{L}/\hbar}, \tag{10.96}$$

By using the fact (1.264) that a matrix obeys its characteristic equation, one may show (exercise 10.17) that the 3×3 matrix $D(\boldsymbol{\theta})$ that represents a right-handed rotation of θ radians about the axis $\boldsymbol{\theta}$ is

$$D_{ij}(\boldsymbol{\theta}) = \cos\theta\, \delta_{ij} - \sin\theta\, \epsilon_{ijk}\, \theta_k/\theta + (1 - \cos\theta)\, \theta_i\theta_j/\theta^2, \tag{10.97}$$

in which a sum over $k = 1, 2, 3$ is understood.

Example 10.17 (Demonstration of commutation relations) Take a big sphere with a distinguished point and orient the sphere so that the point lies in the y-direction from the center of the sphere. Now rotate the sphere by a small angle, say 15 degrees or $\epsilon = \pi/12$, right-handedly about the x-axis, then right-handedly about the y-axis by the same angle, then left-handedly about the x-axis and then left-handedly about the y-axis. These rotations amount to a smaller, left-handed rotation about the (vertical) z-axis in accordance with equation (10.77) with $\hbar t_a = L_1 = L_x$, $\hbar t_b = L_2 = L_y$, and $\hbar f_{abc} t_c = \epsilon_{12c} L_c = L_3 = L_z$

$$e^{i\epsilon L_y/\hbar}\, e^{i\epsilon L_x/\hbar}\, e^{-i\epsilon L_y/\hbar}\, e^{-i\epsilon L_x/\hbar} \approx e^{i\epsilon^2 L_z/\hbar}. \tag{10.98}$$

The magnitude of that rotation should be about $\epsilon^2 = (\pi/12)^2 \approx 0.069$ or about 3.9 degrees. Photographs of an actual demonstration are displayed in Fig. 10.1.

By expanding both sides of the demonstrated equation (10.98) in powers of ϵ and keeping only the biggest terms that don't cancel, you may show (exercise 10.16) that the generators L_x and L_y satisfy the commutation relation

$$[L_x, L_y] = i\hbar L_z \tag{10.99}$$

of the rotation group. □

Physical demonstration of the commutation relations

Figure 10.1 Demonstration of equation (10.98) and the commutation relation (10.99). Upper left: black ball with a white stick pointing in the y-direction; the x-axis is to the reader's left, the z-axis is vertical. Upper right: ball after a small right-handed rotation about the x-axis. Center left: ball after that rotation is followed by a small right-handed rotation about the y-axis. Center right: ball after these rotations are followed by a small left-handed rotation about the x-axis. Bottom: ball after these rotations are followed by a small left-handed rotation about the y-axis. The net effect is approximately a small left-handed rotation about the z-axis.

10.17 The Lie algebra and representations of $SU(2)$

The three generators of $SU(2)$ in its 2×2 defining representation are the Pauli matrices divided by 2, $t_a = \sigma_a/2$. The structure constants of $SU(2)$ are $f_{abc} = \epsilon_{abc}$, which is totally antisymmetric with $\epsilon_{123} = 1$

$$[t_a, t_b] = i f_{abc} t_c = \left[\frac{\sigma_a}{2}, \frac{\sigma_b}{2}\right] = i\epsilon_{abc}\frac{\sigma_c}{2}. \tag{10.100}$$

For every half-integer

$$j = \frac{n}{2} \quad \text{for} \quad n = 0, 1, 2, 3, \ldots \tag{10.101}$$

there is an irreducible representation of $SU(2)$

$$D^{(j)}(\boldsymbol{\theta}) = e^{-i\boldsymbol{\theta}\cdot\boldsymbol{J}^{(j)}}, \tag{10.102}$$

in which the three generators $t_a^{(j)} \equiv J_a^{(j)}$ are $(2j+1) \times (2j+1)$ square hermitian matrices. In a basis in which $J_3^{(j)}$ is diagonal, the matrix elements of the complex linear combinations $J_\pm^{(j)} \equiv J_1^{(j)} \pm iJ_2^{(j)}$ are

$$\left[J_1^{(j)} \pm iJ_2^{(j)}\right]_{s',s} = \delta_{s',s\pm1}\sqrt{(j \mp s)(j \pm s + 1)}, \tag{10.103}$$

where s and s' run from $-j$ to j in integer steps, and those of $J_3^{(j)}$ are

$$\left[J_3^{(j)}\right]_{s',s} = s\,\delta_{s',s}. \tag{10.104}$$

The sum of the squares of the three generators $J_a^{(j)}$ is a multiple of the $(2j+1) \times (2j+1)$ identity matrix

$$\left(J_a^{(j)}\right)^2 = j(j+1)\,I. \tag{10.105}$$

Combinations of generators that are a multiple of the identity are called **Casimir operators**.

Example 10.18 (Spin-two) For $j = 2$, the spin-two matrices $J_+^{(2)}$ and $J_3^{(2)}$ are

$$J_+^{(2)} = \begin{pmatrix} 0 & 2 & 0 & 0 & 0 \\ 0 & 0 & \sqrt{6} & 0 & 0 \\ 0 & 0 & 0 & \sqrt{6} & 0 \\ 0 & 0 & 0 & 0 & 2 \\ 0 & 0 & 0 & 0 & 0 \end{pmatrix} \quad \text{and} \quad J_3^{(2)} = \begin{pmatrix} 2 & 0 & 0 & 0 & 0 \\ 0 & 1 & 0 & 0 & 0 \\ 0 & 0 & 0 & 0 & 0 \\ 0 & 0 & 0 & -1 & 0 \\ 0 & 0 & 0 & 0 & -2 \end{pmatrix} \tag{10.106}$$

and $J_- = \left(J_+^{(2)}\right)^\dagger$. □

The tensor product of any two irreducible representations $D^{(j)}$ and $D^{(k)}$ of $SU(2)$ is equivalent to the direct sum of all the irreducible representations D^ℓ for $|j - k| \le \ell \le j + k$

$$D^{(j)} \otimes D^{(k)} = \bigoplus_{\ell=|j-k|}^{j+k} D^{\ell}, \tag{10.107}$$

each D^ℓ occurring once.

Under a rotation R, a field $\psi_\ell(x)$ that transforms under the $D^{(j)}$ representation of $SU(2)$ responds as

$$U(R)\, \psi_\ell(x)\, U^{-1}(R) = D^{(j)}_{\ell\ell'}(R^{-1})\, \psi_{\ell'}(Rx). \tag{10.108}$$

Example 10.19 (Spin and statistics) Suppose $|a,m\rangle$ and $|b,m\rangle$ are any eigenstates of the rotation operator J_3 with eigenvalue m (in units with $\hbar = c = 1$). Let u and v be any two points whose separation $u - v$ is space-like $(u-v)^2 > 0$. Then, in some Lorentz frame, the two points are at the same time t, and we may chose our coordinate system so that $u' = (t, x, 0, 0)$ and $v' = (t, -x, 0, 0)$. Let U be the unitary operator that represents a right-handed rotation by π about the 3-axis or z-axis of this Lorentz frame. Then

$$U|a,m\rangle = e^{-im\pi}|a,m\rangle \quad \text{and} \quad \langle b,m|U^{-1} = \langle b,m|e^{im\pi}. \tag{10.109}$$

And by (10.108), U transforms a field ψ of spin j with $\boldsymbol{x} \equiv (x,0,0)$ to

$$U(R)\, \psi_\ell(t,\boldsymbol{x})\, U^{-1}(R) = D^{(j)}_{\ell\ell'}(R^{-1})\, \psi_{\ell'}(t,-\boldsymbol{x}) = e^{i\pi\ell}\psi_\ell(t,-\boldsymbol{x}). \tag{10.110}$$

Thus by inserting the identity operator in the form $I = U^{-1}U$ and using both (10.109) and (10.108), we find, since the phase factors $\exp(-im\pi)$ and $\exp(im\pi)$ cancel,

$$\begin{aligned}\langle b,m|\psi_\ell(t,\boldsymbol{x})\, \psi_\ell(t,-\boldsymbol{x})|a,m\rangle &= \langle b,m|U\psi_\ell(t,\boldsymbol{x})U^{-1}\,U\psi_\ell(t,-\boldsymbol{x})U^{-1}|a,m\rangle \\ &= e^{2i\pi\ell}\,\langle b,m|\psi_\ell(t,-\boldsymbol{x})\psi_\ell(t,\boldsymbol{x})|a,m\rangle. \end{aligned} \tag{10.111}$$

Now if j is an integer, then so is ℓ, and the phase factor $\exp(2i\pi\ell) = 1$ is unity. In this case, we find that the mean value of the equal-time commutator vanishes

$$\langle b,m|[\psi_\ell(t,\boldsymbol{x}), \psi_\ell(t,-\boldsymbol{x})]|a,m\rangle = 0. \tag{10.112}$$

On the other hand, if j is half an odd integer, that is, $j = (2n+1)/2$, where n is an integer, then the phase factor $\exp(2i\pi\ell)$ is -1. In this case, the mean value of the equal-time anticommutator vanishes

$$\langle b,m|\{\psi_\ell(t,\boldsymbol{x}), \psi_\ell(t,-\boldsymbol{x})\}|a,m\rangle = 0. \tag{10.113}$$

While not a proof of the spin-statistics theorem, this argument shows that the behavior of fields under rotations does determine their statistics. □

10.18 The defining representation of $SU(2)$

The smallest positive value of angular momentum is $\hbar/2$. The spin-one-half angular momentum operators are represented by three 2×2 matrices

$$S_a = \frac{\hbar}{2} \sigma_a, \tag{10.114}$$

in which the σ_a are the **Pauli matrices**

$$\sigma_1 = \begin{pmatrix} 0 & 1 \\ 1 & 0 \end{pmatrix}, \quad \sigma_2 = \begin{pmatrix} 0 & -i \\ i & 0 \end{pmatrix}, \quad \text{and} \quad \sigma_3 = \begin{pmatrix} 1 & 0 \\ 0 & -1 \end{pmatrix}, \tag{10.115}$$

which obey the multiplication law

$$\sigma_i \sigma_j = \delta_{ij} + i \sum_{k=1}^{3} \epsilon_{ijk} \, \sigma_k. \tag{10.116}$$

The Pauli matrices divided by 2 satisfy the commutation relations (10.93) of the rotation group

$$\left[\frac{1}{2}\sigma_a, \frac{1}{2}\sigma_b \right] = i\epsilon_{abc} \, \frac{1}{2}\sigma_c \tag{10.117}$$

and generate the elements of the group $SU(2)$

$$\exp\left(i \, \boldsymbol{\theta} \cdot \frac{\boldsymbol{\sigma}}{2} \right) = I \, \cos\frac{\theta}{2} + i \hat{\boldsymbol{\theta}} \cdot \boldsymbol{\sigma} \, \sin\frac{\theta}{2}, \tag{10.118}$$

in which I is the 2×2 identity matrix, $\theta = \sqrt{\boldsymbol{\theta}^2}$ and $\hat{\boldsymbol{\theta}} = \boldsymbol{\theta}/\theta$.

It follows from (10.117) that the spin operators satisfy

$$[S_a, S_b] = i\hbar\epsilon_{abc} \, S_c. \tag{10.119}$$

The raising and lowering operators

$$S_\pm = S_1 \pm i S_2 \tag{10.120}$$

have simple commutators with S_3

$$[S_3, S_\pm] = \pm \hbar S_\pm. \tag{10.121}$$

This relation implies that if the state $|j, m\rangle$ is an eigenstate of S_3 with eigenvalue $\hbar m$, then the states $S_\pm |j, m\rangle$ either vanish or are eigenstates of S_3 with eigenvalues $\hbar(m \pm 1)$

$$S_3 S_\pm |j, m\rangle = S_\pm S_3 |j, m\rangle \pm \hbar S_\pm |j, m\rangle = \hbar(m \pm 1) S_\pm |j, m\rangle. \tag{10.122}$$

Thus the raising and lowering operators raise and lower the eigenvalues of S_3. When $j = 1/2$, the possible values of m are $m = \pm 1/2$, and so with the usual sign and normalization conventions

$$S_+|-\rangle = \hbar|+\rangle \quad \text{and} \quad S_-|+\rangle = \hbar|-\rangle \tag{10.123}$$

while

$$S_+|+\rangle = 0 \quad \text{and} \quad S_-|-\rangle = 0. \tag{10.124}$$

The square of the total spin operator is simply related to the raising and lowering operators and to S_3

$$\boldsymbol{S}^2 = S_1^2 + S_2^2 + S_3^2 = \frac{1}{2}S_+S_- + \frac{1}{2}S_-S_+ + S_3^2. \tag{10.125}$$

But the squares of the Pauli matrices are unity, and so $S_a^2 = (\hbar/2)^2$ for all three values of a. Thus

$$\boldsymbol{S}^2 = \frac{3}{4}\hbar^2 \tag{10.126}$$

is a Casimir operator (10.105) for a spin-one-half system.

Example 10.20 (Two spin-one-half systems) Consider two spin operators $\boldsymbol{S}^{(1)}$ and $\boldsymbol{S}^{(2)}$ as defined by (10.114) acting on two spin-one-half systems. Let the tensor-product states

$$|\pm,\pm\rangle = |\pm\rangle_1|\pm\rangle_2 = |\pm\rangle_1 \otimes |\pm\rangle_2 \tag{10.127}$$

be eigenstates of $S_3^{(1)}$ and $S_3^{(2)}$ so that

$$\begin{aligned} S_3^{(1)}|+,\pm\rangle &= \frac{\hbar}{2}|+,\pm\rangle \quad \text{and} \quad S_3^{(2)}|\pm,+\rangle = \frac{\hbar}{2}|\pm,+\rangle, \\ S_3^{(1)}|-,\pm\rangle &= -\frac{\hbar}{2}|-,\pm\rangle \quad \text{and} \quad S_3^{(2)}|\pm,-\rangle = -\frac{\hbar}{2}|\pm,-\rangle. \end{aligned} \tag{10.128}$$

The total spin of the system is the sum of the two spins $\boldsymbol{S} = \boldsymbol{S}^{(1)} + \boldsymbol{S}^{(2)}$, so

$$\boldsymbol{S}^2 = \left(\boldsymbol{S}^{(1)} + \boldsymbol{S}^{(2)}\right)^2 \quad \text{and} \quad S_3 = S_3^{(1)} + S_3^{(2)}. \tag{10.129}$$

The state $|+,+\rangle$ is an eigenstate of S_3 with eigenvalue $\hbar$

$$\begin{aligned} S_3|+,+\rangle &= S_3^{(1)}|+,+\rangle + S_3^{(2)}|+,+\rangle \\ &= \frac{\hbar}{2}|+,+\rangle + \frac{\hbar}{2}|+,+\rangle = \hbar|+,+\rangle. \end{aligned} \tag{10.130}$$

So the state of angular momentum $\hbar$ in the 3-direction is $|1,1\rangle = |+,+\rangle$. Similarly, the state $|-,-\rangle$ is an eigenstate of S_3 with eigenvalue $-\hbar$

$$\begin{aligned} S_3|-,-\rangle &= S_3^{(1)}|-,-\rangle + S_3^{(2)}|-,-\rangle \\ &= -\frac{\hbar}{2}|-,-\rangle - \frac{\hbar}{2}|-,-\rangle = -\hbar|-,-\rangle \end{aligned} \tag{10.131}$$

and so the state of angular momentum $\hbar$ in the negative 3-direction is $|1,-1\rangle = |-,-\rangle$. The states $|+,-\rangle$ and $|-,+\rangle$ are eigenstates of S_3 with eigenvalue 0

$$S_3|+,-\rangle = S_3^{(1)}|+,-\rangle + S_3^{(2)}|+,-\rangle = \frac{\hbar}{2}|+,-\rangle - \frac{\hbar}{2}|+,-\rangle = 0,$$
$$S_3|-,+\rangle = S_3^{(1)}|-,+\rangle + S_3^{(2)}|-,+\rangle = -\frac{\hbar}{2}|-,+\rangle + \frac{\hbar}{2}|-,+\rangle = 0. \quad (10.132)$$

To see which states are eigenstates of $\boldsymbol{S}^2$, we use the lowering operator for the combined system $S_- = S_-^{(1)} + S_-^{(2)}$ and the rules (10.103, 10.123, & 10.124) to lower the state $|1,1\rangle$

$$S_-|+,+\rangle = \left(S_-^{(1)} + S_-^{(2)}\right)|+,+\rangle = \hbar\left(|-,+\rangle + |+,-\rangle\right) = \hbar\sqrt{2}\,|1,0\rangle.$$

Thus the state $|1,0\rangle$ is

$$|1,0\rangle = \frac{1}{\sqrt{2}}\left(|+,-\rangle + |-,+\rangle\right). \quad (10.133)$$

The orthogonal and normalized combination of $|+,-\rangle$ and $|-,+\rangle$ must be the state of spin zero

$$|0,0\rangle = \frac{1}{\sqrt{2}}\left(|+,-\rangle - |-,+\rangle\right) \quad (10.134)$$

with the usual sign convention.

To check that the states $|1,0\rangle$ and $|0,0\rangle$ really are eigenstates of $\boldsymbol{S}^2$, we use (10.125 & 10.126) to write $\boldsymbol{S}^2$ as

$$\boldsymbol{S}^2 = \left(\boldsymbol{S}^{(1)} + \boldsymbol{S}^{(2)}\right)^2 = \frac{3}{2}\hbar^2 + 2\boldsymbol{S}^{(1)}\cdot\boldsymbol{S}^{(2)}$$
$$= \frac{3}{2}\hbar^2 + S_+^{(1)}S_-^{(2)} + S_-^{(1)}S_+^{(2)} + 2S_3^{(1)}S_3^{(2)}. \quad (10.135)$$

Now the sum $S_+^{(1)}S_-^{(2)} + S_-^{(1)}S_+^{(2)}$ merely interchanges the states $|+,-\rangle$ and $|-,+\rangle$ and multiplies them by $\hbar^2$, so

$$\boldsymbol{S}^2|1,0\rangle = \frac{3}{2}\hbar^2|1,0\rangle + \hbar^2|1,0\rangle - \frac{2}{4}\hbar^2|1,0\rangle$$
$$= 2\hbar^2|1,0\rangle = s(s+1)\hbar^2|1,0\rangle, \quad (10.136)$$

which confirms that $s = 1$. Because of the relative minus sign in formula (10.134) for the state $|0,0\rangle$, we have

$$\boldsymbol{S}^2|0,0\rangle = \frac{3}{2}\hbar^2|0,0\rangle - \hbar^2|1,0\rangle - \frac{1}{2}\hbar^2|1,0\rangle$$
$$= 0\hbar^2|1,0\rangle = s(s+1)\hbar^2|1,0\rangle, \quad (10.137)$$

which confirms that $s = 0$. □

10.19 The Jacobi identity

Any three square matrices A, B, and C satisfy the commutator-product rule

$$[A, BC] = ABC - BCA = ABC - BAC + BAC - BCA$$
$$= [A, B]C + B[A, C]. \tag{10.138}$$

Interchanging B and C gives

$$[A, CB] = [A, C]B + C[A, B]. \tag{10.139}$$

Subtracting the second equation from the first, we get the Jacobi identity

$$[A, [B, C]] = [[A, B], C] + [B, [A, C]] \tag{10.140}$$

and its equivalent cyclic form

$$[A, [B, C]] + [B, [C, A]] + [C, [A, B]] = 0. \tag{10.141}$$

Another Jacobi identity uses the anticommutator $\{A, B\} \equiv AB + BA$

$$\{[A, B], C\} + \{[A, C], B\} + [\{B, C\}, A] = 0. \tag{10.142}$$

10.20 The adjoint representation

Any three generators t_a, t_b, and t_c satisfy the Jacobi identity (10.141)

$$[t_a, [t_b, t_c]] + [t_b, [t_c, t_a]] + [t_c, [t_a, t_b]] = 0. \tag{10.143}$$

By using the structure-constant formula (10.80), we may express each of these double commutators as a linear combination of the generators

$$[t_a, [t_b, t_c]] = [t_a, if^d_{bc} t_d] = -f^d_{bc} f^e_{ad} t_e,$$
$$[t_b, [t_c, t_a]] = [t_b, if^d_{ca} t_d] = -f^d_{ca} f^e_{bd} t_e,$$
$$[t_c, [t_a, t_b]] = [t_c, if^d_{ab} t_d] = -f^d_{ab} f^e_{cd} t_e. \tag{10.144}$$

So the Jacobi identity (10.143) implies that

$$\left(f^d_{bc} f^e_{ad} + f^d_{ca} f^e_{bd} + f^d_{ab} f^e_{cd}\right) t_e = 0 \tag{10.145}$$

or since the generators are linearly independent

$$f^d_{bc} f^e_{ad} + f^d_{ca} f^e_{bd} + f^d_{ab} f^e_{cd} = 0. \tag{10.146}$$

If we define a set of matrices T_a by

$$(T_b)_{ac} = i f^c_{ab} \tag{10.147}$$

then, since the structure constants are antisymmetric in their lower indices, we may write the three terms in the preceding equation (10.146) as

$$f^d_{bc} f^e_{ad} = f^d_{cb} f^e_{da} = (-T_b T_a)_{ce}, \tag{10.148}$$

$$f^d_{ca} f^e_{bd} = -f^d_{ca} f^e_{db} = (T_a T_b)_{ce}, \tag{10.149}$$

and

$$f^d_{ab} f^e_{cd} = -if^d_{ab} (T_d)_{ce} \tag{10.150}$$

or in matrix notation

$$[T_a, T_b] = if^c_{ab} T_c. \tag{10.151}$$

So the matrices T_a, which we made out of the structure constants by the rule $(T_b)_{ac} = if_{abc}$ (10.147), obey the same algebra (10.63) as do the generators t_a. They are the **generators in the adjoint representation** of the Lie algebra. If the Lie algebra has N generators t_a, then the N generators T_a in the adjoint representation are $N \times N$ matrices.

10.21 Casimir operators

For any compact Lie algebra, the sum of the squares of all the generators

$$C = \sum_{n=1}^{N} t^a t^a \equiv t^a t^a \tag{10.152}$$

commutes with every generator t^b

$$\begin{aligned} [C, t^b] &= [t^a t^a, t^b] = [t^a, t^b] t^a + t^a [t^a, t^b] \\ &= if_{abc} t^c t^a + t^a if_{abc} t^c = i\,(f_{abc} + f_{cba})\, t^c t^a = 0 \end{aligned} \tag{10.153}$$

because of the total antisymmetry (10.75) of the structure constants. This sum, called a **Casimir operator**, commutes with every matrix

$$[C, D(\alpha)] = [C, \exp(i\alpha_a t^a)] = 0 \tag{10.154}$$

of the representation generated by the t^as. Thus by part 2 of Schur's lemma (section 10.7), it must be a multiple of the identity matrix

$$C = t^a t^a = cI. \tag{10.155}$$

The constant c depends upon the representation $D(\alpha)$ and is called the **quadratic Casimir**.

The generators of some noncompact groups come in pairs t^a and it^a, and so the sum of the squares of these generators vanishes, $C = t^a t^a - t^a t^a = 0$.

10.22 Tensor operators for the rotation group

Suppose $A_m^{(j)}$ is a set of $2j+1$ operators whose commutation relations with the generators J_i of rotations are

$$[J_i, A_m^{(j)}] = A_\ell^{(j)} (J_i^{(j)})_{\ell m} \tag{10.156}$$

in which the sum over ℓ runs from $-j$ to j. Then $A^{(j)}$ is said to be a **spin-j tensor operator** for the group $SU(2)$.

Example 10.21 (A spin-one tensor operator) For instance, if $j = 1$, then $(J_i^{(1)})_{\ell m} = i\hbar\epsilon_{\ell i m}$, and so a spin-one tensor operator of $SU(2)$ is a vector $A_m^{(1)}$ that transforms as

$$[J_i, A_m^{(1)}] = A_\ell^{(1)} i\hbar\epsilon_{\ell i m} = i\hbar\epsilon_{i m \ell} A_\ell^{(1)} \tag{10.157}$$

under rotations. □

Let's rewrite the definition (10.156) as

$$J_i A_m^{(j)} = A_\ell^{(j)} (J_i^{(j)})_{\ell m} + A_m^{(j)} J_i \tag{10.158}$$

and specialize to the case $i = 3$ so that $(J_3^{(j)})_{\ell m}$ is diagonal, $(J_3^{(j)})_{\ell m} = \hbar m \delta_{\ell m}$

$$J_3 A_m^{(j)} = A_\ell^{(j)} (J_3^{(j)})_{\ell m} + A_m^{(j)} J_3 = A_\ell^{(j)} \hbar m \delta_{\ell m} + A_m^{(j)} J_3 = A_m^{(j)} (\hbar m + J_3)\,. \tag{10.159}$$

Thus if the state $|j, m', E\rangle$ is an eigenstate of J_3 with eigenvalue $\hbar m'$, then the state $A_m^{(j)}|j, m', E\rangle$ is an eigenstate of J_3 with eigenvalue $\hbar(m + m')$

$$J_3 A_m^{(j)} |j, m', E\rangle = A_m^{(j)} (\hbar m + J_3)\, |j, m', E\rangle = \hbar\left(m + m'\right) A_m^{(j)} |j, m', E\rangle. \tag{10.160}$$

The J_3 eigenvalues of the tensor operator $A_m^{(j)}$ and the state $|j, m', E\rangle$ add.

10.23 Simple and semisimple Lie algebras

An **invariant subalgebra** is a set of generators $t_a^{(i)}$ whose commutator with every generator t_b of the group is a linear combination of the $t_c^{(i)}$s

$$[t_a^{(i)}, t_b] = i f_{abc} t_c^{(i)}. \tag{10.161}$$

The whole algebra and the null algebra are trivial invariant subalgebras.

An algebra with no nontrivial invariant subalgebras is a **simple** algebra. A simple algebra generates a **simple group**. An algebra that has no nontrivial abelian invariant subalgebras is a **semisimple** algebra. A semisimple algebra generates a **semisimple group**.

Example 10.22 (Some simple Lie groups) The groups of unitary matrices of unit determinant $SU(2)$, $SU(3)$, ... are simple. So are the groups of orthogonal matrices of unit determinant $SO(2)$, $SO(3)$, ... and the groups of symplectic matrices $Sp(2)$, $Sp(4)$, ... (section 10.28). □

Example 10.23 (Unification and grand unification) The symmetry group of the **standard model of particle physics** is a **direct product** of an $SU(3)$ group that acts on colored fields, an $SU(2)$ group that acts on **left-handed** quark and lepton fields, and a $U(1)$ group that acts on fields that carry hypercharge. Each of these three groups is an invariant subgroup of the full symmetry group $SU(3)_c \otimes SU(2)_\ell \otimes U(1)_Y$, and the last one is abelian. Thus the symmetry group of the standard model is neither simple nor semisimple. A simple symmetry group relates all its quantum numbers, and so physicists have invented **grand unification** in which a simple symmetry group G contains the symmetry group of the standard model. Georgi and Glashow suggested the group $SU(5)$ in 1976 (Howard Georgi, 1947–; Sheldon Glashow, 1932–). Others have proposed $SO(10)$ and even bigger groups. □

10.24 $SU(3)$

The Gell-Mann matrices are

$$\lambda_1 = \begin{pmatrix} 0 & 1 & 0 \\ 1 & 0 & 0 \\ 0 & 0 & 0 \end{pmatrix}, \quad \lambda_2 = \begin{pmatrix} 0 & -i & 0 \\ i & 0 & 0 \\ 0 & 0 & 0 \end{pmatrix}, \quad \lambda_3 = \begin{pmatrix} 1 & 0 & 0 \\ 0 & -1 & 0 \\ 0 & 0 & 0 \end{pmatrix},$$

$$\lambda_4 = \begin{pmatrix} 0 & 0 & 1 \\ 0 & 0 & 0 \\ 1 & 0 & 0 \end{pmatrix}, \quad \lambda_5 = \begin{pmatrix} 0 & 0 & -i \\ 0 & 0 & 0 \\ i & 0 & 0 \end{pmatrix}, \quad \lambda_6 = \begin{pmatrix} 0 & 0 & 0 \\ 0 & 0 & 1 \\ 0 & 1 & 0 \end{pmatrix},$$

$$\lambda_7 = \begin{pmatrix} 0 & 0 & 0 \\ 0 & 0 & -i \\ 0 & i & 0 \end{pmatrix}, \quad \text{and} \quad \lambda_8 = \frac{1}{\sqrt{3}} \begin{pmatrix} 1 & 0 & 0 \\ 0 & 1 & 0 \\ 0 & 0 & -2 \end{pmatrix}. \tag{10.162}$$

The generators t_a of the 3×3 defining representation of $SU(3)$ are these Gell-Mann matrices divided by 2

$$t_a = \lambda_a / 2 \tag{10.163}$$

(Murray Gell-Mann, 1929–).

The eight generators t_a are orthogonal with $k = 1/2$

$$\mathrm{Tr}\,(t_a t_b) = \frac{1}{2}\delta_{ab} \tag{10.164}$$

and satisfy the commutation relation

$$[t_a, t_b] = if_{abc}\, t_c. \tag{10.165}$$

The trace formula (10.65) gives us the ***SU*(3) structure constants** as

$$f_{abc} = -2i\mathrm{Tr}\left([t_a, t_b]t_c\right). \tag{10.166}$$

They are real and totally antisymmetric with $f_{123} = 1, f_{458} = f_{678} = \sqrt{3}/2$, and $f_{147} = -f_{156} = f_{246} = f_{257} = f_{345} = -f_{367} = 1/2$.

While no two generators of $SU(2)$ commute, two generators of $SU(3)$ do. In the representation (10.162, 10.163), t_3 and t_8 are diagonal and so commute

$$[t_3, t_8] = 0. \tag{10.167}$$

They generate the **Cartan subalgebra** (section 10.26) of $SU(3)$.

10.25 *SU*(3) and quarks

The generators defined by equations (10.163 & 10.162) give us the 3×3 representation

$$D(\alpha) = \exp\left(i\alpha_a t_a\right) \tag{10.168}$$

in which the sum $a = 1, 2, \ldots, 8$ is over the eight generators t_a. This representation acts on complex 3-vectors and is called the **3**.

Note that if

$$D(\alpha_1)D(\alpha_2) = D(\alpha_3) \tag{10.169}$$

then the complex conjugates of these matrices obey the same multiplication rule

$$D^*(\alpha_1)D^*(\alpha_2) = D^*(\alpha_3) \tag{10.170}$$

and so form another representation of $SU(3)$. It turns out that (unlike in $SU(2)$) this representation is inequivalent to the **3**; it is the $\overline{\mathbf{3}}$.

There are three quarks with masses less than about 100 MeV/c^2 – the u, d, and s quarks. The other three quarks c, b, and t are more massive by factors of 12, 45, and 173. Nobody knows why. Gell-Mann suggested that the low-energy strong interactions were approximately invariant under unitary transformations of the three light quarks, which he represented by a **3**, and of the three light antiquarks, which he represented by a $\overline{\mathbf{3}}$. He imagined that the eight light pseudo-scalar mesons, that is, the three pions π^-, π^0, π^+, the neutral η, and the four kaons K^0, K^+, K^-, $\overline{K}^0$, were composed of a quark and an antiquark. So they should transform as the tensor product

$$\mathbf{3} \otimes \overline{\mathbf{3}} = \mathbf{8} \oplus \mathbf{1}. \tag{10.171}$$

He put the eight pseudo-scalar mesons into an **8**.

He imagined that the eight light baryons – the two nucleons N and P, the three sigmas Σ^-, Σ^0, Σ^+, the neutral lambda Λ, and the two cascades Ξ^- and Ξ^0 – were each made of three quarks. They should transform as the tensor product

$$\mathbf{3} \otimes \mathbf{3} \otimes \mathbf{3} = \mathbf{10} \oplus \mathbf{8} \oplus \mathbf{8} \oplus \mathbf{1}. \tag{10.172}$$

He put the eight light baryons into one of these **8**s. When he was writing these papers, there were nine spin-3/2 resonances with masses somewhat heavier than 1200 MeV/c^2 – four Δs, three Σ^*s, and two Ξ^*s. He put these into the ten and predicted the tenth and its mass. When a tenth spin-3/2 resonance, the Ω^-, was found with a mass close to his prediction of 1680 MeV/c^2, his $SU(3)$ theory became wildly popular among high-energy physicists. Within a few years, a SLAC team had discovered quarks, and Gell-Mann had won the Nobel prize.

10.26 Cartan subalgebra

In any Lie group, the maximum set of mutually commuting generators H_a generates the **Cartan subalgebra**

$$[H_a, H_b] = 0, \tag{10.173}$$

which is an abelian subalgebra. The number of generators in the Cartan subalgebra is the **rank** of the Lie algebra. The Cartan generators H_a can be simultaneously diagonalized, and their eigenvalues or diagonal elements are the **weights**

$$H_a|\mu, x, D\rangle = \mu_a|\mu, x, D\rangle, \tag{10.174}$$

in which D labels the representation and x whatever other variables are needed to specify the state. The vector μ is the **weight vector**. The **roots** are the weights of the adjoint representation.

10.27 Quaternions

If z and w are any two complex numbers, then the 2×2 matrix

$$q = \begin{pmatrix} z & w \\ -w^* & z^* \end{pmatrix} \tag{10.175}$$

is a quaternion. The quaternions are closed under addition and multiplication and under multiplication by a real number (exercise 10.21), but not under multiplication by an arbitrary complex number. The squared norm of q is its determinant

$$\|q\|^2 = |z|^2 + |w|^2 = \det q. \tag{10.176}$$

The matrix products $q^\dagger q$ and $q\,q^\dagger$ are the squared norm $\|q\|^2$ multiplied by the 2×2 identity matrix

$$q^\dagger q = q\,q^\dagger = \|q\|^2\, I. \tag{10.177}$$

The 2×2 matrix

$$i\sigma_2 = \begin{pmatrix} 0 & 1 \\ -1 & 0 \end{pmatrix} \tag{10.178}$$

provides another expression for $\|q\|^2$ in terms of q and its transpose q^{T}

$$q^{\mathsf{T}}\, i\sigma_2\, q = \|q\|^2\, i\sigma_2. \tag{10.179}$$

Clearly $\|q\| = 0$ implies $q = 0$. The norm of a product of quaternions is the product of their norms

$$\|q_1 q_2\| = \sqrt{\det(q_1 q_2)} = \sqrt{\det q_1 \det q_2} = \|q_1\|\|q_2\|. \tag{10.180}$$

The quaternions therefore form an **associative division algebra** (over the real numbers); the only others are the real numbers and the complex numbers; the **octonions** are a nonassociative division algebra.

One may use the Pauli matrices to define for any real 4-vector x a **quaternion** $q(x)$ as

$$\begin{aligned} q(x) &= x^0 - i\sigma_k x^k = x^0 - i\boldsymbol{\sigma} \cdot \boldsymbol{x} \\ &= \begin{pmatrix} x^0 - ix^3 & -x^2 - ix^1 \\ x^2 - ix^1 & x^0 + ix^3 \end{pmatrix}. \end{aligned} \tag{10.181}$$

The product rule (10.116) for the Pauli matrices tells us that the product of two quaternions is

$$\begin{aligned} q(x)\,q(y) &= (x^0 - i\boldsymbol{\sigma} \cdot \boldsymbol{x})(y^0 - i\boldsymbol{\sigma} \cdot \boldsymbol{y}) \\ &= x^0 y^0 - i\boldsymbol{\sigma} \cdot (y^0 \boldsymbol{x} + x^0 \boldsymbol{y}) - i(\boldsymbol{x} \times \boldsymbol{y}) \cdot \boldsymbol{\sigma} - \boldsymbol{x} \cdot \boldsymbol{y} \end{aligned} \tag{10.182}$$

so their commutator is

$$[q(x), q(y)] = -2i(\boldsymbol{x} \times \boldsymbol{y}) \cdot \boldsymbol{\sigma}. \tag{10.183}$$

Example 10.24 (Lack of analyticity) One may define a function $f(q)$ of a quaternionic variable and then ask what functions are analytic in the sense that the (one-sided) derivative

$$f'(q) = \lim_{q' \to 0} [f(q + q') - f(q)]q'^{-1} \tag{10.184}$$

exists and is independent of the direction through which $q' \to 0$. This space of functions is extremely limited and does not even include the function $f(q) = q^2$ (exercise 10.22). □

10.28 The symplectic group $Sp\,(2n)$

The symplectic group $Sp(2n)$ consists of $2n \times 2n$ matrices W that map n-tuples q of quaternions into n-tuples $q' = Wq$ of quaternions with the same value of the quadratic quaternionic form

$$\|q'\|^2 = \|q'_1\|^2 + \|q'_2\|^2 + \cdots + \|q'_n\|^2 = \|q_1\|^2 + \|q_2\|^2 + \cdots + \|q_n\|^2 = \|q\|^2. \tag{10.185}$$

By (10.177), the quadratic form $\|q'\|^2$ times the 2×2 identity matrix I is equal to the hermitian form $q'^\dagger q'$

$$\|q'\|^2 I = q'^\dagger q' = q_1'^\dagger q_1' + \cdots + q_n'^\dagger q_n' = q^\dagger W^\dagger W q \tag{10.186}$$

and so any matrix W that is both a $2n \times 2n$ unitary matrix and an $n \times n$ matrix of quaternions keeps $\|q'\|^2 = \|q\|^2$

$$\|q'\|^2 I = q^\dagger W^\dagger W q = q^\dagger q = \|q\|^2 I. \tag{10.187}$$

The group $Sp(2n)$ thus consists of all $2n \times 2n$ unitary matrices that also are $n \times n$ matrices of quaternions. (This last requirement is needed so that $q' = Wq$ is an n-tuple of quaternions.)

The generators t_a of the symplectic group $Sp(2n)$ are $2n \times 2n$ direct-product matrices of the form

$$I \otimes A, \quad \sigma_1 \otimes S_1, \quad \sigma_2 \otimes S_2, \quad \text{and} \quad \sigma_3 \otimes S_3, \tag{10.188}$$

in which I is the 2×2 identity matrix, the three σ_is are the Pauli matrices, A is an imaginary $n \times n$ antisymmetric matrix, and the S_i are $n \times n$ real symmetric matrices. These generators t_a close under commutation

$$[t_a, t_b] = i f_{abc} t_c. \tag{10.189}$$

Any imaginary linear combination $i\alpha_a t_a$ of these generators is not only a $2n \times 2n$ antihermitian matrix but also an $n \times n$ matrix of quaternions. Thus the matrices

$$D(\alpha) = e^{i\alpha_a t_a} \tag{10.190}$$

are both unitary $2n \times 2n$ matrices and $n \times n$ quaternionic matrices and so are elements of the group $Sp(2n)$.

Example 10.25 ($Sp(2) = SU(2)$) There is no 1×1 antisymmetric matrix, and there is only one 1×1 symmetric matrix. So the generators t_a of the group $Sp(2)$ are the Pauli matrices $t_a = \sigma_a$, and $Sp(2) = SU(2)$. The elements $g(\alpha)$ of $SU(2)$ are quaternions of unit norm (exercise 10.20), and so the product $g(\alpha)q$ is a quaternion

$$\|g(\alpha)q\|^2 = \det(g(\alpha)q) = \det(g(\alpha)) \det q = \det q = \|q\|^2 \tag{10.191}$$

with the same squared norm. □

Example 10.26 ($Sp(4) = SO(5)$) Apart from scale factors, there are three real symmetric 2×2 matrices $S_1 = \sigma_1$, $S_2 = I$, and $S_3 = \sigma_3$ and one imaginary antisymmetric 2×2 matrix $A = \sigma_2$. So there are ten generators of $Sp(4) = SO(5)$

$$t_1 = I \otimes \sigma_2 = \begin{pmatrix} 0 & -iI \\ iI & 0 \end{pmatrix}, \quad t_{k1} = \sigma_k \otimes \sigma_1 = \begin{pmatrix} 0 & \sigma_k \\ \sigma_k & 0 \end{pmatrix}$$

$$t_{k2} = \sigma_k \otimes I = \begin{pmatrix} \sigma_k & 0 \\ 0 & \sigma_k \end{pmatrix}, \quad t_{k3} = \sigma_k \otimes \sigma_3 = \begin{pmatrix} \sigma_k & 0 \\ 0 & -\sigma_k \end{pmatrix} \tag{10.192}$$

where k runs from 1 to 3. □

We may see $Sp(2n)$ from a different viewpoint if we use (10.179) to write the quadratic form $\|q\|^2$ in terms of a $2n \times 2n$ matrix J that has n copies of $i\sigma_2$ on its 2×2 diagonal

$$J = \begin{pmatrix} i\sigma_2 & 0 & 0 & 0 & \dots & 0 \\ 0 & i\sigma_2 & 0 & 0 & \dots & 0 \\ 0 & 0 & i\sigma_2 & 0 & \dots & 0 \\ 0 & 0 & 0 & \ddots & \dots & 0 \\ \vdots & \vdots & \vdots & \vdots & \ddots & 0 \\ 0 & 0 & 0 & 0 & 0 & i\sigma_2 \end{pmatrix} \tag{10.193}$$

(and zeros elsewhere) as

$$\|q\|^2 J = q^{\mathsf{T}} J q. \tag{10.194}$$

Thus any $n \times n$ matrix of quaternions W that satisfies

$$W^{\mathsf{T}} J W = J \tag{10.195}$$

also satisfies

$$\| Wq\|^2 J = q^{\mathsf{T}}\, W^{\mathsf{T}} J W q = q^{\mathsf{T}} J q = \|q\|^2 J \tag{10.196}$$

and so leaves invariant the quadratic form (10.185). The group $Sp(2n)$ therefore consists of all $2n \times 2n$ matrices W that satisfy (10.195) and that also are $n \times n$ matrices of quaternions.

The symplectic group is something of a physics orphan. Its best-known application is in classical mechanics, and that application uses the **noncompact** symplectic group $Sp(2n, R)$, not the compact symplectic group $Sp(2n)$. The elements of $Sp(2n, R)$ are all real $2n \times 2n$ matrices T that satisfy $T^{\mathsf{T}} J T = J$ with the J of (10.193); those near the identity are of the form $T = \exp(JS)$ in which S is a $2n \times 2n$ real symmetric matrix (exercise 10.24).

Example 10.27 ($Sp(2,R)$) The matrices (exercise 10.25)

$$T = \pm \begin{pmatrix} \cosh\theta & \sinh\theta \\ \sinh\theta & \cosh\theta \end{pmatrix} \tag{10.197}$$

are elements of the **noncompact** symplectic group $Sp(2,R)$. □

A dynamical **map** $\mathcal{M}$ takes the phase-space $2n$-tuple $z = (q_1, p_1, \dots, q_n, p_n)$ from $z(t_1)$ to $z(t_2)$. One may show that $\mathcal{M}$'s **jacobian** matrix

$$M_{ab} = \frac{\partial z_a(t_2)}{\partial z_b(t_1)} \tag{10.198}$$

is in $Sp(2n,R)$ if and only if its dynamics are **hamiltonian**

$$\dot{q}_a = \frac{\partial H}{\partial p_a} \quad \text{and} \quad \dot{p}_a = -\frac{\partial H}{\partial q_a} \tag{10.199}$$

(Carl Jacobi, 1804–1851; William Hamilton, 1805–1865, inventor of quaternions).

10.29 Compact simple Lie groups

Élie Cartan (1869–1951) showed that all compact, simple Lie groups fall into four infinite classes and five discrete cases. For $n = 1, 2, \dots$, his four classes are

- $A_n = SU(n+1)$, which are $(n+1) \times (n+1)$ unitary matrices with unit determinant,
- $B_n = SO(2n+1)$, which are $(2n+1) \times (2n+1)$ orthogonal matrices with unit determinant,
- $C_n = Sp(2n)$, which are $2n \times 2n$ symplectic matrices, and
- $D_n = SO(2n)$, which are $2n \times 2n$ orthogonal matrices with unit determinant.

The five discrete cases are the **exceptional groups** G_2, F_4, E_6, E_7, and E_8.

The exceptional groups are associated with the **octonians**

$$a + b_\alpha i_\alpha \tag{10.200}$$

where the α-sum runs from 1 to 7; the eight numbers a and b_α are real; and the seven i_αs obey the multiplication law

$$i_\alpha \, i_\beta = -\delta_{\alpha\beta} + g_{\alpha\beta\gamma} \, i_\gamma, \tag{10.201}$$

in which $g_{\alpha\beta\gamma}$ is totally antisymmetric with

$$g_{123} = g_{247} = g_{451} = g_{562} = g_{634} = g_{375} = g_{716} = 1. \tag{10.202}$$

Like the quaternions and the complex numbers, the octonians form a **division algebra** with an absolute value

$$|a + b_\alpha i_\alpha| = \left(a^2 + b_\alpha^2\right)^{1/2} \tag{10.203}$$

that satisfies

$$|AB| = |A||B| \tag{10.204}$$

but they lack associativity.

The group G_2 is the subgroup of $SO(7)$ that leaves the $g_{\alpha\beta\gamma}$s of (10.201) invariant.

10.30 Group integration

Suppose we need to integrate some function $f(g)$ over a group. Naturally, we want to do so in a way that gives equal weight to every element of the group. In particular, if g' is any group element, we want the integral of the shifted function $f(g'g)$ to be the same as the integral of $f(g)$

$$\int f(g)\, dg = \int f(g'g)\, dg. \tag{10.205}$$

Such a measure dg is said to be **left invariant** (Creutz, 1983, chap. 8).

Let's use the letters $a = a_1, \ldots, a_n$, $b = b_1, \ldots, b_n$, and so forth to label the elements $g(a)$, $g(b)$, so that an integral over the group is

$$\int f(g)\, dg = \int f(g(a))\, m(a)\, d^n a, \tag{10.206}$$

in which $m(a)$ is the left-invariant measure and the integration is over the n-space of as that label all the elements of the group.

To find the left-invariant measure $m(a)$, we use the multiplication law of the group

$$g(a(c,b)) \equiv g(c)\, g(b) \tag{10.207}$$

and impose the requirement (10.205) of left invariance with $g' \equiv g(c)$

$$\int f(g(b))\, m(b)\, d^n b = \int f(g(c)g(b))\, m(b)\, d^n b = \int f(g(a(c,b)))\, m(b)\, d^n b. \tag{10.208}$$

We change variables from b to $a = a(c,b)$ by using the jacobian $\det(\partial b/\partial a)$, which gives us $d^n b = \det(\partial b/\partial a)\, d^n a$

$$\int f(g(b))\, m(b)\, d^n b = \int f(g(a))\, \det(\partial b/\partial a)\, m(b)\, d^n a. \tag{10.209}$$

Replacing b by $a = a(c, b)$ on the left-hand side of this equation, we find

$$m(a) = \det(\partial b/\partial a)\, m(b) \tag{10.210}$$

or since $\det(\partial b/\partial a) = 1/\det(\partial a(c, b)/\partial b)$

$$m(a(c, b)) = m(b)/\det(\partial a(c, b)/\partial b). \tag{10.211}$$

So if we let $g(b) \to g(0) = e$, the identity element of the group, and set $m(e) = 1$, then we find for the measure

$$m(a) = m(c) = m(a(c, b))|_{b=0} = 1/\det(\partial a(c, b)/\partial b)|_{b=0}\,. \tag{10.212}$$

Example 10.28 (The invariant measure for $SU(2)$) A general element of the group $SU(2)$ is given by (10.118) as

$$\exp\left(i\, \boldsymbol{\theta} \cdot \frac{\boldsymbol{\sigma}}{2}\right) = I \cos\frac{\theta}{2} + i\hat{\boldsymbol{\theta}} \cdot \boldsymbol{\sigma} \sin\frac{\theta}{2}. \tag{10.213}$$

Setting $a_0 = \cos(\theta/2)$ and $\boldsymbol{a} = \hat{\boldsymbol{\theta}} \sin(\theta/2)$, we have

$$g(a) = a_0 + i\, \boldsymbol{a} \cdot \boldsymbol{\sigma}, \tag{10.214}$$

in which $a^2 \equiv a_0^2 + \boldsymbol{a} \cdot \boldsymbol{a} = 1$. Thus, the parameter space for $SU(2)$ is the unit sphere S_3 in four dimensions. Its invariant measure is

$$\int \delta(1 - a^2)\, d^4a = \int \delta(1 - a_0^2 - \boldsymbol{a}^2)\, d^4a = \int (1 - \boldsymbol{a}^2)^{-1/2}\, d^3a \tag{10.215}$$

or

$$m(\boldsymbol{a}) = (1 - \boldsymbol{a}^2)^{-1/2}. \tag{10.216}$$

We also can write the arbitrary element (10.214) of $SU(2)$ as

$$g(\boldsymbol{a}) = \pm\sqrt{1 - \boldsymbol{a}^2} + i\, \boldsymbol{a} \cdot \boldsymbol{\sigma} \tag{10.217}$$

and the group-multiplication law (10.207) as

$$\sqrt{1 - \boldsymbol{a}^2} + i\, \boldsymbol{a} \cdot \boldsymbol{\sigma} = \left(\sqrt{1 - \boldsymbol{c}^2} + i\, \boldsymbol{c} \cdot \boldsymbol{\sigma}\right)\left(\sqrt{1 - \boldsymbol{b}^2} + i\, \boldsymbol{b} \cdot \boldsymbol{\sigma}\right). \tag{10.218}$$

Thus, by multiplying both sides of this equation by σ_i and taking the trace, we find (exercise 10.26) that the parameters $\boldsymbol{a}(\boldsymbol{c}, \boldsymbol{b})$ that describe the product $g(\boldsymbol{c})\, g(\boldsymbol{b})$ are

$$\boldsymbol{a}(\boldsymbol{c}, \boldsymbol{b}) = \sqrt{1 - \boldsymbol{c}^2}\, \boldsymbol{b} + \sqrt{1 - \boldsymbol{b}^2}\, \boldsymbol{c} - \boldsymbol{c} \times \boldsymbol{b}. \tag{10.219}$$

To compute the jacobian of our formula (10.212) for the invariant measure, we differentiate this expression (10.219) at $\boldsymbol{b} = \boldsymbol{0}$ and so find (exercise 10.27)

$$m(\boldsymbol{a}) = 1/\det(\partial a(c, b)/\partial b)|_{b=0} = (1 - \boldsymbol{a}^2)^{-1/2} \tag{10.220}$$

as the left-invariant measure in agreement with (10.216). □

10.31 The Lorentz group

The Lorentz group $O(3,1)$ is the set of all linear transformations L that leave invariant the Minkowski inner product

$$xy \equiv \boldsymbol{x} \cdot \boldsymbol{y} - x^0 y^0 = x^{\mathsf{T}} \eta y \tag{10.221}$$

in which η is the diagonal matrix

$$\eta = \begin{pmatrix} -1 & 0 & 0 & 0 \\ 0 & 1 & 0 & 0 \\ 0 & 0 & 1 & 0 \\ 0 & 0 & 0 & 1 \end{pmatrix}. \tag{10.222}$$

So L is in $O(3,1)$ if for all 4-vectors x and y

$$(Lx)^{\mathsf{T}} \, \eta L y = x^{\mathsf{T}} L^{\mathsf{T}} \, \eta \, L y = x^{\mathsf{T}} \eta \, y. \tag{10.223}$$

Since x and y are arbitrary, this condition amounts to

$$L^{\mathsf{T}} \eta \, L = \eta. \tag{10.224}$$

Taking the determinant of both sides and using the transpose (1.194) and product (1.207) rules, we have

$$(\det L)^2 = 1. \tag{10.225}$$

So $\det L = \pm 1$, and every Lorentz transformation L has an inverse. Multiplying (10.224) by η, we find

$$\eta L^{\mathsf{T}} \eta L = \eta^2 = I, \tag{10.226}$$

which identifies L^{-1} as

$$L^{-1} = \eta L^{\mathsf{T}} \eta. \tag{10.227}$$

The subgroup of $O(3,1)$ with $\det L = 1$ is the proper Lorentz group $SO(3,1)$.

To find its Lie algebra, we consider a Lorentz matrix $L = I + \omega$ that differs from the identity matrix I by a tiny matrix ω and require it to satisfy the condition (10.224) for membership in the Lorentz group

$$\left(I + \omega^{\mathsf{T}}\right) \eta \, (I + \omega) = \eta + \omega^{\mathsf{T}} \eta + \eta \, \omega + \omega^{\mathsf{T}} \omega = \eta. \tag{10.228}$$

Neglecting $\omega^{\mathsf{T}} \omega$, we have $\omega^{\mathsf{T}} \eta = -\eta \, \omega$ or since $\eta^2 = I$

$$\omega^{\mathsf{T}} = -\, \eta \, \omega \, \eta. \tag{10.229}$$

This equation says (exercise 10.29) that under transposition the time-time and space-space elements of ω change sign, while the time-space and space-time elements do not. That is, the tiny matrix ω must be for infinitesimal $\boldsymbol{\theta}$ and $\boldsymbol{\lambda}$ a linear combination

$$\omega = \boldsymbol{\theta} \cdot \boldsymbol{R} + \boldsymbol{\lambda} \cdot \boldsymbol{B} \tag{10.230}$$

of the six matrices

$$R_1 = \begin{pmatrix} 0 & 0 & 0 & 0 \\ 0 & 0 & 0 & 0 \\ 0 & 0 & 0 & -1 \\ 0 & 0 & 1 & 0 \end{pmatrix}, \quad R_2 = \begin{pmatrix} 0 & 0 & 0 & 0 \\ 0 & 0 & 0 & 1 \\ 0 & 0 & 0 & 0 \\ 0 & -1 & 0 & 0 \end{pmatrix}, \quad R_3 = \begin{pmatrix} 0 & 0 & 0 & 0 \\ 0 & 0 & -1 & 0 \\ 0 & 1 & 0 & 0 \\ 0 & 0 & 0 & 0 \end{pmatrix} \tag{10.231}$$

and

$$B_1 = \begin{pmatrix} 0 & 1 & 0 & 0 \\ 1 & 0 & 0 & 0 \\ 0 & 0 & 0 & 0 \\ 0 & 0 & 0 & 0 \end{pmatrix}, \quad B_2 = \begin{pmatrix} 0 & 0 & 1 & 0 \\ 0 & 0 & 0 & 0 \\ 1 & 0 & 0 & 0 \\ 0 & 0 & 0 & 0 \end{pmatrix}, \quad B_3 = \begin{pmatrix} 0 & 0 & 0 & 1 \\ 0 & 0 & 0 & 0 \\ 0 & 0 & 0 & 0 \\ 1 & 0 & 0 & 0 \end{pmatrix}, \tag{10.232}$$

which satisfy condition (10.229). The three R_j are 4×4 versions of the rotation generators (10.88); the three B_j generate Lorentz boosts.

If we write $L = I + \omega$ as

$$L = I - i\theta_\ell iR_\ell - i\lambda_j iB_j \equiv I - i\theta_\ell J_\ell - i\lambda_j K_j \tag{10.233}$$

then the three matrices $J_\ell = iR_\ell$ are imaginary and antisymmetric, and therefore hermitian. But the three matrices $K_j = iB_j$ are imaginary and symmetric, and so are antihermitian. Thus, the 4×4 matrix L is **not unitary**. The reason is that the Lorentz group is **not compact**.

One may verify (exercise 10.30) that the six generators J_ℓ and K_j satisfy three sets of commutation relations:

$$[J_i, J_j] = i\epsilon_{ijk} J_k, \tag{10.234}$$

$$[J_i, K_j] = i\epsilon_{ijk} K_k, \tag{10.235}$$

$$[K_i, K_j] = -\, i\epsilon_{ijk} J_k. \tag{10.236}$$

The first (10.234) says that the three J_ℓ generate the rotation group $SO(3)$; the second (10.235) says that the three boost generators transform as a 3-vector under $SO(3)$; and the third (10.236) implies that four canceling infinitesimal boosts can amount to a rotation. These three sets of commutation relations form the Lie algebra of the Lorentz group $SO(3,1)$. Incidentally, one may show (exercise 10.31) that if $\boldsymbol{J}$ and $\boldsymbol{K}$ satisfy these commutation relations (10.234–10.236), then so do

$$\boldsymbol{J} \text{ and } -\boldsymbol{K}. \tag{10.237}$$

The infinitesimal Lorentz transformation (10.233) is the 4×4 matrix

$$L = I + \omega = I + \theta_\ell R_\ell + \lambda_j B_j = \begin{pmatrix} 1 & \lambda_1 & \lambda_2 & \lambda_3 \\ \lambda_1 & 1 & -\theta_3 & \theta_2 \\ \lambda_2 & \theta_3 & 1 & -\theta_1 \\ \lambda_3 & -\theta_2 & \theta_1 & 1 \end{pmatrix}. \tag{10.238}$$

It moves any 4-vector x to $x' = Lx$ or in components $x'^a = L^a{}_b x^b$

$$\begin{aligned} x'^0 &= x^0 + \lambda_1 x^1 + \lambda_2 x^2 + \lambda_3 x^3, \\ x'^1 &= \lambda_1 x^0 + x^1 - \theta_3 x^2 + \theta_2 x^3, \\ x'^2 &= \lambda_2 x^0 + \theta_3 x^1 + x^2 - \theta_1 x^3, \\ x'^3 &= \lambda_3 x^0 - \theta_2 x^1 + \theta_1 x^2 + x^3. \end{aligned} \tag{10.239}$$

More succinctly with $t = x^0$, this is

$$\begin{aligned} t' &= t + \boldsymbol{\lambda} \cdot \boldsymbol{x}, \\ \boldsymbol{x}' &= \boldsymbol{x} + t\boldsymbol{\lambda} + \boldsymbol{\theta} \wedge \boldsymbol{x}, \end{aligned} \tag{10.240}$$

in which $\wedge \equiv \times$ means cross-product.

For arbitrary real $\boldsymbol{\theta}$ and $\boldsymbol{\lambda}$, the matrices

$$L = e^{-i\theta_\ell J_\ell - i\lambda_j K_j} \tag{10.241}$$

form the subgroup of $SO(3,1)$ that is connected to the identity matrix I. This subgroup preserves the sign of the time of any time-like vector, that is, if $x^2 < 0$, and $y = Lx$, then $y^0 x^0 > 0$. It is called the proper orthochronous Lorentz group. The rest of the (homogeneous) Lorentz group can be obtained from it by space $\mathcal{P}$, time $\mathcal{T}$, and space-time $\mathcal{PT}$ reflections.

The task of finding all the finite-dimensional irreducible representations of the proper orthochronous homogeneous Lorentz group becomes vastly simpler when we write the commutation relations (10.234–10.236) in terms of the nonhermitian matrices

$$J^\pm_\ell = \frac{1}{2}\left(J_\ell \pm i K_\ell\right), \tag{10.242}$$

which generate two independent rotation groups

$$\begin{aligned} [J^+_i, J^+_j] &= i\epsilon_{ijk} J^+_k, \\ [J^-_i, J^-_j] &= i\epsilon_{ijk} J^-_k, \\ [J^+_i, J^-_j] &= 0. \end{aligned} \tag{10.243}$$

Thus the Lie algebra of the Lorentz group is equivalent to two copies of the Lie algebra (10.100) of $SU(2)$. Its finite-dimensional irreducible representations are the direct products

$$D^{(j,j')}(\theta, \lambda) = e^{-i\theta_\ell J_\ell - i\lambda_\ell K_\ell} = e^{(-i\theta_\ell - \lambda_\ell) J^+_\ell}\, e^{(-i\theta_\ell + \lambda_\ell) J^-_\ell} \tag{10.244}$$

of the nonunitary representations $D^{(j)}(\theta,\lambda) = e^{(-i\theta_\ell - \lambda_\ell)J_\ell^+}$ and $D^{(j')}(\theta,\lambda) = e^{(-i\theta_\ell + \lambda_\ell)J_\ell^-}$ generated by the three $(2j+1)\times(2j+1)$ matrices J_ℓ^+ and by the three $(2j'+1)\times(2j'+1)$ matrices J_ℓ^-. Under a Lorentz transformation L, a field $\psi_{m,m'}^{(j,j')}(x)$ that transforms under the $D^{(j,j')}$ representation of the Lorentz group responds as

$$U(L)\,\psi_{m,m'}^{(j,j')}(x)\,U^{-1}(L) = D_{mm''}^{(j)}(L^{-1})\,D_{m'm'''}^{(j')}(L^{-1})\,\psi_{m'',m'''}^{(j,j')}(Lx). \tag{10.245}$$

Although these representations are not unitary, the $SO(3)$ subgroup of the Lorentz group is represented unitarily by the hermitian matrices

$$\boldsymbol{J} = \boldsymbol{J}^+ + \boldsymbol{J}^-. \tag{10.246}$$

Thus, the representation $D^{(j,j')}$ describes objects of the spins s that can arise from the direct product of spin-j with spin-j' (Weinberg, 1995, p. 231)

$$s = j+j',\, j+j'-1, \ldots, |j-j'|. \tag{10.247}$$

For instance, $D^{(0,0)}$ describes a spinless field or particle, while $D^{(1/2,0)}$ and $D^{(0,1/2)}$ respectively describe right-handed and left-handed spin-1/2 fields or particles. The representation $D^{(1/2,1/2)}$ describes objects of spin 1 and spin 0 – the spatial and time components of a 4-vector.

The generators K_j of the Lorentz boosts are related to $\boldsymbol{J}^{\pm}$ by

$$\boldsymbol{K} = -i\boldsymbol{J}^+ + i\boldsymbol{J}^-, \tag{10.248}$$

which like (10.246) follows from the definition (10.242).

The interchange of $\boldsymbol{J}^+$ and $\boldsymbol{J}^-$ replaces the generators $\boldsymbol{J}$ and $\boldsymbol{K}$ with $\boldsymbol{J}$ and $-\boldsymbol{K}$, a substitution that we know (10.237) is legitimate.

10.32 Two-dimensional representations of the Lorentz group

The generators of the representation $D^{(1/2,0)}$ with $j = 1/2$ and $j' = 0$ are given by (10.246 & 10.248) with $\boldsymbol{J}^+ = \boldsymbol{\sigma}/2$ and $\boldsymbol{J}^- = 0$. They are

$$\boldsymbol{J} = \frac{1}{2}\boldsymbol{\sigma} \quad \text{and} \quad \boldsymbol{K} = -i\frac{1}{2}\boldsymbol{\sigma}. \tag{10.249}$$

The 2×2 matrix $D^{(1/2,0)}$ that represents the Lorentz transformation (10.241)

$$L = e^{-i\theta_\ell J_\ell - i\lambda_j K_j} \tag{10.250}$$

is

$$D^{(1/2,0)}(\boldsymbol{\theta},\boldsymbol{\lambda}) = \exp\left(-i\boldsymbol{\theta}\cdot\boldsymbol{\sigma}/2 - \boldsymbol{\lambda}\cdot\boldsymbol{\sigma}/2\right). \tag{10.251}$$

And so the generic $D^{(1/2,0)}$ matrix is

$$D^{(1/2,0)}(\boldsymbol{\theta}, \boldsymbol{\lambda}) = e^{-\boldsymbol{z}\cdot\boldsymbol{\sigma}/2} \tag{10.252}$$

with $\boldsymbol{\lambda} = \text{Re}\boldsymbol{z}$ and $\boldsymbol{\theta} = \text{Im}\boldsymbol{z}$. It is nonunitary and of unit determinant; it is a member of the group $SL(2, C)$ of complex unimodular 2×2 matrices. The group $SL(2, C)$ relates to the Lorentz group $SO(3, 1)$ as $SU(2)$ relates to the rotation group $SO(3)$.

Example 10.29 (The standard left-handed boost) For a particle of mass $m > 0$, the "standard" boost that takes the 4-vector $k = (m, \boldsymbol{0})$ to $p = (p^0, \boldsymbol{p})$, where $p^0 = \sqrt{m^2 + \boldsymbol{p}^2}$, is a boost in the $\hat{\boldsymbol{p}}$ direction

$$B(p) = R(\hat{\boldsymbol{p}})\, B_3(p^0)\, R^{-1}(\hat{\boldsymbol{p}}) = \exp\left(\alpha\, \hat{\boldsymbol{p}} \cdot \boldsymbol{B}\right) \tag{10.253}$$

in which $\cosh\alpha = p^0/m$ and $\sinh\alpha = |\boldsymbol{p}|/m$, as one may show by expanding the exponential (exercise 10.33).

For $\boldsymbol{\lambda} = \alpha\hat{\boldsymbol{p}}$, one may show (exercise 10.34) that the matrix $D^{(1/2,0)}(\boldsymbol{0}, \boldsymbol{\lambda})$ is

$$\begin{aligned} D^{(1/2,0)}(\boldsymbol{0}, \alpha\,\hat{\boldsymbol{p}}) &= e^{-\alpha\hat{\boldsymbol{p}}\cdot\boldsymbol{\sigma}/2} = I\,\cosh(\alpha/2) - \hat{\boldsymbol{p}} \cdot \boldsymbol{\sigma}\,\sinh(\alpha/2) \\ &= I\sqrt{(p^0 + m)/(2m)} - \hat{\boldsymbol{p}} \cdot \boldsymbol{\sigma}\sqrt{(p^0 - m)/(2m)} \\ &= \frac{p^0 + m - \boldsymbol{p}\cdot\boldsymbol{\sigma}}{\sqrt{2m(p^0 + m)}} \end{aligned} \tag{10.254}$$

in the third line of which the 2×2 identity matrix I is suppressed. □

Under $D^{(1/2,0)}$, the vector $(-I, \boldsymbol{\sigma})$ transforms like a 4-vector. For tiny $\boldsymbol{\theta}$ and $\boldsymbol{\lambda}$, one may show (exercise 10.36) that the vector $(-I, \boldsymbol{\sigma})$ transforms as

$$\begin{aligned} D^{\dagger(1/2,0)}(\boldsymbol{\theta}, \boldsymbol{\lambda})(-I)D^{(1/2,0)}(\boldsymbol{\theta}, \boldsymbol{\lambda}) &= -I + \boldsymbol{\lambda} \cdot \boldsymbol{\sigma}, \\ D^{\dagger(1/2,0)}(\boldsymbol{\theta}, \boldsymbol{\lambda})\, \boldsymbol{\sigma}\, D^{(1/2,0)}(\boldsymbol{\theta}, \boldsymbol{\lambda}) &= \boldsymbol{\sigma} + (-I)\boldsymbol{\lambda} + \boldsymbol{\theta} \wedge \boldsymbol{\sigma}, \end{aligned} \tag{10.255}$$

which is how the 4-vector $(t, \boldsymbol{x})$ transforms (10.240). Under a finite Lorentz transformation L the 4-vector $S^a \equiv (-I, \boldsymbol{\sigma})$ becomes

$$D^{\dagger(1/2,0)}(L)\, S^a\, D^{(1/2,0)}(L) = L^a{}_b S^b. \tag{10.256}$$

A field $\xi(x)$ that responds to a unitary Lorentz transformation $U(L)$ like

$$U(L)\,\xi(x)\,U^{-1}(L) = D^{(1/2,0)}(L^{-1})\,\xi(Lx) \tag{10.257}$$

is called a **left-handed Weyl spinor**. We will see in example 10.30 why the action density for such spinors

$$\mathcal{L}_\ell(x) = i\,\xi^\dagger(x)\,(\partial_0 I - \nabla \cdot \boldsymbol{\sigma})\,\xi(x) \tag{10.258}$$

is Lorentz covariant, that is

$$U(L)\,\mathcal{L}_\ell(x)\,U^{-1}(L) = \mathcal{L}_\ell(Lx). \tag{10.259}$$

Example 10.30 (Why $\mathcal{L}_\ell$ is Lorentz covariant) We first note that the derivatives ∂'_b in $\mathcal{L}_\ell(Lx)$ are with respect to $x' = Lx$. Since the inverse matrix L^{-1} takes x' back to $x = L^{-1}x'$ or in tensor notation $x^a = L^{-1a}{}_b\, x'^b$, the derivative ∂'_b is

$$\partial'_b = \frac{\partial}{\partial x'^b} = \frac{\partial x^a}{\partial x'^b}\frac{\partial}{\partial x^a} = L^{-1a}{}_b \frac{\partial}{\partial x^a} = \partial_a\, L^{-1a}{}_b. \tag{10.260}$$

Now using the abbreviation $\partial_0 I - \nabla \cdot \boldsymbol{\sigma} \equiv -\,\partial_a S^a$ and the transformation laws (10.256 & 10.257), we have

$$\begin{aligned} U(L)\,\mathcal{L}_\ell(x)\,U^{-1}(L) &= i\,\xi^\dagger(Lx) D^{(1/2,0)\dagger}(L^{-1})(-\,\partial_a S^a) D^{(1/2,0)}(L^{-1})\,\xi(Lx) \\ &= i\,\xi^\dagger(Lx)(-\,\partial_a L^{-1a}{}_b S^b)\,\xi(Lx) \\ &= i\,\xi^\dagger(Lx)(-\,\partial'_b S^b)\,\xi(Lx) = \mathcal{L}_\ell(Lx), \end{aligned} \tag{10.261}$$

which shows that $\mathcal{L}_\ell$ is Lorentz covariant. □

Incidentally, the rule (10.260) ensures, among other things, that the divergence $\partial_a V^a$ is invariant

$$\left(\partial_a V^a\right)' = \partial'_a V'^a = \partial_b\, L^{-1b}{}_a\, L^a{}_c V^c = \partial_b\, \delta^b{}_c\, V^c = \partial_b\, V^b. \tag{10.262}$$

Example 10.31 (Why ξ is left-handed) The space-time integral S of the action density $\mathcal{L}_\ell$ is stationary when $\xi(x)$ satisfies the wave equation

$$(\partial_0 I - \nabla \cdot \boldsymbol{\sigma})\,\xi(x) = 0 \tag{10.263}$$

or in momentum space

$$(E + \boldsymbol{p} \cdot \boldsymbol{\sigma})\,\xi(p) = 0. \tag{10.264}$$

Multiplying from the left by $(E - \boldsymbol{p} \cdot \boldsymbol{\sigma})$, we see that the energy of a particle created or annihilated by the field ξ is the same as its momentum $E = |\boldsymbol{p}|$ in accord with the absence of a mass term in the action density $\mathcal{L}_\ell$. And because the spin of the particle is represented by the matrix $\boldsymbol{J} = \boldsymbol{\sigma}/2$, the momentum-space relation (10.264) says that $\xi(p)$ is an eigenvector of $\hat{\boldsymbol{p}} \cdot \boldsymbol{J}$

$$\hat{\boldsymbol{p}} \cdot \boldsymbol{J}\,\xi(p) = -\,\frac{1}{2}\,\xi(p) \tag{10.265}$$

with eigenvalue $-\,1/2$. A particle whose spin is opposite to its momentum is said to have **negative helicity** or to be **left-handed**. Nearly massless neutrinos are nearly left-handed. □

One may add to this action density the **Majorana mass term**

$$\mathcal{L}_{\rm M}(x) = \tfrac{1}{2}\left[m\,\xi^{\mathsf T}(x)\,\sigma_2\,\xi(x) + \left(m\,\xi^{\mathsf T}(x)\,\sigma_2\,\xi(x)\right)^\dagger\right], \tag{10.266}$$

which is Lorentz covariant because the matrices σ_1 and σ_3 anticommute with σ_2, which is antisymmetric (exercise 10.39). Since charge is conserved, only neutral fields like neutrinos can have Majorana mass terms.

The generators of the representation $D^{(0,1/2)}$ with $j = 0$ and $j' = 1/2$ are given by (10.246 & 10.248) with $\boldsymbol{J}^+ = 0$ and $\boldsymbol{J}^- = \boldsymbol{\sigma}/2$; they are

$$\boldsymbol{J} = \frac{1}{2}\boldsymbol{\sigma} \quad \text{and} \quad \boldsymbol{K} = i\frac{1}{2}\boldsymbol{\sigma}. \tag{10.267}$$

Thus the 2×2 matrix $D^{(0,1/2)}(\boldsymbol{\theta},\boldsymbol{\lambda})$ that represents the Lorentz transformation (10.241)

$$L = e^{-i\theta_\ell J_\ell - i\lambda_j K_j} \tag{10.268}$$

is

$$D^{(0,1/2)}(\boldsymbol{\theta},\boldsymbol{\lambda}) = \exp\left(-i\boldsymbol{\theta}\cdot\boldsymbol{\sigma}/2 + \boldsymbol{\lambda}\cdot\boldsymbol{\sigma}/2\right) = D^{(1/2,0)}(\boldsymbol{\theta},\,-\boldsymbol{\lambda}), \tag{10.269}$$

which differs from $D^{(1/2,0)}(\boldsymbol{\theta},\boldsymbol{\lambda})$ only by the sign of $\boldsymbol{\lambda}$. The generic $D^{(0,1/2)}$ matrix is the complex unimodular 2×2 matrix

$$D^{(0,1/2)}(\boldsymbol{\theta},\boldsymbol{\lambda}) = e^{\boldsymbol{z}^*\cdot\boldsymbol{\sigma}/2} \tag{10.270}$$

with $\boldsymbol{\lambda} = \mathrm{Re}\boldsymbol{z}$ and $\boldsymbol{\theta} = \mathrm{Im}\boldsymbol{z}$.

Example 10.32 (The standard right-handed boost) For a particle of mass $m > 0$, the "standard" boost (10.253) that transforms $k = (m, \boldsymbol{0})$ to $p = (p^0, \boldsymbol{p})$ is the 4×4 matrix $B(p) = \exp\left(\alpha\,\hat{\boldsymbol{p}}\cdot\boldsymbol{B}\right)$ in which $\cosh\alpha = p^0/m$ and $\sinh\alpha = |\boldsymbol{p}|/m$. This Lorentz transformation with $\boldsymbol{\theta} = \boldsymbol{0}$ and $\boldsymbol{\lambda} = \alpha\,\hat{\boldsymbol{p}}$ is represented by the matrix (exercise 10.35)

$$\begin{aligned} D^{(0,1/2)}(\boldsymbol{0},\alpha\,\hat{\boldsymbol{p}}) &= e^{\alpha\hat{\boldsymbol{p}}\cdot\boldsymbol{\sigma}/2} = I\,\cosh(\alpha/2) + \hat{\boldsymbol{p}}\cdot\boldsymbol{\sigma}\,\sinh(\alpha/2) \\ &= I\sqrt{(p^0+m)/(2m)} + \hat{\boldsymbol{p}}\cdot\boldsymbol{\sigma}\sqrt{(p^0-m)/(2m)} \\ &= \frac{p^0 + m + \boldsymbol{p}\cdot\boldsymbol{\sigma}}{\sqrt{2m(p^0+m)}}, \end{aligned} \tag{10.271}$$

in the third line of which the 2×2 identity matrix I is suppressed. □

Under $D^{(0,1/2)}$, the vector $(I, \boldsymbol{\sigma})$ transforms as a 4-vector; for tiny $\boldsymbol{z}$

$$\begin{aligned} D^{\dagger(0,1/2)}(\boldsymbol{\theta},\boldsymbol{\lambda})\,I\,D^{(0,1/2)}(\boldsymbol{\theta},\boldsymbol{\lambda}) &= I + \boldsymbol{\lambda}\cdot\boldsymbol{\sigma}, \\ D^{\dagger(0,1/2)}(\boldsymbol{\theta},\boldsymbol{\lambda})\,\boldsymbol{\sigma}\,D^{(0,1/2)}(\boldsymbol{\theta},\boldsymbol{\lambda}) &= \boldsymbol{\sigma} + I\boldsymbol{\lambda} + \boldsymbol{\theta}\wedge\boldsymbol{\sigma} \end{aligned} \tag{10.272}$$

as in (10.240).

A field $\zeta(x)$ that responds to a unitary Lorentz transformation $U(L)$ as

$$U(L)\,\zeta(x)\,U^{-1}(L) = D^{(0,1/2)}(L^{-1})\,\zeta(Lx) \tag{10.273}$$

is called a **right-handed Weyl spinor**. One may show (exercise 10.37) that the action density

$$\mathcal{L}_{\rm r}(x) = i\,\zeta^{\dagger}(x)\,(\partial_0 I + \nabla\cdot\boldsymbol{\sigma})\,\zeta(x) \tag{10.274}$$

is Lorentz covariant

$$U(L)\,\mathcal{L}_{\rm r}(x)\,U^{-1}(L) = \mathcal{L}_{\rm r}(Lx). \tag{10.275}$$

Example 10.33 (Why ζ is right-handed) An argument like that of example (10.31) shows that the field $\zeta(x)$ satisfies the wave equation

$$(\partial_0 I + \nabla\cdot\boldsymbol{\sigma})\,\zeta(x) = 0 \tag{10.276}$$

or in momentum space

$$(E - \boldsymbol{p}\cdot\boldsymbol{\sigma})\,\zeta(p) = 0. \tag{10.277}$$

Thus, $E = |\boldsymbol{p}|$, and $\zeta(p)$ is an eigenvector of $\hat{\boldsymbol{p}}\cdot\boldsymbol{J}$

$$\hat{\boldsymbol{p}}\cdot\boldsymbol{J}\,\zeta(p) = \frac{1}{2}\,\zeta(p) \tag{10.278}$$

with eigenvalue 1/2. A particle whose spin is parallel to its momentum is said to have **positive helicity** or to be **right-handed**. Nearly massless antineutrinos are nearly right-handed. □

The Majorana mass term

$$\mathcal{L}_{\rm M}(x) = \tfrac{1}{2}\left[im\,\zeta^{\sf T}(x)\,\sigma_2\,\zeta(x) + \left(im\,\zeta^{\sf T}(x)\,\sigma_2\,\zeta(x)\right)^{\dagger}\right] \tag{10.279}$$

like (10.266) is Lorentz covariant.

10.33 The Dirac representation of the Lorentz group

Dirac's representation of $SO(3,1)$ is the direct sum $D^{(1/2,0)} \oplus D^{(0,1/2)}$ of $D^{(1/2,0)}$ and $D^{(0,1/2)}$. Its generators are the 4×4 matrices

$$\boldsymbol{J} = \frac{1}{2}\begin{pmatrix}\boldsymbol{\sigma} & 0\\ 0 & \boldsymbol{\sigma}\end{pmatrix} \quad\text{and}\quad \boldsymbol{K} = \frac{i}{2}\begin{pmatrix}-\boldsymbol{\sigma} & 0\\ 0 & \boldsymbol{\sigma}\end{pmatrix}. \tag{10.280}$$

Dirac's representation uses the **Clifford algebra** of the gamma matrices γ^a, which satisfy the anticommutation relation

$$\{\gamma^a, \gamma^b\} \equiv \gamma^a\gamma^b + \gamma^b\gamma^a = 2\eta^{ab} I, \tag{10.281}$$

in which η is the 4×4 diagonal matrix (10.222) with $\eta^{00} = -1$ and $\eta^{jj} = 1$ for $j = 1, 2,$ and 3 and I is the 4×4 identity matrix.

Remarkably, the generators of the Lorentz group

$$J^{ij} = \epsilon_{ijk} J_k \quad \text{and} \quad J^{0j} = K_j \tag{10.282}$$

may be represented as commutators of gamma matrices

$$J^{ab} = -\frac{i}{4}[\gamma^a, \gamma^b]. \tag{10.283}$$

They transform the gamma matrices as a 4-vector

$$[J^{ab}, \gamma^c] = -i\gamma^a \eta^{bc} + i\gamma^b \eta^{ac} \tag{10.284}$$

(exercise 10.40) and satisfy the commutation relations

$$i[J^{ab}, J^{cd}] = \eta^{bc} J^{ad} - \eta^{ac} J^{bd} - \eta^{da} J^{cb} + \eta^{db} J^{ca} \tag{10.285}$$

of the Lorentz group (Weinberg, 1995, pp. 213–217) (exercise 10.41).

The gamma matrices γ^a are not unique; if S is any 4×4 matrix with an inverse, then the matrices $\gamma'^a \equiv S\gamma^a S^{-1}$ also satisfy the definition (10.281). The choice

$$\gamma^0 = -i\begin{pmatrix} 0 & 1 \\ 1 & 0 \end{pmatrix} \quad \text{and} \quad \boldsymbol{\gamma} = -i\begin{pmatrix} 0 & \boldsymbol{\sigma} \\ -\boldsymbol{\sigma} & 0 \end{pmatrix} \tag{10.286}$$

is useful in high-energy physics because it lets us assemble a left-handed spinor and a right-handed spinor into a 4-component Majorana spinor

$$\psi_{\mathrm{M}} = \begin{pmatrix} \xi \\ \zeta \end{pmatrix}. \tag{10.287}$$

If two Majorana spinors $\psi_{\mathrm{M}}^{(1)}$ and $\psi_{\mathrm{M}}^{(2)}$ have the same mass, then one may combine them into a Dirac spinor

$$\psi_{\mathrm{D}} = \frac{1}{\sqrt{2}}\left(\psi_{\mathrm{M}}^{(1)} + i\psi_{\mathrm{M}}^{(2)}\right) = \frac{1}{\sqrt{2}}\begin{pmatrix} \xi^{(1)} + i\xi^{(2)} \\ \zeta^{(1)} + i\zeta^{(2)} \end{pmatrix} = \begin{pmatrix} \xi_{\mathrm{D}} \\ \zeta_{\mathrm{D}} \end{pmatrix}. \tag{10.288}$$

The action for a Majorana or Dirac 4-spinor often is written as

$$\mathcal{L} = -\overline{\psi}\left(\gamma^a \partial_a + m\right)\psi \equiv -\overline{\psi}\left(\partial\!\!\!/ + m\right)\psi, \tag{10.289}$$

in which

$$\overline{\psi} \equiv i\psi^\dagger \gamma^0 = \psi^\dagger \begin{pmatrix} 0 & 1 \\ 1 & 0 \end{pmatrix} = \begin{pmatrix} \zeta^\dagger & \xi^\dagger \end{pmatrix}. \tag{10.290}$$

The kinetic part is the sum of the left-handed $\mathcal{L}_\ell$ and right-handed $\mathcal{L}_{\mathrm{r}}$ action densities (10.258 & 10.274)

$$-\overline{\psi}\gamma^a \partial_a \psi = i\xi^\dagger\left(\partial_0 I - \nabla \cdot \boldsymbol{\sigma}\right)\xi + i\zeta^\dagger\left(\partial_0 I + \nabla \cdot \boldsymbol{\sigma}\right)\zeta. \tag{10.291}$$

The **Dirac mass term**

$$-m\overline{\psi}\psi = -m\left(\zeta^\dagger\xi + \xi^\dagger\zeta\right) \tag{10.292}$$

conserves charge even if ψ is a charged Dirac 4-spinor ψ_D

$$\psi_D = \frac{1}{\sqrt{2}}\left(\psi^{(1)} + i\psi^{(2)}\right), \tag{10.293}$$

in which case it is

$$\begin{aligned}-m\overline{\psi}_D\psi_D &= -m\left(\zeta_D^\dagger\xi_D + \xi_D^\dagger\zeta_D\right)\\ &= -\frac{m}{2}\left[\left(\zeta^{(1)\dagger} - i\zeta^{(2)\dagger}\right)\left(\xi^{(1)} + i\xi^{(2)}\right)\right.\\ &\qquad\left.+\left(\xi^{(1)\dagger} - i\xi^{(2)\dagger}\right)\left(\zeta^{(1)} + i\zeta^{(2)}\right)\right].\end{aligned} \tag{10.294}$$

One may show (exercise 10.42) that if ξ is a left-handed spinor transforming as (10.257), then the spinor

$$\zeta = \sigma_2\,\xi^* \equiv \begin{pmatrix}0 & -i\\ i & 0\end{pmatrix}\begin{pmatrix}\xi_1^\dagger\\ \xi_2^\dagger\end{pmatrix} \tag{10.295}$$

transforms as a right-handed spinor (10.273), that is

$$e^{z^*\cdot\sigma/2}\sigma_2\,\xi^* = \sigma_2\left(e^{-z\cdot\sigma/2}\xi\right)^*. \tag{10.296}$$

Similarly, $\xi = \sigma_2\,\zeta^*$ is left-handed if ζ is right-handed. Thus

$$\zeta^\dagger\xi = \xi^\top\sigma_2\,\xi = \zeta^\dagger\sigma_2\,\zeta^*. \tag{10.297}$$

One therefore can write a Dirac mass term (10.298) as a specific combination of Majorana mass terms

$$\begin{aligned}-m\overline{\psi}_D\psi_D = -\frac{m}{2}&\left[\left(\xi^{(1)\top} - i\xi^{(2)\top}\right)\sigma_2\left(\xi^{(1)} + i\xi^{(2)}\right)\right.\\ &\left.+\left(\zeta^{(1)\top} - i\zeta^{(2)\top}\right)\sigma_2\left(\zeta^{(1)} + i\zeta^{(2)}\right)\right]\end{aligned} \tag{10.298}$$

or entirely in terms of either left-handed ξ or right-handed ζ spinors.

10.34 The Poincaré group

The elements of the Poincaré group are products of Lorentz transformations and translations in space and time. The Lie algebra of the Poincaré group therefore includes the generators $\boldsymbol{J}$ and $\boldsymbol{K}$ of the Lorentz group as well as the hamiltonian H and the momentum operator $\boldsymbol{P}$, which respectively generate translations in time and space.

Suppose $T(y)$ is a translation that takes a 4-vector x to $x+y$ and $T(z)$ is a translation that takes a 4-vector x to $x+z$. Then $T(z)T(y)$ and $T(y)T(z)$ both take x to $x+y+z$. So if a translation $T(y) = T(t, \boldsymbol{y})$ is represented by a unitary operator $U(t, \boldsymbol{y}) = \exp(iHt - i\boldsymbol{P} \cdot \boldsymbol{y})$, then the hamiltonian H and the momentum operator $\boldsymbol{P}$ commute with each other

$$[H, P^j] = 0 \quad \text{and} \quad [P^i, P^j] = 0. \tag{10.299}$$

We can figure out the commutation relations of H and $\boldsymbol{P}$ with the angular-momentum $\boldsymbol{J}$ and boost $\boldsymbol{K}$ operators by realizing that $P^a = (H, \boldsymbol{P})$ is a 4-vector. Let

$$U(\boldsymbol{\theta}, \boldsymbol{\lambda}) = e^{-i\boldsymbol{\theta}\cdot\boldsymbol{J} - i\boldsymbol{\lambda}\cdot\boldsymbol{K}} \tag{10.300}$$

be the (infinite-dimensional) unitary operator that represents (in Hilbert space) the infinitesimal Lorentz transformation

$$L = I + \boldsymbol{\theta} \cdot \boldsymbol{R} + \boldsymbol{\lambda} \cdot \boldsymbol{B} \tag{10.301}$$

where $\boldsymbol{R}$ and $\boldsymbol{B}$ are the six 4×4 matrices (10.231 & 10.232). Then because P is a 4-vector under Lorentz transformations, we have

$$U^{-1}(\boldsymbol{\theta}, \boldsymbol{\lambda}) P U(\boldsymbol{\theta}, \boldsymbol{\lambda}) = e^{+i\boldsymbol{\theta}\cdot\boldsymbol{J} + i\boldsymbol{\lambda}\cdot\boldsymbol{K}} P e^{-i\boldsymbol{\theta}\cdot\boldsymbol{J} - i\boldsymbol{\lambda}\cdot\boldsymbol{K}} = (I + \boldsymbol{\theta} \cdot \boldsymbol{R} + \boldsymbol{\lambda} \cdot \boldsymbol{B})\, P \tag{10.302}$$

or using (10.272)

$$\begin{aligned} (I + i\boldsymbol{\theta} \cdot \boldsymbol{J} + i\boldsymbol{\lambda} \cdot \boldsymbol{K})\, H\, (I - i\boldsymbol{\theta} \cdot \boldsymbol{J} - i\boldsymbol{\lambda} \cdot \boldsymbol{K}) &= H + \boldsymbol{\lambda} \cdot \boldsymbol{P}, \\ (I + i\boldsymbol{\theta} \cdot \boldsymbol{J} + i\boldsymbol{\lambda} \cdot \boldsymbol{K})\, \boldsymbol{P}\, (I - i\boldsymbol{\theta} \cdot \boldsymbol{J} - i\boldsymbol{\lambda} \cdot \boldsymbol{K}) &= \boldsymbol{P} + H\boldsymbol{\lambda} + \boldsymbol{\theta} \wedge \boldsymbol{P}. \end{aligned} \tag{10.303}$$

Thus, one finds (exercise 10.42) that H is invariant under rotations, while $\boldsymbol{P}$ transforms as a 3-vector

$$[J_i, H] = 0 \quad \text{and} \quad [J_i, P_j] = i\epsilon_{ijk} P_k \tag{10.304}$$

and that

$$[K_i, H] = -iP_i \quad \text{and} \quad [K_i, P_j] = -i\delta_{ij} H. \tag{10.305}$$

By combining these equations with (10.285), one may write (exercise 10.44) the Lie algebra of the Poincaré group as

$$\begin{aligned} i[J^{ab}, J^{cd}] &= \eta^{bc} J^{ad} - \eta^{ac} J^{bd} - \eta^{da} J^{cb} + \eta^{db} J^{ca}, \\ i[P^a, J^{bc}] &= \eta^{ab} P^c - \eta^{ac} P^b, \\ [P^a, P^b] &= 0. \end{aligned} \tag{10.306}$$

Further reading

The classic *Lie Algebras in Particle Physics* (Georgi, 1999), which inspired much of this chapter, is outstanding.

Exercises

10.1 Show that all $n \times n$ (real) orthogonal matrices O leave invariant the quadratic form $x_1^2 + x_2^2 + \cdots + x_n^2$, that is, that if $x' = Ox$, then $x'^2 = x^2$.

10.2 Show that the set of all $n \times n$ orthogonal matrices forms a group.

10.3 Show that all $n \times n$ unitary matrices U leave invariant the quadratic form $|x_1|^2 + |x_2|^2 + \cdots + |x_n|^2$, that is, that if $x' = Ux$, then $|x'|^2 = |x|^2$.

10.4 Show that the set of all $n \times n$ unitary matrices forms a group.

10.5 Show that the set of all $n \times n$ unitary matrices with unit determinant forms a group.

10.6 Show that the matrix $D^{(j)}_{m'm}(g) = \langle j, m'|U(g)|j, m\rangle$ is unitary because the rotation operator $U(g)$ is unitary $\langle j, m'|U^\dagger(g)U(g)|j, m\rangle = \delta_{m'm}$.

10.7 Invent a group of order 3 and compute its multiplication table. For extra credit, prove that the group is unique.

10.8 Show that the relation (10.20) between two equivalent representations is an isomorphism.

10.9 Suppose that D_1 and D_2 are equivalent, finite-dimensional, irreducible representations of a group G so that $D_2(g) = SD_1(g)S^{-1}$ for all $g \in G$. What can you say about a matrix A that satisfies $D_2(g)\, A = A\, D_1(g)$ for all $g \in G$?

10.10 Find all components of the matrix $\exp(i\alpha A)$ in which

$$A = \begin{pmatrix} 0 & 0 & -i \\ 0 & 0 & 0 \\ i & 0 & 0 \end{pmatrix}. \tag{10.307}$$

10.11 If $[A, B] = B$, find $e^{i\alpha A} B e^{-i\alpha A}$. Hint: what are the α-derivatives of this expression?

10.12 Show that the tensor-product matrix (10.31) of two representations D_1 and D_2 is a representation.

10.13 Find a 4×4 matrix S that relates the tensor-product representation $D_{1/2 \otimes 1/2}$ to the direct sum $D_1 \oplus D_0$.

10.14 Find the generators in the adjoint representation of the group with structure constants $f_{abc} = \epsilon_{abc}$ where a, b, c run from 1 to 3. Hint: the answer is three 3×3 matrices t_a, often written as L_a.

10.15 Show that the generators (10.90) satisfy the commutation relations (10.93).

10.16 Show that the demonstrated equation (10.98) implies the commutation relation (10.99).

10.17 Use the Cayley–Hamilton theorem (1.264) to show that the 3×3 matrix (10.96) that represents a right-handed rotation of θ radians about the axis $\boldsymbol{\theta}$ is given by (10.97).

10.18 Verify the mixed Jacobi identity (10.142).

10.19 For the group $SU(3)$, find the structure constants f_{123} and f_{231}.

10.20 Show that every 2×2 unitary matrix of unit determinant is a quaternion of unit norm.

10.21 Show that the quaternions as defined by (10.175) are closed under addition and multiplication and that the product xq is a quaternion if x is real and q is a quaternion.

10.22 Show that the one-sided derivative $f'(q)$ (10.184) of the quaternionic function $f(q) = q^2$ depends upon the direction along which $q' \to 0$.

10.23 Show that the generators (10.188) of $Sp(2n)$ obey commutation relations of the form (10.189) for some real structure constants f_{abc} and a suitably extended set of matrices $A, A', \ldots$ and $S_k, S'_k, \ldots$.

10.24 Show that for $0 < \epsilon \ll 1$, the real $2n \times 2n$ matrix $T = \exp(\epsilon JS)$ in which S is symmetric satisfies $T^{\mathsf{T}} J T = J$ (at least up to terms of order ϵ^2) and so is in $Sp(2n, R)$.

10.25 Show that the matrix T of (10.197) is in $Sp(2, R)$.

10.26 Using the parametrization (10.217) of the group $SU(2)$, show that the parameters $\boldsymbol{a}(\boldsymbol{c}, \boldsymbol{b})$ that describe the product $g(\boldsymbol{a}(\boldsymbol{c}, \boldsymbol{b})) = g(\boldsymbol{c})\, g(\boldsymbol{b})$ are those of (10.219).

10.27 Use formulas (10.219) and (10.212) to show that the left-invariant measure for $SU(2)$ is given by (10.220).

10.28 In tensor notation, which is explained in chapter 11, the condition (10.229) that $I + \omega$ be an infinitesimal Lorentz transformation reads $\left(\omega^{\mathsf{T}}\right)_b{}^a = \omega^a{}_b = -\,\eta_{bc}\, \omega^c{}_d\, \eta^{da}$ in which sums over c and d from 0 to 3 are understood. In this notation, the matrix η_{ef} lowers indices and η^{gh} raises them, so that $\omega_b{}^a = -\,\omega_{bd}\, \eta^{da}$. (Both η_{ef} and η^{gh} are numerically equal to the matrix η displayed in equation (10.222).) Multiply both sides of the condition (10.229) by $\eta_{ae} = \eta_{ea}$ and use the relation $\eta^{da}\, \eta_{ae} = \eta^d{}_e \equiv \delta^d{}_e$ to show that the matrix ω_{ab} with both indices lowered (or raised) is antisymmetric, that is,

$$\omega_{ba} = -\,\omega_{ab} \quad \text{and} \quad \omega^{ba} = -\,\omega^{ab}. \tag{10.308}$$

10.29 Show that the six matrices (10.231) and (10.232) satisfy the $SO(3,1)$ condition (10.229).

10.30 Show that the six generators $\boldsymbol{J}$ and $\boldsymbol{K}$ obey the commutations relations (10.234–10.236).

10.31 Show that if $\boldsymbol{J}$ and $\boldsymbol{K}$ satisfy the commutation relations (10.234–10.236) of the Lie algebra of the Lorentz group, then so do $\boldsymbol{J}$ and $-\,\boldsymbol{K}$.

10.32 Show that if the six generators $\boldsymbol{J}$ and $\boldsymbol{K}$ obey the commutation relations (10.234–10.236), then the six generators $\boldsymbol{J}^+$ and $\boldsymbol{J}^-$ obey the commutation relations (10.243).

10.33 Relate the parameter α in the definition (10.253) of the standard boost $B(p)$ to the 4-vector p and the mass m.

10.34 Derive the formulas for $D^{(1/2,0)}(\boldsymbol{0}, \alpha\, \hat{\boldsymbol{p}})$ given in equation (10.254).

10.35 Derive the formulas for $D^{(0,1/2)}(\boldsymbol{0}, \alpha\, \hat{\boldsymbol{p}})$ given in equation (10.271).

10.36 For infinitesimal complex $\boldsymbol{z}$, derive the 4-vector properties (10.255 & 10.272) of $(-I, \boldsymbol{\sigma})$ under $D^{(1/2,0)}$ and of $(I, \boldsymbol{\sigma})$ under $D^{(0,1/2)}$.

10.37 Show that under the unitary Lorentz transformation (10.257), the action density (10.258) is Lorentz covariant (10.259).

10.38 Show that under the unitary Lorentz transformation (10.273), the action density (10.274) is Lorentz covariant (10.275).

10.39 Show that under the unitary Lorentz transformations (10.257 & 10.273), the Majorana mass terms (10.266 & 10.279) are Lorentz covariant.

10.40 Show that the definitions of the gamma matrices (10.281) and of the generators (10.283) imply that the gamma matrices transform as a 4-vector under Lorentz transformations (10.284).

10.41 Show that (10.283) and (10.284) imply that the generators J^{ab} satisfy the commutation relations (10.285) of the Lorentz group.

10.42 Show that the spinor $\zeta = \sigma_2 \xi^*$ defined by (10.295) is right-handed (10.273) if ξ is left-handed (10.257).

10.43 Use (10.303) to get (10.304 & 10.305).

10.44 Derive (10.306) from (10.285, 10.299, 10.304 & 10.305).

11

Tensors and local symmetries

11.1 Points and coordinates

A point on a curved surface or in a curved space also is a point in a higher-dimensional flat space called an embedding space. For instance, a point on a sphere also is a point in three-dimensional euclidean space and in four-dimensional space-time. One always can add extra dimensions, but it's simpler to use as few as possible, three in the case of a sphere.

On a sufficiently small scale, any reasonably smooth space locally looks like n-dimensional euclidean space. Such a space is called a **manifold**. Incidentally, according to Whitney's embedding theorem, every n-dimensional connected, smooth manifold can be embedded in $2n$-dimensional euclidean space $\mathbb{R}^{2n}$. So the embedding space for such spaces in general relativity has no more than eight dimensions.

We use coördinates to label points. For example, we can choose a polar axis and a meridian and label a point on the sphere by its polar and azimuthal angles (θ, ϕ) with respect to that axis and meridian. If we use a different axis and meridian, then the coordinates (θ', ϕ') for the same point will change. **Points are physical, coordinates are metaphysical. When we change our system of coordinates, the points don't change, but their coordinates do.**

Most points p have unique coordinates $x^i(p)$ and $x'^i(p)$ in their coordinate systems. For instance, polar coordinates (θ, ϕ) are unique for all points on a sphere – except the north and south poles which are labeled by $\theta = 0$ and $\theta = \pi$ and all $0 \leq \phi < 2\pi$. By using more than one coordinate system, one usually can arrange to label every point uniquely. In the flat three-dimensional space in which the sphere is a surface, each point of the sphere has unique coordinates, $\vec{p} = (x, y, z)$.

We will use coordinate systems that represent points on the manifold uniquely and smoothly at least in local patches, so that the maps

$$x'^i = x'^i(p) = x'^i(p(x)) = x'^i(x) \tag{11.1}$$

and

$$x^i = x^i(p) = x^i(p(x')) = x^i(x') \tag{11.2}$$

are well defined, differentiable, and one to one in the patches. We'll often group the n coordinates x^i together and write them collectively as x without a superscript. Since the coordinates $x(p)$ label the point p, we sometimes will call them "the point x." But p and x are different. The point p is unique with infinitely many coordinates $x, x', x'', \ldots$ in infinitely many coordinate systems.

11.2 Scalars

A **scalar** is a quantity B that is the same in all coordinate systems

$$B' = B. \tag{11.3}$$

If it also depends upon the coordinates x of the space-time point p, and

$$B'(x') = B(x), \tag{11.4}$$

then it is a **scalar field**.

11.3 Contravariant vectors

The change dx'^i due to changes in the unprimed coordinates is

$$dx'^i = \sum_j \frac{\partial x'^i}{\partial x^j} dx^j. \tag{11.5}$$

This rule defines **contravariant vectors**: a quantity A^i is a contravariant vector if it transforms like dx^i

$$A'^i = \sum_j \frac{\partial x'^i}{\partial x^j} A^j. \tag{11.6}$$

The coordinate differentials dx^i form a contravariant vector. A contravariant vector $A^i(x)$ that depends on the coordinates x and transforms as

$$A'^i(x') = \sum_j \frac{\partial x'^i}{\partial x^j} A^j(x) \tag{11.7}$$

is a **contravariant vector field**.

11.4 Covariant vectors

The chain rule for partial derivatives

$$\frac{\partial}{\partial x'^i} = \sum_j \frac{\partial x^j}{\partial x'^i} \frac{\partial}{\partial x^j} \tag{11.8}$$

defines **covariant vectors**: a vector C_i that transforms as

$$C'_i = \sum_j \frac{\partial x^j}{\partial x'^i} C_j \tag{11.9}$$

is a **covariant vector**. If it also is a function of x, then it is a **covariant vector field** and

$$C'_i(x') = \sum_j \frac{\partial x^j}{\partial x'^i} C_j(x). \tag{11.10}$$

Example 11.1 (Gradient of a scalar) The derivatives of a scalar field form a covariant vector field. For by using the chain rule to differentiate the equation $B'(x') = B(x)$ that defines a scalar field, one finds

$$\frac{\partial B'(x')}{\partial x'^i} = \frac{\partial B(x)}{\partial x'^i} = \sum_j \frac{\partial x^j}{\partial x'^i} \frac{\partial B(x)}{\partial x^j}, \tag{11.11}$$

which shows that the gradient $\partial B(x)/\partial x^j$ is a covariant vector field. □

11.5 Euclidean space in euclidean coordinates

If we use euclidean coordinates to describe points in euclidean space, then covariant and contravariant vectors are the same.

Euclidean space has a natural inner product (section 1.6), the usual dot-product, which is real and symmetric. In a euclidean space of n dimensions, we may choose any n fixed, orthonormal basis vectors e_i

$$(e_i, e_j) \equiv e_i \cdot e_j = \sum_{k=1}^{n} e_i^k \, e_j^k = \delta_{ij} \tag{11.12}$$

and use them to represent any point p as the linear combination

$$p = \sum_{i=1}^{n} e_i \, x^i. \tag{11.13}$$

The coefficients x^i are the euclidean coordinates in the e_i basis. Since the basis vectors e_i are orthonormal, each x^i is an inner product or dot-product

$$x^i = e_i \cdot p = \sum_{j=1}^{n} e_i \cdot e_j \, x^j = \sum_{j=1}^{n} \delta_{ij} \, x^j. \tag{11.14}$$

The dual vectors e^i are defined as those vectors whose inner products with the e_j are $(e^i, e_j) = \delta^i_j$. In this section, they are the same as the vectors e_i, and so we shall not bother to distinguish e^i from $e_i = e^i$.

If we use different orthonormal vectors e'_i as a basis

$$p = \sum_{i=1}^{n} e'_i \, x'^i \tag{11.15}$$

then we get new euclidean coordinates $x'_i = e'_i \cdot p$ for the same point p. These two sets of coordinates are related by the equations

$$x'^i = e'_i \cdot p = \sum_{j=1}^{n} e'_i \cdot e_j \, x^j,$$
$$x^j = e_j \cdot p = \sum_{k=1}^{n} e_j \cdot e'_k \, x'^k. \tag{11.16}$$

Because the basis vectors e and e' are all **independent of** x, the coefficients $\partial x'^i / \partial x^j$ of the transformation laws for contravariant (11.6) and covariant (11.9) vectors are

$$\text{contravariant} \quad \frac{\partial x'^i}{\partial x^j} = e'_i \cdot e_j \quad \text{and} \quad \frac{\partial x^j}{\partial x'^i} = e_j \cdot e'_i \quad \text{covariant}. \tag{11.17}$$

But the dot-product (1.82) is symmetric, and so these are the same:

$$\frac{\partial x'^i}{\partial x^j} = e'_i \cdot e_j = e_j \cdot e'_i = \frac{\partial x^j}{\partial x'^i}. \tag{11.18}$$

Contravariant and covariant vectors transform the same way in euclidean space with euclidean coordinates.

The relations between x'^i and x^j imply that

$$x'^i = \sum_{j,k=1}^{n} \left(e'_i \cdot e_j\right) \left(e_j \cdot e'_k\right) x'^k. \tag{11.19}$$

Since this holds for all coordinates x'^i, we have

$$\sum_{j=1}^{n} \left(e'_i \cdot e_j\right) \left(e_j \cdot e'_k\right) = \delta_{ik}. \tag{11.20}$$

The coefficients $e'_i \cdot e_j$ form an orthogonal matrix, and the linear operator

$$\sum_{i=1}^{n} e_i e_i'^{\mathsf{T}} = \sum_{i=1}^{n} |e_i\rangle\langle e'_i| \tag{11.21}$$

is an orthogonal (real, unitary) transformation. The change $x \to x'$ is a rotation plus a possible reflection (exercise 11.2).

Example 11.2 (A euclidean space of two dimensions) In two-dimensional euclidean space, one can describe the same point by euclidean (x, y) and polar (r, θ) coordinates. The derivatives

$$\frac{\partial r}{\partial x} = \frac{x}{r} = \frac{\partial x}{\partial r} \quad \text{and} \quad \frac{\partial r}{\partial y} = \frac{y}{r} = \frac{\partial y}{\partial r} \tag{11.22}$$

respect the symmetry (11.18), but (exercise 11.1) these derivatives

$$\frac{\partial \theta}{\partial x} = -\frac{y}{r^2} \neq \frac{\partial x}{\partial \theta} = -y \quad \text{and} \quad \frac{\partial \theta}{\partial y} = \frac{x}{r^2} \neq \frac{\partial y}{\partial \theta} = x \tag{11.23}$$

do not. □

11.6 Summation conventions

When a given index is repeated in a product, that index usually is being summed over. So to avoid distracting summation symbols, one writes

$$A_i\, B_i \equiv \sum_{i=1}^{n} A_i\, B_i. \tag{11.24}$$

The sum is understood to be over the relevant range of indices, usually from 0 or 1 to 3 or n. Where the distinction between covariant and contravariant indices matters, an index that appears twice in the same monomial, once as a subscript and once as a superscript, is a dummy index that is summed over as in

$$A_i\, B^i \equiv \sum_{i=1}^{n} A_i\, B^i. \tag{11.25}$$

These summation conventions make tensor notation almost as compact as matrix notation. They make equations easier to read and write.

Example 11.3 (The Kronecker delta) The summation convention and the chain rule imply that

$$\frac{\partial x'^i}{\partial x^k}\frac{\partial x^k}{\partial x'^j} = \frac{\partial x'^i}{\partial x'^j} = \delta^i_j = \begin{cases} 1 & \text{if } i = j, \\ 0 & \text{if } i \neq j. \end{cases} \tag{11.26}$$

The repeated index k has disappeared in this **contraction**. □

11.7 Minkowski space

Minkowski space has one time dimension, labeled by $k = 0$, and n space dimensions. In special relativity $n = 3$, and the Minkowski metric η

$$\eta_{kl} = \eta^{kl} = \begin{cases} -1 & \text{if } k = l = 0, \\ 1 & \text{if } 1 \le k = l \le 3\,, \\ 0 & \text{if } k \neq l \end{cases} \tag{11.27}$$

defines an inner product between points p and q with coordinates x^k_p and x^ℓ_q as

$$(p, q) = p \cdot q = p^k\,\eta_{kl}\,q^l = (q, p). \tag{11.28}$$

If one time component vanishes, the Minkowski inner product reduces to the euclidean dot-product (1.82).

We can use different sets $\{e_i\}$ and $\{e'_i\}$ of $n + 1$ **Lorentz-orthonormal** basis vectors

$$(e_i, e_j) = e_i \cdot e_j = e^k_i\,\eta_{kl}\,e^l_j = \eta_{ij} = e'_i \cdot e'_j = (e'_i, e'_j) \tag{11.29}$$

to represent any point p in the space either as a linear combination of the vectors e_i with coefficients x^i or as a linear combination of the vectors e'_i with coefficients x'^i

$$p = e_i\,x^i = e'_i\,x'^i. \tag{11.30}$$

The dual vectors, which carry upper indices, are defined as

$$e^i = \eta^{ij}\,e_j \quad \text{and} \quad e'^i = \eta^{ij}\,e'_j. \tag{11.31}$$

They are orthonormal to the vectors e_i and e'_i because

$$(e^i, e_j) = e^i \cdot e_j = \eta^{ik}\,e_k \cdot e_j = \eta^{ik}\,\eta_{kj} = \delta^i_j \tag{11.32}$$

and similarly $(e'^i, e'_j) = e'^i \cdot e'_j = \delta^i_j$. Since the square of the matrix η is the identity matrix $\eta_{\ell i}\eta^{ij} = \delta^j_\ell$, it follows that

$$e_i = \eta_{ij}\,e^j \quad \text{and} \quad e'_i = \eta_{ij}\,e'^j. \tag{11.33}$$

The metric η raises (11.31) and lowers (11.33) the index of a basis vector.

The component x'^i is related to the components x^j by the linear map

$$x'^i = e'^i \cdot p = e'^i \cdot e_j \, x^j. \tag{11.34}$$

Such a map from a 4-vector x to a 4-vector x' is a **Lorentz transformation**

$$x'^i = L^i_j \, x^j \quad \text{with matrix} \quad L^i_j = e'^i \cdot e_j. \tag{11.35}$$

The inner product (p, q) of two points $p = e_i \, x^i = e'_i \, x'^i$ and $q = e_k \, y^k = e'_k \, y'^k$ is **physical** and so is **invariant under Lorentz transformations**

$$(p, q) = x^i \, y^k \, e_i \cdot e_k = \eta_{ik} \, x^i \, y^k = x'^i \, y'^k \, e'_i \cdot e'_k = \eta_{ik} \, x'^i \, y'^k. \tag{11.36}$$

With $x'^i = L^i{}_r \, x^r$ and $y'^k = L^k{}_s \, x^s$, this invariance is

$$\eta_{rs} \, x^r y^s = \eta_{ik} \, L^i{}_r \, x^r \, L^k{}_s \, y^s \tag{11.37}$$

or since x^r and y^s are arbitrary

$$\eta_{rs} = \eta_{ik} \, L^i{}_r \, L^k{}_s = L^i{}_r \, \eta_{ik} \, L^k{}_s. \tag{11.38}$$

In matrix notation, a left index labels a row, and a right index labels a column. Transposition interchanges rows and columns $L^i{}_r = L^{\mathsf{T}i}{}_r$, so

$$\eta_{rs} = L^{\mathsf{T}i}{}_r \, \eta_{ik} \, L^k{}_s \quad \text{or} \quad \eta = L^{\mathsf{T}} \, \eta \, L \tag{11.39}$$

in matrix notation. In such matrix products, the height of an index – whether it is up or down – determines whether it is contravariant or covariant but does not affect its place in its matrix.

Example 11.4 (A boost) The matrix

$$L = \begin{pmatrix} \gamma & \sqrt{\gamma^2 - 1} & 0 & 0 \\ \sqrt{\gamma^2 - 1} & \gamma & 0 & 0 \\ 0 & 0 & 1 & 0 \\ 0 & 0 & 0 & 1 \end{pmatrix} \tag{11.40}$$

where $\gamma = 1/\sqrt{1 - v^2/c^2}$ represents a Lorentz transformation that is a boost in the x-direction. Boosts and rotations are Lorentz transformations. Working with 4 × 4 matrices can get tedious, so students are advised to think in terms of scalars, like $p \cdot x = p^i \eta_{ij} x^j = \boldsymbol{p} \cdot \boldsymbol{x} - Et$ whenever possible. □

If the basis vectors e and e' are independent of p and of x, then the coefficients of the transformation law (11.6) for contravariant vectors are

$$\frac{\partial x'^i}{\partial x^j} = e'^i \cdot e_j. \tag{11.41}$$

Similarly, the component x^j is $x^j = e^j \cdot p = e^j \cdot e'_i x'^i$, so the coefficients of the transformation law (11.9) for covariant vectors are

$$\frac{\partial x^j}{\partial x'^i} = e^j \cdot e'_i. \tag{11.42}$$

Using η to raise and lower the indices in the formula (11.41) for the coefficients of the transformation law (11.6) for contravariant vectors, we find

$$\frac{\partial x'^i}{\partial x^j} = e'^i \cdot e_j = \eta^{ik} \, \eta_{j\ell} \, e'_k \cdot e^\ell = \eta^{ik} \, \eta_{j\ell} \, \frac{\partial x^\ell}{\partial x'^k}, \tag{11.43}$$

which is $\pm \, \partial x^j / \partial x'^i$. So if we use coordinates associated with fixed basis vectors e_i in Minkowski space, then the coefficients for the two kinds of transformation laws differ only by occasional minus signs.

So if A^i is a contravariant vector

$$A'^i = \frac{\partial x'^i}{\partial x^j} \, A^j \tag{11.44}$$

then the relation (11.43) between the two kinds of coefficient implies that

$$\eta_{si} \, A'^i = \eta_{si} \, \frac{\partial x'^i}{\partial x^j} \, A^j = \eta_{si} \, \eta^{ik} \, \eta_{j\ell} \, \frac{\partial x^\ell}{\partial x'^k} \, A^j = \delta^k_s \, \frac{\partial x^\ell}{\partial x'^k} \, \eta_{j\ell} \, A^j = \frac{\partial x^\ell}{\partial x'^s} \, \eta_{\ell j} \, A^j, \tag{11.45}$$

which shows that $A_\ell = \eta_{\ell j} \, A^j$ transforms covariantly

$$A'_s = \frac{\partial x^\ell}{\partial x'^s} A_\ell. \tag{11.46}$$

The metric η turns a contravariant vector into a covariant one. It also switches a covariant vector A_ℓ back to its contravariant form A^k

$$\eta^{k\ell} A_\ell = \eta^{k\ell} \eta_{\ell j} A^j = \delta^k_j A^j = A^k. \tag{11.47}$$

In Minkowski space, one uses η to raise and lower indices

$$A_i = \eta_{ij} \, A^j \text{ and } A^i = \eta^{ij} \, A_j. \tag{11.48}$$

In general relativity, the space-time metric g raises and lowers indices.

11.8 Lorentz transformations

In section 11.7, Lorentz transformations arose as linear maps of the coordinates due to a change of basis. They also are linear maps of the basis vectors e_i that preserve the inner products

$$(e_i, e_j) = e_i \cdot e_j = \eta_{ij} = e'_i \cdot e'_j = (e'_i, e'_j). \tag{11.49}$$

The vectors e_i are four linearly independent four-dimensional vectors, and so they span four-dimensional Minkowski space and can represent the vectors e'_i as

$$e'_i = \Lambda_i{}^k \, e_k \tag{11.50}$$

where the coefficients $\Lambda_i{}^k$ are real numbers. The requirement that the new basis vectors e'_i are Lorentz orthonormal gives

$$\eta_{ij} = e'_i \cdot e'_j = \Lambda_i{}^k \, e_k \cdot \Lambda_j{}^\ell \, e_\ell = \Lambda_i{}^k \, e_k \cdot e_\ell \, \Lambda_j{}^\ell = \Lambda_i{}^k \, \eta_{k\ell} \, \Lambda_j{}^\ell \tag{11.51}$$

or in matrix notation

$$\eta = \Lambda \, \eta \, \Lambda^{\mathsf{T}} \tag{11.52}$$

where Λ^{T} is the transpose $(\Lambda^{\mathsf{T}})^\ell{}_j = \Lambda_j{}^\ell$. Evidently Λ^{T} satisfies the definition (11.39) of a Lorentz transformation. What Lorentz transformation is it? The point p must remain invariant, so by (11.35 & 11.50) one has

$$p = e'_i \, x'^i = \Lambda_i{}^k \, e_k \, L^i{}_j x^j = \delta^k_j \, e_k \, x^j = e_j \, x^j \tag{11.53}$$

whence $\Lambda_i{}^k \, L^i{}_j = \delta^k_j$ or $\Lambda^{\mathsf{T}} L = I$. So $\Lambda^{\mathsf{T}} = L^{-1}$.

By multiplying condition (11.52) by the metric η first from the left and then from the right and using the fact that $\eta^2 = I$, we find

$$1 = \eta^2 = \eta \, \Lambda \, \eta \, \Lambda^{\mathsf{T}} = \Lambda \, \eta \, \Lambda^{\mathsf{T}} \, \eta, \tag{11.54}$$

which gives us the inverse matrices

$$\Lambda^{-1} = \eta \, \Lambda^{\mathsf{T}} \, \eta = L^{\mathsf{T}} \quad \text{and} \quad (\Lambda^{\mathsf{T}})^{-1} = \eta \, \Lambda \, \eta = L. \tag{11.55}$$

In special relativity, contravariant vectors transform as

$$dx'^i = L^i{}_j \, dx^j \tag{11.56}$$

and since $x^j = L^{-1j}{}_i \, x'^i$, the covariant ones transform as

$$\frac{\partial}{\partial x'^i} = \frac{\partial x^j}{\partial x'^i} \frac{\partial}{\partial x^j} = L^{-1j}{}_i \, \frac{\partial}{\partial x^j} = \Lambda_i{}^j \, \frac{\partial}{\partial x^j}. \tag{11.57}$$

By taking the determinant of both sides of (11.52) and using the transpose (1.194) and product (1.207) rules for determinants, we find that $\det \Lambda = \pm 1$.

11.9 Special relativity

The space-time of special relativity is flat, four-dimensional Minkowski space. The inner product $(p - q) \cdot (p - q)$ of the interval $p - q$ between two points is physical and independent of the coordinates and therefore invariant. If the points p and q are close neighbors with coordinates $x^i + dx^i$ for p and x^i for q, then that invariant inner product is

$$(p-q)\cdot(p-q) = e_i\,dx^i \cdot e_j\,dx^j = dx^i\,\eta_{ij}\,dx^j = \mathbf{dx}^2 - (dx^0)^2 \tag{11.58}$$

with $dx^0 = c\,dt$. (At some point in what follows, we'll measure distance in light-seconds so that $c = 1$.) If the points p and q are on the trajectory of a massive particle moving at velocity $\boldsymbol{v}$, then this invariant quantity is the square of the **invariant distance**

$$ds^2 = \mathbf{dx}^2 - c^2 dt^2 = \left(\boldsymbol{v}^2 - c^2\right) dt^2, \tag{11.59}$$

which always is negative since $v < c$. The time in the rest frame of the particle is the **proper time**. The square of its differential element is

$$d\tau^2 = -\,ds^2/c^2 = \left(1 - v^2/c^2\right) dt^2. \tag{11.60}$$

A particle of mass zero moves at the speed of light, and so its proper time is zero. But for a particle of mass $m > 0$ moving at speed v, the element of proper time $d\tau$ is smaller than the corresponding element of laboratory time dt by the factor $\sqrt{1 - v^2/c^2}$. The proper time is the time in the rest frame of the particle, $d\tau = dt$ when $v = 0$. So if $T(0)$ is the lifetime of a particle (at rest), then the apparent lifetime $T(v)$ when the particle is moving at speed v is

$$T(v) = dt = \frac{d\tau}{\sqrt{1 - v^2/c^2}} = \frac{T(0)}{\sqrt{1 - v^2/c^2}}, \tag{11.61}$$

which is longer – an effect known as **time dilation**.

Example 11.5 (Time dilation in muon decay) A muon at rest has a mean life of $T(0) = 2.2 \times 10^{-6}$ seconds. Cosmic rays hitting nitrogen and oxygen nuclei make pions high in the Earth's atmosphere. The pions rapidly decay into muons in 2.6×10^{-8} s. A muon moving at the speed of light from 10 km takes at least $t = 10\,\text{km}/300{,}000\,(\text{km/sec}) = 3.3 \times 10^{-5}$ s to hit the ground. Were it not for time dilation, the probability P of such a muon reaching the ground as a muon would be

$$P = e^{-t/T(0)} = \exp(-33/2.2) = e^{-15} = 2.6 \times 10^{-7}. \tag{11.62}$$

The (rest) mass of a muon is 105.66 MeV. So a muon of energy $E = 749$ MeV has by (11.69) a time-dilation factor of

$$\frac{1}{\sqrt{1 - v^2/c^2}} = \frac{E}{mc^2} = \frac{749}{105.7} = 7.089 = \frac{1}{\sqrt{1 - (0.99)^2}}. \tag{11.63}$$

So a muon moving at a speed of $v = 0.99\,c$ has an apparent mean life $T(v)$ given by equation (11.61) as

$$T(v) = \frac{E}{mc^2}\,T(0) = \frac{T(0)}{\sqrt{1 - v^2/c^2}} = \frac{2.2 \times 10^{-6}\,\mathrm{s}}{\sqrt{1 - (0.99)^2}} = 1.6 \times 10^{-5}\,\mathrm{s}. \qquad (11.64)$$

The probability of survival with time dilation is

$$P = e^{-t/T(v)} = \exp(-33/16) = 0.12 \qquad (11.65)$$

so that 12% survive. Time dilation increases the chance of survival by a factor of 460,000 – no small effect. □

11.10 Kinematics

From the scalar $d\tau$, and the contravariant vector dx^i, we can make the 4-vector

$$u^i = \frac{dx^i}{d\tau} = \frac{dt}{d\tau}\left(\frac{dx^0}{dt}, \frac{d\boldsymbol{x}}{dt}\right) = \frac{1}{\sqrt{1 - v^2/c^2}}\,(c, \boldsymbol{v}), \qquad (11.66)$$

in which $u^0 = c\,dt/d\tau = c/\sqrt{1 - v^2/c^2}$ and $\boldsymbol{u} = u^0\,\boldsymbol{v}/c$. The product mu^i is the **energy–momentum 4-vector** p^i

$$p^i = m\,u^i = m\,\frac{dx^i}{d\tau} = m\,\frac{dt}{d\tau}\frac{dx^i}{dt} = \frac{m}{\sqrt{1 - v^2/c^2}}\,\frac{dx^i}{dt}$$

$$= \frac{m}{\sqrt{1 - v^2/c^2}}\,(c, \boldsymbol{v}) = \left(\frac{E}{c}, \boldsymbol{p}\right). \qquad (11.67)$$

Its invariant inner product is a constant characteristic of the particle and proportional to the square of its mass

$$c^2\,p^i\,p_i = mc\,u^i\,mc\,u_i = -E^2 + c^2\,\boldsymbol{p}^2 = -m^2\,c^4. \qquad (11.68)$$

Note that the time-dilation factor is the ratio of the energy of a particle to its rest energy

$$\frac{1}{\sqrt{1 - v^2/c^2}} = \frac{E}{mc^2} \qquad (11.69)$$

and the velocity of the particle is its momentum divided by its equivalent mass E/c^2

$$\boldsymbol{v} = \frac{\boldsymbol{p}}{E/c^2}. \qquad (11.70)$$

The analog of $\boldsymbol{F} = m\,\boldsymbol{a}$ is

$$m\,\frac{d^2x^i}{d\tau^2} = m\,\frac{du^i}{d\tau} = \frac{dp^i}{d\tau} = f^i, \qquad (11.71)$$

in which $p^0 = E$, and f^i is a 4-vector force.

Example 11.6 (Time dilation and proper time) In the frame of a laboratory, a particle of mass m with 4-momentum $p^i_{\text{lab}} = (E/c, p, 0, 0)$ travels a distance L in a time t for a 4-vector displacement of $x^i_{\text{lab}} = (ct, L, 0, 0)$. In its own rest frame, the particle's 4-momentum and 4-displacement are $p^i_{\text{rest}} = (mc, 0, 0, 0)$ and $x^i_{\text{rest}} = (c\tau, 0, 0, 0)$. Since the Minkowski inner product of two 4-vectors is Lorentz invariant, we have

$$\left(p^i x_i\right)_{\text{rest}} = \left(p^i x_i\right)_{\text{lab}} \quad \text{or} \quad Et - pL = mc^2\tau = mc^2 t\sqrt{1 - v^2/c^2} \tag{11.72}$$

so a massive particle's phase $\exp(-ip^i x_i/\hbar)$ is $\exp(imc^2\tau/\hbar)$. □

Example 11.7 ($p + \pi \to \Sigma + K$) What is the minimum energy that a beam of pions must have to produce a sigma hyperon and a kaon by striking a proton at rest? Conservation of the energy–momentum 4-vector gives $p_p + p_\pi = p_\Sigma + p_K$. We set $c = 1$ and use this equality in the invariant form $(p_p + p_\pi)^2 = (p_\Sigma + p_K)^2$. We compute $(p_p + p_\pi)^2$ in the $p_p = (m_p, \mathbf{0})$ frame and set it equal to $(p_\Sigma + p_K)^2$ in the frame in which the spatial momenta of the Σ and the K cancel:

$$\begin{aligned}(p_p + p_\pi)^2 &= p_p^2 + p_\pi^2 + 2p_p \cdot p_\pi = -m_p^2 - m_\pi^2 - 2m_p E_\pi \\ &= (p_\Sigma + p_K)^2 = -(m_\Sigma + m_K)^2 .\end{aligned} \tag{11.73}$$

Thus, since the relevant masses (in MeV) are $m_{\Sigma^+} = 1189.4$, $m_{K^+} = 493.7$, $m_p = 938.3$, and $m_{\pi^+} = 139.6$, the minimum total energy of the pion is

$$E_\pi = \frac{(m_\Sigma + m_K)^2 - m_p^2 - m_\pi^2}{2m_p} \approx 1030 \quad \text{MeV}, \tag{11.74}$$

of which 890 MeV is kinetic. □

11.11 Electrodynamics

In electrodynamics and in MKSA (SI) units, the three-dimensional vector potential $\boldsymbol{A}$ and the scalar potential ϕ form a covariant 4-vector potential

$$A_i = \left(\frac{-\phi}{c}, \boldsymbol{A}\right). \tag{11.75}$$

The contravariant 4-vector potential is $A^i = (\phi/c, \boldsymbol{A})$. The **magnetic induction** is

$$\boldsymbol{B} = \nabla \times \boldsymbol{A} \quad \text{or} \quad B_i = \epsilon_{ijk}\partial_j A_k, \tag{11.76}$$

in which $\partial_j = \partial/\partial x^j$, the sum over the repeated indices j and k runs from 1 to 3, and ϵ_{ijk} is totally antisymmetric with $\epsilon_{123} = 1$. The electric field is

$$E_i = c\left(\frac{\partial A_0}{\partial x^i} - \frac{\partial A_i}{\partial x^0}\right) = -\frac{\partial \phi}{\partial x^i} - \frac{\partial A_i}{\partial t} \tag{11.77}$$

where $x^0 = ct$. In 3-vector notation, $\boldsymbol{E}$ is given by the gradient of ϕ and the time-derivative of $\boldsymbol{A}$

$$\boldsymbol{E} = -\nabla\phi - \dot{\boldsymbol{A}}. \tag{11.78}$$

In terms of the second-rank, antisymmetric Faraday field-strength tensor

$$F_{ij} = \frac{\partial A_j}{\partial x^i} - \frac{\partial A_i}{\partial x^j} = -F_{ji} \tag{11.79}$$

the electric field is $E_i = c\,F_{i0}$ and the magnetic field B_i is

$$B_i = \frac{1}{2}\epsilon_{ijk}\,F_{jk} = \frac{1}{2}\epsilon_{ijk}\left(\frac{\partial A_k}{\partial x^j} - \frac{\partial A_j}{\partial x^k}\right) = (\nabla \times \boldsymbol{A})_i \tag{11.80}$$

where the sum over repeated indices runs from 1 to 3. The inverse equation $F_{jk} = \epsilon_{jki}B_i$ for spatial j and k follows from the Levi-Civita identity (1.449)

$$\begin{aligned}\epsilon_{jki}B_i &= \frac{1}{2}\epsilon_{jki}\epsilon_{inm}\,F_{nm} = \frac{1}{2}\epsilon_{ijk}\epsilon_{inm}\,F_{nm}\\ &= \frac{1}{2}\left(\delta_{jn}\,\delta_{km} - \delta_{jm}\,\delta_{kn}\right)F_{nm} = \frac{1}{2}\left(F_{jk} - F_{kj}\right) = F_{jk}.\end{aligned} \tag{11.81}$$

In 3-vector notation and MKSA = SI units, Maxwell's equations are a ban on magnetic monopoles and **Faraday's law**, both homogeneous,

$$\nabla\cdot\boldsymbol{B} = 0 \quad\text{and}\quad \nabla\times\boldsymbol{E} + \dot{\boldsymbol{B}} = 0 \tag{11.82}$$

and **Gauss's law** and the **Maxwell-Ampère law**, both inhomogeneous,

$$\nabla\cdot\boldsymbol{D} = \rho_f \quad\text{and}\quad \nabla\times\boldsymbol{H} = \boldsymbol{j}_f + \dot{\boldsymbol{D}}. \tag{11.83}$$

Here ρ_f is the density of **free** charge and $\boldsymbol{j}_f$ is the **free current density**. By *free*, we understand charges and currents that do not arise from polarization and are not restrained by chemical bonds. The divergence of $\nabla\times\boldsymbol{H}$ vanishes (like that of any curl), and so the Maxwell–Ampère law and Gauss's law imply that free charge is conserved

$$0 = \nabla\cdot(\nabla\times\boldsymbol{H}) = \nabla\cdot\boldsymbol{j}_f + \nabla\cdot\dot{\boldsymbol{D}} = \nabla\cdot\boldsymbol{j}_f + \dot{\rho}_f. \tag{11.84}$$

If we use this **continuity equation** to replace $\nabla\cdot\boldsymbol{j}_f$ with $-\dot{\rho}_f$ in its middle form $0 = \nabla\cdot\boldsymbol{j}_f + \nabla\cdot\dot{\boldsymbol{D}}$, then we see that the Maxwell–Ampère law preserves the Gauss law constraint in time

$$0 = \nabla\cdot\boldsymbol{j}_f + \nabla\cdot\dot{\boldsymbol{D}} = \frac{\partial}{\partial t}\left(-\rho_f + \nabla\cdot\boldsymbol{D}\right). \tag{11.85}$$

Similarly, Faraday's law preserves the constraint $\nabla\cdot\boldsymbol{B} = 0$

$$0 = -\nabla\cdot(\nabla\times\boldsymbol{E}) = \frac{\partial}{\partial t}\nabla\cdot\boldsymbol{B} = 0. \tag{11.86}$$

In a **linear, isotropic** medium, the **electric displacement** $\boldsymbol{D}$ is related to the electric field $\boldsymbol{E}$ by the **permittivity** $\boldsymbol{D} = \epsilon \boldsymbol{E}$ and the **magnetic** or **magnetizing** field $\boldsymbol{H}$ differs from the magnetic induction $\boldsymbol{B}$ by the **permeability** $\boldsymbol{H} = \boldsymbol{B}/\mu$.

On a sub-nanometer scale, the microscopic form of Maxwell's equations applies. On this scale, the homogeneous equations (11.82) are unchanged, but the inhomogeneous ones are

$$\nabla \cdot \boldsymbol{E} = \frac{\rho}{\epsilon_0} \quad \text{and} \quad \nabla \times \boldsymbol{B} = \mu_0 \boldsymbol{j} + \epsilon_0 \mu_0 \dot{\boldsymbol{E}} = \mu_0 \boldsymbol{j} + \frac{\dot{\boldsymbol{E}}}{c^2}, \tag{11.87}$$

in which ρ and $\boldsymbol{j}$ are the total charge and current densities, and $\epsilon_0 = 8.854 \times 10^{-12}$ F/m and $\mu_0 = 4\pi \times 10^{-7}$ N/A^2 are the **electric** and **magnetic constants**, whose product is the inverse of the square of the speed of light, $\epsilon_0 \mu_0 = 1/c^2$. Gauss's law and the Maxwell–Ampère law (11.87) imply (exercise 11.6) that the microscopic (total) current 4-vector $j = (c\rho, \boldsymbol{j})$ obeys the continuity equation $\dot{\rho} + \nabla \cdot \boldsymbol{j} = 0$. Electric charge is conserved.

In vacuum, $\rho = \boldsymbol{j} = 0$, $\boldsymbol{D} = \epsilon_0 \boldsymbol{E}$, and $\boldsymbol{H} = \boldsymbol{B}/\mu_0$, and Maxwell's equations become

$$\begin{aligned} \nabla \cdot \boldsymbol{B} = 0 \quad &\text{and} \quad \nabla \times \boldsymbol{E} + \dot{\boldsymbol{B}} = 0, \\ \nabla \cdot \boldsymbol{E} = 0 \quad &\text{and} \quad \nabla \times \boldsymbol{B} = \frac{1}{c^2} \dot{\boldsymbol{E}}. \end{aligned} \tag{11.88}$$

Two of these equations $\nabla \cdot \boldsymbol{B} = 0$ and $\nabla \cdot \boldsymbol{E} = 0$ are constraints. Taking the curl of the other two equations, we find

$$\nabla \times (\nabla \times \boldsymbol{E}) = -\frac{1}{c^2} \ddot{\boldsymbol{E}} \quad \text{and} \quad \nabla \times (\nabla \times \boldsymbol{B}) = -\frac{1}{c^2} \ddot{\boldsymbol{B}}. \tag{11.89}$$

One may use the Levi-Civita identity (1.449) to show (exercise 11.8) that

$$\nabla \times (\nabla \times \boldsymbol{E}) = \nabla (\nabla \cdot \boldsymbol{E}) - \Delta \boldsymbol{E} \text{ and } \nabla \times (\nabla \times \boldsymbol{B}) = \nabla (\nabla \cdot \boldsymbol{B}) - \Delta \boldsymbol{B}, \tag{11.90}$$

in which $\triangle \equiv \nabla^2$. Since in vacuum the divergence of $\boldsymbol{E}$ vanishes, and since that of $\boldsymbol{B}$ always vanishes, these identities and the curl–curl equations (11.89) tell us that waves of $\boldsymbol{E}$ and $\boldsymbol{B}$ move at the speed of light

$$\frac{1}{c^2} \ddot{\boldsymbol{E}} - \triangle \boldsymbol{E} = 0 \quad \text{and} \quad \frac{1}{c^2} \ddot{\boldsymbol{B}} - \triangle \boldsymbol{B} = 0. \tag{11.91}$$

We may write the two homogeneous Maxwell equations (11.82) as

$$\begin{aligned} \partial_i F_{jk} + \partial_k F_{ij} + \partial_j F_{ki} = {} & \partial_i \left(\partial_j A_k - \partial_k A_j \right) + \partial_k \left(\partial_i A_j - \partial_j A_i \right) \\ & + \partial_j \left(\partial_k A_i - \partial_i A_k \right) = 0 \end{aligned} \tag{11.92}$$

(exercise 11.9). This relation, known as the **Bianchi identity**, actually is a generally covariant tensor equation

$$\epsilon^{\ell ijk} \partial_i F_{jk} = 0, \tag{11.93}$$

in which $\epsilon^{\ell ijk}$ is totally antisymmetric, as explained in section 11.32. There are four versions of this identity (corresponding to the four ways of choosing three different indices i, j, k from among four and leaving out one, ℓ). The $\ell = 0$ case gives the scalar equation $\nabla \cdot \boldsymbol{B} = 0$, and the three that have $\ell \neq 0$ give the vector equation $\nabla \times \boldsymbol{E} + \dot{\boldsymbol{B}} = 0$.

In tensor notation, the microscopic form of the two inhomogeneous equations (11.87) – the laws of Gauss and Ampère – are

$$\partial_i F^{ki} = \mu_0 j^k, \tag{11.94}$$

in which j^k is the current 4-vector

$$j^k = (c\rho, \boldsymbol{j}) . \tag{11.95}$$

The **Lorentz force law** for a particle of charge q is

$$m\,\frac{d^2x^i}{d\tau^2} = m\,\frac{du^i}{d\tau} = \frac{dp^i}{d\tau} = f^i = q\,F^{ij}\,\frac{dx_j}{d\tau} = q\,F^{ij}\,u_j. \tag{11.96}$$

We may cancel a factor of $dt/d\tau$ from both sides and find for $i = 1, 2, 3$

$$\frac{dp^i}{dt} = q\left(-F^{i0} + \epsilon_{ijk} B_k v_j\right) \quad \text{or} \quad \frac{d\boldsymbol{p}}{dt} = q\,(\boldsymbol{E} + \boldsymbol{v} \times \boldsymbol{B}) \tag{11.97}$$

and for $i = 0$

$$\frac{dE}{dt} = q\,\boldsymbol{E} \cdot \boldsymbol{v}, \tag{11.98}$$

which shows that only the electric field does work. The only special-relativistic correction needed in Maxwell's electrodynamics is a factor of $1/\sqrt{1 - v^2/c^2}$ in these equations. That is, we use $\boldsymbol{p} = m\boldsymbol{u} = m\boldsymbol{v}/\sqrt{1 - v^2/c^2}$ not $\boldsymbol{p} = m\boldsymbol{v}$ in (11.97), and we use the total energy E not the kinetic energy in (11.98). The reason why so little of classical electrodynamics was changed by special relativity is that electric and magnetic effects were accessible to measurement during the 1800s. Classical electrodynamics was almost perfect.

Keeping track of factors of the speed of light is a lot of trouble and a distraction; in what follows, we'll often use units with $c = 1$.

11.12 Tensors

Tensors are structures that transform like products of vectors. A first-rank tensor is a covariant or a contravariant vector. Second-rank tensors also are distinguished by how they transform under changes of coordinates:

$$\text{contravariant} \quad M'^{ij} = \frac{\partial x'^i}{\partial x^k} \frac{\partial x'^j}{\partial x^l} M^{kl},$$
$$\text{mixed} \quad N'^i_j = \frac{\partial x'^i}{\partial x^k} \frac{\partial x^l}{\partial x'^j} N^k_l,$$
$$\text{covariant} \quad F'_{ij} = \frac{\partial x^k}{\partial x'^i} \frac{\partial x^l}{\partial x'^j} F_{kl}. \tag{11.99}$$

We can define tensors of higher rank by extending these definitions to quantities with more indices.

Example 11.8 (Some second-rank tensors) If A_k and B_ℓ are covariant vectors, and C^m and D^n are contravariant vectors, then the product $C^m D^n$ is a second-rank contravariant tensor, and all four products $A_k C^m$, $A_k D^n$, $B_k C^m$, and $B_k D^n$ are second-rank mixed tensors, while $C^m D^n$ as well as $C^m C^n$ and $D^m D^n$ are second-rank contravariant tensors. □

Since the transformation laws that define tensors are linear, any linear combination of tensors of a given rank and kind is a tensor of that rank and kind. Thus if F_{ij} and G_{ij} are both second-rank covariant tensors, then so is their sum

$$H_{ij} = F_{ij} + G_{ij}. \tag{11.100}$$

A covariant tensor is **symmetric** if it is independent of the order of its indices. That is, if $S_{ik} = S_{ki}$, then S is symmetric. Similarly, a contravariant tensor is symmetric if permutations of its indices leave it unchanged. Thus A is symmetric if $A^{ik} = A^{ki}$.

A covariant or contravariant tensor is **antisymmetric** if it changes sign when any two of its indices are interchanged. So A_{ik}, B^{ik}, and C_{ijk} are antisymmetric if

$$A_{ik} = -A_{ki} \quad \text{and} \quad B^{ik} = -B^{ki}, \quad \text{and}$$
$$C_{ijk} = C_{jki} = C_{kij} = -C_{jik} = -C_{ikj} = -C_{kji}. \tag{11.101}$$

Example 11.9 (Three important tensors) The Maxwell field strength $F_{kl}(x)$ is a second-rank covariant tensor; so is the metric of space-time $g_{ij}(x)$. The Kronecker delta δ^i_j is a mixed second-rank tensor; it transforms as

$$\delta'^i_j = \frac{\partial x'^i}{\partial x^k} \frac{\partial x^l}{\partial x'^j} \delta^k_l = \frac{\partial x'^i}{\partial x^k} \frac{\partial x^k}{\partial x'^j} = \frac{\partial x'^i}{\partial x'^j} = \delta^i_j. \tag{11.102}$$

So it is **invariant** under changes of coordinates. □

Example 11.10 (Contractions) Although the product $A_k\, C^\ell$ is a mixed second-rank tensor, the product $A_k\, C^k$ transforms as a scalar because

$$A'_k\, C'^k = \frac{\partial x^\ell}{\partial x'^k}\frac{\partial x'^k}{\partial x^m} A_\ell\, C^m = \frac{\partial x^\ell}{\partial x^m} A_\ell\, C^m = \delta^\ell_m A_\ell\, C^m = A_\ell\, C^\ell. \tag{11.103}$$

A sum in which an index is repeated once covariantly and once contravariantly is a **contraction** as in the Kronecker-delta equation (11.26). In general, the rank of a tensor is the number of uncontracted indices. □

11.13 Differential forms

By (11.10 & 11.5), a covariant vector field contracted with contravariant coordinate differentials is invariant under arbitrary coordinate transformations

$$A' = A'_i\, dx'^i = \frac{\partial x^j}{\partial x'^i} A_j \frac{\partial x'^i}{\partial x^k} dx^k = \delta^j_k\, A_j\, dx^k = A_k\, dx^k = A. \tag{11.104}$$

This invariant quantity $A = A_k\, dx^k$ is a called a **1-form** in the language of **differential forms** introduced about a century ago by Élie Cartan (1869–1951, son of a blacksmith).

The **wedge product** $dx \wedge dy$ of two coordinate differentials is the directed area spanned by the two differentials and is defined to be antisymmetric

$$dx \wedge dy = -\, dy \wedge dx \quad \text{and} \quad dx \wedge dx = dy \wedge dy = 0 \tag{11.105}$$

so as to transform correctly under a change of coordinates. In terms of the coordinates $u = u(x, y)$ and $v = v(x, y)$, the new element of area is

$$du \wedge dv = \left(\frac{\partial u}{\partial x}dx + \frac{\partial u}{\partial y}dy\right) \wedge \left(\frac{\partial v}{\partial x}dx + \frac{\partial v}{\partial y}dy\right). \tag{11.106}$$

Labeling partial derivatives by subscripts (6.20) and using the antisymmetry (11.105), we see that the new element of area $du \wedge dv$ is the old area $dx \wedge dy$ multiplied by the Jacobian $J(u, v; x, y)$ of the transformation $x, y \to u, v$

$$\begin{aligned} du \wedge dv &= \left(u_x dx + u_y dy\right) \wedge \left(v_x dx + v_y dy\right) \\ &= u_x\, v_x\, dx \wedge dx + u_x\, v_y\, dx \wedge dy + u_y\, v_x\, dy \wedge dx + u_y\, v_y\, dy \wedge dy \\ &= \left(u_x\, v_y - u_y v_x\right)\, dx \wedge dy \\ &= \begin{vmatrix} u_x & u_y \\ v_x & v_y \end{vmatrix} dx \wedge dy = J(u, v; x, y)\, dx \wedge dy. \end{aligned} \tag{11.107}$$

A contraction $H = \frac{1}{2} H_{ik}\, dx^i \wedge dx^k$ of a second-rank covariant tensor with a wedge product of two differentials is a 2-form. A ***p*-form** is a rank-p covariant tensor contracted with a wedge product of p differentials

$$K = \frac{1}{p!} K_{i_1 \ldots i_p} dx^{i_1} \wedge \ldots dx^{i_p}. \tag{11.108}$$

The **exterior derivative** d differentiates and adds a differential; it turns a p-form into a $(p+1)$-form. It converts a function or a **0-form** f into a 1-form

$$df = \frac{\partial f}{\partial x^i} dx^i \tag{11.109}$$

and a 1-form $A = A_j \, dx^j$ into a 2-form $dA = d(A_j \, dx^j) = (\partial_i A_j) \, dx^i \wedge dx^j$.

Example 11.11 (The curl) The exterior derivative of the 1-form

$$A = A_x \, dx + A_y \, dy + A_z \, dz \tag{11.110}$$

is the 2-form

$$\begin{aligned} dA &= \partial_y A_x \, dy \wedge dx + \partial_z A_x \, dz \wedge dx \\ &\quad + \partial_x A_y \, dx \wedge dy + \partial_z A_y \, dz \wedge dy \\ &\quad + \partial_x A_z \, dx \wedge dz + \partial_y A_z \, dy \wedge dz \\ &= \left(\partial_y A_z - \partial_z A_y\right) dy \wedge dz \\ &\quad + (\partial_z A_x - \partial_x A_z) \, dz \wedge dx \\ &\quad + \left(\partial_x A_y - \partial_y A_x\right) dx \wedge dy \\ &= (\nabla \times A)_x \, dy \wedge dz + (\nabla \times A)_y \, dz \wedge dx + (\nabla \times A)_z \, dx \wedge dy, \end{aligned} \tag{11.111}$$

in which we recognize the curl (6.39) of A. □

The exterior derivative of the 1-form $A = A_j \, dx^j$ is the 2-form

$$dA = dA_j \wedge dx^j = \partial_i A_j \, dx^i \wedge dx^j = \tfrac{1}{2} F_{ij} \, dx^i \wedge dx^j = F, \tag{11.112}$$

in which $\partial_i = \partial/\partial x^i$. So d turns the electromagnetic 1-form A – the 4-vector potential or gauge field A_j – into the Faraday 2-form – the tensor F_{ij}. Its square vanishes: dd applied to any p-form Q is zero

$$ddQ_{i\ldots} dx^i \wedge \cdots = d(\partial_r Q_{i\ldots}) \wedge dx^r \wedge dx^i \wedge \cdots = (\partial_s \partial_r Q_{i\ldots}) dx^s \wedge dx^r \wedge dx^i \wedge \cdots = 0 \tag{11.113}$$

because $\partial_s \partial_r Q$ is symmetric in r and s while $dx^s \wedge dx^r$ is antisymmetric.

Some writers drop the wedges and write $dx^i \wedge dx^j$ as $dx^i dx^j$ while keeping the rules of antisymmetry $dx^i dx^j = -dx^j dx^i$ and $\left(dx^i\right)^2 = 0$. But this economy prevents one from using invariant quantities like $S = \frac{1}{2} S_{ik} \, dx^i dx^k$, in which S_{ik} is a second-rank covariant symmetric tensor. If M_{ik} is a covariant second-rank tensor with no particular symmetry, then (exercise 11.7) only its antisymmetric part contributes to the 2-form $M_{ik} \, dx^i \wedge dx^k$ and only its symmetric part contributes to the quantity $M_{ik} \, dx^i dx^k$.

The exterior derivative d applied to the Faraday 2-form $F = dA$ gives

$$dF = ddA = 0, \tag{11.114}$$

which is the Bianchi identity (11.93). A p-form H is **closed** if $dH = 0$. By (11.114), the Faraday 2-form is closed, $dF = 0$.

A p-form H is **exact** if there is a $(p+1)$-form K whose differential is $H = dK$. The identity (12.64) or $dd = 0$ implies that **every exact form is closed**. The lemma of Poincaré shows that **every closed form is locally exact**.

If the A_i in the 1-form $A = A_i dx^i$ commute with each other, then the 2-form $A^2 = 0$. But if the A_i don't commute because they are matrices or operators or Grassmann variables, then A^2 need not vanish.

Example 11.12 (A static electric field is closed and locally exact) If $\dot{\boldsymbol{B}} = 0$, then by Faraday's law (11.82) the curl of the electric field vanishes, $\nabla \times \boldsymbol{E} = 0$. Writing the electrostatic field as the 1-form $E = E_i\, dx^i$ for $i = 1, 2, 3$, we may express the vanishing of its curl as

$$dE = \partial_j E_i\, dx^j\, dx^i = \frac{1}{2}\left(\partial_j E_i - \partial_i E_j\right) dx^j\, dx^i = 0, \tag{11.115}$$

which says that E is closed. We can define a quantity $V_P(\boldsymbol{x})$ as a line integral of the 1-form E along a path P to $\boldsymbol{x}$ from some starting point $\boldsymbol{x}_0$

$$V_P(\boldsymbol{x}) = -\int_{P,\boldsymbol{x}_0}^{\boldsymbol{x}} E_i\, dx^i = -\int_P E \tag{11.116}$$

and so $V_P(\boldsymbol{x})$ will depend on the path P as well as on $\boldsymbol{x}_0$ and $\boldsymbol{x}$. But if $\nabla \times \boldsymbol{E} = 0$ in some ball (or neighborhood) around $\boldsymbol{x}$ and $\boldsymbol{x}_0$, then within that ball the dependence on the path P drops out because the difference $V_{P'}(\boldsymbol{x}) - V_P(\boldsymbol{x})$ is the line integral of E around a closed loop in the ball, which by Stokes's theorem (6.44) is an integral of the vanishing curl $\nabla \times \boldsymbol{E}$ over any surface S in the ball whose boundary ∂S is the closed curve $P' - P$

$$V_{P'}(\boldsymbol{x}) - V_P(\boldsymbol{x}) = \oint_{P'-P} E_i\, dx^i = \int_S (\nabla \times \boldsymbol{E}) \cdot d\boldsymbol{a} = 0 \tag{11.117}$$

or

$$V_{P'}(\boldsymbol{x}) - V_P(\boldsymbol{x}) = \int_{\partial S} E = \int_S dE = 0 \tag{11.118}$$

in the language of forms (George Stokes, 1819–1903). Thus the potential $V_P(\boldsymbol{x}) = V(\boldsymbol{x})$ is independent of the path, $\boldsymbol{E} = -\nabla V(\boldsymbol{x})$, and the 1-form $E = E_i\, dx^i = -\partial_i V\, dx^i = -dV$ is exact. □

The general form of Stokes's theorem is that the integral of any p-form H over the boundary ∂R of any $(p+1)$-dimensional, simply connected, orientable region R is equal to the integral of the $(p+1)$-form dH over R

$$\int_{\partial R} H = \int_R dH, \tag{11.119}$$

which for $p = 1$ gives (6.44).

Example 11.13 (Stokes's theorem for 0-forms) Here $p = 0$, the region $R = [a, b]$ is one-dimensional, H is a 0-form, and Stokes's theorem is

$$H(b) - H(a) = \int_{\partial R} H = \int_R dH = \int_a^b dH(x) = \int_a^b H'(x)\, dx, \tag{11.120}$$

familiar from elementary calculus. □

Example 11.14 (Exterior derivatives anticommute with differentials) The exterior derivative acting on two 1-forms $A = A_i dx^i$ and $B = B_j dx^j$ is

$$\begin{aligned} d(A\,B) &= d(A_i dx^i \wedge B_j dx^j) = \partial_k(A_i B_j)\, dx^k \wedge dx^i \wedge dx^j \\ &= (\partial_k A_i) B_j\, dx^k \wedge dx^i \wedge dx^j + A_i(\partial_k B_j)\, dx^k \wedge dx^i \wedge dx^j \\ &= (\partial_k A_i) B_j\, dx^k \wedge dx^i \wedge dx^j - A_i(\partial_k B_j)\, dx^i \wedge dx^k \wedge dx^j \\ &= (\partial_k A_i) dx^k \wedge dx^i \wedge B_j dx^j - A_i dx^i \wedge (\partial_k B_j)\, dx^k \wedge dx^j \\ &= dA \wedge B - A \wedge dB. \end{aligned} \tag{11.121}$$

If A is a p-form, then $d(A \wedge B) = dA \wedge B + (-1)^p A \wedge dB$ (exercise 11.10). □

11.14 Tensor equations

Maxwell's equations (11.93 & 11.94) relate the derivatives of the field-strength tensor to the current density

$$\frac{\partial F^{ik}}{\partial x^k} = \mu_0 j^i \tag{11.122}$$

and the derivatives of the field-strength tensor to each other

$$0 = \partial_i F_{jk} + \partial_k F_{ij} + \partial_j F_{ki}. \tag{11.123}$$

They are generally covariant **tensor equations** (sections 11.31 & 11.32). We also can write Maxwell's equations in terms of invariant forms; his homogeneous equations are simply the Bianchi identity (11.114)

$$dF = ddA = 0 \tag{11.124}$$

and we'll write his inhomogeneous ones in terms of forms in section 11.26.

If we can write a physical law in one coordinate system as a tensor equation

$$K^{kl} = 0 \tag{11.125}$$

then in any other coordinate system, the corresponding tensor equation

$$K'^{ij} = 0 \tag{11.126}$$

also is valid since

$$K'^{ij} = \frac{\partial x'^i}{\partial x^k} \frac{\partial x'^j}{\partial x^l} K^{kl} = 0. \tag{11.127}$$

Similarly, physical laws remain the same when expressed in terms of invariant forms. Thus **by writing a theory in terms of tensors or forms, one gets a theory that is true in all coordinate systems if it is true in any**. Only such "covariant" theories have a chance at being right in our coordinate system, which is not special. One way to make a covariant theory is to start with an action that is invariant under all coordinate transformations.

11.15 The quotient theorem

Suppose that the product BA of a quantity B (with unknown transformation properties) with an arbitrary tensor A (of a given rank and kind) is a tensor. Then B is itself a tensor. The simplest example is when $B_i A^i$ is a scalar for all contravariant vectors A^i

$$B'_i A'^i = B_j A^j. \tag{11.128}$$

Then since A^i is a contravariant vector

$$B'_i A'^i = B'_i \frac{\partial x'^i}{\partial x^j} A^j = B_j A^j \tag{11.129}$$

or

$$\left(B'_i \frac{\partial x'^i}{\partial x^j} - B_j \right) A^j = 0. \tag{11.130}$$

Since this equation holds for all vectors A, we may promote it to the level of a vector equation

$$B'_i \frac{\partial x'^i}{\partial x^j} - B_j = 0. \tag{11.131}$$

Multiplying both sides by $\partial x^j / \partial x'^k$ and summing over j

$$B'_i \frac{\partial x'^i}{\partial x^j} \frac{\partial x^j}{\partial x'^k} = B_j \frac{\partial x^j}{\partial x'^k} \tag{11.132}$$

we see that the unknown quantity B_i does transform as a covariant vector

$$B'_k = \frac{\partial x^j}{\partial x'^k} B_j. \tag{11.133}$$

The quotient rule works for unknowns B and tensors A of arbitrary rank and kind. The proof in each case is very similar to the one given here.

11.16 The metric tensor

So far we have been considering coordinate systems with constant basis vectors e_i that do not vary with the physical point p. Now we shall assume only that we can write the change in the point $p(x)$ due to an infinitesimal change $dx^i(p)$ in its coordinates $x^i(p)$ as

$$dp(x) = e_i(x)\, dx^i. \tag{11.134}$$

In a different system of coordinates x', this displacement is $dp = e'_i(x')\, dx'^i$. The basis vectors e_i and e'_i are partial derivatives of the point p

$$e_i(x) = \frac{\partial p}{\partial x^i} \quad \text{and} \quad e'_i(x') = \frac{\partial p}{\partial x'^i}. \tag{11.135}$$

They are linearly related to each other, transforming as covariant vectors

$$e'_i(x') = \frac{\partial p}{\partial x'^i} = \frac{\partial x^j}{\partial x'^i}\frac{\partial p}{\partial x^j} = \frac{\partial x^j}{\partial x'^i}\, e_j(x). \tag{11.136}$$

They also are vectors in the n-dimensional embedding space with inner product

$$e_i(x) \cdot e_j(x) = \sum_{a=1}^{n}\sum_{b=1}^{n} e_i^a(x)\, \eta_{ab}\, e_j^b(x), \tag{11.137}$$

which will be positive-definite (1.75) if all the eigenvalues of the real symmetric matrix η are positive. For instance, the eigenvalues are positive in euclidean 3-space with cylindrical or spherical coordinates but not in Minkowski 4-space where η is a diagonal matrix with main diagonal $(-1, 1, 1, 1)$.

The basis vectors $e_i(x)$ constitute a **moving frame**, a concept introduced by Élie Cartan. In general, they are not normalized or orthogonal. Their inner products define the metric of the manifold or of space-time

$$g_{ij}(x) = e_i(x) \cdot e_j(x). \tag{11.138}$$

An inner product by definition (1.73) satisfies $(f, g) = (g, f)^*$ and so a real inner product is symmetric. For real coordinates on a real manifold the basis vectors are real, so the metric tensor is real and symmetric

$$g_{ij} = g_{ji}. \tag{11.139}$$

The basis vectors $e'_j(x')$ of a different coordinate system define the metric in that coordinate system $g'_{ij}(x') = e'_i(x') \cdot e'_j(x')$. Since the basis vectors e_i are covariant vectors, the metric g_{ij} is a second-rank covariant tensor

$$g'_{ij}(x') = e'_i(x') \cdot e'_j(x') = \frac{\partial x^k}{\partial x'^i}\, e_k(x) \cdot \frac{\partial x^\ell}{\partial x'^j}\, e_\ell(x) = \frac{\partial x^k}{\partial x'^i}\frac{\partial x^\ell}{\partial x'^j}\, g_{k\ell}(x). \tag{11.140}$$

Example 11.15 (The sphere) Let the point $\boldsymbol{p}$ be a euclidean 3-vector representing a point on the two-dimensional surface of a sphere of radius r. The spherical coordinates (θ, ϕ) label the point $\boldsymbol{p}$, and the basis vectors are

$$\boldsymbol{e}_\theta = \frac{\partial \boldsymbol{p}}{\partial \theta} = r\hat{\theta} \quad \text{and} \quad \boldsymbol{e}_\phi = \frac{\partial \boldsymbol{p}}{\partial \phi} = r \sin\theta\, \hat{\phi}. \tag{11.141}$$

Their inner products are the components (11.138) of the sphere's metric tensor, which is the matrix

$$\begin{pmatrix} g_{\theta\theta} & g_{\theta\phi} \\ g_{\phi\theta} & g_{\phi\phi} \end{pmatrix} = \begin{pmatrix} \boldsymbol{e}_\theta \cdot \boldsymbol{e}_\theta & \boldsymbol{e}_\theta \cdot \boldsymbol{e}_\phi \\ \boldsymbol{e}_\phi \cdot \boldsymbol{e}_\theta & \boldsymbol{e}_\phi \cdot \boldsymbol{e}_\phi \end{pmatrix} = \begin{pmatrix} r^2 & 0 \\ 0 & r^2 \sin^2\theta \end{pmatrix} \tag{11.142}$$

with determinant $r^4 \sin^2 \phi$. □

11.17 A basic axiom

Points are physical, coordinate systems metaphysical. So p, q, $p - q$, and $(p - q) \cdot (p - q)$ are all invariant quantities. When p and $q = p + dp$ both lie on the (space-time) manifold and are infinitesimally close to each other, the vector $dp = e_i\, dx^i$ is the sum of the basis vectors multiplied by the changes in the coordinates x^i. Both dp and the inner product $dp \cdot dp$ are physical and so are independent of the coordinates. The (squared) distance dp^2 is the same in one coordinate system

$$dp^2 \equiv dp \cdot dp = (e_i\, dx^i) \cdot (e_j\, dx^j) = g_{ij}\, dx^i dx^j \tag{11.143}$$

as in another

$$dp^2 \equiv dp \cdot dp = (e'_i\, dx'^i) \cdot (e'_j\, dx'^j) = g'_{ij}\, dx'^i dx'^j. \tag{11.144}$$

This invariance and the quotient rule provide a second reason why g_{ij} is a second-rank covariant tensor.

We want dp to be infinitesimal so that it is tangent to the manifold.

11.18 The contravariant metric tensor

The **inverse g^{ik} of the covariant metric tensor** g_{kj} satisfies

$$g'^{ik} g'_{kj} = \delta^i{}_j = g^{ik} g_{kj} \tag{11.145}$$

in all coordinate systems. To see how it transforms, we use the transformation law (11.140) of g_{kj}

$$\delta^i{}_j = g'^{ik} g'_{kj} = g'^{ik} \frac{\partial x^t}{\partial x'^k}\, g_{tu}\, \frac{\partial x^u}{\partial x'^j}. \tag{11.146}$$

Thus in matrix notation, we have as $I = g'^{-1} H g H$, which implies $g'^{-1} = H^{-1} g^{-1} H^{-1}$ or in tensor notation

$$g'^{i\ell} = \frac{\partial x'^i}{\partial x^v} \frac{\partial x'^\ell}{\partial x^w} g^{vw}. \tag{11.147}$$

Thus the inverse g^{ik} of the covariant metric tensor is a second-rank contravariant tensor called the **contravariant metric tensor**.

11.19 Raising and lowering indices

The contraction of a contravariant vector A^i with any rank-2 covariant tensor gives a covariant vector. We reserve the symbol A_i for the covariant vector that is the contraction of A^j with the metric tensor

$$A_i = g_{ij} A^j. \tag{11.148}$$

This operation is called **lowering the index** on A^j.

Similarly the contraction of a covariant vector B_j with any rank-2 contravariant tensor is a contravariant vector. But we reserve the symbol B^i for contravariant vector that is the contraction

$$B^i = g^{ij} B_j \tag{11.149}$$

of B_j with the inverse of the metric tensor. This is called **raising the index** on B_j.

The vectors e^i, for instance, are given by

$$e^i = g^{ij} e_j. \tag{11.150}$$

They are therefore orthonormal or dual to the basis vectors e_i

$$e_i \cdot e^j = e_i \cdot g^{jk} e_k = g^{jk} e_i \cdot e_k = g^{jk} g_{ik} = g^{jk} g_{ki} = \delta_i^j. \tag{11.151}$$

11.20 Orthogonal coordinates in euclidean n-space

In flat n-dimensional euclidean space, it is convenient to use **orthogonal basis vectors** and **orthogonal coordinates**. A change dx^i in the coordinates moves the point p by (11.134)

$$dp = e_i \, dx^i. \tag{11.152}$$

The metric g_{ij} is the inner product (11.138)

$$g_{ij} = e_i \cdot e_j. \tag{11.153}$$

Since the vectors e_i are orthogonal, the metric is diagonal

$$g_{ij} = e_i \cdot e_j = h_i^2 \delta_{ij}. \tag{11.154}$$

The inverse metric

$$g^{ij} = h_i^{-2} \delta_{ij} \tag{11.155}$$

raises indices. For instance, the dual vectors

$$e^i = g^{ij}\, e_j = h_i^{-2}\, e_i \quad \text{satisfy} \quad e^i \cdot e_k = \delta^i_k. \tag{11.156}$$

The invariant squared distance dp^2 between nearby points (11.143) is

$$dp^2 = dp \cdot dp = g_{ij}\, dx^i\, dx^j = h_i^2\, (dx^i)^2 \tag{11.157}$$

and the invariant volume element is

$$dV = d^n p = h_1 \dots h_n\, dx^1 \wedge \dots \wedge dx^n = g\, dx^1 \wedge \dots \wedge dx^n = g\, d^n x, \tag{11.158}$$

in which $g = \sqrt{\det g_{ij}}$ is the square-root of the positive determinant of g_{ij}.

The important special case in which all the scale factors h_i are unity is cartesian coordinates in euclidean space (section 11.5).

We also can use basis vectors $\hat{e}_i$ that are **orthonormal**. By (11.154 & 11.156), these vectors

$$\hat{e}_i = e_i / h_i = h_i\, e^i \quad \text{satisfy} \quad \hat{e}_i \cdot \hat{e}_j = \delta_{ij}. \tag{11.159}$$

In terms of them, a physical and invariant vector V takes the form

$$V = e_i\, V^i = h_i\, \hat{e}_i\, V^i = e^i\, V_i = h_i^{-1}\, \hat{e}_i\, V_i = \hat{e}_i\, \overline{V}_i \tag{11.160}$$

where

$$\overline{V}_i \equiv h_i\, V^i = h_i^{-1}\, V_i \quad \text{(no sum)}. \tag{11.161}$$

The dot-product is then

$$V \cdot U = g_{ij}\, V^i\, V^j = \overline{V}_i\, \overline{U}_i. \tag{11.162}$$

In euclidean n-space, we even can choose coordinates x^i so that the vectors e_i defined by $dp = e_i\, dx^i$ are orthonormal. The metric tensor is then the $n \times n$ identity matrix $g_{ik} = e_i \cdot e_k = I_{ik} = \delta_{ik}$. But since this is euclidean n-space, we also can expand the n fixed orthonormal cartesian unit vectors $\hat{\ell}$ in terms of the $e_i(x)$ which vary with the coordinates as $\hat{\ell} = e_i(x)(e_i(x) \cdot \hat{\ell})$.

11.21 Polar coordinates

In polar coordinates in flat 2-space, the change dp in a point p due to changes in its coordinates is $dp = \hat{r}\, dr + \hat{\theta}\, r\, d\theta$ so $dp = e_r\, dr + e_\theta\, d\theta$ with $e_r = \hat{e}_r = \hat{r}$ and $e_\theta = r\, \hat{e}_\theta = r\, \hat{\theta}$. The metric tensor for polar coordinates is

$$(g_{ij}) = (e_i \cdot e_j) = \begin{pmatrix} 1 & 0 \\ 0 & r^2 \end{pmatrix}. \tag{11.163}$$

The contravariant basis vectors are $e^r = \hat{r}$ and $e^\theta = \hat{e}_\theta/r$. A physical vector V is $V = V^i e_i = V_i e^i = \overline{V}_r \hat{r} + \overline{V}_\theta \hat{\theta}$.

11.22 Cylindrical coordinates

For cylindrical coordinates in flat 3-space, the change $\boldsymbol{dp}$ in a point $\boldsymbol{p}$ due to changes in its coordinates is

$$\boldsymbol{dp} = \hat{\boldsymbol{\rho}}\, d\rho + \hat{\boldsymbol{\phi}}\, \rho\, d\phi + \hat{\boldsymbol{z}}\, dz = \boldsymbol{e}_\rho\, d\rho + \boldsymbol{e}_\phi\, d\phi + \boldsymbol{e}_z\, dz \tag{11.164}$$

with $\boldsymbol{e}_\rho = \hat{\boldsymbol{e}}_\rho = \hat{\boldsymbol{\rho}}$, $\boldsymbol{e}_\phi = \rho\, \hat{\boldsymbol{e}}_\phi = \rho\, \hat{\boldsymbol{\phi}}$, and $\boldsymbol{e}_z = \hat{\boldsymbol{e}}_z = \hat{\boldsymbol{z}}$. The metric tensor for cylindrical coordinates is

$$(g_{ij}) = (\boldsymbol{e}_i \cdot \boldsymbol{e}_j) = \begin{pmatrix} 1 & 0 & 0 \\ 0 & \rho^2 & 0 \\ 0 & 0 & 1 \end{pmatrix} \tag{11.165}$$

with determinant $\det g_{ij} \equiv g = \rho^2$. The invariant volume element is

$$dV = \rho\, dx^1 \wedge dx^2 \wedge dx^3 = \sqrt{g}\, d\rho d\phi dz = \rho\, d\rho d\phi dz. \tag{11.166}$$

The contravariant basis vectors are $\boldsymbol{e}^\rho = \hat{\boldsymbol{\rho}}$, $\boldsymbol{e}^\phi = \hat{\boldsymbol{e}}_\phi/\rho$, and $\boldsymbol{e}^z = \hat{\boldsymbol{z}}$. A physical vector V is

$$V = V^i \boldsymbol{e}_i = V_i \boldsymbol{e}^i = \overline{V}_\rho\, \hat{\boldsymbol{\rho}} + \overline{V}_\phi\, \hat{\boldsymbol{\phi}} + \overline{V}_z\, \hat{\boldsymbol{z}}. \tag{11.167}$$

Incidentally, since

$$\boldsymbol{p} = (\rho \cos\phi, \rho \sin\phi, z) \tag{11.168}$$

the formulas for the basis vectors of cylindrical coordinates in terms of those of rectangular coordinates are (exercise 11.13)

$$\begin{aligned} \hat{\boldsymbol{\rho}} &= \cos\phi\, \hat{\boldsymbol{x}} + \sin\phi\, \hat{\boldsymbol{y}}, \\ \hat{\boldsymbol{\phi}} &= -\sin\phi\, \hat{\boldsymbol{x}} + \cos\phi\, \hat{\boldsymbol{y}}, \\ \hat{\boldsymbol{z}} &= \hat{\boldsymbol{z}}. \end{aligned} \tag{11.169}$$

11.23 Spherical coordinates

For spherical coordinates in flat 3-space, the change $\boldsymbol{dp}$ in a point $\boldsymbol{p}$ due to changes in its coordinates is

$$\boldsymbol{dp} = \hat{\boldsymbol{r}}\, dr + \hat{\boldsymbol{\theta}}\, r\, d\theta + \hat{\boldsymbol{\phi}}\, r \sin\theta\, d\phi = \boldsymbol{e}_r\, dr + \boldsymbol{e}_\theta\, d\theta + \boldsymbol{e}_\phi\, d\phi \tag{11.170}$$

so $\boldsymbol{e}_r = \hat{\boldsymbol{r}}$, $\boldsymbol{e}_\theta = r\hat{\boldsymbol{\theta}}$, and $\boldsymbol{e}_\phi = r \sin\theta\, \hat{\boldsymbol{\phi}}$. The metric tensor for spherical coordinates is

$$(g_{ij}) = (e_i \cdot e_j) = \begin{pmatrix} 1 & 0 & 0 \\ 0 & r^2 & 0 \\ 0 & 0 & r^2 \sin^2\theta \end{pmatrix} \tag{11.171}$$

with determinant $\det g_{ij} \equiv g = r^4 \sin^2\theta$. The invariant volume element is

$$dV = r^2 \sin^2\theta \, dx^1 \wedge dx^2 \wedge dx^3 = \sqrt{g}\, dr d\theta d\phi = r^2 \sin\theta \, dr d\theta d\phi. \tag{11.172}$$

The orthonormal basis vectors are $\hat{\boldsymbol{e}}_r = \hat{\boldsymbol{r}}$, $\hat{\boldsymbol{e}}_\theta = \hat{\boldsymbol{\theta}}$, and $\hat{\boldsymbol{e}}_\phi = \hat{\boldsymbol{\phi}}$. The contravariant basis vectors are $\boldsymbol{e}^r = \hat{\boldsymbol{r}}$, $\boldsymbol{e}^\theta = \hat{\boldsymbol{\theta}}/r$, $\boldsymbol{e}^\phi = \hat{\boldsymbol{\phi}}/r\sin\theta$. A physical vector V is

$$V = V^i \boldsymbol{e}_i = V_i \boldsymbol{e}^i = \overline{V}_r \hat{\boldsymbol{r}} + \overline{V}_\theta \hat{\boldsymbol{\theta}} + \overline{V}_\phi \hat{\boldsymbol{\phi}}. \tag{11.173}$$

Incidentally, since

$$\boldsymbol{p} = (r\sin\theta\cos\phi, r\sin\theta\sin\phi, r\cos\theta) \tag{11.174}$$

the formulas for the basis vectors of spherical coordinates in terms of those of rectangular coordinates are (exercise 11.14)

$$\begin{aligned} \hat{\boldsymbol{r}} &= \sin\theta\cos\phi\,\hat{\boldsymbol{x}} + \sin\theta\sin\phi\,\hat{\boldsymbol{y}} + \cos\theta\,\hat{\boldsymbol{z}}, \\ \hat{\boldsymbol{\theta}} &= \cos\theta\cos\phi\,\hat{\boldsymbol{x}} + \cos\theta\sin\phi\,\hat{\boldsymbol{y}} - \sin\theta\,\hat{\boldsymbol{z}}, \\ \hat{\boldsymbol{\phi}} &= -\sin\phi\,\hat{\boldsymbol{x}} + \cos\phi\,\hat{\boldsymbol{y}}. \end{aligned} \tag{11.175}$$

11.24 The gradient of a scalar field

If $f(x)$ is a scalar field, then the difference between it and $f(x+dx)$ defines the **gradient** ∇f as (6.26)

$$df(x) = f(x+dx) - f(x) = \frac{\partial f(x)}{\partial x^i}\, dx^i = \nabla f(x) \cdot dp. \tag{11.176}$$

Since $dp = e_j\, dx^j$, the **invariant** form

$$\nabla f = e^i \frac{\partial f}{\partial x^i} = \frac{\hat{e}_i}{h_i} \frac{\partial f}{\partial x^i} \tag{11.177}$$

satisfies this definition (11.176) of the gradient

$$\nabla f \cdot dp = \frac{\partial f}{\partial x^i} e^i \cdot e_j dx^j = \frac{\partial f}{\partial x^i}\, \delta^i_j\, dx^j = \frac{\partial f}{\partial x^i}\, dx^i = df. \tag{11.178}$$

In two polar coordinates, the gradient is

$$\nabla f = e^i \frac{\partial f}{\partial x^i} = \frac{\hat{e}_i}{h_i} \frac{\partial f}{\partial x^i} = \hat{r} \frac{\partial f}{\partial r} + \frac{\hat{\theta}}{r} \frac{\partial f}{\partial \theta}. \tag{11.179}$$

In three cylindrical coordinates, it is (6.27)

$$\nabla f = e^i \frac{\partial f}{\partial x^i} = \frac{\hat{e}_i}{h_i} \frac{\partial f}{\partial x^i} = \frac{\partial f}{\partial \rho}\hat{\rho} + \frac{1}{\rho}\frac{\partial f}{\partial \phi}\hat{\phi} + \frac{\partial f}{\partial z}\hat{z} \tag{11.180}$$

and in three spherical coordinates it is (6.28)

$$\nabla f = \frac{\partial f}{\partial x^i} e^i = \frac{\hat{e}_i}{h_i} \frac{\partial f}{\partial x^i} = \frac{\partial f}{\partial r} \hat{r} + \frac{1}{r} \frac{\partial f}{\partial \theta} \hat{\boldsymbol{\theta}} + \frac{1}{r \sin\theta} \frac{\partial f}{\partial \phi} \hat{\phi}. \tag{11.181}$$

11.25 Levi-Civita's tensor

In three dimensions, Levi-Civita's **symbol** $\epsilon_{ijk} \equiv \epsilon^{ijk}$ is totally antisymmetric with $\epsilon_{123} = 1$ in all coordinate systems.

We can turn his symbol into something that transforms as a tensor by multiplying it by the square-root of the determinant of a rank-2 covariant tensor. A natural choice is the metric tensor. Thus the Levi-Civita **tensor** η_{ijk} is the totally antisymmetric rank-3 covariant (pseudo-)tensor

$$\eta_{ijk} = \sqrt{g}\, \epsilon_{ijk} \tag{11.182}$$

in which $g = |\det g_{mn}|$ is the absolute value of the determinant of the metric tensor g_{mn}. The determinant's definition (1.184) and product rule (1.207) imply that Levi-Civita's tensor η_{ijk} transforms as

$$\begin{aligned}
\eta'_{ijk} = \sqrt{g'}\, \epsilon'_{ijk} = \sqrt{g'}\, \epsilon_{ijk} &= \sqrt{\left|\det\left(\frac{\partial x^t}{\partial x'^m} \frac{\partial x^u}{\partial x'^n} g_{tu}\right)\right|}\, \epsilon_{ijk} \\
&= \sqrt{\left|\det\left(\frac{\partial x^t}{\partial x'^m}\right) \det\left(\frac{\partial x^u}{\partial x'^n}\right) \det\left(g_{tu}\right)\right|}\, \epsilon_{ijk} \\
&= \left|\det\left(\frac{\partial x}{\partial x'}\right)\right| \sqrt{g}\, \epsilon_{ijk} = \sigma \det\left(\frac{\partial x}{\partial x'}\right) \sqrt{g}\, \epsilon_{ijk} \\
&= \sigma \frac{\partial x^t}{\partial x'^i} \frac{\partial x^u}{\partial x'^j} \frac{\partial x^v}{\partial x'^k} \sqrt{g}\, \epsilon_{tuv} = \sigma \frac{\partial x^t}{\partial x'^i} \frac{\partial x^u}{\partial x'^j} \frac{\partial x^v}{\partial x'^k} \eta_{tuv}
\end{aligned} \tag{11.183}$$

in which σ is the sign of the Jacobian $\det(\partial x/\partial x')$. Levi-Civita's tensor is a **pseudo-tensor** because it doesn't change sign under the parity transformation $x'^i = -x^i$.

We get η with upper indices by using the inverse g^{nm} of the metric tensor

$$\begin{aligned}
\eta^{ijk} &= g^{it} g^{ju} g^{kv}\, \eta_{tuv} = g^{it} g^{ju} g^{kv} \sqrt{g}\, \epsilon_{tuv} \\
&= \sqrt{g}\, \epsilon_{ijk} / \det(g_{mn}) = s\epsilon_{ijk}/\sqrt{g} = s\epsilon^{ijk}/\sqrt{g},
\end{aligned} \tag{11.184}$$

in which s is the sign of the determinant $\det g_{ij} = sg$.

Similarly in four dimensions, Levi-Civita's **symbol** $\epsilon_{ijk\ell} \equiv \epsilon^{ijk\ell}$ is totally antisymmetric with $\epsilon_{0123} = 1$ in all coordinate systems. No meaning attaches to whether the indices of the Levi-Civita symbol are up or down; some authors even use the notation $\epsilon(ijk\ell)$ or $\epsilon[ijk\ell]$ to emphasize this fact.

In four dimensions, the Levi-Civita pseudo-**tensor** is

$$\eta_{ijk\ell} = \sqrt{g}\,\epsilon_{ijk\ell}. \tag{11.185}$$

It transforms as

$$\begin{aligned}\eta'_{ijk\ell} &= \sqrt{g'}\,\epsilon_{ijk\ell} = \left|\det\left(\frac{\partial x}{\partial x'}\right)\right| \sqrt{g}\,\epsilon_{ijk\ell} = \sigma \det\left(\frac{\partial x}{\partial x'}\right) \sqrt{g}\,\epsilon_{ijk\ell} \\ &= \sigma\, \frac{\partial x^t}{\partial x'^i}\frac{\partial x^u}{\partial x'^j}\frac{\partial x^v}{\partial x'^k}\frac{\partial x^w}{\partial x'^\ell} \sqrt{g}\,\epsilon_{tuvw} = \sigma\, \frac{\partial x^t}{\partial x'^i}\frac{\partial x^u}{\partial x'^j}\frac{\partial x^v}{\partial x'^k}\frac{\partial x^w}{\partial x'^\ell}\, \eta_{tuvw}\end{aligned} \tag{11.186}$$

where σ is the sign of the Jacobian $\det(\partial x/\partial x')$.

Raising the indices on η with $\det g_{ij} = sg$ we have

$$\begin{aligned}\eta^{ijk\ell} &= g^{it}\, g^{ju}\, g^{kv}\, g^{\ell w}\, \eta_{tuvw} = g^{it}\, g^{ju}\, g^{kv}\, g^{\ell w}\, \sqrt{g}\,\epsilon_{tuvw} \\ &= \sqrt{g}\,\epsilon_{ijk\ell}/\det(g_{mn}) = s\,\epsilon_{ijk\ell}/\sqrt{g} \equiv s\,\epsilon^{ijk\ell}/\sqrt{g}.\end{aligned} \tag{11.187}$$

In n dimensions, one may define Levi-Civita's symbol $\epsilon(i_1 \ldots i_n)$ as totally antisymmetric with $\epsilon(1 \ldots n) = 1$ and his tensor as $\eta_{i_1 \ldots i_n} = \sqrt{g}\,\epsilon(i_1 \ldots i_n)$.

11.26 The Hodge star

In three cartesian coordinates, the Hodge dual turns 1-forms into 2-forms

$$*\,dx = dy \wedge dz, \qquad *\,dy = dz \wedge dx, \qquad *\,dz = dx \wedge dy \tag{11.188}$$

and 2-forms into 1-forms

$$*\,(dx \wedge dy) = dz, \qquad *\,(dy \wedge dz) = dx, \qquad *\,(dz \wedge dx) = dy. \tag{11.189}$$

It also maps the 0-form 1 and the volume 3-form into each other

$$*\,1 = dx \wedge dy \wedge dz, \qquad *\,(dx \wedge dy \wedge dz) = 1 \tag{11.190}$$

(William Vallance Douglas Hodge, 1903–1975). More generally in 3-space, we define the Hodge dual, also called the Hodge star, as

$$\begin{aligned} &*1 = \tfrac{1}{3!}\eta_{\ell jk} dx^\ell \wedge dx^j \wedge dx^k, \quad *\,(dx^\ell \wedge dx^j \wedge dx^k) = g^{\ell t} g^{ju} g^{kv} \eta_{tuv}, \\ &*\,dx^i = \tfrac{1}{2}\, g^{i\ell}\, \eta_{\ell jk}\, dx^j \wedge dx^k, \quad *\,(dx^i \wedge dx^j) = g^{ik}\, g^{j\ell}\, \eta_{k\ell m}\, dx^m \end{aligned} \tag{11.191}$$

and so if the sign of $\det g_{ij}$ is $s = +1$, then $**1 = 1$, $**dx^i = dx^i$, $**(dx^i \wedge dx^k) = dx^i \wedge dx^k$, and $**\,(dx^i \wedge dx^j \wedge dx^k) = dx^i \wedge dx^j \wedge dx^k$.

Example 11.16 (Divergence and Laplacian) The dual of the 1-form

$$df = \frac{\partial f}{\partial x}\, dx + \frac{\partial f}{\partial y}\, dy + \frac{\partial f}{\partial z}\, dz \tag{11.192}$$

is the 2-form

$$* \, df = \frac{\partial f}{\partial x} \, dy \wedge dz + \frac{\partial f}{\partial y} \, dz \wedge dx + \frac{\partial f}{\partial z} \, dx \wedge dy \tag{11.193}$$

and its exterior derivative is the Laplacian

$$d * df = \left(\frac{\partial^2 f}{\partial x^2} + \frac{\partial^2 f}{\partial y^2} + \frac{\partial^2 f}{\partial z^2} \right) dx \wedge dy \wedge dz \tag{11.194}$$

multiplied by the volume 3-form.

Similarly, the dual of the 1-form

$$A = A_x \, dx + A_y \, dy + A_z \, dz \tag{11.195}$$

is the 2-form

$$* \, A = A_x \, dy \wedge dz + A_y \, dz \wedge dx + A_z \, dx \wedge dy \tag{11.196}$$

and its exterior derivative is the divergence

$$d * A = \left(\frac{\partial A_x}{\partial x} + \frac{\partial A_y}{\partial y} + \frac{\partial A_z}{\partial z} \right) dx \wedge dy \wedge dz \tag{11.197}$$

times $dx \wedge dy \wedge dz$. □

In flat Minkowski 4-space with $c = 1$, the Hodge dual turns 1-forms into 3-forms

$$\begin{aligned} * \, dt &= - \, dx \wedge dy \wedge dz, & * \, dx &= - \, dy \wedge dz \wedge dt, \\ * \, dy &= - \, dz \wedge dx \wedge dt, & * \, dz &= - \, dx \wedge dy \wedge dt, \end{aligned} \tag{11.198}$$

2-forms into 2-forms

$$\begin{aligned} * \, (dx \wedge dt) &= dy \wedge dz, & * \, (dx \wedge dy) &= - \, dz \wedge dt, \\ * \, (dy \wedge dt) &= dz \wedge dx, & * \, (dy \wedge dz) &= - \, dx \wedge dt, \\ * \, (dz \wedge dt) &= dx \wedge dy, & * \, (dz \wedge dx) &= - \, dy \wedge dt, \end{aligned} \tag{11.199}$$

3-forms into 1-forms

$$\begin{aligned} * \, (dx \wedge dy \wedge dz) &= - \, dt, & * \, (dy \wedge dz \wedge dt) &= - \, dx, \\ * \, (dz \wedge dx \wedge dt) &= - \, dy, & * \, (dx \wedge dy \wedge dt) &= - \, dz, \end{aligned} \tag{11.200}$$

and interchanges 0-forms and 4-forms

$$* \, 1 = dt \wedge dx \wedge dy \wedge dz, \qquad * \, (dt \wedge dx \wedge dy \wedge dz) = - \, 1. \tag{11.201}$$

More generally in four dimensions, we define the Hodge star as

$$
\begin{aligned}
*\,1 &= \frac{1}{4!}\,\eta_{k\ell mn}\,dx^k \wedge dx^\ell \wedge dx^m \wedge dx^n, \\
*\,dx^i &= \frac{1}{3!}\,g^{ik}\,\eta_{k\ell mn}\,dx^\ell \wedge dx^m \wedge dx^n, \\
*\,(dx^i \wedge dx^j) &= \frac{1}{2}\,g^{ik}\,g^{j\ell}\,\eta_{k\ell mn}\,dx^m \wedge dx^n, \\
*\,(dx^i \wedge dx^j \wedge dx^k) &= g^{it}\,g^{ju}\,g^{kv}\,\eta_{tuvw}\,dx^w, \\
*\left(dx^i \wedge dx^j \wedge dx^k \wedge dx^\ell\right) &= g^{it}\,g^{ju}\,g^{kv}\,g^{\ell w}\eta_{tuvw} = \eta^{ijk\ell}.
\end{aligned} \tag{11.202}
$$

Thus (exercise 11.16) if the determinant $\det g_{ij}$ of the metric is negative, then

$$
\begin{aligned}
&**\; dx^i = dx^i, \qquad **\;(dx^i \wedge dx^j) = -\,dx^i \wedge dx^j, \\
&**\;(dx^i \wedge dx^j \wedge dx^k) = dx^i \wedge dx^j \wedge dx^k, \qquad **\,1 = -\,1.
\end{aligned} \tag{11.203}
$$

In n dimensions, the Hodge star turns p-forms into $(n-p)$-forms

$$
*\left(dx^{i_1} \wedge \ldots \wedge dx^{i_p}\right) = g^{i_1 k_1} \ldots g^{i_p k_p}\,\frac{\eta_{k_1 \ldots k_p \ell_1 \ldots \ell_{n-p}}}{(n-p)!}\,dx^{\ell_1} \wedge \ldots \wedge dx^{\ell_{n-p}}. \tag{11.204}
$$

Example 11.17 (The inhomogeneous Maxwell equations) Since the homogeneous Maxwell equations are

$$
dF = ddA = 0 \tag{11.205}
$$

we first form the dual $*F = *dA$

$$
*\,F = \tfrac{1}{2}F_{ij} * \left(dx^i \wedge dx^j\right) = \tfrac{1}{4}F_{ij}g^{ik}g^{j\ell}\eta_{k\ell mn}dx^m \wedge dx^n = \tfrac{1}{4}F^{k\ell}\eta_{k\ell mn}dx^m \wedge dx^n
$$

and then apply the exterior derivative

$$
d * F = \tfrac{1}{4}d\left(F^{k\ell}\eta_{k\ell mn}dx^m \wedge dx^n\right) = \tfrac{1}{4}\partial_p\left(F^{k\ell}\eta_{k\ell mn}\right)dx^p \wedge dx^m \wedge dx^n.
$$

To get back to a 1-form like $j = j_k dx^k$, we apply a second Hodge star

$$
\begin{aligned}
*\,d * F &= \tfrac{1}{4}\,\partial_p\left(F^{k\ell}\,\eta_{k\ell mn}\right) * \left(dx^p \wedge dx^m \wedge dx^n\right) \\
&= \tfrac{1}{4}\,\partial_p\left(F^{k\ell}\,\eta_{k\ell mn}\right) g^{ps}g^{mt}g^{nu}\eta_{stuv}\,dx^v \\
&= \tfrac{1}{4}\,\partial_p\left(\sqrt{g}\,F^{k\ell}\right)\epsilon_{k\ell mn}\,g^{ps}g^{mt}g^{nu}\sqrt{g}\,\epsilon_{stuv}\,dx^v \\
&= \tfrac{1}{4}\,\partial_p\left(\sqrt{g}\,F^{k\ell}\right)\epsilon_{k\ell mn}\,g^{ps}g^{mt}g^{nu}g^{wv}\,\epsilon_{stuv}\,\sqrt{g}\,dx \\
&= \tfrac{1}{4}\,\partial_p\left(\sqrt{g}\,F^{k\ell}\right)\epsilon_{k\ell mn}\,\epsilon_{pmnw}\,\frac{\sqrt{g}}{\det g_{ij}}\,dx_w,
\end{aligned} \tag{11.206}
$$

in which we used the definition (1.184) of the determinant. Levi-Civita's 4-symbol obeys the identity (exercise 11.17)

$$\epsilon_{k\ell mn}\,\epsilon^{pwmn} = 2\left(\delta^p_k\,\delta^w_\ell - \delta^w_k\,\delta^p_\ell\right). \tag{11.207}$$

Applying it to $* d * F$, we get

$$* \, d * F = \frac{s}{2\sqrt{g}}\partial_p\left(\sqrt{g}\,F^{k\ell}\right)\left(\delta^p_k\,\delta^w_\ell - \delta^w_k\,\delta^p_\ell\right)dx_w = -\frac{s}{\sqrt{g}}\,\partial_p\left(\sqrt{g}\,F^{kp}\right)\,dx_k.$$

In our space-time $s = -1$. Setting $*d*F$ equal to $j = j_k\,dx^k = j^k\,dx_k$ multiplied by the permeability μ_0 of the vacuum, we arrive at expressions for the microscopic inhomogeneous Maxwell equations in terms of both tensors and forms

$$\partial_p\left(\sqrt{g}\,F^{kp}\right) = \mu_0\,\sqrt{g}\,j^k \quad \text{and} \quad * \, d * F = \mu_0 j. \tag{11.208}$$

They and the homogeneous Bianchi identity (11.93, 11.114, & 11.247)

$$\epsilon^{ijk\ell}\,\partial_\ell F_{jk} = dF = d\,dA = 0 \tag{11.209}$$

are invariant under general coordinate transformations. □

11.27 Derivatives and affine connections

If $F(x)$ is a vector field, then its invariant description in terms of space-time dependent basis vectors $e_i(x)$ is

$$F(x) = F^i(x)\,e_i(x). \tag{11.210}$$

Since the basis vectors $e_i(x)$ vary with x, the derivative of $F(x)$ contains two terms

$$\frac{\partial F}{\partial x^\ell} = \frac{\partial F^i}{\partial x^\ell}\,e_i + F^i\,\frac{\partial e_i}{\partial x^\ell}. \tag{11.211}$$

In general, the derivative of a vector e_i is not a linear combination of the basis vectors e_k. For instance, on the two-dimensional surface of a sphere in three dimensions, the derivative

$$\frac{\partial e_\theta}{\partial \theta} = -\hat{r} \tag{11.212}$$

points to the sphere's center and isn't a linear combination of e_θ and e_ϕ.

The inner product of a derivative $\partial e_i/\partial x^\ell$ with a dual basis vector e^k is the **Levi-Civita affine connection**

$$\Gamma^k_{\ell i} = e^k \cdot \frac{\partial e_i}{\partial x^\ell}, \tag{11.213}$$

which relates spaces that are tangent to the manifold at infinitesimally separated points. It is called an *affine* connection because the different tangent spaces lack a common origin.

In terms of the affine connection (11.213), the inner product of the derivative of (11.211) with e^k is

$$e^k \cdot \frac{\partial F}{\partial x^\ell} = e^k \cdot \frac{\partial F^i}{\partial x^\ell} e_i + F^i e^k \cdot \frac{\partial e_i}{\partial x^\ell} = \frac{\partial F^k}{\partial x^\ell} + \Gamma^k_{\ell i} F^i \tag{11.214}$$

a combination that is called a **covariant derivative** (section 11.30)

$$D_\ell F^k \equiv \nabla_\ell F^k \equiv \frac{\partial F^k}{\partial x^\ell} + \Gamma^k_{\ell i} F^i. \tag{11.215}$$

Some physicists write the affine connection $\Gamma^k_{i\ell}$ as

$$\left\{ \begin{matrix} k \\ i\ell \end{matrix} \right\} = \Gamma^k_{i\ell} \tag{11.216}$$

and call it a **Christoffel symbol of the second kind**.

The vectors e_i are the space-time derivatives (11.135) of the point p, and so the affine connection (11.213) is a double derivative of p

$$\Gamma^k_{\ell i} = e^k \cdot \frac{\partial e_i}{\partial x^\ell} = e^k \cdot \frac{\partial^2 p}{\partial x^\ell \partial x^i} = e^k \cdot \frac{\partial^2 p}{\partial x^i \partial x^\ell} = e^k \cdot \frac{\partial e_\ell}{\partial x^i} = \Gamma^k_{i\ell} \tag{11.217}$$

and thus is symmetric in its two lower indices

$$\Gamma^k_{i\ell} = \Gamma^k_{\ell i}. \tag{11.218}$$

Affine connections are **not** tensors. Tensors transform homogeneously; connections transform inhomogeneously. The connection $\Gamma^k_{i\ell}$ transforms as

$$\begin{aligned} \Gamma'^k_{i\ell} &= e'^k \cdot \frac{\partial e'_\ell}{\partial x'^i} = \frac{\partial x'^k}{\partial x^p} e^p \cdot \frac{\partial x^m}{\partial x'^i} \frac{\partial}{\partial x^m} \left(\frac{\partial x^n}{\partial x'^\ell} e_n \right) \\ &= \frac{\partial x'^k}{\partial x^p} \frac{\partial x^m}{\partial x'^i} \frac{\partial x^n}{\partial x'^\ell} e^p \cdot \frac{\partial e_n}{\partial x^m} + \frac{\partial x'^k}{\partial x^p} \frac{\partial^2 x^p}{\partial x'^i \partial x'^\ell} \\ &= \frac{\partial x'^k}{\partial x^p} \frac{\partial x^m}{\partial x'^i} \frac{\partial x^n}{\partial x'^\ell} \Gamma^p_{mn} + \frac{\partial x'^k}{\partial x^p} \frac{\partial^2 x^p}{\partial x'^i \partial x'^\ell}. \end{aligned} \tag{11.219}$$

The electromagnetic field $A_i(x)$ and other gauge fields are connections.

Since the Levi-Civita connection $\Gamma^k_{i\ell}$ is symmetric in i and ℓ, in four-dimensional space-time, there are ten of them for k, or 40 in all. The ten correspond to three rotations, three boosts, and four translations.

Einstein–Cartan theories do not assume that the space-time manifold is embedded in a flat space of higher dimension. So their basis vectors need not be partial derivatives of a point in the embedding space, and their affine connections Γ^a_{bc} need not be symmetric in their lower indices. The antisymmetric part is the **torsion tensor**

$$T^a_{bc} = \Gamma^a_{bc} - \Gamma^a_{cb}. \tag{11.220}$$

11.28 Parallel transport

The movement of a vector along a curve on a manifold so that its direction in successive tangent spaces does not change is called **parallel transport**. If the vector is $F = F^i e_i$, then we want $e^k \cdot dF$ to vanish along the curve. But this is just the condition that the covariant derivative of F should vanish along the curve

$$e^k \cdot \frac{\partial F}{\partial x^\ell} = e^k \cdot \frac{\partial F^i}{\partial x^\ell} e_i + F^i e^k \cdot \frac{\partial e_i}{\partial x^\ell} = \frac{\partial F^k}{\partial x^\ell} + \Gamma^k_{\ell i} F^i = D_\ell F^k = 0. \tag{11.221}$$

Example 11.18 (Parallel transport on a sphere) The tangent space on a 2-sphere is spanned by the unit basis vectors

$$\begin{aligned} \hat{\boldsymbol{\theta}} &= (\cos\theta\,\cos\phi,\ \cos\theta\,\sin\phi,\ -\sin\theta)\,, \\ \hat{\boldsymbol{\phi}} &= (-\sin\phi,\ \cos\phi,\ 0)\,. \end{aligned} \tag{11.222}$$

We can parallel-transport the vector $\hat{\boldsymbol{\phi}}$ down from the north pole along the meridian $\phi = 0$ to the equator; all along this path $\hat{\boldsymbol{\phi}} = (0, 1, 0)$. Then we can parallel-transport it along the equator to $\phi = \pi/2$ where it is $(-1, 0, 0)$. Then we can parallel-transport it along the meridian $\phi = \pi/2$ up to the north pole where it is $(-1, 0, 0)$ as it was on the equator. The change from $(0, 1, 0)$ to $(-1, 0, 0)$ is due to the curvature of the sphere. □

11.29 Notations for derivatives

We have various notations for derivatives. We can use the variables x, y, and so forth as subscripts to label derivatives

$$f_x = \partial_x f = \frac{\partial f}{\partial x} \quad \text{and} \quad f_y = \partial_y f = \frac{\partial f}{\partial y}. \tag{11.223}$$

If we use indices to label variables, then we can use commas

$$f_{,i} = \partial_i f = \frac{\partial f}{\partial x^i} \quad \text{and} \quad f_{,ik} = \partial_k \partial_i f = \frac{\partial^2 f}{\partial x^k \partial x^i} \tag{11.224}$$

and $f_{,k'} = \partial f / \partial x'^k$. For instance, we may write part of (11.217) as $e_{i,\ell} = e_{\ell,i}$.

11.30 Covariant derivatives

In comma notation, the derivative of a contravariant vector field $F = F^i\, e_i$ is

$$F_{,\ell} = F^i_{,\ell}\, e_i + F^i\, e_{i,\ell}, \tag{11.225}$$

which in general lies outside the space spanned by the basis vectors e_i. So we use the affine connections (11.213) to form the inner product

$$e^k \cdot F_{,\ell} = e^k \cdot \left(F^i_{,\ell} e_i + F^i e_{i,\ell}\right) = F^i_{,\ell}\, \delta^k_i + F^i\, \Gamma^k_{\ell i} = F^k_{,\ell} + \Gamma^k_{\ell i}\, F^i. \tag{11.226}$$

This **covariant derivative of a contravariant vector field** often is written with a semicolon

$$F^k_{;\ell} = e^k \cdot F_{,\ell} = F^k_{,\ell} + \Gamma^k_{\ell i}\, F^i. \tag{11.227}$$

It transforms as a mixed second-rank tensor. The invariant change dF projected onto e^k is

$$e^k \cdot dF = e^k \cdot F_{,\ell}\, dx^\ell = F^k_{;\ell}\, dx^\ell. \tag{11.228}$$

In terms of its covariant components, the derivative of a vector V is

$$V_{,\ell} = (V_k\, e^k)_{,\ell} = V_{k,\ell}\, e^k + V_k\, e^k_{,\ell}. \tag{11.229}$$

To relate the derivatives of the vectors e^i to the affine connections $\Gamma^k_{i\ell}$, we differentiate the orthonormality relation

$$\delta^k_i = e^k \cdot e_i, \tag{11.230}$$

which gives us

$$0 = e^k_{,\ell} \cdot e_i + e^k \cdot e_{i,\ell} \quad \text{or} \quad e^k_{,\ell} \cdot e_i = -e^k \cdot e_{i,\ell} = -\Gamma^k_{i\ell}. \tag{11.231}$$

Since $e_i \cdot e^k_{,\ell} = -\Gamma^k_{i\ell}$, the inner product of e_i with the derivative of V is

$$e_i \cdot V_{,\ell} = e_i \cdot \left(V_{k,\ell} e^k + V_k\, e^k_{,\ell}\right) = V_{i,\ell} - V_k \Gamma^k_{i\ell}. \tag{11.232}$$

This **covariant derivative of a covariant vector field** also is often written with a semicolon

$$V_{i;\ell} = e_i \cdot V_{,\ell} = V_{i,\ell} - V_k \Gamma^k_{i\ell}. \tag{11.233}$$

It transforms as a rank-2 covariant tensor. Note the minus sign in $V_{i;\ell}$ and the plus sign in $F^k_{;\ell}$. The change $e_i \cdot dV$ is

$$e_i \cdot dV = e_i \cdot V_{,\ell}\, dx^\ell = V_{i;\ell}\, dx^\ell. \tag{11.234}$$

Since dV is invariant, e_i covariant, and dx^ℓ contravariant, the quotient rule (section 11.15) confirms that the covariant derivative $V_{i;\ell}$ of a covariant vector V_i is a rank-2 covariant tensor.

11.31 The covariant curl

Because the connection $\Gamma^k_{i\ell}$ is symmetric (11.218) in its lower indices, the covariant curl of a covariant vector V_i is simply its ordinary curl

$$V_{\ell;i} - V_{i;\ell} = V_{\ell,i} - V_k \Gamma^k_{\ell i} - V_{i,\ell} + V_k \Gamma^k_{i\ell} = V_{\ell,i} - V_{i,\ell}. \tag{11.235}$$

Thus the Faraday field-strength tensor $F_{i\ell}$ which is defined as the curl of the covariant vector field A_i

$$F_{i\ell} = A_{\ell,i} - A_{i,\ell} \tag{11.236}$$

is a generally covariant second-rank tensor.

In orthogonal coordinates, the curl is defined (6.39, 11.111) in terms of the totally antisymmetric Levi-Civita symbol ϵ^{ijk} (with $\epsilon_{123} = \epsilon^{123} = 1$), as

$$\nabla \times \overline{V} = \sum_{i=1}^{3} (\nabla \times \overline{V})_i \, \hat{e}_i = \frac{1}{h_1 h_2 h_3} \sum_{ijk=1}^{3} \boldsymbol{e}_i \, \epsilon^{ijk} \, V_{k;j}, \tag{11.237}$$

which, in view of (11.235) and the antisymmetry of ϵ^{ijk}, is

$$\nabla \times \overline{V} = \sum_{i=1}^{3} (\nabla \times \overline{V})_i \, \hat{e}_i = \sum_{ijk=1}^{3} \frac{1}{h_i h_j h_k} \boldsymbol{e}_i \, \epsilon^{ijk} \, V_{k,j} \tag{11.238}$$

or by (11.159 & 11.161)

$$\nabla \times \overline{V} = \sum_{ijk=1}^{3} \frac{1}{h_i h_j h_k} h_i \hat{e}_i \, \epsilon^{ijk} \, V_{k,j} = \sum_{ijk=1}^{3} \frac{1}{h_i h_j h_k} h_i \hat{e}_i \, \epsilon^{ijk} \, (h_k \overline{V}_k)_{,j}. \tag{11.239}$$

Often one writes this as a determinant

$$\nabla \times \overline{V} = \frac{1}{h_1 h_2 h_3} \begin{vmatrix} \boldsymbol{e}_1 & \boldsymbol{e}_2 & \boldsymbol{e}_3 \\ \partial_1 & \partial_2 & \partial_3 \\ V_1 & V_2 & V_3 \end{vmatrix} = \frac{1}{h_1 h_2 h_3} \begin{vmatrix} h_1 \hat{e}_1 & h_2 \hat{e}_2 & h_3 \hat{e}_3 \\ \partial_1 & \partial_2 & \partial_3 \\ h_1 \overline{V}_1 & h_2 \overline{V}_2 & h_3 \overline{V}_3 \end{vmatrix}. \tag{11.240}$$

In cylindrical coordinates, the curl is

$$\nabla \times \overline{V} = \frac{1}{\rho} \begin{vmatrix} \hat{\rho} & \rho \hat{\phi} & \hat{z} \\ \partial_\rho & \partial_\phi & \partial_z \\ \overline{V}_\rho & \rho \overline{V}_\phi & \overline{V}_z \end{vmatrix}. \tag{11.241}$$

In spherical coordinates, it is

$$\nabla \times \overline{V} = \frac{1}{r^2 \sin\theta} \begin{vmatrix} \hat{r} & r\hat{\theta} & r \sin\theta \, \hat{\phi} \\ \partial_r & \partial_\theta & \partial_\phi \\ \overline{V}_r & r\overline{V}_\theta & r \sin\theta \, \overline{V}_\phi \end{vmatrix}. \tag{11.242}$$

In more formal language, the curl is

$$dV = d\left(V_k dx^k\right) = V_{k,i}\, dx^i \wedge dx^k = \frac{1}{2}\left(V_{k,i} - V_{i,k}\right) dx^i \wedge dx^k. \tag{11.243}$$

11.32 Covariant derivatives and antisymmetry

By applying our rule (11.233) for the covariant derivative of a covariant vector to a second-rank tensor $A_{i\ell}$, we get

$$A_{i\ell;k} = A_{i\ell,k} - \Gamma^m_{ik} A_{m\ell} - \Gamma^m_{\ell k} A_{im}. \tag{11.244}$$

Suppose now that our tensor is antisymmetric

$$A_{i\ell} = -A_{\ell i}. \tag{11.245}$$

Then by adding together the three cyclic permutations of the indices $i\ell k$ we find that the antisymmetry of the tensor and the symmetry (11.218) of the affine connection conspire to cancel the nonlinear terms

$$\begin{aligned} A_{i\ell;k} + A_{ki;\ell} + A_{\ell k;i} &= A_{i\ell,k} - \Gamma^m_{ik} A_{m\ell} - \Gamma^m_{\ell k} A_{im} \\ &\quad + A_{ki,\ell} - \Gamma^m_{k\ell} A_{mi} - \Gamma^m_{i\ell} A_{km} \\ &\quad + A_{\ell k,i} - \Gamma^m_{\ell i} A_{mk} - \Gamma^m_{ki} A_{\ell m} \\ &= A_{i\ell,k} + A_{ki,\ell} + A_{\ell k,i}, \end{aligned} \tag{11.246}$$

an identity named after Luigi Bianchi (1856–1928).

The Maxwell field-strength tensor $F_{i\ell}$ is antisymmetric by construction ($F_{i\ell} = A_{\ell,i} - A_{i,\ell}$), and so the homogeneous Maxwell equations

$$\epsilon^{ijk\ell}\, F_{jk,\ell} = F_{jk,\ell} + F_{k\ell,j} + F_{\ell j,k} = 0 \tag{11.247}$$

are tensor equations valid in all coordinate systems. This is another example of how amazingly right Maxwell was in the middle of the nineteenth century.

11.33 Affine connection and metric tensor

To relate the affine connection $\Gamma^m_{\ell i}$ to the derivatives of the metric tensor $g_{k\ell}$, we lower the contravariant index m to get

$$\Gamma_{k\ell i} = g_{km}\, \Gamma^m_{\ell i} = g_{km}\, \Gamma^m_{i\ell} = \Gamma_{ki\ell}, \tag{11.248}$$

which is symmetric in its last two indices and which some call a **Christoffel symbol of the first kind**, written $[\ell i, k]$. One can raise the index k back up by using the inverse of the metric tensor

$$g^{mk}\, \Gamma_{k\ell i} = g^{mk}\, g_{kn}\, \Gamma^n_{\ell i} = \delta^m_n\, \Gamma^n_{\ell i} = \Gamma^m_{\ell i}. \tag{11.249}$$

Although we can raise and lower these indices, the connections $\Gamma^m_{\ell i}$ and $\Gamma_{k\ell i}$ are not tensors.

The definition (11.213) of the affine connection tells us that

$$\Gamma_{k\ell i} = g_{km}\, \Gamma^m_{\ell i} = g_{km}\, e^m \cdot e_{\ell,i} = e_k \cdot e_{\ell,i} = \Gamma_{ki\ell} = e_k \cdot e_{i,\ell}. \tag{11.250}$$

By differentiating the definition $g_{i\ell} = e_i \cdot e_\ell$ of the metric tensor, we find

$$g_{i\ell,k} = e_{i,k} \cdot e_\ell + e_i \cdot e_{\ell,k} = e_\ell \cdot e_{i,k} + e_i \cdot e_{\ell,k} = \Gamma_{\ell ik} + \Gamma_{i\ell k}. \tag{11.251}$$

Permuting the indices cyclicly, we have

$$\begin{aligned} g_{ki,\ell} &= \Gamma_{ik\ell} + \Gamma_{ki\ell}, \\ g_{\ell k,i} &= \Gamma_{k\ell i} + \Gamma_{\ell ki}. \end{aligned} \tag{11.252}$$

If we now subtract relation (11.251) from the sum of the two formulas (11.252) keeping in mind the symmetry $\Gamma_{abc} = \Gamma_{acb}$, then we find that four of the six terms cancel

$$g_{ki,\ell} + g_{\ell k,i} - g_{i\ell,k} = \Gamma_{ik\ell} + \Gamma_{ki\ell} + \Gamma_{k\ell i} + \Gamma_{\ell ki} - \Gamma_{\ell ik} - \Gamma_{i\ell k} = 2\Gamma_{k\ell i} \tag{11.253}$$

leaving a formula for $\Gamma_{k\ell i}$

$$\Gamma_{k\ell i} = \tfrac{1}{2}\left(g_{ki,\ell} + g_{\ell k,i} - g_{i\ell,k}\right). \tag{11.254}$$

Thus the connection is three derivatives of the metric tensor

$$\Gamma^s_{i\ell} = g^{sk}\Gamma_{k\ell i} = \tfrac{1}{2} g^{sk}\left(g_{ki,\ell} + g_{\ell k,i} - g_{i\ell,k}\right). \tag{11.255}$$

11.34 Covariant derivative of the metric tensor

Covariant derivatives of second-rank and higher-rank tensors are formed by iterating our formulas for the covariant derivatives of vectors. For instance, the covariant derivative of the metric tensor is

$$g_{i\ell;k} \equiv g_{i\ell,k} - \Gamma^m_{ik}\, g_{m\ell} - \Gamma^n_{k\ell}\, g_{in}. \tag{11.256}$$

One way to derive this formula is to proceed as in section 11.30 by differentiating the invariant metric tensor $g_{i\ell}\, e^i \otimes e^\ell$ in which the vector product $e^i \otimes e^\ell$ is a kind of direct product

$$g_{,k} = (g_{i\ell} e^i \otimes e^\ell)_{,k} = g_{i\ell,k}\, e^i \otimes e^\ell + g_{i\ell}\, e^i_{,k} \otimes e^\ell + g_{i\ell}\, e^i \otimes e^\ell_{,k}. \tag{11.257}$$

We now take the inner product of this derivative with $e_m \otimes e_n$

$$(e_m \otimes e_n, g_{,k}) = g_{i\ell,k}\, e_m \cdot e^i\, e_n \cdot e^\ell + g_{i\ell}\, e_m \cdot e^i_{,k}\, e_n \cdot e^\ell + g_{i\ell}\, e_m \cdot e^i\, e_n \cdot e^\ell_{,k} \tag{11.258}$$

and use the rules $e_m \cdot e^i = \delta^i_m$ and $e_m \cdot e^i_{,k} = -\Gamma^i_{mk}$ (11.231) to write

$$(e_m \otimes e_n, g_{,k}) = g_{mn;k} = g_{mn,k} - g_{i\ell}\Gamma^i_{mk}\delta^\ell_n - g_{i\ell}\delta^i_m\Gamma^\ell_{nk} \tag{11.259}$$

or

$$g_{mn;k} = g_{mn,k} - \Gamma^{i}_{mk} g_{in} - \Gamma^{\ell}_{nk} g_{m\ell}, \tag{11.260}$$

which is (11.256) inasmuch as both $g_{i\ell}$ and $\Gamma^{k}_{i\ell}$ are symmetric in their two lower indices.

If we now substitute our formula (11.255) for the connections Γ^{l}_{ik} and $\Gamma^{n}_{k\ell}$

$$g_{i\ell;k} = g_{i\ell,k} - \tfrac{1}{2} g^{ms}\left(g_{is,k} + g_{sk,i} - g_{ik,s}\right) g_{m\ell} - \tfrac{1}{2} g^{ns}\left(g_{\ell s,k} + g_{sk,\ell} - g_{\ell k,s}\right) g_{in} \tag{11.261}$$

and use the fact (11.145) that the metric tensors $g_{i\ell}$ and $g^{\ell k}$ are mutually inverse, then we find

$$\begin{aligned} g_{i\ell;k} &= g_{i\ell,k} - \tfrac{1}{2}\delta^{s}_{\ell}\left(g_{is,k} + g_{sk,i} - g_{ik,s}\right) - \tfrac{1}{2}\delta^{s}_{i}\left(g_{\ell s,k} + g_{sk,\ell} - g_{\ell k,s}\right) \\ &= g_{i\ell,k} - \tfrac{1}{2}\left(g_{i\ell,k} + g_{\ell k,i} - g_{ik,\ell}\right) - \tfrac{1}{2}\left(g_{\ell i,k} + g_{ik,\ell} - g_{\ell k,i}\right) \\ &= 0. \end{aligned} \tag{11.262}$$

The covariant derivative of the metric tensor vanishes. This result follows from our choice of the Levi-Civita connection (11.213); it is not true for some other connections.

11.35 Divergence of a contravariant vector

The contraction of the covariant derivative of a contravariant vector is a **scalar** known as the **divergence**,

$$\nabla \cdot V = V^{i}_{;i} = V^{i}_{,i} + \Gamma^{i}_{ik} V^{k}. \tag{11.263}$$

Because two indices in the connection

$$\Gamma^{i}_{ik} = \tfrac{1}{2} g^{im}\left(g_{im,k} + g_{km,i} - g_{ik,m}\right) \tag{11.264}$$

are contracted, its last two terms cancel because they differ only by the interchange of the dummy indices i and m

$$g^{im} g_{km,i} = g^{mi} g_{km,i} = g^{im} g_{ki,m} = g^{im} g_{ik,m}. \tag{11.265}$$

So the contracted connection collapses to

$$\Gamma^{i}_{ik} = \tfrac{1}{2} g^{im} g_{im,k}. \tag{11.266}$$

There is a nice formula for this last expression involving the absolute value of the determinant $\det \underline{g} \equiv \det g_{mn}$ of the metric tensor considered as a matrix $\underline{g} \equiv g_{mn}$. To derive it, we recall that like any determinant, the determinant $\det(\underline{g})$ of the metric tensor is given by the cofactor sum (1.195)

$$\det(\underline{g}) = \sum_{\ell} g_{i\ell} C_{i\ell} \tag{11.267}$$

along any row or column, that is, over ℓ for fixed i or over i for fixed ℓ, where $C_{i\ell}$ is the cofactor defined as $(-1)^{i+\ell}$ times the determinant of the reduced matrix consisting of the matrix $g_{i\ell}$ with row i and column ℓ omitted. Thus the partial derivative of $\det \underline{g}$ with respect to the $i\ell$th element $g_{i\ell}$ is

$$\frac{\partial \det(\underline{g})}{\partial g_{i\ell}} = C_{i\ell}, \tag{11.268}$$

in which we consider $g_{i\ell}$ and $g_{\ell i}$ to be independent variables for the purposes of this differentiation. The inverse $g^{i\ell}$ of the metric tensor g, like the inverse (1.197) of any matrix, is the transpose of the cofactor matrix divided by its determinant $\det g$,

$$g^{i\ell} = \frac{C_{\ell i}}{\det(\underline{g})} = \frac{1}{\det(\underline{g})} \frac{\partial \det(\underline{g})}{\partial g_{\ell i}}. \tag{11.269}$$

The chain rule gives us the derivative of the determinate $\det(\underline{g})$ as

$$\det(\underline{g})_{,k} = g_{i\ell,k} \frac{\partial \det(\underline{g})}{\partial g_{i\ell}} = g_{i\ell,k} \det(\underline{g})\, g^{\ell i} \tag{11.270}$$

and so, since $g_{i\ell} = g_{\ell i}$, the contracted connection (11.266) is

$$\Gamma^i_{ik} = \tfrac{1}{2} g^{im} g_{im,k} = \frac{\det(\underline{g})_{,k}}{2\det(\underline{g})} = \frac{|\det(\underline{g})|_{,k}}{2|\det(\underline{g})|} = \frac{g_{,k}}{2g} = \frac{(\sqrt{g})_{,k}}{\sqrt{g}}, \tag{11.271}$$

in which $g \equiv \left|\det(\underline{g})\right|$ is the absolute value of the determinant of the metric tensor.

Thus from (11.263), we arrive at our formula for the covariant divergence of a contravariant vector:

$$\nabla \cdot V = V^i_{;i} = V^i_{,i} + \Gamma^i_{ik} V^k = V^k_{,k} + \frac{(\sqrt{g})_{,k}}{\sqrt{g}} V^k = \frac{(\sqrt{g}\, V^k)_{,k}}{\sqrt{g}}. \tag{11.272}$$

More formally, the Hodge dual (11.202) of the 1-form $V = V_i\, dx^i$ is

$$\begin{aligned} * V = V_i * dx^i &= V_i \frac{1}{3!} g^{ik}\, \eta_{k\ell mn}\, dx^\ell \wedge dx^m \wedge dx^n \\ &= \frac{1}{3!} \sqrt{g}\, V^k\, \epsilon_{k\ell mn}\, dx^\ell \wedge dx^m \wedge dx^n, \end{aligned} \tag{11.273}$$

in which g is the absolute value of the determinant of the metric tensor g_{ij}. The exterior derivative now gives

$$d * V = \frac{1}{3!} \left(\sqrt{g}\, V^k\right)_{,p} \epsilon_{k\ell mn}\, dx^p \wedge dx^\ell \wedge dx^m \wedge dx^n. \tag{11.274}$$

So using (11.202) to apply a second Hodge star, we get (exercise 11.19)

$$
\begin{aligned}
* \, d * V &= \frac{1}{3!} \left(\sqrt{g}\, V^k\right)_{,p} \epsilon_{k\ell mn} * \left(dx^p \wedge dx^\ell \wedge dx^m \wedge dx^n\right) \\
&= \frac{1}{3!} \left(\sqrt{g}\, V^k\right)_{,p} \epsilon_{k\ell mn}\, g^{pt}\, g^{\ell u}\, g^{mv}\, g^{nw} \eta_{tuvw} \\
&= \frac{1}{3!} \left(\sqrt{g}\, V^k\right)_{,p} \epsilon_{k\ell mn}\, g^{pt}\, g^{\ell u}\, g^{mv}\, g^{nw} \epsilon_{tuvw} \sqrt{g} \\
&= \frac{1}{3!} \left(\sqrt{g}\, V^k\right)_{,p} \epsilon_{k\ell mn} \frac{\sqrt{g}}{\det g_{ij}}\, \epsilon^{p\ell mn} \\
&= \frac{s}{\sqrt{g}} \left(\sqrt{g}\, V^k\right)_{,p} \delta^p_k = \frac{s}{\sqrt{g}} \left(\sqrt{g}\, V^k\right)_{,k}.
\end{aligned} \tag{11.275}
$$

So in our space-time with $\det g_{ij} = -g$

$$
- * d * V = \frac{1}{\sqrt{g}} \left(\sqrt{g}\, V^k\right)_{,k}. \tag{11.276}
$$

In 3-space the Hodge star (11.191) of a 1-form $V = V_i\, dx^i$ is

$$
* \; V = V_i * dx^i = V_i \frac{1}{2} g^{i\ell}\, \eta_{\ell jk}\, dx^j \wedge dx^k = \frac{1}{2} \sqrt{g}\, V^\ell\, \epsilon_{\ell jk}\, dx^j \wedge dx^k. \tag{11.277}
$$

Applying the exterior derivative, we get the invariant form

$$
d * V = \frac{1}{2} \left(\sqrt{g}\, V^\ell\right)_{,p} \epsilon_{\ell jk}\, dx^p \wedge dx^j \wedge dx^k. \tag{11.278}
$$

We add a star by using the definition (11.191) of the Hodge dual in a 3-space in which the determinant $\det g_{ij}$ is positive and the identity (exercise 11.18)

$$
\epsilon_{\ell jk}\, \epsilon^{pjk} = 2\, \delta^p_\ell \tag{11.279}
$$

as well as the definition (1.184) of the determinant

$$
\begin{aligned}
* \; d \; * \; V &= \frac{1}{2} \left(\sqrt{g}\, V^\ell\right)_{,p} \epsilon_{\ell jk} * \left(dx^p \wedge dx^j \wedge dx^k\right) \\
&= \frac{1}{2} \left(\sqrt{g}\, V^\ell\right)_{,p} \epsilon_{\ell jk}\, g^{pt} g^{ju} g^{kv} \eta_{tuv} \\
&= \frac{1}{2} \left(\sqrt{g}\, V^\ell\right)_{,p} \epsilon_{\ell jk}\, g^{pt} g^{ju} g^{kv} \epsilon_{tuv} \sqrt{g} \\
&= \frac{1}{2} \left(\sqrt{g}\, V^\ell\right)_{,p} \epsilon_{\ell jk}\, \epsilon^{pjk} \frac{\sqrt{g}}{\det g_{ij}} \\
&= \frac{1}{\sqrt{g}} \left(\sqrt{g}\, V^\ell\right)_{,p} \delta^p_\ell = \frac{1}{\sqrt{g}} (\sqrt{g}\, V^p)_{,p}.
\end{aligned} \tag{11.280}
$$

Example 11.19 (Divergence in orthogonal coordinates) In two orthogonal coordinates, equations (11.154 & 11.161) imply that $\sqrt{g} = h_1 h_2$ and $V^k = \overline{V}_k / h_k$, and so the divergence of a vector $\overline{V}$ is

$$\nabla \cdot V = \frac{1}{h_1 h_2} \sum_{k=1}^{2} \left(\frac{h_1 h_2}{h_k} \overline{V}_k \right)_{,k}, \tag{11.281}$$

which in polar coordinates (section 11.21), with $h_r = 1$ and $h_\theta = r$, is

$$\nabla \cdot V = \frac{1}{r} \left[\left(r \overline{V}_r \right)_{,r} + \left(\overline{V}_\theta \right)_{,\theta} \right] = \frac{1}{r} \left[\left(r \overline{V}_r \right)_{,r} + \overline{V}_{\theta,\theta} \right]. \tag{11.282}$$

In three orthogonal coordinates, equations (11.154 & 11.161) give $\sqrt{g} = h_1 h_2 h_3$ and $V^k = \overline{V}_k / h_k$, and so the divergence of a vector V is (6.29)

$$\nabla \cdot V = \frac{1}{h_1 h_2 h_3} \sum_{k=1}^{3} \left(\frac{h_1 h_2 h_3}{h_k} \overline{V}_k \right)_{,k}. \tag{11.283}$$

In cylindrical coordinates (section 11.22), $h_\rho = 1$, $h_\phi = \rho$, and $h_z = 1$; so

$$\begin{aligned} \nabla \cdot V &= \frac{1}{\rho} \left[\left(\rho \overline{V}_\rho \right)_{,\rho} + \left(\overline{V}_\phi \right)_{,\phi} + \left(\rho \overline{V}_z \right)_{,z} \right] \\ &= \frac{1}{\rho} \left[\left(\rho \overline{V}_\rho \right)_{,\rho} + \overline{V}_{\phi,\phi} + \rho \overline{V}_{z,z} \right]. \end{aligned} \tag{11.284}$$

In spherical coordinates (section 11.23), $h_r = 1$, $h_\theta = r$, $h_\phi = r \sin\theta$, $g = |\det \underline{g}| = r^4 \sin^2\theta$ and the inverse g^{ij} of the metric tensor is

$$(g^{ij}) = \begin{pmatrix} 1 & 0 & 0 \\ 0 & r^{-2} & 0 \\ 0 & 0 & r^{-2} \sin^{-2}\theta \end{pmatrix}. \tag{11.285}$$

So our formula (11.281) gives us

$$\begin{aligned} \nabla \cdot V &= \frac{1}{r^2 \sin\theta} \left[\left(r^2 \sin\theta \, \overline{V}_r \right)_{,r} + \left(r \sin\theta \, \overline{V}_\theta \right)_{,\theta} + \left(r \overline{V}_\phi \right)_{,\phi} \right] \\ &= \frac{1}{r^2 \sin\theta} \left[\sin\theta \left(r^2 \overline{V}_r \right)_{,r} + r \left(\sin\theta \, \overline{V}_\theta \right)_{,\theta} + r \overline{V}_{\phi,\phi} \right] \end{aligned} \tag{11.286}$$

as the divergence $\nabla \cdot V$. □

11.36 The covariant Laplacian

In flat 3-space, we write the Laplacian as $\nabla \cdot \nabla = \nabla^2$ or as $\triangle$. In euclidean coordinates, both mean $\partial_x^2 + \partial_y^2 + \partial_z^2$. In flat Minkowski space, one often turns the triangle into a square and writes the 4-Laplacian as $\Box = \triangle - \partial_t^2$.

Since the gradient of a scalar field f is a covariant vector, we may use the inverse metric tensor g^{ij} to write the Laplacian $\Box f$ of a scalar f as the covariant divergence of the contravariant vector $g^{ik} f_{,k}$

$$\Box f = (g^{ik} f_{,k})_{;i}. \tag{11.287}$$

The divergence formula (11.272) now expresses the **invariant Laplacian** as

$$\Box f = \frac{(\sqrt{g}\, g^{ik} f_{,k})_{,i}}{\sqrt{g}} = \frac{(\sqrt{g} f^{,i})_{,i}}{\sqrt{g}}. \tag{11.288}$$

To find the Laplacian $\Box f$ in terms of forms, we apply the exterior derivative to the Hodge dual (11.202) of the 1-form $df = f_{,i} dx^i$

$$\begin{aligned} d * df &= d\left(f_{,i} * dx^i\right) = d\left(\frac{1}{3!} f_{,i}\, g^{ik}\, \eta_{k\ell mn}\, dx^\ell \wedge dx^m \wedge dx^n\right) \\ &= \frac{1}{3!}\left(f^{,k}\sqrt{g}\right)_{,p} \epsilon_{k\ell mn}\, dx^p \wedge dx^\ell \wedge dx^m \wedge dx^n \end{aligned} \tag{11.289}$$

and then add a star using (11.202)

$$\begin{aligned} * \, d * df &= \frac{1}{3!}\left(f^{,k}\sqrt{g}\right)_{,p} \epsilon_{k\ell mn} * \left(dx^p \wedge dx^\ell \wedge dx^m \wedge dx^n\right) \\ &= \frac{1}{3!}\left(f^{,k}\sqrt{g}\right)_{,p} \epsilon_{k\ell mn}\, g^{pt} g^{\ell u} g^{mv} g^{nw} \sqrt{g}\, \epsilon_{tuvw}. \end{aligned} \tag{11.290}$$

The definition (1.184) of the determinant now gives (exercise 11.19)

$$\begin{aligned} * \, d * df &= \frac{1}{3!}\left(f^{,k}\sqrt{g}\right)_{,p} \epsilon_{k\ell mn}\, \epsilon^{p\ell mn} \frac{\sqrt{g}}{\det g} \\ &= \left(f^{,k}\sqrt{g}\right)_{,p} \delta^p_k \frac{s}{\sqrt{g}} = \frac{s}{\sqrt{g}}\left(f^{,k}\sqrt{g}\right)_{,k}. \end{aligned} \tag{11.291}$$

In our space-time $\det g_{ij} = sg = -g$, and so the Laplacian is

$$\Box f = - * d * df = \frac{1}{\sqrt{g}}\left(f^{,k}\sqrt{g}\right)_{,k}. \tag{11.292}$$

Example 11.20 (Invariant Laplacians) In two orthogonal coordinates, equations (11.154 & 11.155) imply that $\sqrt{g} = \sqrt{|\det(g_{ij})|} = h_1 h_2$ and that $f^{,i} = g^{ik} f_{,k} = h_i^{-2} f_{,i}$, and so the Laplacian of a scalar f is

$$\Delta f = \frac{1}{h_1 h_2}\left(\sum_{i=1}^{2} \frac{h_1 h_2}{h_i^2} f_{,i}\right)_{,i}. \tag{11.293}$$

In polar coordinates, where $h_1 = 1$, $h_2 = r$, and $g = r^2$, the Laplacian is

$$\Delta f = \frac{1}{r}\left[\left(r f_{,r}\right)_{,r} + \left(r^{-1} f_{,\theta}\right)_{,\theta}\right] = f_{,rr} + r^{-1} f_{,r} + r^{-2} f_{,\theta\theta}. \tag{11.294}$$

In three orthogonal coordinates, equations (11.154 & 11.155) imply that $\sqrt{g} = \sqrt{|\det(g_{ij})|} = h_1 h_2 h_3$ and that $f^{,i} = g^{ik} f_{,k} = h_i^{-2} f_{,i}$, and so the Laplacian of a scalar f is (6.33)

$$\triangle f = \frac{1}{h_1 h_2 h_3} \left(\sum_{i=1}^{3} \frac{h_1 h_2 h_3}{h_i^2} f_{,i} \right)_{,i}. \tag{11.295}$$

In cylindrical coordinates (section 11.22), $h_\rho = 1$, $h_\phi = \rho$, $h_z = 1$, $g = \rho^2$, and the Laplacian is

$$\triangle f = \frac{1}{\rho} \left[\left(\rho f_{,\rho}\right)_{,\rho} + \frac{1}{\rho} f_{,\phi\phi} + \rho f_{,zz} \right] = f_{,\rho\rho} + \frac{1}{\rho} f_{,\rho} + \frac{1}{\rho^2} f_{,\phi\phi} + f_{,zz}. \tag{11.296}$$

In spherical coordinates (section 11.23), $h_r = 1$, $h_\theta = r$, $h_\phi = r \sin\theta$, and $g = |\det \underline{g}| = r^4 \sin^2\theta$. So (11.295) gives us the Laplacian of f as

$$\begin{aligned} \triangle f &= \frac{\left(r^2 \sin\theta f_{,r}\right)_{,r} + \left(\sin\theta f_{,\theta}\right)_{,\theta} + \left(f_{,\phi}/\sin\theta\right)_{,\phi}}{r^2 \sin\theta} \\ &= \frac{\left(r^2 f_{,r}\right)_{,r}}{r^2} + \frac{\left(\sin\theta f_{,\theta}\right)_{,\theta}}{r^2 \sin\theta} + \frac{f_{,\phi\phi}}{r^2 \sin^2\theta}. \end{aligned} \tag{11.297}$$

If the function f is a function only of the radial variable r, then the Laplacian is simply

$$\triangle f(r) = \frac{1}{r^2} \left[r^2 f'(r) \right]' = \frac{1}{r} [r f(r)]'' = f''(r) + \frac{2}{r} f'(r), \tag{11.298}$$

in which the primes denote r-derivatives. □

11.37 The principle of stationary action

It follows from a path-integral formulation of quantum mechanics that the classical motion of a particle is given by the **principle of stationary action** $\delta S = 0$. In the simplest case of a free nonrelativistic particle, the lagrangian is $L = m\dot{x}^2/2$ and the action is

$$S = \int_{t_1}^{t_2} \frac{m}{2} \dot{\boldsymbol{x}}^2 \, dt. \tag{11.299}$$

The classical trajectory is the one that when varied slightly by $\delta\boldsymbol{x}$ (with $\delta\boldsymbol{x}(t_1) = \delta\boldsymbol{x}(t_2) = \boldsymbol{0}$) does not change the action to first order in $\delta\boldsymbol{x}$. We first note that the change $\delta\dot{\boldsymbol{x}}$ in the velocity is the time derivative of the change in the path

$$\delta\dot{\boldsymbol{x}} = \dot{\boldsymbol{x}}' - \dot{\boldsymbol{x}} = \frac{d}{dt}(\boldsymbol{x}' - \boldsymbol{x}) = \frac{d}{dt}\delta\boldsymbol{x}. \tag{11.300}$$

So since $\delta \boldsymbol{x}(t_1) = \delta \boldsymbol{x}(t_2) = \boldsymbol{0}$, the stationary path satisfies

$$
\begin{aligned}
0 = \delta S = \int_{t_1}^{t_2} m\dot{\boldsymbol{x}} \cdot \delta \dot{\boldsymbol{x}}\, dt &= \int_{t_1}^{t_2} m\dot{\boldsymbol{x}} \cdot \frac{d\delta \boldsymbol{x}}{dt}\, dt \\
= \int_{t_1}^{t_2} \left[m\frac{d}{dt}(\dot{\boldsymbol{x}} \cdot \delta \boldsymbol{x}) - m\ddot{\boldsymbol{x}} \cdot \delta \boldsymbol{x} \right] dt \\
= m\,[\dot{\boldsymbol{x}} \cdot \delta \boldsymbol{x}]_{t_1}^{t_2} - m \int_{t_1}^{t_2} \ddot{\boldsymbol{x}} \cdot \delta \boldsymbol{x}\, dt &= -m \int_{t_1}^{t_2} \ddot{\boldsymbol{x}} \cdot \delta \boldsymbol{x}\, dt. \qquad (11.301)
\end{aligned}
$$

If the first-order change in the action is to vanish for arbitrary small variations $\delta \boldsymbol{x}$ in the path, then the acceleration must vanish

$$\ddot{\boldsymbol{x}} = \boldsymbol{0}, \qquad (11.302)$$

which is the classical equation of motion for a free particle.

If the particle is moving under the influence of a potential $V(\boldsymbol{x})$, then the action is

$$S = \int_{t_1}^{t_2} \left(\frac{m}{2}\dot{\boldsymbol{x}}^2 - V(\boldsymbol{x}) \right) dt. \qquad (11.303)$$

Since $\delta V(\boldsymbol{x}) = \nabla V(\boldsymbol{x}) \cdot \delta \boldsymbol{x}$, the principle of stationary action requires that

$$0 = \delta S = \int_{t_1}^{t_2} (-m\ddot{\boldsymbol{x}} - \nabla V) \cdot \delta \boldsymbol{x}\, dt \qquad (11.304)$$

or

$$m\ddot{\boldsymbol{x}} = -\nabla V, \qquad (11.305)$$

which is the classical equation of motion for a particle of mass m in a potential V.

The action for a free particle of mass m in special relativity is

$$S = -m \int_{\tau_1}^{\tau_2} d\tau = -\int_{t_1}^{t_2} m\sqrt{1 - \dot{\boldsymbol{x}}^2}\, dt \qquad (11.306)$$

where $c = 1$ and $\dot{\boldsymbol{x}} = d\boldsymbol{x}/dt$. The requirement of stationary action is

$$0 = \delta S = -\delta \int_{t_1}^{t_2} m\sqrt{1 - \dot{\boldsymbol{x}}^2}\, dt = m \int_{t_1}^{t_2} \frac{\dot{\boldsymbol{x}} \cdot \delta \dot{\boldsymbol{x}}}{\sqrt{1 - \dot{\boldsymbol{x}}^2}}\, dt. \qquad (11.307)$$

But $1/\sqrt{1 - \dot{\boldsymbol{x}}^2} = dt/d\tau$ and so

$$
\begin{aligned}
0 = \delta S &= m \int_{t_1}^{t_2} \frac{d\boldsymbol{x}}{dt} \cdot \frac{d\delta \boldsymbol{x}}{dt} \frac{dt}{d\tau}\, dt = m \int_{\tau_1}^{\tau_2} \frac{d\boldsymbol{x}}{dt} \cdot \frac{d\delta \boldsymbol{x}}{dt} \frac{dt}{d\tau} \frac{dt}{d\tau}\, d\tau \\
&= m \int_{\tau_1}^{\tau_2} \frac{d\boldsymbol{x}}{d\tau} \cdot \frac{d\delta \boldsymbol{x}}{d\tau}\, d\tau. \qquad (11.308)
\end{aligned}
$$

So, integrating by parts, keeping in mind that $\delta\boldsymbol{x}(\tau_2) = \delta\boldsymbol{x}(\tau_1) = \boldsymbol{0}$, we have

$$0 = \delta S = m \int_{\tau_1}^{\tau_2} \left[\frac{d}{d\tau} (\dot{\boldsymbol{x}} \cdot \delta\boldsymbol{x}) - \frac{d^2\boldsymbol{x}}{d\tau^2} \cdot \delta\boldsymbol{x} \right] d\tau = -m \int_{\tau_1}^{\tau_2} \frac{d^2\boldsymbol{x}}{d\tau^2} \cdot \delta\boldsymbol{x}\, d\tau. \quad (11.309)$$

To have this hold for arbitrary $\delta\boldsymbol{x}$, we need

$$\frac{d^2\boldsymbol{x}}{d\tau^2} = \boldsymbol{0}, \quad (11.310)$$

which is the equation of motion for a free particle in special relativity.

What about a charged particle in an electromagnetic field A_i? Its action is

$$S = -m \int_{\tau_1}^{\tau_2} d\tau + q \int_{x_1}^{x_2} A_i(x)\, dx^i = \int_{\tau_1}^{\tau_2} \left(-m + qA_i(x) \frac{dx^i}{d\tau} \right) d\tau. \quad (11.311)$$

We now treat the first term in a four-dimensional manner

$$\delta d\tau = \delta\sqrt{-\eta_{ik}dx^i dx^k} = \frac{-\eta_{ik}dx^i \delta dx^k}{\sqrt{-\eta_{ik}dx^i dx^k}} = -u_k \delta dx^k = -u_k d\delta x^k, \quad (11.312)$$

in which $u_k = dx_k/d\tau$ is the 4-velocity (11.66) and η is the Minkowski metric (11.27) of flat space-time. The variation of the other term is

$$\delta \left(A_i\, dx^i \right) = (\delta A_i)\; dx^i + A_i\, \delta dx^i = A_{i,k}\delta x^k\, dx^i + A_i\, d\delta x^i. \quad (11.313)$$

Putting them together, we get for δS

$$\delta S = \int_{\tau_1}^{\tau_2} \left(m u_k \frac{d\delta x^k}{d\tau} + qA_{i,k}\delta x^k \frac{dx^i}{d\tau} + qA_i \frac{d\delta x^i}{d\tau} \right) d\tau. \quad (11.314)$$

After integrating by parts the last term, dropping the boundary terms, and changing a dummy index, we get

$$\begin{aligned} \delta S &= \int_{\tau_1}^{\tau_2} \left(- m\frac{du_k}{d\tau}\delta x^k + qA_{i,k}\delta x^k \frac{dx^i}{d\tau} - q\frac{dA_k}{d\tau}\delta x^k \right) d\tau \\ &= \int_{\tau_1}^{\tau_2} \left[- m\frac{du_k}{d\tau} + q\left(A_{i,k} - A_{k,i}\right) \frac{dx^i}{d\tau} \right] \delta x^k\, d\tau. \end{aligned} \quad (11.315)$$

If this first-order variation of the action is to vanish for arbitrary δx^k, then the particle must follow the path

$$0 = -m\frac{du_k}{d\tau} + q\left(A_{i,k} - A_{k,i}\right) \frac{dx^i}{d\tau} \quad \text{or} \quad \frac{dp^k}{d\tau} = qF^{ki}u_i, \quad (11.316)$$

which is the Lorentz force law (11.96).

11.38 A particle in a gravitational field

The invariant action for a particle of mass m moving along a path $x^i(t)$ is

$$S = -m\int_{\tau_1}^{\tau_2} d\tau = -m\int\left(-g_{i\ell}dx^i dx^\ell\right)^{\frac{1}{2}}. \tag{11.317}$$

Proceeding as in equation (11.312), we compute the variation $\delta d\tau$ as

$$\begin{aligned}\delta d\tau &= \delta\sqrt{-g_{i\ell}dx^i dx^\ell} = \frac{-\delta(g_{i\ell})dx^i dx^\ell - 2g_{i\ell}dx^i \delta dx^\ell}{2\sqrt{-g_{i\ell}dx^i dx^\ell}}\\ &= -\tfrac{1}{2}g_{i\ell,k}\delta x^k u^i u^\ell d\tau - g_{i\ell}u^i \delta dx^\ell\\ &= -\tfrac{1}{2}g_{i\ell,k}\delta x^k u^i u^\ell d\tau - g_{i\ell}u^i d\delta x^\ell,\end{aligned} \tag{11.318}$$

in which $u^\ell = dx^\ell/d\tau$ is the 4-velocity (11.66). The condition of stationary action then is

$$0 = \delta S = -m\int_{\tau_1}^{\tau_2} \delta d\tau = m\int_{\tau_1}^{\tau_2}\left(\tfrac{1}{2}g_{i\ell,k}\delta x^k u^i u^\ell + g_{i\ell}u^i \frac{d\delta x^\ell}{d\tau}\right)d\tau, \tag{11.319}$$

which we integrate by parts keeping in mind that $\delta x^\ell(\tau_2) = \delta x^\ell(\tau_1) = 0$

$$\begin{aligned}0 &= m\int_{\tau_1}^{\tau_2}\left(\tfrac{1}{2}g_{i\ell,k}\delta x^k u^i u^\ell - \frac{d(g_{i\ell}u^i)}{d\tau}\delta x^\ell\right)d\tau\\ &= m\int_{\tau_1}^{\tau_2}\left(\tfrac{1}{2}g_{i\ell,k}\delta x^k u^i u^\ell - g_{i\ell,k}u^i u^k \delta x^\ell - g_{i\ell}\frac{du^i}{d\tau}\delta x^\ell\right)d\tau.\end{aligned} \tag{11.320}$$

Now interchanging the dummy indices ℓ and k on the second and third terms, we have

$$0 = m\int_{\tau_1}^{\tau_2}\left(\tfrac{1}{2}g_{i\ell,k}u^i u^\ell - g_{ik,\ell}u^i u^\ell - g_{ik}\frac{du^i}{d\tau}\right)\delta x^k d\tau \tag{11.321}$$

or since δx^k is arbitrary

$$0 = \tfrac{1}{2}g_{i\ell,k}u^i u^\ell - g_{ik,\ell}u^i u^\ell - g_{ik}\frac{du^i}{d\tau}. \tag{11.322}$$

If we multiply this equation of motion by g^{rk} and note that $g_{ik,\ell}u^i u^\ell = g_{\ell k,i}u^i u^\ell$, then we find

$$0 = \frac{du^r}{d\tau} + \tfrac{1}{2}g^{rk}\left(g_{ik,\ell} + g_{\ell k,i} - g_{i\ell,k}\right)u^i u^\ell. \tag{11.323}$$

So, using the symmetry $g_{i\ell} = g_{\ell i}$ and the formula (11.255) for $\Gamma^r_{i\ell}$, we get

$$0 = \frac{du^r}{d\tau} + \Gamma^r_{i\ell}\,u^i u^\ell \quad \text{or} \quad 0 = \frac{d^2x^r}{d\tau^2} + \Gamma^r_{i\ell}\,\frac{dx^i}{d\tau}\frac{dx^\ell}{d\tau}, \tag{11.324}$$

which is the geodesic equation. In empty space, particles fall along geodesics **independently of their masses.**

The right-hand side of the geodesic equation (11.324) is a contravariant vector because (Weinberg, 1972) under general coordinate transformations, the inhomogeneous terms arising from $\ddot{x}^r$ cancel those from $\Gamma^r_{i\ell}\dot{x}^i\dot{x}^\ell$. Here and often in what follows we'll use dots to mean proper-time derivatives.

The action for a particle of mass m and charge q in a gravitational field $\Gamma^r_{i\ell}$ and an electromagnetic field A_i is

$$S = -m \int \left(- g_{i\ell} dx^i dx^\ell \right)^{\frac{1}{2}} + q \int_{\tau_1}^{\tau_2} A_i(x)\, dx^i \tag{11.325}$$

because the interaction $q \int A_i dx^i$ is invariant under general coordinate transformations. By (11.315 & 11.321), the first-order change in S is

$$\delta S = m \int_{\tau_1}^{\tau_2} \left[\tfrac{1}{2} g_{i\ell,k} u^i u^\ell - g_{ik,\ell} u^i u^\ell - g_{ik} \frac{du^i}{d\tau} + q \left(A_{i,k} - A_{k,i} \right) u^i \right] \delta x^k d\tau \tag{11.326}$$

and so by combining the Lorentz force law (11.316) and the geodesic equation (11.324) and by writing $F^{ri}\dot{x}_i$ as $F^r{}_i\dot{x}^i$, we have

$$0 = \frac{d^2x^r}{d\tau^2} + \Gamma^r_{i\ell} \frac{dx^i}{d\tau}\frac{dx^\ell}{d\tau} - \frac{q}{m} F^r{}_i \frac{dx^i}{d\tau} \tag{11.327}$$

as the equation of motion of a particle of mass m and charge q. It is striking how nearly perfect the electromagnetism of Faraday and Maxwell is.

11.39 The principle of equivalence

The **principle of equivalence** says that in any gravitational field, one may choose free-fall coordinates in which all physical laws take the same form as in special relativity without acceleration or gravitation – at least over a suitably small volume of space-time. Within this volume and in these coordinates, things behave as they would at rest deep in empty space far from any matter or energy. The volume must be small enough so that the gravitational field is constant throughout it.

Example 11.21 (Elevators) When a modern elevator starts going down from a high floor, it accelerates downward at something less than the local acceleration of gravity. One feels less pressure on one's feet; one feels lighter. (This is as close to free fall as I like to get.) After accelerating downward for a few seconds, the elevator assumes a constant downward speed, and then one feels the normal pressure of one's weight on one's feet. The elevator seems to be slowing down for a stop, but actually it has just stopped accelerating downward.

If in those first few seconds the elevator really were falling, then the physics in it would be the same as if it were at rest in empty space far from any gravitational field. A clock in it would tick as fast as it would at rest in the absence of gravity. □

The transformation from arbitrary coordinates x^k to free-fall coordinates y^i changes the metric $g_{j\ell}$ to the diagonal metric η_{ik} of flat space-time $\eta = \text{diag}(-1, 1, 1, 1)$, which has two indices and is not a Levi-Civita tensor. Algebraically, this transformation is a congruence (1.308)

$$\eta_{ik} = \frac{\partial x^j}{\partial y^i} g_{j\ell} \frac{\partial x^\ell}{\partial y^k}. \tag{11.328}$$

The geodesic equation (11.324) follows from the **principle of equivalence** (Weinberg, 1972; Hobson *et al.*, 2006). Suppose a particle is moving under the influence of gravitation alone. Then one may choose free-fall coordinates $y(x)$ so that the particle obeys the force-free equation of motion

$$\frac{d^2 y^i}{d\tau^2} = 0 \tag{11.329}$$

with $d\tau$ the proper time $d\tau^2 = -\eta_{ik}\, dy^i dy^k$. The chain rule applied to $y^i(x)$ in (11.329) gives

$$\begin{aligned} 0 &= \frac{d}{d\tau}\left(\frac{\partial y^i}{\partial x^k}\frac{dx^k}{d\tau}\right) \\ &= \frac{\partial y^i}{\partial x^k}\frac{d^2 x^k}{d\tau^2} + \frac{\partial^2 y^i}{\partial x^k \partial x^\ell}\frac{dx^k}{d\tau}\frac{dx^\ell}{d\tau}. \end{aligned} \tag{11.330}$$

We multiply by $\partial x^m / \partial y^i$ and use the identity

$$\frac{\partial x^m}{\partial y^i}\frac{\partial y^i}{\partial x^k} = \delta^m_k \tag{11.331}$$

to get the equation of motion (11.329) in the x-coordinates

$$\frac{d^2 x^m}{d\tau^2} + \Gamma^m_{k\ell}\frac{dx^k}{d\tau}\frac{dx^\ell}{d\tau} = 0, \tag{11.332}$$

in which the affine connection is

$$\Gamma^m_{k\ell} = \frac{\partial x^m}{\partial y^i}\frac{\partial^2 y^i}{\partial x^k \partial x^\ell}. \tag{11.333}$$

So the principle of equivalence tells us that a particle in a gravitational field obeys the geodesic equation (11.324).

11.40 Weak, static gravitational fields

Slow motion in a weak, static gravitational field is an important example. Because the motion is slow, we neglect u^i compared to u^0 and simplify the geodesic equation (11.324) to

$$0 = \frac{du^r}{d\tau} + \Gamma^r_{00}\,(u^0)^2. \tag{11.334}$$

Because the gravitational field is static, we neglect the time derivatives $g_{k0,0}$ and $g_{0k,0}$ in the connection formula (11.255) and find for Γ^r_{00}

$$\Gamma^r_{00} = \tfrac{1}{2}\, g^{rk}\,\left(g_{0k,0} + g_{0k,0} - g_{00,k}\right) = -\tfrac{1}{2}\, g^{rk}\, g_{00,k} \tag{11.335}$$

with $\Gamma^0_{00} = 0$. Because the field is weak, the metric can differ from η_{ij} by only a tiny tensor $g_{ij} = \eta_{ij} + h_{ij}$ so that to first order in $|h_{ij}| \ll 1$ we have $\Gamma^r_{00} = -\frac{1}{2}\, h_{00,r}$ for $r = 1, 2, 3$. With these simplifications, the geodesic equation (11.324) reduces to

$$\frac{d^2x^r}{d\tau^2} = \tfrac{1}{2}\,(u^0)^2\, h_{00,r} \qquad \text{or} \qquad \frac{d^2x^r}{d\tau^2} = \frac{1}{2}\left(\frac{dx^0}{d\tau}\right)^2 h_{00,r}. \tag{11.336}$$

So for slow motion, the ordinary acceleration is described by Newton's law

$$\frac{d^2\boldsymbol{x}}{dt^2} = \frac{c^2}{2}\,\nabla h_{00}. \tag{11.337}$$

If ϕ is his potential, then for slow motion in weak static fields

$$g_{00} = -1 + h_{00} = -1 - 2\phi/c^2 \qquad \text{and so} \qquad h_{00} = -\,2\phi/c^2. \tag{11.338}$$

Thus, if the particle is at a distance r from a mass M, then $\phi = -\,GM/r$ and $h_{00} = -2\phi/c^2 = 2GM/rc^2$ and so

$$\frac{d^2\boldsymbol{x}}{dt^2} = -\,\nabla\phi = \nabla\frac{GM}{r} = -\,GM\frac{\boldsymbol{r}}{r^3}. \tag{11.339}$$

How weak are the static gravitational fields we know about? The dimensionless ratio ϕ/c^2 is 10^{-39} on the surface of a proton, 10^{-9} on the Earth, 10^{-6} on the surface of the Sun, and 10^{-4} on the surface of a white dwarf.

11.41 Gravitational time dilation

Suppose we have a system of coordinates x^i with a metric g_{ik} and a clock at rest in this system. Then the proper time $d\tau$ between ticks of the clock is

$$d\tau = (1/c)\sqrt{-g_{ij}\,dx^i\,dx^j} = \sqrt{-g_{00}}\,dt \tag{11.340}$$

where dt is the time between ticks in the x^i coordinates, which is the laboratory frame in the gravitational field g_{00}. By the principle of equivalence (section 11.39), the proper time $d\tau$ between ticks is the same as the time between ticks when the same clock is at rest deep in empty space.

If the clock is in a weak static gravitational field due to a mass M at a distance r, then

$$- g_{00} = 1 + 2\phi/c^2 = 1 - 2GM/c^2 r \tag{11.341}$$

is a little less than unity, and the interval of proper time between ticks

$$d\tau = \sqrt{-g_{00}}\, dt = \sqrt{1 - 2GM/c^2 r\, dt} \tag{11.342}$$

is slightly less than the interval dt between ticks in the coordinate system of an observer at x in the rest frame of the clock and the mass, and in its gravitational field. Since $dt > d\tau$, the laboratory time dt between ticks is greater than the proper or intrinsic time $d\tau$ between ticks of the clock unaffected by any gravitational field. Clocks near big masses run slow.

Now suppose we have two identical clocks at different heights above sea level. The time T_ℓ for the lower clock to make N ticks will be longer than the time T_u for the upper clock to make N ticks. The ratio of the clock times will be

$$\frac{T_\ell}{T_u} = \frac{\sqrt{1 - 2GM/c^2(r+h)}}{\sqrt{1 - 2GM/c^2 r}} \approx 1 + \frac{gh}{c^2}. \tag{11.343}$$

Now imagine that a photon going down passes the upper clock, which measures its frequency as ν_u, and then passes the lower clock, which measures its frequency as ν_ℓ. The slower clock will measure a higher frequency. The ratio of the two frequencies will be the same as the ratio of the clock times

$$\frac{\nu_\ell}{\nu_u} = 1 + \frac{gh}{c^2}. \tag{11.344}$$

As measured by the lower, slower clock, the photon is blue shifted.

Example 11.22 (Pound, Rebka, and Mössbauer) Pound and Rebka in 1960 used the Mössbauer effect to measure the blue shift of light falling down a 22.6 m shaft. They found

$$\frac{\nu_\ell - \nu_u}{\nu} = \frac{gh}{c^2} = 2.46 \times 10^{-15} \tag{11.345}$$

(Robert Pound, 1919–2010; Glen Rebka, 1931–; Rudolf Mössbauer, 1929–2011). □

Example 11.23 (Redshift of the Sun) A photon emitted with frequency ν_0 at a distance r from a mass M would be observed at spatial infinity to have frequency ν

$$\nu = \nu_0 \sqrt{-g_{00}} = \nu_0 \sqrt{1 - 2MG/c^2 r} \tag{11.346}$$

for a redshift of $\Delta\nu = \nu_0 - \nu$. Since the Sun's dimensionless potential $\phi_\odot/c^2$ is $-MG/c^2 r = -2.12 \times 10^{-6}$ at its surface, sunlight is shifted to the red by 2 parts per million. □

11.42 Curvature

The **curvature tensor** or **Riemann tensor** is

$$R^i_{\ mnk} = \Gamma^i_{\ mn,k} - \Gamma^i_{\ mk,n} + \Gamma^i_{\ kj}\,\Gamma^j_{\ nm} - \Gamma^i_{\ nj}\,\Gamma^j_{\ km}, \tag{11.347}$$

which we may write as the commutator

$$\begin{aligned} R^i_{\ mnk} &= (R_{nk})^i_{\ m} = [\partial_k + \Gamma_k, \partial_n + \Gamma_n]^i_{\ m} \\ &= \left(\Gamma_{n,k} - \Gamma_{k,n} + \Gamma_k\,\Gamma_n - \Gamma_n\,\Gamma_k\right)^i_{\ m}, \end{aligned} \tag{11.348}$$

in which the Γs are treated as matrices

$$\Gamma_k = \begin{pmatrix} \Gamma^0_{\ k0} & \Gamma^0_{\ k1} & \Gamma^0_{\ k2} & \Gamma^0_{\ k3} \\ \Gamma^1_{\ k0} & \Gamma^1_{\ k1} & \Gamma^1_{\ k2} & \Gamma^1_{\ k3} \\ \Gamma^2_{\ k0} & \Gamma^2_{\ k1} & \Gamma^2_{\ k2} & \Gamma^2_{\ k3} \\ \Gamma^3_{\ k0} & \Gamma^3_{\ k1} & \Gamma^3_{\ k2} & \Gamma^3_{\ k3} \end{pmatrix} \tag{11.349}$$

with $(\Gamma_k\,\Gamma_n)^i_{\ m} = \Gamma^i_{\ kj}\,\Gamma^j_{\ nm}$ and so forth. Just as there are two conventions for the Faraday tensor F_{ik}, which differ by a minus sign, so too there are two conventions for the curvature tensor $R^i_{\ mnk}$. Weinberg (1972) uses the definition (11.347); Carroll (2003) uses an extra minus sign.

The Ricci tensor is a contraction of the curvature tensor

$$R_{mk} = R^n_{\ mnk} \tag{11.350}$$

and the curvature scalar is a further contraction

$$R = g^{mk}\,R_{mk}. \tag{11.351}$$

Example 11.24 (Curvature of a sphere) While in four-dimensional space-time indices run from 0 to 3, on the sphere they are just θ and ϕ. There are only eight possible affine connections, and because of the symmetry (11.218) in their lower indices $\Gamma^i_{\ \theta\phi} = \Gamma^i_{\ \phi\theta}$, only six are independent.

The point $\boldsymbol{p}$ on a sphere of radius r has cartesian coordinates

$$\boldsymbol{p} = r\,(\sin\theta\cos\phi,\ \sin\theta\sin\phi,\ \cos\theta) \tag{11.352}$$

so the two 3-vectors are

$$\boldsymbol{e}_\theta = \frac{\partial \boldsymbol{p}}{\partial \theta} = r\,(\cos\theta\,\cos\phi,\,\cos\theta\,\sin\phi,\,-\sin\theta) = r\,\hat{\boldsymbol{\theta}}$$
$$\boldsymbol{e}_\phi = \frac{\partial \boldsymbol{p}}{\partial \phi} = r\sin\theta\,(-\sin\phi,\,\cos\phi,\,0) = r\sin\theta\,\hat{\boldsymbol{\phi}} \tag{11.353}$$

and the metric $g_{ij} = \boldsymbol{e}_i \cdot \boldsymbol{e}_j$ is

$$(g_{ij}) = \begin{pmatrix} r^2 & 0 \\ 0 & r^2\sin^2\theta \end{pmatrix}. \tag{11.354}$$

Differentiating the vectors $\boldsymbol{e}_\theta$ and $\boldsymbol{e}_\phi$, we find

$$e_{\theta,\theta} = -r\,(\sin\theta\cos\phi,\,\sin\theta\sin\phi,\,\cos\theta) = -r\,\hat{\boldsymbol{r}}, \tag{11.355}$$
$$e_{\theta,\phi} = r\cos\theta\,(-\sin\phi,\,\cos\phi,\,0) = r\cos\theta\,\hat{\boldsymbol{\phi}}, \tag{11.356}$$
$$e_{\phi,\theta} = e_{\theta,\phi}, \tag{11.357}$$
$$e_{\phi,\phi} = -r\sin\theta\,(\cos\phi,\,\sin\phi,\,0)\,. \tag{11.358}$$

The metric with upper indices (g^{ij}) is the inverse of the metric (g_{ij})

$$(g^{ij}) = \begin{pmatrix} r^{-2} & 0 \\ 0 & r^{-2}\sin^{-2}\theta \end{pmatrix} \tag{11.359}$$

so the dual vectors $\boldsymbol{e}^i$ are

$$\boldsymbol{e}^\theta = r^{-1}\,(\cos\theta\,\cos\phi,\,\cos\theta\,\sin\phi,\,-\sin\theta) = r^{-1}\hat{\boldsymbol{\theta}},$$
$$\boldsymbol{e}^\phi = = \frac{1}{r\sin\theta}\,(-\sin\phi,\,\cos\phi,\,0) = \frac{1}{r\sin\theta}\,\hat{\boldsymbol{\phi}}. \tag{11.360}$$

The affine connections are given by (11.213) as

$$\Gamma^i_{jk} = \Gamma^i_{kj} = \boldsymbol{e}^i \cdot \boldsymbol{e}_{j,k}. \tag{11.361}$$

Since both $\boldsymbol{e}^\theta$ and $\boldsymbol{e}^\phi$ are perpendicular to $\hat{r}$, the affine connections $\Gamma^\theta_{\theta\theta}$ and $\Gamma^\phi_{\theta\theta}$ both vanish. Also, $e_{\phi,\phi}$ is orthogonal to $\hat{\phi}$, so $\Gamma^\phi_{\phi\phi} = 0$ as well. Similarly, $e_{\theta,\phi}$ is perpendicular to $\hat{\theta}$, so $\Gamma^\theta_{\theta\phi} = \Gamma^\theta_{\phi\theta}$ also vanishes.

The two nonzero affine connections are

$$\Gamma^\phi_{\theta\phi} = \boldsymbol{e}^\phi \cdot e_{\theta,\phi} = r^{-1}\sin^{-1}\theta\,\hat{\boldsymbol{\phi}} \cdot r\cos\theta\,\hat{\boldsymbol{\phi}} = \cot\theta \tag{11.362}$$

and

$$\begin{aligned}\Gamma^\theta_{\phi\phi} &= \boldsymbol{e}^\theta \cdot e_{\phi,\phi} \\ &= -\sin\theta\,(\cos\theta\,\cos\phi,\,\cos\theta\,\sin\phi,\,-\sin\theta)\cdot(\cos\phi,\,\sin\phi,\,0) \\ &= -\sin\theta\cos\theta.\end{aligned} \tag{11.363}$$

In terms of the two nonzero affine connections $\Gamma^\phi_{\theta\phi} = \Gamma^\phi_{\phi\theta} = \cot\theta$ and $\Gamma^\theta_{\phi\phi} = -\sin\theta\cos\theta$, the two Christoffel matrices (11.349) are

$$\Gamma_\theta = \begin{pmatrix} 0 & 0 \\ 0 & \Gamma^\phi_{\theta\phi} \end{pmatrix} = \begin{pmatrix} 0 & 0 \\ 0 & \cot\theta \end{pmatrix} \tag{11.364}$$

and

$$\Gamma_\phi = \begin{pmatrix} 0 & \Gamma^\theta_{\phi\phi} \\ \Gamma^\phi_{\phi\theta} & 0 \end{pmatrix} = \begin{pmatrix} 0 & -\sin\theta\cos\theta \\ \cot\theta & 0 \end{pmatrix}. \tag{11.365}$$

Their commutator is

$$[\Gamma_\theta, \Gamma_\phi] = \begin{pmatrix} 0 & \cos^2\theta \\ \cot^2\theta & 0 \end{pmatrix} = -[\Gamma_\phi, \Gamma_\theta] \tag{11.366}$$

and both $[\Gamma_\theta, \Gamma_\theta]$ and $[\Gamma_\phi, \Gamma_\phi]$ vanish.

So the commutator formula (11.348) gives for Riemann's curvature tensor

$$\begin{aligned} R^\theta_{\ \theta\theta\theta} &= [\partial_\theta + \Gamma_\theta, \partial_\theta + \Gamma_\theta]^\theta{}_\theta = 0, \\ R^\phi_{\ \theta\phi\theta} &= [\partial_\theta + \Gamma_\theta, \partial_\phi + \Gamma_\phi]^\phi{}_\theta = \left(\Gamma_{\phi,\theta}\right)^\phi{}_\theta + [\Gamma_\theta, \Gamma_\phi]^\phi{}_\theta \\ &= (\cot\theta)_{,\theta} + \cot^2\theta = -1, \\ R^\theta_{\ \phi\theta\phi} &= [\partial_\phi + \Gamma_\phi, \partial_\theta + \Gamma_\theta]^\theta{}_\phi = -\left(\Gamma_{\phi,\theta}\right)^\theta{}_\phi + [\Gamma_\phi, \Gamma_\theta]^\theta{}_\phi \\ &= \cos^2\theta - \sin^2\theta - \cos^2\theta = -\sin^2\theta, \\ R^\phi_{\ \phi\phi\phi} &= [\partial_\phi + \Gamma_\phi, \partial_\phi + \Gamma_\phi]^\phi{}_\phi = 0. \end{aligned} \tag{11.367}$$

The Ricci tensor (11.350) is the contraction $R_{mk} = R^n_{\ mnk}$, and so

$$\begin{aligned} R_{\theta\theta} &= R^\theta_{\ \theta\theta\theta} + R^\phi_{\ \theta\phi\theta} = -1, \\ R_{\phi\phi} &= R^\theta_{\ \phi\theta\phi} + R^\phi_{\ \phi\phi\phi} = -\sin^2\theta. \end{aligned} \tag{11.368}$$

The curvature scalar (11.351) is the contraction $R = g^{km} R_{mk}$, and so since $g^{\theta\theta} = r^{-2}$ and $g^{\phi\phi} = r^{-2}\sin^{-2}\theta$, it is

$$\begin{aligned} R &= g^{\theta\theta} R_{\theta\theta} + g^{\phi\phi} R_{\phi\phi} \\ &= -r^{-2} - \sin^2\theta\, r^{-2}\sin^{-2}\theta = -\frac{2}{r^2} \end{aligned} \tag{11.369}$$

for a 2-sphere of radius r.

Gauss invented a formula for the curvature K of a surface; for all two-dimensional surfaces, his $K = -R/2$. □

11.43 Einstein's equations

The source of the gravitational field is the **energy–momentum tensor** T_{ij}. In many astrophysical and most cosmological models, the energy–momentum tensor is assumed to be that of a **perfect fluid**, which is isotropic in its rest frame, does not conduct heat, and has zero viscosity. For a perfect fluid of pressure p

and density ρ with 4-velocity u^i (defined by (11.66)), the energy–momentum or **stress–energy** tensor T_{ij} is

$$T_{ij} = p\, g_{ij} + (p + \rho)\, u_i\, u_j, \tag{11.370}$$

in which g_{ij} is the space-time metric.

An important special case is the energy–momentum tensor due to a nonzero value of the energy density of the vacuum. In this case $p = -\rho$ and the energy–momentum tensor is

$$T_{ij} = -\rho\, g_{ij}, \tag{11.371}$$

in which ρ is the (presumably constant) value of the energy density of the ground state of the theory. This energy density ρ is a plausible candidate for the **dark-energy** density. It is equivalent to a **cosmological constant** $\Lambda = 8\pi G\rho$.

Whatever its nature, the energy–momentum tensor usually is defined so as to satisfy the conservation law

$$0 = \left(T^i{}_j\right)_{;i} = \partial_i T^i{}_j + \Gamma^i_{ic} T^c{}_j - T^i{}_c \Gamma^c_{ji}. \tag{11.372}$$

Einstein's equations relate the Ricci tensor (11.350) and the scalar curvature (11.351) to the energy–momentum tensor

$$R_{ij} - \tfrac{1}{2}\, g_{ij}\, R = -\frac{8\pi\, G}{c^4}\, T_{ij}, \tag{11.373}$$

in which $G = 6.7087 \times 10^{-39}\, \hbar c\, (\text{GeV}/c^2)^{-2} = 6.6742 \times 10^{-11}\, \text{m}^3\, \text{kg}^{-1}\, \text{s}^{-2}$ is Newton's constant. Taking the trace and using $g^{ji}\, g_{ij} = \delta^j_j = 4$, we relate the scalar curvature to the trace $T = T^i{}_i$ of the energy–momentum tensor

$$R = \frac{8\pi\, G}{c^4}\, T. \tag{11.374}$$

So another form of Einstein's equations (11.373) is

$$R_{ij} = -\frac{8\pi\, G}{c^4}\left(T_{ij} - \frac{T}{2} g_{ij}\right). \tag{11.375}$$

On small scales, such as that of our Solar System, one may neglect dark energy. So in empty space and on small scales, the energy–momentum tensor vanishes $T_{ij} = 0$ along with its trace and the scalar curvature $T = 0 = R$, and Einstein's equations are

$$R_{ij} = 0. \tag{11.376}$$

11.44 The action of general relativity

If we make an action that is a scalar, invariant under general coordinate transformations, and then apply to it the principle of stationary action, we will get tensor field equations that are invariant under general coordinate transformations. If the metric of space-time is among the fields of the action, then the resulting theory will be a possible theory of gravity. If we make the action as simple as possible, it will be Einstein's theory.

To make the action of the gravitational field, we need a scalar. Apart from the volume 4-form $\sqrt{g}\, d^4x$, the only scalar we can form from the metric tensor and its first and second derivatives is the scalar curvature R, which gives us the **Einstein–Hilbert action**

$$S_{\rm EH} = -\frac{c^4}{16\pi G}\int *R = -\frac{c^4}{16\pi G}\int R\sqrt{g}\, d^4x. \tag{11.377}$$

If $\delta g^{ik}(x)$ is a tiny change in the inverse metric that vanishes as any coordinate $x_j \to \pm\infty$, then one may write the first-order change in the action $S_{\rm EH}$ as

$$\delta S_{\rm EH} = -\frac{c^4}{16\pi G}\int \left(R_{ik} - \frac{1}{2} g_{ik} R\right)\sqrt{g}\, \delta g^{ik}\, d^4x. \tag{11.378}$$

The principle of least action $\delta S_{\rm EH} = 0$ now leads to Einstein's equations

$$G_{ik} = R_{ik} - \frac{1}{2} g_{ik}\, R = 0 \tag{11.379}$$

for empty space in which G_{ik} is Einstein's tensor.

The stress–energy tensor T_{ik} is defined so that the change in the action of the matter fields due to a tiny change $\delta g^{ik}(x)$ (vanishing at infinity) in the metric is

$$\delta S_{\rm m} = -\frac{1}{2}\int T_{ik}\sqrt{g}\, \delta g^{ik}\, d^4x. \tag{11.380}$$

So the principle of least action $\delta S = \delta S_{\rm EH} + \delta S_{\rm m} = 0$ implies Einstein's equations (11.373, 11.375) in the presence of matter and energy

$$R_{ik} - \frac{1}{2} g_{ik}\, R = -\frac{8\pi G}{c^4} T_{ij} \quad \text{or} \quad R_{ij} = -\frac{8\pi G}{c^4}\left(T_{ij} - \frac{T}{2} g_{ij}\right). \tag{11.381}$$

11.45 Standard form

Tensor equations are independent of the choice of coordinates, so it's wise to choose coordinates that simplify one's work. For a **static** and **isotropic** gravitational field, this choice is the **standard form** (Weinberg, 1972, ch. 8)

$$d\tau^2 = B(r)\, dt^2 - A(r)\, dr^2 - r^2\left(d\theta^2 + \sin^2\theta\, d\phi^2\right), \tag{11.382}$$

in which $c = 1$, and $B(r)$ and $A(r)$ are functions that one may find by solving the field equations (11.373). Since $d\tau^2 = -ds^2 = -g_{ij}\,dx^i dx^j$, the nonzero components of the metric tensor are $g_{rr} = A(r)$, $g_{\theta\theta} = r^2$, $g_{\phi\phi} = r^2 \sin^2\theta$, and $g_{00} = -B(r)$, and those of its inverse are $g^{rr} = A^{-1}(r)$, $g^{\theta\theta} = r^{-2}$, $g^{\phi\phi} = r^{-2}\sin^{-2}\theta$, and $g^{00} = -B^{-1}(r)$. By differentiating the metric tensor and using (11.255), one gets the components of the connection $\Gamma^i_{k\ell}$, such as $\Gamma^{\theta}_{\phi\phi} = -\sin\theta\cos\theta$, and the components (11.350) of the Ricci tensor R_{ij}, such as (Weinberg, 1972, ch. 8)

$$R_{rr} = \frac{B''(r)}{2B(r)} - \frac{1}{4}\left(\frac{B'(r)}{B(r)}\right)\left(\frac{A'(r)}{A(r)} + \frac{B'(r)}{B(r)}\right) - \frac{1}{r}\left(\frac{A'(r)}{A(r)}\right), \tag{11.383}$$

in which the primes mean d/dr.

11.46 Schwarzschild's solution

If one ignores the small dark-energy parameter Λ, one may solve Einstein's field equations (11.376) in empty space

$$R_{ij} = 0 \tag{11.384}$$

outside a mass M for the standard form of the Ricci tensor. One finds (Weinberg, 1972) that $A(r)\,B(r) = 1$ and that $r\,B(r) = r$ plus a constant, and one determines the constant by invoking the Newtonian limit $g_{00} = -B \to -1 + 2MG/c^2 r$ as $r \to \infty$. In 1916, Schwarzschild found the solution

$$d\tau^2 = \left(1 - \frac{2MG}{c^2 r}\right) c^2 dt^2 - \left(1 - \frac{2MG}{c^2 r}\right)^{-1} dr^2 - r^2\left(d\theta^2 + \sin^2\theta\, d\phi^2\right), \tag{11.385}$$

which one can use to analyze orbits around a star. The singularity in

$$g_{rr} = \left(1 - \frac{2MG}{c^2 r}\right)^{-1} \tag{11.386}$$

at the Schwarzschild radius $r = 2MG/c^2$ is an artifact of the coordinates; the scalar curvature R and other invariant curvatures are not singular at the Schwarzschild radius. Moreover, for the Sun, the Schwarzschild radius $r = 2MG/c^2$ is only 2.95 km, far less than the radius of the Sun, which is 6.955×10^5 km. So the surface at $r = 2MG/c^2$ is far from the empty space in which Schwarzschild's solution applies (Karl Schwarzschild, 1873–1916).

11.47 Black holes

Suppose an uncharged, spherically symmetric star of mass M has collapsed within a sphere of radius r_b less than its Schwarzschild radius $r = 2MG/c^2$.

Then for $r > r_b$, the Schwarzschild metric (11.385) is correct. By (11.340), the apparent time dt of a process of proper time $d\tau$ at $r \geq 2MG/c^2$ is

$$dt = d\tau / \sqrt{-g_{00}} = d\tau / \sqrt{1 - \frac{2MG}{c^2 r}}. \tag{11.387}$$

The apparent time dt becomes infinite as $r \to 2MG/c^2$. To outside observers, the star seems frozen in time.

Due to the gravitational redshift (11.346), light of frequency ν_p emitted at $r \geq 2MG/c^2$ will have frequency ν

$$\nu = \nu_p \sqrt{-g_{00}} = \nu_p \sqrt{1 - \frac{2MG}{c^2 r}} \tag{11.388}$$

when observed at great distances. Light coming from the surface at $r = 2MG/c^2$ is redshifted to zero frequency $\nu = 0$. The star is black. It is a black hole with a surface or horizon at its Schwarzschild radius $r = 2MG/c^2$, although there is no singularity there. If the radius of the Sun were less than its Schwarzschild radius of 2.95 km, then the Sun would be a black hole. The radius of the Sun is 6.955×10^5 km.

Black holes are not really black. Stephen Hawking (1942–) has shown that the intense gravitational field of a black hole of mass M radiates at temperature

$$T = \frac{\hbar c^3}{8\pi k G M}, \tag{11.389}$$

in which $k = 8.617343 \times 10^{-5}$ eV K^{-1} is Boltzmann's constant, and $\hbar$ is Planck's constant $h = 6.6260693 \times 10^{-34}$ J s divided by 2π, $\hbar = h/(2\pi)$.

The black hole is entirely converted into radiation after a time

$$t = \frac{5120\pi G^2}{\hbar c^4} M^3 \tag{11.390}$$

proportional to the cube of its mass.

11.48 Cosmology

Astrophysical observations tell us that on the largest observable scales, space is **flat** or very nearly flat; that the visible Universe contains at least 10^{90} particles; and that the cosmic microwave background radiation is isotropic to one part in 10^5 apart from a Doppler shift due the motion of the Earth. These and other observations suggest that potential energy expanded our Universe by $\exp(60) = 10^{26}$ during an **era of inflation** that could have been as brief as 10^{-35} s. The potential energy that powered inflation became the radiation of the **Big Bang**.

During the first three minutes, some of that radiation became hydrogen, helium, neutrinos, and **dark matter**. But for 65,000 years after the Big Bang, most of the energy of the visible Universe was radiation. Because the momentum of a particle but not its mass falls with the expansion of the Universe, this **era of radiation** gradually gave way to an **era of matter**. This transition happened when the temperature kT of the Universe fell to 1.28 eV.

The era of matter lasted for 8.8 billion years. After 380,000 years, the Universe had cooled to $kT = 0.26$ eV, and fewer than 1% of the atoms were ionized. Photons no longer scattered off a plasma of electrons and ions. The Universe became **transparent**. The photons that last scattered just before this **initial transparency** became the **cosmic microwave background radiation** or **CMBR** that now surrounds us, red shifted to 2.7255 ± 0.0006 K.

The era of matter was followed by the current **era of dark energy** during which the energy of the visible Universe is mostly a potential energy called **dark energy** (something like a **cosmological constant**). Dark energy has been accelerating the expansion of the Universe for the past 5 billion years and may continue to do so forever.

It is now 13.817 ± 0.048 billion years after the Big Bang, and the dark-energy density is $\rho_{\rm de} = 5.827 \times 10^{-30}\, c^2\,\text{g cm}^{-3}$ or 68.5 percent ($\pm$ 1.8%) of the **critical energy density** $\rho_c = 3H_0^2/8\pi G = 1.87837\, h^2 \times 10^{-29} f^2\, c^2\,\text{g cm}^{-3}$ needed to make the Universe flat. Here $H_0 = 100\, h\,\text{km s}^{-1}\,\text{Mpc}^{-1} = 1.022/(10^{10}\text{yr})$ is the **Hubble constant**, one parsec is 3.262 light-years, the Hubble time is $1/H_0 = 9.788\, h^{-1} \times 10^9$ years, and $h = 0.673 \pm 0.012$ is not to be confused with Planck's constant.

Matter makes up 31.5 ± 1.8% of the critical density, and baryons only 4.9 ± 0.06% of it. Baryons are 15% of the total matter in the visible Universe. The other 85% does not interact with light and is called **dark matter**.

Einstein's equations (11.373) are second-order, nonlinear partial differential equations for ten unknown functions $g_{ij}(x)$ in terms of the energy–momentum tensor $T_{ij}(x)$ throughout the Universe, which of course we don't know. The problem is not quite hopeless, however. The ability to choose arbitrary coordinates, the appeal to symmetry, and the choice of a reasonable form for T_{ij} all help.

Hubble showed us that the Universe is expanding. The cosmic microwave background radiation looks the same in all spatial directions (apart from a Doppler shift due to the motion of the Earth relative to the local super-cluster of galaxies). Observations of clusters of galaxies reveal a Universe that is homogeneous on suitably large scales of distance. So it is plausible that the Universe is **homogeneous** and **isotropic** in space, but not in time. One may show (Carroll, 2003) that for a universe of such symmetry, the line element in **comoving coordinates** is

$$ds^2 = -dt^2 + a^2 \left[\frac{dr^2}{1 - k\, r^2} + r^2 \left(d\theta^2 + \sin^2\theta\, d\phi^2 \right) \right]. \tag{11.391}$$

Whitney's embedding theorem tells us that any smooth four-dimensional manifold can be embedded in a flat space of eight dimensions with a suitable **signature**. We need only four or five dimensions to embed the space-time described by the line element (11.391). If the Universe is closed, then the signature is $(-1, 1, 1, 1, 1)$, and our three-dimensional space is the **3-sphere**, which is the surface of a four-dimensional sphere in four space dimensions. The points of the Universe then are

$$\boldsymbol{p} = (t,\ a \sin\chi\, \sin\theta \cos\phi,\ a \sin\chi\, \sin\theta \sin\phi,\ a \sin\chi\, \cos\theta,\ a \cos\chi), \quad (11.392)$$

in which $0 \le \chi \le \pi$, $0 \le \theta \le \pi$, and $0 \le \phi \le 2\pi$. If the Universe is flat, then the embedding space is flat, four-dimensional Minkowski space with points

$$\boldsymbol{p} = (t,\ ar \sin\theta \cos\phi,\ ar \sin\theta \sin\phi,\ ar \cos\theta), \quad (11.393)$$

in which $0 \le \theta \le \pi$ and $0 \le \phi \le 2\pi$. If the Universe is open, then the embedding space is a flat five-dimensional space with signature $(-1, 1, 1, 1, -1)$, and our three-dimensional space is a hyperboloid in a flat Minkowski space of four dimensions. The points of the Universe then are

$$\boldsymbol{p} = (t,\ a \sinh\chi\, \sin\theta \cos\phi,\ a \sinh\chi\, \sin\theta \sin\phi,\ a \sinh\chi\, \cos\theta,\ a \cosh\chi), \quad (11.394)$$

in which $0 \le \chi \le \infty$, $0 \le \theta \le \pi$, and $0 \le \phi \le 2\pi$.

In all three cases, the corresponding **Robertson–Walker metric** is

$$g_{ij} = \begin{pmatrix} -1 & 0 & 0 & 0 \\ 0 & a^2/(1-kr^2) & 0 & 0 \\ 0 & 0 & a^2 r^2 & 0 \\ 0 & 0 & 0 & a^2 r^2 \sin^2\theta \end{pmatrix}, \quad (11.395)$$

in which the coordinates (t, r, θ, ϕ) are numbered $(0, 1, 2, 3)$, the speed of light is $c = 1$, and k is a constant. One always may choose coordinates (exercise 11.30) such that k is either 0 or ± 1. This constant determines whether the spatial Universe is **open** $k = -1$, **flat** $k = 0$, or **closed** $k = 1$. The **scale factor** a, which in general is a function of time $a(t)$, tells us how space expands and contracts. These coordinates are called **comoving** because a point at rest (fixed r, θ, ϕ) sees the same Doppler shift in all directions.

The metric (11.395) is diagonal; its inverse g^{ij} also is diagonal; and so we may use our formula (11.255) to compute the affine connections $\Gamma^k_{i\ell}$, such as

$$\Gamma^0_{\ell\ell} = \tfrac{1}{2} g^{0k} \left(g_{\ell k,\ell} + g_{\ell k,\ell} - g_{\ell\ell,k} \right) = \tfrac{1}{2} g^{00} \left(g_{\ell 0,\ell} + g_{\ell 0,\ell} - g_{\ell\ell,0} \right) = \tfrac{1}{2} g_{\ell\ell,0} \quad (11.396)$$

so that

$$\Gamma^0_{11} = \frac{a\dot{a}}{1 - kr^2} \qquad \Gamma^0_{22} = a\dot{a}\, r^2 \quad \text{and} \quad \Gamma^0_{22} = a\dot{a}\, r^2 \sin^2\theta, \quad (11.397)$$

in which a dot means a time-derivative. The other Γ^0_{ij}s vanish. Similarly, for *fixed* $\ell = 1, 2$, or 3

$$\begin{aligned}\Gamma^{\ell}_{0\ell} &= \tfrac{1}{2}g^{\ell k}\left(g_{0k,\ell} + g_{\ell k,0} - g_{0\ell,k}\right)\\ &= \tfrac{1}{2}g^{\ell\ell}\left(g_{0\ell,\ell} + g_{\ell\ell,0} - g_{0\ell,\ell}\right)\\ &= \tfrac{1}{2}g^{\ell\ell}\,g_{\ell\ell,0} = \frac{\dot a}{a} = \Gamma^{\ell}_{\ell 0}, \quad \text{no sum over } \ell. \end{aligned} \tag{11.398}$$

The other nonzero Γs are

$$\Gamma^1_{22} = -r(1-kr^2), \qquad \Gamma^1_{33} = -r(1-kr^2)\sin^2\theta, \tag{11.399}$$

$$\Gamma^2_{12} = \Gamma^3_{13} = \frac{1}{r} = \Gamma^2_{21} = \Gamma^3_{31}, \tag{11.400}$$

$$\Gamma^2_{33} = -\sin\theta\,\cos\theta, \qquad \Gamma^3_{23} = \cot\theta = \Gamma^3_{32}. \tag{11.401}$$

Our formulas (11.350 & 11.348) for the Ricci and curvature tensors give

$$R_{00} = R^n_{\ 0n0} = [\partial_0 + \Gamma_0, \partial_n + \Gamma_n]^n{}_0. \tag{11.402}$$

Clearly the commutator of Γ_0 with itself vanishes, and one may use the formulas (11.397–11.401) for the other connections to check that

$$[\Gamma_0, \Gamma_n]^n{}_0 = \Gamma^n_{0k}\,\Gamma^k_{n0} - \Gamma^n_{nk}\,\Gamma^k_{00} = 3\left(\frac{\dot a}{a}\right)^2 \tag{11.403}$$

and that

$$\partial_0\,\Gamma^n_{n\,0} = 3\,\partial_0\left(\frac{\dot a}{a}\right) = 3\,\frac{\ddot a}{a} - 3\left(\frac{\dot a}{a}\right)^2 \tag{11.404}$$

while $\partial_n\Gamma^n_{0\,0} = 0$. So the 00-component of the Ricci tensor is

$$R_{00} = 3\,\frac{\ddot a}{a}. \tag{11.405}$$

Similarly, one may show that the other nonzero components of Ricci's tensor are

$$R_{11} = -\frac{A}{1-kr^2}, \qquad R_{22} = -r^2 A, \quad \text{and} \quad R_{33} = -r^2 A\sin^2\theta, \tag{11.406}$$

in which $A = a\ddot a + 2\dot a^2 + 2k$. The scalar curvature (11.351) is

$$R = g^{ab}R_{ba} = -\frac{6}{a^2}\left(a\ddot a + \dot a^2 + k\right). \tag{11.407}$$

In comoving coordinates such as those of the Robertson–Walker metric (11.395) $u_i = (1, 0, 0, 0)$, and so the energy–momentum tensor (11.370) is

$$T_{ij} = \begin{pmatrix} \rho & 0 & 0 & 0 \\ 0 & p\,g_{11} & 0 & 0 \\ 0 & 0 & p\,g_{22} & 0 \\ 0 & 0 & 0 & p\,g_{33} \end{pmatrix}. \tag{11.408}$$

Its trace is

$$T = g^{ij}\, T_{ij} = -\rho + 3p. \tag{11.409}$$

Thus, using our formula (11.395) for $g_{00} = -1$, (11.405) for R_{00}, (11.408) for T_{ij}, and (11.409) for T, we find that the 00 Einstein equation (11.375) becomes the second-order equation

$$\frac{\ddot{a}}{a} = -\frac{4\pi G}{3}\,(\rho + 3p)\,, \tag{11.410}$$

which is nonlinear because ρ and $3p$ depend upon a. The sum $\rho + 3p$ determines the acceleration $\ddot{a}$ of the scale factor $a(t)$. When it is negative, it accelerates the expansion of the Universe.

Because of the isotropy of the metric, the three nonzero spatial Einstein equations (11.375) give us only one relation

$$\frac{\ddot{a}}{a} + 2\left(\frac{\dot{a}}{a}\right)^2 + 2\frac{k}{a^2} = 4\pi G\,(\rho - p)\,. \tag{11.411}$$

Using the 00-equation (11.410) to eliminate the second derivative $\ddot{a}$, we have

$$\left(\frac{\dot{a}}{a}\right)^2 = \frac{8\pi G}{3}\,\rho - \frac{k}{a^2}, \tag{11.412}$$

which is a first-order nonlinear equation. It and the second-order equation (11.410) are known as the **Friedmann equations**.

The LHS of the first-order Friedmann equation (11.412) is the square of the **Hubble rate**

$$H = \frac{\dot{a}}{a}, \tag{11.413}$$

which is an inverse time. Its present value H_0 is the **Hubble constant**. In terms of H, Friedmann's first-order equation (11.412) is

$$H^2 = \frac{8\pi G}{3}\,\rho - \frac{k}{a^2}. \tag{11.414}$$

The energy density of a flat universe with $k = 0$ is the **critical energy density**

$$\rho_c = \frac{3H^2}{8\pi G}. \tag{11.415}$$

The ratio of the energy density ρ to the critical energy density is called Ω

$$\Omega = \frac{\rho}{\rho_c} = \frac{8\pi G}{3H^2}\,\rho. \tag{11.416}$$

From (11.414), we see that Ω is

$$\Omega = 1 + \frac{k}{(aH)^2} = 1 + \frac{k}{\dot{a}^2}. \tag{11.417}$$

Thus $\Omega = 1$ both in a flat universe ($k = 0$) and as $aH \to \infty$. One use of inflation is to expand a by 10^{26} so as to force Ω to almost exactly unity.

Something like inflation is needed because in a universe in which the energy density is due to matter and/or radiation, the present value of Ω

$$\Omega_0 = 1.000 \pm 0.036 \tag{11.418}$$

is unlikely. To see why, we note that conservation of energy ensures that a^3 times the matter density ρ_m is constant. Radiation red shifts by a, so energy conservation implies that a^4 times the radiation density ρ_r is constant. So with $n = 3$ for matter and 4 for radiation, $\rho a^n \equiv 3F^2/8\pi G$ is a constant. In terms of F and n, Friedmann's first-order equation (11.412) is

$$\dot{a}^2 = \frac{8\pi G}{3}\rho a^2 - k = \frac{F^2}{a^{n-2}} - k. \tag{11.419}$$

In small-a limit of the early Universe, we have

$$\dot{a} = F/a^{(n-2)/2} \qquad \text{or} \qquad a^{(n-2)/2} da = F dt, \tag{11.420}$$

which we integrate to $a \sim t^{2/n}$ so that $\dot{a} \sim t^{2/n-1}$. Now (11.417) says that

$$|\Omega - 1| = \frac{1}{\dot{a}^2} \propto t^{2-4/n} = \begin{cases} t & \text{radiation,} \\ t^{2/3} & \text{matter.} \end{cases} \tag{11.421}$$

Thus, Ω deviated from unity faster than $t^{2/3}$ during the eras of radiation and matter. At this rate, the inequality $|\Omega_0 - 1| < 0.036$ could last 13.8 billion years only if Ω at $t = 1$ second had been unity to within six parts in 10^{14}. The only *known* explanation for such early flatness is inflation.

Manipulating our relation (11.417) between Ω and aH, we see that

$$(aH)^2 = \frac{k}{\Omega - 1}. \tag{11.422}$$

So $\Omega > 1$ implies $k = 1$, and $\Omega < 1$ implies $k = -1$, and as $\Omega \to 1$ the product $aH \to \infty$, which is the essence of flatness since curvature vanishes as the scale factor $a \to \infty$. Imagine blowing up a balloon.

Staying for the moment with a universe without inflation and with an energy density composed of radiation and/or matter, we note that the first-order equation (11.419) in the form $\dot{a}^2 = F^2/a^{n-2} - k$ tells us that for a closed ($k = 1$) universe, in the limit $a \to \infty$ we'd have $\dot{a}^2 \to -1$, which is impossible. Thus a closed universe eventually collapses, which is incompatible with the flatness (11.422) implied by the present value $\Omega_0 = 1.000 \pm 0.036$.

The first-order Friedmann equation (11.412) tells us that $\rho a^2 \geq 3k/8\pi G$. So in a closed universe ($k = 1$), the energy density ρ is positive and increases without limit as $a \to 0$ as in a collapse. In open ($k < 0$) and flat ($k = 0$) universes, the same Friedmann equation (11.412) in the form $\dot{a}^2 = 8\pi G\rho a^2/3 - k$ tells us that if ρ is positive, then $\dot{a}^2 > 0$, which means that $\dot{a}$ never vanishes. Hubble told us that $\dot{a} > 0$ now. So if our Universe is open or flat, then it always expands.

Due to the expansion of the Universe, the wave-length of radiation grows with the scale factor $a(t)$. A photon emitted at time t and scale factor $a(t)$ with wave-length $\lambda(t)$ will be seen now at time t_0 and scale factor $a(t_0)$ to have a longer wave-length $\lambda(t_0)$

$$\frac{\lambda(t_0)}{\lambda(t)} = \frac{a(t_0)}{a(t)} = z + 1, \tag{11.423}$$

in which the **redshift** z is the ratio

$$z = \frac{\lambda(t_0) - \lambda(t)}{\lambda(t)} = \frac{\Delta\lambda}{\lambda}. \tag{11.424}$$

Now $H = \dot{a}/a = da/(adt)$ implies $dt = da/(aH)$, and $z = a_0/a - 1$ implies $dz = -a_0 da/a^2$, so we find

$$dt = -\frac{dz}{(1+z)H(z)}, \tag{11.425}$$

which relates time intervals to redshift intervals. An on-line calculator is available for macroscopic intervals (Wright, 2006).

11.49 Model cosmologies

The 0-component of the energy–momentum conservation law (11.372) is

$$\begin{aligned} 0 = \left(T^a{}_0\right)_{;a} &= \partial_a T^a{}_0 + \Gamma^a_{ac} T^c{}_0 - T^a{}_c \Gamma^c_{0a} \\ &= -\partial_0 T_{00} - \Gamma^a_{a0} T_{00} - g^{cc} T_{cc} \Gamma^c_{0c} \\ &= -\dot{\rho} - 3\frac{\dot{a}}{a}\rho - 3p\frac{\dot{a}}{a} = -\dot{\rho} - 3\frac{\dot{a}}{a}(\rho + p) \end{aligned} \tag{11.426}$$

or

$$\frac{d\rho}{da} = -\frac{3}{a}(\rho + p). \tag{11.427}$$

The energy density ρ is composed of fractions ρ_k each contributing its own partial pressure p_k according to its own **equation of state**

$$p_k = w_k \rho_k, \tag{11.428}$$

in which w_k is a constant. In terms of these components, the energy–momentum conservation law (11.427) is

$$\sum_k \frac{d\rho_k}{da} = -\frac{3}{a} \sum_k (1 + w_k)\, \rho_k \tag{11.429}$$

with solution

$$\rho = \sum_k \overline{\rho_k} \left(\frac{\overline{a}}{a}\right)^{3(1+w_k)} = \sum_k \overline{\rho_k} \left(\frac{\overline{a}}{a}\right)^{3(1+\overline{p_k}/\overline{\rho_k})}. \tag{11.430}$$

Simple cosmological models take the energy density and pressure each to have a single component with $p = w\rho$, and in this case

$$\rho = \overline{\rho} \left(\frac{\overline{a}}{a}\right)^{3(1+w)} = \overline{\rho} \left(\frac{\overline{a}}{a}\right)^{3(1+\overline{p}/\overline{\rho})}. \tag{11.431}$$

Example 11.25 ($w = -1/3$, no acceleration) If $w = -1/3$, then $p = w\rho = -\rho/3$ and $\rho + 3p = 0$. The second-order Friedmann equation (11.410) then tells us that $\ddot{a} = 0$. The scale factor does not accelerate.

To find its constant speed, we use its equation of state (11.431)

$$\rho = \overline{\rho} \left(\frac{\overline{a}}{a}\right)^{3(1+w)} = \overline{\rho} \left(\frac{\overline{a}}{a}\right)^{2}. \tag{11.432}$$

Now all the terms in Friedmann's first-order equation (11.412) have a common factor of $1/a^2$, which cancels leaving us with the square of the constant speed

$$\dot{a}^2 = \frac{8\pi G}{3} \overline{\rho}\, \overline{a}^2 - k. \tag{11.433}$$

Incidentally, $\overline{\rho}\, \overline{a}^2$ must exceed $3k/8\pi G$. The scale factor grows linearly with time as

$$a(t) = \left(\frac{8\pi G}{3} \overline{\rho}\, \overline{a}^2 - k\right)^{1/2} (t - t_0) + a(t_0). \tag{11.434}$$

Setting $t_0 = 0$ and $a(0) = 0$, we use the definition of the Hubble parameter $H = \dot{a}/a$ to write the constant linear growth $\dot{a}$ as aH and the time as

$$t = \int_0^a da'/a'H = (1/aH) \int_0^a da' = 1/H. \tag{11.435}$$

So in a universe without acceleration, the age of the universe is the inverse of the Hubble rate. For our Universe, the present Hubble time is $1/H_0 = 14.5$ billion years, which isn't far from the actual age of 13.817 ± 0.048 billion years. Presumably, a slower Hubble rate during the era of matter compensates for the higher rate during the era of dark energy. □

Example 11.26 ($w = -1$, inflation) Inflation occurs when the ground state of the theory has a positive and constant energy density $\rho > 0$ that dwarfs the energy densities of the matter and radiation. The **internal energy** of the Universe then is proportional to its volume $U = \rho V$, and the pressure p as given by the thermodynamic relation

$$p = -\frac{\partial U}{\partial V} = -\rho \tag{11.436}$$

is **negative**. The equation of state (11.428) tells us that in this case $w = -1$. The second-order Friedmann equation (11.410) becomes

$$\frac{\ddot{a}}{a} = -\frac{4\pi G}{3}(\rho + 3p) = \frac{8\pi G\rho}{3} \equiv g^2. \tag{11.437}$$

By it and the first-order Friedmann equation (11.412) and by choosing $t = 0$ as the time at which the scale factor a is minimal, one may show (exercise 11.37) that in a closed ($k = 1$) universe

$$a(t) = \frac{\cosh g\,t}{g}. \tag{11.438}$$

Similarly, in an open ($k = -1$) universe with $a(0) = 0$, we have

$$a(t) = \frac{\sinh g\,t}{g}. \tag{11.439}$$

Finally, in a flat ($k = 0$) expanding universe, the scale factor is

$$a(t) = a(0)\,\exp(g\,t). \tag{11.440}$$

Studies of the cosmic microwave background radiation suggest that inflation did occur in the **very early** Universe, possibly on a time scale as short as 10^{-35} s. What is the origin of the vacuum energy density ρ that drove inflation? Current theories attribute it to the assumption by at least one scalar field ϕ of a mean value $\langle\phi\rangle$ different from the one $\langle 0|\phi|0\rangle$ that minimizes the energy density of the vacuum. When $\langle\phi\rangle$ settled to $\langle 0|\phi|0\rangle$, the vacuum energy was released as radiation and matter in a **Big Bang**. □

Example 11.27 ($w = 1/3$, the era of radiation) Until a redshift of $z = 3400$ or 50,000 years after inflation, our Universe was dominated by **radiation** (Frieman *et al.*, 2008). During *The First Three Minutes* (Weinberg, 1988) of the era of radiation, the quarks and gluons formed hadrons, which decayed into protons and neutrons. As the neutrons decayed ($\tau = 885.7$ s), they and the protons formed the light elements – principally hydrogen, deuterium, and helium – in a process called **big-bang nucleosynthesis**.

We can guess the value of w for radiation by noticing that the energy–momentum tensor of the electromagnetic field (in suitable units)

$$T^{ab} = F^a{}_c F^{bc} - \frac{1}{4} g^{ab} F_{cd} F^{cd} \tag{11.441}$$

is traceless

$$T = T^a{}_a = F^a{}_c F_a{}^c - \frac{1}{4} \delta^a_a F_{cd} F^{cd} = 0. \tag{11.442}$$

But by (11.409) its trace must be $T = 3p - \rho$. So for radiation $p = \rho/3$ and $w = 1/3$. The relation (11.431) between the energy density and the scale factor then is

$$\rho = \overline{\rho} \left(\frac{\overline{a}}{a} \right)^4. \tag{11.443}$$

The energy drops both with the volume a^3 and with the scale factor a due to a redshift; so it drops as $1/a^4$. Thus the quantity

$$f^2 \equiv \frac{8\pi G \rho a^4}{3} \tag{11.444}$$

is a constant. The Friedmann equations (11.410 & 11.411) now are

$$\frac{\ddot{a}}{a} = -\frac{4\pi G}{3}(\rho + 3p) = -\frac{8\pi G \rho}{3} \quad \text{or} \quad \ddot{a} = -\frac{f^2}{a^3} \tag{11.445}$$

and

$$\dot{a}^2 + k = \frac{f^2}{a^2}. \tag{11.446}$$

With calendars chosen so that $a(0) = 0$, this last equation (11.446) tells us that for a flat universe ($k = 0$)

$$a(t) = (2f\,t)^{1/2} \tag{11.447}$$

while for a closed universe ($k = 1$)

$$a(t) = \sqrt{f^2 - (t - f)^2} \tag{11.448}$$

and for an open universe ($k = -1$)

$$a(t) = \sqrt{(t + f)^2 - f^2} \tag{11.449}$$

as we saw in (6.422). The scale factor (11.448) of a closed universe of radiation has a maximum $a = f$ at $t = f$ and falls back to zero at $t = 2f$. □

Example 11.28 ($w = 0$, the era of matter) A universe composed only of **dust** or **nonrelativistic collisionless matter** has no pressure. Thus $p = w\rho = 0$ with $\rho \neq 0$, and so $w = 0$. Conservation of energy (11.430), or equivalently (11.431), implies that the energy density falls with the volume

$$\rho = \overline{\rho} \left(\frac{\overline{a}}{a} \right)^3 . \tag{11.450}$$

As the scale factor $a(t)$ increases, the matter energy density, which falls as $1/a^3$, eventually dominates the radiation energy density, which falls as $1/a^4$. This happened in our Universe about 50,000 years after inflation at a temperature of $T = 9{,}400\,K$ or $kT = 0.81$ eV. Were baryons most of the matter, the era of radiation dominance would have lasted for a few hundred thousand years. But the kind of matter that we know about, which interacts with photons, is only about 17% of the total; the rest – an unknown substance called **dark matter** – shortened the era of radiation dominance by nearly 2 million years.

Since $\rho \propto 1/a^3$, the quantity

$$m^2 = \frac{4\pi G \rho a^3}{3} \tag{11.451}$$

is a constant. For a matter-dominated universe, the Friedmann equations (11.410 & 11.411) then are

$$\frac{\ddot{a}}{a} = -\frac{4\pi G}{3} (\rho + 3p) = -\frac{4\pi G \rho}{3} \quad \text{or} \quad \ddot{a} = -\frac{m^2}{a^2} \tag{11.452}$$

and

$$\dot{a}^2 + k = 2m^2/a. \tag{11.453}$$

For a flat universe, $k = 0$, we get

$$a(t) = \left[\frac{3m}{\sqrt{2}} \, t \right]^{2/3} . \tag{11.454}$$

For a closed universe, $k = 1$, we use example 6.46 to integrate

$$\dot{a} = \sqrt{2m^2/a - 1} \tag{11.455}$$

to

$$t - t_0 = -\sqrt{a(2m^2 - a)} - m^2 \arcsin(1 - a/m^2). \tag{11.456}$$

With a suitable calendar and choice of t_0, one may parametrize this solution in terms of the **development angle** $\phi(t)$ as

$$\begin{aligned} a(t) &= m^2 \left[1 - \cos\phi(t) \right], \\ t &= m^2 \left[\phi(t) - \sin\phi(t) \right]. \end{aligned} \tag{11.457}$$

For an open universe, $k = -1$, we use example 6.47 to integrate

$$\dot{a} = \sqrt{2m^2/a + 1} \tag{11.458}$$

to

$$t - t_0 = \left[a(2m^2 + a)\right]^{1/2} - m^2 \ln \left\{2\left[a(2m^2 + a)\right]^{1/2} + 2a + 2m^2\right\}. \quad (11.459)$$

The conventional parametrization is

$$\begin{aligned} a(t) &= m^2 \left[\cosh \phi(t) - 1\right], \\ t &= m^2 \left[\sinh \phi(t) - \phi(t)\right]. \end{aligned} \quad (11.460)$$

Transparency Some 380,000 years after inflation at a redshift of about $z = 1090$, the Universe had cooled to about $T = 3000\,K$ or $kT = 0.26\,\text{eV}$ – a temperature at which less than 1% of the hydrogen is ionized. Ordinary matter became a gas of neutral atoms rather than a plasma of ions and electrons, and the Universe suddenly became **transparent** to light. Some scientists call this moment of last scattering or first transparency **recombination**. □

Example 11.29 ($w = -1$, the era of dark energy) About 10.3 billion years after inflation at a redshift of $z = 0.30$, the matter density falling as $1/a^3$ dropped below the very small but positive value of the energy density $\rho_v = (2.23\,\text{meV})^4$ of the vacuum. The present time is 13.817 billion years after inflation. So for the past 3 billion years, this constant energy density, called **dark energy**, has accelerated the expansion of the Universe approximately as (11.439)

$$a(t) = a(t_\text{m}) \exp\left((t - t_\text{m})\sqrt{8\pi G\rho_v/3}\right), \quad (11.461)$$

in which $t_\text{m} = 10.3 \times 10^9$ years. □

Observations and measurements on the largest scales indicate that the Universe is flat: $k = 0$. So the evolution of the scale factor $a(t)$ is given by the $k = 0$ equations (11.440, 11.447, 11.454, & 11.461) for a flat universe. During the brief era of inflation, the scale factor $a(t)$ grew as (11.440)

$$a(t) = a(0) \exp\left(t\sqrt{8\pi G\rho_\text{i}/3}\right), \quad (11.462)$$

in which ρ_i is the positive energy density that drove inflation.

During the 50,000-year era of radiation, $a(t)$ grew as $\sqrt{t}$ as in (11.447)

$$a(t) = \left(2\,(t - t_i)\sqrt{8\pi G\rho(t_\text{r}')a^4(t_\text{r}')/3}\right)^{1/2} + a(t_\text{i}) \quad (11.463)$$

where t_i is the time at the end of inflation, and t_r' is any time during the era of radiation. During this era, the energy of highly relativistic particles dominated the energy density, and $\rho a^4 \propto T^4 a^4$ was approximately constant, so that $T(t) \propto 1/a(t) \propto 1/\sqrt{t}$. When the temperature was in the range $10^{12} > T > 10^{10}\,\text{K}$

or $m_\mu c^2 > kT > m_e c^2$, where m_μ is the mass of the muon and m_e that of the electron, the radiation was mostly electrons, positrons, photons, and neutrinos, and the relation between the time t and the temperature T was (Weinberg, 2010, ch. 3)

$$t = 0.994 \ \text{sec} \times \left[\frac{10^{10}\,\text{K}}{T}\right]^2 + \text{constant}. \tag{11.464}$$

By 10^9 K, the positrons had annihilated with electrons, and the neutrinos fallen out of equilibrium. Between 10^9 K and 10^6 K, when the energy density of nonrelativistic particles became relevant, the time–temperature relation was (Weinberg, 2010, ch. 3)

$$t = 1.78 \ \text{sec} \times \left[\frac{10^{10}\,\text{K}}{T}\right]^2 + \text{constant}'. \tag{11.465}$$

During the 10.3 billion years of the matter era, $a(t)$ grew as (11.454)

$$a(t) = \left[(t - t_\text{r})\sqrt{3\pi G\rho(t'_\text{m})a(t'_\text{m})} + a^{3/2}(t_\text{r})\right]^{2/3} + a(t_r), \tag{11.466}$$

where t_r is the time at the end of the radiation era, and t'_m is any time in the matter era. By 380,000 years, the temperature had dropped to 3000 K, the Universe had become transparent, and the CMBR had begun to travel freely.

Over the past 3 billion years of the era of vacuum dominance, $a(t)$ has been growing exponentially (11.461)

$$a(t) = a(t_\text{m}) \exp\left((t - t_\text{m})\sqrt{8\pi G\rho_\text{v}/3}\right), \tag{11.467}$$

in which t_m is the time at the end of the matter era, and ρ_v is the density of dark energy, which, while vastly less than the energy density ρ_i that drove inflation, currently amounts to 68.5% of the total energy density.

11.50 Yang–Mills theory

The gauge transformation of an **abelian** gauge theory like electrodynamics multiplies a *single* charged field by a space-time-dependent *phase factor* $\phi'(x) = \exp(iq\theta(x))\,\phi(x)$. Yang and Mills generalized this gauge transformation to one that multiplies a *vector* ϕ of matter fields by a space-time dependent *unitary matrix* $U(x)$

$$\phi'_a(x) = \sum_{b=1}^{n} U_{ab}(x)\,\phi_b(x) \quad \text{or} \quad \phi'(x) = U(x)\,\phi(x) \tag{11.468}$$

and showed how to make the action of the theory invariant under such **nonabelian** gauge transformations. (The fields ϕ are scalars for simplicity.)

Since the matrix U is unitary, inner products like $\phi^\dagger(x)\,\phi(x)$ are automatically invariant

$$\left(\phi^\dagger(x)\,\phi(x)\right)' = \phi^\dagger(x)U^\dagger(x)U(x)\phi(x) = \phi^\dagger(x)\phi(x). \tag{11.469}$$

But inner products of derivatives $\partial^i\phi^\dagger\,\partial_i\phi$ are not invariant because the derivative acts on the matrix $U(x)$ as well as on the field $\phi(x)$.

Yang and Mills made derivatives $D_i\phi$ that transform like the fields ϕ

$$(D_i\phi)' = U\,D_i\phi. \tag{11.470}$$

To do so, they introduced **gauge-field matrices** A_i that play the role of the connections Γ_i in general relativity and set

$$D_i = \partial_i + A_i, \tag{11.471}$$

in which A_i, like ∂_i, is antihermitian. They required that under the gauge transformation (11.468), the gauge-field matrix A_i transform to A'_i in such a way as to make the derivatives transform as in (11.470)

$$(D_i\phi)' = \left(\partial_i + A'_i\right)\phi' = \left(\partial_i + A'_i\right)U\phi = U\,D_i\phi = U\left(\partial_i + A_i\right)\phi. \tag{11.472}$$

So they set

$$\left(\partial_i + A'_i\right)U\phi = U\left(\partial_i + A_i\right)\phi \quad \text{or} \quad \left(\partial_i U\right)\phi + A'_i\,U\phi = UA_i\,\phi \tag{11.473}$$

and made the gauge-field matrix A_i transform as

$$A'_i = UA_iU^{-1} - \left(\partial_i U\right)U^{-1}. \tag{11.474}$$

Thus under the gauge transformation (11.468), the derivative $D_i\phi$ transforms as in (11.470), like the vector ϕ in (11.468), and the inner product of covariant derivatives

$$\left[\left(D^i\phi\right)^\dagger D_i\phi\right]' = \left(D^i\phi\right)^\dagger U^\dagger UD_i\phi = \left(D^i\phi\right)^\dagger D_i\phi \tag{11.475}$$

remains invariant.

To make an invariant action density for the gauge-field matrices A_i, they used the transformation law (11.472), which implies that $D'_i\,U\phi = UD_i\,\phi$ or $D'_i = UD_i\,U^{-1}$. So they defined their generalized Faraday tensor as

$$F_{ik} = [D_i, D_k] = \partial_i A_k - \partial_k A_i + [A_i, A_k], \tag{11.476}$$

which transforms covariantly

$$F'_{ik} = UF_{ik}U^{-1}. \tag{11.477}$$

They then generalized the action density $F_{ik}F^{ik}$ of electrodynamics to the trace $\mathrm{Tr}\left(F_{ik}F^{ik}\right)$ of the square of the Faraday matrices which is invariant under gauge transformations since

$$\mathrm{Tr}\left(UF_{ik}U^{-1}\,UF^{ik}U^{-1}\right) = \mathrm{Tr}\left(UF_{ik}F^{ik}U^{-1}\right) = \mathrm{Tr}\left(F_{ik}F^{ik}\right). \tag{11.478}$$

As an action density for fermionic matter fields, they replaced the ordinary derivative in Dirac's formula $\overline{\psi}(\gamma^i\partial_i + m)\psi$ by the covariant derivative (11.471) to get $\overline{\psi}(\gamma^i D_i + m)\psi$ (Chen-Ning Yang, 1922–; Robert L. Mills, 1927–1999).

In an abelian gauge theory, the square of the 1-form $A = A_i\,dx^i$ vanishes $A^2 = A_i\,A_k\,dx^i \wedge dx^k = 0$, but in a nonabelian gauge theory the gauge fields are matrices, and $A^2 \neq 0$. The sum $dA + A^2$ is the Faraday 2-form

$$\begin{aligned} F = dA + A^2 &= (\partial_i A_k + A_i A_k)\; dx^i \wedge dx^k \\ &= \tfrac{1}{2}\left(\partial_i A_k - \partial_k A_i + [A_i\,A_k]\right)\; dx^i \wedge dx^k \\ &= \tfrac{1}{2} F_{ik}\,dx^i \wedge dx^k. \end{aligned} \tag{11.479}$$

The scalar matter fields ϕ may have self-interactions described by a potential $V(\phi)$ such as $V(\phi) = \lambda(\phi^\dagger\phi - m^2/\lambda)^2$, which is positive unless $\phi^\dagger\phi = m^2/\lambda$. The kinetic action of these fields is $(D^i\phi)^\dagger D_i\phi$. At low temperatures, these scalar fields assume mean values $\langle 0|\phi|0\rangle = \phi_0$ in the vacuum with $\phi_0^\dagger\phi_0 = m^2/\lambda$ so as to minimize their potential energy density $V(\phi)$. Their kinetic action $(D^i\phi)^\dagger D_i\phi = (\partial^i\phi + A^i\phi)^\dagger(\partial_i\phi + A_i\phi)$ then is in effect $\phi_0^\dagger\,A^i\,A_i\phi_0$. The gauge-field matrix $A^i_{ab} = i\,t^\alpha_{ab}A^i_\alpha$ is a linear combination of the generators t^α of the gauge group. So the action of the scalar fields contains the term $\phi_0^\dagger\,A^i\,A_i\,\phi_0 = -M^2_{\alpha\beta}\,A^i_\alpha\,A_{i\beta}$ in which the mass-squared matrix for the gauge fields is $M^2_{\alpha\beta} = \phi_0^{*a}\,t^\alpha_{ab}\,t^\beta_{bc}\,\phi_0^c$. This **Higgs mechanism** gives masses to those linear combinations $b_{\beta i}\,A_\beta$ of the gauge fields for which $M^2_{\alpha\beta}\,b_{\beta i} = m_i^2 b_{\alpha i} \neq 0$.

The Higgs mechanism also gives masses to the fermions. The mass term m in the Yang–Mills–Dirac action is replaced by something like $c\,\phi$ in which c is a constant different for each fermion. In the vacuum and at low temperatures, each fermion in effect acquires as its mass $c\,\phi_0$. On 4 July 2012, physicists at CERN's Large Hadron Collider announced the discovery of a Higgs-like particle with a mass near 126 GeV/c^2 (Peter Higgs, 1929–).

11.51 Gauge theory and vectors

This section is optional on a first reading.

We can formulate Yang–Mills theory in terms of vectors as we did relativity. To accommodate noncompact groups, we will generalize the unitary matrices $U(x)$ of the Yang–Mills gauge group to nonsingular matrices $V(x)$ that act on n matter fields $\psi^a(x)$ as

$$\psi'^a(x) = \sum_{a=1}^{n} V^a{}_b(x)\,\psi^b(x). \tag{11.480}$$

The field

$$\Psi(x) = \sum_{a=1}^{n} e_a(x)\,\psi^a(x) \tag{11.481}$$

will be gauge invariant $\Psi'(x) = \Psi(x)$ if the vectors $e_a(x)$ transform as

$$e'_a(x) = \sum_{b=1}^{n} e_b(x)\,V^{-1b}{}_a(x). \tag{11.482}$$

In what follows, we will sum over repeated indices from 1 to n and often will suppress explicit mention of the space-time coordinates. In this compressed notation, the field Ψ is gauge invariant because

$$\Psi' = e'_a\,\psi'^a = e_b\,V^{-1b}{}_a\,V^a{}_c\,\psi^c = e_b\,\delta^b{}_c\,\psi^c = e_b\,\psi^b = \Psi, \tag{11.483}$$

which is $e'^{\mathsf{T}}\psi' = e^{\mathsf{T}}\,V^{-1}\,V\psi = e^{\mathsf{T}}\,\psi$ in matrix notation.

The inner product of two basis vectors is an internal "metric tensor"

$$e^*_a \cdot e_b = \sum_{\alpha=1}^{N}\sum_{\beta=1}^{N} e^{\alpha *}_a\,\eta_{\alpha\beta} e^{\alpha}_b = \sum_{\alpha=1}^{N} e^{\alpha *}_a\,e^{\alpha}_b = g_{ab}, \tag{11.484}$$

in which for simplicity I used the N-dimensional identity matrix for the metric η. As in relativity, we'll assume the matrix g_{ab} to be nonsingular. We then can use its inverse to construct dual vectors $e^a = g^{ab}e_b$ that satisfy $e^{a\dagger}\cdot e_b = \delta^a_b$.

The free Dirac action density of the invariant field Ψ

$$\overline{\Psi}(\gamma^i\partial_i + m)\Psi = \overline{\psi}_a e^{a\dagger}(\gamma^i\partial_i + m)e_b\psi^b = \overline{\psi}_a\left[\gamma^i(\delta^a{}_b\partial_i + e^{a\dagger}\cdot e_{b,i}) + m\delta^a{}_b\right]\psi^b \tag{11.485}$$

is the full action of the component fields ψ^b

$$\overline{\Psi}(\gamma^i\partial_i + m)\Psi = \overline{\psi}_a(\gamma^i D^a_{i\,b} + m\,\delta^a{}_b)\psi^b = \overline{\psi}_a\left[\gamma^i(\delta^a{}_b\partial_i + A^a_{i\,b}) + m\,\delta^a{}_b\right]\psi^b \tag{11.486}$$

if we identify the gauge-field matrix as $A^a_{i\,b} = e^{a\dagger}\cdot e_{b,i}$ in harmony with the definition (11.213) of the affine connection $\Gamma^k_{i\ell} = e^k \cdot e_{\ell,i}$.

Under the gauge transformation $e'_a = e_b\,V^{-1b}{}_a$, the metric matrix transforms as

$$g'_{ab} = V^{-1c*}{}_a\,g_{cd}\,V^{-1d}{}_b \qquad \text{or as} \qquad g' = V^{-1\dagger}\,g\,V^{-1} \tag{11.487}$$

in matrix notation. Its inverse goes as $g'^{-1} = V\,g^{-1}\,V^{\dagger}$.

The gauge-field matrix $A^a_{i\,b} = e^{a\dagger} \cdot e_{b,i} = g^{ac} e^{\dagger}_c \cdot e_{b,i}$ transforms as

$$A'^a_{i\,b} = g'^{ac} e'^{\dagger}_a \cdot e'_{b,i} = V^a{}_c A^c_{id} V^{-1d}{}_b + V^a{}_c V^{-1c}{}_{b,i} \tag{11.488}$$

or as $A'_i = VA_i V^{-1} + V\partial_i V^{-1} = VA_i V^{-1} - (\partial_i V)\, V^{-1}$.

By using the identity $e^{a\dagger} \cdot e_{c,i} = -\, e^{a\dagger}_{,i} \cdot e_c$, we may write (exercise 11.44) the Faraday tensor as

$$F^a_{ijb} = [D_i, D_j]^a{}_b = e^{a\dagger}_{,i} \cdot e_{b,j} - e^{a\dagger}_{,i} \cdot e_c\; e^{c\dagger} \cdot e_{b,j} - e^{a\dagger}_{,j} \cdot e_{b,i} + e^{a\dagger}_{,j} \cdot e_c\; e^{c\dagger} \cdot e_{b,i}. \tag{11.489}$$

If $n = N$, then

$$\sum_{c=1}^{n} e^\alpha_c\; e^{\beta c *} = \delta^{\alpha\beta} \quad \text{and} \quad F^a_{ijb} = 0. \tag{11.490}$$

The Faraday tensor vanishes when $n = N$ because the dimension of the embedding space is too small to allow the tangent space to have different orientations at different points x of space-time. The Faraday tensor, which represents internal curvature, therefore must vanish. One needs at least three dimensions in which to bend a sheet of paper. The embedding space must have $N > 2$ dimensions for $SU(2)$, $N > 3$ for $SU(3)$, and $N > 5$ for $SU(5)$.

The covariant derivative of the internal metric matrix

$$g_{;i} = g_{,i} - gA_i - A^{\dagger}_i g \tag{11.491}$$

does not vanish and transforms as $\left(g_{;i}\right)' = V^{-1\dagger} g_{,i} V^{-1}$. A suitable action density for it is the trace $\mathrm{Tr}(g_{;i} g^{-1} g^{;i} g^{-1})$. If the metric matrix assumes a (constant, hermitian) mean value g_0 in the vacuum at low temperatures, then its action is

$$m^2 \mathrm{Tr}\left[(g_0 A_i + A^{\dagger}_i g_0) g_0^{-1} (g_0 A^i + A^{i\dagger} g_0) g_0^{-1}\right], \tag{11.492}$$

which is a mass term for the matrix of gauge bosons

$$W_i = g_0^{1/2}\, A_i\, g_0^{-1/2} + g_0^{-1/2}\, A^{\dagger}_i\, g_0^{1/2}. \tag{11.493}$$

This mass mechanism also gives masses to the fermions. To see how, we write the Dirac action density (11.486) as

$$\overline{\psi}_a \left[\gamma^i (\delta^a{}_b \partial_i + A^a_{i\,b}) + m\, \delta^a{}_b\right] \psi^b = \overline{\psi}^a \left[\gamma^i (g_{ab} \partial_i + g_{ac} A^c_{i\,b}) + m\, g_{ab}\right] \psi^b. \tag{11.494}$$

Each fermion now gets a mass $m\, c_i$ proportional to an eigenvalue c_i of the hermitian matrix g_0.

This mass mechanism does not leave behind scalar bosons. Whether Nature uses it is unclear.

11.52 Geometry

This section is optional on a first reading.

In gauge theory, what plays the role of space-time? Could it be the group manifold? Let us consider the gauge group $SU(2)$ whose group manifold is the 3-sphere in flat euclidean 4-space. A point on the 3-sphere is

$$p = \left(\pm\sqrt{1-\boldsymbol{r}^2},\, r^1,\, r^2,\, r^3\right) \tag{11.495}$$

as explained in example 10.28. The coordinates $r^a = r_a$ are not vectors. The three basis vectors are

$$e_a = \frac{\partial p}{\partial r^a} = \left(\mp\frac{r_a}{\sqrt{1-\boldsymbol{r}^2}},\, \delta^1_a,\, \delta^2_a,\, \delta^3_a\right) \tag{11.496}$$

and so the metric $g_{ab} = e_a \cdot e_b$ is

$$g_{ab} = \frac{r_a\, r_b}{1-\boldsymbol{r}^2} + \delta_{ab} \tag{11.497}$$

or

$$\| g \| = \frac{1}{1-\boldsymbol{r}^2}\begin{pmatrix} 1-r_2^2-r_3^2 & r_1\, r_2 & r_1\, r_3 \\ r_2\, r_1 & 1-r_1^2-r_3^2 & r_2\, r_3 \\ r_3\, r_1 & r_3\, r_2 & 1-r_1^2-r_2^2 \end{pmatrix}. \tag{11.498}$$

The inverse matrix is

$$g^{bc} = \delta_{bc} - r_b\, r_c. \tag{11.499}$$

The dual vectors

$$e^b = g^{bc} e_c = \left(\mp r_b\sqrt{1-\boldsymbol{r}^2},\, \delta^b_1 - r_b r_1,\, \delta^b_2 - r_b r_2,\, \delta^b_3 - r_b r_3\right) \tag{11.500}$$

satisfy $e^b \cdot e_a = \delta^b_a$.

There are two kinds of affine connection $e^b \cdot e_{a,c}$ and $e^b \cdot e_{a,i}$. If we differentiate e_a with respect to an $SU(2)$ coordinate r_c, then

$$E^b_{ca} = e^b \cdot e_{a,c} = r_b\left(\delta_{ac} + \frac{r_a\, r_c}{1-\boldsymbol{r}^2}\right), \tag{11.501}$$

in which we used E (for Einstein) instead of Γ for the affine connection. If we differentiate e_a with respect to a space-time coordinate x^i, then

$$E^b_{ia} = e^b \cdot e_{a,i} = e^b \cdot e_{a,c}\, r^c_{,i} = r_b\, r^c_{,i}\left(\delta_{ac} + \frac{r_a\, r_c}{1-\boldsymbol{r}^2}\right). \tag{11.502}$$

But if the group coordinates r_a are functions of the space-time coordinates x^i, then there are four new basis 4-vectors $e_i = e_a r_{a,i}$. The metric then is a 7×7

matrix $\| g \|$ with entries $g_{a,b} = e_a \cdot e_b$, $g_{a,k} = e_a \cdot e_k$, $g_{i,b} = e_i \cdot e_b$, and $g_{i,k} = e_i \cdot e_k$ or

$$\| g \| = \begin{pmatrix} g_{a,b} & g_{a,b}\, r_{b,k} \\ g_{a,b}\, r_{a,i} & g_{a,b}\, r_{a,i}\, r_{b,k}. \end{pmatrix}. \tag{11.503}$$

Further reading

The classics *Gravitation and Cosmology* (Weinberg, 1972), *Gravitation* (Misner *et al.*, 1973), and *Cosmology* (Weinberg, 2010) as well as the terse *General Theory of Relativity* (Dirac, 1996) and the very accessible *Spacetime and Geometry* (Carroll, 2003) are of special interest, as is Daniel Finley's website (panda.unm.edu/Courses/Finley/p570.html).

Exercises

11.1 Compute the derivatives (11.22 & 11.23).

11.2 Show that the transformation $x \to x'$ defined by (11.16) is a rotation and a reflection.

11.3 Show that the matrix (11.40) satisfies the Lorentz condition (11.39).

11.4 If $\eta = L\,\eta\,L^{\mathsf{T}}$, show that $\Lambda = L^{-1}$ satisfies the definition (11.39) of a Lorentz transformation $\eta = \Lambda^{\mathsf{T}}\eta\,\Lambda$.

11.5 The LHC is designed to collide 7 TeV protons against 7 TeV protons for a total collision energy of 14 TeV. Suppose one used a linear accelerator to fire a beam of protons at a target of protons at rest at one end of the accelerator. What energy would you need to see the same physics as at the LHC?

11.6 Use Gauss's law and the Maxwell–Ampère law (11.87) to show that the microscopic (total) current 4-vector $j = (c\rho, \boldsymbol{j})$ obeys the continuity equation $\dot{\rho} + \nabla \cdot \boldsymbol{j} = 0$.

11.7 Show that if M_{ik} is a covariant second-rank tensor with no particular symmetry, then only its antisymmetric part contributes to the 2-form $M_{ik}\, dx^i \wedge dx^k$ and only its symmetric part contributes to the quantity $M_{ik}\, dx^i dx^k$.

11.8 In rectangular coordinates, use the Levi-Civita identity (1.449) to derive the curl–curl equations (11.90).

11.9 Derive the Bianchi identity (11.92) from the definition (11.79) of the Faraday field-strength tensor, and show that it implies the two homogeneous Maxwell equations (11.82).

11.10 Show that if A is a p-form, then $d(AB) = dA \wedge B + (-1)^p A \wedge dB$.

11.11 Show that if $\omega = a_{ij} dx^i \wedge dx^j/2$ with $a_{ij} = -a_{ji}$, then

$$d\omega = \frac{1}{3!}\left(\partial_k a_{ij} + \partial_i a_{jk} + \partial_j a_{ki}\right) dx^i \wedge dx^j \wedge dx^k. \tag{11.504}$$

11.12 Using tensor notation throughout, derive (11.147) from (11.145 & 11.146).

11.13 Use the flat-space formula (11.168) to compute the change $\boldsymbol{dp}$ due to $d\rho$, $d\phi$, and dz, and so derive the expressions (11.169) for the orthonormal basis vectors $\hat{\boldsymbol{\rho}}$, $\hat{\boldsymbol{\phi}}$, and $\hat{\boldsymbol{z}}$.

11.14 Similarly, derive (11.175) from (11.174).

11.15 Use the definition (11.191) to show that in flat 3-space, the dual of the Hodge dual is the identity: $**\, dx^i = dx^i$ and $**\,(dx^i \wedge dx^k) = dx^i \wedge dx^k$.

11.16 Use the definition of the Hodge star (11.202) to derive (a) two of the four identities (11.203) and (b) the other two.

11.17 Show that Levi-Civita's 4-symbol obeys the identity (11.207).

11.18 Show that $\epsilon_{\ell mn}\, \epsilon^{pmn} = 2\, \delta^p_\ell$.

11.19 Show that $\epsilon_{k\ell mn}\, \epsilon^{p\ell mn} = 3!\, \delta^p_k$.

11.20 (a) Using the formulas (11.175) for the basis vectors of spherical coordinates in terms of those of rectangular coordinates, compute the derivatives of the unit vectors $\hat{\boldsymbol{r}}$, $\hat{\boldsymbol{\theta}}$, and $\hat{\boldsymbol{\phi}}$ with respect to the variables r, θ, and ϕ and express them in terms of the basis vectors $\hat{\boldsymbol{r}}$, $\hat{\boldsymbol{\theta}}$, and $\hat{\boldsymbol{\phi}}$. (b) Using the formulas of (a) and our expression (6.28) for the gradient in spherical coordinates, derive the formula (11.297) for the Laplacian $\nabla \cdot \nabla$.

11.21 Consider the torus with coordinates θ, ϕ labeling the arbitrary point

$$\boldsymbol{p} = (\cos\phi(R + r\sin\theta), \sin\phi(R + r\sin\theta), r\cos\theta) \tag{11.505}$$

in which $R > r$. Both θ and ϕ run from 0 to 2π. (a) Find the basis vectors e_θ and e_ϕ. (b) Find the metric tensor and its inverse.

11.22 For the same torus, (a) find the dual vectors e^θ and e^ϕ and (b) find the nonzero connections Γ^i_{jk} where i, j, and k take the values θ and ϕ.

11.23 For the same torus, (a) find the two Christoffel matrices Γ_θ and Γ_ϕ, (b) find their commutator $[\Gamma_\theta, \Gamma_\phi]$, and (c) find the elements $R^\theta_{\theta\theta\theta}$, $R^\phi_{\theta\phi\theta}$, $R^\theta_{\phi\theta\phi}$, and $R^\phi_{\phi\phi\phi}$ of the curvature tensor.

11.24 Find the curvature scalar R of the torus with points (11.505). Hint: in these four problems, you may imitate the corresponding calculation for the sphere in section 11.42.

11.25 By differentiating the identity $g^{ik}\, g_{k\ell} = \delta^i_\ell$, show that $\delta g^{ik} = -\, g^{is} g^{kt} \delta g_{st}$ or equivalently that $dg^{ik} = -g^{is} g^{kt} dg_{st}$.

11.26 Just to get an idea of the sizes involved in black holes, imagine an isolated sphere of matter of uniform density ρ that as an initial condition is all at rest within a radius r_b. Its radius will be less than its Schwarzschild radius if

$$r_\text{b} < \frac{2MG}{c^2} = 2\left(\frac{4}{3}\pi r_b^3 \rho\right)\frac{G}{c^2}. \tag{11.506}$$

If the density ρ is that of water under standard conditions (1 gram per cc), for what range of radii r_b might the sphere be or become a black hole? Same question if ρ is the density of dark energy.

11.27 For the points (11.392), derive the metric (11.395) with $k = 1$. Don't forget to relate $d\chi$ to dr.

11.28 For the points (11.393), derive the metric (11.395) with $k = 0$.

11.29 For the points (11.394), derive the metric (11.395) with $k = -1$. Don't forget to relate $d\chi$ to dr.

11.30 Suppose the constant k in the Robertson–Walker metric (11.391 or 11.395) is some number other than 0 or ± 1. Find a coordinate transformation such that in the new coordinates, the Robertson–Walker metric has $k = k/|k| = \pm 1$. Hint: You can also change the scale factor a.

11.31 Derive the affine connections in equation (11.399).

11.32 Derive the affine connections in equation (11.400).

11.33 Derive the affine connections in equation (11.401).

11.34 Derive the spatial Einstein equation (11.411) from (11.375, 11.395, 11.406, 11.408, & 11.409).

11.35 Assume there had been no inflation, no era of radiation, and no dark energy. In this case, the magnitude of the difference $|\Omega - 1|$ would have increased as $t^{2/3}$ over the past 13.8 billion years. Show explicitly how close to unity Ω would have had to have been at $t = 1$ s so as to satisfy the observational constraint $|\Omega_0 - 1| < 0.036$ on the present value of Ω.

11.36 Derive the relation (11.431) between the energy density ρ and the Robertson–Walker scale factor $a(t)$ from the conservation law (11.427) and the equation of state $p = w\rho$.

11.37 Use the Friedmann equations (11.410 & 11.412) for constant with $\rho = -p$ and $k = 1$ to derive (11.438) subject to the boundary condition that $a(t)$ has its minimum at $t = 0$.

11.38 Use the Friedmann equations (11.410 & 11.412) with $w = -1$, ρ constant, and $k = -1$ to derive (11.439) subject to the boundary condition that $a(0) = 0$.

11.39 Use the Friedmann equations (11.410 & 11.412) with $w = -1$, ρ constant, and $k = 0$ to derive (11.440). Show why a linear combination of the two solutions (11.440) does not work.

11.40 Use the conservation equation (11.444) and the Friedmann equations (11.410 & 11.412) with $w = 1/3$ and $k = 0$ to derive (11.447) subject to the boundary condition that $a(0) = 0$.

11.41 Show that if the matrix $U(x)$ is nonsingular, then

$$(\partial_i U)\, U^{-1} = -\, U\, \partial_i\, U^{-1}. \tag{11.507}$$

11.42 The gauge-field matrix is a linear combination $A_k = -i\, g\, t^b A^b_k$ of the generators t^b of a representation of the gauge group. The generators obey the commutation relations

$$[t^a, t^b] = i f_{abc} t^c, \tag{11.508}$$

in which the f_{abc} are the structure constants of the gauge group. Show that under a gauge transformation (11.474)

$$A'_i = U A_i U^{-1} - (\partial_i U)\, U^{-1} \tag{11.509}$$

by the unitary matrix $U = \exp(-ig\lambda^a t^a)$ in which λ^a is infinitesimal, the gauge-field matrix A_i transforms as

$$-igA_i'^a t^a = -igA_i^a t^a - ig^2 f_{abc}\lambda^a A_i^b t^c + ig\partial_i\lambda^a t^a. \tag{11.510}$$

Show further that the gauge field transforms as

$$A_i'^a = A_i^a - \partial_i\lambda^a - gf_{abc}A_i^b\lambda^c. \tag{11.511}$$

11.43 Show that if the vectors $e_a(x)$ are orthonormal, then $e^{a\dagger} \cdot e_{c,i} = -e^{a\dagger}_{,i} \cdot e_c$.

11.44 Use the identity of exercise 11.43 to derive the formula (11.489) for the nonabelian Faraday tensor.

12

Forms

12.1 Exterior forms

1-forms **A 1-form** is a linear function ω that maps vectors into numbers. Thus, if A and B are vectors in $\mathbb{R}^n$ and z and w are numbers, then

$$\omega(zA + wB) = z\,\omega(A) + w\,\omega(B). \tag{12.1}$$

The n coordinates $x_1, \ldots, x_n$ are 1-forms; they map a vector A into its coordinates: $x_1(A) = A_1, \ldots, x_n(A) = A_n$. Every 1-form may be expanded in terms of these **basic** 1-forms as

$$\omega = B_1 x_1 + \cdots + B_n x_n \tag{12.2}$$

so that

$$\begin{aligned} \omega(A) &= B_1 x_1(A) + \cdots + B_n x_n(A) \\ &= B_1 A_1 + \cdots + B_n A_n \\ &= (B, A) = \boldsymbol{B} \cdot \boldsymbol{A}. \end{aligned} \tag{12.3}$$

Thus, every 1-form is associated with a (dual) vector, in this case B.

2-forms **A 2-form** is a function that maps pairs of vectors into numbers linearly and skew-symmetrically. Thus, if A, B, and C are vectors in $\mathbb{R}^n$ and z and w are numbers, then

$$\begin{aligned} \omega^2(zA + wB, C) &= z\,\omega^2(A, C) + w\,\omega^2(B, C), \\ \omega^2(A, B) &= -\,\omega^2(B, A). \end{aligned} \tag{12.4}$$

One often drops the superscript and writes the addition of two 2-forms as

$$(\omega_1 + \omega_2)(A, B) = \omega_1(A, B) + \omega_2(A, B). \tag{12.5}$$

Example 12.1 (Parallelogram) The **oriented area** of the parallelogram defined by two 2-vectors A and B is the determinant

$$\omega(A,B) = \begin{vmatrix} A_1 & A_2 \\ B_1 & B_2 \end{vmatrix}. \tag{12.6}$$

This 2-form maps the ordered pair of vectors (A, B) into the oriented area ($\pm$ the usual area) of the parallelogram they describe. To check that this 2-form gives the area to within a sign, rotate the coordinates so that the 2-vector A runs from the origin along the x-axis. Then $A_2 = 0$, and the 2-form gives $A_1 B_2$ which is the base A_1 of the parallelogram times its height B_2. □

Example 12.2 (Parallelepiped) The **triple scalar product** of three 3-vectors

$$\omega_A^2(B,C) = \boldsymbol{A} \cdot \boldsymbol{B} \times \boldsymbol{C} = \begin{vmatrix} A_1 & A_2 & A_3 \\ B_1 & B_2 & B_3 \\ C_1 & C_2 & C_3 \end{vmatrix} = \omega^3(A,B,C) \tag{12.7}$$

is both a 2-form that depends upon the vector A and also a **3-form** that maps the triplet of vectors A, B, C into the signed volume of their parallelepiped. □

k*-forms** **A *k*-form** (or an **exterior form of degree *k) is a linear function of k vectors that is antisymmetric. For vectors $A_1, \ldots, A_k$ and numbers z and w

$$\omega(zA_1' + wA_1'', A_2, \ldots, A_k) = z\,\omega(A_1', A_2, \ldots, A_k) + w\,\omega(A_1'', A_2, \ldots, A_k) \tag{12.8}$$

and the interchange of any two vectors makes a minus sign

$$\omega(A_2, A_1, \ldots, A_k) = -\omega(A_1, A_2, \ldots, A_k). \tag{12.9}$$

Exterior product of two 1-forms The 1-form ω_1 maps the vectors A and B into the numbers $\omega_1(A)$ and $\omega_1(B)$, and the 1-form ω_2 does the same thing with $1 \to 2$. The value of the **exterior product** $\omega_1 \wedge \omega_2$ on the two vectors A and B is the 2-form defined by the 2×2 determinant

$$\omega_1 \wedge \omega_2(A,B) = \begin{vmatrix} \omega_1(A) & \omega_2(A) \\ \omega_1(B) & \omega_2(B) \end{vmatrix} = \omega_1(A)\omega_2(B) - \omega_2(A)\omega_1(B) \tag{12.10}$$

or more formally

$$\omega_1 \wedge \omega_2 = \omega_1 \otimes \omega_2 - \omega_2 \otimes \omega_1. \tag{12.11}$$

The **most general 2-form** on $\mathbb{R}^n$ is a linear combination of the basic 2-forms $x_i \wedge x_j$

$$\omega^2 = \sum_{1 \le i < k \le n} a_{ik}\, x_i \wedge x_k. \tag{12.12}$$

If the unit vectors in the n orthogonal directions of $\mathbb{R}^n$ are $e_1, \ldots, e_n$, then $x_i(e_k) = \delta_{ik}$ and so

$$\omega^2(e_i, e_k) = a_{ik} \begin{vmatrix} x_i(e_i) & x_k(e_i) \\ x_i(e_k) & x_k(e_k) \end{vmatrix} = a_{ik} \begin{vmatrix} 1 & 0 \\ 0 & 1 \end{vmatrix} = a_{ik}. \tag{12.13}$$

Exterior product of k 1-forms The **exterior product** of k 1-forms $\omega_1, \ldots, \omega_k$ maps the k n-vectors $A_1, A_2, \ldots, A_k$ to the determinant

$$\omega_1 \wedge \omega_2 \wedge \cdots \wedge \omega_k(A_1, A_2, \ldots, A_k) = \begin{vmatrix} \omega_1(A_1) & \ldots & \omega_k(A_1) \\ \vdots & \ddots & \vdots \\ \omega_1(A_k) & \ldots & \omega_k(A_k) \end{vmatrix}. \tag{12.14}$$

The **most general k-form** on $\mathbb{R}^n$ is a linear combination of the various exterior products of k basic 1-forms $x_{i_1} \wedge \cdots \wedge x_{i_k}$

$$\omega^k = \sum_{1 \le i_1 < \ldots i_k \le n} a_{i_1 \ldots i_k} \, x_{i_1} \wedge \cdots \wedge x_{i_k}. \tag{12.15}$$

Exterior multiplication The **exterior multiplication** of a k-form with an ℓ-form is linear, associative, and antisymmetric

$$\omega^k \wedge \omega^\ell = (-1)^{k\ell} \omega^\ell \wedge \omega^k. \tag{12.16}$$

Restriction of forms A p-form ω^p is a map from the product $V \times \cdots \times V$ of p copies of some vector space V into the real numbers. The restriction $\omega^p|_U(A_1, A_2, \ldots, A_p)$ of the p-form ω^p to a subspace $U \subset V$ is the same p-form ω^p but with its domain restricted to vectors $A_i \in U$.

12.2 Differential forms

A **manifold** is a set of points that can be labeled locally by coordinates in $\mathbb{R}^n$ in such a way that the coordinates make sense when the local regions overlap. The k-dimensional surface S^k of the unit sphere in $\mathbb{R}^{k+1}$

$$\sum_{i=1}^{k+1} y_i^2 = 1 \tag{12.17}$$

is an example of a manifold. A **smooth** function $f(x_1, \ldots, x_n)$ is one that is infinitely differentiable with respect to all combinations of its arguments $x_1, \ldots, x_n$.

There are two ways of thinking about differential forms. The Russian literature views a manifold as embedded in $\mathbb{R}^n$ and so is somewhat more straightforward. We will discuss it first.

The Russian way Suppose $x(t)$ is a curve with $x(0) = x$ on some **manifold** M, and $f(x(t))$ is a smooth function $f : \mathbb{R}^n \to \mathbb{R}$ that maps points $x(t)$ into numbers. Then the **differential** $df(\dot{x}(t))$ maps $\dot{x}(t)$ at x into

$$df\left(\frac{d}{dt}x(t)\right) \equiv \frac{d}{dt}f(x(t)) = \sum_{j=1}^{n} \dot{x}(t)_j \frac{\partial f(x(t))}{\partial x_j} = \dot{x}(t) \cdot \nabla f(x(t)) \tag{12.18}$$

all at $t = 0$. As physicists, we think of df as a number – the change in the function $f(x)$ when its argument x is changed by dx. Russian mathematicians think of df as a linear map of tangent vectors $\dot{x}$ at x into numbers. Since this map is linear, we may multiply the definition (12.18) by dt and arrive at the more familiar formula

$$dt\, df\left(\frac{d}{dt}x(t)\right) = df\left(dt\frac{d}{dt}x(t)\right) = df\,(dx(t)) = dx(t) \cdot \nabla f(x(t)) \tag{12.19}$$

all at $t = 0$. So

$$df(dx) = dx \cdot \nabla f \tag{12.20}$$

is the physicist's df.

Since the differential df is a linear map of vectors $\dot{x}(0)$ into numbers, it is a 1-form; since it is defined on vectors like $\dot{x}(0)$, it is a **differential 1-form**. The term *differential 1-form* underscores the fact that the actual value of the differential df depends upon the vector $\dot{x}(0)$ and the point $x = x(0)$. Mathematicians call the space of vectors $\dot{x}(0)$ at the point $x = x(0)$ the **tangent space** TM_x. They say df is a smooth map of the **tangent bundle** TM, which is the union of the tangent spaces for all points x in the manifold M, to the real line, so $df : TM \to \mathbb{R}$.

Consider a curve $x(t)$ on a smooth surface with coordinates x_i. In the special case in which $f(x) = x_i(x) = x_i$, the differential $dx_i(\dot{x}(t))$ by (12.18) is

$$dx_i(\dot{x}(t)) = \sum_{j=1}^{n} \dot{x}_j(t) \frac{\partial x_i(x)}{\partial x_j} = \sum_{j=1}^{n} \dot{x}_j(t) \frac{\partial x_i}{\partial x_j} = \sum_{j=1}^{n} \dot{x}(t)_j\, \delta_{ij} = \dot{x}_i(t). \tag{12.21}$$

These dx_is are the **basic differentials**. Using A for the vector $\dot{x}(t)$, we find from our definition (12.18) that

$$dx_i(A) = \sum_{j=1}^{n} A_j \frac{\partial x_i}{\partial x_j} = \sum_{j=1}^{n} A_j \delta_{ij} = A_i \tag{12.22}$$

as well as

$$df(A) = \sum_{j=1}^{n} A_j \frac{\partial f(x)}{\partial x_j} = \sum_{j=1}^{n} \frac{\partial f(x)}{\partial x_j}\, dx_j(A) \tag{12.23}$$

or

$$df = \sum_{j=1}^{n} \frac{\partial f(x)}{\partial x_j}\, dx_j. \tag{12.24}$$

Example 12.3 (dr^2) If $r^2 = x_1^2 + x_2^2$, then the differential 1-form dr^2 is

$$dr^2 = 2x_1\, dx_1 + 2x_2\, dx_2. \tag{12.25}$$

It takes the 2-vector A into the number

$$dr^2(A) = 2x_1\, dx_1(A) + 2x_2\, dx_2(A) = 2x_1\, A_1 + 2x_2\, A_2. \tag{12.26}$$

So if $A = (\epsilon_1, \epsilon_2)$, then $dr^2(A) = 2x_1\,\epsilon_1 + 2x_2\,\epsilon_2$. □

The other way Most American, French, and English mathematicians use a more abstract approach. They abstract from the basic definition (12.18) the rule

$$df\left(\frac{d}{dt}\right) = \frac{d}{dt}f \tag{12.27}$$

or more generally

$$df\left(\frac{\partial}{\partial x_k}\right) = \frac{\partial}{\partial x_k}f. \tag{12.28}$$

In particular, if $f(x) = x_i$, then

$$dx_i\left(\frac{\partial}{\partial x_k}\right) = \frac{\partial}{\partial x_k}x_i = \delta_{ik}. \tag{12.29}$$

In this frequently used notation, the idea is that the derivatives

$$\partial_k \equiv \frac{\partial}{\partial x_k} \tag{12.30}$$

form a set of orthonormal vectors to which the forms dx_i are dual

$$dx_i(\partial_k) = \delta_{ik}. \tag{12.31}$$

In the nonRussian literature, equations (12.27–12.31) *define* how the basic 1-forms dx_i act on the vectors ∂_k.

Change of variables Suppose that $x_1, \ldots, x_n$ and $y_1, \ldots, y_n$ are two systems of coordinates on $\mathbb{R}^n$, and that $dx_1, \ldots, dx_n$ and $dy_1, \ldots, dy_n$ are two sets of basic differentials. Then by applying the formula (12.24) to the function $y_k(x)$, we get

$$dy_k = \sum_{j=1}^{n} \frac{\partial y_k(x)}{\partial x_j}\, dx_j, \tag{12.32}$$

which is the familiar rule for changing variables.

The **most general differential 1-form** ω on the space $\mathbb{R}^n$ with coordinates $x_1, \ldots, x_n$ is a linear combination of the basic differentials dx_i with coefficients $a_i(x)$ that are smooth functions of $x = (x_1, \ldots, x_n)$

$$\omega = a_1(x)\, dx_1 + \cdots + a_n(x)\, dx_n. \tag{12.33}$$

The **basic differential 2-forms** are $dx_i \wedge dx_k$ defined as

$$dx_i \wedge dx_k(A, B) = \begin{vmatrix} dx_i(A) & dx_k(A) \\ dx_i(B) & dx_k(B) \end{vmatrix} = \begin{vmatrix} A_i & A_k \\ B_i & B_k \end{vmatrix} = A_i B_k - A_k B_i. \tag{12.34}$$

So in particular

$$dx_i \wedge dx_i = 0. \tag{12.35}$$

The **basic differential *k*-forms** $dx_1 \wedge \cdots \wedge dx_k$ are defined as

$$dx_1 \wedge \cdots \wedge dx_k(A_1, \ldots A_k) = \begin{vmatrix} dx_1(A_1) & \ldots & dx_k(A_1) \\ \vdots & \ddots & \vdots \\ dx_1(A_k) & \ldots & dx_k(A_k) \end{vmatrix} = \begin{vmatrix} A_{11} & \ldots & A_{1k} \\ \vdots & \ddots & \vdots \\ A_{k1} & \ldots & A_{kk} \end{vmatrix}. \tag{12.36}$$

Example 12.4 ($dx_3 \wedge dr^2$) If $r^2 = x_1^2 + x_2^2 + x_3^2$, then dr^2 is

$$dr^2 = 2(x_1 dx_1 + x_2 dx_2 + x_3 dx_3) \tag{12.37}$$

and the differential 2-form $\omega = dx_3 \wedge dr^2$ is

$$\omega = dx_3 \wedge 2(x_1 dx_1 + x_2 dx_2 + x_3 dx_3) = 2x_1 dx_3 \wedge dx_1 + 2x_2 dx_3 \wedge dx_2 \tag{12.38}$$

since in view of (12.35) $dx_3 \wedge dx_3 = 0$. So the value of the 2-form ω on the vectors $A = (1, 2, 3)$ and $B = (2, 1, 1)$ at the point $x = (3, 0, 3)$ is

$$\omega(A, B) = 2x_1 dx_3 \wedge dx_1(A, B) = 6 \begin{vmatrix} dx_3(A) & dx_1(A) \\ dx_3(B) & dx_1(B) \end{vmatrix} = 6 \begin{vmatrix} 3 & 1 \\ 1 & 2 \end{vmatrix} = 30. \tag{12.39}$$

On the vectors, $C = (1, 0, 0)$ and $D = (0, 0, 1)$ at $x = (2, 3, 4)$, this 2-form has the value $\omega(C, D) = -4$. □

The most general **differential *k*-form** ω^k on the space $\mathbb{R}^n$ with coordinates $x_1, \ldots, x_n$ is

$$\omega^k = \sum_{1 \le i_1 < \ldots i_k \le n} a_{i_1 \ldots i_k}(x)\, dx_{i_1} \wedge \cdots \wedge dx_{i_k}, \tag{12.40}$$

in which the functions $a_{i_1 \ldots i_k}(x)$ are smooth on $\mathbb{R}^n$.

Example 12.5 (Change of variables) If x_1, x_2, x_3 and y_1, y_2, y_3 are two coordinate systems on $\mathbb{R}^3$, then in terms of the basic 1-forms dy_k, the 2-form $\omega = X dx_2 \wedge dx_3$ is by (12.32)

$$\omega = X dx_2 \wedge dx_3 = X \left(\sum_{k=1}^{3} \frac{\partial x_2}{\partial y_j} dy_j \right) \wedge \left(\sum_{k=1}^{3} \frac{\partial x_3}{\partial y_k} dy_k \right), \tag{12.41}$$

in which jacobians appear such as

$$\frac{\partial(x_2, x_3)}{\partial(y_1, y_2)} = \frac{\partial x_2}{\partial y_1} \frac{\partial x_3}{\partial y_2} - \frac{\partial x_2}{\partial y_2} \frac{\partial x_3}{\partial y_1}. \tag{12.42}$$

In terms of these jacobians, the 2-form $\omega = X dx_2 \wedge dx_3$ is (exercise 12.2)

$$\omega = X \left(\frac{\partial(x_2, x_3)}{\partial(y_1, y_2)} dy_1 \wedge dy_2 + \frac{\partial(x_2, x_3)}{\partial(y_2, y_3)} dy_2 \wedge dy_3 + \frac{\partial(x_2, x_3)}{\partial(y_3, y_1)} dy_3 \wedge dy_1 \right). \tag{12.43}$$

On the vectors of example 12.4, both forms of ω give $\omega(A, B) = -X$. □

Example 12.6 (Euclidean 3-space) Let us recall the formula (11.157) for the square ds^2 of the length of a vector $\boldsymbol{dx}$ in orthogonal coordinates

$$ds^2 = h_1^2 \, dx_1^2 + h_2^2 \, dx_2^2 + h_3^2 \, dx_3^2 \tag{12.44}$$

as well as our formula (11.177) for the gradient

$$\nabla f = \frac{1}{h_i} \frac{\partial f}{\partial x_i} \hat{e}_i. \tag{12.45}$$

Then in cylindrical coordinates (ρ, ϕ, z), we have $h_\rho = 1$, $h_\phi = \rho$, and $h_z = 1$, while in spherical coordinates (r, θ, ϕ), we have $h_r = 1$, $h_\theta = r$, and $h_\phi = r \sin\theta$. The value of the form dx_i on the unit vector $\hat{\boldsymbol{e}}_j$ is by (12.20)

$$dx_k(\hat{\boldsymbol{e}}_j) = \hat{\boldsymbol{e}}_j \cdot \nabla x_k = \hat{\boldsymbol{e}}_j \cdot \frac{1}{h_i} \frac{\partial x_k}{\partial x_i} \hat{e}_i = \frac{1}{h_j} \frac{\partial x_k}{\partial x_j} = \frac{\delta_{kj}}{h_j}. \tag{12.46}$$

Thus $d\rho(\hat{\boldsymbol{e}}_\rho) = 1$, $d\phi(\hat{\boldsymbol{e}}_\phi) = 1/\rho$, and $dz(\hat{\boldsymbol{e}}_z) = 1$. □

Example 12.7 (Three-dimensional vectors and their forms) Any three-dimensional vector $\boldsymbol{A}$ defines a 1-form as the dot-product

$$\omega_A^1(\boldsymbol{U}) = \boldsymbol{A} \cdot \boldsymbol{U} \tag{12.47}$$

and a 2-form as the triple cross-product

$$\omega_A^2(\boldsymbol{U}, \boldsymbol{V}) = \boldsymbol{A} \cdot (\boldsymbol{U} \times \boldsymbol{V}). \tag{12.48}$$

Here we assume that we have a right-handed set of basis vectors $\hat{\boldsymbol{e}}_1$, $\hat{\boldsymbol{e}}_2$, and $\hat{\boldsymbol{e}}_3$ with $\hat{\boldsymbol{e}}_1 \times \hat{\boldsymbol{e}}_2 = \hat{\boldsymbol{e}}_3$ so as to define the cross-product $\boldsymbol{U} \times \boldsymbol{V}$. Such a manifold is said to be **oriented**.

The quantity

$$A = A_1\hat{e}_1 + A_2\hat{e}_2 + A_3\hat{e}_3 \tag{12.49}$$

is a vector field $A(x)$. So if we use (12.44) for the squared length ds^2, then we can write the 1-form (12.47) as

$$\omega_A^1 = A_1\, h_1\, dx_1 + A_2\, h_2\, dx_2 + A_3\, h_3\, dx_3 \tag{12.50}$$

because by (12.46) and summing over i and k we get

$$\omega_A^1(U) = A_i\, h_i\, dx_i\left(U_k\hat{e}_k\right) = A_i\, h_i\, U_k\, \delta_{ik}/h_k = A_i U_i = A \cdot U. \tag{12.51}$$

Similarly (exercise 12.6), the 2-form (12.48) is

$$\omega_A^2 = A_1\, h_2 h_3\, dx_2 \wedge dx_3 + A_2\, h_3 h_1\, dx_3 \wedge dx_1 + A_3\, h_1 h_2\, dx_1 \wedge dx_2. \tag{12.52}$$

By analogy with the definition (12.50), the gradient 1-form $\omega_{\nabla f}^1$ is

$$\omega_{\nabla f}^1 = (\nabla f)_k\; h_k\, dx_k \tag{12.53}$$

summed over repeated indices. The relation $\omega_{\nabla f}^1 = df$ gives

$$\omega_{\nabla f}^1 = (\nabla f)_k\; h_k\, dx_k = df = \frac{\partial f}{\partial x_k} dx_k \tag{12.54}$$

according to the definition (12.24) of df. So the vector field ∇f is

$$\nabla f = \sum_{k=1}^{3} \frac{1}{h_k} \frac{\partial f}{\partial x_k} \hat{e}_k, \tag{12.55}$$

which in cylindrical and spherical coordinates is

$$\nabla f = \frac{\partial f}{\partial \rho}\hat{e}_\rho + \frac{1}{\rho}\frac{\partial f}{\partial \phi}\hat{e}_\phi + \frac{\partial f}{\partial z}\hat{e}_z = \frac{\partial f}{\partial r}\hat{e}_r + \frac{1}{r}\frac{\partial f}{\partial \theta}\hat{e}_\theta + \frac{1}{r\sin\theta}\frac{\partial f}{\partial \phi}\hat{e}_\phi \tag{12.56}$$

in agreement with (11.180–11.181). □

12.3 Exterior differentiation

Exterior differentiation is nifty. The differential (12.24)

$$df = \sum_{k=1}^{n} \frac{\partial f}{\partial x_k}\, dx_k \tag{12.57}$$

is the **exterior derivative** of the function $f(x)$, itself a 0-form. The operator

$$d = \sum_{k=1}^{n} \frac{\partial}{\partial x_k}\, dx_k \tag{12.58}$$

turns the 0-form f into the differential 1-form df.

Applied to the 1-form

$$\omega^1 = \sum_{i=1}^{n} a_i(x)\, dx_i \tag{12.59}$$

the exterior derivative d generates the 2-form

$$d\omega^1 = d\left(\sum_{i=1}^{n} a_i(x)\, dx_i\right) = \sum_{i,k=1}^{n} a_{i,k}(x)\, dx_k \wedge dx_i, \tag{12.60}$$

in which $a_{i,k} = \partial_k a_i$. But a second application of d gives zero:

$$\begin{aligned} dd\omega^1 &= dd\left(\sum_{i=1}^{n} a_i\, dx_i\right) = d\left(\sum_{i,k=1}^{n} a_{i,k}\, dx_k \wedge dx_i\right) \\ &= \sum_{i,k,\ell=1}^{n} a_{i,k\ell}\, dx_\ell \wedge dx_k \wedge dx_i = 0 \end{aligned} \tag{12.61}$$

because the double partial derivative $a_{i,k\ell} = \partial_\ell \partial_k a_i$ is symmetric in k and ℓ while the wedge product $dx_\ell \wedge dx_k$ is antisymmetric in these indices.

We have seen (12.40) that the most general differential k-form is

$$\omega^k = \sum_{1 \le i_1 < \cdots i_k \le n} a_{i_1 \ldots i_k}(x)\, dx_{i_1} \wedge \cdots \wedge dx_{i_k} \tag{12.62}$$

in which the functions $a_{i_1 \ldots i_k}(x)$ are smooth on $\mathbb{R}^n$. The exterior derivative operator d turns ω^k into the $(k+1)$-form

$$\begin{aligned} d\omega^k &= \sum_{1 \le i_1 < \cdots i_k \le n} d\left(a_{i_1 \ldots i_k}(x)\, dx_{i_1} \wedge \cdots \wedge dx_{i_k}\right) \\ &= \sum_{1 \le \ell, i_1 < \cdots i_k \le n} a_{i_1 \ldots i_k, \ell}(x)\, dx_\ell \wedge dx_{i_1} \wedge \cdots \wedge dx_{i_k}. \end{aligned} \tag{12.63}$$

Once again $d\,d\,\omega^k = 0$ so quite generally

$$d\,d = 0. \tag{12.64}$$

If ω is the wedge product of two 1-forms $\omega_a = a_i\, dx_i$ and $\omega_b = b_k\, dx_k$

$$\omega = \omega_a \wedge \omega_b = a_i\, dx_i \wedge b_k\, dx_k \tag{12.65}$$

then d maps it to

$$\begin{aligned} d\,\omega &= d\,(a_i\,dx_i \wedge b_k\,dx_k) = d\,(a_i\,b_k\,dx_i \wedge dx_k) \\ &= (a_i\,b_k)_{,\ell}\,dx_\ell \wedge dx_i \wedge dx_k \\ &= (a_{i,\ell}\,b_k + a_i\,b_{k,\ell})\,dx_\ell \wedge dx_i \wedge dx_k \\ &= a_{i,\ell}\,dx_\ell \wedge dx_i \wedge b_k\,dx_k + a_i\,b_{k,\ell}\,dx_\ell \wedge dx_i \wedge dx_k \\ &= (a_{i,\ell}\,dx_\ell \wedge dx_i) \wedge b_k\,dx_k - a_i\,dx_i \wedge (b_{k,\ell}\,dx_\ell \wedge dx_k) \\ &= d\,\omega_a \wedge \omega_b - \omega_a \wedge d\,\omega_b. \end{aligned} \tag{12.66}$$

More generally, the exterior derivative operator d maps the wedge product of a k-form ω^k and a p-form ω^p to

$$d\left(\omega^k \wedge \omega^p\right) = \left(d\omega^k\right) \wedge \omega^p + (-1)^k \omega^k \wedge \left(d\omega^p\right). \tag{12.67}$$

Example 12.8 (Phase space) If ω^1 is the 1-form

$$\omega^1 = p_1 dq_1 + \cdots + p_n dq_n = p \cdot dq \tag{12.68}$$

with coordinates $p_1, \ldots, p_n, q_1, \ldots, q_n$, then $d\,\omega^1$ is the 2-form

$$\begin{aligned} d\,\omega^1 &= d\left(\sum_{i=1}^{n} p_i\,dq_i\right) = \sum_{i,k=1}^{n} (\partial_{p_k} p_i)\,dp_k \wedge dq_i = \sum_{i,k=1}^{n} \delta_{ik}\,dp_k \wedge dq_i \\ &= dp_1 \wedge dq_1 + \cdots + dp_n \wedge dq_n = dp \wedge dq. \end{aligned} \tag{12.69}$$

It follows that $d\,(dp \wedge dq) = d\,d\,\omega^1 = 0$. □

Example 12.9 (A Poincaré invariant) The 1-form $\omega^1 = p \cdot dq$ maps a tiny piece $\partial_s q\,ds$ of a phase-space trajectory into a small element of action $p \cdot dq(\partial_s q\,ds) = p \cdot \partial_s q\,ds$. The sum of these pieces along a *closed* trajectory

$$A = \int_{\partial S} \omega^1 = \oint p \cdot dq = \oint \sum_{i=1}^{n} p_i\,dq_i = \oint p \cdot \frac{\partial q}{\partial s}\,ds \tag{12.70}$$

is a Poincaré invariant. Because the trajectory is a loop, we may integrate the second term in its time derivative by parts

$$\dot{A} = \oint \left(\dot{p} \cdot \frac{\partial q}{\partial s} + p \cdot \frac{\partial^2 q}{\partial s \partial t}\right) ds = \oint \left(\dot{p} \cdot \frac{\partial q}{\partial s} - \frac{\partial p}{\partial s} \cdot \dot{q}\right) ds \tag{12.71}$$

without acquiring an extra term. The trajectory is physical, and so we can use Hamilton's equations

$$\dot{p}_i = -\frac{\partial H}{\partial q_i} \quad \text{and} \quad \dot{q}_i = \frac{\partial H}{\partial p_i} \quad \text{for} \quad i = 1, \ldots, n \tag{12.72}$$

to write $\dot{A}$ as

$$\dot{A} = \oint \left(-\frac{\partial H}{\partial q} \cdot \frac{\partial q}{\partial s} - \frac{\partial p}{\partial s} \cdot \frac{\partial H}{\partial p} \right) ds = -\oint \frac{\partial H}{\partial s}\, ds = -\oint dH = 0 \tag{12.73}$$

because the trajectory is closed. □

Example 12.10 (The Bohr model) In 1912, Bohr considered an electron in a circular orbit around a proton, set Poincaré's invariant equal to an integral multiple of Planck's constant

$$A = \oint p dq = 2\pi r p = nh \tag{12.74}$$

and so quantized the orbital angular momentum as $L = rp = n\hbar$. One can derive the energy levels of the hydrogen atom from this rule (exercise 12.12). In 1924, Arnold Sommerfeld applied this trick to a more general orbit of a relativistic electron about a proton and got the energy levels that Dirac would four years later. □

Example 12.11 (Constant area) Consider three nearby points in phase space (p, q), $(p + \delta p, q + \delta q)$, and $(p + \Delta p, q + \Delta q)$ that move according to Hamilton's equations (12.72). The time derivatives of the tiny displacements δp_i and δq_i are (exercise 12.9)

$$\begin{aligned} \frac{d}{dt}\delta p_i &= \delta \dot{p}_i = \sum_{k=1}^{n} -\frac{\partial^2 H}{\partial q_i \partial q_k}\delta q_k - \frac{\partial^2 H}{\partial q_i \partial p_k}\delta p_k \\ \frac{d}{dt}\delta q_i &= \delta \dot{q}_i = \sum_{k=1}^{n} \frac{\partial^2 H}{\partial p_i \partial q_k}\delta q_k + \frac{\partial^2 H}{\partial p_i \partial p_k}\delta p_k. \end{aligned} \tag{12.75}$$

Similar equations give the derivatives of the small differences Δp_i and Δq_i.

The 2-form (12.69) maps the n pairs of 2-vectors $(\delta p, \delta q)$ and $(\Delta p, \Delta q)$ into a sum of areas of parallelograms

$$d\omega^1(\delta p, \delta q; \Delta p, \Delta q) = \begin{vmatrix} \delta p_1 & \delta q_1 \\ \Delta p_1 & \Delta q_1 \end{vmatrix} + \cdots + \begin{vmatrix} \delta p_n & \delta q_n \\ \Delta p_n & \Delta q_n \end{vmatrix}. \tag{12.76}$$

By using the time derivatives (12.75) of δp_i and δq_i and those of Δp_i and Δq_i, one may show (exercise 12.10) that this sum of areas remains constant

$$\frac{d}{dt} d\omega^1(\delta p, \delta q; \Delta p, \Delta q) = 0 \tag{12.77}$$

along the trajectories in phase space (Gutzwiller, 1990, chap. 7). □

Example 12.12 (The curl) We saw in example 12.7 that the 1-form (12.50) of a vector field A is $\omega_A = A_1\, h_1\, dx_1 + A_2\, h_2\, dx_2 + A_3\, h_3\, dx_3$ in which the h_ks are those that determine (12.44) the squared length $ds^2 = h_k^2\, dx_k^2$ of the triply orthogonal coordinate system with unit vectors $\hat{e}_1$, $\hat{e}_2$, $\hat{e}_3$. So the exterior derivative of the 1-form ω_A is

$$
\begin{aligned}
d\omega_A &= \sum_{i,k=1}^{3} \partial_k (A_i\, h_i)\, dx_k \wedge dx_i \\
&= \left(\frac{\partial (A_3\, h_3)}{\partial x_2} - \frac{\partial (A_2\, h_2)}{\partial x_3} \right) dx_2 \wedge dx_3 \\
&\quad + \left(\frac{\partial (A_2\, h_2)}{\partial x_1} - \frac{\partial (A_1\, h_1)}{\partial x_2} \right) dx_1 \wedge dx_2 \\
&\quad + \left(\frac{\partial A_1\, h_1}{\partial x_3} - \frac{\partial (A_3\, h_3)}{\partial x_1} \right) dx_3 \wedge dx_1 \equiv \omega_{\nabla \times \boldsymbol{A}}.
\end{aligned} \tag{12.78}
$$

Comparison with equation (12.52) shows that the curl of $\boldsymbol{A}$ is

$$
\begin{aligned}
\boldsymbol{\nabla} \times \boldsymbol{A} &= \frac{1}{h_2\, h_3} \left(\frac{\partial A_3\, h_3}{\partial x_2} - \frac{\partial A_2\, h_2}{\partial_3} \right) dx_2 \wedge dx_3\, \hat{\boldsymbol{e}}_1 + \cdots \\
&= \frac{1}{h_1\, h_2\, h_3} \begin{vmatrix} h_1 \hat{\boldsymbol{e}}_1 & h_2 \hat{\boldsymbol{e}}_2 & h_3 \hat{\boldsymbol{e}}_3 \\ \partial_1 & \partial_2 & \partial_3 \\ A_1 h_1 & A_2 h_2 & A_3 h_3 \end{vmatrix} \\
&= \frac{1}{h_1 h_2 h_3} \sum_{i,j,k=1}^{3} \epsilon_{ijk} h_i\, \hat{\boldsymbol{e}}_i\, \frac{\partial (A_k h_k)}{\partial x_j}
\end{aligned} \tag{12.79}
$$

as we saw in (11.240). This formula gives our earlier expressions for the curl in cylindrical and spherical coordinates (11.241 & 11.242). □

Example 12.13 (The divergence) We have seen in equations (12.48, 12.49, & 12.52) that the 2-form $\omega_A(U, V) = A \cdot (U \times V)$ of the vector field $\boldsymbol{A} = A_1 \hat{\boldsymbol{e}}_1 + A_2 \hat{\boldsymbol{e}}_2 + A_3 \hat{\boldsymbol{e}}_3$ is

$$
\omega_A^2 = A_1\, h_2\, h_3\; dx_2 \wedge dx_3 + A_2\, h_3\, h_1\; dx_3 \wedge dx_1 + A_3\, h_1\, h_2\; dx_1 \wedge dx_2. \tag{12.80}
$$

The exterior derivative of this 2-form is

$$
\begin{aligned}
d\,\omega_A &= \sum_{k=1}^{3} \frac{\partial}{\partial x_k}\, \omega_A \\
&= \frac{\partial A_1\, h_2 h_3}{\partial x_1}\, dx_1 \wedge dx_2 \wedge dx_3 + \frac{\partial A_2\, h_3\, h_1}{\partial x_2}\, dx_2 \wedge dx_3 \wedge dx_1 \\
&\quad + \frac{\partial A_3\, h_1\, h_2}{\partial x_3}\, dx_3 \wedge dx_1 \wedge dx_2 \\
&= \left(\sum_{k=1}^{3} \frac{\partial (A_k\, h_1\, h_2\, h_3 / h_k)}{\partial x_k} \right) dx_1 \wedge dx_2 \wedge dx_3.
\end{aligned} \tag{12.81}
$$

If one defines the divergence $\boldsymbol{\nabla} \cdot \boldsymbol{A}$ as

$$
d\omega_A = (\boldsymbol{\nabla} \cdot \boldsymbol{A})\; h_1\, h_2\, h_3\, dx_1 \wedge dx_2 \wedge dx_3 \tag{12.82}
$$

then $\nabla \cdot A$ must be

$$\nabla \cdot A = \frac{1}{h_1 h_2 h_3} \left(\sum_{k=1}^{3} \frac{\partial (A_k h_1 h_2 h_3 / h_k)}{\partial x_k} \right) \tag{12.83}$$

in agreement with (11.283) from which the specific formulas for cylindrical (11.284) and spherical (11.286) coordinates follow. □

Example 12.14 (The divergence of a gradient) By combining our expression (12.83) for the divergence with our formula (12.55) for the gradient of a function f, we find that its Laplacian Δf in orthogonal coordinates is

$$\Delta f(\boldsymbol{x}) \equiv \nabla \cdot \nabla f(\boldsymbol{x}) = \frac{1}{h_1 h_2 h_3} \left[\sum_{k=1}^{3} \frac{\partial}{\partial x_k} \left(\frac{h_1 h_2 h_3}{h_k^2} \frac{\partial f(\boldsymbol{x})}{\partial x_k} \right) \right], \tag{12.84}$$

which agrees with (11.295) and so yields (11.296) for cylindrical coordinates and (11.297) for spherical ones. □

12.4 Integration of forms

Let's follow the Russian approach at first. Let $\gamma(t)$ be a smooth map from the unit interval [0, 1] into some manifold $M \subset \mathbb{R}^n$. We divide this interval into tiny segments $[t_i, t_{i+1}]$ of length $dt = t_{i+1} - t_i$, which γ maps into vectors $d\gamma(dt_i) = \dot{\gamma}(t_i)\, dt$ that are tangent to the manifold at the point $\gamma(t_i)$. The integral of a 1-form ω along the curve γ is then the usual Riemann sum

$$\int_\gamma \omega = \lim_{dt \to 0} \sum_i \omega(\dot{\gamma}(t_i))\, dt. \tag{12.85}$$

If for example, the 1-form is $\omega_A(U) = A \cdot U$, then

$$\int_\gamma \omega = \int_\gamma \omega_A(\dot{\gamma}(t_i))\, dt = \int A \cdot \dot{\gamma}(t_i)\, dt = \int A \cdot d\gamma. \tag{12.86}$$

And if $\omega = a_k(x)\, dx_k$, then since $dt\, dx_k(\dot{\gamma}) = d\gamma_k$, the integral

$$\int_\gamma \omega = \lim_{dt \to 0} \sum_{i=1}^{n} \omega(\dot{\gamma}(t_i)\, dt) = \lim_{dt \to 0} \sum_{i=1}^{n} a_k(x)\, dx_k(\dot{\gamma}(t_i))\, dt = \int a_k(x)\, d\gamma_k \tag{12.87}$$

is a line integral on the manifold

$$\int_\gamma \omega = \int a_k\, dx_k = \int a_k\, d\gamma_k. \tag{12.88}$$

Suppose now that our 1-form ω is **exact**, that is, that

$$\omega = d\alpha = \alpha_{,k}\, dx_k \tag{12.89}$$

where $\alpha(x)$ is a 0-form, that is a function, defined on the manifold. Then by (12.88) the integral of $d\alpha$ is

$$\int_\gamma d\alpha = \int \alpha_{,k}\, dx_k = \alpha(\gamma(1)) - \alpha(\gamma(0)). \tag{12.90}$$

The signed endpoints $\gamma(1)$ and $-\gamma(0)$ are the **boundary** of the curve $\gamma(t)$, which one writes as $\partial\gamma$. In this notation, we have

$$\int_\gamma d\alpha = \int_{\partial\gamma} \alpha. \tag{12.91}$$

Example 12.15 (Green's theorem) Let $\boldsymbol{\xi}$ and $\boldsymbol{\eta}$ be two infinitesimal vectors that form a parallelogram Π in the tangent space. We will compute the line integral of the 1-form $\omega = a_1(x_1, x_2)\, dx_1 + a_2(x_1, x_2)\, dx_2$ around the boundary $\partial\Pi$ this parallelogram. This boundary $\partial\Pi$ is a **chain** of four maps $t \to t\boldsymbol{\xi}$, $t \to \boldsymbol{\xi} + t\boldsymbol{\eta}$, $t \to \boldsymbol{\eta} + t\boldsymbol{\xi}$, and $t \to t\boldsymbol{\eta}$ of the unit interval $0 \le t \le 1$ into the plane defined by the two vectors $\boldsymbol{\xi}$ and $\boldsymbol{\eta}$. We assign multiplicities 1, 1, -1, and -1 to these four maps, that is, the full chain runs from a point that we'll call the origin to the point $\boldsymbol{\xi}$, then from $\boldsymbol{\xi}$ to $\boldsymbol{\xi} + \boldsymbol{\eta}$, and then from $\boldsymbol{\xi} + \boldsymbol{\eta}$ to $\boldsymbol{\eta}$, and then from $\boldsymbol{\eta}$ back to the origin. On the curve $\gamma(t) = t\boldsymbol{\xi}$, the differentials dx_k map $\dot\gamma(t)\, dt$ into $dx_k(\dot\gamma(t)\, dt) = dx_k(\boldsymbol{\xi}\, dt) = \xi_k\, dt$. Similarly, on the curve $\gamma(t) = t\boldsymbol{\eta}$, we have $dx_k(\dot\gamma(t)\, dt) = \eta_k\, dt$. So summing over $k = 1, 2$, we find

$$\int_{\partial\Pi} \omega = \int_0^1 \{[a_k(t\boldsymbol{\xi}) - a_k(t\boldsymbol{\xi} + \boldsymbol{\eta})]\, \xi_k - [a_k(t\boldsymbol{\eta}) - a_k(t\boldsymbol{\eta} + \boldsymbol{\xi})]\, \eta_k\}\, dt. \tag{12.92}$$

Since the tangent vectors $\boldsymbol{\xi}$ and $\boldsymbol{\eta}$ are infinitesimal, the square brackets are

$$a_k(t\boldsymbol{\xi}) - a_k(t\boldsymbol{\xi} + \boldsymbol{\eta}) = -\eta_j \frac{\partial a_k}{\partial x_j},$$
$$a_k(t\boldsymbol{\eta}) - a_k(t\boldsymbol{\eta} + \boldsymbol{\xi}) = -\xi_j \frac{\partial a_k}{\partial x_j} \tag{12.93}$$

and so we have

$$\int_{\partial\Pi} \omega = \int_0^1 \left(-\eta_j \frac{\partial a_k}{\partial x_j} \xi_k + \xi_j \frac{\partial a_k}{\partial x_j} \eta_k \right) dt = \frac{\partial a_k}{\partial x_j} \left(\xi_j\, \eta_k - \xi_k\, \eta_j \right). \tag{12.94}$$

But the exterior derivative of ω is

$$d\,\omega = d\,(a_k\, dx_k) = \frac{\partial a_k}{\partial x_j}\, dx_j \wedge dx_k \tag{12.95}$$

so the last term in (12.94) is just the 2-form $d\,\omega$ applied to the tangent vectors $\boldsymbol{\xi}$ and $\boldsymbol{\eta}$

$$d\,\omega(\boldsymbol{\xi}, \boldsymbol{\eta}) = \frac{\partial a_k}{\partial x_j}\, dx_j \wedge dx_k(\boldsymbol{\xi}, \boldsymbol{\eta}) = \frac{\partial a_k}{\partial x_j} \left(\xi_j \eta_k - \xi_k \eta_j \right). \tag{12.96}$$

And $dx_j \wedge dx_k(\boldsymbol{\xi}, \boldsymbol{\eta}) = \xi_j \eta_k - \xi_k \eta_j$ is the area of the tiny parallelogram Π. So this last expression (12.96) is $d\omega$ integrated over the tiny parallelogram Π formed by the tangent vectors $\boldsymbol{\xi}$ and $\boldsymbol{\eta}$, and we have

$$\int_{\Pi} d\omega = \int_{\partial\Pi} \omega \tag{12.97}$$

for an infinitesimal parallelogram.

It is easy to extend this identity to an arbitrary surface S of finite extent. To do this, we tile the surface S with infinitesimal parallelograms Π_α with boundaries $\partial\Pi_\alpha$. The integral over the finite surface S is then the sum of the integrals over the Π_α that tile S

$$\int_S d\omega = \sum_\alpha \int_{\Pi_\alpha} d\omega, \tag{12.98}$$

which by (12.96) is a sum of integrals over the boundaries $\partial\Pi_\alpha$

$$\int_S d\omega = \sum_\alpha \int_{\Pi_\alpha} d\omega = \sum_\alpha \int_{\partial\Pi_\alpha} \omega. \tag{12.99}$$

In the sum of the integrals over the boundaries $\partial\Pi_\alpha$, the internal boundaries all cancel, leaving us with the integral over the boundary ∂S of the surface S. Thus we have

$$\int_S d\omega = \sum_\alpha \int_{\Pi_\alpha} d\omega = \sum_\alpha \int_{\partial\Pi_\alpha} \omega = \int_{\partial S} \omega \tag{12.100}$$

or more simply

$$\int_S d\omega = \int_{\partial S} \omega, \tag{12.101}$$

which generalizes the identity (12.91) from 0-forms to 1-forms.

In the notation of ordinary vector calculus, this relation is

$$\int_S (\nabla \times \boldsymbol{A}) \cdot d\boldsymbol{S} = \oint_{\partial S} \boldsymbol{A} \cdot d\boldsymbol{x} \tag{12.102}$$

in accord (exercise 12.13) with the curl formulas (12.78–12.79). □

Example 12.16 (How electric motors and generators work) Since by (11.80) the curl of the vector potential $\boldsymbol{A}$ is the magnetic field $\boldsymbol{B}$, this last identity (12.102) implies that the magnetic flux Φ through a surface S is the line integral of the vector potential $\boldsymbol{A}$ around the edge of the surface

$$\Phi = \int_S \boldsymbol{B} \cdot d\boldsymbol{S} = \int_S (\nabla \times \boldsymbol{A}) \cdot d\boldsymbol{S} = \oint_{\partial S} \boldsymbol{A} \cdot d\boldsymbol{x}. \tag{12.103}$$

If we take the time derivative of this relation and remember (11.78) that the time derivative of the vector potential is $\dot{\boldsymbol{A}} = -\boldsymbol{E} - \nabla\phi$, then we find that the rate of change of the magnetic flux through a surface is the negative of the line integral of the electric field along the boundary of the surface

$$\dot{\Phi} = \int_S \dot{\boldsymbol{B}} \cdot \boldsymbol{dS} = \int_S (\nabla \times \dot{\boldsymbol{A}}) \cdot \boldsymbol{dS} = -\oint_{\partial S} \boldsymbol{E} \cdot \boldsymbol{dx} \tag{12.104}$$

or minus the voltage ($-\nabla\phi$ drops out because its curl vanishes). □

Example 12.17 (Stokes's theorem) Suppose $\boldsymbol{\xi}$, $\boldsymbol{\eta}$, and $\boldsymbol{\zeta}$ form a triplet of infinitesimal vectors oriented so as to form a right-handed coordinate system, $\boldsymbol{\xi} \times \boldsymbol{\eta} \cdot \boldsymbol{\zeta} > 0$. These vectors form a tiny parallelepiped Π. We want to integrate the 2-form $\omega = a_{jk} dx_j \wedge dx_k$ over the surface $\partial\Pi$ of this tiny parallelepiped Π. We find

$$\begin{aligned}
\int_{\partial\Pi} \omega = & \int_0^1 dt \int_0^1 ds \left\{ \left[a_{jk}(t\boldsymbol{\xi} + s\boldsymbol{\eta} + \boldsymbol{\zeta}) - a_{jk}(t\boldsymbol{\xi} + s\boldsymbol{\eta} \right] \right\} dx_j \wedge dx_k(\boldsymbol{\xi}, \boldsymbol{\eta}) \\
& + \int_0^1 dt \int_0^1 ds \left\{ \left[a_{jk}(t\boldsymbol{\eta} + s\boldsymbol{\zeta} + \boldsymbol{\xi}) - a_{jk}(t\boldsymbol{\eta} + s\boldsymbol{\zeta} \right] \right\} dx_j \wedge dx_k(\boldsymbol{\eta}, \boldsymbol{\zeta}) \\
& + \int_0^1 dt \int_0^1 ds \left\{ \left[a_{jk}(t\boldsymbol{\zeta} + s\boldsymbol{\xi} + \boldsymbol{\eta}) - a_{jk}(t\boldsymbol{\zeta} + s\boldsymbol{\xi} \right] \right\} dx_j \wedge dx_k(\boldsymbol{\zeta}, \boldsymbol{\xi}) \\
= & \ a_{jk,\ell} \zeta_\ell \left(\xi_j \eta_k - \eta_j \xi_k \right) + a_{jk,\ell} \xi_\ell \left(\eta_j \zeta_k - \zeta_j \eta_k \right) + a_{jk,\ell} \eta_\ell \left(\zeta_j \xi_k - \xi_j \zeta_k \right) \\
= & \ a_{jk,\ell} \, dx_\ell \wedge dx_j \wedge dx_k(\boldsymbol{\xi}, \boldsymbol{\eta}, \boldsymbol{\zeta}) = d\omega(\boldsymbol{\xi}, \boldsymbol{\eta}, \boldsymbol{\zeta}) = \int_\Pi d\omega
\end{aligned} \tag{12.105}$$

for an infinitesimal parallelepiped Π.

It is easy to extend this identity to an arbitrary volume V of finite extent. To do this, we tile the volume V with infinitesimal parallelepipeds Π_α with boundaries $\partial\Pi_\alpha$. The integral over the finite volume V is then the sum of the integrals over the Π_α

$$\int_V d\omega = \sum_\alpha \int_{\Pi_\alpha} d\omega, \tag{12.106}$$

which by (12.105) is a sum of the integrals over the boundaries $\partial\Pi_\alpha$

$$\int_V d\omega = \sum_\alpha \int_{\Pi_\alpha} d\omega = \sum_\alpha \int_{\partial\Pi} \omega. \tag{12.107}$$

In this sum over surface integrals, the internal boundaries all cancel, leaving us with the integral over the boundary ∂V of the volume V, so that

$$\int_V d\omega = \sum_\alpha \int_{\Pi_\alpha} d\omega = \sum_\alpha \int_{\partial\Pi_\alpha} \omega = \int_{\partial V} \omega. \tag{12.108}$$

Thus we have

$$\int_V d\omega = \int_{\partial V} \omega, \tag{12.109}$$

which generalizes the identity (12.91) from 1-forms to 2-forms.

Before leaving this example, it may be instructive to examine the value of the integral of ω on one of the faces of the infinitesimal parallelepiped Π as well as that of the integral of $d\omega$ on the infinitesimal parallelepiped Π. The 2-form ω on the upper $\boldsymbol{\xi}, \boldsymbol{\eta}$ face of the surface $\partial\Pi$ of Π may be seen from equation (12.105) to be

$$\omega(\boldsymbol{\xi}, \boldsymbol{\eta}) = a_{jk}\, dx_j \wedge dx_k(\boldsymbol{\xi}, \boldsymbol{\eta}) = a_{jk}\, \left(\xi_j \eta_k - \eta_j \xi_k\right). \tag{12.110}$$

The wedge $dx_j \wedge dx_k(\boldsymbol{\xi}, \boldsymbol{\eta})$ defines an area vector $\boldsymbol{S} = \boldsymbol{\xi} \times \boldsymbol{\eta}$ with components $S_i = \epsilon_{ijk}\xi_j\eta_k$. In terms of $\boldsymbol{S}$, the 2-form ω is

$$\omega(\boldsymbol{\xi}, \boldsymbol{\eta}) = (a_{23} - a_{32})\, S_1 + (a_{31} - a_{13})\, S_2 + (a_{12} - a_{21})\, S_3, \tag{12.111}$$

which suggests defining the vector field $A_k = \epsilon_{kij} a_{ij}$. In terms of $\boldsymbol{S}$ and $\boldsymbol{A}$, the 2-form ω on $\boldsymbol{\xi}, \boldsymbol{\eta}$ is $\omega(\boldsymbol{\xi}, \boldsymbol{\eta}) = \boldsymbol{A} \cdot \boldsymbol{S}$.

The integral of the 3-form $d\omega$ on the infinitesimal parallelepiped Π is

$$\int_\Pi d\omega = d\omega(\boldsymbol{\xi}, \boldsymbol{\eta}, \boldsymbol{\zeta}) = a_{jk,\ell}\, dx_\ell \wedge dx_j \wedge dx_k(\boldsymbol{\xi}, \boldsymbol{\eta}, \boldsymbol{\zeta}). \tag{12.112}$$

The wedge product is $\epsilon_{\ell jk}$ times the determinant $\det(\boldsymbol{\xi}, \boldsymbol{\eta}, \boldsymbol{\zeta})$ of the 3×3 matrix that has $\boldsymbol{\xi}$ as its first row, $\boldsymbol{\eta}$ as its second, and $\boldsymbol{\zeta}$ as its third row

$$dx_\ell \wedge dx_j \wedge dx_k(\boldsymbol{\xi}, \boldsymbol{\eta}, \boldsymbol{\zeta}) = \epsilon_{\ell jk}\, \det(\boldsymbol{\xi}, \boldsymbol{\eta}, \boldsymbol{\zeta}). \tag{12.113}$$

So $d\omega(\boldsymbol{\xi}, \boldsymbol{\eta}, \boldsymbol{\zeta}) = a_{jk,\ell}\, \epsilon_{\ell jk}\, \det(\boldsymbol{\xi}, \boldsymbol{\eta}, \boldsymbol{\zeta})$ or more explicitly

$$d\omega(\boldsymbol{\xi}, \boldsymbol{\eta}, \boldsymbol{\zeta}) = \left[(a_{23} - a_{32})_{,1} + (a_{31} - a_{13})_{,2} + (a_{12} - a_{21})_{,3}\right] \det(\boldsymbol{\xi}, \boldsymbol{\eta}, \boldsymbol{\zeta}),$$

which we recognize as the divergence of the vector field $A_k = \epsilon_{ijk} a_{ij}$

$$d\omega(\boldsymbol{\xi}, \boldsymbol{\eta}, \boldsymbol{\zeta}) = \nabla \cdot \boldsymbol{A}\, \det(\boldsymbol{\xi}, \boldsymbol{\eta}, \boldsymbol{\zeta}). \tag{12.114}$$

Thus we have rediscovered the vector identity

$$\int_V \nabla \cdot \boldsymbol{A}\, dV = \oint_{\partial V} \boldsymbol{A} \cdot \boldsymbol{dS}. \tag{12.115}$$

Example 12.18 (Gauss's law) The divergence of the electric displacement $\boldsymbol{D}$ is the density ρ_f of free charge, $\nabla \cdot \boldsymbol{D} = \rho_f$, and so this last identity (12.115) gives

$$Q_{fV} = \int_V \rho_f\, dV = \int_V \nabla \cdot \boldsymbol{D}\, dV = \oint_{\partial V} \boldsymbol{D} \cdot \boldsymbol{dS}, \tag{12.116}$$

which is the integral form of Gauss's law. □

One may generalize these examples to what has been called the Newton–Leibniz–Gauss–Green–Ostrogradskii–Stokes–Poincaré theorem

$$\int_{\partial C} \omega = \int_C d\omega, \tag{12.117}$$

in which ω is a k-form and C is any $(k+1)$-chain on a manifold. □

Example 12.19 (Poincaré's invariant action) In example 12.9, we saw that Poincaré's action

$$A = \int_{\partial S} \omega^1 = \oint p \cdot dq = \oint \sum_{i=1}^{n} p_i \wedge dq_i \tag{12.118}$$

does not change with time, $\dot{A} = 0$. In example 12.11, we learned that the element of area $d\omega^1$ and therefore its surface integral

$$I = \int_S d\omega^1 \tag{12.119}$$

does not change with time. The identity (12.117) in the form

$$A = \int_{\partial S} \omega^1 = \int_S d\omega^1 = I \tag{12.120}$$

relates these two examples. □

12.5 Are closed forms exact?

A form ω is said to be **closed** if its exterior derivative vanishes

$$d\omega = 0. \tag{12.121}$$

A form ω is **exact** if it's the exterior derivative of another form ψ

$$\omega = d\psi. \tag{12.122}$$

We have seen in (12.64) that the exterior derivative of any exterior derivative vanishes; in effect, $dd = 0$. Thus the exterior derivative of any exact form ω must be zero

$$d\omega = dd\psi = 0. \tag{12.123}$$

So every exact form is closed.

But are closed forms exact? Poincaré's lemma provides the answer. A form that is defined and closed on a **simply connected** part of a manifold is exact there. More technically, if a form ω is defined and closed on a region U of a manifold M and if U can be mapped by a one-to-one differentiable map onto the interior of the unit ball in $\mathbb{R}^n$, then there is a form ψ such that $d\psi = \omega$ on U. The unit ball in $\mathbb{R}^n$ is the interior of the sphere S^{n-1} defined by $x_1^2 + \cdots + x_n^2 = 1$. You may find a proof of this result in section 4.19 of Schutz's book (Schutz, 1980).

Example 12.20 (Two dimensions) Suppose that the 1-form $\omega = fdx + gdy$ is closed

$$d\omega = f_{,y}\, dy \wedge dx + g_{,x}\, dx \wedge dy = \left(g_{,x} - f_{,y}\right)\, dx \wedge dy = 0 \tag{12.124}$$

on the real plane $\mathbb{R}^2$. Since $\mathbb{R}^2$ is simply connected, Poincaré's lemma tells us that there exists a 0-form h whose exterior derivative is ω

$$\omega = fdx + gdy = dh = h_{,x}dx + h_{,y}dy. \tag{12.125}$$

We may construct such a function $h(x, y)$ as the line integral

$$h(x,y) = \int_0^1 [f(u(t), v(t))\, \dot{u}(t) + g(u(t), v(t))\, \dot{v}(t)]\, dt \tag{12.126}$$

along any differentiable curve $\gamma(t) = (u(t), v(t))$ that goes from $(u(0), v(0)) = (x_0, y_0)$ to $(u(1), v(1)) = (x, y)$. Clearly the exterior derivative of this 0-form is

$$dh = h_{,x} dx + h_{,y} dy = f(x,y)\, dx + g(x,y)\, dy = \omega. \tag{12.127}$$

So the real issue here is whether the line integral (12.126)

$$h = \int_\gamma \omega \tag{12.128}$$

defines a function $h(x, y)$ that is the *same* for any two curves $\gamma_1(t)$ and $\gamma_2(t)$ that both go from (x_0, y_0) to (x, y). The difference $h_1(x, y) - h_2(x, y)$ is an integral of ω along a closed curve $\Gamma = \gamma_1 - \gamma_2$ that runs from (x_0, y_0) to (x, y) along $\gamma_1(t)$ and then from (x, y) to (x_0, y_0) backwards along $\gamma_2(t)$. The closed curve Γ is the boundary ∂S of the (plane) surface S that it encloses. Thus by Stokes's theorem (12.117)

$$h_1 - h_2 = \int_\Gamma \omega = \int_{\partial S} \omega = \int_S d\omega = 0, \tag{12.129}$$

in which we used the fact that ω is closed so that $d\omega = 0$. The two curves $\gamma_1(t)$ and $\gamma_2(t)$ define the same function $h(x, y)$, and $\omega = dh$ is exact.

What if the region in which $d\omega = 0$ is not simply connected? Consider for example the 1-form

$$\omega = -\frac{y}{x^2 + y^2}\, dx + \frac{x}{x^2 + y^2}\, dy, \tag{12.130}$$

which is well defined except at the origin. The plane $\mathbb{R}^2$ minus the origin is not simply connected. One may check that ω is closed

$$d\omega = 0 \tag{12.131}$$

except at the origin. But ω is not exact. In fact, by writing it as

$$\omega = \arctan(y/x)_{,y}\, dy \wedge dx + \arctan(y/x)_{,x}\, dx \wedge dy \tag{12.132}$$

we see that ω is almost exact. It is the exterior derivative of

$$\theta = \arctan(y/x), \tag{12.133}$$

which would be a 0-form if it were single valued. □

Example 12.21 (Some exact forms) It's easy to make lots of exact forms; one just applies the exterior derivative d to any form. For instance, taking the exterior derivative of the 0-form $\omega^0 = x\, y^2 \exp(zw)$, we get the 1-form

$$d\omega^0 = y^2\, e^{zw}\, dx + 2x\, y\, e^{zw}\, dy + x\, y^2\, w\, e^{zw}\, dz + x\, y^2\, z\, e^{zw}\, dw, \tag{12.134}$$

which is exact (and closed). Applying d to the 1-form $\omega^1 = y^2 z dx + x^3 dy$, we get the exact 2-form $d\omega^1 = (3x^2 - 2yz)\, dx \wedge dy - y^2 dx \wedge dz$. Incidentally, any n-form in n variables, such as $f(x, y, z)\, dx \wedge dy \wedge dz$, is closed. □

12.6 Complex differential forms

Any function $f(x, y)$ of two real variables also is a function of the two complex variables $z = x + iy$ and $\bar{z} = x - iy$. For instance, $4xy = -i(z + \bar{z})(z - \bar{z})$ and $x^2 + y^2 = z\bar{z}$. We can write any 1-form $\omega = adx + bdy$ with complex coefficients a and b in terms of the complex differentials $dz = dx + idy$ and $d\bar{z} = dx - idy$ as

$$\omega = \tfrac{1}{2}(a - ib)dz + \tfrac{1}{2}(a + ib)\, d\bar{z}. \tag{12.135}$$

A 1-form of the variables $z_1, \ldots, z_n$ and $\bar{z}_1, \ldots, \bar{z}_n$ is a sum of their differentials $\omega = a_j dz_j + b_j d\bar{z}_j$. The expression

$$\omega^{1,1} = a\, dz_1 \wedge d\bar{z}_1 + b\, dz_1 \wedge d\bar{z}_2 + c\, dz_2 \wedge d\bar{z}_1 + d\, dz_2 \wedge d\bar{z}_2 \tag{12.136}$$

is a 1,1-form in z_1 and z_2, while 2,0- and 0,2-forms look like

$$\omega^{2,0} = e\, dz_1 \wedge dz_2 \quad \text{and} \quad \omega^{0,2} = f\, d\bar{z}_1 \wedge d\bar{z}_2. \tag{12.137}$$

The complex differentials anticommute: $d\bar{z}_j \wedge dz_k = -dz_k \wedge d\bar{z}_j$ as well as $dz_j \wedge dz_k = -dz_k \wedge dz_j$ and $d\bar{z}_j \wedge d\bar{z}_k = -d\bar{z}_k \wedge d\bar{z}_j$.

There are two exterior derivatives ∂ and $\bar{\partial}$ defined by

$$\partial = \sum_{j=1}^{n} \frac{\partial}{\partial z_j}\, dz_j \wedge \quad \text{and} \quad \bar{\partial} = \sum_{j=1}^{n} \frac{\partial}{\partial \bar{z}_j}\, d\bar{z}_j \wedge . \tag{12.138}$$

Their sum is the ordinary exterior derivative $\partial + \bar{\partial} = d$, and one has

$$\partial^2 = \bar{\partial}^2 = \partial\bar{\partial} + \bar{\partial}\partial = 0. \tag{12.139}$$

Example 12.22 ($\partial + \bar{\partial} = d$) We illustrate the rule $\partial + \bar{\partial} = d$ for the 1-form $\omega = z\bar{z}dz = (x^2 + y^2)(dx + idy)$. The sum $\partial + \bar{\partial}$ acting on ω gives

$$\begin{aligned}\left(\partial + \bar{\partial}\right) z\bar{z}\, dz &= \bar{\partial} z\bar{z}\, d\bar{z} \wedge dz = z\, d\bar{z} \wedge dz \\ &= (x + iy)(dx - idy) \wedge (dx + idy) = 2i(x + iy)dx \wedge dy\end{aligned} \tag{12.140}$$

while $d\omega$ is

$$d(x^2 + y^2)(dx + idy) = 2xidx \wedge dy + 2ydy \wedge dx = 2i(x + iy)dx \wedge dy, \tag{12.141}$$

which is the same as $\left(\partial + \bar{\partial}\right) \omega$. □

12.7 Frobenius's theorem

This section, optional on a first reading, begins with some definitions.

- If ω is a k-form that maps all vectors V of the tangent space T_P at the point P into numbers $\omega(V_1, \ldots, V_k)$, and $S \subset T_P$ is a subspace of the tangent space T_P, then the **restriction** ω_S of the form ω to S maps the vectors $S_i \in S$ into the numbers $\omega(S_1, \ldots, S_k)$.
- The **annihilator** (actually the annihilated) of a set of forms β_i at a point P of a manifold is the subspace of vectors X_P that every β_i maps to zero. (The vectors X_P are in the tangent space T_P.)
- The **complete ideal** of a set B of forms β_i is the set of all forms at P whose restriction to B's annihilator X_P vanishes. Note that for *any* form α and any vectors X_ℓ in the annihilator X_P, the wedge product $\alpha \wedge \beta_i(X_1, \ldots, X_n)$ vanishes, so $\alpha \wedge \beta_i$ is in the complete ideal of the set B of forms β_i.
- The complete ideal of a set B of forms β_i has a set α_i of linearly independent 1-forms that **generates** it, that is, whose complete ideal is the same as that of the set B.
- The complete ideal of a set B of fields β_i is the set of fields that map the annihilator X_P of B to zero at each point P of the manifold.
- If the exterior derivative $d\alpha_i$ is in an ideal whenever α_i is, then the ideal is a **differential ideal**.
- A set A of 1-forms α_i has a **closed ideal** if every $d\alpha_i$ is in the complete ideal generated by the α_is. (Some authors call such a set A of 1-forms closed, but this terminology can be confusing.)

Example 12.23 (A rank-2 annihilator) Consider a manifold with coordinates $x_1, \ldots, x_n$, tangent vectors $\partial_1, \ldots, \partial_n$, and 1-forms $dx_1, \ldots, dx_n$ with $dx_i(\partial_k) = \delta_{ik}$ as in (12.31). The subspace $\{c_1\partial_1 + c_2\partial_2 \,|\, c_1, c_2 \text{ real}\}$ is the annihilator of the set $\{dx_3, \ldots, dx_n\}$ of 1-forms. Any linear combination of these 1-forms

$$\omega = \sum_{k=3}^{n} a_k(x)\, dx_k \tag{12.142}$$

is in the complete ideal of the 1-forms $dx_3, \ldots, dx_n$. So is any 2-form

$$\omega^2 = \sum_{i,k=3}^{n} a_{ik}(x)\, dx_i \wedge dx_k. \tag{12.143}$$

For any forms α_k the linear combination

$$\omega = \sum_{k=3}^{n} \alpha_k \wedge dx_k \tag{12.144}$$

is in the complete ideal of the set of 1-forms $dx_1, \ldots, dx_n$. In fact, each member of this complete ideal is such a linear combination. □

Frobenius's theorem Let $\omega_1, \ldots, \omega_n$ be a linearly independent set of 1-form fields in an open region U of a k-dimensional manifold M. Then there exist functions $P_{\ell j}$ and Q_j for $i, j = 1, \ldots, n$ that express the n 1-forms ω_ℓ as

$$\omega_\ell = \sum_{j=1}^{n} P_{\ell j} dQ_j \tag{12.145}$$

if and only if every $d\omega_\ell$ is in the complete ideal generated by the ω_ℓ's (Schutz, 1980, secs. 3.8 & 4.26).

Further reading

Three good discussions of differential forms are *Mathematical Methods of Classical Mechanics* (Arnold, 1989), *Geometrical Methods of Mathematical Physics* (Schutz, 1980), and *Classical Mechanics* (Matzner and Shepley, 1991). They inspired this chapter.

Exercises

12.1 Why do you think mathematicians use the definition (12.18) rather than our (12.19)?

12.2 Show explicitly that the 2-form $\omega = Xdx_2 \wedge dx_3$ is given by (12.43) in terms of the 1-forms dy_k.

12.3 Show explicitly that for the two 3-vectors of example 12.4, the 2-form $\omega(A, B) = Xdx_2 \wedge dx_3(A, B) = -X$.

12.4 In example 12.5, let $x_1 = y_1 + y_2$, $x_2 = y_1 - y_2$, and $x_3 = y_1 - y_3$. Show explicitly that the y-version (12.43) of the 2-form $\omega(A, B) = Xdx_2 \wedge dx_3$ maps the two 3-vectors of example 12.4 into the same real number $-X$ as its x-version. Hint: in the y-version, first express $dy_k(A)$ and $dy_k(B)$ in terms of $dx_j(A)$ and $dx_j(B)$.

12.5 Compute $dr(\hat{e}_r)$, $d\theta(\hat{e}_\theta)$, and $d\phi(\hat{e}_\phi)$ as well as the six off-diagonal ones $dr(\hat{e}_\theta)$, $dr(\hat{e}_\phi)$, and so forth.

12.6 Show that the 2-form (12.52) applied to the vectors $\boldsymbol{U}$ and $\boldsymbol{V}$ gives the triple scalar product (12.48).

12.7 Show that $d\,d\,\omega^k = 0$ for the general k-form (12.62).

12.8 Show that if α and β are both 2-forms, then $d(\alpha \wedge \beta) = (d\alpha) \wedge \beta + \alpha \wedge d\beta$.

12.9 Use Hamilton's equations (12.72) to derive the formula (12.75) for the time derivatives $\delta\dot{p}_i$ and $\delta\dot{q}_i$.

12.10 Use Hamilton's equations (12.72) to compute the time derivatives of the n pairs of tiny displacements $\Delta p_j, \Delta q_j$. Then use your resulting formulas and those (12.75) for the time derivatives of the n pairs of small differences $\delta p_j, \delta q_j$ to show that the time derivative of the sum (12.76) of areas of tiny parallelograms vanishes (12.77).

12.11 In the early days of quantum mechanics, Bohr and Sommerfeld set action integrals like those of example 12.9 equal to a multiple of Planck's constant, $A = \oint p \cdot dq = nh$. Why do you think they chose invariant quantities to quantize in this way?

12.12 Use Bohr's quantization of angular momentum $L = rp = n\hbar$ to find the energy levels of an electron in a circular orbit about a proton. Take the energy as $E = p^2/2m - Ze^2/4\pi\epsilon_0 r$ and balance radial forces $mv^2/r = Ze^2/4\pi\epsilon_0 r^2$ where $p = mv$.

12.13 Work out the details of using the curl formulas (12.78–12.79) to derive the curl formula (12.102) from the general identity (12.101). Do this in rectangular, cylindrical, and spherical coordinates.

12.14 Is the 1-form $y^2\, e^{zw}\, dx + 2x\, y\, e^{zw}\, dy + x\, y^2\, w\, e^{zw}\, dz + x\, y^2\, z\, e^{zw}\, dw$ closed? Why? Why not?

12.15 Is the 1-form $ze^y/xw + ze^y \ln x/w + e^y \ln x/w - ze^y \ln x/w^2$ closed? Why? Why not?

12.16 Show that ∂ and $\bar{\partial}$ satisfy (12.139).

13

Probability and statistics

13.1 Probability and Thomas Bayes

The probability $P(A)$ of an outcome in a set A is the sum of the probabilities P_j of all the different (mutually exclusive) outcomes j in A

$$P(A) = \sum_{j \in A} P_j. \tag{13.1}$$

For instance, if one throws two fair dice, then the probability that the sum is 2 is $P(1, 1) = 1/36$, while the probability that the sum is 3 is $P(1, 2)+P(2, 1) = 1/18$.

If A and B are two sets of possible outcomes, then the probability of an outcome in the **union** $A \cup B$ is the sum of the probabilities $P(A)$ and $P(B)$ minus that of their **intersection** $A \cap B$

$$P(A \cup B) = P(A) + P(B) - P(A \cap B). \tag{13.2}$$

If the outcomes are mutually exclusive, then $P(A \cap B) = 0$, and the probability of the union is the sum $P(A \cup B) = P(A) + P(B)$. The **joint probability** $P(A, B) \equiv P(A \cap B)$ is the probability of an outcome that is in both sets A and B. If the joint probability is the product $P(A, B) = P(A)\,P(B)$, then the outcomes in sets A and B are **statistically independent**.

The probability that a result in set B also is in set A is the **conditional probability** $P(A|B)$, the probability of A given B

$$P(A|B) = \frac{P(A \cap B)}{P(B)}. \tag{13.3}$$

Also $P(B|A) = P(A \cap B)/P(A)$. The substitution $B \to B \cap C$ in (13.3) gives $P(A|B,C) = P(A \cap B \cap C)/P(B \cap C)$. If we multiply (13.3) by $P(B)$, we get

$$P(A,B) = P(A \cap B) = P(B|A)\,P(A) = P(A|B)\,P(B). \tag{13.4}$$

Combination of (13.3 & 13.4) gives **Bayes's theorem** (Riley *et al.*, 2006, p. 1132)

$$P(A|B) = \frac{P(B|A)\,P(A)}{P(B)} \tag{13.5}$$

(Thomas Bayes, 1702–1761).

If there are N mutually exclusive **theories**, **causes**, or **ways** A_j that B can happen, then we must sum over them

$$P(B) = \sum_{j=1}^{N} P(B|A_j)\,P(A_j). \tag{13.6}$$

The probabilities $P(A_j)$ are called ***a priori*** probabilities. In this case, Bayes's theorem is (Roe, 2001, p. 119)

$$P(A_k|B) = \frac{P(B|A_k)\,P(A_k)}{\sum_{j=1}^{N} P(B|A_j)\,P(A_j)}. \tag{13.7}$$

If there are several Bs, then a third form of Bayes's theorem is

$$P(A_k|B_\ell) = \frac{P(B_\ell|A_k)\,P(A_k)}{\sum_{j=1}^{N} P(B_\ell|A_j)\,P(A_j)}. \tag{13.8}$$

Example 13.1 (The low-base-rate problem) Suppose the incidence of a rare disease in a population is $P(D) = 0.001$. Suppose a test for the disease has a **sensitivity** of 99%, that is, the probability that a carrier will test positive is $P(+|D) = 0.99$. Suppose the test also is highly **selective** with a false-positive rate of only $P(+|N) = 0.005$. Then the probability that a random person in the population would test positive is by (13.6)

$$P(+) = P(+|D)\,P(D) + P(+|N)\,P(N) = 0.005993. \tag{13.9}$$

And by Bayes's theorem (13.5), the probability that a person who tests positive actually has the disease is only

$$P(D|+) = \frac{P(+|D)\,P(D)}{P(+)} = \frac{0.99 \times 0.001}{0.005993} = 0.165 \tag{13.10}$$

and the probability that a person testing positive actually is healthy is $P(N|+) = 1 - P(D|+) = 0.835$.

Even with an excellent test, screening for rare diseases is problematic. Similarly, screening for rare behaviors, such as drug use in the CIA or disloyalty in the army, is difficult with a good test and absurd with a poor one like a polygraph. □

Example 13.2 (The three-door problem) A prize lies behind one of three closed doors. A contestant gets to pick which door to open, but before the chosen door is opened, a door that does not lead to the prize and was not picked by the contestant swings open. Should the contestant switch and choose a different door?

We note that a contestant who picks the wrong door and switches always wins, so $P(W|Sw, WD) = 1$, while one who picks the right door and switches never does $P(W|Sw, RD) = 0$. Since the probability of picking the wrong door is $P(WD) = 2/3$, the probability of winning if one switches is

$$P(W|Sw) = P(W|Sw, WD)\, P(WD) + P(W|Sw, RD)\, P(RD) = 2/3. \tag{13.11}$$

The probability of picking the right door is $P(RD) = 1/3$, and the probability of winning if one picks the right door and stays put is $P(W|Sp, RD) = 1$. So the probability of winning if one stays put is

$$P(W|Sp) = P(W|Sp, RD)\, P(RD) + P(W|Sp, WD)\, P(WD) = 1/3. \tag{13.12}$$

Thus, one should switch after the door opens. □

If the set A is the interval $(x - dx/2, x + dx/2)$ of the real line, then $P(A) = P(x)\, dx$, and the second version of Bayes's theorem (13.7) says

$$P(x|B) = \frac{P(B|x)\, P(x)}{\int_{-\infty}^{\infty} P(B|x')\, P(x')\, dx'}. \tag{13.13}$$

Example 13.3 (A tiny poll) We ask four people if they will vote for Nancy Pelosi, and three say *yes*. If the probability that a random voter will vote for her is y, then the probability that three in our sample of four will is

$$P(3|y) = 4\, y^3\, (1 - y). \tag{13.14}$$

We don't know the **prior** probability distribution $P(y)$, so we set it equal to unity on the interval $(0, 1)$. Then the continuous form of Bayes's theorem (13.13) and our cheap poll give the probability distribution of the fraction y who will vote for her as

$$\begin{aligned} P(y|3) &= \frac{P(3|y)\, P(y)}{\int_0^1 P(3|y')\, P(y')\, dy'} = \frac{P(3|y)}{\int_0^1 P(3|y')\, dy'} \\ &= \frac{4\, y^3\, (1 - y)}{\int_0^1 4\, y'^3\, (1 - y')\, dy'} = 20\, y^3\, (1 - y). \end{aligned} \tag{13.15}$$

Our best guess then for the probability that she will win the election is

$$\int_{1/2}^{1} P(y|3)\, dy = \int_{1/2}^{1} 20\, y^3\, (1-y)\, dy = \frac{13}{16}, \tag{13.16}$$

which is slightly higher that the naive estimate of 3/4. □

13.2 Mean and variance

In roulette and many other games, N outcomes x_j can occur with probabilities P_j that sum to unity

$$\sum_{j=1}^{N} P_j = 1. \tag{13.17}$$

The **expected value** $E[x]$ of the outcome x is its **mean** μ or **average** value $\langle x \rangle = \overline{x}$

$$E[x] = \mu = \langle x \rangle = \overline{x} = \sum_{j=1}^{N} x_j\, P_j. \tag{13.18}$$

The **expected value** $E[x]$ also is called the **expectation** of x or **expectation value** of x.

The **ℓth moment** is

$$E[x^\ell] = \mu_\ell = \langle x^\ell \rangle = \sum_{j=1}^{N} x_j^\ell P_j \tag{13.19}$$

and the **ℓth central moment** is

$$E[(x-\mu)^\ell] = \nu_\ell = \sum_{j=1}^{N} (x_j - \mu)^\ell P_j \tag{13.20}$$

where always $\mu_0 = \nu_0 = 1$ and $\nu_1 = 0$ (exercise 13.2).

The **variance** $V[x]$ is the second central moment

$$V[x] \equiv E[(x - \langle x \rangle)^2] = \sum_{j=1}^{N} \left(x_j - \langle x \rangle\right)^2 P_j, \tag{13.21}$$

which one may write as (exercise 13.4)

$$V[x] = \langle x^2 \rangle - \langle x \rangle^2 \tag{13.22}$$

and the standard deviation σ is its square-root

$$\sigma = \sqrt{V[x]}. \tag{13.23}$$

If the values of x are distributed continuously according to a **probability distribution** or **density** $P(x)$ normalized to unity

$$\int P(x)\,dx = 1 \tag{13.24}$$

then the mean value is

$$E[x] = \mu = \langle x \rangle = \int x\,P(x)\,dx \tag{13.25}$$

and the ℓth moment is

$$E[x^\ell] = \mu_\ell = \langle x^\ell \rangle = \int x^\ell\,P(x)\,dx. \tag{13.26}$$

The ℓth central moment is

$$E[(x-\mu)^\ell] = \nu_\ell = \int (x-\mu)^\ell\,P(x)\,dx. \tag{13.27}$$

The variance of the distribution is the second central moment

$$V[x] = \nu_2 = \int (x - \langle x \rangle)^2\,P(x)\,dx = \mu_2 - \mu^2 \tag{13.28}$$

and the standard deviation σ is its square-root $\sigma = \sqrt{V[x]}$.

Many authors use $f(x)$ for the probability distribution $P(x)$ and $F(x)$ for the cumulative probability $\Pr(-\infty, x)$ of an outcome in the interval $(-\infty, x)$

$$F(x) \equiv \Pr(-\infty, x) = \int_{-\infty}^{x} P(x')\,dx' = \int_{-\infty}^{x} f(x')\,dx', \tag{13.29}$$

a function that is necessarily **monotonic**

$$F'(x) = \Pr{}'(-\infty, x) = f(x) = P(x) \geq 0. \tag{13.30}$$

Some mathematicians reserve the term probability **distribution** for probabilities like $\Pr(-\infty, x)$ and P_j and call a continuous distribution $P(x)$ a **probability density function**. But usage of the Maxwell–Boltzmann distribution is too widespread in physics for me to observe this distinction.

Although a probability distribution $P(x)$ is normalized (13.24), it can have **fat tails**, which are important in financial applications (Bouchaud and Potters, 2003). Fat tails can make the variance and even the **mean absolute deviation**

$$E_{\text{abs}} \equiv \int |x - \mu|\,P(x)\,dx \tag{13.31}$$

diverge.

Example 13.4 (Heisenberg's uncertainty principle) In quantum mechanics, the absolute-value squared $|\psi(x)|^2$ of a wave-function $\psi(x)$ is the probability distribution $P(x) = |\psi(x)|^2$ of the position x of the particle, and $P(x)\,dx$ is the probability that the particle is found between $x - dx/2$ and $x + dx/2$. The variance

$\langle (x - \langle x \rangle)^2 \rangle$ of the position operator x is written as the square $(\Delta x)^2$ of the standard deviation $\sigma = \Delta x$, which is the **uncertainty** in the position of the particle. Similarly, the square of the uncertainty in the momentum $(\Delta p)^2$ is the variance $\langle (p - \langle p \rangle)^2 \rangle$ of the momentum.

For the wave-function (3.70)

$$\psi(x) = \left(\frac{2}{\pi}\right)^{1/4} \frac{1}{\sqrt{a}} \, e^{-(x/a)^2} \tag{13.32}$$

these uncertainties are $\Delta x = a/2$ and $\Delta p = \hbar/a$. They provide a (saturated) example $\Delta x \, \Delta p = \hbar/2$ of Heisenberg's uncertainty principle

$$\Delta x \, \Delta p \geq \frac{\hbar}{2}. \tag{13.33}$$

□

If x and y are two random variables that occur with a **joint distribution** $P(x, y)$, then the expected value of the linear combination $ax^n y^m + bx^p y^q$ is

$$\begin{aligned} E[ax^n y^m + bx^p y^q] &= \int (ax^n y^m + bx^p y^q) \, P(x, y) \, dxdy \\ &= a \int x^n y^m \, P(x, y) \, dxdy + b \int x^p y^q \, P(x, y) \, dxdy \\ &= a \, E[x^n y^m] + b \, E[x^p y^q]. \end{aligned} \tag{13.34}$$

This result and its analog for discrete probability distributions show that **expected values are linear**.

The **correlation coefficient** or **covariance** of two variables x and y that occur with a **joint distribution** $P(x, y)$ is

$$C[x, y] \equiv \int P(x, y)(x - \bar{x})(y - \bar{y}) \, dxdy = \langle (x - \bar{x})(y - \bar{y}) \rangle = \langle x \, y \rangle - \langle x \rangle \langle y \rangle. \tag{13.35}$$

The variables x and y are said to be **independent** if

$$P(x, y) = P(x) \, P(y). \tag{13.36}$$

Independence implies that the covariance vanishes, but $C[x, y] = 0$ does not guarantee that x and y are independent (Roe, 2001, p. 9).

The variance of $x + y$

$$\langle (x + y)^2 \rangle - \langle x + y \rangle^2 = \langle x^2 \rangle - \langle x \rangle^2 + \langle y^2 \rangle - \langle y \rangle^2 + 2 \left(\langle x \, y \rangle - \langle x \rangle \langle y \rangle \right) \tag{13.37}$$

is the sum

$$V[x + y] = V[x] + V[y] + 2 \, C[x, y]. \tag{13.38}$$

It follows (exercise 13.6) that for any constants a and b the variance of $ax + by$ is

$$V[ax + by] = a^2 \, V[x] + b^2 \, V[y] + 2 \, ab \, C[x, y]. \tag{13.39}$$

More generally (exercise 13.7), the variance of the sum $a_1 x_1 + a_2 x_2 + \cdots + a_N x_N$ is

$$V[a_1 x_1 + \cdots + a_N x_N] = \sum_{j=1}^{N} a_j^2 V[x_j] + \sum_{j,k=1, j<k}^{N} 2 a_j a_k C[x_j, x_k]. \tag{13.40}$$

If the variables x_j and x_k are independent for $j \neq k$, then their covariances vanish $C[x_j, x_k] = 0$, and the variance of the sum $a_1 x_1 + \cdots + a_N x_N$ is

$$V[a_1 x_1 + \cdots + a_N x_N] = \sum_{j=1}^{N} a_j^2 \, V[x_j]. \tag{13.41}$$

13.3 The binomial distribution

If the probability of success is p on each try, then we expect that in N tries the mean number of successes will be

$$\langle n \rangle = N p. \tag{13.42}$$

The probability of failure on each try is $q = 1 - p$. So the probability of a particular sequence of successes and failures, such as n successes followed by $N - n$ failures is $p^n \, q^{N-n}$. There are $N!/n!\,(N-n)!$ different sequences of n successes and $N - n$ failures, all with the same probability $p^n \, q^{N-n}$. So the probability of n successes (and $N - n$ failures) in N tries is

$$P_{\mathrm{B}}(n, p, N) = \frac{N!}{n!\,(N-n)!} p^n \, q^{N-n} = \binom{N}{n} p^n \,(1-p)^{N-n}. \tag{13.43}$$

This **binomial distribution** also is called **Bernoulli's distribution** (Jacob Bernoulli, 1654–1705).

The sum of the probabilities $P_{\mathrm{B}}(n, p, N)$ for all possible values of n is unity

$$\sum_{n=0}^{N} P_{\mathrm{B}}(n, p, N) = (p + 1 - p)^N = 1. \tag{13.44}$$

In Fig. 13.1, the probabilities $P_{\mathrm{B}}(n, p, N)$ for $0 \le n \le 250$ and $p = 0.2$ are plotted for $N = 125, 250, 500$, and 1000 tries.

The mean number of successes

$$\mu = \langle n \rangle_{\mathrm{B}} = \sum_{n=0}^{N} n \, P_{\mathrm{B}}(n, p, N) = \sum_{n=0}^{N} n \binom{N}{n} p^n q^{N-n} \tag{13.45}$$

is a partial derivative with respect to p with q held fixed

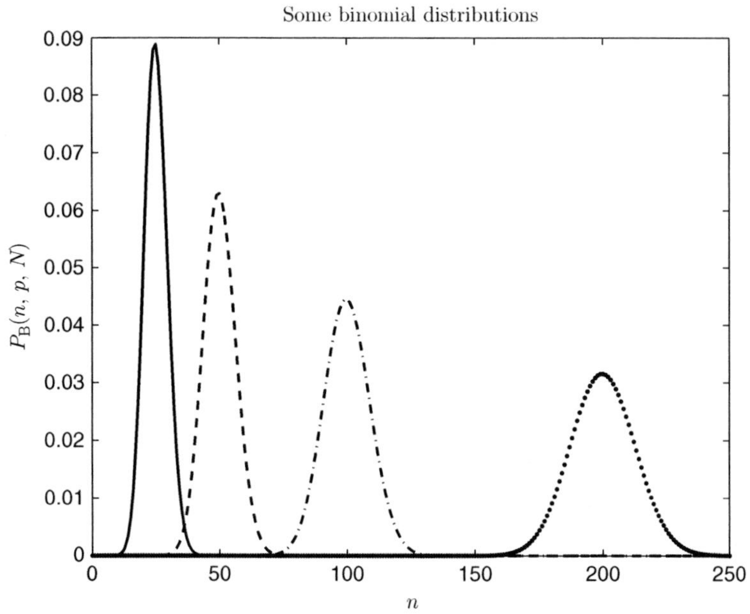

Figure 13.1 If the probability of success on any try is p, then the probability $P_B(n,p,N)$ of n successes in N tries is given by equation (13.43). For $p = 0.2$, this binomial probability distribution $P_B(n,p,N)$ is plotted against n for $N = 125$ (solid), 250 (dashes), 500 (dot dash), and 1000 tries (dots).

$$\begin{aligned} \langle n \rangle_B &= p \frac{\partial}{\partial p} \sum_{n=0}^{N} \binom{N}{n} p^n q^{N-n} \\ &= p \frac{\partial}{\partial p} (p+q)^N = Np\,(p+q)^N = Np, \end{aligned} \tag{13.46}$$

which verifies the estimate (13.42).

One may show (exercise 13.9) that the variance (13.21) of the binomial distribution is

$$V_B = \langle (n - \langle n \rangle)^2 \rangle = p\,(1-p)\,N. \tag{13.47}$$

Its standard deviation (13.23) is

$$\sigma_B = \sqrt{V_B} = \sqrt{p\,(1-p)\,N}. \tag{13.48}$$

The ratio of the width to the mean

$$\frac{\sigma_B}{\langle n \rangle_B} = \frac{\sqrt{p\,(1-p)\,N}}{Np} = \sqrt{\frac{1-p}{Np}} \tag{13.49}$$

decreases with N as $1/\sqrt{N}$.

Example 13.5 (Avogadro's number) A mole of gas is Avogadro's number $N_A = 6 \times 10^{23}$ of molecules. If the gas is in a cubical box, then the chance that each molecule will be in the left half of the cube is $p = 1/2$. The mean number of molecules there is $\langle n \rangle_B = pN_A = 3 \times 10^{23}$, and the uncertainty in n is $\sigma_B = \sqrt{p(1-p)N} = \sqrt{3 \times 10^{23}/4} = 3 \times 10^{11}$. So the numbers of gas molecules in the two halves of the box are equal to within $\sigma_B/\langle n \rangle_B = 10^{-12}$ or to 1 part in 10^{12}. □

Because $N!$ increases very rapidly with N, the rule

$$P_B(n+1, p, N) = \frac{p}{1-p} \frac{N-n}{n+1} P_B(n, p, N) \tag{13.50}$$

is helpful when N is big. But when N exceeds a few hundred, the formula (13.43) for $P_B(n, p, N)$ becomes unmanageable even in quadruple precision. One way of computing $P_B(n, p, N)$ for large N is to use Srinivasa Ramanujan's correction to Stirling's formula $N! \approx \sqrt{2\pi N}(N/e)^N$

$$N! \approx \sqrt{2\pi N} \left(\frac{N}{e}\right)^N \left(1 + \frac{1}{2N} + \frac{1}{8N^2}\right)^{1/6}. \tag{13.51}$$

When N and $N - n$, but not n, are big, one may use (13.51) for $N!$ and $(N-n)!$ in the formula (13.43) for $P_B(n, p, N)$ and so may show (exercise 13.11) that

$$P_B(n, p, N) \approx \frac{(pN)^n}{n!} q^{N-n} R_2(n, N), \tag{13.52}$$

in which

$$\begin{aligned} R_2(n, N) = &\left(1 - \frac{n}{N}\right)^{n-1/2} \left(1 + \frac{1}{2N} + \frac{1}{8N^2}\right)^{1/6} \\ &\times \left[1 + \frac{1}{2(N-n)} + \frac{1}{8(N-n)^2}\right]^{-1/6} \end{aligned} \tag{13.53}$$

tends to unity as $N \to \infty$ for any fixed n.

When all three factorials in $P_B(n, p, N)$ are huge, one may use Ramanujan's approximation (13.51) to show (exercise 13.12) that

$$P_B(n, p, N) \approx \sqrt{\frac{N}{2\pi n(N-n)}} \left(\frac{pN}{n}\right)^n \left(\frac{qN}{N-n}\right)^{N-n} R_3(n, N) \tag{13.54}$$

where

$$R_3(n,N) = \left(1+\frac{1}{2n}+\frac{1}{8n^2}\right)^{-1/6}\left(1+\frac{1}{2N}+\frac{1}{8N^2}\right)^{1/6}$$
$$\times\left[1+\frac{1}{2(N-n)}+\frac{1}{8(N-n)^2}\right]^{-1/6} \tag{13.55}$$

tends to unity as $N \to \infty$, $N - n \to \infty$, and $n \to \infty$.

Another way of coping with the unwieldy factorials in the binomial formula $P_B(n, p, N)$ is to use limiting forms of (13.43) due to Poisson and to Gauss.

13.4 The Poisson distribution

Poisson took the two limits $N \to \infty$ and $p = \langle n \rangle / N \to 0$. So we let N and $N - n$, but not n, tend to infinity, and use (13.52) for the binomial distribution (13.43). Since $R_2(n, N) \to 1$ as $N \to \infty$, we get

$$P_B(n,p,N) \approx \frac{(pN)^n}{n!}\, q^{N-n} = \frac{\langle n \rangle^n}{n!}\, q^{N-n}. \tag{13.56}$$

Now $q = 1 - p = 1 - \langle n \rangle / N$, and so for any fixed n we have

$$\lim_{N\to\infty} q^{N-n} = \lim_{N\to\infty}\left(1-\frac{\langle n \rangle}{N}\right)^N\left(1-\frac{\langle n \rangle}{N}\right)^{-n} = e^{-\langle n \rangle}. \tag{13.57}$$

Thus as $N \to \infty$ with pN fixed at $\langle n \rangle$, the binomial distribution becomes the Poisson distribution

$$P_P(n, \langle n \rangle) = \frac{\langle n \rangle^n}{n!}\, e^{-\langle n \rangle}. \tag{13.58}$$

(Siméon-Denis Poisson, 1781–1840. Incidentally, *poisson* means *fish* and sounds like pwahsahn.)

The Poisson distribution is normalized to unity

$$\sum_{n=0}^{\infty} P_P(n, \langle n \rangle) = \sum_{n=0}^{\infty} \frac{\langle n \rangle^n}{n!}\, e^{-\langle n \rangle} = e^{\langle n \rangle}\, e^{-\langle n \rangle} = 1. \tag{13.59}$$

Its mean μ is the parameter $\langle n \rangle = pN$ of the binomial distribution

$$\mu = \sum_{n=0}^{\infty} n\, P_P(n, \langle n \rangle) = \sum_{n=1}^{\infty} n\, \frac{\langle n \rangle^n}{n!}\, e^{-\langle n \rangle} = \langle n \rangle \sum_{n=1}^{\infty} \frac{\langle n \rangle^{(n-1)}}{(n-1)!}\, e^{-\langle n \rangle}$$
$$= \langle n \rangle \sum_{n=0}^{\infty} \frac{\langle n \rangle^n}{n!}\, e^{-\langle n \rangle} = \langle n \rangle. \tag{13.60}$$

As $N \to \infty$ and $p \to 0$ with $pN = \langle n \rangle$ fixed, the variance (13.47) of the binomial distribution tends to the limit

$$V_{\rm P} = \lim_{\substack{N\to\infty \\ p\to 0}} V_{\rm B} = \lim_{\substack{N\to\infty \\ p\to 0}} p(1-p)\,N = \langle n \rangle. \tag{13.61}$$

Thus the mean and the variance of a Poisson distribution are equal

$$V_{\rm P} = \langle (n - \langle n \rangle)^2 \rangle = \langle n \rangle = \mu \tag{13.62}$$

as one may show directly (exercise 13.13).

Example 13.6 (Coherent states) The **coherent state** $|\alpha\rangle$ introduced in equation (2.138)

$$|\alpha\rangle = e^{-|\alpha|^2/2} e^{\alpha a^\dagger} |0\rangle = e^{-|\alpha|^2/2} \sum_{n=0}^{\infty} \frac{\alpha^n}{\sqrt{n!}} |n\rangle \tag{13.63}$$

is an eigenstate $a|\alpha\rangle = \alpha|\alpha\rangle$ of the annihilation operator a with eigenvalue α. The probability $P(n)$ of finding n quanta in the state $|\alpha\rangle$ is the square of the absolute value of the inner product $\langle n|\alpha\rangle$

$$P(n) = |\langle n|\alpha\rangle|^2 = \frac{|\alpha|^{2n}}{n!} e^{-|\alpha|^2}, \tag{13.64}$$

which is a Poisson distribution $P(n) = P_P(n, |\alpha|^2)$ with mean and variance $\mu = \langle n \rangle = V(\alpha) = |\alpha|^2$. □

13.5 The Gaussian distribution

Gauss considered the binomial distribution in the limit $N \to \infty$ with the probability p fixed. In this limit, the binomial probability

$$P_{\rm B}(n, p, N) = \frac{N!}{n!\,(N-n)!}\, p^n\, q^{N-n} \tag{13.65}$$

is very tiny unless n is near pN, which means that $n \approx pN$ and $N - n \approx (1 - p)N = qN$ are comparable. So the limit $N \to \infty$ effectively is one in which n and $N - n$ also tend to infinity. The approximation (13.54)

$$P_{\rm B}(n, p, N) \approx \sqrt{\frac{N}{2\pi n(N-n)}} \left(\frac{pN}{n}\right)^n \left(\frac{qN}{N-n}\right)^{N-n} R_3(n, N) \tag{13.66}$$

applies in which $R_3(n, N) \to 1$ as N, $N - n$, and n all increase without limit.

Because the probability $P_{\rm B}(n, p, N)$ is negligible unless $n \approx pN$, we set $y = n - pN$ and treat y/n as small. Since $n = pN + y$ and $N - n = (1-p)N + pN - n = qN - y$, we may write the square-root as

$$\sqrt{\frac{N}{2\pi\, n\,(N-n)}} = \frac{1}{\sqrt{2\pi\, N\,(pN+y)/N\,(qN-y)/N}}$$
$$= \frac{1}{\sqrt{2\pi\, pqN\,(1+y/pN)\,(1-y/qN)}}. \tag{13.67}$$

Since y remains finite as $N \to \infty$, we get in this limit

$$\lim_{N\to\infty}\sqrt{\frac{N}{2\pi\, n\,(N-n)}} = \frac{1}{\sqrt{2\pi\, pqN}}. \tag{13.68}$$

Substituting $pN+y$ for n and $qN-y$ for $N-n$ in (13.66), we find

$$P_{\rm B}(n,p,N) \approx \frac{1}{\sqrt{2\pi\, pqN}}\left(\frac{pN}{pN+y}\right)^{pN+y}\left(\frac{qN}{qN-y}\right)^{qN-y}$$
$$= \frac{1}{\sqrt{2\pi\, pqN}}\left(1+\frac{y}{pN}\right)^{-(pN+y)}\left(1-\frac{y}{qN}\right)^{-(qN-y)}, \tag{13.69}$$

which implies

$$\ln\left[P_{\rm B}(n,p,N)\sqrt{2\pi\, pqN}\right] \approx -(pN+y)\ln\left[1+\frac{y}{pN}\right] - (qN-y)\ln\left[1-\frac{y}{qN}\right]. \tag{13.70}$$

The first two terms of the power series (4.88) for $\ln(1+\epsilon)$ are

$$\ln(1+\epsilon) \approx \epsilon - \frac{1}{2}\epsilon^2. \tag{13.71}$$

So, using this expansion for $\ln(1+y/pN)$ and also for $\ln(1-y/qN)$, we get

$$\ln\left(P_{\rm B}(n,p,N)\sqrt{2\pi\, pqN}\right) \approx -(pN+y)\left[\frac{y}{pN}-\frac{1}{2}\left(\frac{y}{pN}\right)^2\right]$$
$$-(qN-y)\left[-\frac{y}{qN}-\frac{1}{2}\left(\frac{y}{qN}\right)^2\right]$$
$$\approx -\frac{y^2}{2pqN}. \tag{13.72}$$

Gauss's approximation to the binomial probability distribution thus is

$$P_{\rm BG}(n,p,N) = \frac{1}{\sqrt{2\pi pqN}}\exp\left(-\frac{(n-pN)^2}{2pqN}\right), \tag{13.73}$$

in which we've replaced y by $n-pN$ and $1-p$ by q.

Extending the integer n to a continuous variable x, we have

$$P_G(x,p,N) = \frac{1}{\sqrt{2\pi pqN}} \exp\left(-\frac{(x-pN)^2}{2pqN}\right), \tag{13.74}$$

which is (exercise 13.14) a normalized probability distribution with mean $\langle x \rangle = \mu = pN$ and variance $\langle (x-\mu)^2 \rangle = \sigma^2 = pqN$. Replacing pN by μ and pqN by σ^2, we get the standard form of **Gauss's distribution**

$$P_G(x,\mu,\sigma) = \frac{1}{\sigma\sqrt{2\pi}} \exp\left(-\frac{(x-\mu)^2}{2\sigma^2}\right). \tag{13.75}$$

This distribution occurs so often in mathematics and in Nature that it is often called **the normal distribution.** Its odd central moments all vanish $\nu_{2n+1} = 0$, and its even ones are $\nu_{2n} = (2n-1)!!\,\sigma^{2n}$ (exercise 13.16).

Example 13.7 (Single-molecule super-resolution microscopy) If the wave-length of visible light were a nanometer, microscopes would yield much sharper images. Each photon from a (single-molecule) fluorophore entering the lens of a microscope would follow ray optics and be focused within a tiny circle of about a nanometer on a detector. Instead, a photon arrives not at $\boldsymbol{x} = (x_1, x_2)$ but at $\boldsymbol{y}_i = (y_{1i}, y_{2i})$ with gaussian probability

$$P(\boldsymbol{y}_i) = \frac{1}{2\pi\sigma^2} e^{-(\boldsymbol{y}_i-\boldsymbol{x})^2/2\sigma^2} \tag{13.76}$$

where $\sigma \approx 150$ nm is about a quarter of a wave-length. What to do?

In the **centroid** method, one collects $N \approx 500$ points $\boldsymbol{y}_i$ and finds the point $\boldsymbol{x}$ that maximizes the joint probability of the N image points

$$P = \prod_{i=1}^{N} P(\boldsymbol{y}_i) = d^N \prod_{i=1}^{N} e^{-(\boldsymbol{y}_i-\boldsymbol{x})^2/(2\sigma^2)} = d^N \exp\left[-\sum_{i=1}^{N} (\boldsymbol{y}_i-\boldsymbol{x})^2/(2\sigma^2)\right] \tag{13.77}$$

where $d = 1/2\pi\sigma^2$, by solving for $k = 1$ and 2 the equations

$$\frac{\partial P}{\partial x_k} = 0 = P\frac{\partial P}{\partial x_k}\left[-\sum_{i=1}^{N} (\boldsymbol{y}_i-\boldsymbol{x})^2/(2\sigma^2)\right] = \frac{P}{\sigma^2}\sum_{i=1}^{N} (y_{ik}-x_k). \tag{13.78}$$

This **maximum-likelihood** estimate of the image point $\boldsymbol{x}$ is the average of the observed points $\boldsymbol{y}_i$

$$\boldsymbol{x} = \frac{1}{N}\sum_{i=1}^{N} \boldsymbol{y}_i. \tag{13.79}$$

This method is an improvement, but it is biased by auto-fluorescence and out-of-focus fluorophores. Fang Huang and Keith Lidke use **direct stochastic**

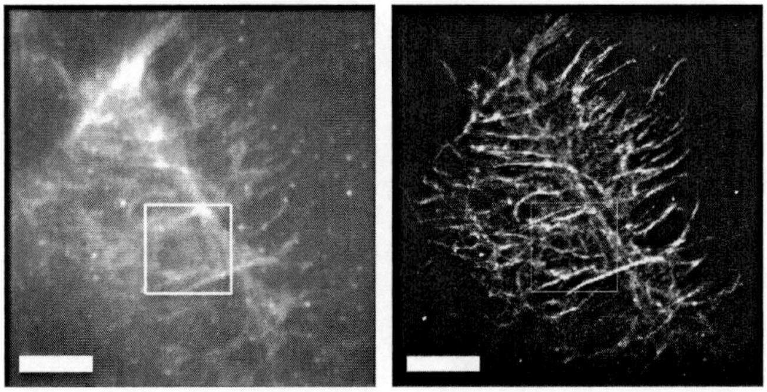
Actin fibers in HELA cells

Figure 13.2 Conventional (left, fuzzy) and dSTORM (right, sharp) images of actin fibers in HELA cells. The actin is labeled with Alexa Fluor 647 Phalloidin. The white rectangles are 5 microns in length. Images courtesy of Fang Huang and Keith Lidke.

optical reconstruction microscopy (dSTORM) to locate the image point $\boldsymbol{x}$ of the fluorophore in ways that account for the finite accuracy of their pixilated detector and the randomness of photo-detection.

Actin filaments are double helices of the protein actin some 5–9 nm wide. They occur throughout a eukaryotic cell but are concentrated near its surface and determine its shape. Together with tubulin and intermediate filaments, they form a cell's cytoskeleton. Figure 13.2 shows conventional (left, fuzzy) and dSTORM (right, sharp) images of actin filaments. The finite size of the fluorophore and the motion of the molecules of living cells limit dSTORM's improvement in resolution to a factor of 10 to 20. □

13.6 The error function erf

The probability that a random variable x distributed according to Gauss's distribution (13.75) has a value between $\mu - \delta$ and $\mu + \delta$ is

$$P(|x-\mu| < \delta) = \int_{\mu-\delta}^{\mu+\delta} P_G(x,\mu,\sigma)\, dx = \frac{1}{\sigma\sqrt{2\pi}} \int_{\mu-\delta}^{\mu+\delta} \exp\left(-\frac{(x-\mu)^2}{2\sigma^2}\right) dx$$
$$= \frac{1}{\sigma\sqrt{2\pi}} \int_{-\delta}^{\delta} \exp\left(-\frac{x^2}{2\sigma^2}\right) dx = \frac{2}{\sqrt{\pi}} \int_0^{\delta/\sigma\sqrt{2}} e^{-t^2}\, dt. \quad (13.80)$$

The last integral is the error function

$$\text{erf}(x) = \frac{2}{\sqrt{\pi}} \int_0^x e^{-t^2} dt, \tag{13.81}$$

so in terms of it the probability that x lies within δ of the mean μ is

$$P(|x - \mu| < \delta) = \text{erf}\left(\frac{\delta}{\sigma\sqrt{2}}\right). \tag{13.82}$$

In particular, the probabilities that x falls within one, two, or three standard deviations of μ are

$$\begin{aligned} P(|x - \mu| < \sigma) &= \text{erf}(1/\sqrt{2}) = 0.6827, \\ P(|x - \mu| < 2\sigma) &= \text{erf}(2/\sqrt{2}) = 0.9545, \\ P(|x - \mu| < 3\sigma) &= \text{erf}(3/\sqrt{2}) = 0.9973. \end{aligned} \tag{13.83}$$

The error function erf(x) is plotted in Fig. 13.3 in which the vertical lines are at $x = \delta/(\sigma\sqrt{2})$ for $\delta = \sigma, 2\sigma$, and 3σ.

The probability that x falls between a and b is (exercise 13.17)

$$P(a < x < b) = \frac{1}{2}\left[\text{erf}\left(\frac{b-\mu}{\sigma\sqrt{2}}\right) - \text{erf}\left(\frac{a-\mu}{\sigma\sqrt{2}}\right)\right]. \tag{13.84}$$

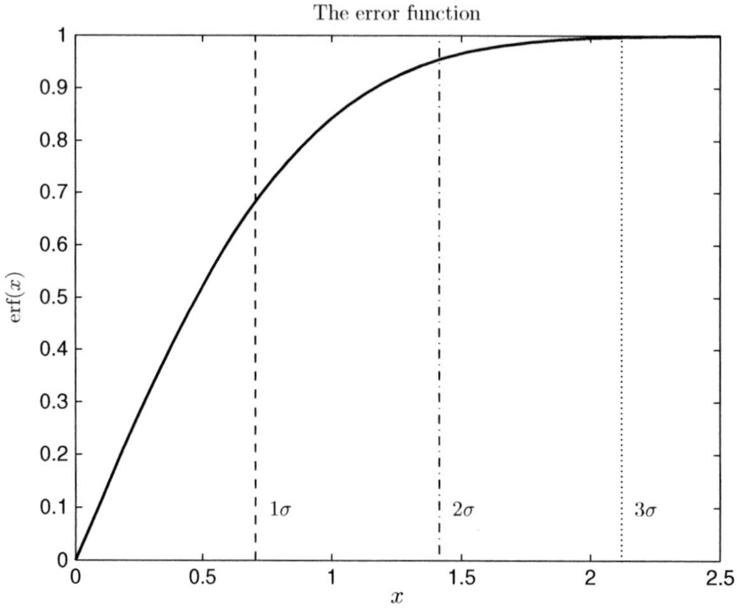

Figure 13.3 The error function erf(x) is plotted for $0 < x < 2.5$. The vertical lines are at $x = \delta/(\sigma\sqrt{2})$ for $\delta = \sigma, 2\sigma$, and 3σ with $\sigma = 1/\sqrt{2}$.

In particular, the cumulative probability $P(-\infty, x)$ that the random variable is less than x is for $\mu = 0$ and $\sigma = 1$

$$P(-\infty, x) = \frac{1}{2}\left[\operatorname{erf}\left(\frac{x}{\sqrt{2}}\right) - \operatorname{erf}\left(\frac{-\infty}{\sqrt{2}}\right)\right] = \frac{1}{2}\left[\operatorname{erf}\left(\frac{x}{\sqrt{2}}\right) + 1\right]. \tag{13.85}$$

The complement erfc of the error function is defined as

$$\operatorname{erfc}(x) = \frac{2}{\sqrt{\pi}} \int_x^\infty e^{-t^2}\, dt = 1 - \operatorname{erf}(x) \tag{13.86}$$

and is numerically useful for large x where round-off errors may occur in subtracting erf(x) from unity. Both erf and erfc are intrinsic functions in FORTRAN available without any effort on the part of the programmer.

Example 13.8 (Summing binomial probabilities) To add up several binomial probabilities when the factorials in $P_B(n, p, N)$ are too big to handle, we first use Gauss's approximation (13.73)

$$P_B(n, p, N) = \frac{N!}{n!\,(N-n)!}\, p^n\, q^{N-n} \approx \frac{1}{\sqrt{2\pi pqN}} \exp\left(-\frac{(n-pN)^2}{2pqN}\right). \tag{13.87}$$

Then, using (13.84) with $\mu = pN$, we find (exercise 13.15)

$$P_B(n, p, N) \approx \frac{1}{2}\left[\operatorname{erf}\left(\frac{n + \frac{1}{2} - pN}{\sqrt{2pqN}}\right) - \operatorname{erf}\left(\frac{n - \frac{1}{2} - pN}{\sqrt{2pqN}}\right)\right], \tag{13.88}$$

which we can sum over the integer n to get

$$\sum_{n=n_1}^{n_2} P_B(n, p, N) \approx \frac{1}{2}\left[\operatorname{erf}\left(\frac{n_2 + \frac{1}{2} - pN}{\sqrt{2pqN}}\right) - \operatorname{erf}\left(\frac{n_1 - \frac{1}{2} - pN}{\sqrt{2pqN}}\right)\right], \tag{13.89}$$

which is easy to evaluate. □

Example 13.9 (Polls) Suppose in a poll of 1000 likely voters, 600 have said they would vote for Nancy Pelosi. Repeating the analysis of example 13.3, we see that if the probability that a random voter will vote for her is y, then the probability that 600 in our sample of 1000 will is by (13.87)

$$\begin{aligned} P(600|y) = P_B(600, y) &= \binom{1000}{600} y^{600}\, (1-y)^{400} \\ &\approx \frac{1}{10\sqrt{20\pi\, y(1-y)}} \exp\left(-\frac{20(3-5y)^2}{y(1-y)}\right) \end{aligned} \tag{13.90}$$

and so the probability density that a fraction y of the voters will vote for her is

$$P(y|600) = \frac{P(600|y)}{\int_0^1 P(600, y')\, dy'}$$

$$= \frac{[y(1-y)]^{-1/2} \exp\left(-\frac{20(3-5y)^2}{y(1-y)}\right)}{\int_0^1 [y'(1-y')]^{-1/2} \exp\left(-\frac{20(3-5y')^2}{y'(1-y')}\right) dy'}. \tag{13.91}$$

This normalized probability distribution is negligible except for y near $3/5$ (exercise 13.18), where it is approximately Gauss's distribution

$$P(y|600) \approx \frac{1}{\sigma\sqrt{2\pi}} \exp\left(-\frac{(y-3/5)^2}{2\sigma^2}\right) \tag{13.92}$$

with mean $\mu = 3/5$ and variance

$$\sigma^2 = \frac{3}{12500} = 2.4 \times 10^{-4}. \tag{13.93}$$

The probability that $y > 1/2$ then is by (13.84)

$$P(a < x < b) = \frac{1}{2}\left[\operatorname{erf}\left(\frac{1-\mu}{\sigma\sqrt{2}}\right) - \operatorname{erf}\left(\frac{1/2-\mu}{\sigma\sqrt{2}}\right)\right]$$

$$= \frac{1}{2}\left[\operatorname{erf}\left(\frac{20}{\sqrt{1.2}}\right) - \operatorname{erf}\left(\frac{-5}{\sqrt{1.2}}\right)\right] \approx 1. \tag{13.94}$$

The probability that $y < 1/2$ is 5.4×10^{-11}. □

13.7 The Maxwell–Boltzmann distribution

It is a small jump from Gauss's distribution (13.75) to the Maxwell–Boltzmann distribution of velocities of molecules in a gas. We start in one dimension and focus on a single molecule that is being hit fore and aft with equal probabilities by other molecules. If each hit increases or decreases its speed by dv, then after n aft hits and $N-n$ fore hits, the speed v_x of a molecule initially at rest would be

$$v_x = ndv - (N-n)dv = (2n-N)dv. \tag{13.95}$$

The probability of this speed is given by Gauss's approximation (13.73) to the binomial distribution $P_{\rm B}(n, \frac{1}{2}, N)$ as

$$P_{\rm BG}(n, \tfrac{1}{2}, N) = \sqrt{\frac{2}{\pi N}} \exp\left(-\frac{(2n-N)^2}{2N}\right) = \sqrt{\frac{2}{\pi N}} \exp\left(-\frac{v_x^2}{2Ndv^2}\right). \tag{13.96}$$

This argument applies to any physical variable subject to unbiased random fluctuations. It is why Gauss's distribution describes statistical errors and why it occurs so often in Nature as to be called the normal distribution.

We now write the argument of the exponential in terms of the temperature T and Boltzmann's constant k by setting $N = kT/(m\,dv^2)$ so that

$$-\frac{\frac{1}{2}v_x^2}{Ndv^2} = -\frac{\frac{1}{2}mv_x^2}{mNdv^2} = -\frac{\frac{1}{2}mv_x^2}{kT}. \tag{13.97}$$

Then with $dv_x = 2dv$, we have

$$P_G(v_x)dv_x = \sqrt{\frac{2m}{\pi kT}}\,dv\,\exp\left(-\frac{\frac{1}{2}mv_x^2}{kT}\right) = \sqrt{\frac{m}{2\pi kT}}\,dv_x\,\exp\left(-\frac{\frac{1}{2}mv_x^2}{kT}\right). \tag{13.98}$$

Gauss's distribution is normalized to unity because it is the limit of the binomial distribution (13.44)

$$\int_{-\infty}^{\infty}\sqrt{\frac{m}{2\pi kT}}\,\exp\left(-\frac{\frac{1}{2}mv_x^2}{kT}\right)\,dv_x = 1 \tag{13.99}$$

as you may verify by explicit integration.

In three space dimensions, the Maxwell–Boltzmann distribution $P_{MB}(\boldsymbol{v})$ is the product

$$P_{MB}(\boldsymbol{v})d^3v = P_G(v_x)\,P_G(v_y)\,P_G(v_z)d^3v = \left(\frac{m}{2\pi kT}\right)^{3/2} e^{-\frac{1}{2}mv^2/(kT)}4\pi v^2 dv. \tag{13.100}$$

The mean value of the velocity of a Maxwell–Boltzmann gas vanishes

$$\langle \boldsymbol{v}\rangle = \int \boldsymbol{v}\,P_{MB}(\boldsymbol{v})d^3v = \boldsymbol{0} \tag{13.101}$$

but the mean value of the square of the velocity $\boldsymbol{v}^2 = \boldsymbol{v}\cdot\boldsymbol{v}$ is the sum of the three variances $\sigma_x^2 = \sigma_y^2 = \sigma_z^2 = kT/m$

$$\langle v^2\rangle = V[\boldsymbol{v}^2] = \int \boldsymbol{v}^2\,P_{MB}(\boldsymbol{v})\,d^3v = 3kT/m \tag{13.102}$$

which is the familiar statement

$$\frac{1}{2}m\langle v^2\rangle = \frac{3}{2}kT \tag{13.103}$$

that each degree of freedom gets $kT/2$ of energy.

13.8 Diffusion

We may apply the same reasoning as in the preceding section (13.7) to the diffusion of a gas of particles treated as a random walk with step size dx. In one dimension, after n steps forward and $N - n$ steps backward, a particle starting at $x = 0$ is at $x = (2n - N)dx$. Thus, as in (13.96), the probability of being at x is given by Gauss's approximation (13.73) to the binomial distribution $P_B(n, \frac{1}{2}, N)$ as

$$P_{BG}(n, \tfrac{1}{2}, N) = \sqrt{\frac{2}{\pi N}}\, \exp\left(-\frac{(2n-N)^2}{2N}\right) = \sqrt{\frac{2}{\pi N}}\, \exp\left(-\frac{x^2}{2Ndx^2}\right). \tag{13.104}$$

In terms of the diffusion constant

$$D = \frac{Ndx^2}{2t} \tag{13.105}$$

this distribution is

$$P_G(x) = \left(\frac{1}{4\pi Dt}\right)^{1/2} \exp\left(-\frac{x^2}{4Dt}\right) \tag{13.106}$$

when normalized to unity on $(-\infty, \infty)$.

In three dimensions, this gaussian distribution is the product

$$P(\boldsymbol{r}, t) = P_G(x)\, P_G(y)\, P_G(z) = \left(\frac{1}{4\pi Dt}\right)^{3/2} \exp\left(-\frac{\boldsymbol{r}^2}{4Dt}\right). \tag{13.107}$$

The variance $\sigma^2 = 2Dt$ gives the average of the squared displacement of each of the three coordinates. Thus the mean of the squared displacement $\langle \boldsymbol{r}^2 \rangle$ rises **linearly** with the time as

$$\langle \boldsymbol{r}^2 \rangle = V[\boldsymbol{r}] = 3\sigma^2 = \int \boldsymbol{r}^2\, P(\boldsymbol{r}, t)\, d^3r = 6\, D\, t. \tag{13.108}$$

The distribution $P(\boldsymbol{r}, t)$ satisfies the **diffusion equation**

$$\dot{P}(\boldsymbol{r}, t) = D\, \nabla^2 P(\boldsymbol{r}, t), \tag{13.109}$$

in which the dot means time derivative.

13.9 Langevin's theory of brownian motion

Einstein made the first theory of brownian motion in 1905, but Langevin's approach (Langevin, 1908) is simpler. A tiny particle of colloidal size and mass m in a fluid is buffeted by a force $\boldsymbol{F}(t)$ due to the 10^{21} collisions per second it suffers with the molecules of the surrounding fluid. Its equation of motion is

$$m\, \frac{d\boldsymbol{v}(t)}{dt} = \boldsymbol{F}(t). \tag{13.110}$$

Langevin suggested that the force $\boldsymbol{F}(t)$ is the sum of a viscous drag $-\boldsymbol{v}(t)/B$ and a rapidly fluctuating part $\boldsymbol{f}(t)$

$$\boldsymbol{F}(t) = -\, \boldsymbol{v}(t)/B + \boldsymbol{f}(t) \tag{13.111}$$

so that

$$m \frac{d\boldsymbol{v}(t)}{dt} = -\frac{\boldsymbol{v}(t)}{B} + \boldsymbol{f}(t). \tag{13.112}$$

The parameter B is called the **mobility**. The **ensemble average** (the average over the set of particles) of the fluctuating force $\boldsymbol{f}(t)$ is zero

$$\langle \boldsymbol{f}(t) \rangle = \boldsymbol{0}. \tag{13.113}$$

Thus the ensemble average of the velocity satisfies

$$m \frac{d\langle \boldsymbol{v} \rangle}{dt} = -\frac{\langle \boldsymbol{v} \rangle}{B} \tag{13.114}$$

whose solution with $\tau = mB$ is

$$\langle \boldsymbol{v}(t) \rangle = \langle \boldsymbol{v}(0) \rangle \, e^{-t/\tau}. \tag{13.115}$$

The instantaneous equation (13.112) divided by the mass m is

$$\frac{d\boldsymbol{v}(t)}{dt} = -\frac{\boldsymbol{v}(t)}{\tau} + \boldsymbol{a}(t), \tag{13.116}$$

in which $\boldsymbol{a}(t) = \boldsymbol{f}(t)/m$ is the acceleration. The ensemble average of the scalar product of the position vector $\boldsymbol{r}$ with this equation is

$$\left\langle \boldsymbol{r} \cdot \frac{d\boldsymbol{v}}{dt} \right\rangle = -\frac{\langle \boldsymbol{r} \cdot \boldsymbol{v} \rangle}{\tau} + \langle \boldsymbol{r} \cdot \boldsymbol{a} \rangle. \tag{13.117}$$

But since the ensemble average $\langle \boldsymbol{r} \cdot \boldsymbol{a} \rangle$ of the scalar product of the position vector $\boldsymbol{r}$ with the random, fluctuating part $\boldsymbol{a}$ of the acceleration vanishes, we have

$$\left\langle \boldsymbol{r} \cdot \frac{d\boldsymbol{v}}{dt} \right\rangle = -\frac{\langle \boldsymbol{r} \cdot \boldsymbol{v} \rangle}{\tau}. \tag{13.118}$$

Now

$$\frac{1}{2} \frac{d\boldsymbol{r}^2}{dt} = \frac{1}{2} \frac{d}{dt} (\boldsymbol{r} \cdot \boldsymbol{r}) = \boldsymbol{r} \cdot \boldsymbol{v} \tag{13.119}$$

and so

$$\frac{1}{2} \frac{d^2 \boldsymbol{r}^2}{dt^2} = \boldsymbol{r} \cdot \frac{d\boldsymbol{v}}{dt} + \boldsymbol{v}^2. \tag{13.120}$$

The ensemble average of this equation gives us

$$\frac{d^2 \langle \boldsymbol{r}^2 \rangle}{dt^2} = 2 \left\langle \boldsymbol{r} \cdot \frac{d\boldsymbol{v}}{dt} \right\rangle + 2 \langle \boldsymbol{v}^2 \rangle \tag{13.121}$$

or in view of (13.118)

$$\frac{d^2 \langle \boldsymbol{r}^2 \rangle}{dt^2} = -2 \frac{\langle \boldsymbol{r} \cdot \boldsymbol{v} \rangle}{\tau} + 2 \langle \boldsymbol{v}^2 \rangle. \tag{13.122}$$

We now use (13.119) to replace $\langle \boldsymbol{r} \cdot \boldsymbol{v} \rangle$ with half the first time derivative of $\langle \boldsymbol{r}^2 \rangle$ so that we have

$$\frac{d^2 \langle \boldsymbol{r}^2 \rangle}{dt^2} = -\frac{1}{\tau}\frac{d\langle \boldsymbol{r}^2 \rangle}{dt} + 2\langle \boldsymbol{v}^2 \rangle. \tag{13.123}$$

If the fluid is in equilibrium, then the ensemble average of $\boldsymbol{v}^2$ is given by the Maxwell–Boltzmann value (13.103)

$$\langle \boldsymbol{v}^2 \rangle = \frac{3kT}{m} \tag{13.124}$$

and so the acceleration (13.123) is

$$\frac{d^2 \langle \boldsymbol{r}^2 \rangle}{dt^2} + \frac{1}{\tau}\frac{d\langle \boldsymbol{r}^2 \rangle}{dt} = \frac{6kT}{m}, \tag{13.125}$$

which we can integrate.

The general solution (6.13) to a second-order linear inhomogeneous differential equation is the sum of any particular solution to the inhomogeneous equation plus the general solution of the homogeneous equation. The function $\langle \boldsymbol{r}^2(t) \rangle_{\mathrm{pi}} = 6kTt\tau/m$ is a particular solution of the inhomogeneous equation. The general solution to the homogeneous equation is $\langle \boldsymbol{r}^2(t) \rangle_{\mathrm{gh}} = U + W \exp(-t/\tau)$ where U and W are constants. So $\langle \boldsymbol{r}^2(t) \rangle$ is

$$\langle \boldsymbol{r}^2(t) \rangle = U + W e^{-t/\tau} + 6kT\tau t/m \tag{13.126}$$

where U and W make $\langle \boldsymbol{r}^2(t) \rangle$ fit the boundary conditions. If the individual particles start out at the origin $\boldsymbol{r} = \boldsymbol{0}$, then one boundary condition is

$$\langle \boldsymbol{r}^2(0) \rangle = 0, \tag{13.127}$$

which implies that

$$U + W = 0. \tag{13.128}$$

And since the particles start out at $\boldsymbol{r} = \boldsymbol{0}$ with an isotropic distribution of initial velocities, the formula (13.119) for $\dot{r}^2$ implies that at $t = 0$

$$\left.\frac{d\,\langle \boldsymbol{r}^2 \rangle}{dt}\right|_{t=0} = 2\langle \boldsymbol{r}(0) \cdot \boldsymbol{v}(0) \rangle = 0. \tag{13.129}$$

This boundary condition means that our solution (13.126) must satisfy

$$\left.\frac{d\,\langle \boldsymbol{r}^2(t) \rangle}{dt}\right|_{t=0} = -\frac{W}{\tau} + \frac{6kT\tau}{m} = 0. \tag{13.130}$$

Thus $W = -U = 6kT\tau^2/m$, and so our solution (13.126) is

$$\langle \boldsymbol{r}^2(t) \rangle = \frac{6kT\tau^2}{m}\left[\frac{t}{\tau} + e^{-t/\tau} - 1\right]. \tag{13.131}$$

At times short compared to τ, the first two terms in the power series for the exponential $\exp(-t/\tau)$ cancel the terms $-1 + t/\tau$, leaving

$$\langle r^2(t)\rangle = \frac{6kT\tau^2}{m}\left[\frac{t^2}{2\tau^2}\right] = \frac{3kT}{m}\,t^2 = \langle v^2\rangle\, t^2. \tag{13.132}$$

But at times long compared to τ, the exponential vanishes, leaving

$$\langle r^2(t)\rangle = \frac{6kT\tau}{m}\,t = 6\,BkT\,t. \tag{13.133}$$

The **diffusion constant** D is defined by

$$\langle r^2(t)\rangle = 6\,D\,t \tag{13.134}$$

and so we arrive at **Einstein's relation**

$$D = BkT, \tag{13.135}$$

which often is written in terms of the **viscous-friction coefficient** ζ

$$\zeta \equiv \frac{1}{B} = \frac{m}{\tau} \tag{13.136}$$

as

$$\zeta\, D = kT. \tag{13.137}$$

This equation expresses Boltzmann's constant k in terms of three quantities ζ, D, and T that were accessible to measurement in the first decade of the twentieth century. It enabled scientists to measure Boltzmann's constant k for the first time. And since Avogadro's number N_A was the known gas constant R divided by k, the number of molecules in a mole was revealed to be $N_A = 6.022 \times 10^{23}$. Chemists could then divide the mass of a mole of any pure substance by 6.022×10^{23} and find the mass of the molecules that composed it. Suddenly the masses of the molecules of chemistry became known, and molecules were recognized as real particles and not tricks for balancing chemical equations.

13.10 The Einstein–Nernst relation

If a particle of mass m carries an electric charge q and is exposed to an electric field $\boldsymbol{E}$, then in addition to viscosity $-\boldsymbol{v}/B$ and random buffeting $\boldsymbol{f}$, the constant force $q\boldsymbol{E}$ acts on it

$$m\,\frac{d\boldsymbol{v}}{dt} = -\frac{\boldsymbol{v}}{B} + q\boldsymbol{E} + \boldsymbol{f}. \tag{13.138}$$

The mean value of its velocity will then satisfy the differential equation

$$\left\langle\frac{d\boldsymbol{v}}{dt}\right\rangle = -\frac{\langle\boldsymbol{v}\rangle}{\tau} + \frac{q\boldsymbol{E}}{m} \tag{13.139}$$

where $\tau = mB$. A particular solution of this inhomogeneous equation is

$$\langle \boldsymbol{v}(t) \rangle_{\rm pi} = \frac{q\tau \boldsymbol{E}}{m} = qB\boldsymbol{E}. \tag{13.140}$$

The general solution of its homogeneous version is $\langle \boldsymbol{v}(t) \rangle_{\rm gh} = \boldsymbol{A} \exp(-t/\tau)$ in which the constant $\boldsymbol{A}$ is chosen to give $\langle \boldsymbol{v}(0) \rangle$ at $t = 0$. So by (6.13), the general solution $\langle \boldsymbol{v}(t) \rangle$ to equation (13.139) is (exercise 13.19) the sum of $\langle \boldsymbol{v}(t) \rangle_{\rm pi}$ and $\langle \boldsymbol{v}(t) \rangle_{\rm gh}$

$$\langle \boldsymbol{v}(t) \rangle = qB\boldsymbol{E} + [\langle \boldsymbol{v}(0) \rangle - qB\boldsymbol{E}]\, e^{-t/\tau}. \tag{13.141}$$

By applying the tricks of the previous section (13.9), one may show (exercise 13.20) that the variance of the position $\boldsymbol{r}$ about its mean $\langle \boldsymbol{r}(t) \rangle$ is

$$\left\langle (\boldsymbol{r} - \langle \boldsymbol{r}(t) \rangle)^2 \right\rangle = \frac{6kT\tau^2}{m} \left(\frac{t}{\tau} - 1 + e^{-t/\tau} \right). \tag{13.142}$$

where $\langle \boldsymbol{r}(t) \rangle = (q\tau^2 \boldsymbol{E}/m)\,(t/\tau - 1 + e^{-t/\tau})$ if $\langle \boldsymbol{r}(0) \rangle = \langle \boldsymbol{v}(0) \rangle = 0$. So for times $t \gg \tau$, this variance is

$$\left\langle (\boldsymbol{r} - \langle \boldsymbol{r}(t) \rangle)^2 \right\rangle = \frac{6kT\tau t}{m}. \tag{13.143}$$

Since the diffusion constant D is defined by (13.134) as

$$\left\langle (\boldsymbol{r} - \langle \boldsymbol{r}(t) \rangle)^2 \right\rangle = 6\,D\,t \tag{13.144}$$

we arrive at the Einstein–Nernst relation

$$D = BkT = \frac{qB}{q} kT = \frac{\mu}{q} kT, \tag{13.145}$$

in which the electric mobility is $\mu = qB$.

13.11 Fluctuation and dissipation

Let's look again at Langevin's equation (13.116) but with u as the independent variable

$$\frac{d\boldsymbol{v}(u)}{du} + \frac{\boldsymbol{v}(u)}{\tau} = \boldsymbol{a}(u). \tag{13.146}$$

If we multiply both sides by the exponential $\exp(u/\tau)$

$$\left(\frac{d\boldsymbol{v}}{du} + \frac{\boldsymbol{v}}{\tau} \right) e^{u/\tau} = \frac{d}{du} \left(\boldsymbol{v}\, e^{u/\tau} \right) = \boldsymbol{a}(u)\, e^{u/\tau} \tag{13.147}$$

and integrate from 0 to t

$$\int_0^t \frac{d}{du} \left(\boldsymbol{v}\, e^{u/\tau} \right) du = \boldsymbol{v}(t)\, e^{t/\tau} - \boldsymbol{v}(0) = \int_0^t \boldsymbol{a}(u)\, e^{u/\tau}\, du \tag{13.148}$$

then we get

$$\boldsymbol{v}(t) = e^{-t/\tau}\, \boldsymbol{v}(0) + e^{-t/\tau} \int_0^t \boldsymbol{a}(u)\, e^{u/\tau}\, du. \tag{13.149}$$

Thus the ensemble average of the square of the velocity is

$$\begin{aligned} \langle \boldsymbol{v}^2(t) \rangle = {} & e^{-2t/\tau} \langle \boldsymbol{v}^2(0) \rangle + 2e^{-2t/\tau} \int_0^t \langle \boldsymbol{a}(u) \rangle\, e^{u/\tau}\, du \\ & + e^{-2t/\tau} \int_0^t \int_0^t \langle \boldsymbol{a}(u_1) \cdot \boldsymbol{a}(u_2) \rangle\, e^{(u_1+u_2)/\tau}\, du_1 du_2. \end{aligned} \tag{13.150}$$

The second term on the RHS is zero, so we have

$$\langle \boldsymbol{v}^2(t) \rangle = e^{-2t/\tau} \langle \boldsymbol{v}^2(0) \rangle + e^{-2t/\tau} \int_0^t \int_0^t \langle \boldsymbol{a}(u_1) \cdot \boldsymbol{a}(u_2) \rangle\, e^{(u_1+u_2)/\tau}\, du_1 du_2. \tag{13.151}$$

The ensemble average

$$C(u_1, u_2) = \langle \boldsymbol{a}(u_1) \cdot \boldsymbol{a}(u_2) \rangle \tag{13.152}$$

is an example of an **autocorrelation function.**

All autocorrelation functions have some simple properties, which are easy to prove (Pathria, 1972, p. 458).

1 If the system is independent of time, then its autocorrelation function for any given variable $\boldsymbol{A}(t)$ depends only upon the time delay s:

$$C(t, t+s) = \langle \boldsymbol{A}(t) \cdot \boldsymbol{A}(t+s) \rangle \equiv C(s). \tag{13.153}$$

2 The autocorrelation function for $s = 0$ is necessarily nonnegative

$$C(t, t) = \langle \boldsymbol{A}(t) \cdot \boldsymbol{A}(t) \rangle = \langle \boldsymbol{A}(t)^2 \rangle \geq 0. \tag{13.154}$$

If the system is time independent, then $C(t, t) = C(0) \geq 0$.

3 The absolute value of $C(t_1, t_2)$ is never greater than the average of $C(t_1, t_1)$ and $C(t_2, t_2)$ because

$$\langle |\boldsymbol{A}(t_1) \pm \boldsymbol{A}(t_2)|^2 \rangle = \langle \boldsymbol{A}(t_1)^2 \rangle + \langle \boldsymbol{A}(t_2)^2 \rangle \pm 2 \langle \boldsymbol{A}(t_1) \cdot \boldsymbol{A}(t_2) \rangle \geq 0, \tag{13.155}$$

which implies that

$$- C(t_1, t_2) \leq \frac{1}{2} \left(C(t_1, t_1) + C(t_2, t_2) \right) \geq C(t_1, t_2) \tag{13.156}$$

or

$$|C(t_1, t_2)| \leq \frac{1}{2} \left(C(t_1, t_1) + C(t_2, t_2) \right). \tag{13.157}$$

For a time-independent system, this inequality is $|C(s)| \leq C(0)$ for every time delay s.

4 If the variables $A(t_1)$ and $A(t_2)$ commute, then their autocorrelation function is symmetric

$$C(t_1, t_2) = \langle A(t_1) \cdot A(t_2) \rangle = \langle A(t_2) \cdot A(t_1) \rangle = C(t_2, t_1). \tag{13.158}$$

For a time-independent system, this symmetry is $C(s) = C(-s)$.

5 If the variable $A(t)$ is randomly fluctuating with zero mean, then we expect both that its ensemble average vanishes

$$\langle A(t) \rangle = \mathbf{0} \tag{13.159}$$

and that there is some characteristic time scale T beyond which the correlation function falls to zero:

$$\langle A(t_1) \cdot A(t_2) \rangle \to \langle A(t_1) \rangle \cdot \langle A(t_2) \rangle = 0 \tag{13.160}$$

when $|t_1 - t_2| \gg T$.

In terms of the autocorrelation function $C(u_1, u_2) = \langle \boldsymbol{a}(u_1) \cdot \boldsymbol{a}(u_2) \rangle$ of the acceleration, the variance of the velocity (13.152) is

$$\langle \boldsymbol{v}^2(t) \rangle = e^{-2t/\tau} \langle \boldsymbol{v}^2(0) \rangle + e^{-2t/\tau} \int_0^t \int_0^t C(u_1, u_2) \, e^{(u_1+u_2)/\tau} \, du_1 du_2. \tag{13.161}$$

Since $C(u_1, u_2)$ is big only for tiny values of $|u_2 - u_1|$, it makes sense to change variables to

$$s = u_2 - u_1 \quad \text{and} \quad w = \frac{1}{2}(u_1 + u_2). \tag{13.162}$$

The element of area then is by (12.6–12.14)

$$du_1 \wedge du_2 = dw \wedge ds \tag{13.163}$$

and the limits of integration are $-2w \le s \le 2w$ for $0 \le w \le t/2$ and $-2(t-w) \le s \le 2(t-w)$ for $t/2 \le w \le t$. So $\langle \boldsymbol{v}^2(t) \rangle$ is

$$\begin{aligned} \langle \boldsymbol{v}^2(t) \rangle = {} & e^{-2t/\tau} \langle \boldsymbol{v}^2(0) \rangle + e^{-2t/\tau} \int_0^{t/2} e^{2w/\tau} dw \int_{-2w}^{2w} C(s) \, ds \\ & + e^{-2t/\tau} \int_{t/2}^{t} e^{2w/\tau} dw \int_{-2(t-w)}^{2(t-w)} C(s) \, ds. \end{aligned} \tag{13.164}$$

Since by (13.160) the autocorrelation function $C(s)$ vanishes outside a narrow window of width $2T$, we may approximate each of the s-integrals by

$$C = \int_{-\infty}^{\infty} C(s) \, ds. \tag{13.165}$$

It follows then that

$$\begin{aligned}\langle \boldsymbol{v}^2(t)\rangle &= e^{-2t/\tau}\,\langle \boldsymbol{v}^2(0)\rangle + C\,e^{-2t/\tau}\int_0^t e^{2w/\tau}\,dw \\ &= e^{-2t/\tau}\,\langle \boldsymbol{v}^2(0)\rangle + C\,e^{-2t/\tau}\,\frac{\tau}{2}\left(e^{2t/\tau}-1\right) \\ &= e^{-2t/\tau}\,\langle \boldsymbol{v}^2(0)\rangle + C\,\frac{\tau}{2}\left(1-e^{-2t/\tau}\right).\end{aligned} \tag{13.166}$$

As $t \to \infty$, $\langle \boldsymbol{v}^2(t)\rangle$ must approach its equilibrium value of $3kT/m$, and so

$$\lim_{t\to\infty}\langle \boldsymbol{v}^2(t)\rangle = C\,\frac{\tau}{2} = \frac{3kT}{m}, \tag{13.167}$$

which implies that

$$C = \frac{6kT}{m\tau} \quad \text{or} \quad \frac{1}{B} = \frac{m^2 C}{6kT}. \tag{13.168}$$

Our final formula for $\langle \boldsymbol{v}^2(t)\rangle$ then is

$$\langle \boldsymbol{v}^2(t)\rangle = e^{-2t/\tau}\,\langle \boldsymbol{v}^2(0)\rangle + \frac{3kT}{m}\left(1-e^{-2t/\tau}\right). \tag{13.169}$$

Referring back to the definition (13.136) of the viscous-friction coefficient $\zeta = 1/B$, we see that ζ is related to the integral

$$\zeta = \frac{1}{B} = \frac{m^2}{6kT}\,C = \frac{m^2}{6kT}\int_{-\infty}^{\infty}\langle \boldsymbol{a}(0)\cdot\boldsymbol{a}(s)\rangle ds = \frac{1}{6kT}\int_{-\infty}^{\infty}\langle \boldsymbol{f}(0)\cdot\boldsymbol{f}(s)\rangle ds \tag{13.170}$$

of the autocorrelation function of the random acceleration $\boldsymbol{a}(t)$ or equivalently of the random force $\boldsymbol{f}(t)$. This equation relates the dissipation of viscous friction to the random fluctuations. It is an example of a **fluctuation–dissipation theorem**.

If we substitute our formula (13.169) for $\langle \boldsymbol{v}^2(t)\rangle$ into the expression (13.123) for the acceleration of $\langle \boldsymbol{r}^2\rangle$, then we get

$$\frac{d^2\langle \boldsymbol{r}^2(t)\rangle}{dt^2} = -\frac{1}{\tau}\frac{d\langle \boldsymbol{r}^2(t)\rangle}{dt} + 2e^{-2t/\tau}\,\langle \boldsymbol{v}^2(0)\rangle + \frac{6kT}{m}\left(1-e^{-2t/\tau}\right). \tag{13.171}$$

The solution with both $\langle \boldsymbol{r}^2(0)\rangle = 0$ and $d\langle \boldsymbol{r}^2(0)\rangle/dt = 0$ is (exercise 13.21)

$$\langle \boldsymbol{r}^2(t)\rangle = \langle \boldsymbol{v}^2(0)\rangle\,\tau^2\left(1-e^{-t/\tau}\right)^2 - \frac{3kT}{m}\,\tau^2\left(1-e^{-t/\tau}\right)\left(3-e^{-t/\tau}\right) + \frac{6kT\tau}{m}\,t. \tag{13.172}$$

13.12 Characteristic and moment-generating functions

The Fourier transform (3.9) of a probability distribution $P(x)$ is its **characteristic function** $\hat{P}(k)$, sometimes written as $\chi(k)$

$$\hat{P}(k) \equiv \chi(k) \equiv E[e^{ikx}] = \int e^{ikx}\, P(x)\, dx. \tag{13.173}$$

The probability distribution $P(x)$ is the inverse Fourier transform (3.9)

$$P(x) = \int e^{-ikx}\, \hat{P}(k)\, \frac{dk}{2\pi}. \tag{13.174}$$

Example 13.10 (Gauss) The characteristic function of the gaussian

$$P_G(x, \mu, \sigma) = \frac{1}{\sigma\sqrt{2\pi}} \exp\left(-\frac{(x-\mu)^2}{2\sigma^2}\right) \tag{13.175}$$

is by (3.18)

$$\begin{aligned} \hat{P}_G(k, \mu, \sigma) &= \frac{1}{\sigma\sqrt{2\pi}} \int \exp\left(ikx - \frac{(x-\mu)^2}{2\sigma^2}\right) dx \\ &= \frac{e^{ik\mu}}{\sigma\sqrt{2\pi}} \int \exp\left(ikx - \frac{x^2}{2\sigma^2}\right) dx = \exp\left(i\mu k - \frac{1}{2}\sigma^2 k^2\right). \end{aligned} \tag{13.176}$$

□

For a discrete probability distribution P_n the characteristic function is

$$\chi(k) \equiv E[e^{ikx}] = \sum_n e^{ikn}\, P_n. \tag{13.177}$$

The normalization of both continuous and discrete probability distributions implies that their characteristic functions satisfy $\hat{P}(0) = \chi(0) = 1$.

Example 13.11 (Poisson) The Poisson distribution (13.58)

$$P_P(n, \langle n \rangle) = \frac{\langle n \rangle^n}{n!}\, e^{-\langle n \rangle} \tag{13.178}$$

has the characteristic function

$$\chi(k) = \sum_{n=0}^{\infty} e^{ikn}\, \frac{\langle n \rangle^n}{n!}\, e^{-\langle n \rangle} = e^{-\langle n \rangle} \sum_{n=0}^{\infty} \frac{(\langle n \rangle e^{ik})^n}{n!} = \exp\left[\langle n \rangle \left(e^{ik} - 1\right)\right]. \tag{13.179}$$

□

The **moment-generating function** is the characteristic function evaluated at an imaginary argument

$$M(k) \equiv E[e^{kx}] = \hat{P}(-ik) = \chi(-ik). \tag{13.180}$$

For a continuous probability distribution $P(x)$, it is

$$M(k) = E[e^{kx}] = \int e^{kx}\, P(x)\, dx \tag{13.181}$$

and for a discrete probability distribution P_n it is

$$M(k) = E[e^{kx}] = \sum_n e^{kx_n}\, P_n. \tag{13.182}$$

In both cases, the normalization of the probability distribution implies that $M(0) = 1$.

Derivatives of the moment-generating function and of the characteristic function give the moments

$$E[x^n] = \mu_n = \left.\frac{d^n M(k)}{dk^n}\right|_{k=0} = (-i)^n \left.\frac{d^n \hat{P}(k)}{dk^n}\right|_{k=0}. \tag{13.183}$$

Example 13.12 (Gauss and Poisson) The moment-generating functions for the distributions of Gauss (13.175) and Poisson (13.178) are

$$M_{\rm G}(k,\mu,\sigma) = \exp\left(\mu k + \frac{1}{2}\sigma^2 k^2\right) \quad \text{and} \quad M_{\rm P}(k,\langle n\rangle) = \exp\left[\langle n\rangle\left(e^k - 1\right)\right]. \tag{13.184}$$

They give as the first three moments of these distributions

$$\mu_{\rm G0} = 1, \quad \mu_{\rm G1} = \mu, \quad \mu_{\rm G2} = \mu^2 + \sigma^2, \tag{13.185}$$

$$\mu_{\rm P0} = 1, \quad \mu_{\rm P1} = \langle n\rangle, \quad \mu_{\rm P2} = \langle n\rangle + \langle n\rangle^2 \tag{13.186}$$

(exercise 13.22). □

Since the characteristic and moment-generating functions have derivatives (13.183) proportional to the moments μ_n, their Taylor series are

$$\hat{P}(k) = E[e^{ikx}] = \sum_{n=0}^{\infty} \frac{(ik)^n}{n!} E[x^n] = \sum_{n=0}^{\infty} \frac{(ik)^n}{n!} \mu_n \tag{13.187}$$

and

$$M(k) = E[e^{kx}] = \sum_{n=0}^{\infty} \frac{k^n}{n!} E[x^n] = \sum_{n=0}^{\infty} \frac{k^n}{n!} \mu_n. \tag{13.188}$$

The **cumulants** c_n of a probability distribution are the derivatives of the logarithm of its moment-generating function

$$c_n = \left.\frac{d^n \ln M(k)}{dk^n}\right|_{k=0} = (-i)^n \left.\frac{d^n \ln \hat{P}(k)}{dk^n}\right|_{k=0}. \tag{13.189}$$

One may show (exercise 13.24) that the first five cumulants of an arbitrary probability distribution are

$$c_0 = 0, \quad c_1 = \mu, \quad c_2 = \sigma^2, \quad c_3 = \nu_3, \quad \text{and} \quad c_4 = \nu_4 - 3\sigma^4 \tag{13.190}$$

where the νs are its central moments (13.27). The third and fourth **normalized cumulants** are the **skewness** $\zeta = c_3/\sigma^3 = \nu_3/\sigma^3$ and the **kurtosis** $\kappa = c_4/\sigma^4 = \nu_4/\sigma^4 - 3$.

Example 13.13 (Gaussian cumulants) The logarithm of the moment-generating function (13.184) of Gauss's distribution is $\mu k + \sigma^2 k^2/2$. Thus by (13.189), $P_G(x, \mu, \sigma)$ has no skewness or kurtosis, its cumulants vanish $c_{Gn} = 0$ for $n > 2$, and its fourth central moment is $\nu_4 = 3\sigma^4$. □

13.13 Fat tails

The gaussian probability distribution $P_G(x, \mu, \sigma)$ falls off for $|x - \mu| \gg \sigma$ very fast – as $\exp\left(-(x-\mu)^2/2\sigma^2\right)$. Many other probability distributions fall off more slowly; they have **fat tails**. Rare "black-swan" events – wild fluctuations, market bubbles, and crashes – lurk in their fat tails.

Gosset's distribution, which is known as **Student's t-distribution** with ν degrees of freedom

$$P_S(x, \nu, a) = \frac{1}{\sqrt{\pi}} \frac{\Gamma((1+\nu)/2)}{\Gamma(\nu/2)} \frac{a^\nu}{(a^2 + x^2)^{(1+\nu)/2}}, \tag{13.191}$$

has **power-law tails**. Its even moments are

$$\mu_{2n} = (2n-1)!! \, \frac{\Gamma(\nu/2 - n)}{\Gamma(\nu/2)} \left(\frac{a^2}{2}\right)^n \tag{13.192}$$

for $2n < \nu$ and infinite otherwise. For $\nu = 1$, it coincides with the Breit–Wigner or Cauchy distribution

$$P_S(x, 1, a) = \frac{1}{\pi} \frac{a}{a^2 + x^2}, \tag{13.193}$$

in which $x = E - E_0$ and $a = \Gamma/2$ is the half-width at half-maximum.

Two representative cumulative probabilities are (Bouchaud and Potters, 2003, p.15–16)

$$\Pr(x,\infty) = \int_x^\infty P_S(x',3,1)\,dx' = \frac{1}{2} - \frac{1}{\pi}\left[\arctan x + \frac{x}{1+x^2}\right], \tag{13.194}$$

$$\Pr(x,\infty) = \int_x^\infty P_S(x',4,\sqrt{2})\,dx' = \frac{1}{2} - \frac{3}{4}u + \frac{1}{4}u^3 \tag{13.195}$$

where $u = x/\sqrt{2+x^2}$ and a is picked so $\sigma^2 = 1$. William Gosset (1876–1937), who worked for Guinness, wrote as Student because Guinness didn't let its employees publish.

The **log-normal** probability distribution on $(0,\infty)$

$$P_{\ln}(x) = \frac{1}{\sigma x\sqrt{2\pi}}\exp\left[-\frac{\ln^2(x/x_0)}{2\sigma^2}\right] \tag{13.196}$$

describes distributions of rates of return (Bouchaud and Potters, 2003, p. 9). Its moments are (exercise 13.27)

$$\mu_n = x_0^n\, e^{n^2\sigma^2/2}. \tag{13.197}$$

The **exponential distribution** on $[0,\infty)$

$$P_e(x) = \alpha e^{-\alpha x} \tag{13.198}$$

has (exercise 13.28) mean $\mu = 1/\alpha$ and variance $\sigma^2 = 1/\alpha^2$. The sum of n independent exponentially and identically distributed random variables $x = x_1 + \cdots + x_n$ is distributed on $[0,\infty)$ as (Feller, 1966, p.10)

$$P_{n,e}(x) = \alpha\frac{(\alpha x)^{n-1}}{(n-1)!}e^{-\alpha x}. \tag{13.199}$$

The sum of the squares $x^2 = x_1^2 + \cdots + x_n^2$ of n independent normally and identically distributed random variables of zero mean and variance σ^2 gives rise to Pearson's **chi-squared distribution** on $(0,\infty)$

$$P_{n,G}(x,\sigma)dx = \frac{\sqrt{2}}{\sigma}\frac{1}{\Gamma(n/2)}\left(\frac{x}{\sigma\sqrt{2}}\right)^{n-1} e^{-x^2/(2\sigma^2)}dx, \tag{13.200}$$

which for $x = v$, $n = 3$, and $\sigma^2 = kT/m$ is (exercise 13.29) the Maxwell–Boltzmann distribution (13.100). In terms of $\chi = x/\sigma$, it is

$$P_{n,G}(\chi^2/2)\,d\chi^2 = \frac{1}{\Gamma(n/2)}\left(\frac{\chi^2}{2}\right)^{n/2-1} e^{-\chi^2/2}d\left(\chi^2/2\right). \tag{13.201}$$

It has mean and variance

$$\mu = n \quad \text{and} \quad \sigma^2 = 2n \tag{13.202}$$

and is used in the chi-squared test (Pearson, 1900).

Personal income, the amplitudes of catastrophes, the price changes of financial assets, and many other phenomena occur on both small and large scales. **Lévy** distributions describe such multi-scale phenomena. The characteristic function for a symmetric Lévy distribution is for $\nu \le 2$

$$\hat{L}_\nu(k, a_\nu) = \exp\left(- a_\nu |k|^\nu \right). \tag{13.203}$$

Its inverse Fourier transform (13.174) is for $\nu = 1$ (exercise 13.30) the **Cauchy** or **Lorentz** distribution

$$L_1(x, a_1) = \frac{a_1}{\pi(x^2 + a_1^2)} \tag{13.204}$$

and for $\nu = 2$ the gaussian

$$L_2(x, a_2) = P_G(x, 0, \sqrt{2a_2}) = \frac{1}{2\sqrt{\pi a_2}} \exp\left(- \frac{x^2}{4a_2} \right) \tag{13.205}$$

but for other values of ν no simple expression for $L_\nu(x, a_\nu)$ is available. For $0 < \nu < 2$ and as $x \to \pm\infty$, it falls off as $|x|^{-(1+\nu)}$, and for $\nu > 2$ it assumes negative values, ceasing to be a probability distribution (Bouchaud and Potters, 2003, pp. 10–13).

13.14 The central limit theorem and Jarl Lindeberg

We have seen in sections 13.7 and 13.8 that unbiased fluctuations tend to distribute the position and velocity of molecules according to Gauss's distribution (13.75). Gaussian distributions occur very frequently. The **central limit theorem** suggests why they occur so often.

Let $x_1, \ldots, x_N$ be N **independent** random variables described by probability distributions $P_1(x_1), \ldots, P_N(x_N)$ with finite means μ_j and finite variances σ_j^2. The P_js may be all different. The central limit theorem says that as $N \to \infty$ the probability distribution $P^{(N)}(y)$ for the average of the x_js

$$y = \frac{1}{N}(x_1 + x_2 + \cdots + x_N) \tag{13.206}$$

tends to a gaussian in y quite independently of what the underlying probability distributions $P_j(x_j)$ happen to be.

Because expected values are linear (13.34), the mean value of the average y is the average of the N means

$$\begin{aligned}\mu_y = E[y] = E[(x_1 + \cdots + x_N)/N] &= \frac{1}{N}(E[x_1] + \cdots + E[x_N]) \\ &= \frac{1}{N}(\mu_1 + \cdots + \mu_N).\end{aligned} \tag{13.207}$$

Similarly, our rule (13.41) for the variance of a linear combination of *independent* variables tells us that the variance of the average y is

$$\sigma_y^2 = V[(x_1 + \cdots + x_N)/N] = \frac{1}{N^2}\left(\sigma_1^2 + \cdots + \sigma_N^2\right). \tag{13.208}$$

The independence of the random variables $x_1, x_2, \ldots, x_N$ implies (13.36) that their joint probability distribution factorizes

$$P(x_1, \ldots, x_N) = P_1(x_1)P_2(x_2)\cdots P_N(x_N). \tag{13.209}$$

We can use a delta function (3.36) to write the probability distribution $P^{(N)}(y)$ for the average $y = (x_1 + x_2 + \cdots + x_N)/N$ of the x_js as

$$P^{(N)}(y) = \int P(x_1, \ldots, x_N)\,\delta((x_1 + x_2 + \cdots + x_N)/N - y)\,d^N x \tag{13.210}$$

where $d^N x = dx_1 \ldots dx_N$. Its characteristic function

$$\begin{aligned}
\hat{P}^{(N)}(k) &= \int e^{iky}\,P^{(N)}(y)\,dy \\
&= \int e^{iky} \int P(x_1, \ldots, x_N)\,\delta((x_1 + x_2 + \cdots + x_N)/N - y)\,d^N x\,dy \\
&= \int \exp\left[\frac{ik}{N}(x_1 + x_2 + \cdots + x_N)\right] P(x_1, \ldots, x_N)\,d^N x \\
&= \int \exp\left[\frac{ik}{N}(x_1 + x_2 + \cdots + x_N)\right] P_1(x_1)P_2(x_2)\cdots P_N(x_N)\,d^N x
\end{aligned} \tag{13.211}$$

is then the product

$$\hat{P}^{(N)}(k) = \hat{P}_1(k/N)\,\hat{P}_2(k/N)\,\cdots\,\hat{P}_N(k/N) \tag{13.212}$$

of the characteristic functions

$$\hat{P}_j(k/N) = \int e^{ikx_j/N}\,P_j(x_j)\,dx \tag{13.213}$$

of the probability distributions $P_1(x_1), \ldots, P_N(x_N)$.

The Taylor series (13.187) for each characteristic function is

$$\hat{P}_j(k/N) = \sum_{n=0}^{\infty} \frac{(ik)^n}{n!\,N^n}\,\mu_{nj} \tag{13.214}$$

and so for big N we can use the approximation

$$\hat{P}_j(k/N) \approx 1 + \frac{ik}{N}\,\mu_j - \frac{k^2}{2N^2}\,\mu_{2j}, \tag{13.215}$$

in which $\mu_{2j} = \sigma_j^2 + \mu_j^2$ by the formula (13.22) for the variance. So we have

$$\hat{P}_j(k/N) \approx 1 + \frac{ik}{N}\mu_j - \frac{k^2}{2N^2}\left(\sigma_j^2 + \mu_j^2\right) \tag{13.216}$$

or for large N

$$\hat{P}_j(k/N) \approx \exp\left(\frac{ik}{N}\mu_j - \frac{k^2}{2N^2}\sigma_j^2\right). \tag{13.217}$$

Thus as $N \to \infty$, the characteristic function (13.212) for the variable y converges to

$$\begin{aligned} \hat{P}^{(N)}(k) &= \prod_{j=1}^{N} \hat{P}_j(k/N) = \prod_{j=1}^{N} \exp\left(\frac{ik}{N}\mu_j - \frac{k^2}{2N^2}\sigma_j^2\right) \\ &= \exp\left[\sum_{j=1}^{N}\left(\frac{ik}{N}\mu_j - \frac{k^2}{2N^2}\sigma_j^2\right)\right] = \exp\left(i\mu_y k - \frac{1}{2}\sigma_y^2 k^2\right), \end{aligned} \tag{13.218}$$

which is the characteristic function (13.176) of a gaussian (13.175) with mean and variance

$$\mu_y = \frac{1}{N}\sum_{j=1}^{N}\mu_j \quad \text{and} \quad \sigma_y^2 = \frac{1}{N^2}\sum_{j=1}^{N}\sigma_j^2. \tag{13.219}$$

The inverse Fourier transform (13.174) now gives the probability distribution $P^{(N)}(y)$ for the average $y = (x_1 + x_2 + \cdots + x_N)/N$ as

$$P^{(N)}(y) = \int_{-\infty}^{\infty} e^{-iky}\, \hat{P}^{(N)}(k)\, \frac{dk}{2\pi}, \tag{13.220}$$

which in view of (13.218) and (13.176) tends as $N \to \infty$ to Gauss's distribution $P_G(y, \mu_y, \sigma_y)$

$$\begin{aligned} \lim_{N\to\infty} P^{(N)}(y) &= \int_{-\infty}^{\infty} e^{-iky} \lim_{N\to\infty} \hat{P}^{(N)}(k)\, \frac{dk}{2\pi} \\ &= \int_{-\infty}^{\infty} e^{-iky} \exp\left(i\mu_y k - \frac{1}{2}\sigma_y^2 k^2\right) \frac{dk}{2\pi} \\ &= P_G(y, \mu_y, \sigma_y) = \frac{1}{\sigma_y\sqrt{2\pi}} \exp\left[-\frac{(y-\mu_y)^2}{2\sigma_y^2}\right] \end{aligned} \tag{13.221}$$

with mean μ_y and variance σ_y^2 as given by (13.219). The sense in which $P^{(N)}(y)$ converges to $P_G(y, \mu_y, \sigma_y)$ is that for all a and b the probability $\Pr_N(a < y < b)$ that y lies between a and b as determined by $P^{(N)}(y)$ converges as $N \to \infty$

to the probability that y lies between a and b as determined by the gaussian $P_G(y, \mu_y, \sigma_y)$

$$\lim_{N\to\infty} \Pr_N(a < y < b) = \lim_{N\to\infty} \int_a^b P^{(N)}(y)\, dy = \int_a^b P_G(y, \mu_y, \sigma_y)\, dy. \quad (13.222)$$

This type of convergence is called **convergence in probability** (Feller, 1966, pp. 231, 241–248).

For the special case in which all the means and variances are the same, with $\mu_j = \mu$ and $\sigma_j^2 = \sigma^2$, the definitions in (13.219) imply that $\mu_y = \mu$ and $\sigma_y^2 = \sigma^2/N$. In this case, one may show (exercise 13.32) that in terms of the variable

$$u \equiv \frac{\sqrt{N}(y-\mu)}{\sigma} = \frac{\left(\sum_{n=1}^{N} x_j\right) - N\mu}{\sqrt{N}\,\sigma} \quad (13.223)$$

$P^{(N)}(y)$ converges to a distribution that is normal

$$\lim_{N\to\infty} P^{(N)}(y)\, dy = \frac{1}{\sqrt{2\pi}} e^{-u^2/2}\, du. \quad (13.224)$$

To get a clearer idea of when the **central limit theorem** holds, let us write the sum of the N variances as

$$S_N \equiv \sum_{j=1}^{N} \sigma_j^2 = \sum_{j=1}^{N} \int_{-\infty}^{\infty} (x_j - \mu_j)^2\, P_j(x_j)\, dx_j \quad (13.225)$$

and the part of this sum due to the regions within δ of the means μ_j as

$$S_N(\delta) \equiv \sum_{j=1}^{N} \int_{\mu_j-\delta}^{\mu_j+\delta} (x_j - \mu_j)^2\, P_j(x_j)\, dx_j. \quad (13.226)$$

In terms of these definitions, Jarl Lindeberg (1876–1932) showed that $P^{(N)}(y)$ converges (in probability) to the gaussian (13.221) as long as the part $S_N(\delta)$ is most of S_N in the sense that for every $\epsilon > 0$

$$\lim_{N\to\infty} \frac{S_N\left(\epsilon\sqrt{S_N}\right)}{S_N} = 1. \quad (13.227)$$

This is **Lindeberg's condition** (Feller, 1968, p. 254; Feller, 1966, pp. 252–259; Gnedenko, 1968, p. 304).

Because we dropped all but the first three terms of the series (13.214) for the characteristic functions $\hat{P}_j(k/N)$, we may infer that the convergence of the distribution $P^{(N)}(y)$ to a gaussian is quickest near its mean μ_y. If the higher moments μ_{nj} are big, then for finite N the distribution $P^{(N)}(y)$ can have tails that are fatter than those of the limiting gaussian $P_G(y, \mu_y, \sigma_y)$.

Example 13.14 (Illustration of the central limit theorem) The simplest probability distribution is a random number x uniformly distributed on the interval $(0, 1)$. The probability distribution $P^{(2)}(y)$ of the mean of two such random numbers is the integral

$$P^{(2)}(y) = \int_0^1 dx_1 \int_0^1 dx_2 \; \delta((x_1 + x_2)/2 - y). \tag{13.228}$$

Letting $u_1 = x_1/2$ and $u_2 = x_2/2$, we find

$$P^{(2)}(y) = 4 \int_{\max(0,y-\frac{1}{2})}^{\min(y,\frac{1}{2})} \theta(\tfrac{1}{2}+u_1-y)\, du_1 = 4y\, \theta(\tfrac{1}{2}-y)+4(1-y)\, \theta(y-\tfrac{1}{2}), \tag{13.229}$$

which is the dot-dashed triangle in Fig. 13.4. The probability distribution $P^{(4)}(y)$ is the dashed somewhat gaussian curve in the figure, while $P^{(8)}(y)$ is the solid, nearly gaussian curve.

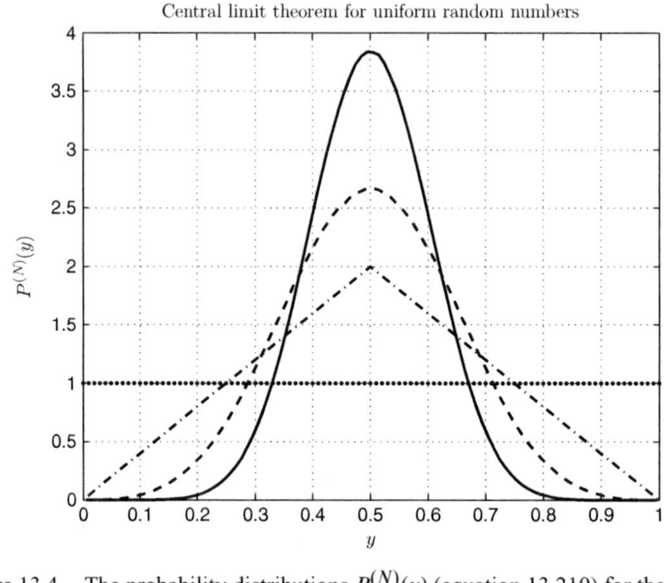

Figure 13.4 The probability distributions $P^{(N)}(y)$ (equation 13.210) for the mean $y = (x_1 + \cdots + x_N)/N$ of N random variables drawn from the uniform distribution are plotted for $N = 1$ (dots), 2 (dot dash), 4 (dashes), and 8 (solid). The distributions $P^{(N)}(y)$ rapidly approach gaussians with the same mean $\mu_y = 1/2$ but with shrinking variances $\sigma^2 = 1/12N$.

□

To work through a more complicated example of the central limit theorem, we first need to learn how to generate random numbers that follow an arbitrary distribution.

13.15 Random-number generators

To generate truly random numbers, one might use decaying nuclei or an electronic device that makes white noise. But people usually settle for **pseudo-random numbers** computed by a mathematical algorithm. Such algorithms are deterministic, so the numbers they generate are not truly random. But for most purposes, they are random enough.

The easiest way to generate pseudo-random numbers is to use a random-number algorithm that is part of one's favorite FORTRAN, C, or C^{++} compiler. To run it, one first gives it a random starting point called a **seed**, which is a number or a vector. For instance, to start the GNU or Intel FORTRAN90 compiler, one includes in the code the line

```
call random_seed()
```

before using the line

```
call random_number(x)
```

to generate a random number x uniformly distributed on the interval $0 < x < 1$, or an array of such random numbers.

Some applications require random numbers of very high quality. For such applications, one might use Lüscher's RANLUX (Lüscher, 1994; James, 1994).

Most random-number generators are periodic with very long periods. The **Mersenne twister** (Saito and Matsumoto, 2007) has the exceptionally long period $2^{19937} - 1 \gtrsim 4.3 \times 10^{6001}$. Matlab uses it.

Random-number generators distribute random numbers uniformly on the interval $(0, 1)$. How do we make them follow an arbitrary distribution $P(x)$? If the distribution is strictly positive $P(x) > 0$ on the relevant interval (a, b), then its integral

$$F(x) = \int_a^x P(x')\, dx' \tag{13.230}$$

is a strictly increasing function on (a, b), that is, $a < x < y < b$ implies $F(x) < F(y)$. Moreover, the function $F(x)$ rises from $F(a) = 0$ to $F(b) = 1$ and takes on every value $0 < y < 1$ for exactly one x in the interval (a, b). Thus the inverse function $F^{-1}(y)$

$$x = F^{-1}(y) \quad \text{if and only if} \quad y = F(x) \tag{13.231}$$

is well defined on the interval (0, 1).

Our random-number generator gives us random numbers u that are uniform on (0, 1). We want a random variable r whose probability $\Pr(r < x)$ of being less than x is $F(x)$. The trick (Knuth, 1981, p. 116) is to set

$$r = F^{-1}(u) \tag{13.232}$$

so that $\Pr(r < x) = \Pr(F^{-1}(u) < x)$. For by (13.231) $F^{-1}(u) < x$ if and only if $u < F(x)$. So $\Pr(r < x) = \Pr(F^{-1}(u) < x) = \Pr(u < F(x)) = F(x)$. The trick works.

Example 13.15 ($P(r) = 3r^2$) To turn a distribution of random numbers u uniform on (0, 1) into a distribution $P(r) = 3r^2$ of random numbers r, we integrate and find

$$F(x) = \int_0^x P(x')\,dx' = \int_0^x 3x'^2\,dx' = x^3. \tag{13.233}$$

We then set $r = F^{-1}(u) = u^{1/3}$. □

13.16 Illustration of the central limit theorem

To make things simple, we'll take all the probability distributions $P_j(x)$ to be the same and equal to $P_j(x_j) = 3x_j^2$ on the interval (0, 1) and zero elsewhere. Our random-number generator gives us random numbers u that are uniformly distributed on (0, 1), so by the example (13.15) the variable $r = u^{1/3}$ is distributed as $P_j(x) = 3x^2$.

The central limit theorem tells us that the distribution

$$P^{(N)}(y) = \int 3x_1^2\, 3x_2^2 \ldots 3x_N^2\, \delta((x_1 + x_2 + \cdots + x_N)/N - y)\, d^N x \tag{13.234}$$

of the mean $y = (x_1 + \cdots + x_N)/N$ tends as $N \to \infty$ to Gauss's distribution

$$\lim_{N\to\infty} P^{(N)}(y) = \frac{1}{\sigma_y\sqrt{2\pi}} \exp\left(-\frac{(x-\mu_y)^2}{2\sigma_y^2}\right) \tag{13.235}$$

with mean μ_y and variance σ_y^2 given by (13.219). Since the P_js are all the same, they all have the same mean

$$\mu_y = \mu_j = \int_0^1 3x^3 dx = \frac{3}{4} \tag{13.236}$$

and the same variance

$$\sigma_j^2 = \int_0^1 3x^4 dx - \left(\frac{3}{4}\right)^2 = \frac{3}{5} - \frac{9}{16} = \frac{3}{80}. \tag{13.237}$$

By (13.219), the variance of the mean y is then $\sigma_y^2 = 3/80N$. Thus as N increases, the mean y tends to a gaussian with mean $\mu_y = 3/4$ and ever narrower peaks.

For $N = 1$, the probability distribution $P^{(1)}(y)$ is

$$P^{(1)}(y) = \int 3x_1^2 \, \delta(x_1 - y) \, dx_1 = 3y^2, \tag{13.238}$$

which is the probability distribution we started with. In Fig. 13.5, this is the quadratic, dotted curve.

For $N = 2$, the probability distribution $P^{(1)}(y)$ is (exercise 13.31)

$$\begin{aligned} P^{(2)}(y) &= \int 3x_1^2 \, 3x_2^2 \, \delta((x_1 + x_2)/2 - y) \, dx_1 \, dx_2 \\ &= \theta(\tfrac{1}{2} - y) \frac{96}{5} y^5 + \theta(y - \tfrac{1}{2}) \left(\frac{36}{5} - \frac{96}{5} y^5 + 48y^2 - 36y \right). \end{aligned} \tag{13.239}$$

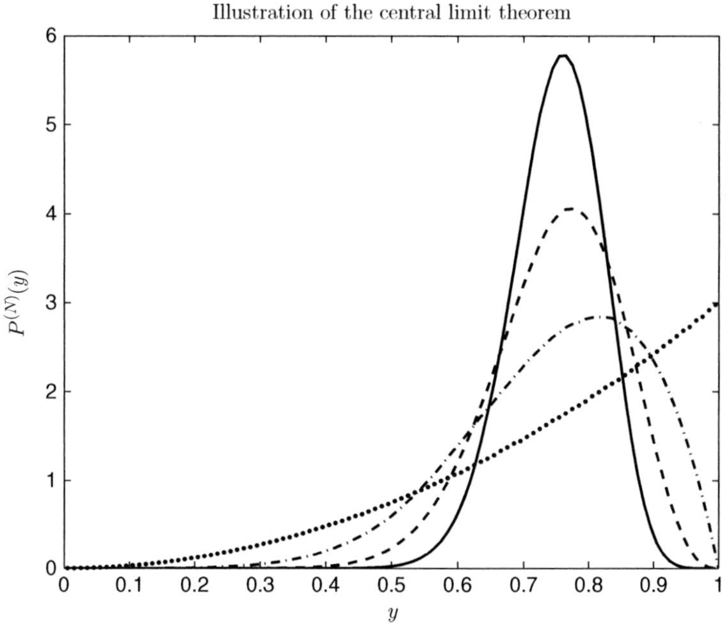

Figure 13.5 The probability distributions $P^{(N)}(y)$ (equation 13.234) for the mean $y = (x_1 + \cdots + x_N)/N$ of N random variables drawn from the quadratic distribution $P(x) = 3x^2$ are plotted for $N = 1$ (dots), 2 (dot dash), 4 (dashes), and 8 (solid). The four distributions $P^{(N)}(y)$ rapidly approach gaussians with the same mean $\mu_y = 3/4$ but with shrinking variances $\sigma_y^2 = 3/80N$.

The probability distributions $P^{(N)}(y)$ for $N = 2^j$ can be obtained by running the FORTAN95 program

```
program clt
  implicit none ! avoids typos
  character(len=1)::ch_i1
  integer,parameter::dp = kind(1.d0) !define double
    precision
  integer::j,k,n,m
  integer,dimension(100)::plot = 0
  real(dp)::y
  real(dp),dimension(100)::rplot
  real(dp),allocatable,dimension(:)::r,u
  real(dp),parameter::onethird = 1.d0/3.d0
  write(6,*)'What is j?'; read(5,*) j
  allocate(u(2**j),r(2**j))
  call init_random_seed() ! set new seed, see below
  do k = 1, 10000000  ! Make the N = 2**j plot
     call random_number(u)
     r = u**onethird
     y = sum(r)/2**j
     n = 100*y + 1
     plot(n) = plot(n) + 1
  end do
  rplot = 100*real(plot)/sum(plot)
  write(ch_i1,"(i1)") j ! turns integer j into
    character ch_i1
  open(7,file='plot'//ch_i1) ! opens and names files
  do m = 1, 100
     write(7,*) 0.01d0*(m-0.5), rplot(m)
  end do
end program clt
subroutine init_random_seed()
  implicit none
  integer i, n, clock
  integer, dimension(:), allocatable :: seed
  call random_seed(size = n) ! find size of seed
  allocate(seed(n))
  call system_clock(count=clock)!get time of processor
    clock
  seed = clock + 37 * (/ (i-1, i=1, n) /) ! make seed
  call random_seed(put=seed) ! set seed
```

```
  deallocate(seed)
end subroutine init_random_seed
```

The distributions $P^{(N)}(y)$ for $N = 1, 2, 4,$ and 8 are plotted in Fig. 13.5. $P^{(1)}(y) = 3y^2$ is the original distribution. $P^{(2)}(y)$ is trying to be a gaussian, while $P^{(4)}(y)$ and $P^{(8)}(y)$ have almost succeeded. The variance $\sigma_y^2 = 3/80N$ shrinks with N.

Although FORTRAN95 is an ideal language for computation, C++ is more versatile, more modular, and more suited to large projects involving many programmers. An equivalent C++ code written by Sean Cahill is:

```
#include <stdlib.h>
#include <time.h>
#include <math.h>
#include <string>
#include <iostream>
#include <fstream>
#include <sstream>
#include <iomanip>
#include <valarray>

using namespace std;

// Fills the array val with random numbers between
0 and 1 void rand01(valarray<double>& val)
{
  // Records the size
  unsigned int size = val.size();

  // Loops through the size
  unsigned int i=0;
  for (i=0; i<size; i++)
  {
    // Generates a random number between 0 and 1
    val[i] = static_cast<double>(rand()) / RAND_MAX;
  }
}

void clt ()
{
  // Declares local constants
  const int PLOT_SIZE      = 100;
```

```
const int LOOP_CALC_ITR   = 10000000;
const double ONE_THIRD    = 1.0 / 3.0;

// Inits local variables
double y=0;
int i=0, j=0, n=0;

// Gets the value of J
cout << "What is J? ";
cin >> j;

// Bases the vec size on J
const int VEC_SIZE = static_cast<int>(pow(2.0,j));

// Inits vectors
valarray<double> plot(PLOT_SIZE);
valarray<double> rplot(PLOT_SIZE);
valarray<double> r(VEC_SIZE);

// Seeds random number generator
srand ( time(NULL) );

// Performs the calculations
for (i=0; i<LOOP_CALC_ITR; i++)
{
  rand01(r);
  r = pow(r, ONE_THIRD);
  y = r.sum() / VEC_SIZE;
  n = static_cast<int>(100 * y);
  plot[n]++;
}

// Normalizes RPLOT
rplot = plot * (100.0 / plot.sum());

// Opens a data file
ostringstream fileName;
fileName << "plot_" << j << ".txt";
ofstream fileHandle;
fileHandle.open (fileName.str().c_str());

// Sets precision
```

```
  fileHandle.setf(ios::fixed,ios::floatfield);
  fileHandle.precision(7);

  // Writes the data to a file
  for (i=1; i<=PLOT_SIZE; i++)
   fileHandle << 0.01*(i-0.5) << "     " << rplot[i-1]
     << endl;

  // Closes the data file
  fileHandle.close();
}
```

13.17 Measurements, estimators, and Friedrich Bessel

A probability distribution $P(x;\boldsymbol{\theta})$ for a stochastic variable x may depend upon one or more unknown parameters $\boldsymbol{\theta} = (\theta_1, \dots, \theta_m)$ such as the mean μ and the variance σ^2.

Experimenters seek to determine the unknown parameters $\boldsymbol{\theta}$ by collecting data in the form of values $\boldsymbol{x} = x_1, \dots, x_N$ of the stochastic variable x. They assume that the probability distribution for the sequence $\boldsymbol{x} = (x_1, \dots, x_N)$ is the product of N factors of the physical distribution $P(x;\boldsymbol{\theta})$

$$P(\boldsymbol{x};\boldsymbol{\theta}) = \prod_{j=1}^{N} P(x_j;\boldsymbol{\theta}). \tag{13.240}$$

They approximate the unknown value of a parameter θ_ℓ as the mean value of an **estimator** $u_\ell^{(N)}(\boldsymbol{x})$ of θ_ℓ

$$E[u_\ell^{(N)}] = \int u_\ell^{(N)}(\boldsymbol{x})\, P(\boldsymbol{x};\boldsymbol{\theta})\, d^N x = \theta_\ell + b_\ell^{(N)}, \tag{13.241}$$

in which the **bias** $b_\ell^{(N)}$ depends upon $\boldsymbol{\theta}$ and N. If as $N \to \infty$, the bias $b_\ell^{(N)} \to 0$, then the estimator $u_\ell^{(N)}(\boldsymbol{x})$ is **consistent**.

Inasmuch as the mean (13.25) is the integral of the physical distribution

$$\mu = \int x\, P(x;\boldsymbol{\theta})\, dx \tag{13.242}$$

a natural estimator for the mean is

$$u_\mu^{(N)}(\boldsymbol{x}) = (x_1 + \cdots + x_N)/N. \tag{13.243}$$

Its expected value is

$$E[u_\mu^{(N)}] = \int u_\mu^{(N)}(\boldsymbol{x})\, P(\boldsymbol{x};\boldsymbol{\theta})\, d^N x = \int \frac{x_1 + \cdots + x_N}{N}\, P(\boldsymbol{x};\boldsymbol{\theta})\, d^N x$$

$$= \frac{1}{N} \sum_{k=1}^{N} \int x_k \, P(x_k;\boldsymbol{\theta}) \, dx_k \prod_{k \neq j=1}^{N} \int P(x_j;\boldsymbol{\theta}) \, dx_j$$

$$= \frac{1}{N} \sum_{k=1}^{N} \mu = \mu. \tag{13.244}$$

Thus the natural estimator $u_\mu^{(N)}(\boldsymbol{x})$ of the mean (13.243) has $b_\ell^{(N)} = 0$, and so it is a consistent and unbiased estimator for the mean.

Since the variance (13.28) of the probability distribution $P(x;\boldsymbol{\theta})$ is the integral

$$\sigma^2 = \int (x - \mu)^2 \, P(x;\boldsymbol{\theta}) \, dx, \tag{13.245}$$

that of the estimator u_μ^N is

$$V[u_\mu^N] = \int \left(u_\mu^{(N)}(\boldsymbol{x}) - \mu \right)^2 P(\boldsymbol{x};\boldsymbol{\theta}) \, d^N x = \int \left[\frac{1}{N} \sum_{j=1}^{N} (x_j - \mu) \right]^2 P(\boldsymbol{x};\boldsymbol{\theta}) \, d^N x$$

$$= \frac{1}{N^2} \sum_{j,k=1}^{N} \int (x_j - \mu) \, (x_k - \mu) \, P(\boldsymbol{x};\boldsymbol{\theta}) \, d^N x$$

$$= \frac{1}{N^2} \sum_{j,k=1}^{N} \delta_{jk} \int (x_j - \mu)^2 \, P(\boldsymbol{x};\boldsymbol{\theta}) \, d^N x = \frac{1}{N^2} \sum_{j,k=1}^{N} \delta_{jk} \, \sigma^2 = \frac{\sigma^2}{N}, \tag{13.246}$$

in which σ^2 is the variance (13.245) of the physical distribution $P(x;\boldsymbol{\theta})$. We'll learn in the next section that no estimator of the mean can have a lower variance than this.

A natural estimator for the variance of the probability distribution $P(x;\boldsymbol{\theta})$ is

$$u_{\sigma^2}^{(N)}(x) = B \sum_{j=1}^{N} \left(x_j - u_\mu^{(N)}(\boldsymbol{x}) \right)^2, \tag{13.247}$$

in which $B = B(N)$ is a constant of proportionality. The naive choice $B(N) = 1/N$ leads to a biased estimator. To find the correct value of B, we set the expected value $E[u_{\sigma^2}^{(N)}]$ equal to σ^2

$$E[u_{\sigma^2}^{(N)}] = \int B \sum_{j=1}^{N} \left(x_j - u_\mu^{(N)}(\boldsymbol{x}) \right)^2 P(\boldsymbol{x};\boldsymbol{\theta}) \, d^N x = \sigma^2 \tag{13.248}$$

and solve for B. Subtracting the mean μ from both x_j and $u_\mu^{(N)}(\boldsymbol{x})$, we express σ^2/B as the sum of three terms

$$\frac{\sigma^2}{B} = \sum_{j=1}^{N} \int \left[x_j - \mu - \left(u_\mu^{(N)}(\boldsymbol{x}) - \mu\right)\right]^2 P(\boldsymbol{x};\boldsymbol{\theta})\, d^N x = S_{jj}+S_{j\mu}+S_{\mu\mu} \quad (13.249)$$

the first of which is

$$S_{jj} = \sum_{j=1}^{N} \int (x_j - \mu)^2\, P(\boldsymbol{x};\boldsymbol{\theta})\, d^N x = N\sigma^2. \quad (13.250)$$

The cross-term $S_{j\mu}$ is

$$\begin{aligned} S_{j\mu} &= -2\sum_{j=1}^{N} \int (x_j - \mu)\left(u_\mu^{(N)}(\boldsymbol{x}) - \mu\right) P(\boldsymbol{x};\boldsymbol{\theta})\, d^N x \quad (13.251) \\ &= -\frac{2}{N}\sum_{j=1}^{N} \int (x_j - \mu) \sum_{k=1}^{N} (x_k - \mu)\, P(\boldsymbol{x};\boldsymbol{\theta})\, d^N x = -2\sigma^2. \end{aligned}$$

The third term is related to the variance (13.246)

$$S_{\mu\mu} = \sum_{j=1}^{N} \int \left(u_\mu^{(N)}(\boldsymbol{x}) - \mu\right)^2 P(\boldsymbol{x};\boldsymbol{\theta})\, d^N x = NV[u_\mu^N] = \sigma^2. \quad (13.252)$$

Thus the factor B must satisfy

$$\sigma^2/B = N\sigma^2 - 2\sigma^2 + \sigma^2 = (N-1)\sigma^2, \quad (13.253)$$

which tells us that $B = 1/(N-1)$, which is **Bessel's correction**. Our estimator for the variance of the probability distribution $P(x;\boldsymbol{\theta})$ then is

$$u_{\sigma^2}^{(N)}(\boldsymbol{x}) = \frac{1}{N-1}\sum_{j=1}^{N}\left(x_j - u_\mu^{(N)}(\boldsymbol{x})\right)^2 = \frac{1}{N-1}\sum_{j=1}^{N}\left(x_j - \frac{1}{N}\sum_{k=1}^{N} x_k\right)^2. \quad (13.254)$$

It is consistent and unbiased since $E[u_{\sigma^2}^{(N)}] = \sigma^2$ by construction (13.248). It gives for the variance σ^2 of a single measurement the undefined ratio 0/0, as it should, whereas the naive choice $B = 1/N$ absurdly gives zero.

On the basis of N measurements $x_1, \ldots, x_N$ we can estimate the mean of the unknown probability distribution $P(x;\boldsymbol{\theta})$ as $\mu_N = (x_1 + \cdots + x_N)/N$. And we can use Bessel's formula (13.254) to estimate the variance σ_N^2 of the unknown distribution $P(x;\boldsymbol{\theta})$. Our formula (13.246) for the variance $\sigma^2(\mu_N)$ of the mean μ_N then gives

$$\sigma^2(\mu_N) = \frac{\sigma_N^2}{N} = \frac{1}{N(N-1)} \sum_{j=1}^{N} \left(x_j - \frac{1}{N} \sum_{k=1}^{N} x_k \right)^2 . \tag{13.255}$$

Thus we can use N measurements x_j to estimate the mean μ to within a standard error or standard deviation of

$$\sigma(\mu_N) = \sqrt{\frac{\sigma_N^2}{N}} = \sqrt{\frac{1}{N(N-1)} \sum_{j=1}^{N} \left(x_j - \frac{1}{N} \sum_{k=1}^{N} x_k \right)^2} . \tag{13.256}$$

Few formulas have seen so much use.

13.18 Information and Ronald Fisher

The **Fisher information matrix** of a distribution $P(\boldsymbol{x};\boldsymbol{\theta})$ is the mean of products of its partial logarithmic derivatives

$$\begin{aligned} F_{k\ell}(\boldsymbol{\theta}) &\equiv E\left[\frac{\partial \ln P(\boldsymbol{x};\boldsymbol{\theta})}{\partial \theta_k} \frac{\partial \ln P(\boldsymbol{x};\boldsymbol{\theta})}{\partial \theta_\ell} \right] \\ &= \int \frac{\partial \ln P(\boldsymbol{x};\boldsymbol{\theta})}{\partial \theta_k} \frac{\partial \ln P(\boldsymbol{x};\boldsymbol{\theta})}{\partial \theta_\ell} P(\boldsymbol{x};\boldsymbol{\theta})\, d^N x \end{aligned} \tag{13.257}$$

(Ronald Fisher, 1890–1962). Fisher's matrix (exercise 13.33) is symmetric $F_{k\ell} = F_{\ell k}$ and nonnegative (1.38), and when it is positive (1.39), it has an inverse. By differentiating the normalization condition

$$\int P(\boldsymbol{x};\boldsymbol{\theta})\, d^N x = 1 \tag{13.258}$$

we have

$$0 = \int \frac{\partial P(\boldsymbol{x};\boldsymbol{\theta})}{\partial \theta_k}\, d^N x = \int \frac{\partial \ln P(\boldsymbol{x};\boldsymbol{\theta})}{\partial \theta_k} P(\boldsymbol{x};\boldsymbol{\theta})\, d^N x, \tag{13.259}$$

which says that the mean value of the logarithmic derivative of the probability distribution, a quantity called the **score**, vanishes. Using the notation $P_{,k} \equiv \partial P/\partial \theta_k$ and $(\ln P)_{,k} \equiv \partial \ln P/\partial \theta_k$ and differentiating again, one has (exercise 13.34)

$$\int (\ln P)_{,k} (\ln P)_{,\ell}\, P\, d^N x = -\int (\ln P)_{,k,\ell}\, P\, d^N x \tag{13.260}$$

so that another form of Fisher's information matrix is

$$F_{k\ell}(\boldsymbol{\theta}) = -E\left[(\ln P)_{,k,\ell} \right] = -\int (\ln P)_{,k,\ell}\, P\, d^N x. \tag{13.261}$$

Cramér and Rao used Fisher's information matrix to form a lower bound on the covariance (13.35) matrix $C[u_k, u_\ell]$ of any two estimators. To see how this works, we use the vanishing (13.259) of the mean of the score to write the covariance of the kth score $V_k \equiv (\ln P(\boldsymbol{x}; \boldsymbol{\theta}))_{,k}$ with the ℓth estimator $u_\ell(\boldsymbol{x})$ as a derivative $\langle u_\ell \rangle_{,k}$ of the mean $\langle u_\ell \rangle$

$$\begin{aligned} C[V_k, u_\ell] &= \int (\ln P)_{,k} (u_\ell - b_\ell - \theta_\ell) \, P \, d^N x = \int (\ln P)_{,k} \, u_\ell \, P \, d^N x \\ &= \int P_{,k}(\boldsymbol{x}; \boldsymbol{\theta}) \, u_\ell(\boldsymbol{x}) \, d^N x = \langle u_\ell \rangle_{,k}. \end{aligned} \tag{13.262}$$

Thus for any two sets of constants y_k and w_ℓ, we have with $P = \sqrt{P}\sqrt{P}$

$$\sum_{\ell,k=1}^{m} y_k \, \partial_k \langle u_\ell \rangle \, w_\ell = \int \sum_{\ell,k=1}^{m} y_k \, (\ln P)_{,k} \sqrt{P} \, (u_\ell - b_\ell - \theta_\ell) \, w_\ell \, \sqrt{P} \, d^N x. \tag{13.263}$$

We can suppress some indices by grouping the y_js, the w_js, and so forth into the vectors $Y^{\mathsf{T}} = (y_1, \ldots, y_m)$, $W^{\mathsf{T}} = (w_1, \ldots, w_m)$, $U^{\mathsf{T}} = (u_1, \ldots, u_m)$, $B^{\mathsf{T}} = (b_1, \ldots, b_m)$, and $\Theta^{\mathsf{T}} = (\theta_1, \ldots, \theta_m)$, and by grouping the $\partial_k \langle u_\ell \rangle$s into a matrix $(\nabla \overline{U})_{kl}$, which by (13.241) is

$$(\nabla \overline{U})_{kl} \equiv \partial_k \langle u_\ell \rangle = \partial_k \, (\theta_\ell + b_\ell) = \delta_{kl} + \partial_k b_\ell. \tag{13.264}$$

In this compact notation, our relation (13.263) is

$$Y^{\mathsf{T}} \, \nabla \overline{U} \, W = \int Y^{\mathsf{T}} (\nabla \ln P) \sqrt{P} (U^{\mathsf{T}} - B^{\mathsf{T}} - \Theta^{\mathsf{T}}) W \sqrt{P} \, d^N x. \tag{13.265}$$

Squaring, we apply Schwarz's inequality (6.379)

$$\begin{aligned} \left[Y^{\mathsf{T}} \, \nabla \overline{U} \, W \right]^2 &= \left[\int Y^{\mathsf{T}} (\nabla \ln P) \sqrt{P} (U^{\mathsf{T}} - B^{\mathsf{T}} - \Theta^{\mathsf{T}}) W \sqrt{P} \, d^N x \right]^2 \\ &\le \int \left[Y^{\mathsf{T}} (\nabla \ln P) \sqrt{P} \right]^2 d^N x \int \left[(U^{\mathsf{T}} - B^{\mathsf{T}} - \Theta^{\mathsf{T}}) W \sqrt{P} \right]^2 d^N x \\ &= \int \left[Y^{\mathsf{T}} \nabla \ln P \right]^2 \, P \, d^N x \int \left[(U^{\mathsf{T}} - B^{\mathsf{T}} - \Theta^{\mathsf{T}}) W \right]^2 \, P \, d^N x. \end{aligned} \tag{13.266}$$

In the last line, we recognize the first integral as $Y^{\mathsf{T}} F Y$, where F is Fisher's matrix (13.257), and the second as $W^{\mathsf{T}} C W$ in which C is the covariance of the estimators

$$C_{k\ell} \equiv C[U, U]_{k\ell} = C[u_k - b_k - \theta_k, u_\ell - b_\ell - \theta_\ell]. \tag{13.267}$$

So (13.266) says

$$\left(Y^{\mathsf{T}}\nabla\overline{U}W\right)^2 \leq Y^{\mathsf{T}}FY\ W^{\mathsf{T}}CW. \tag{13.268}$$

Thus as long as the symmetric nonnegative matrix F is positive and so has an inverse, we can set the arbitrary constant vector $Y = F^{-1}\nabla\overline{U}\,W$ and get

$$(W^{\mathsf{T}}\nabla\overline{U}^{\mathsf{T}}F^{-1}\nabla\overline{U}W)^2 \leq W^{\mathsf{T}}\nabla\overline{U}^{\mathsf{T}}F^{-1}\nabla\overline{U}W\ W^{\mathsf{T}}CW. \tag{13.269}$$

Canceling a common factor, we obtain the **Cramér–Rao inequality**

$$W^{\mathsf{T}}C\,W \geq W^{\mathsf{T}}\nabla\overline{U}^{\mathsf{T}}F^{-1}\,\nabla\overline{U}\,W \tag{13.270}$$

often written as

$$C \geq \nabla\overline{U}^{\mathsf{T}}\,F^{-1}\,\nabla\overline{U}. \tag{13.271}$$

By (13.264), the matrix $\nabla\overline{U}$ is the identity matrix I plus the gradient of the bias B

$$\nabla\overline{U} = I + \nabla B. \tag{13.272}$$

Thus another form of the Cramér–Rao inequality is

$$C \geq (I + \nabla B)^{\mathsf{T}}\,F^{-1}\,(I + \nabla B) \tag{13.273}$$

or in terms of the arbitrary vector W

$$W^{\mathsf{T}}C\,W \geq W^{\mathsf{T}}\,(I + \nabla B)^{\mathsf{T}}\,F^{-1}\,(I + \nabla B)\,W. \tag{13.274}$$

Letting the arbitrary vector W be $W_j = \delta_{jk}$, one arrives at (exercise 13.35) the **Cramér–Rao lower bound on the variance** $V[u_k] = C[u_k, u_k]$

$$V[u_k] \geq (F^{-1})_{kk} + \sum_{\ell=1}^{m} 2(F^{-1})_{k\ell}\partial_l b_k + \sum_{\ell,j=1}^{m}(F^{-1})_{\ell j}\partial_l b_k \partial_j b_k. \tag{13.275}$$

If the estimator $u_k(\boldsymbol{x})$ is unbiased, then this lower bound simplifies to

$$V[u_k] \geq (F^{-1})_{kk}. \tag{13.276}$$

Example 13.16 (Cramér–Rao bound for a gaussian) The elements of Fisher's information matrix for the mean μ and variance σ^2 of Gauss's distribution for N data points $x_1, \ldots, x_N$

$$P_{\mathrm{G}}^{(N)}(\boldsymbol{x}, \mu, \sigma) = \prod_{j=1}^{N} P_{\mathrm{G}}(x_j; \mu, \sigma) = \left(\frac{1}{\sigma\sqrt{2\pi}}\right)^N \exp\left(-\sum_{j=1}^{N}\frac{(x_j-\mu)^2}{2\sigma^2}\right) \tag{13.277}$$

are

$$F_{\mu\mu} = \int \left[\left(\ln P_G^{(N)}(\boldsymbol{x}, \mu, \sigma) \right)_{,\mu} \right]^2 P_G^{(N)}(\boldsymbol{x}, \mu, \sigma)\, d^N x$$
$$= \sum_{i,j=1}^{N} \int \left(\frac{x_i - \mu}{\sigma^2} \right) \left(\frac{x_j - \mu}{\sigma^2} \right) P_G^{(N)}(\boldsymbol{x}, \mu, \sigma)\, d^N x$$
$$= \sum_{i=1}^{N} \int \left(\frac{x_i - \mu}{\sigma^2} \right)^2 P_G^{(N)}(\boldsymbol{x}, \mu, \sigma)\, d^N x = \frac{N}{\sigma^2}, \tag{13.278}$$
$$F_{\mu\sigma^2} = \int (\ln P_G^{(N)}(\boldsymbol{x}, \mu, \sigma))_{,\mu} \, (\ln P_G^{(N)}(\boldsymbol{x}, \mu, \sigma))_{,\sigma^2} \, P_G^{(N)}(\boldsymbol{x}, \mu, \sigma)\, d^N x$$
$$= \sum_{i,j=1}^{N} \int \left[\frac{x_i - \mu}{\sigma^2} \right] \left[\frac{(x_j - \mu)^2}{2\sigma^4} - \frac{1}{2\sigma^2} \right] P_G^{(N)}(\boldsymbol{x}, \mu, \sigma)\, d^N x = 0,$$

$F_{\sigma^2\mu} = F_{\mu\sigma^2} = 0$, and

$$F_{\sigma^2\sigma^2} = \int \left[(\ln P_G^{(N)}(\boldsymbol{x}, \mu, \sigma))_{,\sigma^2} \right]^2 P_G^{(N)}(\boldsymbol{x}, \mu, \sigma)\, d^N x$$
$$= \sum_{i,j=1}^{N} \int \left[\frac{(x_i - \mu)^2}{2\sigma^4} - \frac{1}{2\sigma^2} \right] \left[\frac{(x_j - \mu)^2}{2\sigma^4} - \frac{1}{2\sigma^2} \right] P_G^{(N)}(\boldsymbol{x}, \mu, \sigma)\, d^N x$$
$$= \frac{N}{2\sigma^4}. \tag{13.279}$$

The inverse of Fisher's matrix then is diagonal with $(F^{-1})_{\mu\mu} = \sigma^2/N$ and $(F^{-1})_{\sigma^2\sigma^2} = 2\sigma^4/N$.

The variance of any unbiased estimator $u_\mu(x)$ of the mean must exceed its Cramér–Rao lower bound (13.276), and so $V[u_\mu] \geq (F^{-1})_{\mu\mu} = \sigma^2/N$. The variance $V[u_\mu^{(N)}]$ of the natural estimator of the mean $u_\mu^{(N)}(\boldsymbol{x}) = (x_1 + \cdots + x_N)/N$ is σ^2/N by (13.246), and so it respects and saturates the lower bound (13.276)

$$V[u_\mu^{(N)}] = E[(u_\mu^{(N)} - \mu)^2] = \sigma^2/N = (F^{-1})_{\mu\mu}. \tag{13.280}$$

One may show (exercise 13.36) that the variance $V[u_{\sigma^2}^{(N)}]$ of Bessel's estimator (13.254) of the variance is (Riley *et al.*, 2006, p. 1248)

$$V[u_{\sigma^2}^{(N)}] = \frac{1}{N} \left(\nu_4 - \frac{N-3}{N-1} \sigma^4 \right) \tag{13.281}$$

where ν_4 is the fourth central moment (13.26) of the probability distribution. For the gaussian $P_G(\boldsymbol{x}; \mu, \sigma)$ one may show (exercise 13.37) that this moment is $\nu_4 = 3\sigma^4$, and so for it

$$V_G[u_{\sigma^2}^{(N)}] = \frac{2}{N-1} \sigma^4. \tag{13.282}$$

Thus the variance of Bessel's estimator of the variance respects but does not saturate its Cramér–Rao lower bound (13.276, 13.279)

$$V_G[u_{\sigma^2}^{(N)}] = \frac{2}{N-1}\,\sigma^4 > \frac{2}{N}\,\sigma^4. \tag{13.283}$$

□

Estimators that saturate their Cramér–Rao lower bounds are **efficient**. The natural estimator $u_\mu^{(N)}(\boldsymbol{x})$ of the mean is efficient as well as consistent and unbiased, and Bessel's estimator $u_{\sigma^2}^{(N)}(\boldsymbol{x})$ of the variance is consistent and unbiased but not efficient.

13.19 Maximum likelihood

Suppose we measure some quantity x at various values of another variable t and find the values $x_1, x_2, \ldots, x_N$ at the known points $t_1, t_2, \ldots, t_N$. We might want to fit these measurements to a curve $x = f(t; \boldsymbol{\alpha})$ where $\boldsymbol{\alpha} = \alpha_1, \ldots, \alpha_M$ is a set of $M < N$ parameters. In view of the central limit theorem, we'll assume that the points x_j fall in Gauss's distribution about the values $x_j = f(t_j; \boldsymbol{\alpha})$ with some known variance σ^2. The probability of getting the N values $x_1, \ldots, x_N$ then is

$$P(\boldsymbol{x}) = \prod_{j=1}^{N} P(x_j, t_j, \sigma) = \left(\frac{1}{\sigma\sqrt{2\pi}}\right)^N \exp\left(-\sum_{j=1}^{N} \frac{(x_j - f(t_j; \boldsymbol{\alpha}))^2}{2\sigma^2}\right). \tag{13.284}$$

To find the M parameters $\boldsymbol{\alpha}$, we maximize the likelihood $P(\boldsymbol{x})$ by minimizing the argument of its exponential

$$0 = \frac{\partial}{\partial \alpha_\ell} \sum_{j=1}^{N} \left(x_j - f(t_j; \boldsymbol{\alpha})\right)^2 = -2 \sum_{j=1}^{N} \left(x_j - f(t_j; \boldsymbol{\alpha})\right) \frac{\partial f(t_j; \boldsymbol{\alpha})}{\partial \alpha_\ell}. \tag{13.285}$$

If the function $f(t; \boldsymbol{\alpha})$ depends nonlinearly upon the parameters $\boldsymbol{\alpha}$, then we may need to use numerical methods to solve this **least-squares** problem.

But if the function $f(t; \boldsymbol{\alpha})$ depends **linearly** upon the M parameters $\boldsymbol{\alpha}$

$$f(t; \boldsymbol{\alpha}) = \sum_{k=1}^{M} g_k(t)\,\alpha_k \tag{13.286}$$

then the equations (13.285) that determine these parameters $\boldsymbol{\alpha}$ are linear

$$0 = \sum_{j=1}^{N} \left(x_j - \sum_{k=1}^{M} g_k(t_j)\,\alpha_k\right) g_\ell(t_j). \tag{13.287}$$

In matrix notation with G the $N \times M$ rectangular matrix with entries $G_{jk} = g_k(t_j)$, they are

$$G^{\mathsf{T}} \boldsymbol{x} = G^{\mathsf{T}} G \boldsymbol{\alpha}. \tag{13.288}$$

The basis functions $g_k(t)$ may depend nonlinearly upon the independent variable t. If one chooses them to be sufficiently different that the columns of G are linearly independent, then the rank of G is M, and the nonnegative matrix $G^{\mathsf{T}} G$ has an inverse. The matrix G then has a pseudoinverse (1.397)

$$G^{+} = \left(G^{\mathsf{T}} G\right)^{-1} G^{\mathsf{T}} \tag{13.289}$$

and it maps the N-vector $\boldsymbol{x}$ into our parameters $\boldsymbol{\alpha}$

$$\boldsymbol{\alpha} = G^{+} \boldsymbol{x}. \tag{13.290}$$

The product $G^{+} G = I_M$ is the $M \times M$ identity matrix, while

$$G G^{+} = P \tag{13.291}$$

is an $N \times N$ projection operator (exercise 13.38) onto the $M \times M$ subspace for which $G^{+} G = I_M$ is the identity operator. Like all projection operators, P satisfies $P^2 = P$.

13.20 Karl Pearson's chi-squared statistic

The argument of the exponential (13.284) in $P(\boldsymbol{x})$ is (the negative of) Karl Pearson's chi-squared statistic (Pearson, 1900)

$$\chi^2 \equiv \sum_{j=1}^{N} \frac{(x_j - f(t_j; \boldsymbol{\alpha}))^2}{2\sigma^2}. \tag{13.292}$$

When the function $f(t; \boldsymbol{\alpha})$ is linear (13.286) in $\boldsymbol{\alpha}$, the N-vector $f(t_j; \boldsymbol{\alpha})$ is $f = G\boldsymbol{\alpha}$. Pearson's χ^2 then is

$$\chi^2 = (\boldsymbol{x} - G\boldsymbol{\alpha})^2 / 2\sigma^2. \tag{13.293}$$

Now (13.290) tells us that $\boldsymbol{\alpha} = G^{+} \boldsymbol{x}$, and so in terms of the projection operator $P = G G^{+}$, the vector $\boldsymbol{x} - G\boldsymbol{\alpha}$ is

$$\boldsymbol{x} - G\boldsymbol{\alpha} = \boldsymbol{x} - G G^{+} \boldsymbol{x} = \left(I - G G^{+}\right) \boldsymbol{x} = (I - P) \boldsymbol{x}. \tag{13.294}$$

So χ^2 is proportional to the squared length

$$\chi^2 = \tilde{\boldsymbol{x}}^2 / 2\sigma^2 \tag{13.295}$$

of the vector

$$\tilde{\boldsymbol{x}} \equiv (I - P) \boldsymbol{x}. \tag{13.296}$$

Thus if the matrix G has rank M, and the vector $\boldsymbol{x}$ has N independent components, then the vector $\tilde{\boldsymbol{x}}$ has only $N - M$ independent components.

Example 13.17 (Two position measurements) Suppose we measure a position twice with error σ and get x_1 and x_2. Then the single parameter α is their average $\alpha = (x_1 + x_2)/2$, and χ^2 is

$$\begin{aligned}\chi^2 &= \left\{[x_1 - (x_1 + x_2)/2]^2 + [x_2 - (x_1 + x_2)/2]^2\right\} \Big/ 2\sigma^2 \\ &= \left\{[(x_1 - x_2)/2]^2 + [(x_2 - x_1)/2]^2\right\} \Big/ 2\sigma^2 \\ &= \left[(x_1 - x_2)/\sqrt{2}\right]^2 \Big/ 2\sigma^2. \end{aligned} \tag{13.297}$$

Thus instead of having two independent components x_1 and x_2, χ^2 just has one $(x_1 - x_2)/\sqrt{2}$. □

We can see how this happens more generally if we use as basis vectors the $N - M$ orthonormal vectors $|j\rangle$ in the kernel of P (that is, the $|j\rangle$s annihilated by P)

$$P|j\rangle = 0, \quad 1 \le j \le N - M \tag{13.298}$$

and the M that lie in the range of the projection operator P

$$P|k\rangle = |k\rangle, \quad N - M + 1 \le k \le N. \tag{13.299}$$

In terms of these basis vectors, the N-vector $\boldsymbol{x}$ is

$$\boldsymbol{x} = \sum_{j=1}^{N-M} x_j |j\rangle + \sum_{k=N-M+1}^{N} x_k |k\rangle \tag{13.300}$$

and the last M components of the vector $\tilde{\boldsymbol{x}}$ vanish

$$\tilde{\boldsymbol{x}} = (I - P)\boldsymbol{x} = \sum_{j=1}^{N-M} x_j |j\rangle. \tag{13.301}$$

Example 13.18 (N position measurements) Suppose the N values of x_j are the measured values of the position $f(t_j; \alpha) = \alpha$ of some object. Then $M = 1$, and $G_{j1} = g_1(t_j) = 1$ for $j = 1, \ldots, N$. Now $G^{\mathsf{T}} G = N$ is a 1×1 matrix, the number N, and the parameter α is the mean $\overline{x}$

$$\alpha = G^{+} \boldsymbol{x} = \left(G^{\mathsf{T}} G\right)^{-1} G^{\mathsf{T}} \boldsymbol{x} = \frac{1}{N} \sum_{j=1}^{N} x_j = \overline{x} \tag{13.302}$$

of the N position measurements x_j. So the vector $\tilde{\boldsymbol{x}}$ has components $\tilde{x}_j = x_j - \overline{x}$ and is orthogonal to $G^{\mathsf{T}} = (1, 1, \ldots, 1)$

$$G^{\mathsf{T}} \tilde{\boldsymbol{x}} = \left(\sum_{j=1}^{N} x_j \right) - N\overline{x} = 0. \tag{13.303}$$

The matrix G^{T} has rank 1, and the vector $\tilde{\boldsymbol{x}}$ has $N - 1$ independent components. □

Suppose now that we have determined our M parameters $\boldsymbol{\alpha}$ and have a theoretical fit

$$x = f(t; \boldsymbol{\alpha}) = \sum_{k=1}^{M} g_k(t)\, \alpha_k, \tag{13.304}$$

which when we apply it to N measurements x_j gives χ^2 as

$$\chi^2 = (\tilde{\boldsymbol{x}})^2 / 2\sigma^2. \tag{13.305}$$

How good is our fit?

A χ^2 distribution with $N - M$ **degrees of freedom** has by (13.202) mean

$$E[\chi^2] = N - M \tag{13.306}$$

and variance

$$V[\chi^2] = 2(N - M). \tag{13.307}$$

So our χ^2 should be about

$$\chi^2 \approx N - M \pm \sqrt{2(N - M)}. \tag{13.308}$$

If it lies within this range, then (13.304) is a good fit to the data. But if it exceeds $N - M + \sqrt{2(N - M)}$, then the fit isn't so good. On the other hand, if χ^2 is less than $N - M - \sqrt{2(N - M)}$, then we may have used too many parameters. Indeed, by using N parameters with $G\,G^{+} = I_N$, we could get $\chi^2 = 0$ every time.

The probability that χ^2 exceeds χ_0^2 is the integral (13.201)

$$\Pr_n(\chi^2 > \chi_0^2) = \int_{\chi_0^2}^{\infty} P_n(\chi^2/2)\, d\chi^2 = \int_{\chi_0^2}^{\infty} \frac{1}{2\Gamma(n/2)} \left(\frac{\chi^2}{2} \right)^{n/2-1} e^{-\chi^2/2} d\chi^2, \tag{13.309}$$

in which $n = N - M$ is the number of data points minus the number of parameters, and $\Gamma(n/2)$ is the gamma function (5.102, 4.62). So an M-parameter fit to N data points has only a chance of ϵ of being right if its χ^2 is greater than a χ_0^2 for which $\Pr_{N-M}(\chi^2 > \chi_0^2) = \epsilon$. These probabilities $\Pr_{N-M}(\chi^2 > \chi_0^2)$ are

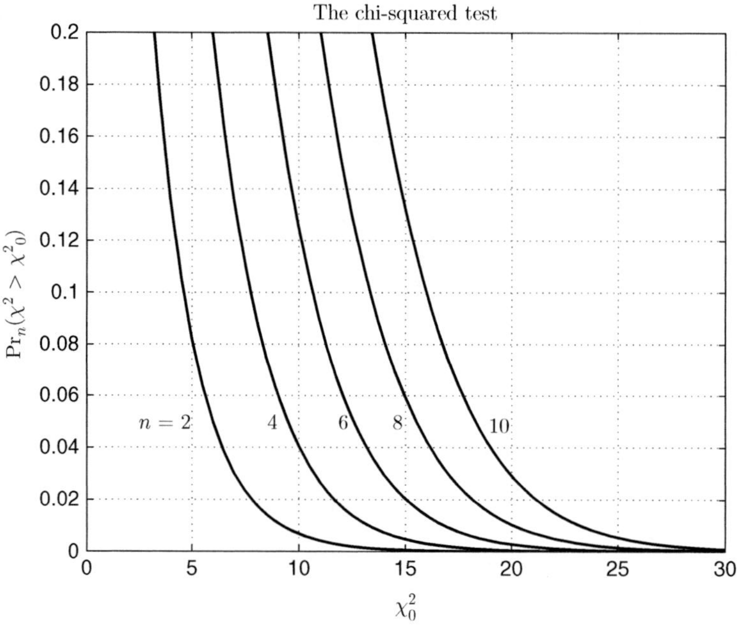

Figure 13.6 The probabilities $\Pr_n(\chi^2 > \chi_0^2)$ are plotted from left to right for $n = N - M = 2, 4, 6, 8$, and 10 degrees of freedom as functions of χ_0^2.

plotted in Fig. 13.6 for $n = N - M = 2, 4, 6, 8$, and 10. In particular, the probability of a value of χ^2 greater than $\chi_0^2 = 20$ respectively is 0.000045, 0.000499, 0.00277, 0.010336, and 0.029253 for $n = N - M = 2, 4, 6, 8$, and 10.

13.21 Kolmogorov's test

Suppose we want to use a sequence of N measurements x_j to determine the probability distribution that they come from. Our empirical probability distribution is

$$P_e^{(N)}(x) = \frac{1}{N}\sum_{j=1}^{N} \delta(x - x_j). \tag{13.310}$$

Our cumulative probability for events less than x then is

$$\Pr_e^{(N)}(-\infty, x) = \int_{-\infty}^{x} P_e^{(N)}(x')\, dx' = \int_{-\infty}^{x} \frac{1}{N}\sum_{j=1}^{N} \delta(x' - x_j)\, dx'. \tag{13.311}$$

So if we label our events in increasing order $x_1 \le x_2 \le \cdots \le x_N$, then the probability of an event less than x is

$$\Pr_e^{(N)}(-\infty, x) = \frac{j}{N} \quad \text{for} \quad x_j < x < x_{j+1}. \tag{13.312}$$

Having approximately and experimentally determined our empirical cumulative probability distribution $\Pr_e^{(N)}(-\infty, x)$, we might want to know whether it comes from some hypothetical, theoretical cumulative probability distribution $\Pr_t(-\infty, x)$. One way to do this is to compute the distance D_N between the two cumulative probability distributions

$$D_N = \sup_{-\infty<x<\infty} \left|\Pr_e^{(N)}(-\infty, x) - \Pr_t(-\infty, x)\right|, \tag{13.313}$$

in which **sup** stands for ***supremum*** and means **least upper bound**. Since cumulative probabilities lie between zero and one, it follows (exercise 13.39) that the Kolmogorov distance is bounded by

$$0 \le D_N \le 1. \tag{13.314}$$

The simpler Smirnov distances

$$D_N^+ = \sup_{-\infty<x<\infty} \left(\Pr_e^{(N)}(-\infty, x) - \Pr_t(-\infty, x)\right),$$

$$D_N^- = \sup_{-\infty<x<\infty} \left(\Pr_t(-\infty, x) - \Pr_e^{(N)}(-\infty, x)\right) \tag{13.315}$$

provide (exercise 13.40) an expression for D_N as the greater of the two

$$D_N = \max(D_N^+, D_N^-). \tag{13.316}$$

Using our explicit expression (13.312) for the empirical cumulative probability $\Pr_e^{(N)}(-\infty, x)$ and the monotonicity (13.30) of cumulative probabilities such as $\Pr_t(-\infty, x)$, one may show (exercise 13.41) that the Smirnov distances are given by

$$D_N^+ = \sup_{1\le j\le N} \left(\frac{j}{N} - \Pr_t(-\infty, x_j)\right)$$

$$D_N^- = \sup_{1\le j\le N} \left(\Pr_t(-\infty, x_j) - \frac{j-1}{N}\right). \tag{13.317}$$

In general, as the number N of data points increases, we expect that our empirical distribution $\Pr_e^{(N)}(-\infty, x)$ should approach the actual empirical distribution $\Pr_e(-\infty, x)$ from which the events x_j came. In this case, the Kolmogorov distance D_N should converge to a limiting value D_∞

$$\lim_{N\to\infty} D_N = D_\infty = \sup_{-\infty<x<\infty} |\Pr_e(-\infty, x) - \Pr_t(-\infty, x)| \in [0, 1]. \tag{13.318}$$

If the empirical distribution $\Pr_e(-\infty, x)$ is the same as the theoretical distribution $\Pr_t(-\infty, x)$, then we expect that $D_\infty = 0$. This expectation is confirmed by a theorem due to Glivenko (Glivenko, 1933; Cantelli, 1933) according to which the probability that the Kolmogorov distance D_N should go to zero as $N \to \infty$ is unity

$$\Pr(D_\infty = 0) = 1. \tag{13.319}$$

The real issue is how fast D_N should decrease with N if our events x_j do come from $\Pr_t(-\infty, x)$. This question was answered by Kolmogorov, who showed (Kolmogorov, 1933) that if the theoretical distribution $\Pr_t(-\infty, x)$ is continuous, then for large N (and for $u > 0$) the probability that $\sqrt{N}\, D_N$ is less than u is given by the **Kolmogorov function** $K(u)$

$$\lim_{N\to\infty} \Pr(\sqrt{N}\, D_N < u) = K(u) \equiv 1 + 2\sum_{k=1}^{\infty}(-1)^k e^{-2k^2u^2}, \tag{13.320}$$

which is **universal and independent of the particular probability distributions** $\Pr_e(-\infty, x)$ and $\Pr_t(-\infty, x)$.

On the other hand, if our events x_j come from a different probability distribution $\Pr_e(-\infty, x)$, then as $N \to \infty$ we should expect that $\Pr_e^{(N)}(-\infty, x) \to \Pr_e(-\infty, x)$, and so that D_N should converge to a positive constant $D_\infty \in (0, 1]$. In this case, we expect that as $N \to \infty$ the quantity $\sqrt{N}\, D_N$ should grow with N as $\sqrt{N}\, D_\infty$.

Example 13.19 (Kolmogorov's test) How do we use (13.320)? As illustrated in Fig. 13.7, Kolmogorov's distribution $K(u)$ rises from zero to unity on $(0, \infty)$, reaching 0.9993 already at $u = 2$. So if our points x_j come from the theoretical distribution, then Kolmogorov's theorem (13.320) tells us that as $N \to \infty$, the probability that $\sqrt{N}\, D_N$ is less than 2 is more than 99.9%. But if the experimental points x_j do not come from the theoretical distribution, then the quantity $\sqrt{N}\, D_N$ should grow as $\sqrt{N}\, D_\infty$ as $N \to \infty$.

To see what this means in practice, I took as the theoretical distribution $P_t(x) = P_G(x, 0, 1)$, which has the cumulative probability distribution (13.85)

$$\Pr_t(-\infty, x) = \frac{1}{2}\left[\operatorname{erf}\left(x/\sqrt{2}\right) + 1\right]. \tag{13.321}$$

I generated $N = 10^m$ points x_j for $m = 1, 2, 3, 4, 5$, and 6 from the theoretical distribution $P_t(x) = P_G(x, 0, 1)$ and computed $u_N = \sqrt{10^m}\, D_{10^m}$ for these points. I found $\sqrt{10^m}\, D_{10^m} = 0.6928, 0.7074, 1.2000, 0.7356, 1.2260$, and 1.0683. All were less than 2, as expected since I had taken the experimental points x_j from the theoretical distribution.

To see what happens when the experimental points do not come from the theoretical distribution $P_t(x) = P_G(x, 0, 1)$, I generated $N = 10^m$ points x_j

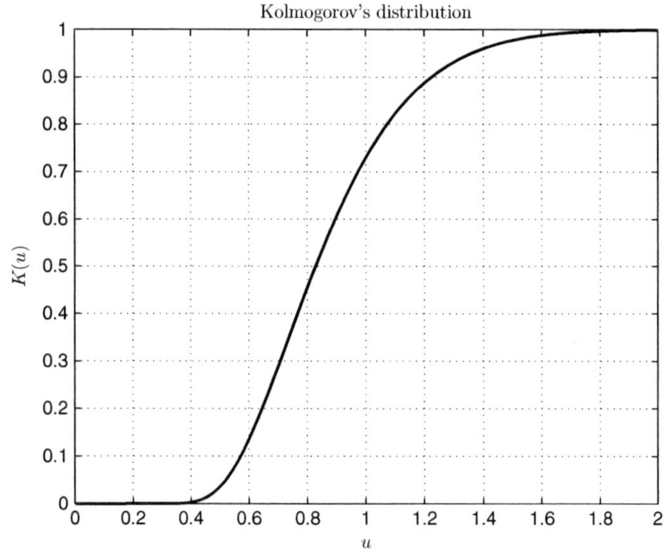

Figure 13.7 Kolmogorov's cumulative probability distribution $K(u)$ defined by (13.320) rises from zero to unity as u runs from zero to about two.

for $m = 1, 2, 3, 4, 5$, and 6 from Gosset's Student's distribution $P_S(x, 3, 1)$ defined by (13.191) with $\nu = 3$ and $a = 1$. Both $P_t(x) = P_G(x, 0, 1)$ and $P_S(x, 3, 1)$ have the same mean $\mu = 0$ and standard deviation $\sigma = 1$, as illustrated in Fig. 13.8. For these points, I computed $u_N = \sqrt{N}\, D_N$ and found $\sqrt{10^m}\, D_{10^m} =$ 0.7741, 1.4522, 3.3837, 9.0478, 27.6414, and 87.8147. Only the first two are less than 2, and the last four grow as $\sqrt{N}$, indicating that the x_j had not come from the theoretical distribution. In fact, we can approximate the limiting value of D_N as $D_\infty \approx u_{10^6}/\sqrt{10^6} = 0.0878$. The exact value is (exercise 13.42) $D_\infty =$ 0.0868552356.

At the risk of overemphasizing this example, I carried it one step further. I generated $\ell = 1, 2, \ldots, 100$ sets of $N = 10^m$ points $x_j^{(\ell)}$ for $m = 2$, 3, and 4 drawn from $P_G(x, 0, 1)$ and from $P_S(x, 3, 1)$ and used them to form 100 empirical cumulative probabilities $\Pr_{e,G}^{(\ell,10^m)}(-\infty, x)$ and $\Pr_{e,S}^{(\ell,10^m)}(-\infty, x)$ as defined by (13.310–13.312). Next, I computed the distances $D_{G,G,10^m}^{(\ell)}$ and $D_{S,G,10^m}^{(\ell)}$ of each of these cumulative probabilities from the gaussian distribution $P_G(x, 0, 1)$. I labeled the two sets of 100 quantities $u_{G,G}^{(\ell,m)} = \sqrt{10^m}\, D_{G,G,10^m}^{(\ell)}$ and $u_{S,G}^{(\ell,m)} = \sqrt{10^m}\, D_{S,G,10^m}^{(\ell)}$ in increasing order as $u_{G,G,1}^{(m)} \le u_{G,G,2}^{(m)} \le \cdots \le u_{G,G,100}^{(m)}$ and $u_{S,G,1}^{(m)} \le u_{S,G,2}^{(m)} \le \cdots \le u_{S,G,100}^{(m)}$. I then used (13.310–13.312) to form the cumulative probabilities

$$\Pr_{e,G,G}^{(m)}(-\infty, u) = \frac{j}{N_s} \quad \text{for} \quad u_{G,G,j}^{(m)} < u < u_{G,G,j+1}^{(m)}, \tag{13.322}$$

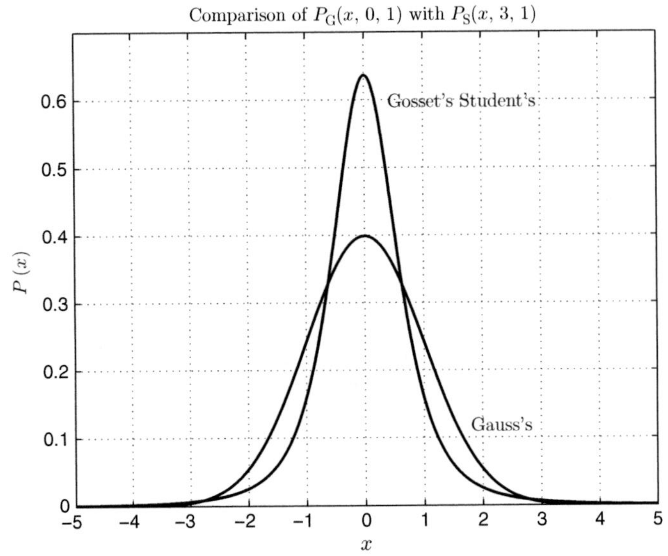

Figure 13.8 The probability distributions of Gauss $P_{\rm G}(x, 0, 1)$ and Gosset/Student $P_{\rm S}(x, 3, 1)$ with zero mean and unit variance.

and

$$\mathrm{Pr}^{(m)}_{\rm e,S,G}(-\infty, u) = \frac{j}{N_s} \quad \text{for} \quad u^{(m)}_{{\rm S,G},j} < u < u^{(m)}_{{\rm S,G},j+1} \tag{13.323}$$

for $N_s = 100$ sets of 10^m points.

I plotted these cumulative probabilities in Fig. 13.9. The thick smooth curve is Kolmogorov's universal cumulative probability distribution $K(u)$ defined by (13.320). The thin jagged curve that clings to $K(u)$ is the cumulative probability distribution $\mathrm{Pr}^{(4)}_{\rm e,G,G}(-\infty, u)$ made from 100 sets of 10^4 points taken from $P_{\rm G}(x, 0, 1)$. As the number of sets increases beyond 100 and the number of points 10^m rises further, the probability distributions $\mathrm{Pr}^{(m)}_{\rm e,G,G}(-\infty, u)$ converge to the universal cumulative probability distribution $K(u)$ and provide a numerical verification of Kolmogorov's theorem. Such curves make poor figures, however, because they hide beneath $K(u)$. The curves labeled $\mathrm{Pr}^{(m)}_{\rm e,S,G}(-\infty, u)$ for $m = 2$ and 3 are made from 100 sets of $N = 10^m$ points taken from $P_{\rm S}(x, 3, 1)$ and tested as to whether they instead come from $P_{\rm G}(x, 0, 1)$. Note that as $N = 10^m$ increases from 100 to 1000, the cumulative probability distribution $\mathrm{Pr}^{(m)}_{\rm e,S,G}(-\infty, u)$ moves farther from Kolmogorov's universal cumulative probability distribution $K(u)$. In fact, the curve $\mathrm{Pr}^{(4)}_{\rm e,S,G}(-\infty, u)$ made from 100 sets of 10^4 points lies beyond $u > 8$, too far to the right to fit in the figure. Kolmogorov's test gets more conclusive as the number of points $N \to \infty$.

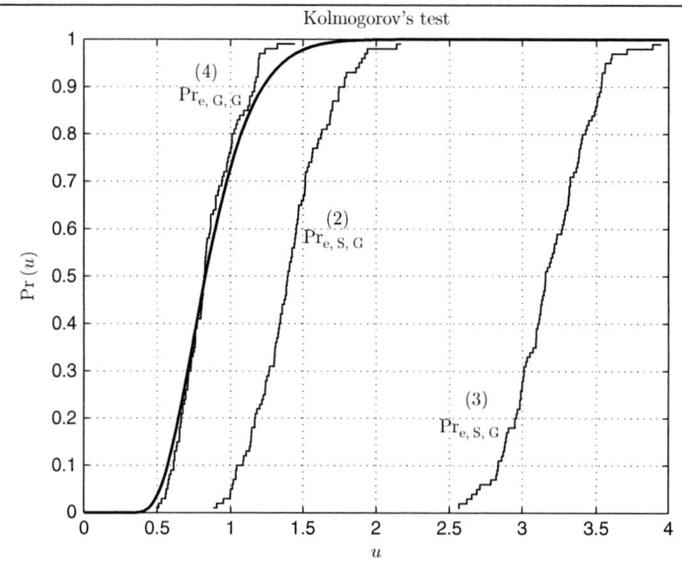

Figure 13.9 Kolmogorov's test is applied to points x_j taken from Gauss's distribution $P_G(x, 0, 1)$ and from Gosset's Student's distribution $P_S(x, 3, 1)$ to see whether the x_j came from $P_G(x, 0, 1)$. The thick smooth curve is Kolmogorov's universal cumulative probability distribution $K(u)$ defined by (13.320). The thin jagged curve that clings to $K(u)$ is the cumulative probability distribution $\mathrm{Pr}^{(4)}_{\mathrm{e,G,G}}(-\infty, u)$ made (13.322) from points taken from $P_G(x, 0, 1)$. The other curves $\mathrm{Pr}^{(m)}_{\mathrm{e,S,G}}(-\infty, u)$ for $m = 2$ and 3 are made (13.323) from 10^m points taken from $P_S(x, 3, 1)$.

□

Warning, mathematical hazard While binned data are ideal for chi-squared fits, they ruin Kolmogorov tests. The reason is that if the data are in bins of width w, then the empirical cumulative probability distribution $\mathrm{Pr}^{(N)}_{\mathrm{e}}(-\infty, x)$ is a staircase function with steps as wide as the bin-width w even in the limit $N \to \infty$. Thus **even if the data come from the theoretical distribution**, the limiting value D_∞ of the Kolmogorov distance will be positive. In fact, one may show (exercise 13.43) that when the data do come from the theoretical probability distribution $P_t(x)$, assumed to be continuous, then the value of D_∞ is

$$D_\infty \approx \sup_{-\infty < x < \infty} \frac{w\, P_t(x)}{2}. \tag{13.324}$$

Thus, in this case, the quantity $\sqrt{N}\, D_N$ would diverge as $\sqrt{N}\, D_\infty$ and lead one to believe that the data had not come from $P_t(x)$.

Suppose we have made some changes in our experimental apparatus and our software, and we want to see whether the new data $x'_1, x'_2, \ldots, x'_{N'}$ we took after the changes are consistent with the old data $x_1, x_2, \ldots, x_N$ we took before the changes. Then, following equations (13.310–13.312), we can make two empirical cumulative probability distributions – one $\Pr_e^{(N)}(-\infty, x)$ made from the N old points x_j and the other $\Pr_e^{(N')}(-\infty, x)$ made from the N' new points x'_j. Next, we compute the distances

$$D^+_{N,N'} = \sup_{-\infty<x<\infty} \left(\Pr_e^{(N)}(-\infty, x) - \Pr_e^{(N')}(-\infty, x) \right),$$
$$D_{N,N'} = \sup_{-\infty<x<\infty} \left| \Pr_e^{(N)}(-\infty, x) - \Pr_e^{(N')}(-\infty, x) \right|, \tag{13.325}$$

which are analogous to (13.313–13.316). Smirnov (Smirnov, 1939; Gnedenko, 1968, p. 453) has shown that as $N, N' \to \infty$ the probabilities that

$$u^+_{N,N'} = \sqrt{\frac{NN'}{N+N'}}\, D^+_{N,N'} \quad \text{and} \quad u_{N,N'} = \sqrt{\frac{NN'}{N+N'}}\, D_{N,N'} \tag{13.326}$$

are less than u are

$$\lim_{N,N'\to\infty} \Pr(u^+_{N,N'} < u) = 1 - e^{-2u^2},$$
$$\lim_{N,N'\to\infty} \Pr(u_{N,N'} < u) = K(u), \tag{13.327}$$

in which $K(u)$ is Kolmogorov's distribution (13.320).

Further reading

Students can learn more about probability and statistics in *Mathematical Methods for Physics and Engineering* (Riley *et al.*, 2006), *An Introduction to Probability Theory and Its Applications I, II* (Feller, 1968, 1966), *Theory of Financial Risk and Derivative Pricing* (Bouchaud and Potters, 2003), and *Probability and Statistics in Experimental Physics* (Roe, 2001).

Exercises

13.1 Find the probabilities that the sum on two thrown fair dice is 4, 5, or 6.

13.2 Show that the zeroth moment μ_0 and the zeroth central moment ν_0 always are unity, and that the first central moment ν_1 always vanishes.

13.3 Compute the variance of the uniform distribution on (0, 1).

13.4 In the formulas (13.21 & 13.28) for the variances of discrete and continuous distributions, show that $E[(x - \langle x \rangle)^2] = \mu_2 - \mu^2$.

13.5 A **convex** function is one that lies above its tangents: if $f(x)$ is convex, then $f(x) \ge f(y) + (x-y)f'(y)$. For instance, $e^x \ge 1 + x$. Show that for any convex function $f(x)$ that $f(x) \ge f(\langle x \rangle) + (x - \langle x \rangle) f'(\langle x \rangle)$ and so that $\langle f(x) \rangle \ge f(\langle x \rangle)$ or $E[f(x)] \ge f(E[x])$ (Johan Jensen, 1859–1925).

13.6 (a) Show that the covariance $\langle (x - \overline{x})(y - \overline{y}) \rangle$ is equal to $\langle x y \rangle - \langle x \rangle \langle y \rangle$ as asserted in (13.35). (b) Derive (13.39) for the variance $V[ax + by]$.

13.7 Derive expression (13.40) for the variance of a sum of N variables.

13.8 Find the range of $pq = p(1 - p)$ for $0 \le p \le 1$.

13.9 Show that the variance of the binomial distribution (13.43) is given by (13.47).

13.10 Redo the polling example (13.14–13.16) for the case of a slightly better poll in which 16 people were asked and 13 said they'd vote for Nancy Pelosi. What's the probability that she'll win the election? (You may use Maple or some other program to do the tedious integral.)

13.11 For the case in which N and $N - n$ are big, derive (13.52 & 13.53) from (13.43 & 13.51).

13.12 For the case in which N, $N - n$, and n are big, derive (13.54 & 13.55) from (13.43 & 13.51).

13.13 Without using the fact that the Poisson distribution is a limiting form of the binomial distribution, show from its definition (13.58) and its mean (13.60) that its variance is equal to its mean, as in (13.62).

13.14 Show that Gauss's approximation (13.74) to the binomial distribution is a normalized probability distribution with mean $\langle x \rangle = \mu = pN$ and variance $V[x] = pqN$.

13.15 Derive the approximations (13.88 & 13.89) for binomial probabilities for large N.

13.16 Compute the central moments (13.27) of the gaussian (13.75).

13.17 Derive formula (13.84) for the probability that a gaussian random variable falls within an interval.

13.18 Show that the expression (13.91) for $P(y|600)$ is negligible on the interval $(0, 1)$ except for y near $3/5$.

13.19 Determine the constant A of the homogeneous solution $\langle \boldsymbol{v}(t) \rangle_{gh}$ and derive expression (13.141) for the general solution $\langle \boldsymbol{v}(t) \rangle$ to (13.139).

13.20 Derive equation (13.142) for the variance of the position $\boldsymbol{r}$ about its mean $\langle \boldsymbol{r}(t) \rangle$. You may assume that $\langle \boldsymbol{r}(0) \rangle = \langle \boldsymbol{v}(0) \rangle = 0$ and that $\langle (\boldsymbol{v} - \langle \boldsymbol{v}(t) \rangle)^2 \rangle = 3kT/m$.

13.21 Derive equation (13.172) for the ensemble average $\langle \boldsymbol{r}^2(t) \rangle$ for the case in which $\langle \boldsymbol{r}^2(0) \rangle = 0$ and $d\langle \boldsymbol{r}^2(0) \rangle / dt = 0$.

13.22 Use (13.183) to derive the lower moments (13.185 & 13.186) of the distributions of Gauss and Poisson.

13.23 Find the third and fourth moments μ_3 and μ_4 for the distributions of Poisson (13.178) and Gauss (13.175).

13.24 Derive formula (13.190) for the first five cumulants of an arbitrary probability distribution.

13.25 Show that, like the characteristic function, the moment-generating function $M(t)$ for an average of several independent random variables factorizes $M(t) = M_1(t/N)\, M_2(t/N) \cdots M_N(t/N)$.

13.26 Derive formula (13.197) for the moments of the log-normal probability distribution (13.196).

13.27 Why doesn't the log-normal probability distribution (13.196) have a sensible power-series about $x = 0$? What are its derivatives there?
13.28 Compute the mean and variance of the exponential distribution (13.198).
13.29 Show that the chi-square distribution $P_{3,G}(v, \sigma)$ with variance $\sigma^2 = kT/m$ is the Maxwell–Boltzmann distribution (13.100).
13.30 Compute the inverse Fourier transform (13.174) of the characteristic function (13.203) of the symmetric Lévy distribution for $\nu = 1$ and 2.
13.31 Show that the integral that defines $P^{(2)}(y)$ gives formula (13.239) with two Heaviside functions. Hint: keep x_1 and x_2 in the interval (0, 1).
13.32 Derive the normal distribution (13.224) in the variable (13.223) from the central limit theorem (13.221) for the case in which all the means and variances are the same.
13.33 Show that Fisher's matrix (13.257) is symmetric $F_{k\ell} = F_{\ell k}$ and nonnegative (1.38), and that when it is positive (1.39), it has an inverse.
13.34 Derive the integral equations (13.259 & 13.260) from the normalization condition $\int P(\boldsymbol{x}; \boldsymbol{\theta})\, d^N x = 1$.
13.35 Derive the Cramér–Rao lower bound (13.275) on the variance $V[t_k]$ from the inequality (13.270).
13.36 Show that the variance $V[u^{(N)}_{\sigma^2}]$ of Bessel's estimator (13.254) is given by (13.281).
13.37 Compute the fourth central moment (13.27) of Gauss's probability distribution $P_G(x; \mu, \sigma^2)$.
13.38 Show that when the real $N \times M$ matrix G has rank M, the matrices $P = G G^+$ and $P_\perp = 1 - P$ are projection operators that are mutually orthogonal $P(I - P) = (I - P)P = 0$.
13.39 Show that Kolmogorov's distance D_N is bounded as in (13.314).
13.40 Show that Kolmogorov's distance D_N is the greater of D_N^+ and D_N^-.
13.41 Derive the formulas (13.317) for D_N^+ and D_N^-.
13.42 Compute the exact limiting value D_∞ of the Kolmogorov distance between $P_G(x, 0, 1)$ and $P_S(x, 3, 1)$. Use the cumulative probabilities (13.321 & 13.194) to find the value of x that maximizes their difference. Using Maple or some other program, you should find $x = 0.6276952185$ and then $D_\infty = 0.0868552356$.
13.43 Show that when the data do come from the theoretical probability distribution (assumed to be continuous) but are in bins of width w, then the limiting value D_∞ of the Kolmogorov distance is given by (13.324).

14

Monte Carlo methods

14.1 The Monte Carlo method

The Monte Carlo method is simple, robust, and useful. It was invented by Enrico Fermi and developed by Metropolis (Metropolis *et al.*, 1953). It has many applications. One can use it for numerical integration. One can use it to decide whether an odd signal is random noise or something to evaluate. One can use it to generate sequences of configurations that are random but occur according to a probability distribution, such as the Boltzmann distribution of statistical mechanics. One even can use it to solve virtually any problem for which one has a criterion to judge the quality of an arbitrary solution and a way of generating a suitably huge space of possible solutions. That's how evolution invented us.

14.2 Numerical integration

Suppose one wants to numerically integrate a function $f(x)$ of a vector $x = (x_1, \ldots, x_n)$ over a region $\mathcal{R}$. One generates a large number N of random values for the n coordinates x within a hypercube of length L that contains the region $\mathcal{R}$, keeps the $N_{\mathcal{R}}$ points $x_k = (x_{1k}, \ldots, x_{nk})$ that fall within the region $\mathcal{R}$, computes the average $\langle f(x_k) \rangle$, and multiplies by the hypervolume $V_{\mathcal{R}}$ of the region

$$\int_{\mathcal{R}} f(x)\, d^n x \approx \frac{V_{\mathcal{R}}}{N_{\mathcal{R}}} \sum_{k=1}^{N_{\mathcal{R}}} f(x_k). \tag{14.1}$$

If the hypervolume $V_{\mathcal{R}}$ is hard to compute, you can have the Monte Carlo code compute it for you. The hypervolume $V_{\mathcal{R}}$ is the volume L^n of the enclosing hypercube multiplied by the number $N_{\mathcal{R}}$ of times the N points fall within the region $\mathcal{R}$

$$V_{\mathcal{R}} = \frac{N_{\mathcal{R}}}{N} L^n. \tag{14.2}$$

The integral formula (14.1) then becomes

$$\int_{\mathcal{R}} f(x)\, d^n x \approx \frac{L^n}{N} \sum_{k=1}^{N_{\mathcal{R}}} f(x_k). \tag{14.3}$$

The utility of the Monte Carlo method of numerical integration rises sharply with the dimension n of the hypervolume.

Example 14.1 (Numerical integration) Suppose one wants to integrate the function

$$f(x, y) = \frac{e^{-2x-3y}}{\sqrt{x^2 + y^2 + 1}} \tag{14.4}$$

over the quarter of the unit disk in which x and y are positive. In this case, $V_{\mathcal{R}}$ is the area $\pi/4$ of the quarter disk.

To generate fresh random numbers, one must set the seed for the code that computes them. The following program sets the seed by using the subroutine init_random_seed defined in a FORTRAN95 program in section 13.16. With some compilers, one can just write "call random_seed()".

```
program integrate
  implicit none ! catches typos
  integer :: k, N
  integer, parameter :: dp = kind(1.0d0)
  real(dp) :: x, y, sum = 0.0d0, f
  real(dp), dimension(2) :: rdn
  real(dp), parameter :: area = atan(1.0d0) ! pi/4
  f(x,y) = exp(-2*x - 3*y)/sqrt(x**2 + y**2 + 1.0d0)
  write(6,*) 'How many points?'
  read(5,*) N
  call init_random_seed() ! set new seed
  do k = 1, N
10    call random_number(rdn); x= rdn(1); y = rdn(2)
      if (x**2+y**2 > 1.0d0) then
         go to 10
      end if
      sum = sum + f(x,y)
  end do
  ! integral = area times mean value < f > of f
  sum = area*sum/real(N,dp)
```

```
  write(6,*) 'The integral is ',sum
end program integrate
```

I ran this code with npoints = 10^ℓ for ℓ = 1, 2, 3, 4, 5, 6, 7, and 8 and found respectively the results 0.059285, 0.113487, 0.119062, 0.115573, 0.118349, 0.117862, 0.117868, and 0.117898. The integral is approximately 0.1179.

An equivalent C++ code by Sean Cahill is:

```
#include <math.h>
#include <iostream>
#include <stdlib.h>

using namespace std;

// The function to integrate
double f(const double& x, const double& y)
{
  double numer = exp(-2*x - 3*y);
  double denom = sqrt(x*x + y*y + 1);
  double retval = numer / denom;

  return retval;
}

void integrate ()
{
  // Declares local constants
  const double area = atan(1); // pi/4

  // Inits local variables
  int n=0;
  double sum=0,x=0,y=0;

  // Seeds random number generator
  srand ( time(NULL) );

  // Gets the value of N
  cout << "What is N? ";
  cin >> n;

  // Loops the given number of times
```

```
  for (int i=0; i<n; i++)
  {
    // Loops until criteria met
    while (true)
    {
      // Generates random points between 0 and 1
      x = static_cast<double>(rand()) / RAND_MAX;
      y = static_cast<double>(rand()) / RAND_MAX;

      // Checks if the points are suitable
      if ((x*x + y*y) <= 1)
      {
        // If so, break out of the while loop
        break;
      }
    }

    // Updates our sum with the given points
    sum += f(x,y);
  }

  // Integral = area times mean value < f > of f
  sum = area * sum / n;

  cout << '' The integral is'' << sum << endl;
}
```

14.3 Applications to experiments

Physicists accumulate vast quantities of data and sometimes must decide whether a particular signal is due to a defect in the detector, to a random fluctuation in the real events that they are measuring, or to a new and unexpected phenomenon. For simplicity, let us assume that the background can be ignored and that the real events arrive randomly in time apart from extraordinary phenomena. One reliable way to evaluate an ambiguous signal is to run a Monte Carlo program that generates the kind of real random events to which one's detector is sensitive and to use these events to compute the probability that the signal occurred randomly.

To illustrate the use of random-event generators, we will consider the work of a graduate student who spent 100 days counting muons produced by atmospheric GeV neutrinos in an underground detector. Each of the very large number N of primary cosmic rays that hit the Earth each day can collide with a nucleus and make a shower of pions which in turn produce atmospheric neutrinos that can make muons in the detector. The probability p that a given cosmic ray will make a muon in the detector is very small, but the number N of primary cosmic rays is very large. In this experiment, their product pN was $\langle n \rangle = 0.1$ muons per day. Since N is huge and p tiny, the probability distribution is Poisson, and so by (13.58) the probability that n muons would be detected on any particular day is

$$P(n, \langle n \rangle) = \frac{\langle n \rangle^n}{n!} e^{-\langle n \rangle} \tag{14.5}$$

in the absence of a failure of the anticoincidence shield or some other problem with the detector – or some hard-to-imagine astrophysical event.

The graduate student might have used the following program to generate 1,000,000 random histories of 100 days of events distributed according to the Poisson distribution (14.5) with $\langle n \rangle = 0.1$:

```
program muons
  implicit none
  interface
     function factorial(n)
       implicit none
       integer, intent(in) :: n
       double precision :: factorial
     end function factorial
  end interface
  integer :: k, m, day, number
  integer, parameter :: N = 1000000 ! number of data
    sets
  integer, dimension(N,100) :: histories
  integer, dimension(0:100) :: maxEvents = 0,
    sumEvents = 0
  double precision :: prob, x, numMuons, totMuons
  double precision, dimension(0:100) :: p
  double precision, parameter :: an = 0.1 ! <n> events
    per day
  prob = exp(-an); p(0) = prob; maxEvents = 0
  ! p(k) is the probability of fewer than k+1 events per
    day
  do k = 1, 100 ! make Poisson distribution
     prob = prob + an**k*exp(-an)/factorial(k)
```

```
    p(k) = prob
  end do
  call init_random_seed() ! sets random seed
  do k = 1, N ! do N histories
     do day = 1, 100 ! do day of kth history
        call random_number(x)
        do m = 100, 0, -1
           if (x < p(m)) then
              number = m
           end if
        end do
        histories(k,day) = number
     end do
     numMuons = maxval(histories(k,:))
     totMuons = sum(histories(k,:))
     maxEvents(numMuons) = maxEvents(numMuons) + 1
     sumEvents(totMuons) = sumEvents(totMuons) + 1
  end do
  open(7,file="maxEvents"); open(8,file="totEvents")
  do k = 0, 100
     write(7,*) k, maxEvents(k); write(8,*) k,
       sumEvents(k)
  end do
end program muons
function factorial(n)  result(fact)
  implicit none
  integer, intent(in) :: n
  integer, parameter :: dp = kind(1.0d0)
  real(dp) :: fact
  fact = 1.0d0
  do i = 1, n
     fact = i*fact
  end do
end function factorial
```

Figure 14.1 plots the results from this simple Monte Carlo of 1,000,000 runs of 100 days each. The boxes show that the maximum number of muons detected on a single day respectively was $n = 1$, 2, and 3 on 62.6%, 35.9%, and 1.5% of the runs – and respectively was $n = 0$, 4, 5, and 6 on only 36, 410, 9, and 1 runs. Thus if the actual run detected no muons at all, that would be by (13.83) about a 4σ event, while a run with more than four muons on a single day would be an event of more than 4σ. Either would be a reason to

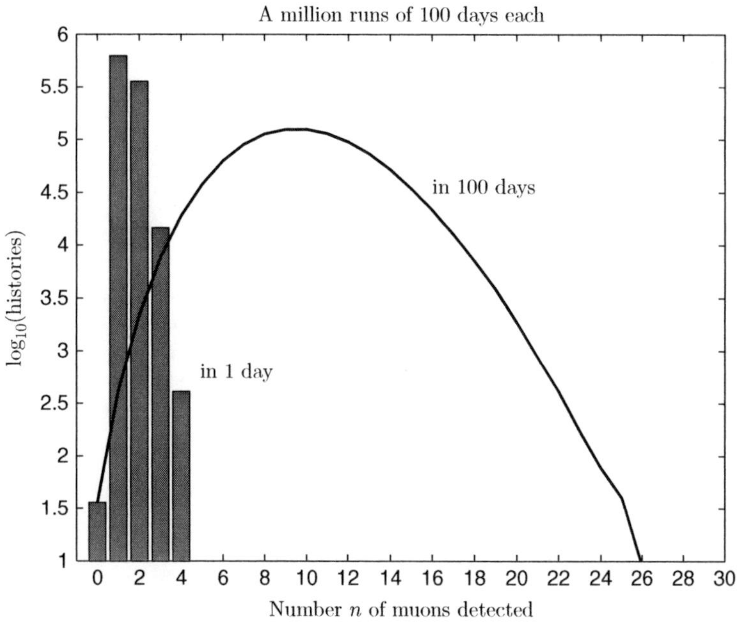

Figure 14.1 The number (out of 1,000,000) of histories of 100 days in which a maximum of n muons is detected on a single day (boxes) and in 100 days (curve).

examine the apparatus or the heavens; the Monte Carlo can't tell us which. The curve shows how many runs had a total of n muons; 125,142 histories had ten muons.

Of course, one could compute the data of Fig. 14.1 by hand without running a Monte Carlo. But suppose one's aging phototubes reduced the mean number of muons detected per day to $\langle n \rangle = 0.1(1 - \alpha d/100)$ on day d? Or suppose one needed the probability of detecting more than one muon on two days separated by one day of zero muons? In such cases, the analytic computation would be difficult and error prone, but the student would need to change only a few lines in the Monte Carlo program.

An equivalent C++ code by Sean Cahill is:

```
#include <stdlib.h>
#include <time.h>
#include <math.h>
#include <iostream>
#include <fstream>
#include <iomanip>
#include <vector>
#include <valarray>
```

```
using namespace std;

// Calculates the factorial of n
double factorial(const int& n)
{
  double f = 1;

  int i=0;
  for(i = 1; i <= n; i++)
  {
    f *= i;
  }
  return f;
}

void muons()
{
  // Declares constants
  const int N = 1000000; // Number of data sets
  const int LOOP_ITR = 101;
  const double AN = 1; // Number of events per day

  // Inits local variables
  int k=0, m=0, day=0, num=0, numMuons=0, totMuons=0;
  int maxEvents[LOOP_ITR];
  int totEvents[LOOP_ITR];
  memset (maxEvents, 0, sizeof(int) * LOOP_ITR);
  memset (totEvents, 0, sizeof(int) * LOOP_ITR);

  // Creates our 2d histories array
  vector<valarray<int> > histories(N, LOOP_ITR);

  double prob=0,tmpProb=0,fact=0, x=0;
  double p[LOOP_ITR];

  // probability of no events
  p[0] = exp(-AN);
  prob = p[0];

  // p(k) is the probability of fewer than k+1 events
     per day
  for (k=1; k<=LOOP_ITR; k++)
```

```
{
  fact = factorial (k);
  tmpProb = k * exp(-AN) / fact;
  prob += pow(AN, tmpProb);
  p[k] = prob;
}

// Random seed
srand ( time(NULL) );

// Goes through all the histories
for (k=0; k<N; k++)
{
  // Goes through all the days
  for (day=1; day<LOOP_ITR; day++)
  {
    // Generates a random number between 0 and 1
    x = static_cast<double>(rand()) / RAND_MAX;

    // Finds an M with p(M) < X
    for (m=100; m>=0; m--)
    {
      if (x < p[m])
      {
        num = m;
      }
    }

    histories[k][day] = num;
  }

  // Calculates max and sum
  numMuons = histories[k].max();
  totMuons = histories[k].sum();

  // Updates our records
  maxEvents[numMuons]++;
  totEvents[totMuons]++;
}
// Opens a data file
ofstream fhMaxEvents, fhSumEvents;
fhMaxEvents.open ("maxEvents.txt");
```

```
  fhSumEvents.open ("totEvents.txt");

  // Sets precision
  fhMaxEvents.setf(ios::fixed,ios::floatfield);
  fhMaxEvents.precision(7);
  fhSumEvents.setf(ios::fixed,ios::floatfield);
  fhSumEvents.precision(7);

  // Writes the data to a file
  for (k=0; k<LOOP_ITR; k++)
  {
    fhMaxEvents << k << "    " << maxEvents[k] << endl;
    fhSumEvents << k << "    " << totEvents[k] << endl;
  }
}
```

14.4 Statistical mechanics

The Metropolis algorithm can generate a sequence of states or configurations of a system distributed according to the Boltzmann probability distribution (1.345). Suppose the state of the system is described by a vector x of many components. For instance, if the system is a protein, the vector x might be the $3N$ spatial coordinates of the N atoms of the protein. A protein composed of 200 amino acids has about 4000 atoms, and so the vector x would have some 12,000 components. Suppose $E(x)$ is the energy of configuration x of the protein in its cellular environment of salty water crowded with macromolecules. How do we generate a sequence of "native states" of the protein at temperature T?

We start with some random or artificial initial configuration x^0 and then make random changes δx in successive configurations x. One way to do this is to make a small, random change δx_i in coordinate x_i and then to test whether to accept this change by comparing the energies $E(x)$ and $E(x')$ of the two configurations x and x', which differ by δx_i in coordinate x_i. (Estimating these energies is not trivial; Gromacs and TINKER can help.) It is important that this random walk be symmetric, that is, the choice of testing whether to go from x to x' when one is at x should be exactly as likely as the choice of testing whether to go from x' to x when one is at x'. Also, the sequences of configurations should be **ergodic**; that is, from any configuration x, one should be able to get to any other configuration x' by a suitable sequence of changes $\delta x_i = x'_i - x_i$.

How do we decide whether to accept or reject δx_i? We use the following **Metropolis step**. If the energy $E' = E(x')$ of the new configuration x' is less than the energy $E(x)$ of the current configuration x, then we accept the new configuration x'. If $E' > E$, then we accept x' with probability

$$P(x \to x') = e^{-(E'-E)/kT}. \tag{14.6}$$

In practice, one generates a pseudo-random number $r \in [0, 1]$ and accepts x' if

$$r < e^{-(E'-E)/kT}. \tag{14.7}$$

If one does not accept x', then the system remains in configuration x.

In FORTRAN90, the Metropolis step might be

```
if ( newE <= oldE ) then ! accept
   x(i) = x(i) + dx
else ! accept conditionally
   call random_number(r)
   if ( r <= exp(-beta*(newE - oldE))) then ! accept
      x(i) = x(i) + dx
   end if
end if
```

in which $\beta = 1/kT$.

One then varies another coordinate, such as x_{i+1}. Once one has varied all of the coordinates, one has finished a **sweep** through the system. After thousands or millions of such sweeps, the system is said to be **thermalized**. Once the system is thermalized, one can start measuring properties of the system. One computes a physical quantity every hundred or every thousand sweeps and takes the average of these measurements. That average is the mean value of the physical quantity at temperature T.

Why does this work? Consider two configurations x and x', which respectively have energies $E = E(x)$ and $E' = E(x')$ and are occupied with probabilities $P_t(x)$ and $P_t(x')$ as the system is thermalizing. If $E' > E$, then the rate $R(x' \to x)$ of going from x' to x is the rate v of choosing to test x when one is at x' times the probability $P_t(x')$ of being at x', that is, $R(x' \to x) = v\, P_t(x')$. The reverse rate is $R(x \to x') = v\, P_t(x)\, e^{-(E'-E)/kT}$ with the same v since the random walk is symmetric. The net rate from $x' \to x$ then is

$$R(x' \to x) - R(x \to x') = v \left(P_t(x') - P_t(x)\, e^{-(E'-E)/kT} \right). \tag{14.8}$$

This net flow of probability from $x' \to x$ is positive if and only if

$$P_t(x')/P_t(x) > e^{-(E'-E)/kT}. \tag{14.9}$$

The probability distribution $P_t(x)$ therefore flows with each sweep toward the Boltzmann distribution $\exp(-E(x)/kT)$. The flow slows and stops when the two rates are equal $R(x' \to x) = R(x \to x')$, a condition called **detailed balance**. At this equilibrium, the distribution $P_t(x)$ satisfies $P_t(x) = P_t(x')\, e^{-(E-E')/kT}$, in which $P_t(x')\, e^{E'/kT}$ is independent of x. So the thermalizing distribution $P_t(x)$ approaches the distribution $P(x) = c\, e^{-E/kT}$, in which c is independent of x. Since the sum of these probabilities must be unity, we have

$$\sum_x P(x) = c \sum_x e^{-E/kT} = 1, \tag{14.10}$$

which means that the constant c is the inverse of the **partition function**

$$Z(T) = \sum_x e^{-E(x)/kT}. \tag{14.11}$$

The thermalizing distribution approaches Boltzmann's distribution (1.345)

$$P_t(x) \to P_B(x) = e^{-E(x)/kT}/Z(T). \tag{14.12}$$

Example 14.2 (Z_2 lattice gauge theory) First, one replaces space-time with a lattice of points in d dimensions. Two nearest neighbor points are separated by the lattice spacing a and joined by a link. Next, one puts an element U of the gauge group on each link. For the Z_2 gauge group (example 10.4), one assigns an action $S_\square$ to each elementary square or *plaquette* of the lattice with vertices 1, 2, 3, and 4

$$S_\square = 1 - U_{1,2}U_{2,3}U_{3,4}U_{4,1}. \tag{14.13}$$

Then, one replaces $E(x)/kT$ with βS, in which the action S is a sum of all the plaquette actions S_p. More details are available at Michael Creutz's website (latticeguy.net/lattice.html). □

Although the generation of configurations distributed according to the Boltzmann probability distribution (1.345) is one of its most useful applications, the Monte Carlo method is much more general. It can generate configurations x distributed according to any probability distribution $P(x)$.

To generate configurations distributed according to $P(x)$, we accept any new configuration x' if $P(x') \geq P(x)$ and also accept x' with probability

$$P(x \to x') = P(x')/P(x) \tag{14.14}$$

if $P(x) > P(x')$.

This works for the same reason that the Boltzmann version works. Consider two configurations x and x'. If the system is thermalized, then the probabilities $P_t(x)$ and $P_t(x')$ have reached equilibrium, and so the rate $R(x \to x')$ from $x \to x'$ must equal that $R(x' \to x)$ from $x' \to x$. If $P(x') < P(x)$, then $R(x' \to x)$ is

$$R(x' \to x) = v\, P_t(x'), \tag{14.15}$$

in which v is the rate of choosing $\delta x = x' - x$, while the rate $R(x \to x')$ is

$$R(x \to x') = v\, P_t(x)\, P(x')/P(x) \tag{14.16}$$

with the same v since the random walk is symmetric. Equating the two rates

$$R(x' \to x) = R(x \to x') \tag{14.17}$$

we find that the flow of probability stops when

$$P_t(x) = P(x)\, P_t(x')/P(x') = cP(x), \tag{14.18}$$

where c is independent of x'. Thus $P_t(x) \to P(x)$.

So far we have assumed that the rate of choosing $x \to x'$ is the same as the rate of choosing $x' \to x$. In **smart Monte Carlo** schemes, physicists arrange the rates $v_{x \to x'}$ and $v_{x' \to x}$ so as to steer the flow and speed-up thermalization. To compensate for this asymmetry, they change the second part of the Metropolis step from $x \to x'$ when $E' = E(x') > E = E(x)$ to accept conditionally with probability

$$P(x \to x') = P(x')\, v_{x' \to x} / [P(x)\, v_{x \to x'}]. \tag{14.19}$$

Now if $P(x') < P(x)$, then $R(x' \to x)$ is

$$R(x' \to x) = v_{x' \to x}\, P_t(x') \tag{14.20}$$

while the rate $R(x \to x')$ is

$$R(x \to x') = v_{x \to x'}\, P_t(x)\, P(x')\, v_{x' \to x} / [P(x)\, v_{x \to x'}]. \tag{14.21}$$

Equating the two rates $R(x' \to x) = R(x \to x')$, we find

$$P_t(x') = P_t(x)\, P(x')/P(x). \tag{14.22}$$

That is $P_t(x) = P(x)\, P_t(x')/P(x')$, which gives

$$P_t(x) = N\, P(x) \tag{14.23}$$

where N is a constant of normalization.

14.5 Solving arbitrary problems

If you know how to generate a suitably large space of trial solutions to a problem, and you also know how to compare the quality of any two of your solutions, then you can use a Monte Carlo method to solve it. The hard parts of this seemingly magical method are characterizing a big enough space of solutions s and constructing a quality function or functional that assigns a number $Q(s)$ to every solution in such a way that if s is a better solution than s', then

$$Q(s) > Q(s'). \tag{14.24}$$

But once one has characterized the space of possible solutions s and has constructed the quality function $Q(s)$, then one simply generates huge numbers of random solutions and selects the one that maximizes the function $Q(s)$ over the space of all solutions.

If one can characterize the solutions as vectors of a certain dimension, $s = (x_1, \ldots, x_n)$, then one may use the Monte Carlo method of the previous section (14.4) by replacing $-E(s)$ with $Q(s)$ and kT with a parameter of the same dimension as $Q(s)$, nominally dimensionless.

14.6 Evolution

The reader may think that the use of Monte Carlo methods to solve arbitrary problems is quite a stretch. Yet Nature has applied them to the problem of evolving species that survive. As a measure of the quality $Q(s)$ of a given solution s, Nature used the time derivative of the logarithm of its population $\dot{P}(t)/P(t)$. The space of solutions is the set of possible genomes. We may idealize each solution or genome as a sequence of nucleotides $s = b_1 b_2 \ldots b_N$ some thousands or billions of bases long, each base b_k being adenine, cytosine, guanine, or thymine (A, C, G, or T). Since there are four choices for each base, the set of solutions is huge. The genome for *homo sapiens* has some 3 billion bases (or base pairs, DNA being double stranded), and so the solution space is a set with

$$\mathcal{N} = 4^{3\times 10^9} = 10^{1.8\times 10^9} \tag{14.25}$$

elements. By comparison, a googol is only 10^{100}.

In evolution, a Metropolis step begins with a random change in the sequence of bases; changes in a germ-line cell can create a new individual. Some of these changes are due to errors in the normal mechanisms by which genomes are copied and repaired. The (holo)enzyme DNA polymerase copies DNA with remarkable fidelity, making one error in every billion base pairs copied. Along a given line of descent, only about one nucleotide pair in a thousand is randomly changed in the germ line every million years. Yet in a population of 10,000 diploid individuals, every possible nucleotide substitution will have been tried out on about 20 occasions during a million years (Alberts *et al.*, 2008).

RNA polymerases transcribe DNA into RNA, and RNAs play many roles: Ribosomes translate messenger RNAs (mRNAs) into proteins, which are sequences of amino acids; ribosomal RNAs (rRNAs) combine with proteins to form ribosomes; long noncoding RNAs (ncRNAs) regulate the rates at which different genes are transcribed; micro RNAs (miRNAs) regulate the rates at which different mRNAs are translated into proteins; and other RNAs have other as yet unknown functions. So a change of one base, e.g. from A to C, might alter a protein or change the expression of a gene or be silent.

Sexual reproduction makes bigger random changes in genomes. In **meiosis**, the paternal and maternal versions of each of our 23 chromosomes are duplicated, and the four versions swap segments of DNA in a process called genetic recombination or crossing-over. The cell then divides twice producing four **haploid** germ cells each with a single paternal, maternal, or mixed version of each chromosome. This second kind of Metropolis step makes evolution more ergodic, which is why most complex modern organisms use sexual reproduction.

Other genomic changes occur when a virus inserts its DNA into that of a cell and when transposable elements (transposons) of DNA move to different sites in a genome.

In evolution, the rest of the Metropolis step is done by the new individual: if he or she survives and multiplies, then the change is accepted; if he or she dies without progeny, then the change is rejected. Evolution is slow, but it has succeeded in turning a soup of simple molecules into humans with brains of 100 billion neurons, each with 1000 connections to other neurons.

John Holland and others have incorporated analogs of these Metropolis steps into Monte Carlo techniques called **genetic algorithms** for solving wide classes of problems (Holland, 1975; Vose, 1999; Schmitt, 2001).

Evolution also occurs at the cellular level when a cell mutates enough to escape the control imposed on its proliferation by its neighbors and transforms into a cancer cell.

Further reading

The classic *Quarks, Gluons, and Lattices* (Creutz, 1983) is a marvelous introduction to the subject; Creutz's Website (latticeguy.net/lattice.html) is an extraordinary resource.

Exercises

14.1 Go to Michael Creutz's website (latticeguy.net/lattice.html) and get his C-code for Z_2 lattice gauge theory. Compile and run it, and make a graph that exhibits strong hysteresis as you raise and lower $\beta = 1/kT$.

14.2 Modify his code and produce a graph showing the coexistence of two phases at the critical coupling $\beta_t = 0.5 \ln(1+\sqrt{2})$. Hint: do a cold start and then 100 updates at β_t, then do a random start and do 100 updates at β_t. Plot the values of the action against the update number $1, 2, 3, \ldots, 100$.

14.3 Modify Creutz's C code for Z_2 lattice gauge theory so as to be able to vary the dimension d of space-time. Show that for $d = 2$, there's no hysteresis loop (there's no phase transition). For $d = 3$, show that any hysteresis loop is minimal (there's a second-order phase transition).

14.4 What happens when $d = 5$?

15

Functional derivatives

15.1 Functionals

A **functional** $G[f]$ is a map from a space of functions to a set of numbers. For instance, the **action** functional $S[q]$ for a particle in one dimension maps the coordinate $q(t)$, which is a function of the time t, into a number – the action of the process. If the particle has mass m and is moving slowly and freely, then for the interval (t_1, t_2) its action is

$$S_0[q] = \int_{t_1}^{t_2} dt \, \frac{m}{2} \left(\frac{dq(t)}{dt} \right)^2 . \tag{15.1}$$

If the particle is moving in a potential $V(q(t))$, then its action is

$$S[q] = \int_{t_1}^{t_2} dt \left[\frac{m}{2} \left(\frac{dq(t)}{dt} \right)^2 - V(q(t)) \right] . \tag{15.2}$$

15.2 Functional derivatives

A **functional derivative** is a functional

$$\delta G[f][h] = \left. \frac{d}{d\epsilon} G[f + \epsilon h] \right|_{\epsilon=0} \tag{15.3}$$

of a functional. For instance, if $G_n[f]$ is the functional

$$G_n[f] = \int dx \, f^n(x) \tag{15.4}$$

then its functional derivative is the functional that maps the pair of functions f, h to the number

$$\begin{aligned}
\delta G_n[f][h] &= \frac{d}{d\epsilon} G_n[f + \epsilon h]\bigg|_{\epsilon=0} \\
&= \frac{d}{d\epsilon} \int dx\, (f(x) + \epsilon h(x))^n \bigg|_{\epsilon=0} \\
&= \int dx\, n f^{n-1}(x) h(x).
\end{aligned} \tag{15.5}$$

Physicists often use the less elaborate notation

$$\frac{\delta G[f]}{\delta f(y)} = \delta G[f][\delta_y], \tag{15.6}$$

in which the function $h(x)$ is $\delta_y(x) = \delta(x - y)$. Thus in the preceding example

$$\frac{\delta G[f]}{\delta f(y)} = \int dx\, n f^{n-1}(x) \delta(x - y) = n f^{n-1}(y). \tag{15.7}$$

Functional derivatives of functionals that involve powers of derivatives also are easily dealt with. Suppose that the functional involves the square of the derivative $f'(x)$

$$G[f] = \int dx\, \left(f'(x)\right)^2. \tag{15.8}$$

Then its functional derivative is

$$\begin{aligned}
\delta G[f][h] &= \frac{d}{d\epsilon} G[f + \epsilon h]\bigg|_{\epsilon=0} \\
&= \frac{d}{d\epsilon} \int dx\, \left(f'(x) + \epsilon h'(x)\right)^2 \bigg|_{\epsilon=0} \\
&= \int dx\, 2 f'(x) h'(x) = -2 \int dx\, f''(x) h(x),
\end{aligned} \tag{15.9}$$

in which we have integrated by parts and used suitable boundary conditions on $h(x)$ to drop the surface terms. In physics notation, we have

$$\frac{\delta G[f]}{\delta f(y)} = -2 \int dx\, f''(x) \delta(x - y) = -2 f''(y). \tag{15.10}$$

Let's now compute the functional derivative of the action (15.2), which involves the square of the time-derivative $\dot{q}(t)$ and the potential energy $V(q(t))$

$$\begin{aligned}\delta S[q][h] &= \frac{d}{d\epsilon} S[q+\epsilon h]\bigg|_{\epsilon=0} \\ &= \frac{d}{d\epsilon}\int dt \left[\frac{m}{2}\left(\dot{q}(t)+\epsilon\dot{h}(t)\right)^2 - V(q(t)+\epsilon h(t))\right]\bigg|_{\epsilon=0} \\ &= \int dt \left[m\dot{q}(t)\dot{h}(t) - V'(q(t))h(t)\right] \\ &= \int dt \left[-m\ddot{q}(t) - V'(q(t))\right] h(t) \end{aligned} \tag{15.11}$$

where we once again have integrated by parts and used suitable boundary conditions to drop the surface terms. In physics notation, this is

$$\frac{\delta S[q]}{\delta q(t)} = \int dt' \left[-m\ddot{q}(t') - V'(q(t'))\right] \delta(t-t') = -m\ddot{q}(t) - V'(q(t)). \tag{15.12}$$

In these terms, the stationarity of the action $S[q]$ is the vanishing of its functional derivative either in the form

$$\delta S[q][h] = 0 \tag{15.13}$$

for arbitrary functions $h(t)$ (that satisfy the boundary conditions) or equivalently in the form

$$\frac{\delta S[q]}{\delta q(t)} = 0, \tag{15.14}$$

which is Lagrange's equation of motion

$$m\ddot{q}(t) = -V'(q(t)). \tag{15.15}$$

Physicists also use the compact notation

$$\frac{\delta^2 Z[j]}{\delta j(y)\delta j(z)} \equiv \frac{\partial^2 Z[j+\epsilon\delta_y+\epsilon'\delta_z]}{\partial\epsilon\,\partial\epsilon'}\bigg|_{\epsilon=\epsilon'=0} \tag{15.16}$$

in which $\delta_y(x) = \delta(x-y)$ and $\delta_z(x) = \delta(x-z)$.

Example 15.1 (Shortest path is a straight line) On a plane, the length of the path $(x, y(x))$ from (x_0, y_0) to (x_1, y_1) is

$$L[y] = \int_{x_0}^{x_1} \sqrt{dx^2 + dy^2} = \int_{x_0}^{x_1} \sqrt{1+y'^2}\, dx. \tag{15.17}$$

The shortest path $y(x)$ minimizes this length $L[y]$

$$\begin{aligned}\frac{\delta L[f]}{\delta y} &= \frac{d}{d\epsilon} L[y+\epsilon h]\bigg|_{\epsilon=0} = \frac{d}{d\epsilon}\int_{x_0}^{x_1} \sqrt{1+(y'+\epsilon h')^2}\, dx\bigg|_{\epsilon=0} \\ &= \int_{x_0}^{x_1} \frac{y'h'}{\sqrt{1+y'^2}}\, dx = -\int_{x_0}^{x_1} h \frac{d}{dx}\frac{y'}{\sqrt{1+y'^2}}\, dx \end{aligned} \tag{15.18}$$

since $h(x)$ satisfies $h(x_0) = h(x_1) = 0$. Differentiating, we set

$$\delta L[y][h] = \int_{x_0}^{x_1} h \frac{y''}{1+y'^2}\, dx = 0, \tag{15.19}$$

which can vanish for arbitrary functions $h(x)$ only if $y'' = 0$, which is to say only if $y(x)$ is a straight line, $y = mx + b$. □

15.3 Higher-order functional derivatives

The second functional derivative is

$$\delta^2 G[f][h] = \frac{d^2}{d\epsilon^2}\, G[f + \epsilon h]|_{\epsilon=0}\,. \tag{15.20}$$

So if $G_N[f]$ is the functional

$$G_N[f] = \int f^N(x) dx \tag{15.21}$$

then

$$\begin{aligned}
\delta^2 G_N[f][h] &= \frac{d^2}{d\epsilon^2}\, G_N[f + \epsilon h]|_{\epsilon=0} \\
&= \frac{d^2}{d\epsilon^2} \int (f(x) + \epsilon h(x))^N\, dx \bigg|_{\epsilon=0} \\
&= \frac{d^2}{d\epsilon^2} \int \binom{N}{2} \epsilon^2 h^2(x) f^{N-2}(x)\, dx \bigg|_{\epsilon=0} \\
&= N(N-1) \int f^{N-2}(x) h^2(x) dx.
\end{aligned} \tag{15.22}$$

Example 15.2 ($\delta^2 S_0$) The second functional derivative of the action $S_0[q]$ (15.1) is

$$\begin{aligned}
\delta^2 S_0[q][h] &= \frac{d^2}{d\epsilon^2} \int_{t_1}^{t_2} dt\, \frac{m}{2} \left(\frac{dq(t)}{dt} + \epsilon \frac{dh(t)}{dt} \right)^2 \bigg|_{\epsilon=0} \\
&= \int_{t_1}^{t_2} dt\, m \left(\frac{dh(t)}{dt} \right)^2 \geq 0
\end{aligned} \tag{15.23}$$

and is positive for all functions $h(t)$. The stationary classical trajectory

$$q(t) = \frac{t - t_1}{t_2 - t_1}\, q(t_2) + \frac{t_2 - t}{t_2 - t_1}\, q(t_1) \tag{15.24}$$

is a **minimum** of the action $S_0[q]$. □

The second functional derivative of the action $S[q]$ (15.2) is

$$\delta^2 S[q][h] = \frac{d^2}{d\epsilon^2} \int_{t_1}^{t_2} dt \left[\frac{m}{2} \left(\frac{dq(t)}{dt} + \epsilon \frac{dh(t)}{dt} \right)^2 - V(q(t) + \epsilon h(t)) \right] \Bigg|_{\epsilon=0}$$

$$= \int_{t_1}^{t_2} dt \left[m \left(\frac{dh(t)}{dt} \right)^2 - 2 \frac{\partial^2 V(q(t))}{\partial q^2(t)} h^2(t) \right] \tag{15.25}$$

and it can be positive, zero, or negative. Chaos sometimes arises in systems of several particles when the second variation of $S[q]$ about a stationary path is negative, $\delta^2 S[q][h] < 0$, while $\delta S[q][h] = 0$.

The nth functional derivative is defined as

$$\delta^n G[f][h] = \frac{d^n}{d\epsilon^n} G[f + \epsilon h]|_{\epsilon=0} . \tag{15.26}$$

The nth functional derivative of the same functional (15.21) is

$$\delta^n G_N[f][h] = \frac{N!}{(N-n)!} \int f^{N-n}(x) h^n(x) dx. \tag{15.27}$$

15.4 Functional Taylor series

It follows from the Taylor-series theorem (section 4.6) that

$$e^{\delta} G[f][h] = \sum_{n=0}^{\infty} \frac{\delta^n}{n!} G[f][h] = \sum_{n=0}^{\infty} \frac{1}{n!} \frac{d^n}{d\epsilon^n} G[f + \epsilon h] \Bigg|_{\epsilon=0} = G[f + \epsilon h], \tag{15.28}$$

which illustrates an advantage of the present mathematical notation.

The functional $S_0[q]$ of Equation (15.1) provides a simple example of the functional Taylor series (15.28):

$$e^{\delta} S_0[q][h] = \left(1 + \frac{d}{d\epsilon} + \frac{1}{2} \frac{d^2}{d\epsilon^2} \right) S_0[q + \epsilon h] \Bigg|_{\epsilon=0}$$

$$= \frac{m}{2} \int_{t_1}^{t_2} \left(1 + \frac{d}{d\epsilon} + \frac{1}{2} \frac{d^2}{d\epsilon^2} \right) \left(\dot{q}(t) + \epsilon \dot{h}(t) \right)^2 dt \Bigg|_{\epsilon=0}$$

$$= \frac{m}{2} \int_{t_1}^{t_2} \left(\dot{q}^2(t) + 2\dot{q}(t)\dot{h}(t) + \dot{h}^2(t) \right) dt$$

$$= \frac{m}{2} \int_{t_1}^{t_2} \left(\dot{q}(t) + \dot{h}(t) \right)^2 dt = S_0[q + h]. \tag{15.29}$$

Note that if the function $q(t)$ makes the action $S_0[q]$ stationary, and if $h(t)$ is smooth and vanishes at the endpoints of the time interval, then by (15.23)

$$S_0[q + h] = S_0[q] + S_0[h], \tag{15.30}$$

in which the functions $q(t)$ and $h(t)$ respectively satisfy the boundary conditions on $h(x)$ $q(t_i) = q_i$ and $h(t_1) = h(t_2) = 0$.

More generally, if $q(t)$ makes the action $S[q]$ stationary, and $h(t)$ is any loop from and to the origin, then

$$S[q+h] = e^{\delta} S[q][h] = S[q] + \sum_{n=2}^{\infty} \frac{1}{n!} \frac{d^n}{d\epsilon^n} S[q+\epsilon h]|_{\epsilon=0} \, . \tag{15.31}$$

If further $S_2[q]$ is purely quadratic in q and $\dot{q}$, like the harmonic oscillator, then

$$S_2[q+h] = S_2[q] + S_2[h]. \tag{15.32}$$

15.5 Functional differential equations

In inner products like $\langle q'|f\rangle$, we represent the momentum operator as

$$p = \frac{\hbar}{i} \frac{d}{dq'} \tag{15.33}$$

because then

$$\langle q'|p\, q|f\rangle = \frac{\hbar}{i} \frac{d}{dq'} \langle q'|q|f\rangle = \frac{\hbar}{i} \frac{d}{dq'} \left(q' \langle q'|f\rangle \right) = \left(\frac{\hbar}{i} + q' \frac{\hbar}{i} \frac{d}{dq'} \right) \langle q'|f\rangle, \tag{15.34}$$

which respects the commutation relation $[q, p] = i\hbar$.

So too in inner products $\langle \phi'|f\rangle$ of eigenstates $|\phi'\rangle$ of $\phi(\boldsymbol{x}, t)$

$$\phi(\boldsymbol{x}, t)|\phi'\rangle = \phi'(\boldsymbol{x})|\phi'\rangle \tag{15.35}$$

we can represent the momentum $\pi(\boldsymbol{x}, t)$ canonically conjugate to the field $\phi(\boldsymbol{x}, t)$ as the functional derivative

$$\pi(\boldsymbol{x}, t) = \frac{\hbar}{i} \frac{\delta}{\delta \phi'(\boldsymbol{x})} \tag{15.36}$$

because then

$$\begin{aligned}
\langle \phi'|\pi(\boldsymbol{x}', t)\phi(\boldsymbol{x}, t)|f\rangle &= \frac{\hbar}{i} \frac{\delta}{\delta \phi'(\boldsymbol{x}')} \langle \phi'|\phi(\boldsymbol{x}, t)|f\rangle \\
&= \frac{\hbar}{i} \frac{\delta}{\delta \phi'(\boldsymbol{x}')} \left(\phi'(\boldsymbol{x}) \langle \phi'|f\rangle \right) \\
&= \frac{\hbar}{i} \frac{\delta}{\delta \phi'(\boldsymbol{x}')} \left(\int \delta(\boldsymbol{x} - \boldsymbol{x}')\, \phi'(\boldsymbol{x}')\, d^3x' \, \langle \phi'|f\rangle \right) \\
&= \frac{\hbar}{i} \left(\delta(\boldsymbol{x} - \boldsymbol{x}') + \phi'(\boldsymbol{x}) \frac{\delta}{\delta \phi'(\boldsymbol{x}')} \right) \langle \phi'|f\rangle \\
&= \langle \phi'| - i\hbar \delta(\boldsymbol{x} - \boldsymbol{x}') + \phi(\boldsymbol{x}, t)\, \pi(\boldsymbol{x}', t)|f\rangle,
\end{aligned} \tag{15.37}$$

which respects the equal-time commutation relation

$$[\phi(\boldsymbol{x},t),\pi(\boldsymbol{x}',t)] = i\hbar\,\delta(\boldsymbol{x}-\boldsymbol{x}'). \tag{15.38}$$

We can use the representation (15.36) for $\pi(x)$ to find the wave-function of the ground state $|0\rangle$ of the hamiltonian

$$H = \frac{1}{2}\int \left[\pi^2 + (\nabla\phi)^2 + m^2\phi^2\right] d^3x \tag{15.39}$$

where we set $\hbar = c = 1$. We will use a trick used to find the ground state $|0\rangle$ of the harmonic-oscillator hamiltonian

$$H_0 = \frac{p^2}{2m} + \frac{m\omega^2 q^2}{2}. \tag{15.40}$$

In that trick, one writes

$$\begin{aligned} H_0 &= \frac{1}{2m}(m\omega q - ip)(m\omega q + ip) + \frac{i\omega}{2}[p,q] \\ &= \frac{1}{2m}(m\omega q - ip)(m\omega q + ip) + \frac{1}{2}\hbar\omega \end{aligned} \tag{15.41}$$

and seeks a state $|0\rangle$ that is annihilated by $m\omega q + ip$

$$\langle q'|m\omega q + ip|0\rangle = \left(m\omega q' + \hbar\frac{d}{dq'}\right)\langle q'|0\rangle = 0. \tag{15.42}$$

The solution to this differential equation

$$\frac{d}{dq'}\langle q'|0\rangle = -\frac{m\omega q'}{\hbar}\langle q'|0\rangle \tag{15.43}$$

is

$$\langle q'|0\rangle = \left(\frac{m\omega}{\pi\hbar}\right)^{1/4}\exp\left(-\frac{m\omega q'^2}{2\hbar}\right), \tag{15.44}$$

in which the prefactor is a constant of normalization.

So extending that trick to the hamiltonian (15.39), we factor H

$$H = \frac{1}{2}\int\left[\sqrt{-\nabla^2+m^2}\,\phi - i\pi\right]\left[\sqrt{-\nabla^2+m^2}\,\phi + i\pi\right] d^3x + C, \tag{15.45}$$

in which C is the (infinite) constant

$$C = \frac{i}{2}\int [\pi, \sqrt{-\nabla^2+m^2}\,\phi]\, d^3x. \tag{15.46}$$

The ground state $|0\rangle$ of H must therefore satisfy the functional differential equation

$$\langle\phi'|\sqrt{-\nabla^2+m^2}\,\phi + i\pi\,|0\rangle = 0 \tag{15.47}$$

or

$$\frac{\delta\langle\phi'|0\rangle}{\delta\phi'(\boldsymbol{x})} = -\sqrt{-\nabla^2 + m^2}\,\phi'(\boldsymbol{x})\,\langle\phi'|0\rangle. \tag{15.48}$$

The solution is

$$\langle\phi'|0\rangle = N\,\exp\left(-\frac{1}{2}\int \phi'(\boldsymbol{x})\,\sqrt{-\nabla^2 + m^2}\,\phi'(\boldsymbol{x})\,d^3x\right), \tag{15.49}$$

in which N is a normalization constant. The spatial Fourier transform $\tilde{\phi}'(\boldsymbol{p})$

$$\phi'(\boldsymbol{x}) = \int e^{i\boldsymbol{p}\cdot\boldsymbol{x}}\,\tilde{\phi}'(\boldsymbol{p})\,\frac{d^3p}{(2\pi)^3} \tag{15.50}$$

satisfies $\tilde{\phi}'(-\boldsymbol{p}) = \tilde{\phi}'^*(\boldsymbol{p})$ since ϕ' is real. In terms of it, the ground-state wave-function is

$$\langle\phi'|0\rangle = N\,\exp\left(-\frac{1}{2}\int |\tilde{\phi}'(\boldsymbol{p})|^2\,\sqrt{\boldsymbol{p}^2 + m^2}\,d^3p\right). \tag{15.51}$$

Exercises

15.1 Compute the action $S_0[q]$ (15.1) for the classical path (15.24).

15.2 Use (15.25) to find a formula for the second functional derivative of the action (15.2) of the harmonic oscillator for which $V(q) = m\omega^2 q^2/2$.

15.3 Derive (15.51) from equations (15.49 & 15.50).

16

Path integrals

16.1 Path integrals and classical physics

Since Richard Feynman invented them over 60 years ago, path integrals have been used with increasing frequency in high-energy and condensed-matter physics, in finance, and in biophysics (Kleinert, 2009). Feynman used them to express matrix elements of the time-evolution operator $\exp(-itH/\hbar)$ in terms of the classical action. Others have used them to compute matrix elements of the Boltzmann operator $\exp(-H/kT)$, which in the limit of zero temperature projects out the ground state $|E_0\rangle$ of the system

$$\lim_{T\to 0} e^{-(H-E_0)/kT} = \lim_{T\to 0} \sum_{n=0}^{\infty} |E_n\rangle e^{-(E_n-E_0)/kT} \langle E_n| = |E_0\rangle\langle E_0|, \tag{16.1}$$

a trick used in lattice gauge theory.

Path integrals magically express the quantum-mechanical probability amplitude for a process as a sum of exponentials $\exp(iS/\hbar)$ of the classical action S of the various ways that process might occur.

16.2 Gaussian integrals

The path integrals we can do are gaussian integrals of infinite order. So we begin by recalling the basic integral formula (5.166)

$$\int_{-\infty}^{\infty} \exp\left[-ia\left(x - \frac{b}{2a}\right)^2\right] dx = \sqrt{\frac{\pi}{ia}}, \tag{16.2}$$

which holds for real a and b, and also the one (5.167)

$$\int_{-\infty}^{\infty} \exp\left[-r\left(x-\frac{c}{2r}\right)^2\right] dx = \sqrt{\frac{\pi}{r}}, \tag{16.3}$$

which is true for positive r and complex c. Equivalent formulas for real a and b, positive r, and complex c are

$$\int_{-\infty}^{\infty} \exp\left(-iax^2 + ibx\right) dx = \sqrt{\frac{\pi}{ia}} \exp\left(i\frac{b^2}{4a}\right), \tag{16.4}$$

$$\int_{-\infty}^{\infty} \exp\left(-rx^2 + cx\right) dx = \sqrt{\frac{\pi}{r}} \exp\left(\frac{c^2}{4r}\right). \tag{16.5}$$

This last formula will be useful with $x = p$, $r = \epsilon/(2m)$, and $c = i\epsilon\dot{q}$

$$\int_{-\infty}^{\infty} \exp\left(-\epsilon\frac{p^2}{2m} + i\epsilon\,\dot{q}p\right) dp = \sqrt{\frac{2\pi m}{\epsilon}} \exp\left(-\epsilon\frac{1}{2}m\dot{q}^2\right) \tag{16.6}$$

as will (16.2) with $x = p$, $a = \epsilon/(2m)$, and $b = \epsilon\dot{q}$

$$\int_{-\infty}^{\infty} \exp\left(-i\epsilon\frac{p^2}{2m} + i\epsilon\,\dot{q}p\right) dp = \sqrt{\frac{2\pi m}{i\epsilon}} \exp\left(i\epsilon\frac{1}{2}m\dot{q}^2\right). \tag{16.7}$$

Doable path integrals are multiple gaussian integrals. One may show (exercise 16.1) that for positive $r_1, \ldots, r_N$ and complex $c_1, \ldots, c_N$, the integral (16.5) leads to

$$\int_{-\infty}^{\infty} \exp\left(\sum_i -r_i x_i^2 + c_i x_i\right) \prod_{i=1}^{N} dx_i = \left(\prod_{i=1}^{N} \sqrt{\frac{\pi}{r_i}}\right) \exp\left(\frac{1}{4}\sum_i \frac{c_i^2}{r_i}\right). \tag{16.8}$$

If R is the $N \times N$ diagonal matrix with positive entries $\{r_1, r_2, \ldots, r_N\}$, and X and C are N-vectors with real $\{x_i\}$ and complex $\{c_i\}$ entries, then this formula (16.8) in matrix notation is

$$\int_{-\infty}^{\infty} \exp\left(-X^{\mathsf{T}}RX + C^{\mathsf{T}}X\right) \prod_{i=1}^{N} dx_i = \sqrt{\frac{\pi^N}{\det(R)}} \exp\left(\frac{1}{4}\,C^{\mathsf{T}}R^{-1}C\right). \tag{16.9}$$

Now every positive symmetric matrix S is of the form $S = ORO^{\mathsf{T}}$ for some positive diagonal matrix R. So inserting $R = O^{\mathsf{T}}SO$ into the previous equation (16.9) and using the invariance of determinants under orthogonal transformations, we find

$$\int_{-\infty}^{\infty} \exp(-X^{\mathsf{T}}O^{\mathsf{T}}SOX + C^{\mathsf{T}}X) \prod_{i=1}^{N} dx_i = \sqrt{\frac{\pi^N}{\det(S)}} \exp\left[\frac{1}{4}\,C^{\mathsf{T}}O^{\mathsf{T}}S^{-1}OC\right]. \tag{16.10}$$

The jacobian of the orthogonal transformations $Y = OX$ and $D = OC$ is unity, and so

$$\int_{-\infty}^{\infty} \exp\left(-\, Y^{\mathsf{T}} S Y + D^{\mathsf{T}} Y\right) \prod_{i=1}^{N} dy_i = \sqrt{\frac{\pi^N}{\det(S)}} \exp\left(\frac{1}{4} D^{\mathsf{T}} S^{-1} D\right), \quad (16.11)$$

in which S is a positive symmetric matrix, and D is a complex vector.

The other basic gaussian integral (16.4) leads for real S and D to (exercise 16.2)

$$\int_{-\infty}^{\infty} \exp\left(-\, iY^{\mathsf{T}} S Y + iD^{\mathsf{T}} Y\right) \prod_{i=1}^{N} dy_i = \sqrt{\frac{\pi^N}{\det(iS)}} \exp\left(\frac{i}{4} D^{\mathsf{T}} S^{-1} D\right). \quad (16.12)$$

The vector $\overline{Y}$ that makes the argument $-\, iY^{\mathsf{T}} S Y + iD^{\mathsf{T}} Y$ of the exponential of this multiple gaussian integral (16.12) stationary is (exercise 16.3)

$$\overline{Y} = \frac{1}{2} S^{-1} D. \quad (16.13)$$

The exponential of that integral evaluated at its stationary point $\overline{Y}$ is

$$\exp\left(-\, i\overline{Y}^{\mathsf{T}} S \overline{Y} + iD^{\mathsf{T}} \overline{Y}\right) = \exp\left(\frac{i}{4} D^{\mathsf{T}} S^{-1} D\right). \quad (16.14)$$

Thus, the multiple gaussian integral (16.12) is equal to its exponential evaluated at its stationary point $\overline{Y}$, apart from a prefactor involving the determinant $\det iS$.

Similarly, the vector $\overline{Y}$ that makes the argument $-\, Y^{\mathsf{T}} S Y + D^{\mathsf{T}} Y$ of the exponential of the multiple gaussian integral (16.11) stationary is $\overline{Y} = S^{-1} D/2$, and that exponential evaluated at $\overline{Y}$ is

$$\exp\left(-\, Y^{\mathsf{T}} S Y + D^{\mathsf{T}} Y\right) = \exp\left(\frac{1}{4} D^{\mathsf{T}} S^{-1} D\right). \quad (16.15)$$

Once again, a multiple gaussian integral is simply its exponential evaluated at its stationary point $\overline{Y}$, apart from a prefactor involving the determinant $\det S$.

16.3 Path integrals in imaginary time

At the imaginary time $t = -i\beta\hbar$, the time-evolution operator $\exp(-itH/\hbar)$ is $\exp(-\beta H)$, in which the inverse temperature $\beta = 1/kT$ is the reciprocal of Boltzmann's constant $k = 8.617 \times 10^{-5}$ eV/K times the absolute temperature T. In the low-temperature limit, $\exp(-\beta H)$ is a projection operator (16.1) on the ground state of the system. These path integrals in imaginary time are called **euclidean** path integrals.

Let us consider a quantum-mechanical system with hamiltonian

$$H = \frac{p^2}{2m} + V(q), \tag{16.16}$$

in which the commutator of the position q and momentum p operators is $[q,p] = i$ in units in which $\hbar = 1$. For tiny ϵ, the corrections to the approximation

$$\exp\left[-\epsilon\left(\frac{p^2}{2m} + V(q)\right)\right] \approx \exp\left(-\epsilon\,\frac{p^2}{2m}\right)\exp\left(-\epsilon\, V(q)\right) + \mathcal{O}(\epsilon^2) \tag{16.17}$$

are of second order in ϵ.

To evaluate the matrix element $\langle q''|\exp(-\epsilon H)|q'\rangle$, we insert the identity operator I in the form of an integral over the momentum eigenstates

$$I = \int_{-\infty}^{\infty} |p'\rangle\langle p'|dp' \tag{16.18}$$

and use the inner product $\langle q''|p'\rangle = \exp(iq''p')/\sqrt{2\pi}$ so as to get as $\epsilon \to 0$

$$\begin{aligned}\langle q''|\exp(-\epsilon H)|q'\rangle &= \int_{-\infty}^{\infty} \langle q''|\exp\left(-\epsilon\,\frac{p^2}{2m}\right)|p'\rangle\langle p'|\exp\left(-\epsilon\, V(q)\right)|q'\rangle\, dp' \\ &= e^{-\epsilon\, V(q')}\int_{-\infty}^{\infty} \exp\left[-\epsilon\,\frac{p'^2}{2m} + i\,p'\,(q''-q')\right]\frac{dp'}{2\pi}.\end{aligned} \tag{16.19}$$

We now adopt the suggestive notation

$$\frac{q''-q'}{\epsilon} = \dot{q}' \tag{16.20}$$

and use the integral formula (16.6) so as to obtain

$$\begin{aligned}\langle q''|\exp(-\epsilon H)|q'\rangle &= \frac{1}{2\pi}\,e^{-\epsilon\, V(q')}\int_{-\infty}^{\infty}\exp\left(-\epsilon\,\frac{p'^2}{2m} + i\,\epsilon\, p'\,\dot{q}'\right)dp' \\ &= \left(\frac{m}{2\pi\epsilon}\right)^{1/2}\exp\left\{-\epsilon\left[\tfrac{1}{2}\,m\,\dot{q}'^2 + V(q')\right]\right\},\end{aligned} \tag{16.21}$$

in which q'' enters through the notation (16.20).

The next step is to link two of these matrix elements together

$$\begin{aligned}\langle q'''|e^{-2\epsilon H}|q'\rangle &= \int_{-\infty}^{\infty}\langle q'''|e^{-\epsilon H}|q''\rangle\langle q''|e^{-\epsilon H}|q'\rangle dq'' \\ &= \frac{m}{2\pi\epsilon}\int_{-\infty}^{\infty}\exp\left\{-\epsilon\left[\tfrac{1}{2}\,m\,\dot{q}''^2 + V(q'') + \tfrac{1}{2}\,m\,\dot{q}'^2 + V(q')\right]\right\}dq''.\end{aligned} \tag{16.22}$$

Linking three of these matrix elements together and using subscripts instead of primes, we have

$$\langle q_3|e^{-3\epsilon H}|q_0\rangle = \iint_{-\infty}^{\infty} \langle q_3|e^{-\epsilon H}|q_2\rangle\langle q_2|e^{-\epsilon H}|q_1\rangle\langle q_1|e^{-\epsilon H}|q_0\rangle\, dq_1 dq_2$$
$$= \left(\frac{m}{2\pi\epsilon}\right)^{3/2} \iint_{-\infty}^{\infty} \exp\left\{ -\epsilon \sum_{j=0}^{2} \left[\tfrac{1}{2} m\, \dot{q}_j^2 + V(q_j) \right] \right\} dq_1 dq_2. \tag{16.23}$$

Boldly passing from 3 to n and suppressing some integral signs, we get

$$\langle q_n|e^{-n\epsilon H}|q_0\rangle = \iiint_{-\infty}^{\infty} \langle q_n|e^{-\epsilon H}|q_{n-1}\rangle \cdots \langle q_1|e^{-\epsilon H}|q_0\rangle\, dq_1 \cdots dq_{n-1}$$
$$= \left(\frac{m}{2\pi\epsilon}\right)^{n/2} \iiint_{-\infty}^{\infty} \exp\left\{ -\epsilon \sum_{j=0}^{n-1} \left[\tfrac{1}{2} m\, \dot{q}_j^2 + V(q_j) \right] \right\} dq_1 \cdots dq_{n-1}. \tag{16.24}$$

Writing dt for ϵ and taking the limits $\epsilon \to 0$ and $n\epsilon \to \beta$, we find that the matrix element $\langle q_\beta|e^{-\beta H}|q_0\rangle$ is a path integral of the exponential of the average energy multiplied by $-\beta$

$$\langle q_\beta|e^{-\beta H}|q_0\rangle = N \int \exp\left[-\int_0^\beta \tfrac{1}{2} m\dot{q}^2(t) + V(q(t))\, dt \right] Dq, \tag{16.25}$$

in which $Dq \equiv (n\,m/2\pi\beta)^{n/2} dq_1 dq_2 \cdots dq_{n-1}$ as $n \to \infty$. We sum over all paths $q(t)$ that go from $q(0) = q_0$ at time 0 to $q(\beta) = q_\beta$ at time β.

In the limit $\beta \to \infty$, the operator $\exp(-\beta H)$ becomes proportional to a projection operator (16.1) on the ground state of the theory.

In three-dimensional space, $\boldsymbol{q}(t)$ replaces $q(t)$ in equation (16.25)

$$\langle \boldsymbol{q}_\beta|e^{-\beta H}|\boldsymbol{q}_0\rangle = N \int \exp\left[-\int_0^\beta \tfrac{1}{2} m\dot{\boldsymbol{q}}^2 + V(\boldsymbol{q})\, dt \right] D\boldsymbol{q}. \tag{16.26}$$

Path integrals in imaginary time are called *euclidean* mainly to distinguish them from *Minkowski* path integrals, which represent matrix elements of the time-evolution operator $\exp(-itH)$ in real time.

16.4 Path integrals in real time

Path integrals in real time represent the time-evolution operator $\exp(-itH)$. Using the integral formula (16.7), we find in the limit $\epsilon \to 0$

$$\begin{aligned}\langle q''|e^{-i\epsilon H}|q'\rangle &= \int_{-\infty}^{\infty}\langle q''|\exp\left[-i\epsilon\frac{p^2}{2m}\right]|p'\rangle\langle p'|\exp[-i\epsilon V(q)]|q'\rangle\, dp' \\ &= \frac{1}{2\pi}e^{-i\epsilon V(q')}\int_{-\infty}^{\infty}\exp\left[-i\epsilon\frac{p'^2}{2m}+ip'(q''-q')\right]dp' \\ &= \frac{1}{2\pi}e^{-i\epsilon V(q')}\int_{-\infty}^{\infty}\exp\left[-i\epsilon\frac{p'^2}{2m}+i\epsilon p'\dot{q}'\right]dp' \\ &= \left(\frac{m}{2\pi i\epsilon}\right)^{1/2}\exp\left[i\epsilon\left(\frac{m\dot{q}'^2}{2}-V(q')\right)\right].\end{aligned} \tag{16.27}$$

When we link together n of these matrix elements, we get the real-time version of (16.24)

$$\langle q_n|e^{-in\epsilon H}|q_0\rangle = \left(\frac{m}{2\pi i\epsilon}\right)^{n/2}\iiint_{-\infty}^{\infty}\exp\left\{i\epsilon\sum_{j=0}^{n-1}\left[\tfrac{1}{2}m\dot{q}_j^2 - V(q_j)\right]\right\}dq_1\cdots dq_{n-1}. \tag{16.28}$$

Writing dt for ϵ and taking the limits $\epsilon\to 0$ and $n\epsilon\to t$, we find that the amplitude $\langle q_t|e^{-itH}|q_0\rangle$ is the path integral

$$\langle q_t|e^{-itH}|q_0\rangle = N\int\exp\left[i\int_0^t \tfrac{1}{2}m\dot{q}^2 - V(q)\,dt'\right]Dq, \tag{16.29}$$

in which Dq differs from the one that appears in euclidean path integrals by the substitution $\beta\to it$:

$$Dq = \lim_{n\to\infty}\left(\frac{nm}{2\pi it}\right)^{n/2}dq_1 dq_2\cdots dq_n. \tag{16.30}$$

The integral in the exponent is the **classical action**

$$S[q] = \int_0^t \tfrac{1}{2}m\dot{q}^2 - V(q)\,dt' \tag{16.31}$$

for a process $q(t')$ that runs from $q(0)=q_0$ to $q(t)=q_t$. We sum over all such processes.

In three-dimensional space

$$\langle \boldsymbol{q}_t|e^{-itH}|\boldsymbol{q}_0\rangle = \int\exp\left[i\int_0^t \tfrac{1}{2}m\dot{\boldsymbol{q}}^2 - V(\boldsymbol{q})\,dt'\right]D\boldsymbol{q} \tag{16.32}$$

replaces (16.29).

The units of action are energy × time, and the argument of the exponential must be dimensionless, so in ordinary units the amplitude (16.29) is

$$\langle q_t|e^{-itH/\hbar}|q_0\rangle = \int e^{iS[q]/\hbar}Dq. \tag{16.33}$$

When is this amplitude big? When is it tiny? Suppose there is a process $q(t) = q_c(t)$ that goes from $q_c(0) = q_0$ to $q_c(t) = q_t$ in time t and that obeys the classical equation of motion (15.14–15.15)

$$\frac{\delta S[q_c]}{\delta q_c} = m\ddot{q}_c + V'(q_c) = 0. \tag{16.34}$$

The action of such a classical process is stationary, that is, $S[q_c + dq]$ differs from $S[q_c]$ only by terms of second order in δq. So there are infinitely many other processes that have the same action to within a fraction of $\hbar$. These processes add with nearly the same phase to the path integral (16.33) and so make a huge contribution to the amplitude $\langle q_t | e^{-itH/\hbar} | q_0 \rangle$.

But if no classical process goes from q_0 to q_t in time t, then the nonclassical processes from q_0 to q_t in time t have actions that differ among themselves by large multiples of $\hbar$. Their amplitudes cancel each other, and so the resulting amplitude is tiny. Thus **the real-time path integral explains the principle of stationary action** (section 11.37).

Does this path integral satisfy Schrödinger's equation? To see that it does, we'll use (16.27) in the more explicit form

$$\langle q'' | e^{-i\epsilon H} | q' \rangle = \left(\frac{m}{2\pi i \epsilon} \right)^{1/2} \exp\left[i \frac{m(q'' - q')^2}{2\epsilon} - i\epsilon \, V(q') \right] \tag{16.35}$$

to write $\psi(q'', t + \epsilon) = \langle q'' | \psi, t + \epsilon \rangle$ as an integral of $\psi(q', t) = \langle q' | \psi, t \rangle$

$$\begin{aligned} \langle q'' | \psi, t + \epsilon \rangle &= \int \langle q'' | e^{-i\epsilon H} | q' \rangle \langle q' | \psi, t \rangle \, dq' \\ &= \left(\frac{m}{2\pi i \epsilon} \right)^{1/2} \int \exp\left[i \frac{m(q'' - q')^2}{2\epsilon} - i\epsilon \, V(q') \right] \langle q' | \psi, t \rangle \, dq'. \end{aligned} \tag{16.36}$$

Keeping only leading terms in ϵ, we have

$$\psi(q'', t + \epsilon) = \left(\frac{m}{2\pi i \epsilon} \right)^{1/2} e^{-i\epsilon V(q'')} \int_{-\infty}^{\infty} \exp\left[i \frac{m(q'' - q')^2}{2\epsilon} \right] \psi(q', t) \, dq'. \tag{16.37}$$

Letting $x = q' - q''$ and $q' = q'' + x$, we find

$$\psi(q'', t + \epsilon) = \left(\frac{m}{2\pi i \epsilon} \right)^{1/2} e^{-i\epsilon V(q'')} \int \exp\left[i \frac{mx^2}{2\epsilon} \right] \psi(q'' + x, t) \, dx. \tag{16.38}$$

We now expand $\psi(q'' + x, t)$ as

$$\psi(q'' + x, t) = \psi(q'', t) + x\psi'(q'', t) + \frac{1}{2} x^2 \psi''(q'', t) + \cdots \tag{16.39}$$

and $\psi(q'', t+\epsilon)$ as

$$\psi(q'', t+\epsilon) = \psi(q'', t) + \epsilon\dot{\psi}(q'', t) + \cdots . \tag{16.40}$$

The integral formula (16.2) implies

$$\int_{-\infty}^{\infty} e^{imx^2/2\epsilon}\, dx = \left(\frac{2\pi i\epsilon}{m}\right)^{1/2} \tag{16.41}$$

and its derivative with respect to $im/2\epsilon$ gives

$$\int_{-\infty}^{\infty} x^2\, e^{imx^2/2\epsilon}\, dx = \frac{i\epsilon}{m}\left(\frac{2\pi i\epsilon}{m}\right)^{1/2} \quad \text{while} \quad \int_{-\infty}^{\infty} x\, e^{imx^2/2\epsilon}\, dx = 0. \tag{16.42}$$

Substituting the expansions (16.39 & 16.40) for $\psi(q''+x, t)$ and $\psi(q'', t+\epsilon)$ into the integral (16.38) and using the integral formulas (16.41 & 16.42), we get

$$\psi(q'', t) + \epsilon\dot{\psi}(q'', t) = \left[1 - i\epsilon V(q'')\right]\left[\psi(q'', t) + \frac{i\epsilon}{m}\psi''(q'', t)\right], \tag{16.43}$$

which is Schrödinger's equation

$$i\dot{\psi} = -\frac{1}{2m}\psi'' + V\psi \tag{16.44}$$

in natural units or $i\hbar\dot{\psi} = -\hbar^2\psi''/2m + V\psi$ in arbitrary units.

16.5 Path integral for a free particle

The amplitude for a free nonrelativistic particle to go from the origin to the point $\boldsymbol{q}$ in time t is the path integral (16.32)

$$\langle \boldsymbol{q}|e^{-itH}|\boldsymbol{q}=\boldsymbol{0}\rangle = \int e^{iS_0[\boldsymbol{q}]}\, D\boldsymbol{q} = \int \exp\left(i\int_0^t \tfrac{1}{2}m\dot{\boldsymbol{q}}^2(t')\, dt'\right) D\boldsymbol{q}. \tag{16.45}$$

The classical path that goes from $\boldsymbol{0}$ to $\boldsymbol{q}$ in time t is $\boldsymbol{q}_c(t') = (t'/t)\,\boldsymbol{q}$. The general path $\boldsymbol{q}(t')$ over which we integrate is $\boldsymbol{q}(t') = \boldsymbol{q}_c(t') + \boldsymbol{\delta q}(t')$. Since both $\boldsymbol{q}(t')$ and $\boldsymbol{q}_c(t')$ go from $\boldsymbol{0}$ to $\boldsymbol{q}$ in time t, the otherwise arbitrary path $\boldsymbol{\delta q}(t')$ must be a loop that goes from $\boldsymbol{\delta q}(0) = 0$ to $\boldsymbol{\delta q}(t) = 0$ in time t. The velocity $\dot{\boldsymbol{q}} = \dot{\boldsymbol{q}}_c + \dot{\boldsymbol{\delta q}}$ is the sum of the constant classical velocity $\dot{\boldsymbol{q}}_c = \boldsymbol{q}/t$ and the loop velocity $\dot{\boldsymbol{\delta q}}$. The first-order change vanishes

$$m\int_{t_1}^{t_2} \dot{\boldsymbol{q}}_c \cdot \frac{d\boldsymbol{\delta q}}{dt}\, dt = m\,\dot{\boldsymbol{q}}_c \cdot \int_{t_1}^{t_2} \frac{d\boldsymbol{\delta q}}{dt}\, dt = m\,\dot{\boldsymbol{q}}_c \cdot [\boldsymbol{\delta q}(t_2) - \boldsymbol{\delta q}(t_1)] = 0 \tag{16.46}$$

and so the action $S_0[\boldsymbol{q}]$ is the classical action plus the loop action

$$S_0[\boldsymbol{q}] = \tfrac{1}{2}m\int_0^t \left(\dot{\boldsymbol{q}}_c + \dot{\boldsymbol{\delta q}}\right)^2 dt' = S_0[\boldsymbol{q}_c] + S_0[\boldsymbol{\delta q}]. \tag{16.47}$$

The path integral therefore factorizes

$$\begin{aligned}\langle \boldsymbol{q}|e^{-itH}|\boldsymbol{0}\rangle &= \int e^{iS_0[\boldsymbol{q}]}\,D\boldsymbol{q} = N\int e^{iS_0[\boldsymbol{q}_c+\delta\boldsymbol{q}]}\,D\delta\boldsymbol{q} \\ &= \int e^{iS_0[\boldsymbol{q}_c]}\,e^{iS_0[\delta\boldsymbol{q}]}\,D\delta\boldsymbol{q} \\ &= e^{iS_0[\boldsymbol{q}_c]}\int e^{iS_0[\delta\boldsymbol{q}]}\,D\delta\boldsymbol{q} \end{aligned} \tag{16.48}$$

into the phase of the classical action times a path integral over the loops. The loop integral L is independent of the spatial points $\boldsymbol{q}$ and $\boldsymbol{0}$ and so can only depend upon the time interval, $L = L(t)$. Thus the amplitude is the product

$$\langle \boldsymbol{q}|e^{-itH}|\boldsymbol{0}\rangle = e^{iS_0[\boldsymbol{q}_c]}\,L(t). \tag{16.49}$$

Since the classical velocity is $\dot{\boldsymbol{q}}_c = \boldsymbol{q}/t$, the classical action is

$$S_0[\boldsymbol{q}_c] = \int_0^T \frac{m}{2}\,\dot{\boldsymbol{q}}_c^2(t)\,dt = \frac{m}{2}\frac{\boldsymbol{q}^2}{t}. \tag{16.50}$$

So the amplitude is

$$\langle \boldsymbol{q}|e^{-i(t_2-t_1)H}|\boldsymbol{0}\rangle = e^{im\boldsymbol{q}^2/2t}\,L(t). \tag{16.51}$$

Since the position eigenstates are orthogonal, this amplitude must reduce to a delta function as $t \to 0$

$$\lim_{t\to 0}\langle \boldsymbol{q}|e^{-itH}|\boldsymbol{0}\rangle = \langle \boldsymbol{q}|\boldsymbol{0}\rangle = \delta^3(\boldsymbol{q}). \tag{16.52}$$

One of the many representations of Dirac's delta function is

$$\delta^3(\boldsymbol{q}) = \lim_{t\to 0}\left(\frac{m}{2\pi i\hbar t}\right)^{3/2} e^{im\boldsymbol{q}^2/2\hbar t}. \tag{16.53}$$

Thus $L(t) = (m/2\pi i t)^{3/2}$ and

$$\langle \boldsymbol{q}|e^{-itH/\hbar}|\boldsymbol{0}\rangle = \left(\frac{m}{2\pi i\hbar t}\right)^{3/2} e^{im\boldsymbol{q}^2/2\hbar t} \tag{16.54}$$

in unnatural units. You can verify (exercise 16.6) this result by inserting a complete set of momentum dyadics $|p\rangle\langle p|$ and doing the resulting Fourier transform.

Example 16.1 (The Bohm–Aharonov effect) From our formula (11.311) for the action of a relativistic particle of mass m and charge q, we infer (exercise 16.7) that the action of a nonrelativistic particle in an electromagnetic field with no scalar potential is

$$S = \int_{x_1}^{x_2} \left[\frac{1}{2} m \boldsymbol{v} + q \boldsymbol{A} \right] \cdot d\boldsymbol{x} \,. \tag{16.55}$$

Now imagine that we shoot a beam of such particles past but not through a narrow cylinder in which a magnetic field is confined. The particles can go either way around the cylinder of area S but can not enter the region of the magnetic field. The difference in the phases of the amplitudes is the loop integral from the source to the detector and back to the source

$$\frac{\Delta S}{\hbar} = \oint \left[\frac{m\boldsymbol{v}}{2} + q\boldsymbol{A} \right] \cdot \frac{d\boldsymbol{x}}{\hbar} = \oint \frac{m\boldsymbol{v} \cdot d\boldsymbol{x}}{2\hbar} + \frac{q}{\hbar} \int_S \boldsymbol{B} \cdot d\boldsymbol{S} = \oint \frac{m\boldsymbol{v} \cdot d\boldsymbol{x}}{2\hbar} + \frac{q\Phi}{\hbar}, \tag{16.56}$$

in which Φ is the magnetic flux through the cylinder. □

16.6 Free particle in imaginary time

If we mimic the steps of the preceding section (16.5) in which the hamiltonian is $H = \boldsymbol{p}^2/2m$, set $\beta = it/\hbar = 1/kT$, and use Dirac's delta function

$$\delta^3(\boldsymbol{q}) = \lim_{t \to 0} \left(\frac{m}{2\pi \hbar t} \right)^{3/2} e^{-m\boldsymbol{q}^2/2\hbar t} \tag{16.57}$$

then we get

$$\langle \boldsymbol{q} | e^{-\beta H} | \boldsymbol{0} \rangle = \left(\frac{m}{2\pi \hbar^2 \beta} \right)^{3/2} \exp\left[-\frac{m\boldsymbol{q}^2}{2\hbar^2 \beta} \right] = \left(\frac{mkT}{2\pi \hbar^2} \right)^{3/2} e^{-mkT\boldsymbol{q}^2/2\hbar^2}. \tag{16.58}$$

To study the ground state of the system, we set $\beta = t/\hbar$ and let $t \to \infty$ in

$$\langle \boldsymbol{q} | e^{-tH/\hbar} | \boldsymbol{0} \rangle = \left(\frac{m}{2\pi \hbar t} \right)^{3/2} \exp\left[-\frac{m}{2} \frac{\boldsymbol{q}^2}{\hbar t} \right], \tag{16.59}$$

which for $D = \hbar/(2m)$ is the solution (3.200 & 13.107) of the diffusion equation.

16.7 Harmonic oscillator in real time

Biologists have mice; physicists have harmonic oscillators with hamiltonian

$$H = \frac{p^2}{2m} + \frac{m\omega^2 q^2}{2}. \tag{16.60}$$

For this hamiltonian, our formula (16.29) for the coordinate matrix elements of the time-evolution operator $\exp(-itH)$ is

$$\langle q'' | e^{-itH} | q' \rangle = \int e^{iS[q]} Dq \tag{16.61}$$

with action

$$S[q] = \int_0^t \tfrac{1}{2} m \dot{q}^2(t') - \tfrac{1}{2} m \omega^2 q^2(t') \, dt'. \tag{16.62}$$

The classical solution $q_c(t) = q' \cos \omega t + \dot{q}_0 \sin(\omega t)/\omega$ in which $q' = q_c(0)$ and $\dot{q}_0 = \dot{q}_c(0)$ are the initial position and velocity makes the action $S[q]$ stationary

$$\frac{d}{d\epsilon} S[q + \epsilon h]\bigg|_{\epsilon=0} = 0 \tag{16.63}$$

and satisfies the classical equation of motion $\ddot{q}_c(t) = -\omega^2 q_c(t)$.

We now apply the trick (16.46–16.48) we used for the free particle. We write an arbitrary process $q(t)$ is the sum of the classical process $q_c(t)$ and a loop $\delta q(t)$ with $\delta q(0) = \delta q(t) = 0$. Since the action $S[q]$ is quadratic in the variables q and $\dot{q}$, the functional Taylor series (15.31) for $S[q_c + \delta q]$ has only two terms

$$S[q_c + \delta q] = S[q_c] + S[\delta q]. \tag{16.64}$$

Thus we can write the path integral (16.61) as

$$\begin{aligned} \langle q''|e^{-itH}|q'\rangle &= \int e^{iS[q]}\, Dq = \int e^{iS[q_c+\delta q]}\, D\delta q \\ &= \int e^{iS[q_c]+iS[\delta q]}\, D\delta q = e^{iS[q_c]} \int e^{iS[\delta q]}\, D\delta q. \end{aligned} \tag{16.65}$$

The remaining path integral over the loops δq does not involve the endpoints q' and q'' and so must be a function $L(t)$ of the time t but not of q' or q''

$$\langle q''|e^{-itH}|q'\rangle = e^{iS[q_c]}\, L(t). \tag{16.66}$$

The action $S[q_c]$ is (exercise 16.8)

$$S[q_c] = \frac{m\omega}{2\sin(\omega t)} \left[\left(q'^2 + q''^2 \right) \cos(\omega t) - 2q'q'' \right]. \tag{16.67}$$

The action $S[\delta q]$ of a loop

$$\delta q(t') = \sum_{j=1}^{n-1} a_j \sin \frac{j\pi t'}{t} \tag{16.68}$$

is (exercise 16.9)

$$S[\delta q] = \sum_{j=1}^{n-1} \frac{mt}{4} a_j^2 \left[\frac{(j\pi)^2}{t^2} - \omega^2 \right]. \tag{16.69}$$

The path integral over the loops is then, apart from a constant jacobian J,

$$\begin{aligned} \int e^{iS[\delta q]}\, D\delta q &= J \left(\frac{nm}{2\pi it} \right)^{n/2} \int \exp \left\{ \sum_{j=1}^{n-1} \frac{imt}{4} a_j^2 \left[\frac{(j\pi)^2}{t^2} - \omega^2 \right] \right\} \sum_{j=1}^{n-1} da_j \\ &= J \left(\frac{nm}{2\pi it} \right)^{n/2} \prod_{j=1}^{n-1} \int_{-\infty}^{\infty} \exp \left\{ \frac{imt}{4} a_j^2 \left[\frac{(j\pi)^2}{t^2} - \omega^2 \right] \right\} da_j. \end{aligned} \tag{16.70}$$

Using the gaussian integral (16.2) and the infinite product (4.140), we get

$$\begin{aligned}\int e^{iS[\delta q]}\, D\delta q &= Jn^{n/2}\sqrt{\frac{m}{2\pi i t}}\prod_{j=1}^{n-1}\frac{\sqrt{2}}{j\pi}\left(1-\frac{\omega^2 t^2}{\pi^2 j^2}\right)^{-1/2} \\ &= \sqrt{\frac{m\omega}{2\pi i \sin \omega t}}\left(\lim_{n\to\infty} Jn^{n/2}\prod_{j=1}^{n-1}\frac{\sqrt{2}}{j\pi}\right).\end{aligned} \tag{16.71}$$

Using (16.66) and (16.67), we see that the number within the parentheses is unity because in that case we have (Feynman and Hibbs, 1965, ch. 3)

$$\langle q''|e^{-itH/\hbar}|q'\rangle = \sqrt{\frac{m\omega}{2\pi i\hbar \sin(\omega t)}}\exp\left[i\frac{m\omega\left[(q'^2+q''^2)\cos(\omega t)-2q'q''\right]}{2\hbar\sin(\omega t)}\right], \tag{16.72}$$

which agrees with the amplitude (16.54) in the limit $t \to 0$ (exercise 16.11).

16.8 Harmonic oscillator in imaginary time

For the harmonic oscillator with hamiltonian (16.60), our formula (16.25) for euclidean path integrals becomes

$$\langle q''|e^{-\beta H}|q'\rangle = N\int \exp\left\{-\int_0^\beta \tfrac{1}{2}m\dot{q}^2(t)+\tfrac{1}{2}m\omega^2 q^2(t)\,dt\right\} Dq. \tag{16.73}$$

The euclidean action, which is a time integral of the energy of the oscillator,

$$S_e[q] = \int_0^\beta \left[\tfrac{1}{2}m\dot{q}^2(t)+\tfrac{1}{2}m\omega^2 q^2(t)\right] dt \tag{16.74}$$

is purely quadratic, and so we may play the trick (15.32) if we can find a path $q_e(t)$ that makes it stationary

$$\begin{aligned}\delta S_e[q_e][h] &= \frac{d}{d\epsilon}\int_0^\beta \tfrac{1}{2}m(\dot{q}_e(t)+\epsilon\dot{h}(t))^2+\tfrac{1}{2}m\omega^2(q_e(t)+\epsilon h(t))^2\,dt\bigg|_{\epsilon=0} \\ &= \int_0^\beta m\dot{q}_e(t)\dot{h}(t)+m\omega^2 q_e(t)h(t)\,dt \\ &= \int_0^\beta\left[-m\ddot{q}_e(t)+m\omega^2 q_e(t)\right]h(t)dt = 0.\end{aligned} \tag{16.75}$$

The path $q_e(t)$ must satisfy the euclidean equation of motion

$$\ddot{q}_e(t) = \omega^2 q_e(t) \tag{16.76}$$

whose general solution is

$$q_e(t) = Ae^{\omega t}+Be^{-\omega t}. \tag{16.77}$$

The path from $q_e(0) = q'$ to $q_e(\beta) = q''$ must have

$$A = \frac{q'' e^{-\omega\beta} - q' e^{-2\omega\beta}}{1 - e^{-2\omega\beta}} \quad \text{and} \quad B = q' - A. \tag{16.78}$$

Its action $S_e[q_e]$ is (exercise 16.12)

$$S_e[q_e] = \tfrac{1}{2} m\omega \left[A^2 \left(e^{2\omega\beta} - 1 \right) - B^2 \left(e^{-2\omega\beta} - 1 \right) \right]. \tag{16.79}$$

Since the action is purely quadratic, the trick (15.32) tells us that the action $S_e[q]$ of the arbitrary path $q(t) = q_e(t) + \delta q(t)$ is the sum

$$S_e[q] = S_e[q_e] + S_e[\delta q], \tag{16.80}$$

in which the action $S_e[\delta q]$ of the loop $\delta q(t)$ depends but upon t but not upon q_β or q_0. It follows then that for some loop function $L(\beta)$ of β alone

$$\begin{aligned} \langle q'' | e^{-\beta H} | q' \rangle &= \exp\left(-S_e[q_e]\right) L(\beta) \\ &= L(\beta) \exp\left\{ -\tfrac{1}{2} m\omega \left[A^2 \left(e^{2\omega\beta} - 1 \right) - B^2 \left(e^{-2\omega\beta} - 1 \right) \right] \right\}. \end{aligned} \tag{16.81}$$

To study the ground state of the harmonic oscillator, we let $\beta \to \infty$ in this equation. Inserting a complete set of eigenstates $H|n\rangle = E_n|n\rangle$, we see that the limit of the left-hand side is

$$\lim_{\beta\to\infty} \langle q'' | e^{-\beta H} | q' \rangle = \lim_{\beta\to\infty} \langle q'' | n \rangle e^{-\beta E_n} \langle n | q' \rangle = e^{-\beta E_0} \langle q'' | 0 \rangle \langle 0 | q' \rangle. \tag{16.82}$$

Our formulas (16.78) for A and B say that $A \to q'' e^{-\omega\beta}$ and $B \to q'$ as $\beta \to \infty$, and so in this limit by (16.81 & 16.82) we have

$$e^{-\beta E_0} \langle q'' | 0 \rangle \langle 0 | q' \rangle = L(\beta) \exp\left[-\tfrac{1}{2} m\omega \left(q''^2 + q'^2 \right) \right], \tag{16.83}$$

from which we may infer our earlier formula (15.44) for the wave-function of the ground state

$$\langle q | 0 \rangle = \left(\frac{m\omega}{\pi\hbar} \right)^{1/4} \exp\left[-\frac{1}{2} \frac{m\omega q^2}{\hbar} \right], \tag{16.84}$$

in which the prefactor ensures the normalization

$$1 = \int_{-\infty}^{\infty} |\langle q | 0 \rangle|^2 dq. \tag{16.85}$$

Euclidean path integrals help one study ground states.

16.9 Euclidean correlation functions

In the Heisenberg picture, the position operator $q(t)$ is

$$q(t) = e^{itH} q\, e^{-itH}, \tag{16.86}$$

in which $q = q(0)$ is the position operator at time $t = 0$ or equivalently the position operator in the Schrödinger picture. The analogous operator in imaginary time is the euclidean position operator $q_e(t)$ defined as

$$q_e(t) = e^{tH} q\, e^{-tH} \tag{16.87}$$

obtained from $q(t)$ by replacing t by $-it$.

The **euclidean** product of two euclidean position operators is

$$\mathcal{T}\,[q_e(t_1)\, q_e(t_2)] = \theta(t_1 - t_2) q_e(t_1)\, q_e(t_2) + \theta(t_2 - t_1) q_e(t_2)\, q_e(t_1), \tag{16.88}$$

in which $\theta(x) = (x + |x|)/2|x|$ is Heaviside's function. We can use the method of section 16.3 to compute the matrix element of the euclidean-time-ordered product $\mathcal{T}\,[q(t_1)q(t_2)]$ sandwiched between two factors of $\exp(-tH)$. For $t_1 \geq t_2$, this matrix element is

$$\langle q_t | e^{-tH} q_e(t_1) q_e(t_2) e^{-tH} | q_{-t} \rangle = \langle q_t | e^{-(t-t_1)H} q e^{-(t_1-t_2)H} q e^{-(t+t_2)H} | q_{-t} \rangle. \tag{16.89}$$

Then instead of the path-integral formula (16.25), we get

$$\langle q_t | e^{-tH} \mathcal{T}\,[q_e(t_1) q_e(t_2)]\, e^{-tH} | q_{-t} \rangle = \int q(t_1) q(t_2) e^{-S_e[q,t,-t]}\, Dq \tag{16.90}$$

where as in (16.25) $S_e[q, t, -t]$ is the **euclidean action**

$$S_e[q, t, -t] = \int_{-t}^{t} \tfrac{1}{2} m \dot{q}^2(t') + V(q(t'))\, dt' \tag{16.91}$$

or the time integral of the energy. As in the path integral (16.25), the integration is over all paths that go from $q(-t) = q_{-t}$ to $q(t) = q_t$. The analog of (16.25) is

$$\langle q_t | e^{-2tH} | q_{-t} \rangle = \int e^{-S_e[q,t,-t]}\, Dq \tag{16.92}$$

and the factors $(nm/2\pi\beta)^{n/2}$ cancel in the ratio of (16.90) to (16.92)

$$\frac{\langle q_t | e^{-tH} \mathcal{T}\,[q_e(t_1) q_e(t_2)]\, e^{-tH} | q_{-t} \rangle}{\langle q_t | e^{-2tH} | q_{-t} \rangle} = \frac{\displaystyle\int q(t_1) q(t_2) e^{-S_e[q,t,-t]}\, Dq}{\displaystyle\int e^{-S_e[q,t,-t]}\, Dq}. \tag{16.93}$$

In the limit $t \to \infty$, the operator $\exp(-tH)$ projects out the ground state $|0\rangle$ of the system

$$\lim_{t\to\infty} e^{-tH}|q_{-t}\rangle = \lim_{t\to\infty} \sum_{n=0}^{\infty} e^{-tH}|n\rangle\langle n|q_{-t}\rangle = \lim_{t\to\infty} e^{-tE_0}|0\rangle\langle 0|q_{-t}\rangle, \tag{16.94}$$

which we assume to be unique and normalized to unity. In the ratio (16.93), most of these factors cancel, leaving us with

$$\langle 0|\mathcal{T}\left[q_e(t_1)q_e(t_2)\right]|0\rangle = \frac{\int q(t_1)q(t_2)e^{-S_e[q,\infty,-\infty]}\,Dq}{\int e^{-S_e[q,\infty,-\infty]}\,Dq}. \tag{16.95}$$

More generally, the mean value in the ground state $|0\rangle$ of *any* euclidean-time-ordered product of position operators $q(t_i)$ is a ratio of path integrals

$$\langle 0|\,\mathcal{T}\left[q(t_1)\cdots q(t_n)\right]|0\rangle = \frac{\int q(t_1)\cdots q(t_n)\,e^{-S_e[q]}\,Dq}{\int e^{-S_e[q]}\,Dq}, \tag{16.96}$$

in which $S_e[q]$ stands for $S_e[q,\infty,-\infty]$. Why do we need the time-ordered product $\mathcal{T}$ on the LHS? Because successive factors of $\exp(-(t_k - t_\ell)H)$ lead to the path integral of $\exp(-S_e[q])$. Why don't we need $\mathcal{T}$ on the RHS? Because the $q(t_i)$s are real numbers which commute with each other.

The result (16.96) is important because it can be generalized to all quantum theories, including field theories.

16.10 Finite-temperature field theory

Matrix elements of the operator $\exp(-\beta H)$ where $\beta = 1/kT$ tell us what a system is like at temperature T. In the low-temperature limit, they describe the ground state of the system.

Quantum mechanics imposes upon n coordinates q_i and conjugate momenta π_k the commutation relations

$$[q_i, p_k] = i\,\delta_{i,k} \quad \text{and} \quad [q_i, q_k] = [p_i, p_k] = 0. \tag{16.97}$$

In quantum field theory, we associate a coordinate $q_{\boldsymbol{x}} \equiv \phi(\boldsymbol{x})$ and a conjugate momentum $p_{\boldsymbol{x}} \equiv \pi(\boldsymbol{x})$ with each point $\boldsymbol{x}$ of space and impose upon them the very similar commutation relations

$$[\phi(\boldsymbol{x}), \pi(\boldsymbol{y})] = i\,\delta(\boldsymbol{x}-\boldsymbol{y}) \quad \text{and} \quad [\phi(\boldsymbol{x}), \phi(\boldsymbol{y})] = [\pi(\boldsymbol{x}), \pi(\boldsymbol{y})] = 0. \tag{16.98}$$

Just as in quantum mechanics the time derivative of a coordinate is its commutator with a hamiltonian $\dot{q}_i = i[H, q_i]$, so too in quantum field theory the time derivative of a field $\dot{\phi}(\boldsymbol{x}, t) \equiv \dot{\phi}(x)$ is $\dot{\phi}(x) = i[H, \phi(x)]$. A typical hamiltonian for a single scalar field ϕ is

$$H = \int \left[\frac{1}{2}\pi^2(\boldsymbol{x}) + \frac{1}{2}\left(\nabla\phi(\boldsymbol{x})\right)^2 + \frac{1}{2}m^2\phi^2(\boldsymbol{x}) + P(\phi(\boldsymbol{x})) \right] d^3x, \tag{16.99}$$

in which P is a quartic polynomial.

Since quantum field theory is just the quantum mechanics of many variables, we can use the methods of sections 16.3 & 16.4 to write matrix elements of $\exp(-\beta H)$ as path integrals. We define a potential

$$V(\phi(\boldsymbol{x})) = \frac{1}{2}\left(\nabla\phi(\boldsymbol{x})\right)^2 + \frac{1}{2}m^2\phi^2(\boldsymbol{x}) + P(\phi(\boldsymbol{x})) \tag{16.100}$$

and write the hamiltonian H as

$$H = \int \left[\frac{1}{2}\pi^2(\boldsymbol{x}) + V(\phi(\boldsymbol{x})) \right] d^3x. \tag{16.101}$$

Like $|q'\rangle$ and $|p'\rangle$, the states $|\phi'\rangle$ and $|\pi'\rangle$ are eigenstates of the hermitian operators $\phi(\boldsymbol{x}, 0)$ and $\pi(\boldsymbol{x}, 0)$

$$\phi(\boldsymbol{x}, 0)|\phi'\rangle = \phi'(\boldsymbol{x})|\phi'\rangle \quad \text{and} \quad \pi(\boldsymbol{x}, 0)|\pi'\rangle = \pi'(\boldsymbol{x})|\pi'\rangle. \tag{16.102}$$

The analog of $\langle q'|p'\rangle$ is

$$\langle\phi'|\pi'\rangle = f \exp\left[i \int \phi'(\boldsymbol{x})\pi'(\boldsymbol{x}) d^3x \right], \tag{16.103}$$

in which f is a factor which eventually will cancel.

Repeating our derivation of equation (16.21) with

$$D\pi' \equiv \prod_{\boldsymbol{x}} d\pi'(\boldsymbol{x}) \tag{16.104}$$

we find in the limit $\epsilon \to 0$

$$\begin{aligned}
\langle\phi''|\exp(-\epsilon H)|\phi'\rangle &= \int \langle\phi''|\exp\left(-\frac{\epsilon}{2}\int \pi^2(\boldsymbol{x}) d^3x\right)|\pi'\rangle \\
&\quad \times \langle\pi'|\exp\left(-\epsilon \int V(\phi(\boldsymbol{x}))\, d^3x\right)|\phi'\rangle\, D\pi' \\
&= |f|^2 \exp\left(-\epsilon \int V(\phi'(\boldsymbol{x}))\, d^3x\right) \\
&\quad \times \int \exp\left[\int -\tfrac{1}{2}\epsilon\pi'^2(\boldsymbol{x}) + i\pi'(\boldsymbol{x})[\phi''(\boldsymbol{x}) - \phi'(\boldsymbol{x})] d^3x\right] D\pi'.
\end{aligned} \tag{16.105}$$

Using the abbreviation

$$\dot{\phi}'(\boldsymbol{x}) \equiv \frac{\phi''(\boldsymbol{x}) - \phi'(\boldsymbol{x})}{\epsilon} \tag{16.106}$$

and the integral formula (16.6), we get

$$\langle\phi''|\exp(-\epsilon H)|\phi'\rangle = f' \exp\left\{-\epsilon\int\left[\tfrac{1}{2}\dot\phi'^2(\boldsymbol{x}) + V(\phi'(\boldsymbol{x}))\right]d^3x\right\}.$$

Putting together $n = \beta/\epsilon$ such terms and integrating over the intermediate states $|\phi'''\rangle\langle\phi'''|$, and absorbing the normalizing factors into $D\phi$, we have

$$\langle\phi_\beta|e^{-\beta H}|\phi_0\rangle = \int_{\phi_0}^{\phi_\beta}\exp\left[-\int_0^\beta\int\tfrac{1}{2}\dot\phi^2(x) + V(\phi(x))\,d^3x dt\right]D\phi. \tag{16.107}$$

Replacing the potential $V(\phi)$ with its definition (16.100), we find

$$\langle\phi_\beta|e^{-\beta H}|\phi_0\rangle = \int_{\phi_0}^{\phi_\beta}\exp\left[-\int_0^\beta\!\!\int\tfrac{1}{2}\left[\dot\phi^2 + (\nabla\phi)^2 + m^2\phi^2\right] + P(\phi)\,d^3x dt\right]D\phi, \tag{16.108}$$

in which the limits ϕ_0 and ϕ_β remind us that we are to integrate over all fields $\phi(\boldsymbol{x}, t)$ that run from $\phi(\boldsymbol{x}, 0) = \phi_0(\boldsymbol{x})$ to $\phi(\boldsymbol{x}, \beta) = \phi_\beta(\boldsymbol{x})$.

In terms of the energy density

$$\mathcal{H}(\phi) \equiv \tfrac{1}{2}\left[(\partial_a\phi)^2 + m^2\phi^2\right] + P(\phi), \tag{16.109}$$

in which a is summed from 0 to 3, the path integral (16.108) is

$$\langle\phi_\beta|e^{-\beta H}|\phi_0\rangle = \int_{\phi_0}^{\phi_\beta}\exp\left[-\int_0^\beta\!\!\int\mathcal{H}(\phi)\,d^3x dt\right]D\phi. \tag{16.110}$$

The partition function $Z(\beta)$ – defined as the trace $Z(\beta) \equiv \mathrm{Tr}\, e^{-\beta H}$ over all states of the system – is an integral over all loop fields (ones for which ϕ_β and ϕ_0 coincide)

$$Z(\beta) \equiv \mathrm{Tr}\, e^{-\beta H} = \int_{\phi_0}^{\phi_0}\langle\phi|e^{-\beta H}|\phi\rangle D\phi = N\int_{\phi_0}^{\phi_0}\exp\left[-\int_0^\beta\!\!\int\mathcal{H}(\phi)\,d^3x dt\right]D\phi. \tag{16.111}$$

Because the four space-time derivatives in $\mathcal{H}(\phi)$ occur with the same sign, finite-temperature field theory is called **euclidean** quantum field theory. The density operator ρ for the system described by the hamiltonian H in equilibrium at temperature T is $\rho = \exp(-\beta H)/Z(\beta)$.

Like the definition (16.87) of the euclidean position operator $q_e(t)$, the euclidean field operator $\phi_e(x)$ is defined as

$$\phi_e(\boldsymbol{x}, t) = e^{tH}\phi(\boldsymbol{x}, 0)e^{-tH}. \tag{16.112}$$

The **euclidean-time-ordered product** (16.88) of two fields is

$$\begin{aligned}\mathcal{T}\left[\phi_e(\boldsymbol{x}_1, t_1)\phi_e(\boldsymbol{x}_2, t_2)\right] &= \theta(t_1 - t_2)e^{t_1H}\phi_e(\boldsymbol{x}_1, 0)e^{-(t_1-t_2)H}\phi_e(\boldsymbol{x}_2, 0)e^{-t_2H}\\ &+ \theta(t_2 - t_1)e^{t_2H}\phi_e(\boldsymbol{x}_2, 0)e^{-(t_2-t_1)H}\phi_e(\boldsymbol{x}_1, 0)e^{-t_1H}.\end{aligned}$$

The logic of equations (16.87–16.96) leads us to write its mean value in a system described by a stationary density operator – one that commutes with the hamiltonian – as the ratio

$$\langle \mathcal{T}\,[\phi_e(x_1)\phi_e(x_2)]\rangle = \mathrm{Tr}\,\{\rho\,\mathcal{T}\,[\phi_e(x_1)\phi_e(x_2)]\}$$
$$= \frac{\mathrm{Tr}\left\{e^{-\beta H}\,\mathcal{T}\,[\phi_e(x_1)\phi_e(x_2)]\right\}}{\mathrm{Tr}\left[e^{-\beta H}\right]}$$
$$= \frac{\int_{\phi_0}^{\phi_0} \phi(x_1)\phi(x_2)\exp\left[-\int_0^\beta\!\int \mathcal{H}(\phi)\,d^3x\,dt\right] D\phi}{\int_{\phi_0}^{\phi_0} \exp\left[-\int_0^\beta\!\int \mathcal{H}(\phi)\,d^3x\,dt\right] D\phi}, \tag{16.113}$$

in which all normalization factors have canceled.

In the zero-temperature ($\beta \to \infty$) limit, the density operator ρ becomes the projection operator $|0\rangle\langle 0|$ on the ground state, and mean-value formulas like (16.113) become

$$\langle 0|\mathcal{T}\,[\phi_e(x_1)\cdots\phi_e(x_n)]\,|0\rangle = \frac{\int \phi(x_1)\cdots\phi(x_n)\,\exp\left[-\int \mathcal{H}(\phi)\,d^4x\right] D\phi}{\int \exp\left[-\int \mathcal{H}(\phi)\,d^4x\right] D\phi}, \tag{16.114}$$

in which Hamilton's density $\mathcal{H}(\phi)$ is integrated over all of euclidean space-time and over all fields that are periodic on the infinite time interval. Statistical field theory and lattice gauge theory use formulas like (16.113) and (16.114).

16.11 Real-time field theory

We now follow the derivation of section 16.10 using the same notation but for real time. In (16.105), we replace $-\epsilon H$ by $-i\epsilon H$ and follow the logic of sections (16.4 & 16.10). We find in the limit $\epsilon \to 0$ with $\dot{\phi}' \equiv (\phi'' - \phi')/\epsilon$

$$\langle \phi''|e^{-i\epsilon H}|\phi'\rangle = \int \langle \phi''|e^{-i\epsilon\int \pi^2/2\,d^3x}|\pi'\rangle\langle \pi'|e^{-i\epsilon\int V(\phi)\,d^3x}|\phi'\rangle D\pi'$$
$$= |f|^2 e^{-i\epsilon\int V(\phi')d^3x}\int e^{-i\epsilon\int \pi'^2/2 + i\pi'(\phi''-\phi')\,d^3x} D\pi'$$
$$= f'\,\exp\left[i\epsilon\int \tfrac{1}{2}\dot{\phi}'^2 - V(\phi')\,d^3x\right]. \tag{16.115}$$

Putting together $2t/\epsilon$ similar factors and integrating over all the intermediate states $|\phi\rangle\langle\phi|$, we arrive at the path integral

$$\langle \phi''|e^{-i2tH}|\phi'\rangle = \int_{\phi'}^{\phi''} \exp\left[i\int \tfrac{1}{2}\dot{\phi}^2(x) - V(\phi(x))\,d^4x\right] D\phi, \tag{16.116}$$

in which we integrate over all fields $\phi(x)$ that run from $\phi'(\boldsymbol{x}, -t)$ to $\phi''(\boldsymbol{x}, t)$. After expanding the definition (16.100) of the potential $V(\phi)$, we have

$$\langle \phi'' | e^{-i2tH} | \phi' \rangle = \int_{\phi'}^{\phi''} \exp\left[i \int \tfrac{1}{2} \left[\dot{\phi}^2 - (\nabla \phi)^2 - m^2 \phi^2 \right] - P(\phi)\, d^4x \right] D\phi. \tag{16.117}$$

This amplitude is a path integral

$$\langle \phi'' | e^{-i2tH} | \phi' \rangle = \int_{\phi'}^{\phi''} e^{iS[\phi]}\, D\phi \tag{16.118}$$

of phases $\exp(iS[\phi])$ that are exponentials of the classical action

$$S[\phi] = \int \tfrac{1}{2} \left[\dot{\phi}^2 - (\nabla \phi)^2 - m^2 \phi^2 \right] - P(\phi)\, d^4x. \tag{16.119}$$

The time dependence of the Heisenberg field operator $\phi(x, t)$ is

$$\phi(\boldsymbol{x}, t) = e^{itH} \phi(\boldsymbol{x}, 0) e^{-itH}. \tag{16.120}$$

The **time-ordered product** of two fields, as in (16.88), is the sum

$$\mathcal{T}[\phi(x_1)\phi(x_2)] = \theta(x_1^0 - x_2^0)\phi(x_1)\phi(x_2) + \theta(x_2^0 - x_1^0)\phi(x_2)\phi(x_1). \tag{16.121}$$

Between two factors of $\exp(-itH)$, it is for $t_1 > t_2$

$$e^{-itH} \mathcal{T}[\phi(x_1)\phi(x_2)]\, e^{-itH} = e^{-i(t-t_1)H} \phi(\boldsymbol{x}_1, 0) e^{-i(t_1-t_2)H} \phi(\boldsymbol{x}_2, 0) e^{-i(t-t_2)H}.$$

So by the logic that led to the path-integral formulas (16.113) and (16.118), we can write a matrix element of the time-ordered product (16.121) as

$$\langle \phi'' | e^{-itH} \mathcal{T}[\phi(x_1)\phi(x_2)]\, e^{-itH} | \phi' \rangle = \int_{\phi'}^{\phi''} \phi(x_1)\, \phi(x_2)\, e^{iS[\phi]}\, D\phi, \tag{16.122}$$

in which we integrate over fields that go from ϕ' at time $-t$ to ϕ'' at time t. The time-ordered product of any combination of fields is then

$$\langle \phi'' | e^{-itH} \mathcal{T}[\phi(x_1) \cdots \phi(x_n)]\, e^{-itH} | \phi' \rangle = \int \phi(x_1) \cdots \phi(x_n)\, e^{iS[\phi]}\, D\phi. \tag{16.123}$$

Like the position eigenstates $|q'\rangle$ of quantum mechanics, the eigenstates $|\phi'\rangle$ are states of infinite energy that overlap most states. Yet we often are interested in the ground state $|0\rangle$ or in states of a few particles. To form such matrix elements, we multiply both sides of equations (16.118 & 16.123) by $\langle 0|\phi''\rangle\langle\phi'|0\rangle$

and integrate over ϕ' and ϕ''. Since the ground state is a normalized eigenstate of the hamiltonian $H|0\rangle = E_0|0\rangle$ with eigenvalue E_0, we find from (16.118)

$$\int \langle 0|\phi''\rangle\langle\phi''|e^{-i2tH}|\phi'\rangle\langle\phi'|0\rangle D\phi'' D\phi' = \langle 0|e^{-i2tH}|0\rangle$$
$$= e^{-i2tE_0} = \int \langle 0|\phi''\rangle e^{iS[\phi]}\langle\phi'|0\rangle \, D\phi D\phi'' D\phi' \tag{16.124}$$

and from (16.123) suppressing the differentials $D\phi'' D\phi'$,

$$e^{-2itE_0}\langle 0|\mathcal{T}[\phi(x_1)\cdots\phi(x_n)]\,|0\rangle = \int \langle 0|\phi''\rangle\phi(x_1)\cdots\phi(x_n)\, e^{iS[\phi]}\langle\phi'|0\rangle\, D\phi. \tag{16.125}$$

The mean value in the ground state of a time-ordered product of field operators is then a ratio of these path integrals

$$\langle 0|\mathcal{T}[\phi(x_1)\cdots\phi(x_n)]\,|0\rangle = \frac{\displaystyle\int \langle 0|\phi''\rangle\, \phi(x_1)\cdots\phi(x_n)\, e^{iS[\phi]}\langle\phi'|0\rangle\, D\phi}{\displaystyle\int \langle 0|\phi''\rangle\, e^{iS[\phi]}\langle\phi'|0\rangle\, D\phi}, \tag{16.126}$$

in which factors involving E_0 have canceled and the integration is over all fields that go from $\phi(\boldsymbol{x}, -t) = \phi'(\boldsymbol{x})$ to $\phi(\boldsymbol{x}, t) = \phi''(\boldsymbol{x})$ and over $\phi'(\boldsymbol{x})$ and $\phi''(\boldsymbol{x})$.

16.12 Perturbation theory

Field theories with hamiltonians that are quadratic in their fields like

$$H_0 = \int \tfrac{1}{2}\left[\pi^2(x) + (\nabla\phi(x))^2 + m^2\phi^2(x)\right] d^3x \tag{16.127}$$

are soluble. Their fields evolve in time as

$$\phi(\boldsymbol{x}, t) = e^{itH_0}\phi(\boldsymbol{x}, 0)e^{-itH_0}. \tag{16.128}$$

The mean value in the ground state of H_0 of a time-ordered product of these fields is by (16.126) a ratio of path integrals

$$\langle 0|\mathcal{T}[\phi(x_1)\cdots\phi(x_n)]\,|0\rangle = \frac{\displaystyle\int \langle 0|\phi''\rangle\, \phi(x_1)\cdots\phi(x_n)\, e^{iS_0[\phi]}\langle\phi'|0\rangle\, D\phi}{\displaystyle\int \langle 0|\phi''\rangle\, e^{iS_0[\phi]}\langle\phi'|0\rangle\, D\phi}, \tag{16.129}$$

in which the action $S_0[\phi]$ is quadratic in the fields

$$S_0[\phi] = \int \tfrac{1}{2}\left[\dot{\phi}^2(x) - (\nabla\phi(x))^2 - m^2\phi^2(x)\right] d^4x$$
$$= \int \tfrac{1}{2}\left[- \partial_a\phi(x)\partial^a\phi(x) - m^2\phi^2(x)\right] d^4x. \tag{16.130}$$

So the path integrals in the ratio (16.129) are gaussian and doable.

The Fourier transforms

$$\tilde{\phi}(p) = \int e^{-ipx}\phi(x)\, d^4x \quad \text{and} \quad \phi(x) = \int e^{ipx}\tilde{\phi}(p)\, \frac{d^4p}{(2\pi)^4} \tag{16.131}$$

turn the space-time derivatives in the action into a quadratic form

$$S_0[\phi] = -\tfrac{1}{2}\int |\tilde{\phi}(p)|^2\, (p^2 + m^2)\, \frac{d^4p}{(2\pi)^4}, \tag{16.132}$$

in which $p^2 = \boldsymbol{p}^2 - p^{02}$, and $\tilde{\phi}(-p) = \tilde{\phi}^*(p)$ by (3.25) since the field ϕ is real.

The initial $\langle\phi'|0\rangle$ and final $\langle 0|\phi''\rangle$ wave-functions produce the $i\epsilon$ in the Feynman propagator (5.233). Although its exact form doesn't matter here, the wave-function $\langle\phi'|0\rangle$ of the ground state of H_0 is the exponential (15.51)

$$\langle\phi'|0\rangle = c\exp\left[-\frac{1}{2}\int |\tilde{\phi}'(\boldsymbol{p})|^2 \sqrt{\boldsymbol{p}^2 + m^2}\, \frac{d^3p}{(2\pi)^3}\right], \tag{16.133}$$

in which $\tilde{\phi}'(\boldsymbol{p})$ is the spatial Fourier transform

$$\tilde{\phi}'(\boldsymbol{p}) = \int e^{-i\boldsymbol{p}\cdot\boldsymbol{x}}\, \phi'(\boldsymbol{x})\, d^3x \tag{16.134}$$

and c is a normalization factor that will cancel in ratios of path integrals.

Apart from $-2i\ln c$, which we will not keep track of, the wave-functions $\langle\phi'|0\rangle$ and $\langle 0|\phi''\rangle$ add to the action $S_0[\phi]$ the term

$$\Delta S_0[\phi] = \frac{i}{2}\int \sqrt{\boldsymbol{p}^2 + m^2}\left(|\tilde{\phi}(\boldsymbol{p},t)|^2 + |\tilde{\phi}(\boldsymbol{p},-t)|^2\right) \frac{d^3p}{(2\pi)^3}, \tag{16.135}$$

in which we envision taking the limit $t \to \infty$ with $\phi(\boldsymbol{x},t) = \phi''(\boldsymbol{x})$ and $\phi(\boldsymbol{x},-t) = \phi'(\boldsymbol{x})$. The identity (Weinberg, 1995, pp. 386–388)

$$f(+\infty) + f(-\infty) = \lim_{\epsilon\to 0+} \epsilon \int_{-\infty}^{\infty} f(t)\, e^{-\epsilon|t|}\, dt \tag{16.136}$$

allows us to write $\Delta S_0[\phi]$ as

$$\Delta S_0[\phi] = \lim_{\epsilon\to 0+} \frac{i\epsilon}{2}\int \sqrt{\boldsymbol{p}^2 + m^2}\int_{-\infty}^{\infty} |\tilde{\phi}(\boldsymbol{p},t)|^2\, e^{-\epsilon|t|}\, dt\, \frac{d^3p}{(2\pi)^3}. \tag{16.137}$$

To first order in ϵ, the change in the action is (exercise 16.15)

$$\Delta S_0[\phi] = \lim_{\epsilon \to 0+} \frac{i\epsilon}{2} \int \sqrt{\boldsymbol{p}^2 + m^2} \int_{-\infty}^{\infty} |\tilde{\phi}(\boldsymbol{p}, t)|^2 \, dt \, \frac{d^3p}{(2\pi)^3}$$
$$= \lim_{\epsilon \to 0+} \frac{i\epsilon}{2} \int \sqrt{\boldsymbol{p}^2 + m^2} \, |\tilde{\phi}(p)|^2 \, \frac{d^4p}{(2\pi)^4}. \tag{16.138}$$

The modified action is therefore

$$S_0[\phi, \epsilon] = S_0[\phi] + \Delta S_0[\phi] = -\frac{1}{2} \int |\tilde{\phi}(p)|^2 \left(p^2 + m^2 - i\epsilon \sqrt{\boldsymbol{p}^2 + m^2} \right) \frac{d^4p}{(2\pi)^4}$$
$$= -\frac{1}{2} \int |\tilde{\phi}(p)|^2 \left(p^2 + m^2 - i\epsilon \right) \frac{d^4p}{(2\pi)^4} \tag{16.139}$$

since the square-root is positive. In terms of the modified action, our formula (16.129) for the time-ordered product is the ratio

$$\langle 0 | \mathcal{T} [\phi(x_1) \cdots \phi(x_n)] | 0 \rangle = \frac{\int \phi(x_1) \cdots \phi(x_n) \, e^{iS_0[\phi, \epsilon]} \, D\phi}{\int e^{iS_0[\phi, \epsilon]} \, D\phi}. \tag{16.140}$$

We can use this formula (16.140) to express the mean value in the vacuum $|0\rangle$ of the time-ordered exponential of a space-time integral of $j(x)\phi(x)$, in which $j(x)$ is a classical (c-number, external) current $j(x)$, as the ratio

$$Z_0[j] \equiv \langle 0 | \mathcal{T} \left\{ \exp \left[i \int j(x) \, \phi(x) \, d^4x \right] \right\} | 0 \rangle$$
$$= \frac{\int \exp \left[i \int j(x) \, \phi(x) \, d^4x \right] e^{iS_0[\phi, \epsilon]} \, D\phi}{\int e^{iS_0[\phi, \epsilon]} \, D\phi}. \tag{16.141}$$

Since the state $|0\rangle$ is normalized, the mean value $Z_0[0]$ is unity,

$$Z_0[0] = 1. \tag{16.142}$$

If we absorb the current into the action

$$S_0[\phi, \epsilon, j] = S_0[\phi, \epsilon] + \int j(x) \, \phi(x) \, d^4x \tag{16.143}$$

then in terms of the current's Fourier transform

$$\tilde{j}(p) = \int e^{-ipx} j(x) \, d^4x \tag{16.144}$$

the modified action $S_0[\phi, \epsilon, j]$ is (exercise 16.15)

$$S_0[\phi,\epsilon,j] = -\tfrac{1}{2}\int \left[|\tilde{\phi}(p)|^2\left(p^2+m^2-i\epsilon\right) - \tilde{j}^*(p)\tilde{\phi}(p) - \tilde{\phi}^*(p)\tilde{j}(p)\right]\frac{d^4p}{(2\pi)^4}. \tag{16.145}$$

Changing variables to

$$\tilde{\psi}(p) = \tilde{\phi}(p) - \tilde{j}(p)/(p^2+m^2-i\epsilon) \tag{16.146}$$

we write the action $S_0[\phi, \epsilon, j]$ as (exercise 16.17)

$$\begin{aligned} S_0[\phi,\epsilon,j] &= -\tfrac{1}{2}\int \left[|\tilde{\psi}(p)|^2\left(p^2+m^2-i\epsilon\right) - \frac{\tilde{j}^*(p)\tilde{j}(p)}{(p^2+m^2-i\epsilon)}\right]\frac{d^4p}{(2\pi)^4} \\ &= S_0[\psi,\epsilon] + \tfrac{1}{2}\int\left[\frac{\tilde{j}^*(p)\tilde{j}(p)}{(p^2+m^2-i\epsilon)}\right]\frac{d^4p}{(2\pi)^4}. \end{aligned} \tag{16.147}$$

And since $D\phi = D\psi$, our formula (16.141) gives simply (exercise 16.18)

$$Z_0[j] = \exp\left(\frac{i}{2}\int\frac{|\tilde{j}(p)|^2}{p^2+m^2-i\epsilon}\frac{d^4p}{(2\pi)^4}\right). \tag{16.148}$$

Going back to position space, one finds (exercise 16.19)

$$Z_0[j] = \exp\left[\frac{i}{2}\int j(x)\,\Delta(x-x')\,j(x')\,d^4x\,d^4x'\right], \tag{16.149}$$

in which $\Delta(x-x')$ is Feynman's **propagator** (5.233)

$$\Delta(x-x') = \int\frac{e^{ip(x-x')}}{p^2+m^2-i\epsilon}\frac{d^4p}{(2\pi)^4}. \tag{16.150}$$

The functional derivative (chapter 15) of $Z_0[j]$, defined by (16.141), is

$$\frac{1}{i}\frac{\delta Z_0[j]}{\delta j(x)} = \langle 0|\,\mathcal{T}\left[\phi(x)\exp\left(i\int j(x')\phi(x')d^4x'\right)\right]|0\rangle \tag{16.151}$$

while that of equation (16.149) is

$$\frac{1}{i}\frac{\delta Z_0[j]}{\delta j(x)} = Z_0[j]\int\Delta(x-x')\,j(x')\,d^4x'. \tag{16.152}$$

Thus the second functional derivative of $Z_0[j]$ evaluated at $j = 0$ gives

$$\langle 0|\,\mathcal{T}\left[\phi(x)\phi(x')\right]|0\rangle = \frac{1}{i^2}\frac{\delta^2 Z_0[j]}{\delta j(x)\delta j(x')}\bigg|_{j=0} = -i\,\Delta(x-x'). \tag{16.153}$$

Similarly, one may show (exercise 16.20) that

$$\langle 0|\, \mathcal{T}\left[\phi(x_1)\phi(x_2)\phi(x_3)\phi(x_4)\right]|0\rangle = \frac{1}{i^4}\frac{\delta^4 Z_0[j]}{\delta j(x_1)\delta j(x_2)\delta j(x_3)\delta j(x_4)}\bigg|_{j=0}$$
$$= -\,\Delta(x_1-x_2)\Delta(x_3-x_4) - \Delta(x_1-x_3)\Delta(x_2-x_4)$$
$$- \Delta(x_1-x_4)\Delta(x_2-x_3). \qquad (16.154)$$

Suppose now that we add a potential $V = P(\phi)$ to the free hamiltonian (16.127). Scattering amplitudes are matrix elements of the time-ordered exponential $\mathcal{T}\exp\left[-i\int P(\phi)\, d^4x\right]$. Our formula (16.140) for the mean value in the ground state $|0\rangle$ of the free hamiltonian H_0 of any time-ordered product of fields leads us to

$$\langle 0|\mathcal{T}\left\{\exp\left[-i\int P(\phi)\, d^4x\right]\right\}|0\rangle = \frac{\displaystyle\int \exp\left[-i\int P(\phi)\, d^4x\right] e^{iS_0[\phi,\epsilon]}\, D\phi}{\displaystyle\int e^{iS_0[\phi,\epsilon]}\, D\phi}. \qquad (16.155)$$

Using (16.153 & 16.154), we can cast this expression into the magical form

$$\langle 0|\mathcal{T}\left\{\exp\left[-i\int P(\phi)\, d^4x\right]\right\}|0\rangle = \exp\left[-i\int P\left(\frac{\delta}{i\delta j(x)}\right) d^4x\right] Z_0[j]\bigg|_{j=0}. \qquad (16.156)$$

The generalization of the path-integral formula (16.140) to the ground state $|\Omega\rangle$ of an interacting theory with action S is

$$\langle \Omega|\mathcal{T}\left[\phi(x_1)\cdots\phi(x_n)\right]|\Omega\rangle = \frac{\displaystyle\int \phi(x_1)\cdots\phi(x_n)\, e^{iS[\phi,\epsilon]}\, D\phi}{\displaystyle\int e^{iS[\phi,\epsilon]}\, D\phi}, \qquad (16.157)$$

in which a term like $i\epsilon\phi^2$ is added to make the modified action $S[\phi,\epsilon]$.

These are some of the techniques one uses to make states of incoming and outgoing particles and to compute scattering amplitudes (Weinberg, 1995, 1996; Srednicki, 2007; Zee, 2010).

16.13 Application to quantum electrodynamics

In the Coulomb gauge $\nabla \cdot \boldsymbol{A} = 0$, the QED hamiltonian is

$$H = H_{\rm m} + \int \left[\tfrac{1}{2}\boldsymbol{\pi}^2 + \tfrac{1}{2}(\nabla \times \boldsymbol{A})^2 - \boldsymbol{A}\cdot\boldsymbol{j}\right] d^3x + V_{\rm C}, \qquad (16.158)$$

in which $H_{\rm m}$ is the matter hamiltonian, and $V_{\rm C}$ is the Coulomb term

$$V_{\rm C} = \frac{1}{2} \int \frac{j^0(\boldsymbol{x},t) j^0(\boldsymbol{y},t)}{4\pi |\boldsymbol{x}-\boldsymbol{y}|} \, d^3x d^3y. \tag{16.159}$$

The operators $\boldsymbol{A}$ and $\boldsymbol{\pi}$ are canonically conjugate, but they satisfy the Coulomb-gauge conditions

$$\nabla \cdot \boldsymbol{A} = 0 \quad \text{and} \quad \nabla \cdot \boldsymbol{\pi} = 0. \tag{16.160}$$

One may show (Weinberg, 1995, pp. 413–418) that in this theory, the analog of equation (16.157) is

$$\langle \Omega | \mathcal{T}[\mathcal{O}_1 \cdots \mathcal{O}_n] | \Omega \rangle = \frac{\displaystyle\int \mathcal{O}_1 \cdots \mathcal{O}_n \, e^{iS_{\rm C}} \, \delta[\nabla \cdot \boldsymbol{A}] \, D\boldsymbol{A} \, D\psi}{\displaystyle\int e^{iS_{\rm C}} \, \delta[\nabla \cdot \boldsymbol{A}] \, D\boldsymbol{A} \, D\psi}, \tag{16.161}$$

in which the Coulomb-gauge action is

$$S_{\rm C} = \int \tfrac{1}{2}\dot{\boldsymbol{A}}^2 - \tfrac{1}{2}(\nabla \times \boldsymbol{A})^2 + \boldsymbol{A} \cdot \boldsymbol{j} + \mathcal{L}_{\rm m} \, d^4x - \int V_{\rm C} \, dt \tag{16.162}$$

and the functional delta function

$$\delta[\nabla \cdot \boldsymbol{A}] = \prod_x \delta(\nabla \cdot \boldsymbol{A}(x)) \tag{16.163}$$

enforces the Coulomb-gauge condition. The term $\mathcal{L}_{\rm m}$ is the action density of the matter field ψ.

Tricks are available. We introduce a new field $A^0(x)$ and consider the factor

$$F = \int \exp\left[i \int \tfrac{1}{2} \left(\nabla A^0 + \nabla \triangle^{-1} j^0 \right)^2 d^4x \right] DA^0, \tag{16.164}$$

which is just a *number* independent of the charge density j^0 since we can cancel the j^0 term by shifting A^0. By $\triangle^{-1}$, we mean $-1/4\pi|\boldsymbol{x}-\boldsymbol{y}|$. By integrating by parts, we can write the number F as (exercise 16.21)

$$\begin{aligned} F &= \int \exp\left[i \int \tfrac{1}{2} \left(\nabla A^0 \right)^2 - A^0 j^0 - \frac{1}{2} j^0 \triangle^{-1} j^0 \, d^4x \right] DA^0 \\ &= \int \exp\left[i \int \tfrac{1}{2} \left(\nabla A^0 \right)^2 - A^0 j^0 \, d^4x + i \int V_{\rm C} \, dt \right] DA^0. \end{aligned} \tag{16.165}$$

So when we multiply the numerator and denominator of the amplitude (16.161) by F, the awkward Coulomb term cancels, and we get

$$\langle \Omega | \mathcal{T}[\mathcal{O}_1 \cdots \mathcal{O}_n] | \Omega \rangle = \frac{\displaystyle\int \mathcal{O}_1 \cdots \mathcal{O}_n \, e^{iS'} \, \delta[\nabla \cdot \boldsymbol{A}] \, D\boldsymbol{A} \, D\psi}{\displaystyle\int e^{iS'} \, \delta[\nabla \cdot \boldsymbol{A}] \, D\boldsymbol{A} \, D\psi} \tag{16.166}$$

where now DA includes all four components A^μ and

$$S' = \int \tfrac{1}{2}\dot{\boldsymbol{A}}^2 - \tfrac{1}{2}(\nabla \times \boldsymbol{A})^2 + \tfrac{1}{2}\left(\nabla A^0\right)^2 + \boldsymbol{A}\cdot\boldsymbol{j} - A^0 j^0 + \mathcal{L}_{\rm m}\, d^4x. \quad (16.167)$$

Since the delta function $\delta[\nabla \cdot \boldsymbol{A}]$ enforces the Coulomb-gauge condition, we can add to the action S' the term $(\nabla \cdot \dot{\boldsymbol{A}})\, A^0$, which is $-\dot{\boldsymbol{A}} \cdot \nabla A^0$, after we integrate by parts and drop the surface term. This extra term makes the action gauge invariant

$$\begin{aligned} S &= \int \tfrac{1}{2}(\dot{\boldsymbol{A}} - \nabla A^0)^2 - \tfrac{1}{2}(\nabla \times \boldsymbol{A})^2 + \boldsymbol{A}\cdot\boldsymbol{j} - A^0 j^0 + \mathcal{L}_{\rm m}\, d^4x \\ &= \int -\tfrac{1}{4}F_{ab}\,F^{ab} + A^b j_b + \mathcal{L}_{\rm m}\; d^4x. \end{aligned} \quad (16.168)$$

Thus at this point we have

$$\langle \Omega | \mathcal{T}\,[\mathcal{O}_1 \cdots \mathcal{O}_n]\,|\Omega\rangle = \frac{\int \mathcal{O}_1 \cdots \mathcal{O}_n\, e^{iS}\, \delta[\nabla \cdot \boldsymbol{A}]\, DA\, D\psi}{\int e^{iS}\, \delta[\nabla \cdot \boldsymbol{A}]\, DA\, D\psi}, \quad (16.169)$$

in which S is the gauge-invariant action (16.168), and the integral is over all fields. The only relic of the Coulomb gauge is the gauge-fixing delta functional $\delta[\nabla \cdot \boldsymbol{A}]$.

We now make the gauge transformation

$$A'_b(x) = A_b(x) + \partial_b \Lambda(x) \quad \text{and} \quad \psi'(x) = e^{iq\Lambda(x)}\psi(x) \quad (16.170)$$

and replace the fields $A_b(x)$ and $\psi(x)$ everywhere in the numerator and (separately) in the denominator in the ratio (16.169) of path integrals by their gauge transforms (16.170) $A'_\mu(x)$ and $\psi'(x)$. This change of variables changes nothing; it's like replacing $\int_{-\infty}^{\infty} f(x)\, dx$ by $\int_{-\infty}^{\infty} f(y)\, dy$, and so

$$\langle \Omega | \mathcal{T}\,[\mathcal{O}_1 \cdots \mathcal{O}_n]\,|\Omega\rangle = \langle \Omega | \mathcal{T}\,[\mathcal{O}_1 \cdots \mathcal{O}_n]\,|\Omega\rangle', \quad (16.171)$$

in which the prime refers to the gauge transformation (16.170).

We've seen that the action S is gauge invariant. So is the measure $DA\, D\psi$, and we now restrict ourselves to operators $\mathcal{O}_1 \ldots \mathcal{O}_n$ that are *gauge invariant*. So in the right-hand side of equation (16.171), the replacement of the fields by their gauge transforms affects only the term $\delta[\nabla \cdot \boldsymbol{A}]$ that enforces the Coulomb-gauge condition

$$\langle \Omega | \mathcal{T}\,[\mathcal{O}_1 \cdots \mathcal{O}_n]\,|\Omega\rangle = \frac{\int \mathcal{O}_1 \cdots \mathcal{O}_n\, e^{iS}\, \delta[\nabla \cdot \boldsymbol{A} + \triangle\Lambda)\, DA\, D\psi}{\int e^{iS}\, \delta[\nabla \cdot \boldsymbol{A} + \triangle\Lambda)\, DA\, D\psi}. \quad (16.172)$$

We now have two choices. If we integrate over all gauge functions $\Lambda(x)$ in both the numerator and the denominator of this ratio (16.172), then apart from over-all constants that cancel, the mean value in the vacuum of the time-ordered product is the ratio

$$\langle\Omega|\mathcal{T}\left[\mathcal{O}_1 \cdots \mathcal{O}_n\right]|\Omega\rangle = \frac{\displaystyle\int \mathcal{O}_1 \cdots \mathcal{O}_n \, e^{iS} \, DA \, D\psi}{\displaystyle\int e^{iS} \, DA \, D\psi}, \tag{16.173}$$

in which we integrate over all matter fields, gauge fields, and gauges. That is, **we do not fix the gauge**.

The analogous formula for the euclidean time-ordered product is

$$\langle\Omega|\mathcal{T}_{\rm e}\left[\mathcal{O}_1 \cdots \mathcal{O}_n\right]|\Omega\rangle = \frac{\displaystyle\int \mathcal{O}_1 \cdots \mathcal{O}_n \, e^{-S_{\rm e}} \, DA \, D\psi}{\displaystyle\int e^{-S_{\rm e}} \, DA \, D\psi}, \tag{16.174}$$

in which the euclidean action $S_{\rm e}$ is the space-time integral of the energy density. This formula is quite general; it holds in nonabelian gauge theories and is important in lattice gauge theory.

Our second choice is to multiply the numerator and the denominator of the ratio (16.172) by the exponential $\exp[-i\frac{1}{2}\alpha \int (-\triangle\Lambda)^2 \, d^4x]$ and then integrate over $\Lambda(x)$ separately in the numerator and denominator. This operation just multiplies the numerator and denominator by the same constant factor, which cancels. But if before integrating over all gauge transformations $\Lambda(x)$, we shift Λ so that $\triangle\Lambda$ decreases by $\dot{A}^0$, then the exponential factor is $\exp[-i\frac{1}{2}\alpha \int (\dot{A}^0 - \triangle\Lambda)^2 \, d^4x]$. Now when we integrate over $\Lambda(x)$, the delta function $\delta(\nabla \cdot \boldsymbol{A} + \triangle\Lambda)$ replaces $\triangle\Lambda$ by $-\nabla \cdot \boldsymbol{A}$ in the inserted exponential, converting it to $\exp[-i\frac{1}{2}\alpha \int (\dot{A}^0 + \nabla \cdot \boldsymbol{A})^2 \, d^4x]$. The result is to replace the gauge-invariant action (16.168) with the gauge-fixed action

$$S_\alpha = \int -\frac{1}{4} F_{ab} F^{ab} - \frac{\alpha}{2}(\partial_b A^b)^2 + A^b j_b + \mathcal{L}_{\rm m} \, d^4x. \tag{16.175}$$

This action is Lorentz invariant and so is much easier to work with than the one (16.162) with the Coulomb term. We can use it to compute scattering amplitudes perturbatively. The mean value of a time-ordered product of operators in the ground state $|0\rangle$ of the free theory is

$$\langle 0|\mathcal{T}\left[\mathcal{O}_1 \cdots \mathcal{O}_n\right]|0\rangle = \frac{\displaystyle\int \mathcal{O}_1 \cdots \mathcal{O}_n \, e^{iS_\alpha} \, DA \, D\psi}{\displaystyle\int e^{iS_\alpha} \, DA \, D\psi}. \tag{16.176}$$

By following steps analogous to those that led to (16.150), one may show (exercise 16.22) that in Feynman's gauge, $\alpha = 1$, the photon propagator is

$$\langle 0|\mathcal{T}\left[A_\mu(x)A_\nu(y)\right]|0\rangle = -i\Delta_{\mu\nu}(x-y) = -i\int \frac{\eta_{\mu\nu}}{q^2 - i\epsilon}\, e^{iq\cdot(x-y)}\, \frac{d^4q}{(2\pi)^4}. \quad (16.177)$$

16.14 Fermionic path integrals

In our brief introduction (1.11–1.12) and (1.43–1.45) to Grassmann variables, we learned that because

$$\theta^2 = 0 \quad (16.178)$$

the most general function $f(\theta)$ of a single Grassmann variable θ is

$$f(\theta) = a + b\,\theta. \quad (16.179)$$

So a complete integral table consists of the integral of this linear function

$$\int f(\theta)\, d\theta = \int a + b\,\theta\, d\theta = a\int d\theta + b\int \theta\, d\theta. \quad (16.180)$$

This equation has two unknowns, the integral $\int d\theta$ of unity and the integral $\int \theta\, d\theta$ of θ. We choose them so that the integral of $f(\theta+\zeta)$

$$\int f(\theta+\zeta)\, d\theta = \int a + b\,(\theta+\zeta)\, d\theta = (a + b\,\zeta)\int d\theta + b\int \theta\, d\theta \quad (16.181)$$

is the same as the integral (16.180) of $f(\theta)$. Thus the integral $\int d\theta$ of unity must vanish, while the integral $\int \theta\, d\theta$ of θ can be any constant, which we choose to be unity. Our complete table of integrals is then

$$\int d\theta = 0 \quad \text{and} \quad \int \theta\, d\theta = 1. \quad (16.182)$$

The anticommutation relations for a fermionic degree of freedom ψ are

$$\{\psi, \psi^\dagger\} \equiv \psi\,\psi^\dagger + \psi^\dagger\psi = 1 \quad \text{and} \quad \{\psi, \psi\} = \{\psi^\dagger, \psi^\dagger\} = 0. \quad (16.183)$$

Because ψ has $\psi^\dagger$, it is conventional to introduce a variable $\theta^* = \theta^\dagger$ that anticommutes with itself and with θ

$$\{\theta^*, \theta^*\} = \{\theta^*, \theta\} = \{\theta, \theta\} = 0. \quad (16.184)$$

The logic that led to (16.182) now gives

$$\int d\theta^* = 0 \quad \text{and} \quad \int \theta^*\, d\theta^* = 1. \quad (16.185)$$

We define the reference state $|0\rangle$ as $|0\rangle \equiv \psi|s\rangle$ for a state $|s\rangle$ that is not annihilated by ψ. Since $\psi^2 = 0$, the operator ψ annihilates the state $|0\rangle$

$$\psi|0\rangle = \psi^2|s\rangle = 0. \tag{16.186}$$

The effect of the operator ψ on the state

$$|\theta\rangle = \exp\left(\psi^\dagger\theta - \tfrac{1}{2}\theta^*\theta\right)|0\rangle = \left(1 + \psi^\dagger\theta - \tfrac{1}{2}\theta^*\theta\right)|0\rangle \tag{16.187}$$

is

$$\psi|\theta\rangle = \psi(1 + \psi^\dagger\theta - \tfrac{1}{2}\theta^*\theta)|0\rangle = \psi\psi^\dagger\theta|0\rangle = (1 - \psi^\dagger\psi)\theta|0\rangle = \theta|0\rangle \tag{16.188}$$

while that of θ on $|\theta\rangle$ is

$$\theta|\theta\rangle = \theta(1 + \psi^\dagger\theta - \tfrac{1}{2}\theta^*\theta)|0\rangle = \theta|0\rangle. \tag{16.189}$$

The state $|\theta\rangle$ therefore is an eigenstate of ψ with eigenvalue θ

$$\psi|\theta\rangle = \theta|\theta\rangle. \tag{16.190}$$

The bra corresponding to the ket $|\zeta\rangle$ is

$$\langle\zeta| = \langle 0|\left(1 + \zeta^*\psi - \frac{1}{2}\zeta^*\zeta\right) \tag{16.191}$$

and the inner product $\langle\zeta|\theta\rangle$ is (exercise 16.23)

$$\begin{aligned}\langle\zeta|\theta\rangle &= \langle 0|\left(1 + \zeta^*\psi - \frac{1}{2}\zeta^*\zeta\right)\left(1 + \psi^\dagger\theta - \frac{1}{2}\theta^*\theta\right)|0\rangle \\ &= \langle 0|1 + \zeta^*\psi\psi^\dagger\theta - \frac{1}{2}\zeta^*\zeta - \frac{1}{2}\theta^*\theta + \frac{1}{4}\zeta^*\zeta\theta^*\theta|0\rangle \\ &= \langle 0|1 + \zeta^*\theta - \frac{1}{2}\zeta^*\zeta - \frac{1}{2}\theta^*\theta + \frac{1}{4}\zeta^*\zeta\theta^*\theta|0\rangle \\ &= \exp\left[\zeta^*\theta - \frac{1}{2}\left(\zeta^*\zeta + \theta^*\theta\right)\right].\end{aligned} \tag{16.192}$$

Example 16.2 (A gaussian integral) For any number c, we can compute the integral of $\exp(c\,\theta^*\theta)$ by expanding the exponential

$$\int e^{c\,\theta^*\theta}\, d\theta^* d\theta = \int (1 + c\,\theta^*\theta)\, d\theta^* d\theta = \int (1 - c\,\theta\,\theta^*)\, d\theta^* d\theta = -c, \tag{16.193}$$

a formula that we'll use over and over. □

The identity operator for the space of states

$$c|0\rangle + d|1\rangle \equiv c|0\rangle + d\psi^\dagger|0\rangle \tag{16.194}$$

is (exercise 16.24) the integral

$$I = \int |\theta\rangle\langle\theta|\, d\theta^* d\theta = |0\rangle\langle 0| + |1\rangle\langle 1|, \tag{16.195}$$

in which the differentials anticommute with each other and with other fermionic variables: $\{d\theta, d\theta^*\} = 0$, $\{d\theta, \theta\} = 0$, $\{d\theta, \psi\} = 0$, and so forth.

The case of several Grassmann variables $\theta_1, \theta_2, \ldots, \theta_n$ and several Fermi operators $\psi_1, \psi_2, \ldots, \psi_n$ is similar. The θ_k anticommute among themselves

$$\{\theta_i, \theta_j\} = \{\theta_i, \theta_j^*\} = \{\theta_i^*, \theta_j^*\} = 0 \tag{16.196}$$

while the ψ_k satisfy

$$\{\psi_k, \psi_\ell^\dagger\} = \delta_{k\ell} \quad \text{and} \quad \{\psi_k, \psi_l\} = \{\psi_k^\dagger, \psi_\ell^\dagger\} = 0. \tag{16.197}$$

The reference state $|0\rangle$ is

$$|0\rangle = \left(\prod_{k=1}^{n} \psi_k \right) |s\rangle, \tag{16.198}$$

in which $|s\rangle$ is any state not annihilated by any ψ_k (so the resulting $|0\rangle$ isn't zero). The direct-product state

$$|\theta\rangle \equiv \exp\left(\sum_{k=1}^{n} \psi_k^\dagger \theta_k - \frac{1}{2} \theta_k^* \theta_k \right) |0\rangle = \left[\prod_{k=1}^{n} \left(1 + \psi_k^\dagger \theta_k - \frac{1}{2} \theta_k^* \theta_k \right) \right] |0\rangle \tag{16.199}$$

is (exercise 16.25) a simultaneous eigenstate of each ψ_k

$$\psi_k |\theta\rangle = \theta_k |\theta\rangle. \tag{16.200}$$

It follows that

$$\psi_\ell \psi_k |\theta\rangle = \psi_\ell \theta_k |\theta\rangle = -\theta_k \psi_\ell |\theta\rangle = -\theta_k \theta_\ell |\theta\rangle = \theta_\ell \theta_k |\theta\rangle \tag{16.201}$$

and so too $\psi_k \psi_\ell |\theta\rangle = \theta_k \theta_\ell |\theta\rangle$. Since the ψs anticommute, their eigenvalues must also

$$\theta_\ell \theta_k |\theta\rangle = \psi_\ell \psi_k |\theta\rangle = -\psi_k \psi_\ell |\theta = -\theta_k \theta_\ell |\theta\rangle, \tag{16.202}$$

which is why they must be Grassmann variables.

The inner product $\langle \zeta | \theta \rangle$ is

$$\begin{aligned} \langle \zeta | \theta \rangle &= \langle 0 | \left[\prod_{k=1}^{n} (1 + \zeta_k^* \psi_k - \frac{1}{2} \zeta_k^* \zeta_k) \right] \left[\prod_{\ell=1}^{n} (1 + \psi_\ell^\dagger \theta_\ell - \frac{1}{2} \theta_\ell^* \theta_\ell) \right] |0\rangle \\ &= \exp\left[\sum_{k=1}^{n} \zeta_k^* \theta_k - \frac{1}{2} \left(\zeta_k^* \zeta_k + \theta_k^* \theta_k \right) \right] = e^{\zeta^\dagger \theta - (\zeta^\dagger \zeta + \theta^\dagger \theta)/2}. \end{aligned} \tag{16.203}$$

The identity operator is

$$I = \int |\theta\rangle \langle \theta| \prod_{k=1}^{n} d\theta_k^* d\theta_k. \tag{16.204}$$

Example 16.3 (Gaussian Grassmann integral) For any 2×2 matrix A, we may compute the gaussian integral

$$g(A) = \int e^{-\theta^\dagger A \theta} \, d\theta_1^* d\theta_1 d\theta_2^* d\theta_2 \tag{16.205}$$

by expanding the exponential. The only terms that survive are the ones that have exactly one of each of the four variables θ_1, θ_2, θ_1^*, and θ_2^*. Thus, the integral is the determinant of the matrix A

$$\begin{aligned} g(A) &= \int \frac{1}{2} \left(\theta_k^\dagger A_{k\ell} \theta_\ell \right)^2 \, d\theta_1^* d\theta_1 d\theta_2^* d\theta_2 \\ &= \int \left(\theta_1^* A_{11} \theta_1 \, \theta_2^* A_{22} \theta_2 + \theta_1^* A_{12} \theta_2 \, \theta_2^* A_{21} \theta_1 \right) \, d\theta_1^* d\theta_1 d\theta_2^* d\theta_2 \\ &= A_{11} A_{22} - A_{12} A_{21} = \det A. \end{aligned} \tag{16.206}$$

The natural generalization to n dimensions

$$\int e^{-\theta^\dagger A \theta} \prod_{k=1}^{n} d\theta_k^* d\theta_k = \det A \tag{16.207}$$

is true for any $n \times n$ matrix A. If A is invertible, then the invariance of Grassmann integrals under translations implies that

$$\begin{aligned} \int e^{-\theta^\dagger A \theta + \theta^\dagger \zeta + \zeta^\dagger \theta} \prod_{k=1}^{n} d\theta_k^* d\theta_k &= \int e^{-\theta^\dagger A (\theta + A^{-1} \zeta) + \theta^\dagger \zeta + \zeta^\dagger (\theta + A^{-1} \zeta)} \prod_{k=1}^{n} d\theta_k^* d\theta_k \\ &= \int e^{-\theta^\dagger A \theta + \zeta^\dagger \theta + \zeta^\dagger A^{-1} \zeta} \prod_{k=1}^{n} d\theta_k^* d\theta_k \\ &= \int e^{-(\theta^\dagger + \zeta^\dagger A^{-1}) A \theta + \zeta^\dagger \theta + \zeta^\dagger A^{-1} \zeta} \prod_{k=1}^{n} d\theta_k^* d\theta_k \\ &= \int e^{-\theta^\dagger A \theta + \zeta^\dagger A^{-1} \zeta} \prod_{k=1}^{n} d\theta_k^* d\theta_k \\ &= \det A \, e^{\zeta^\dagger A^{-1} \zeta}. \end{aligned} \tag{16.208}$$

The values of θ and $\theta^\dagger$ that make the argument $-\theta^\dagger A \theta + \theta^\dagger \zeta + \zeta^\dagger \theta$ of the exponential stationary are $\overline{\theta} = A^{-1} \zeta$ and $\overline{\theta^\dagger} = \zeta^\dagger A^{-1}$. So a gaussian Grassmann integral is equal to its exponential evaluated at its stationary point, apart from a prefactor involving the determinant $\det A$. This result is a fermionic echo of the bosonic results (16.13–16.15). □

One may further extend these definitions to a Grassmann field $\chi_m(x)$ and an associated Dirac field $\psi_m(x)$. The $\chi_m(x)$s anticommute among themselves and with all fermionic variables at all points of space-time

$$\{\chi_m(x), \chi_n(x')\} = \{\chi_m^*(x), \chi_n(x')\} = \{\chi_m^*(x), \chi_n^*(x')\} = 0 \tag{16.209}$$

and the Dirac field $\psi_m(x)$ obeys the equal-time anticommutation relations

$$\begin{aligned} \{\psi_m(\boldsymbol{x}, t), \psi_n^\dagger(\boldsymbol{x}', t)\} &= \delta_{mn}\, \delta(\boldsymbol{x} - \boldsymbol{x}')\, (n, m = 1, \ldots, 4) \\ \{\psi_m(\boldsymbol{x}, t), \psi_n(\boldsymbol{x}', t)\} &= \{\psi_m^\dagger(\boldsymbol{x}, t), \psi_n^\dagger(\boldsymbol{x}', t)\} = 0. \end{aligned} \tag{16.210}$$

As in (16.102 & 16.198), we use eigenstates of the field ψ at $t = 0$. If $|0\rangle$ is defined in terms of a state $|s\rangle$ that is not annihilated by any $\psi_m(\boldsymbol{x}, 0)$ as

$$|0\rangle = \left[\prod_{m,\boldsymbol{x}} \psi_m(\boldsymbol{x}, 0) \right] |s\rangle \tag{16.211}$$

then (exercise 16.26) the state

$$\begin{aligned} |\chi\rangle &= \exp\left(\int \sum_m \psi_m^\dagger(\boldsymbol{x}, 0)\, \chi_m(\boldsymbol{x}) - \frac{1}{2} \chi_m^*(\boldsymbol{x}) \chi_m(\boldsymbol{x})\, d^3x \right) |0\rangle \\ &= \exp\left(\int \psi^\dagger \chi - \tfrac{1}{2} \chi^\dagger \chi\, d^3x \right) |0\rangle \end{aligned} \tag{16.212}$$

is an eigenstate of the operator $\psi_m(\boldsymbol{x}, 0)$ with eigenvalue $\chi_m(\boldsymbol{x})$

$$\psi_m(\boldsymbol{x}, 0)|\chi\rangle = \chi_m(\boldsymbol{x})|\chi\rangle. \tag{16.213}$$

The inner product of two such states is (exercise 16.27)

$$\langle \chi' | \chi \rangle = \exp\left[\int \chi'^\dagger \chi - \frac{1}{2} \chi'^\dagger \chi' - \frac{1}{2} \chi^\dagger \chi\, d^3x \right]. \tag{16.214}$$

The identity operator is the integral

$$I = \int |\chi\rangle\langle\chi|\, D\chi^* D\chi, \tag{16.215}$$

in which

$$D\chi^* D\chi \equiv \prod_{m,\boldsymbol{x}} d\chi_m^*(\boldsymbol{x}) d\chi_m(\boldsymbol{x}). \tag{16.216}$$

The hamiltonian for a free Dirac field ψ of mass m is the spatial integral

$$H_0 = \int \overline{\psi}\, (\boldsymbol{\gamma} \cdot \nabla + m)\, \psi\, d^3x, \tag{16.217}$$

in which $\overline{\psi} \equiv i\psi^\dagger \gamma^0$ and the gamma matrices (10.286) satisfy

$$\{\gamma^a, \gamma^b\} = 2\, \eta^{ab} \tag{16.218}$$

where η is the 4×4 diagonal matrix with entries $(-1,1,1,1)$. Since $\psi|\chi\rangle = \chi|\chi\rangle$ and $\langle\chi'|\psi^\dagger = \langle\chi'|\chi'^\dagger$, the quantity $\langle\chi'|\exp(-i\epsilon H_0)|\chi\rangle$ is by (16.214)

$$\begin{aligned}\langle\chi'|e^{-i\epsilon H_0}|\chi\rangle &= \int\langle\chi'|\chi\rangle\exp\left[-i\epsilon\int\overline{\chi}'\,(\boldsymbol{\gamma}\cdot\nabla+m)\,\chi\;d^3x\right]\\ &= \int\exp\left[\int\tfrac{1}{2}(\chi'^\dagger-\chi^\dagger)\chi-\tfrac{1}{2}\chi'^\dagger(\chi'-\chi)-i\epsilon\overline{\chi}'(\boldsymbol{\gamma}\cdot\nabla+m)\chi d^3x\right]\\ &= \int\exp\left\{\epsilon\int\left[\tfrac{1}{2}\dot{\chi}^\dagger\chi-\tfrac{1}{2}\chi'^\dagger\dot{\chi}-i\overline{\chi}'\,(\boldsymbol{\gamma}\cdot\nabla+m)\,\chi\right]\,d^3x\right\}\lambda,\end{aligned} \tag{16.219}$$

in which $\chi'^\dagger-\chi^\dagger = \epsilon\dot{\chi}^\dagger$ and $\chi'-\chi = \epsilon\dot{\chi}$. Everything within the square brackets is multiplied by ϵ, so we may replace $\chi'^\dagger$ by $\chi^\dagger$ and $\overline{\chi}'$ by $\overline{\chi}$ so as to write to first order in ϵ

$$\langle\chi'|e^{-i\epsilon H_0}|\chi\rangle = \int\exp\left[\epsilon\int\tfrac{1}{2}\dot{\chi}^\dagger\chi-\tfrac{1}{2}\chi^\dagger\dot{\chi}-i\overline{\chi}\,(\boldsymbol{\gamma}\cdot\nabla+m)\,\chi\;d^3x\right], \tag{16.220}$$

in which the dependence upon χ' is through the time derivatives.

Putting together $n = 2t/\epsilon$ such matrix elements, integrating over all intermediate-state dyadics $|\chi\rangle\langle\chi|$, and using our formula (16.215), we find

$$\langle\chi_t|e^{-2itH_0}|\chi_{-t}\rangle = \int\exp\left[\int\tfrac{1}{2}\dot{\chi}^\dagger\chi-\tfrac{1}{2}\chi^\dagger\dot{\chi}-i\overline{\chi}\,(\boldsymbol{\gamma}\cdot\nabla+m)\,\chi d^4x\right]D\chi^*D\chi. \tag{16.221}$$

Integrating $\dot{\chi}^\dagger\chi$ by parts and dropping the surface term, we get

$$\langle\chi_t|e^{-2itH_0}|\chi_{-t}\rangle = \int\exp\left[\int-\chi^\dagger\dot{\chi}-i\overline{\chi}\,(\boldsymbol{\gamma}\cdot\nabla+m)\,\chi\;d^4x\right]D\chi^*D\chi. \tag{16.222}$$

Since $-\chi^\dagger\dot{\chi} = -i\overline{\chi}\gamma^0\dot{\chi}$, the argument of the exponential is

$$i\int-\overline{\chi}\gamma^0\dot{\chi}-\overline{\chi}\,(\boldsymbol{\gamma}\cdot\nabla+m)\,\chi\;d^4x = i\int-\overline{\chi}\left(\gamma^\mu\partial_\mu+m\right)\chi\;d^4x. \tag{16.223}$$

We then have

$$\langle\chi_t|e^{-2itH_0}|\chi_{-t}\rangle = \int\exp\left(i\int\mathcal{L}_0(\chi)\,d^4x\right)D\chi^*D\chi, \tag{16.224}$$

in which $\mathcal{L}_0(\chi) = -\overline{\chi}\left(\gamma^\mu\partial_\mu+m\right)\chi$ is the action density (10.289) for a free Dirac field. Thus the amplitude is a path integral with phases given by the classical action $S_0[\chi]$

$$\langle\chi_t|e^{-2itH_0}|\chi_{-t}\rangle = \int e^{i\int\mathcal{L}_0(\chi)\,d^4x}D\chi^*D\chi = \int e^{iS_0[\chi]}D\chi^*D\chi \tag{16.225}$$

and the integral is over all fields that go from $\chi(\boldsymbol{x},-t) = \chi_{-t}(\boldsymbol{x})$ to $\chi(\boldsymbol{x},t) = \chi_t(\boldsymbol{x})$. Any normalization factor will cancel in ratios of such integrals.

Since Fermi fields anticommute, their time-ordered product has an extra minus sign

$$\mathcal{T}\left[\overline{\psi}(x_1)\psi(x_2)\right] = \theta(x_1^0 - x_2^0)\,\overline{\psi}(x_1)\,\psi(x_2) - \theta(x_2^0 - x_1^0)\,\psi(x_2)\,\overline{\psi}(x_1). \tag{16.226}$$

The logic behind our formulas (16.123) and (16.129) for the time-ordered product of bosonic fields now leads to an expression for the time-ordered product of $2n$ Dirac fields (with $D\chi''$ and $D\chi'$ suppressed)

$$\langle 0|\mathcal{T}\left[\overline{\psi}(x_1)\cdots\psi(x_{2n})\right]|0\rangle = \frac{\int \langle 0|\chi''\rangle\,\overline{\chi}(x_1)\cdots\chi(x_{2n})\,e^{iS_0[\chi]}\langle\chi'|0\rangle\,D\chi^*D\chi}{\int \langle 0|\chi''\rangle\,e^{iS_0[\chi]}\langle\chi'|0\rangle\,D\chi^*D\chi}. \tag{16.227}$$

As in (16.140), the effect of the inner products $\langle 0|\chi''\rangle$ and $\langle\chi'|0\rangle$ is to insert ϵ-terms, which modify the Dirac propagators

$$\langle 0|\mathcal{T}\left[\overline{\psi}(x_1)\cdots\psi(x_{2n})\right]|0\rangle = \frac{\int \overline{\chi}(x_1)\cdots\chi(x_{2n})\,e^{iS_0[\chi,\epsilon]}\,D\chi^*D\chi}{\int e^{iS_0[\chi,\epsilon]}\,D\chi^*D\chi}. \tag{16.228}$$

Imitating (16.141), we introduce a Grassmann external current $\zeta(x)$ and define a fermionic analog of $Z_0[j]$

$$Z_0[\zeta] \equiv \langle 0|\,\mathcal{T}\left[e^{i\int\overline{\zeta}\psi+\overline{\psi}\zeta\,d^4x}\right]|0\rangle = \frac{\int e^{i\int\overline{\zeta}\chi+\overline{\chi}\zeta\,d^4x}e^{iS_0[\chi,\epsilon]}D\chi^*D\chi}{\int e^{iS_0[\chi,\epsilon]}D\chi^*D\chi}. \tag{16.229}$$

16.15 Application to nonabelian gauge theories

The action of a generic nonabelian gauge theory is

$$S = \int -\tfrac{1}{4}F_{a\mu\nu}F_a^{\mu\nu} - \overline{\psi}\left(\gamma^\mu D_\mu + m\right)\psi\;d^4x, \tag{16.230}$$

in which the Maxwell field is

$$F_{a\mu\nu} \equiv \partial_\mu A_{a\nu} - \partial_\nu A_{a\mu} + g f_{abc}\,A_{b\mu}\,A_{c\nu} \tag{16.231}$$

and the covariant derivative is

$$D_\mu\psi \equiv \partial_\mu\psi - ig\,t_a\,A_{a\mu}\,\psi. \tag{16.232}$$

Here g is a coupling constant, f_{abc} is a structure constant (10.63), and t_a is a generator (10.55) of the Lie algebra (section 10.15) of the gauge group.

One may show (Weinberg, 1996, pp. 14–18) that the analog of equation (16.169) for quantum electrodynamics is

$$\langle\Omega|\mathcal{T}\left[\mathcal{O}_1\cdots\mathcal{O}_n\right]|\Omega\rangle = \frac{\int \mathcal{O}_1\cdots\mathcal{O}_n\, e^{iS}\,\delta[A_{a3}]\,DA\,D\psi}{\int e^{iS}\,\delta[A_{a3}]\,DA\,D\psi}, \tag{16.233}$$

in which the functional delta function

$$\delta[A_{a3}] \equiv \prod_{x,b} \delta(A_{a3}(x)) \tag{16.234}$$

enforces the axial-gauge condition, and $D\psi$ stands for $D\psi^* D\psi$.

Initially, physicists had trouble computing nonabelian amplitudes beyond the lowest order of perturbation theory. Then DeWitt showed how to compute to second order (DeWitt, 1967), and Faddeev and Popov, using path integrals, showed how to compute to all orders (Faddeev and Popov, 1967).

16.16 The Faddeev–Popov trick

The path-integral tricks of Faddeev and Popov are described in Weinberg (1996, pp. 19–27). We will use gauge-fixing functions $G_a(x)$ to impose a gauge condition on our nonabelian gauge fields $A^a_\mu(x)$. For instance, we can use $G_a(x) = A^3_a(x)$ to impose an axial gauge or $G_a(x) = i\partial_\mu A^\mu_a(x)$ to impose a Lorentz-invariant gauge.

Under an infinitesimal gauge transformation (11.511)

$$A^\lambda_{a\mu} = A_{a\mu} - \partial_\mu \lambda_a - g f_{abc}\, A_{b\mu}\, \lambda_c \tag{16.235}$$

the gauge fields change, and so the gauge-fixing functions $G_b(x)$, which depend upon them, also change. The jacobian J of that change is

$$J = \det\left(\frac{\delta G^\lambda_a(x)}{\delta \lambda_b(y)}\right)\Bigg|_{\lambda=0} \equiv \frac{DG^\lambda}{D\lambda}\Bigg|_{\lambda=0} \tag{16.236}$$

and it typically involves the delta function $\delta^4(x-y)$.

Let $B[G]$ be any functional of the gauge-fixing functions $G_b(x)$ such as

$$B[G] = \prod_{x,a} \delta(G_a(x)) = \prod_{x,a} \delta(A^3_a(x)) \tag{16.237}$$

in an axial gauge or

$$B[G] = \exp\left[\frac{i}{2}\int (G_a(x))^2\, d^4x\right] = \exp\left[-\frac{i}{2}\int \left(\partial_\mu A^\mu_a(x)\right)^2\, d^4x\right] \tag{16.238}$$

in a Lorentz-invariant gauge.

We want to understand functional integrals like (16.233)

$$\langle\Omega|\mathcal{T}\left[\mathcal{O}_1\cdots\mathcal{O}_n\right]|\Omega\rangle = \frac{\int \mathcal{O}_1\cdots\mathcal{O}_n\, e^{iS}\, B[G]\, J\, DA\, D\psi}{\int e^{iS}\, B[G]\, J\, DA\, D\psi}, \tag{16.239}$$

in which the operators $\mathcal{O}_k$, the action functional $S[A]$, and the differentials $DAD\psi$ (but not the gauge-fixing functional $B[G]$ or the Jacobian J) are gauge invariant. The axial-gauge formula (16.233) is a simple example in which $B[G] = \delta[A_{a3}]$ enforces the axial-gauge condition $A_{a3}(x) = 0$ and the determinant $J = \det\left(\delta_{ab}\partial_3\delta(x-y)\right)$ is a constant that cancels.

If we translate the gauge fields by a gauge transformation Λ, then the ratio (16.239) does not change

$$\langle\Omega|\mathcal{T}\left[\mathcal{O}_1\cdots\mathcal{O}_n\right]|\Omega\rangle = \frac{\int \mathcal{O}_1^\Lambda\cdots\mathcal{O}_n^\Lambda\, e^{iS^\Lambda}\, B[G^\Lambda]\, J^\Lambda\, DA^\Lambda\, D\psi^\Lambda}{\int e^{iS^\Lambda}\, B[G^\Lambda]\, J^\Lambda\, DA^\Lambda\, D\psi^\Lambda} \tag{16.240}$$

any more than $\int f(y)\,dy$ is different from $\int f(x)\,dx$. Since the operators $\mathcal{O}_k$, the action functional $S[A]$, and the differentials $DAD\psi$ are gauge invariant, most of the Λ-dependence goes away

$$\langle\Omega|\mathcal{T}\left[\mathcal{O}_1\cdots\mathcal{O}_n\right]|\Omega\rangle = \frac{\int \mathcal{O}_1\cdots\mathcal{O}_n\, e^{iS}\, B[G^\Lambda]\, J^\Lambda\, DA\, D\psi}{\int e^{iS}\, B[G^\Lambda]\, J^\Lambda\, DA\, D\psi}. \tag{16.241}$$

Let $\Lambda\lambda$ be a gauge transformation Λ followed by an infinitesimal gauge transformation λ. The jacobian J^Λ is a determinant of a product of matrices, which is a product of their determinants

$$\begin{aligned} J^\Lambda &= \det\left(\frac{\delta G_a^{\Lambda\lambda}(x)}{\delta\lambda_b(y)}\right)\bigg|_{\lambda=0} = \det\left(\int \frac{\delta G_a^{\Lambda\lambda}(x)}{\delta\Lambda\lambda_c(z)}\frac{\delta\Lambda\lambda_c(z)}{\delta\lambda_b(y)}\, d^4z\right)\bigg|_{\lambda=0} \\ &= \det\left(\frac{\delta G_a^{\Lambda\lambda}(x)}{\delta\Lambda\lambda_c(z)}\right)\bigg|_{\lambda=0} \det\left(\frac{\delta\Lambda\lambda_c(z)}{\delta\lambda_b(y)}\right)\bigg|_{\lambda=0} \\ &= \det\left(\frac{\delta G_a^{\Lambda}(x)}{\delta\Lambda_c(z)}\right) \det\left(\frac{\delta\Lambda\lambda_c(z)}{\delta\lambda_b(y)}\right)\bigg|_{\lambda=0} \equiv \frac{DG^\Lambda}{D\Lambda}\frac{D\Lambda\lambda}{D\lambda}\bigg|_{\lambda=0}. \end{aligned} \tag{16.242}$$

Now we integrate over the gauge transformation Λ with weight function $\rho(\Lambda) = \left(D\Lambda\lambda/D\lambda|_{\lambda=0}\right)^{-1}$ and find, since the ratio (16.241) is Λ-independent,

$$\langle\Omega|\mathcal{T}\left[\mathcal{O}_1\cdots\mathcal{O}_n\right]|\Omega\rangle = \frac{\int \mathcal{O}_1\cdots\mathcal{O}_n\, e^{iS}\, B[G^\Lambda]\,\frac{DG^\Lambda}{D\Lambda}\, D\Lambda\, DA\, D\psi}{\int e^{iS}\, B[G^\Lambda]\,\frac{DG^\Lambda}{D\Lambda}\, D\Lambda\, DA\, D\psi}$$

$$= \frac{\int \mathcal{O}_1 \cdots \mathcal{O}_n \, e^{iS} \, B[G^\Lambda] \, DG^\Lambda \, DA \, D\psi}{\int e^{iS} \, B[G^\Lambda] \, DG^\Lambda \, DA \, D\psi}$$

$$= \frac{\int \mathcal{O}_1 \cdots \mathcal{O}_n \, e^{iS} \, DA \, D\psi}{\int e^{iS} \, DA \, D\psi}. \tag{16.243}$$

Thus the mean-value in the vacuum of a time-ordered product of gauge-invariant operators is a ratio of path integrals over all gauge fields without any gauge fixing. No matter what gauge condition G or gauge-fixing functional $B[G]$ we use, the resulting gauge-fixed ratio (16.239) is equal to the ratio (16.243) of path integrals over all gauge fields without any gauge fixing. All gauge-fixed ratios (16.239) give the same time-ordered products, and so we can use whatever gauge condition G or gauge-fixing functional $B[G]$ is most convenient.

The analogous formula for the euclidean time-ordered product is

$$\langle \Omega | \mathcal{T}_e \left[\mathcal{O}_1 \cdots \mathcal{O}_n \right] | \Omega \rangle = \frac{\int \mathcal{O}_1 \cdots \mathcal{O}_n \, e^{-S_e} \, DA \, D\psi}{\int e^{-S_e} \, DA \, D\psi} \tag{16.244}$$

where the euclidean action S_e is the space-time integral of the energy density. This formula is the basis for lattice gauge theory.

The path-integral formulas (16.173 & 16.244) derived for quantum electrodynamics therefore also apply to nonabelian gauge theories.

16.17 Ghosts

Faddeev and Popov showed how to do perturbative calculations in which one does fix the gauge. To continue our description of their tricks, we return to the gauge-fixed expression (16.239) for the time-ordered product

$$\langle \Omega | \mathcal{T} \left[\mathcal{O}_1 \cdots \mathcal{O}_n \right] | \Omega \rangle = \frac{\int \mathcal{O}_1 \cdots \mathcal{O}_n \, e^{iS} \, B[G] \, J \, DA \, D\psi}{\int e^{iS} \, B[G] \, J \, DA \, D\psi}, \tag{16.245}$$

set $G_b(x) = -i\partial_\mu A^\mu_b(x)$, and use (16.238) as the gauge-fixing functional $B[G]$

$$B[G] = \exp\left[\frac{i}{2} \int (G_a(x))^2 \, d^4x \right] = \exp\left[-\frac{i}{2} \int \left(\partial_\mu A^\mu_a(x) \right)^2 d^4x \right]. \tag{16.246}$$

This functional adds to the action density the term $-(\partial_\mu A_a^\mu)^2/2$, which leads to a gauge-field propagator like the photon's (16.177)

$$\langle 0|\mathcal{T}\left[A_\mu^a(x)A_\nu^b(y)\right]|0\rangle = -i\delta_{ab}\Delta_{\mu\nu}(x-y) = -i\int \frac{\eta_{\mu\nu}\delta_{ab}}{q^2-i\epsilon}\, e^{iq\cdot(x-y)}\,\frac{d^4q}{(2\pi)^4}. \tag{16.247}$$

What about the determinant J? Under an infinitesimal gauge transformation (16.235), the gauge field becomes

$$A^\lambda_{a\mu} = A_{a\mu} - \partial_\mu\lambda_a - gf_{abc}\,A_{b\mu}\,\lambda_c \tag{16.248}$$

and so $G_a^\lambda(x) = i\partial^\mu A^\lambda_{a\mu}(x)$ is

$$G_a^\lambda(x) = i\partial^\mu A_{a\mu}(x) + i\partial^\mu\int\left[-\delta_{ac}\partial_\mu - gf_{abc}A_{b\mu}(x)\right]\delta^4(x-y)\lambda_c(y)\,d^4y. \tag{16.249}$$

The jacobian J then is the determinant (16.236) of the matrix

$$\left(\frac{\delta G_a^\lambda(x)}{\delta\lambda_c(y)}\right)\Bigg|_{\lambda=0} = -i\delta_{ac}\,\Box\,\delta^4(x-y) - igf_{abc}\frac{\partial}{\partial x^\mu}\left[A_b^\mu(x)\delta^4(x-y)\right], \tag{16.250}$$

that is

$$J = \det\left(-i\delta_{ac}\,\Box\,\delta^4(x-y) - igf_{abc}\frac{\partial}{\partial x^\mu}\left[A_b^\mu(x)\delta^4(x-y)\right]\right). \tag{16.251}$$

But we've seen (16.207) that a determinant can be written as a fermionic path integral

$$\det A = \int e^{-\theta^\dagger A\theta}\prod_{k=1}^n d\theta_k^* d\theta_k. \tag{16.252}$$

So we can write the jacobian J as

$$J = \int \exp\left[\int i\omega_a^*\Box\omega_a + igf_{abc}\omega_a^*\partial_\mu(A_b^\mu\omega_c)\,d^4x\right]D\omega^*D\omega, \tag{16.253}$$

which contributes the terms $-\partial_\mu\omega_a^*\partial^\mu\omega_a$ and

$$-\partial_\mu\omega_a^*\,gf_{abc}\,A_b^\mu\omega_c = \partial_\mu\omega_a^*\,gf_{abc}\,A_c^\mu\,\omega_b \tag{16.254}$$

to the action density.

Thus we can do perturbation theory by using the modified action density

$$\mathcal{L}' = -\tfrac{1}{4}F_{a\mu\nu}F_a^{\mu\nu} - \tfrac{1}{2}\left(\partial_\mu A_a^\mu\right)^2 - \partial_\mu\omega_a^*\partial^\mu\omega_a + \partial_\mu\omega_a^*\,gf_{abc}\,A_c^\mu\,\omega_b - \overline{\psi}\,(\not{D}+m)\,\psi, \tag{16.255}$$

in which $\not{D} \equiv \gamma^\mu D_\mu = \gamma^\mu(\partial_\mu - igt^a A_{a\mu})$. The **ghost** field ω is a mathematical device, not a physical field describing real particles, which would be spinless fermions violating the spin-statistics theorem (example 10.20).

Further reading

The detailed *Quantum Field Theory* (Srednicki, 2007), the classics *The Quantum Theory of Fields I, II, III* (Weinberg, 1995, 1996, 2005), and the delightfully readable *Quantum Field Theory in a Nutshell* (Zee, 2010) all provide excellent treatments of path integrals.

Exercises

16.1 Derive the multiple gaussian integral (16.8) from (5.167).

16.2 Derive the multiple gaussian integral (16.12) from (5.166).

16.3 Show that the vector $\overline{Y}$ that makes the argument of the multiple gaussian integral (16.12) stationary is given by (16.13), and that the multiple gaussian integral (16.12) is equal to its exponential evaluated at its stationary point $\overline{Y}$ apart from a prefactor involving $\det iS$.

16.4 Repeat the previous exercise for the multiple gaussian integral (16.11).

16.5 Compute the double integral (16.23) for the case $V(q_j) = 0$.

16.6 Insert into the LHS of (16.54) a complete set of momentum dyadics $|p\rangle\langle p|$, use the inner product $\langle q|p\rangle = \exp(iqp\hbar)/\sqrt{2\pi\hbar}$, do the resulting Fourier transform, and so verify the free-particle path integral (16.54).

16.7 By taking the nonrelativistic limit of the formula (11.311) for the action of a relativistic particle of mass m and charge q, derive the expression (16.55) for the action of a nonrelativistic particle in an electromagnetic field with no scalar potential.

16.8 Show that for the hamiltonian (16.60) of the simple harmonic oscillator the action $S[q_c]$ of the classical path is (16.67).

16.9 Show that the harmonic-oscillator action of the loop (16.68) is (16.69).

16.10 Show that the harmonic-oscillator amplitude (16.72) for $q' = 0$ and $q'' = q$ reduces as $t \to 0$ to the one-dimensional version of the free-particle amplitude (16.54).

16.11 Work out the path-integral formula for the amplitude for a mass m initially at rest to fall to the ground from height h in a gravitational field of local acceleration g to lowest order and then including loops up to an overall constant. Hint: use the technique of section 16.7.

16.12 Show that the action (16.74) of the stationary solution (16.77) is (16.79).

16.13 Derive formula (16.132) for the action $S_0[\phi]$ from (16.130 & 16.131).

16.14 Derive identity (16.136). Split the time integral at $t = 0$ into two halves, use

$$\epsilon\, e^{\pm\epsilon t} = \pm \frac{d}{dt}\, e^{\pm\epsilon t}, \tag{16.256}$$

and then integrate each half by parts.

16.15 Derive the third term in equation (16.138) from the second term.

16.16 Use (16.143) and the Fourier transform (16.144) of the external current to derive formula (16.145).

16.17 Derive equation (16.147) from equations (16.145 & 16.146).

16.18 Derive the formula (16.148) for $Z_0[j]$ from the expression (16.147) for $S_0[\phi, \epsilon, j]$.

16.19 Derive equations (16.149 & 16.150) from formula (16.148).

16.20 Derive equation (16.154) from the formula (16.149) for $Z_0[j]$.

16.21 Show that the time integral of the Coulomb term (16.159) is the term that is quadratic in j^0 in the number F defined by (16.164).

16.22 By following steps analogous to those that led to (16.150), derive the formula (16.177) for the photon propagator in Feynman's gauge.

16.23 Derive expression (16.192) for the inner product $\langle\zeta|\theta\rangle$.

16.24 Derive the representation (16.195) of the identity operator I for a single fermionic degree of freedom from the rules (16.182 & 16.185) for Grassmann integration and the anticommutation relations (16.178 & 16.184).

16.25 Derive the eigenvalue equation (16.200) from the definition (16.198 & 16.199) of the eigenstate $|\theta\rangle$ and the anticommutation relations (16.196 & 16.197).

16.26 Derive the eigenvalue relation (16.213) for the Fermi field $\psi_m(\boldsymbol{x}, t)$ from the anticommutation relations (16.209 & 16.210) and the definitions (16.211 & 16.212).

16.27 Derive the formula (16.214) for the inner product $\langle\chi'|\chi\rangle$ from the definition (16.212) of the ket $|\chi\rangle$.

17

The renormalization group

17.1 The renormalization group in quantum field theory

Most quantum field theories are nonlinear with infinitely many degrees of freedom and, because they describe point particles, they are rife with infinities. But short-distance effects, probably the finite sizes of the fundamental constituents of matter, mitigate these infinities so that we can cope with them consistently without knowing what happens at very short distances and very high energies. This procedure is called **renormalization**.

For instance, in the theory described by the Lagrange density

$$\mathcal{L} = -\frac{1}{2}\partial_\nu\phi\,\partial^\nu\phi - \frac{1}{2}m^2\phi^2 - \frac{g}{24}\phi^4 \tag{17.1}$$

we can cut off divergent integrals at some high energy Λ. The amplitude for the elastic scattering of two bosons of initial 4-momenta p_1 and p_2 into two of final momenta p_1' and p_2' to one-loop order (Weinberg, 1996, chap. 18) then is proportional to (Zee, 2010, chaps. III & VI)

$$A = g - \frac{g^2}{32\pi^2}\left[\ln\left(\frac{\Lambda^6}{stu}\right) - i\pi + 3\right] \tag{17.2}$$

as long as the absolute values of the Mandelstam variables $s = -(p_1+p_2)^2$, $t = -(p_1-p_1')^2$, and $u = -(p_1-p_2')^2$, which satisfy $stu > 0$ and $s+t+u = 4m^2$, are all much larger than m^2 (Stanley Mandelstam, 1928–). We define the **physical coupling constant** g_μ, as opposed to the bare one g that comes with $\mathcal{L}$, to be the real part of the amplitude A at $s = -t = -u = \mu^2$

$$g_\mu = g - \frac{3g^2}{32\pi^2}\left[\ln\left(\frac{\Lambda^2}{\mu^2}\right) + 1\right]. \tag{17.3}$$

Thus the bare coupling constant is $g = g_\mu + 3g^2\left[\ln(\Lambda^2/\mu^2)+1\right]$, and using this formula, we can write our expression (17.2) for the amplitude A in a form in which the cut off Λ no longer appears

$$A = g_\mu - \frac{g^2}{32\pi^2}\left[\ln\left(\frac{\mu^6}{stu}\right) - i\pi\right]. \tag{17.4}$$

This is the magic of renormalization.

The physical coupling "constant" g_μ is the right coupling at energy μ because when all the Mandelstam variables are near the renormalization point $stu = \mu^6$, the one-loop correction is tiny, and $A \approx g_\mu$.

How does the physical coupling g_μ depend upon the energy μ? The amplitude A must be independent of the renormalization energy μ, and so

$$\frac{dA}{d\mu} = \frac{dg_\mu}{d\mu} - \frac{g^2}{32\pi^2}\frac{6}{\mu} = 0, \tag{17.5}$$

which is a version of the **Callan–Symanzik equation.**

We assume that when the cut off Λ is big but finite, the bare and **running** coupling constants g and g_μ are so tiny that they differ by terms of order g^2 or g_μ^2. Then to lowest order in g and g_μ, we can replace g^2 by g_μ^2 in (17.5) and arrive at the simple differential equation

$$\mu\frac{dg_\mu}{d\mu} \equiv \beta(g_\mu) = \frac{3\,g_\mu^2}{16\pi^2}, \tag{17.6}$$

which we can integrate

$$\ln\frac{E}{M} = \int_M^E \frac{d\mu}{\mu} = \int_{g_M}^{g_E}\frac{dg_\mu}{\beta(g_\mu)} = \frac{16\pi^2}{3}\int_{g_M}^{g_E}\frac{dg_\mu}{g_\mu^2} = \frac{16\pi^2}{3}\left(\frac{1}{g_M} - \frac{1}{g_E}\right) \tag{17.7}$$

to find the running physical coupling constant g_μ at energy $\mu = E$

$$g_E = \frac{g_M}{1 - 3\,g_M\ln(E/M)/16\pi^2}. \tag{17.8}$$

As the energy $E = \sqrt{s}$ rises above M, while staying below the singular value $E = M\exp(16\pi^2/3g_M)$, the running coupling g_E slowly increases. And so does the scattering amplitude, $A \approx g_E$.

Example 17.1 (Quantum electrodynamics) Vacuum polarization makes the amplitude for the scattering of two electrons proportional to (Weinberg, 1995, chap. 11)

$$A(q^2) = e^2\left[1 + \pi(q^2)\right] \tag{17.9}$$

rather than to e^2. Here e is the renormalized charge, $q = p_1' - p_1$ is the 4-momentum transferred to the first electron, and

$$\pi(q^2) = \frac{e^2}{2\pi^2} \int_0^1 x(1-x) \ln\left[1 + \frac{q^2 x(1-x)}{m^2}\right] dx \tag{17.10}$$

represents the polarization of the vacuum. We define the square of the running coupling constant e_μ^2 to be the amplitude (17.9) at $q^2 = \mu^2$

$$e_\mu^2 = A(\mu^2) = e^2 \left[1 + \pi(\mu^2)\right]. \tag{17.11}$$

For $\mu^2 \gg m^2$, the vacuum polarization term $\pi(\mu^2)$ is (exercise 17.1)

$$\pi(\mu^2) \approx \frac{e^2}{6\pi^2}\left[\ln\frac{\mu}{m} - \frac{5}{6}\right]. \tag{17.12}$$

The amplitude (17.9) then is

$$A(q^2) = e_\mu^2 \frac{1 + \pi(q^2)}{1 + \pi(\mu^2)} \tag{17.13}$$

and, since it must be independent of μ, we have

$$0 = \frac{d}{d\mu}\frac{A(q^2)}{1+\pi(q^2)} = \frac{d}{d\mu}\frac{e_\mu^2}{1+\pi(\mu^2)} \approx \frac{d}{d\mu}\left\{e_\mu^2\left[1 - \pi(\mu^2)\right]\right\}. \tag{17.14}$$

So we find

$$0 = 2e_\mu\left(\frac{de_\mu}{d\mu}\right)\left[1 - \pi(\mu^2)\right] - e_\mu^2\frac{d\pi(\mu^2)}{d\mu} = 2e_\mu\left(\frac{de_\mu}{d\mu}\right)\left[1 - \pi(\mu^2)\right] - e_\mu^2\frac{e^2}{6\pi^2\mu}. \tag{17.15}$$

Thus, since by (17.10 & 17.11) $\pi(\mu^2) = \mathcal{O}(e^2)$ and $e_\mu^2 = e^2 + \mathcal{O}(e^4)$, we find to lowest order in e_μ

$$\mu\frac{de_\mu}{d\mu} \equiv \beta(e_\mu) = \frac{e_\mu^3}{12\pi^2}. \tag{17.16}$$

We can integrate this differential equation

$$\ln\frac{E}{M} = \int_M^E \frac{d\mu}{\mu} = \int_{e_M}^{e_E}\frac{de_\mu}{\beta(e_\mu)} = 12\pi^2\int_{e_M}^{e_E}\frac{de_\mu}{e_\mu^3} = 6\pi^2\left(\frac{1}{e_M^2} - \frac{1}{e_E^2}\right) \tag{17.17}$$

and so get for the running coupling constant the formula

$$e_E^2 = \frac{e_M^2}{1 - e_M^2\ln(E/M)/6\pi^2}, \tag{17.18}$$

which shows that it slowly increases with the energy E. Thus, the fine-structure constant $e_\mu^2/4\pi$ rises from $\alpha = 1/137.036$ at m_e to

$$\frac{e^2(45.5\,\text{GeV})}{4\pi} = \frac{\alpha}{1 - 2\alpha\ln(45.5/0.00051)/3\pi} = \frac{1}{134.6} \tag{17.19}$$

at $\sqrt{s} = 91$ GeV. When all light charged particles are included, one finds that the fine-structure constant rises to $\alpha = 1/128.87$ at $E = 91$ GeV. □

Example 17.2 (Quantum chromodynamics) Because of the cubic interaction of the gauge fields of a nonabelian gauge theory, the running coupling constant g_μ can slowly decrease with rising energy. If the gauge group is $SU(3)$, then due to this cubic interaction and that of the ghost fields (16.255), the running coupling constant g_μ is to order g_M^3

$$g_\mu = g_M \left[1 - \frac{11 g_M^2}{16\pi^2} \ln\left(\frac{\mu}{M}\right) \right]. \tag{17.20}$$

It differs from g_M only by terms of order g_M^3 and so satisfies the differential equation

$$\mu \frac{dg_\mu}{d\mu} \equiv \beta(g_\mu) = -\frac{11 g_M^3}{16\pi^2} \approx -\frac{11 g_\mu^3}{16\pi^2}, \tag{17.21}$$

in which the β-function is **negative**. Integrating

$$\ln\frac{E}{M} = \int_M^E \frac{d\mu}{\mu} = \int_{g_M}^{g_G} \frac{dg_\mu}{\beta(g_\mu)} = -\frac{16\pi^2}{11} \int_{g_M}^{g_E} \frac{dg_\mu}{g_\mu^3} = \frac{8\pi^2}{11} \left(\frac{1}{g_M^2} - \frac{1}{g_E^2} \right) \tag{17.22}$$

we find

$$g_E^2 = g_M^2 \left[1 + \frac{11 g_M^2}{8\pi^2} \ln\frac{E}{M} \right]^{-1}, \tag{17.23}$$

which shows that as the energy E of a scattering process increases, the running coupling slowly **decreases**, going to zero at infinite energy, an effect called**asymptotic freedom**.

If the gauge group is $SU(N)$, and the theory has n_f flavors of quarks with masses below μ, then the beta function is

$$\beta(g_\mu) = -\frac{g_\mu^3}{4\pi^2} \left(\frac{11N}{12} - \frac{n_f}{6} \right), \tag{17.24}$$

which remains negative as long as $n_f < 11N/2$. Using this beta function with $N = 3$ and again integrating, we get instead of (17.23)

$$g_E^2 = g_M^2 \left[1 + \frac{(11 - 2n_f/3) g_M^2}{16\pi^2} \ln\frac{E^2}{M^2} \right]^{-1}. \tag{17.25}$$

So with

$$\Lambda^2 \equiv M^2 \exp\left(-\frac{16\pi^2}{(11 - 2n_f/3) g_M^2} \right) \tag{17.26}$$

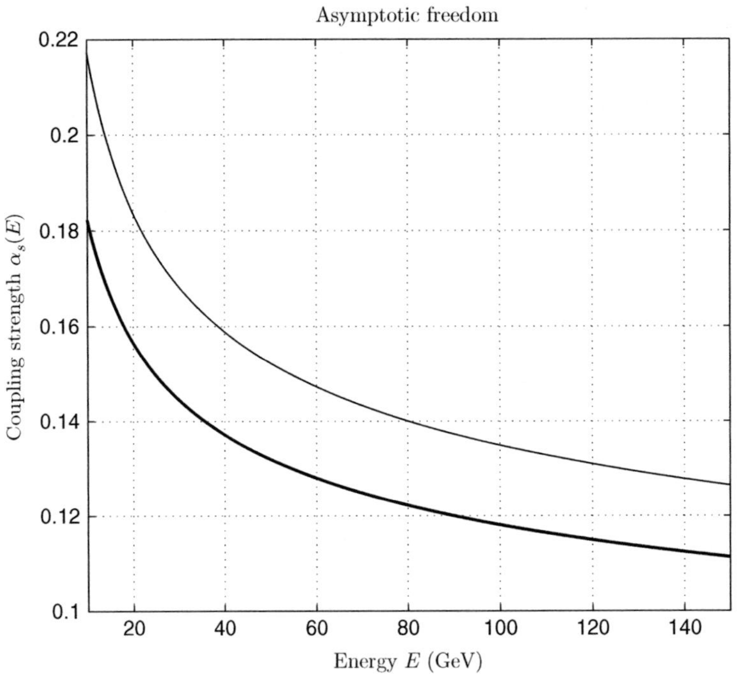

Figure 17.1 The strong-structure constant $\alpha_s(E)$ as given by the one-loop formula (17.27) (thin curve) and by a three-loop formula (thick curve) with $\Lambda = 230$ MeV and $n_f = 5$ is plotted for $m_b \ll E \ll m_t$.

we find (exercise 17.2)

$$\alpha_s(E) \equiv \frac{g^2(E)}{4\pi} = \frac{12\pi}{(33 - 2n_f)\ln(E^2/\Lambda^2)}. \tag{17.27}$$

This formula expresses the dimensionless strong-structure constant $\alpha_s(E)$ appropriate to energy E in terms of a parameter Λ that has the dimension of energy. Some call this **dimensional transmutation**. For $\Lambda = 230$ MeV and $n_f = 5$, Fig. 17.1 displays $\alpha_s(E)$ in the range $4.19 = m_b \ll E \ll m_t = 172$ GeV as given by the one-loop formula (17.27) (thin curve) and a three-loop formula (Weinberg, 1996, p. 156) (thick curve). □

17.2 The renormalization group in lattice field theory

Let us consider a quantum field theory on a lattice (Gattringer and Lang, 2010, chap. 3) in which the strength of the nonlinear interactions depends upon a single dimensionless coupling constant g. The spacing a of the lattice regulates the infinities, which return as $a \to 0$. The value of an observable P computed on this lattice will depend upon the lattice spacing a and on the coupling constant g, and so will be a function $P(a, g)$ of these two parameters. The *right* value of the coupling constant is the value that makes the result of the computation be as close as possible to the physical value P. So the *correct* coupling constant is not a constant at all, but rather a function $g(a)$ that varies with the lattice spacing or cut off a. Thus, as we vary the lattice spacing and go to the continuum limit $a \to 0$, we must adjust the coupling function $g(a)$ so that what we compute, $P(a, g(a))$, is equal to the physical value P. That is, $g(a)$ must vary with a so as to keep $P(a, g(a)) = P$. But then $P(a, g(a))$ must remain constant as a varies, so

$$\frac{dP(a, g(a))}{da} = 0. \tag{17.28}$$

Writing this condition as a dimensionless derivative

$$a \frac{dP(a, g(a))}{da} = \frac{da}{d \ln a} \frac{dP(a, g(a))}{da} = \frac{dP(a, g(a))}{d \ln a} = 0 \tag{17.29}$$

we arrive at the **Callan–Symanzik equation**

$$0 = \frac{dP(a, g(a))}{d \ln a} = \left(\frac{\partial}{\partial \ln a} + \frac{dg}{d \ln a} \frac{\partial}{\partial g} \right) P(a, g(a)). \tag{17.30}$$

The coefficient of the second partial derivative with a minus sign

$$\beta_{\mathrm{L}}(g) \equiv -\frac{dg}{d \ln a} \tag{17.31}$$

is the lattice β-function. Since the lattice spacing a and the energy scale μ are inversely related, the lattice β-function differs from the continuum beta function by a minus sign.

In $SU(N)$ gauge theory, the first two terms of the lattice β-function for small g are

$$\beta_{\mathrm{L}}(g) = -\beta_0 \, g^3 - \beta_1 \, g^5 \tag{17.32}$$

where for n_{f} flavors of light quarks

$$\beta_0 = \frac{1}{(4\pi)^2} \left(\frac{11}{3} N - \frac{2}{3} n_{\mathrm{f}} \right)$$
$$\beta_1 = \frac{1}{(4\pi)^4} \left(\frac{34}{3} N^2 - \frac{10}{3} N n_{\mathrm{f}} - \frac{N^2 - 1}{N} n_{\mathrm{f}} \right). \tag{17.33}$$

In quantum chromodynamics, $N = 3$.

Combining the definition (17.31) of the β-function with its expansion (17.32) for small g, one arrives at the differential equation

$$\frac{dg}{d \ln a} = \beta_0 \, g^3 + \beta_1 \, g^5, \tag{17.34}$$

which one may integrate

$$\int d\ln a = \ln a + c = \int \frac{dg}{\beta_0 g^3 + \beta_1 g^5} = -\frac{1}{2\beta_0 g^2} + \frac{\beta_1}{2\beta_0^2} \ln\left(\frac{\beta_0 + \beta_1 g^2}{g^2}\right) \tag{17.35}$$

to find

$$a(g) = d \left(\frac{\beta_0 + \beta_1 g^2}{g^2}\right)^{\beta_1/2\beta_0^2} e^{-1/2\beta_0 g^2}, \tag{17.36}$$

in which d is a constant of integration. The term $\beta_1 g^2$ is of higher order in g, and if one drops it and absorbs a factor of β_0^2 into a new constant of integration Λ, then one gets

$$a(g) = \frac{1}{\Lambda}\left(\beta_0 g^2\right)^{-\beta_1/2\beta_0^2} e^{-1/2\beta_0 g^2}. \tag{17.37}$$

As $g \to 0$, the lattice spacing $a(g)$ goes to zero *very fast* (as long as $n_f < 17$ for $N = 3$). The inverse of this relation (17.37) is

$$g(a) \approx \left[\beta_0 \ln(a^{-2}\Lambda^{-2}) + (\beta_1/\beta_0) \ln\left(\ln(a^{-2}\Lambda^{-2})\right)\right]^{-1/2}. \tag{17.38}$$

It shows that the coupling constant slowly goes to zero with a, which is a lattice version of **asymptotic freedom**. □

17.3 The renormalization group in condensed-matter physics

The study of condensed matter is concerned mainly with properties that emerge in the bulk, such as the melting point, the boiling point, or the conductivity. So we want to see what happens to the physics when we increase the distance scale many orders of magnitude beyond the size a of an individual molecule or the distance between nearest neighbors.

As a simple example, let's consider a euclidean action in d dimensions

$$S = \int d^d x \left(\frac{1}{2}(\partial\phi)^2 + \sum_n g_n \phi^n\right), \tag{17.39}$$

in which $g_2\phi^2 \equiv m^2\phi^2/2$ is a mass term and $g_4\phi^4 \equiv \lambda\phi^4/24$ is a quartic self-interaction. In terms of an ultraviolet cut off $\Lambda = 1/a$, we may define a partition function

$$Z(\Lambda) = \int_\Lambda e^{-S} D\phi \tag{17.40}$$

to be one in which the field

$$\phi(x) = \int_\Lambda e^{ikx}\phi(k)\,\frac{d^d k}{(2\pi)^d} \tag{17.41}$$

only has Fourier coefficients $\phi(k)$ with $k^2 < \Lambda^2$. Corresponding to each such field $\phi(x)$, we introduce a "stretched" field

$$\phi_L(x) = A(L)\,\phi(x/L) \quad \text{for} \quad L \geq 1, \tag{17.42}$$

in which $A(L)$ is a scale factor that we will use to keep the kinetic part of the action invariant. Since

$$\phi_L(x) = A(L)\,\phi(x/L) = A(L) \int_\Lambda \exp\left(i\frac{kx}{L}\right) \phi(k) \frac{d^d k}{(2\pi)^d} \tag{17.43}$$

the momenta of the stretched field are reduced by the factor $1/L$.

We may define a new partition function in which we integrate over the stretched fields $\phi_L(x)$

$$Z(\Lambda/L) = \int_{\Lambda/L} e^{-S} D\phi \equiv \int_\Lambda e^{-S} D\phi_L. \tag{17.44}$$

The kinetic action of a stretched field is

$$S_\text{k} = \int d^d x \, \frac{A^2(L)}{2} \left(\frac{\partial\phi(x/L)}{\partial x}\right)^2 = \int d^d(x/L)\, L^d \, \frac{A^2(L)}{2} \left(\frac{\partial\phi(x/L)}{L\partial x/L}\right)^2 \tag{17.45}$$

and so if we choose

$$A(L) = L^{-(d-2)/2} \tag{17.46}$$

then letting $x' = x/L$, we find that the kinetic action S_k is invariant

$$S_\text{k} = \int d^d x' \, \frac{1}{2} \left(\frac{\partial\phi(x')}{\partial x'}\right)^2. \tag{17.47}$$

The full action of a stretched field is

$$S(\phi_L) = \int d^d x \left(\frac{1}{2}(\partial\phi)^2 + \sum_n g_{d,n}(L)\phi^n\right), \tag{17.48}$$

in which

$$g_{d,n}(L) = L^d A^n(L)\, g_n = L^{d-n(d-2)/2}\, g_{d,n}. \tag{17.49}$$

The beta function

$$\beta(g_{d,n}) \equiv \frac{L}{g_{d,n}(L)} \frac{dg_{d,n}(L)}{dL} = d - n(d-2)/2 \tag{17.50}$$

is just the exponent of the coupling "constant" $g_{d,n}(L)$. If it is positive, then the coupling constant $g_{d,n}(L)$ gets stronger as $L \to \infty$; such couplings are called **relevant**. Couplings with vanishing exponents are insensitive to changes in L and are **marginal**. Those with negative exponents shrink with increasing L; they are **irrelevant**.

The coupling constant $g_{d,n,p}$ of a term with p derivatives and n powers of the field ϕ in a space of d dimensions varies as

$$g_{d,n,p}(L) = L^d\, A^n(L)\, L^{-p}\, g_{n,p} = L^{d-n(d-2)/2-p}\, g_{d,n,p}. \tag{17.51}$$

Example 17.3 (QCD) In quantum chromodynamics, there is a cubic term $g f_{abc}\, A_0^a A_i^b \partial_0 A_i^c$, which in effect looks like $g f_{abc}\, \phi_a \phi_b \dot{\phi}_c$. Is it relevant? Well, if we stretch space but not time, then the time derivative has no effect, and $d = 3$. So the cubic, $n = 3$, grows as $L^{3/2}$

$$g_{3,3,0}(L) = L^{d-n(d-2)/2}\, g_{3,3,0} = L^{3/2}\, g_{3,3,0}. \tag{17.52}$$

Since this cubic term drives asymptotic freedom, its strengthening as space is stretched by the dimensionless factor L may point to a qualitative explanation of confinement. For if $g_{3,3,0}(L)$ grows with distance as $L^{3/2}$, then $\alpha_s(L) = g_{3,3,0}^2(L)/4\pi$ grows as L^3, and so the strength $\alpha_s(Lr)/(Lr)^2$ of the force between two quarks separated by a distance Lr grows linearly with L

$$F(Lr) = \frac{\alpha_s(Lr)}{(Lr)^2} = \frac{L^3\, \alpha_s(r)}{(Lr)^2} = Lr\, \frac{\alpha_s(r)}{r^2}, \tag{17.53}$$

which may be enough for quark confinement.

On the other hand, if we stretch both space and time, then the cubic $g_{4,3,1}(L)$ and quartic $g_{4,4,0}(L)$ couplings are marginal. □

Exercises

17.1 Show that for $\mu^2 \gg m^2$, the vacuum polarization term (17.10) reduces to (17.12). Hint: use $\ln ab = \ln a + \ln b$ when integrating.

17.2 Show that by choosing the energy scale Λ according to (17.26), one can derive (17.27) from (17.25).

17.3 Show that if we stretch both space and time, then in the notation of (17.51), the cubic $g_{4,3,1}(L)$ and quartic $g_{4,4,0}(L)$ couplings are marginal, that is, are independent of L.

18

Chaos and fractals

18.1 Chaos

Early in the last century, Henri Poincaré studied the three-body problem and found very complicated orbits. In this and other systems, he found that after a transient period, classical motion assumes one of four forms:

1 periodic (a limit cycle),
2 steady or damped or stopped,
3 quasi-periodic (more than one frequency),
4 chaotic.

Example 18.1 (Duffing's equation) If one attaches a thin piece of iron to the end of a rod that moves sinusoidally in the x direction at frequency ω near two magnets then the x coordinate is described by the forced Duffing equation

$$\ddot{x} + a\dot{x} + bx^3 + cx = g\sin(\omega t + \phi). \tag{18.1}$$

For suitable values of a, b, c, g, ω, and ϕ, the coordinate x varies chaotically. □

Example 18.2 (Dripping faucet) Drops from a slowly dripping faucet tend to fall regularly at times t_n separated by a constant interval $\Delta t = t_{n+1} - t_n$. At a slightly higher flow rate, the drops fall separated by intervals that alternate in their durations Δt, ΔT, Δt, ΔT, Δt, ΔT in a **period-two** sequence. At some higher flow rates, no regularity is apparent. □

Example 18.3 (Rayleigh-Benard convection) Consider a fluid in a gravitational field lies above a hot plate and below a cold one. If the difference ΔT is small enough, then steady convective cellular flow occurs. But if ΔT is above the chaotic threshold, the fluid boils chaotically.

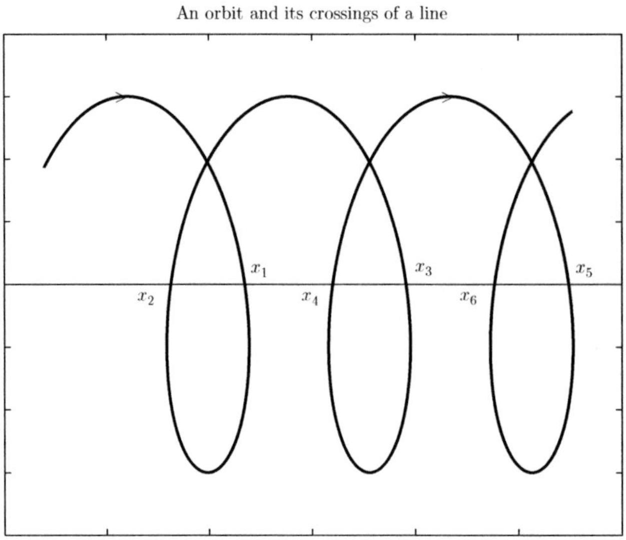

Figure 18.1 The orbit or trajectory of a dynamical system in N dimensions generates a map of crossings in $N - 1$ dimensions.

The N equations

$$\dot{x}^i = F^i(x) \tag{18.2}$$

represent an N-dimensional **dynamical system**; they are said to be **autonomous** because they involve $x(t)$ but not t itself, that is, $\dot{x}^i = F^i(x)$, not $F^i(x, t)$. The crossings of a suitably oriented plane (or more generally of a **Poincaré surface of section**) lead to an **invertible map**

$$x_{n+1} = M(x_n) \tag{18.3}$$

in a space of $N - 1$ dimensions, as shown in Fig. 18.1.

A first-order, autonomous dynamical system like (18.2) can behave chaotically only if

$$N \geq 3. \tag{18.4}$$

□

Example 18.4 (Driven, damped pendulum) The angle θ of a sinusoidally driven, damped pendulum follows the differential equation

$$\ddot{\theta} + f\dot{\theta} + \sin\theta = T\sin(\omega t), \tag{18.5}$$

which is second order and nonautonomous. Will this system exhibit chaos? We put it into autonomous form by defining $x^1 = \dot{\theta}$, $x^2 = \theta$, and $x^3 = \omega t$. In these variables, the pendulum equation (18.5) is the first-order autonomous system

$$\begin{aligned} \dot{x}^1 &= T \sin x^3 - \sin x^2 - f x^1, \\ \dot{x}^2 &= x^1, \\ \dot{x}^3 &= \omega \end{aligned} \tag{18.6}$$

with $N = 3$ dependent variables. So chaos is not excluded and is exhibited by numerical solutions for suitable values of the parameters f, T, and ω.

An invertible map like (18.3)

$$x_n = M^{-1}(x_{n+1}) \tag{18.7}$$

can be chaotic only if it has at least two dimensions. This condition is consistent with condition (18.4) because the Poincaré section of an N-dimensional dynamical system like (18.2) is an (N-1)-dimensional invertible map as in Fig. 18.1. □

Example 18.5 (Logistic map) A map that is not invertible can display chaos even in one dimension. The one-dimensional **logistic map**

$$x_{n+1} = q\, x_n(x_n - 1), \tag{18.8}$$

which is not invertible (because the quadratic equation for x_n in terms of x_{n+1} has two solutions), displays chaos in increasingly striking forms as q exceeds a number slightly greater than 3.57. And at $q = 3.8$, two sequences respectively starting at $x_0 = 0.2$ and $x'_0 = 0.20001$ differ by more than 0.2 after only 21 iterations (Fig. 18.2).

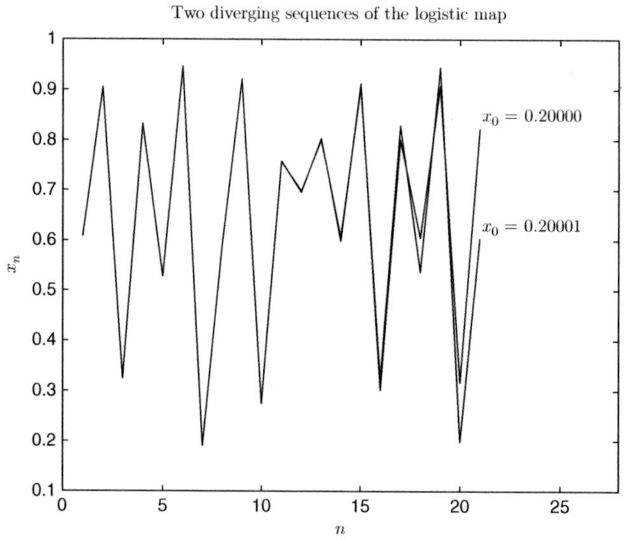

Figure 18.2 At $q = 3.8$, the logistic map (18.8) is so sensitive to initial conditions that after only 21 iterations the sequences starting from $x_0 = 0.2$ and $x_0 = 0.20001$ differ by 0.218.

By $q = 4$, the logistic map (18.8) is totally chaotic and is equivalent to the **tent map**

$$x_{n+1} = 1 - 2|x_n - 1|. \tag{18.9}$$

A similar but simpler chaotic map is the $2x$ **modulo 1 map**

$$x_{n+1} = 2x_n \bmod 1. \tag{18.10}$$

□

Example 18.6 (The Bernoulli shift) The simplest chaotic map is the **Bernoulli shift**, in which the initial point x_0 is an arbitrary number between 0 and 1 with the binary-decimal expansion

$$x_0 = \sum_{j=1}^{\infty} 2^{-j} a_j = 0.a_1a_2a_3a_4 \ldots \tag{18.11}$$

and successive points lack a_1, then a_2, *etc.*:

$$\begin{aligned} x_1 &= 0.a_2a_3a_4a_5 \ldots \\ x_2 &= 0.a_3a_4a_5a_6 \ldots \\ x_3 &= 0.a_4a_5a_6a_7 \ldots . \end{aligned} \tag{18.12}$$

Two unequal irrational numbers x_0 and x_0', no matter how close, generate sequences that roam independently, irregularly, and ergotically over the interval $(0, 1)$. □

Example 18.7 (Hénon's map) The two-dimensional map

$$\begin{aligned} x_{n+1} &= f(x_n) + B y_n, \\ y_{n+1} &= x_n, \end{aligned} \tag{18.13}$$

for $B \neq 0$ is invertible. If $f(x_n) = A - x_n^2$, it is **Hénon's map**, which for $A = 1.4$ and $B = 0.3$ is chaotic and converges to the attractor in Fig. 18.6. □

The **Lyapunov exponent** of a smooth map $x_{n+1} = f(x_n)$ is the limit

$$h(x_1) = \lim_{n \to \infty} \frac{1}{n} \left[\ln |f'(x_1)| + \cdots + \ln |f'(x_n)| \right]. \tag{18.14}$$

A bounded sequence that has a positive Lyapunov exponent and that does not converge to a periodic sequence is **chaotic** (Alligood *et al.*, 1996, p. 110). Other aspects of chaos lead to other definitions.

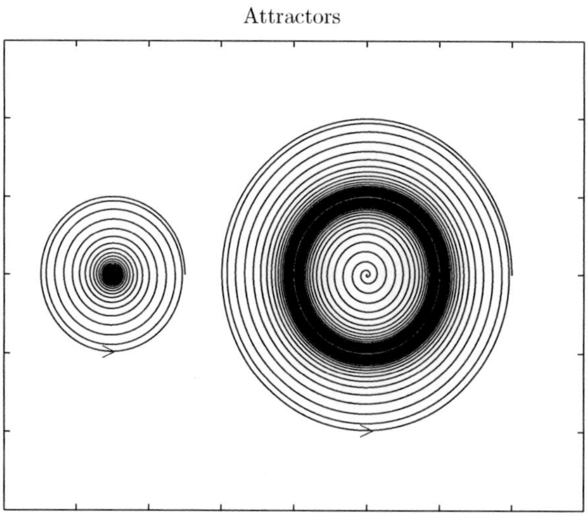

Figure 18.3 On the left, the origin is an attractor; on the right, the circle is a limit cycle.

18.2 Attractors

If the states of a dynamical system converge, as for instance in Fig. 18.3, to a point or set of points, that point or set is called an **attractor**.

Example 18.8 (The van der Pol equation) If the attractor is a loop that the states run around, then the attractor is called a **limit cycle**. The van der Pol equation

$$\ddot{y} + (y^2 - \eta)\dot{y} + \omega^2 y = 0, \tag{18.15}$$

which we may write as the first-order system

$$\begin{aligned} x_1 &= \dot{y}, \quad \dot{x}_1 = -\omega^2 x_2 - (x_2^2 - \eta)x_1, \\ x_2 &= y, \quad \dot{x}_2 = x_1, \end{aligned} \tag{18.16}$$

has a limit cycle. Physicists used the van der Pol equation in the 1920s to study vacuum-tube oscillators. □

18.3 Fractals

A fractal set has a dimension that is not an integer. How can that be?

Felix Hausdorff and Abram Besicovitch showed how to define the dimension of a weird set of points. To compute the **fractal** or **box-counting** dimension of a set, we cover it with line segments, squares, cubes, or n-dimensional "boxes"

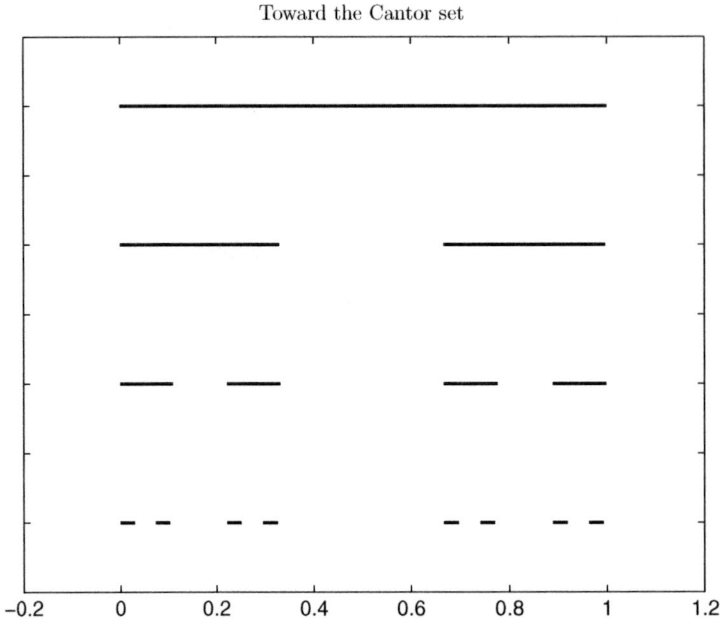

Figure 18.4 The first four approximations to the Cantor set.

of side ϵ. If we need $N(\epsilon)$ boxes, then the fractal dimension D_b is the limit as $\epsilon \to 0$

$$D_b = \lim_{\epsilon \to 0} \frac{\ln(N(\epsilon))}{\ln(1/\epsilon)}. \tag{18.17}$$

For instance, we can cover the interval $[a, b]$ with $N(\epsilon) = (b-a)/\epsilon$ line segments of length ϵ, so the dimension of the segment $[a, b]$ is

$$D_b = \lim_{\epsilon \to 0} \frac{\ln(N(\epsilon))}{\ln(1/\epsilon)} = \lim_{\epsilon \to 0} \frac{\ln((b-a)/\epsilon)}{\ln(1/\epsilon)} = 1 + \frac{\ln(b-a)}{\ln(1/\epsilon)} = 1 \tag{18.18}$$

as it should be.

Example 18.9 (Cantor set) The Cantor set is defined by a limiting process in which the set at the nth stage consists of 2^n line segments each of length $1/3^n$. The first four approximations to the **Cantor set** are sketched in Fig. 18.4. We can cover the nth approximation with $N(\epsilon) = 2^n$ line segments each of length $\epsilon_n = 1/3^n$, and so the fractal dimension is

$$D_{b,C} = \lim_{\epsilon \to 0} \frac{\ln(N(\epsilon))}{\ln(1/\epsilon)} = \lim_{n \to \infty} \frac{\ln(N(\epsilon_n))}{\ln(1/\epsilon_n)} = \lim_{n \to \infty} \frac{\ln(2^n)}{\ln(3^n)} = \frac{\ln 2}{\ln 3} = 0.6309297\ldots, \tag{18.19}$$

which is not an integer or even a rational number. □

Example 18.10 (Koch snowflake) In 1904, the Swedish mathematician Helge von Koch described the Koch curve (or the Koch snowflake), whose construction is shown in Fig. 18.5. With each step, there are four times as many line segments, each one being three times smaller. The length L of the curve at step n is thus

$$L = \left(\frac{4}{3}\right)^n, \tag{18.20}$$

which grows without limit as $n \to \infty$. Its box dimension is

$$D_{\text{b,K}} = \lim_{n\to\infty} \frac{\ln(N(\epsilon_n))}{\ln(1/\epsilon_n)} = \lim_{n\to\infty} \frac{\ln(4^n)}{\ln(3^n)} = \frac{\ln 4}{\ln 3} = 1.2618595\ldots \tag{18.21}$$

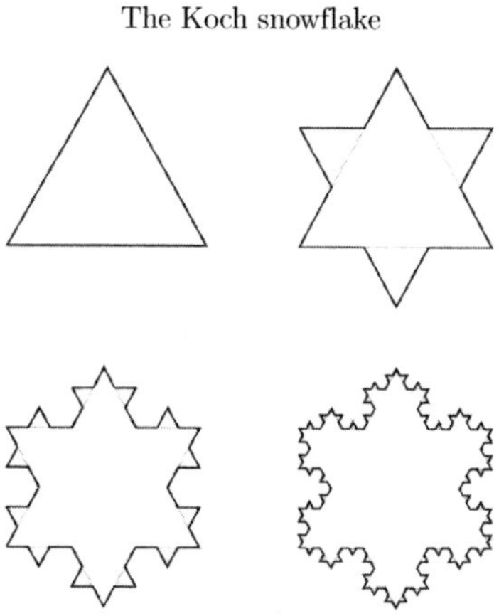

Figure 18.5 Curve of von Koch: steps 0, 1, 2, and 3 of construction (http://en-wikipedia.org/wiki/File:kochFlake.org).

□

Closely related to the box-counting dimension is the **self-similar dimension** D_{s}. To define it, we consider the number of self-similar structures of linear size x needed to cover the figure after n steps and take the limit

$$D_{\text{s}} = \lim_{x\to 0} \frac{\ln N(x)}{\ln 1/x}. \tag{18.22}$$

In the case of von Koch's curve, $x = 1/3^n$ and $N(x) = 4^n$. So the self-similar dimension of von Koch's curve is

$$D_{\text{s,K}} = \lim_{x\to 0} \frac{\ln N(x)}{\ln 1/x} = \lim_{n\to\infty} \frac{\ln 4^n}{\ln 3^n} = \frac{n \ln 4}{n \ln 3} = \frac{\ln 4}{\ln 3} = 1.2618595\ldots \tag{18.23}$$

which is equal to its box dimension $D_{\text{b,K}}$ given by (18.21).

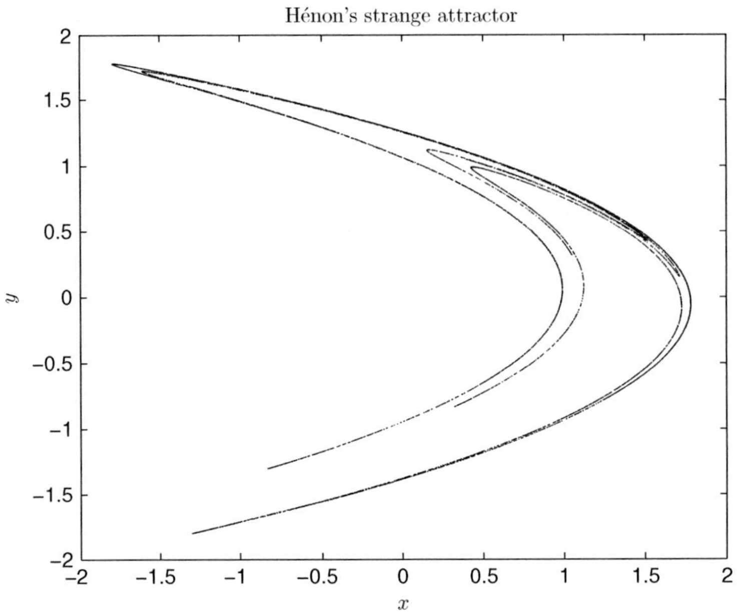

Figure 18.6 Points 20–10,020 of Hénon's map (18.7) with $A = 1.4$ and $B = 0.3$ and $(x_1, y_1) = (0, 0)$.

Attractors of fractal dimension are **strange**. Hénon's map (18.7) with $A = 1.4$ and $B = 0.3$ is chaotic with a strange attractor, as illustrated by Fig. 18.6.Chaotic systems often have strange attractors; but chaotic systems can have nonfractal attractors, and nonchaotic systems can have strange attractors.

Further reading

The book *Chaos: an Introduction to Dynamical Systems* (Alligood *et al.*, 1996) is superb.

Exercises

18.1 A period-one sequence of a map $x_{n+1} = f(x_n)$ is a point p for which $p = f(p)$. Find the period-one sequences of $x_{n+1} = rx_n(1 - x_n/K)$.

18.2 A period-two sequence of a map $x_{n+1} = f(x_n)$ is two different points p and q for which $q = f(p)$ and $p = f(q)$. Estimate the period-two sequences of the logistic map $f(x) = ax(1 - x)$ for $a = 1, 2$, and 3. Hint: graph the functions $f(f(x))$ and $I(x) = x$ on the interval [0,1].

18.3 A period-three sequence of a map $x_{n+1} = f(x_n)$ is three different points p, q, and r for which $q = f(p)$, $r = f(q)$, and $p = f(r)$. Li and Yorke have shown that a map with a period-three sequence is chaotic. Estimate the period-three sequences of the map $f(x) = 4x(1 - x)$.

19

Strings

19.1 The infinities of quantum field theory

Quantum field theory is plagued with infinities. Even exactly soluble theories of free fields have ground-state energies that diverge quarticly, as $+\Lambda^4$ for a theory of bosons and as $-\Lambda^4$ for a theory of fermions as $\Lambda \to \infty$. Theories with the same number of boson and fermion fields are less divergent, and theories with unbroken supersymmetry actually have ground-state energies that vanish.

One may be able to eliminate some of the infinities of a generic quantum field theory without spoiling the symmetries of its action density $\mathcal{L}$ by replacing $\mathcal{L}$ in the path integral by $\mathcal{L}' = \mathcal{L}/(1 - \mathcal{L}/M^4)$ or $\mathcal{L}' = M^4 \left(\exp(\mathcal{L}/M^4) - 1\right)$. In euclidean space, the substitution

$$\mathcal{L}'_e = \frac{\mathcal{L}_e}{1 - \mathcal{L}_e/M^4} \tag{19.1}$$

with the understanding that $\mathcal{L}'_e = \infty$ when $|\mathcal{L}_e| > M^4$ almost certainly removes many infinities, but whether it preserves all the symmetries is less obvious. These substitutions change quantum field theory, but in the limit $M \to \infty$, one recovers standard quantum field theory.

A more physical approach is to represent elementary particles as objects that have finite size. Those that are one-dimensional are called **strings**.

19.2 The Nambu–Goto string action

If we give up the idea of point particle then the next simplest choice is a one-dimensional string. We'll use $0 \le \sigma \le \sigma_1$ and $\tau_i \le \tau \le \tau_f$ to parametrize the space-time coordinates $X^\mu(\sigma, \tau)$ of the string. Nambu and Goto suggested using as the action the area

$$S = -\frac{T_0}{c}\int_{\tau_i}^{\tau_f}\int_0^{\sigma_1}\sqrt{\left(\dot{X}\cdot X'\right)^2 - \left(\dot{X}\right)^2 (X')^2}\, d\tau\, d\sigma, \tag{19.2}$$

in which

$$\dot{X}^{\mu} = \frac{\partial X^{\mu}}{\partial \tau} \quad \text{and} \quad X'^{\mu} = \frac{\partial X^{\mu}}{\partial \sigma} \tag{19.3}$$

and a Lorentz metric $\eta_{\mu\nu} = \text{diag}(-1, 1, 1, \ldots)$ is used to form the inner products like

$$\dot{X}\cdot X' = \dot{X}^{\mu}\eta_{\mu\nu}X^{\nu\prime} \quad \text{etc.} \tag{19.4}$$

This action is the area swept out by a string of length σ_1 in time $\tau_f - \tau_i$.

If $\dot{X}d\tau = dt$ points in the time direction and $X'd\sigma = d\boldsymbol{r}$ points in a spatial direction, then it is easy to see that $\dot{X}\cdot X' = 0$, that $-\left(\dot{X}\right)^2 d\tau^2 = dt^2$, and that $\left(X'\right)^2 d\sigma^2 = d\boldsymbol{r}^2$. So in this simple case, the action (19.2)

$$S = -\frac{T_0}{c}\int_{t_i}^{t_f}\int_0^{r_1} dt\, dr = -\frac{T_0}{c}(t_f - t_i)r_1 \tag{19.5}$$

is the area the string sweeps out. The other term within the square-root ensures that the action is the area swept out for all $\dot{X}$ and X', and that it is invariant under arbitrary reparametrizations $\sigma \to \sigma'$ and $\tau \to \tau'$.

The equation of motion for the relativistic string follows from the requirement that the action (19.2) be stationary, $\delta S = 0$. Since

$$\delta\dot{X}^{\mu} = \delta\frac{\partial X^{\mu}}{\partial \tau} = \frac{\partial(\dot{X}^{\mu} + \delta X^{\mu})}{\partial \tau} - \frac{\partial X^{\mu}}{\partial \tau} = \frac{\partial \delta X^{\mu}}{\partial \tau} \tag{19.6}$$

and similarly

$$\delta X'^{\mu} = \frac{\partial \delta X^{\mu}}{\partial \sigma} \tag{19.7}$$

we may express the change in the action in terms of derivatives of the Lagrange density

$$L = -\frac{T_0}{c}\sqrt{\left(\dot{X}\cdot X'\right)^2 - \left(\dot{X}\right)^2 (X')^2} \tag{19.8}$$

as

$$\delta S = \int_{\tau_i}^{\tau_f}\int_0^{\sigma_1}\left[\frac{\partial L}{\partial \dot{X}^{\mu}}\frac{\partial \delta X^{\mu}}{\partial \tau} + \frac{\partial L}{\partial X'^{\mu}}\frac{\partial \delta X^{\mu}}{\partial \sigma}\right] d\tau\, d\sigma. \tag{19.9}$$

Its derivatives, which we'll call $\mathcal{P}^{\tau}_{\mu}$ and $\mathcal{P}^{\sigma}_{\mu}$, are

$$\mathcal{P}^{\tau}_{\mu} = \frac{\partial L}{\partial \dot{X}^{\mu}} = -\frac{T_0}{c}\frac{(\dot{X}\cdot X')X'_{\mu} - (X')^2\dot{X}_{\mu}}{\sqrt{\left(\dot{X}\cdot X'\right)^2 - \left(\dot{X}\right)^2 (X')^2}} \tag{19.10}$$

and

$$\mathcal{P}^{\sigma}_{\mu} = \frac{\partial L}{\partial X'^{\mu}} = -\frac{T_0}{c} \frac{(\dot{X} \cdot X')\dot{X}_{\mu} - (X')^2 X'_{\mu}}{\sqrt{\left(\dot{X} \cdot X'\right)^2 - \left(\dot{X}\right)^2 (X')^2}}. \tag{19.11}$$

In terms of them, the change in the action is

$$\delta S = \int_{\tau_i}^{\tau_f}\int_0^{\sigma_1} \left[\frac{\partial}{\partial \tau}\left(\delta X^{\mu} \mathcal{P}^{\tau}_{\mu}\right) + \frac{\partial}{\partial \sigma}\left(\delta X^{\mu} \mathcal{P}^{\sigma}_{\mu}\right) - \delta X^{\mu}\left(\frac{\partial \mathcal{P}^{\tau}_{\mu}}{\partial \tau} + \frac{\partial \mathcal{P}^{\sigma}_{\mu}}{\partial \sigma}\right)\right] d\tau\, d\sigma. \tag{19.12}$$

The total τ-derivative integrates to a term involving the variation δX^{μ}, which we require to vanish at the initial and final values of τ. So we drop that term and find that the net change in the action is

$$\delta S = \int_{\tau_i}^{\tau_f} \left[\delta X^{\mu} \mathcal{P}^{\sigma}_{\mu}\right]_0^{\sigma_1} d\tau - \int_{\tau_i}^{\tau_f}\int_0^{\sigma_1} \delta X^{\mu}\left(\frac{\partial \mathcal{P}^{\tau}_{\mu}}{\partial \tau} + \frac{\partial \mathcal{P}^{\sigma}_{\mu}}{\partial \sigma}\right) d\tau\, d\sigma. \tag{19.13}$$

Thus the equations of motion for the string are

$$\frac{\partial \mathcal{P}^{\tau}_{\mu}}{\partial \tau} + \frac{\partial \mathcal{P}^{\sigma}_{\mu}}{\partial \sigma} = 0 \tag{19.14}$$

but the action is stationary only if the boundary condition

$$\delta X^{\mu}(\tau,\sigma_1)\mathcal{P}^{\sigma}_{\mu}(\tau,\sigma_1) - \delta X^{\mu}(\tau,0)\mathcal{P}^{\sigma}_{\mu}(\tau,0) = 0 \tag{19.15}$$

is satisfied for all τ. This is a condition on the ends of open strings; closed strings satisfy it automatically.

Usually the boundary condition (19.15) is interpreted as $2D = 2(d+1)$ conditions – one for each end σ_* of the string and each dimension μ of space-time:

$$\delta X^{\mu}(\tau,\sigma_*)\mathcal{P}^{\sigma}_{\mu}(\tau,\sigma_*) = 0, \quad \text{no sum over } \mu. \tag{19.16}$$

A **Dirichlet boundary condition** fixes a spatial component at an end of the string by

$$\dot{X}^{i}(\tau,\sigma_*) = 0 \tag{19.17}$$

or equivalently by $\delta X^{\mu}(\tau,\sigma_*) = 0$. The time component X^0 can not have a vanishing τ derivative, so it must obey a **free-endpoint-boundary condition**

$$\mathcal{P}^{\sigma}_{\mu}(\tau,\sigma_*) = 0, \tag{19.18}$$

which also may apply to any dimension and any end.

19.3 Regge trajectories

The quantity $\mathcal{P}^\tau_\mu(\tau,\sigma)$ defined as the derivative (19.10) turns out to be the momentum density of the string. The angular momentum M_{12} of a string rigidly rotating in the x, y plane is

$$M_{12}(\tau) = \int_0^{\sigma_1} X_1 \mathcal{P}^\tau_2(\tau,\sigma) - X_2 \mathcal{P}^\tau_1(\tau,\sigma)\, d\sigma. \tag{19.19}$$

In a parametrization of the string with $\tau = t$ and $d\sigma$ proportional to the energy density dE of the string, the x, y coordinates of the string are

$$\vec{X}(t,\sigma) = \frac{\sigma_1}{\pi} \cos\frac{\pi\sigma}{\sigma_1} \left(\cos\frac{\pi c t}{\sigma_1}, \sin\frac{\pi c t}{\sigma_1}\right). \tag{19.20}$$

The x, y components of the momentum density are

$$\vec{\mathcal{P}}^\tau(t,\sigma) = \frac{T_0}{c}\frac{\partial \vec{X}}{\partial t} = \frac{T_0}{c} \cos\frac{\pi\sigma}{\sigma_1} \left(-\sin\frac{\pi c t}{\sigma_1}, \cos\frac{\pi c t}{\sigma_1}\right). \tag{19.21}$$

The angular momentum (19.19) is then given by the integral

$$M_{12} = \frac{\sigma_1}{\pi}\frac{T_0}{c} \int_0^{\sigma_1} \cos^2\frac{\pi\sigma}{\sigma_1} d\sigma = \frac{\sigma_1^2 T_0}{2\pi c}. \tag{19.22}$$

Now the parametrization $d\sigma \propto dE$ implies that $\sigma_1 \propto E$, and in fact the energy of the string is $E = T_0\sigma_1$. Thus the angular momentum $J = |M_{12}|$ of a classical relativistic string is proportional to the square of its total energy

$$J = \frac{E^2}{2\pi T_0 c}. \tag{19.23}$$

This rule is obeyed by many meson and baryon resonances. The nucleon and five baryon resonances fit it with nearly the same value of the string tension

$$T_0 \approx 0.92 \quad \text{GeV/fm} \tag{19.24}$$

as shown by Fig. 19.1, which displays the **Regge trajectories** of the N and Δ resonances on a single curve. Other N and Δ resonances, however, do not fall on this curve.

A string theory of hadrons took off in 1968 when Gabriel Veneziano published his amplitude for $\pi + \pi$ scattering as a sum of three Euler beta functions (Veneziano, 1968). But after eight years of intense work, this effort was largely abandoned with the discovery of quarks at SLAC and the promise of QCD as a theory of the strong interactions. In 1974, Joël Scherk and John H. Schwarz proposed increasing the string tension by 38 orders of magnitude so as to use strings to make a quantum theory that included gravity (Scherk

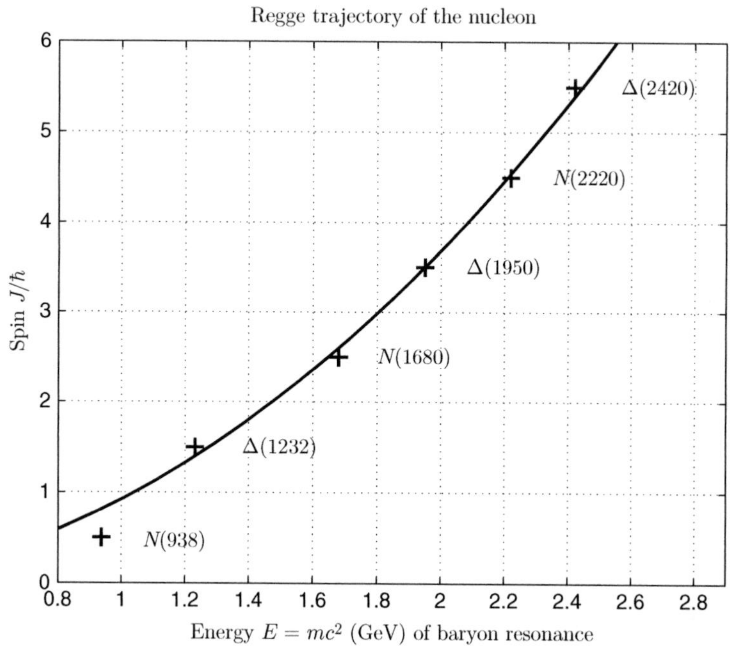

Figure 19.1 The angular momentum and energy of the nucleon and five baryon resonances approximately fit the curve $J/\hbar = T_0 E^2$ with string tension $T_0 = 0.92$ GeV/fm.

and Schwarz, 1974). They identified the graviton as an excitation of the closed string.

19.4 Quantized strings

The coordinates X^μ may be quantized most easily in light-cone coordinates. The resulting relativistic bosonic string must live in a space-time of exactly $D = d + 1 = 26$ dimensions with a tachyon. But if one adds fermionic variables $\psi_1^\mu(\tau, \sigma)$ and $\psi_2^\mu(\tau, \sigma)$ in a supersymmetric way, then the tachyon goes away, and the number of space-time dimensions drops to exactly 10. There are five distinct superstring theories – types I, IIA, and IIB; $E_8 \otimes E_8$ heterotic; and $SO(32)$ heterotic. All five may be related to a single theory in 11 dimensions called **M-theory**, which is not a string theory. M-theory contains membranes (2-branes) and 5-branes, which are not D-branes.

19.5 D-branes

One may satisfy Dirichlet boundary conditions (19.17) by requiring the ends of a string to be attached to a spatial manifold, called a **D-brane** after Dirichlet. If the manifold to which the string is stuck has p dimensions, then it's called a

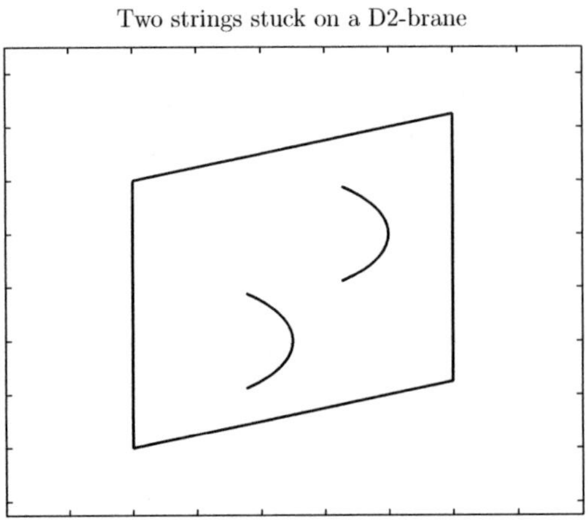
Two strings stuck on a D2-brane

Figure 19.2 Two strings stuck on a D2-brane.

Dp-brane. Figure (19.2) shows a string whose ends are free to move only within a D2-brane.

Dp-branes offer a natural way to explain the extra six dimensions required in a universe of superstrings. One imagines that the ends of all the strings are free to move only in our four-dimensional space-time; the strings are stuck on a D3-brane, which is the three-dimensional space of our physical Universe. The tension of the superstring then keeps it from wandering far enough into the extra six spatial dimensions for us ever to have noticed.

This explanation conflicts, however, with one interpretation of gravity in string theory. In that view, the graviton and the spin-3/2 gravitino are modes of the closed string, are not attached to any D-brane, and propagate in all d space dimensions. Thus if the extra space dimensions were huge, then the gravitational force, which is mainly due to the graviton, would fall off with the distance r as $1/r^{d-1} = 1/r^8$ for superstrings ($1/r^{24}$ for bosonic strings), both much faster than $1/r^2$.

19.6 String–string scattering

Strings interact by joining and by breaking. Figure 19.3 shows two open strings joining to form one open string and then breaking into two open strings. Figure 19.4 shows two closed strings joining to form one closed string and then breaking into two closed strings. The interactions of strings do not occur at points.

Because the fundamental things of string theory are extended objects, string theory is intrinsically free of ultraviolet divergences. It provides a finite theory of quantum gravity.

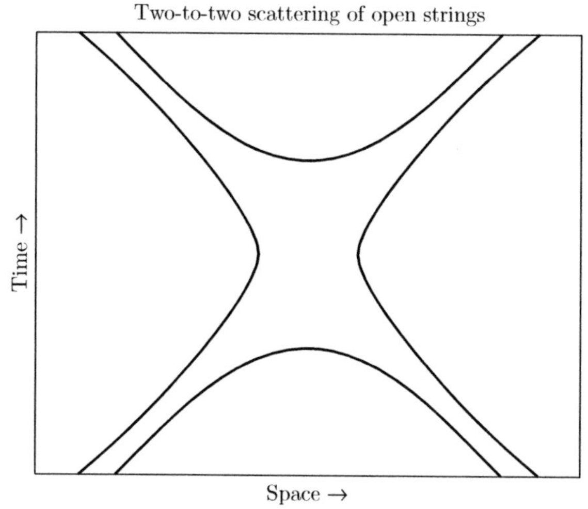

Figure 19.3 A space-time diagram of the scattering of two open strings into two open strings.

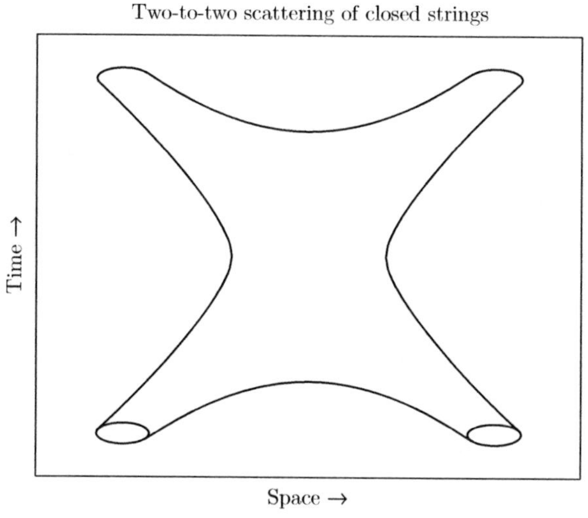

Figure 19.4 A space-time diagram of the scattering of two closed strings into two closed strings.

19.7 Riemann surfaces and moduli

A **homeomorphism** is a map that is one to one and continuous with a continuous inverse. A **Riemann surface** is a two-dimensional real manifold whose open sets U_α are mapped onto open sets of the complex plane $\mathbb{C}$ by **homeomorphisms**

z_α whose transition functions $z_\alpha \circ z_\beta^1$ are analytic on the images of the intersections $U_\alpha \cap U_\beta$. Two Riemann surfaces are **equivalent** if they are related by a continuous analytic map that is one to one and onto.

A parameter that distinguishes a Riemann surface from other, inequivalent Riemann surfaces is called a **modulus**. Some Riemann surfaces have several moduli; others have one modulus; others none at all. Some moduli are continuous parameters; others are discrete.

Further reading

Students who want to learn more about strings should look at the excellent textbook *A First Course in String Theory* (Zwiebach, 2009).

Exercises

19.1 Derive formulas (19.10) and (19.11).
19.2 Derive equation (19.12) from (19.9, 19.10, & 19.11).

References

Aitken, A. C. 1959. *Determinants and Matrices.* Oliver and Boyd.

Alberts, Bruce, Johnson, Alexander, Lewis, Julian, Raff, Martin, Roberts, Keith, and Walter, Peter. 2008. *Molecular Biology of the Cell.* 5th edn. Garland Science. Page 246.

Alligood, Kathleen T., Sauer, Tim D., and Yorke, James A. 1996. *Chaos: an Introduction to Dynamical Systems.* Springer-Verlag.

Arnold, V. I. 1989. *Mathematical Methods of Classical Mechanics.* 2nd edn. Springer. Chapter 7.

Autonne, L. 1915. Sur les Matrices Hypohermitiennes et sur les Matrices Unitaires. *Ann. Univ. Lyon, Nouvelle Série I*, **Fasc. 38**, 1–77.

Bigelow, Matthew S., Lepeshkin, Nick N., and Boyd, Robert W. 2003. Superluminal and slow light propagation in a room-temperature solid. *Science*, **301(5630)**, 200–202.

Bordag, Michael, Klimchitskaya, Galina Leonidovna, Mohideen, Umar, and Mostepanenko, Vladimir Mikhaylovich. 2009. *Advances in the Casimir Effect.* Oxford University Press.

Bouchaud, Jean-Philippe, and Potters, Marc. 2003. *Theory of Financial Risk and Derivative Pricing.* 2nd edn. Cambridge University Press.

Boyd, Robert W. 2000. *Nonlinear Optics.* 2nd edn. Academic Press.

Brillouin, L. 1960. *Wave Propagation and Group Velocity.* Academic Press.

Brunner, N., Scarani, V., Wegmüller, M., Legré, M., and Gisin, N. 2004. Direct measurement of superluminal group velocity and signal velocity in an optical fiber. *Phys. Rev. Lett.*, **93(20)**, 203902.

Cantelli, F. P. 1933. Sulla determinazione empirica di una legge di distribuzione. *Giornale dell'Instituto Italiano degli Attuari*, **4**, 221–424.

Carroll, Sean. 2003. *Spacetime and Geometry: an Introduction to General Relativity.* Benjamin Cummings.

Cohen-Tannoudji, Claude, Diu, Bernard, and Laloë, Frank. 1977. *Quantum Mechanics.* Hermann & John Wiley.

Courant, Richard. 1937. *Differential and Integral Calculus*, Vol. I. Interscience.

Courant, Richard and Hilbert, David. 1955. *Methods of Mathematical Physics, Vol. I.* Interscience.

Creutz, Michael. 1983. *Quarks, Gluons, and Lattices*. Cambridge University Press.

Darden, Tom, York, Darrin, and Pedersen, Lee. 1993. Particle mesh Ewald: an $N\log(N)$ method for Ewald sums in large systems. *J. Chem. Phys.*, **98(12)**, 10089.

DeWitt, Bryce S. 1967. Quantum theory of gravity. II. The manifestly covariant theory. *Phys. Rev.*, **162(5)**, 1195–1239.

Dirac, P. A. M. 1967. *The Principles of Quantum Mechanics*. 4th edn. Oxford University Press.

Dirac, P. A. M. 1996. *General Theory of Relativity*. Princeton University Press.

Faddeev, L. D. and Popov, V. N. 1967. Feynman diagrams for the Yang–Mills field. *Phys. Lett. B*, **25(1)**, 29–30.

Feller, William. 1966. *An Introduction to Probability Theory and Its Applications*. Vol. II. Wiley.

Feller, William. 1968. *An Introduction to Probability Theory and Its Applications*. 3rd edn. Vol. I. Wiley.

Feynman, Richard P. and Hibbs, A. R. 1965. *Quantum Mechanics and Path Integrals*. McGraw-Hill.

Frieman, Joshua A., Turner, Michael S., and Huterer, Dragan. 2008. Dark energy and the accelerating Universe. *Ann. Rev. Astron. Astrophys.*, **46**, 385–432. arXiv:0803.0982v1 [astro-ph].

Gattringer, Christof and Lang, Christian B. 2010. *Quantum Chromodynamics on the Lattice: an Introductory Presentation*. Springer (Lecture Notes in Physics).

Gehring, G. M., Schweinsberg, A., Barsi, C., Kostinski, N., and Boyd, R. W. 2006. Observation of backwards pulse propagation through a medium with a negative group velocity. *Science*, **312(5775)**, 895–897.

Gelfand, Israel M. 1961. *Lectures on Linear Algebra*. Interscience.

Gell-Mann, Murray. 1994. *The Quark and the Jaguar*. W. H. Freeman.

Gell-Mann, Murray. 2008. *Plectics*. Lectures at the University of New Mexico.

Georgi, H. 1999. *Lie Algebras in Particle Physics*. 2nd edn. Perseus Books.

Glauber, Roy J. 1963a. Coherent and incoherent states of the radiation field. *Phys. Rev.*, **131(6)**, 2766–2788.

Glauber, Roy J. 1963b. The quantum theory of optical coherence. *Phys. Rev.*, **130(6)**, 2529–2539.

Glivenko, V. 1933. Sulla determinazione empirica di una legge di distribuzione. *Giornale dell'Instituto Italiano degli Attuari*, **4**, 92–99.

Gnedenko, B. V. 1968. *The Theory of Probability*. Chelsea Publishing Co.

Gutzwiller, Martin C. 1990. *Chaos in Classical and Quantum Mechanics*. Springer.

Hau, L.V., Harris, S. E., Dutton, Z., and Behroozi, C. H. 1999. Light speed reduction to 17 metres per second in an ultracold atomic gas. *Nature*, **397**, 594.

Hobson, M. P., Efstathiou, G. P., and Lasenby, A. N. 2006. *General Relativity: an Introduction for Physicists*. Cambridge University Press.

Holland, John H. 1975. *Adaptation in Natural and Artificial Systems*. University of Michigan Press.

Ince, E. L. 1956. *Integration of Ordinary Differential Equations*. 7th edn. Oliver and Boyd, Ltd. Chapter 1.

James, F. 1994. RANLUX: a Fortran implementation of the high-quality pseudo-random number generator of Lüscher. *Comp. Phys. Comm.*, **79**, 110.

Kleinert, Hagen. 2009. *Path Integrals in Quantum Mechanics, Statistics, Polymer Physics, and Financial Markets*. World Scientific.

Knuth, Donald E. 1981. *The Art of Computer Programming, Volume 2: Seminumerical Algorithms*. 2nd edn. Addison-Wesley.

Kolmogorov, Andrei Nikolaevich. 1933. Sulla determinazione empirica di una legge di distribuzione. *Giornale dell'Instituto Italiano degli Attuari*, **4**, 83–91.

Langevin, Paul. 1908. Sur la théorie du mouvement brownien. *Comptes Rend. Acad. Sci. Paris*, **146**, 530–533.

Larson, D., Dunkley, J., Hinshaw, G., Komatsu, E., Nolta, M.R., *et al.* 2011. Seven-year Wilkinson microwave anisotropy probe (WMAP) observations: power spectra and WMAP-derived parameters. *Astrophys. J. Suppl.*, **192**, 16.

Lifshitz, E. M. 1956. The theory of molecular attractive forces between solids. *Sov. Phys. JETP*, **2**, 73.

Lin, I-Hsiung. 2011. *Classic Complex Analysis*. World Scientific.

Lüscher, M. 1994. A portable high-quality random number generator for lattice field theory simulations. *Comp. Phys. Comm.*, **79**, 100.

Matzner, Richard A. and Shepley, Lawrence C. 1991. *Classical Mechanics*. Prentice Hall.

McCauley, Joseph L. 1994. *Chaos, Dynamics, and Fractals*. Cambridge University Press.

Metropolis, Nicholas, Rosenbluth, Arianna W., Rosenbluth, Marshall N., Teller, Augusta H., and Teller, Edward. 1953. Equation of state calculations by fast computing machines. *J. Chem. Phys.*, **21(6)**, 1087–1092.

Milonni, Peter W. and Shih, M.-L. 1992. Source theory of the Casimir force. *Phys. Rev. A*, **45(7)**, 4241–4253.

Misner, Charles W., Thorne, Kip S., and Wheeler, John Archibald. 1973. *Gravitation*. W. H. Freeman.

Morse, Philip M. and Feshbach, Herman. 1953. *Methods of Theoretical Physics*. Vol. I. McGraw-Hill.

Parsegian, Adrian. 1969. Energy of an ion crossing a low dielectric membrane: solutions to four relevant electrostatic problems. *Nature*, **221**, 844–846.

Pathria, R. K. 1972. *Statistical Mechanics*. Pergamon Press. Chapter 13.

Pearson, Karl. 1900. On the criterion that a given system of deviations from the probable in the case of correlated system of variables is such that it can be reasonably supposed to have arisen from random sampling. *Phil. Mag.*, **50(5)**, 157–175.

Riley, Ken, Hobson, Mike, and Bence, Stephen. 2006. *Mathematical Methods for Physics and Engineering*. 3rd edn. Cambridge University Press.

Roe, Byron P. 2001. *Probability and Statistics in Experimental Physics*. Springer.

Saito, Mutsuo, and Matsumoto, Makoto. 2007. www.math.sci.hiroshima-u.ac.jp/m-mat/MT/emt.html.

Sakurai, J. J. 1982. *Advanced Quantum Mechanics*. 1st edn. Addison Wesley. Pages 62–63.

Scherk, Joël, and Schwarz, John H. 1974. Dual models for non-hadrons. *Nucl. Phys.*, **B81**, 118.

Schmitt, Lothar M. 2001. Theory of genetic algorithms. *Theoretical Computer Science*, **259**, 1–61.

Schutz, Bernard. 1980. *Geometrical Methods of Mathematical Physics*. Cambridge University Press.

Schwinger, Julian, Deraad, Lester, Milton, Kimball A., and Tsai, Wu-yang. 1998. *Classical Electrodynamics*. Westview Press.

Smirnov, N. V. 1939. Estimation of the deviation between empirical distribution curves for two independent random samples. *Bull. Moscow State Univ.*, **2(2)**, 3–14.

Srednicki, Mark. 2007. *Quantum Field Theory*. Cambridge University Press.

Stakgold, Ivar. 1967. *Boundary Value Problems of Mathematical Physics*, Vol. I. Macmillan.

Steinberg, A. M., Kwiat, P. G., and Chiao, R. Y. 1993. Measurement of the single-photon tunneling time. *Phys. Rev. Lett.*, **71(5)**, 708–711.

Stenner, Michael D., Gauthier, Daniel J., and Neifeld, Mark A. 2003. The speed of information in a 'fast-light' optical medium. *Nature*, **425**, 695–698.

Titulaer, U. M., and Glauber, R. J. 1965. Correlation functions for coherent fields. *Phys. Rev.*, **140(3B)**, B676–682.

Veneziano, Gabriel. 1968. Construction of a crossing-symmetric Regge-behaved amplitude for linearly rising regge trajectories. *Nuovo Cim.*, **57A**, 190.

von Foerster, Heinz, Mora, Patricia M., and Amiot, Lawrence W. 1960. Doomsday: Friday, 13 November, A.D. 2026. *Science*, **132**, 1291–1295.

Vose, Michael D. 1999. *The Simple Genetic Algorithm: Foundations and Theory*. MIT Press.

Wang, Yun-ping and Zhang, Dian-lin. 1995. Reshaping, path uncertainty, and superluminal traveling. *Phys. Rev. A*, **52(4)**, 2597–2600.

Watson, George Neville. 1995. *A Treatise on the Theory of Bessel Functions*. Cambridge University Press.

Waxman, David and Peck, Joel R. 1998. Pleiotropy and the preservation of perfection. *Science*, **279**.

Weinberg, Steven. 1972. *Gravitation and Cosmology*. John Wiley & Sons.

Weinberg, Steven. 1988. *The First Three Minutes*. Basic Books.

Weinberg, Steven. 1995. *The Quantum Theory of Fields*. Vol. I: Foundations. Cambridge University Press.

Weinberg, Steven. 1996. *The Quantum Theory of Fields*. Vol. II: Modern applications. Cambridge University Press.

Weinberg, Steven. 2005. *The Quantum Theory of Fields*. Vol. III: Supersymmetry. Cambridge University Press.

Weinberg, Steven. 2010. *Cosmology*. Oxford University Press.

Whittaker, E. T. and Watson, G. N. 1927. *A Course of Modern Analysis*. 4th edn. Cambridge University Press.

Wright, Ned. 2006. A cosmology calculator for the World Wide Web. *Publ. Astron. Soc. Pacific*, **118(850)**, 1711–1715. www.astro.ucla.edu/wright/CosmoCalc.html.
Zee, Anthony. 2010. *Quantum Field Theory in a Nutshell*. 2nd edn. Princeton University Press.
Zwiebach, Barton. 2009. *A First Course in String Theory*. 2nd edn. Cambridge University Press.

Index

adjoint domain, 262
analytic continuation, 179–180
  dimensional regularization, 180
analytic functions, 160–222
  branch of, 194
  definition of, 160
  entire, 161, 177
  essential singularity, 177
  harmonic functions, 170–171
  holomorphic, 177
  isolated singularity, 177
  meromorphic, 177
  multivalued, 194
  pole, 177
  simple pole, 177
angular momentum
  lowering operators, 371
  raising operators, 371
  spin, 371
annihilation and creation operators, 132, 233
arrays, 2–3
associated Legendre functions, 317–323
  Rodrigues's formula for, 318
associated Legendre polynomials, 317–323
  Rodrigues's formula for, 318
asymptotic freedom, 237, 629–634
average value, 117

basis, 7, 16
beats, 77
Bessel functions, 325–347
  and charge near a membrane, 331–333
  and coaxial wave-guides, 342–343
  and cylindrical wave-guides, 333–335
  and scattering off a hard sphere, 344–345
  exercises, 345–347
  Hankel functions, 341–343
  modified, 330–331, 341–343
  Neumann functions, 341–343
  of the first kind, 143, 144, 325–341
  of the second kind, 341–345
  spherical, 145
    and partial waves, 338–340
    and quantum dots, 340–341
    Rayleigh's formula for, 336
  spherical Bessel functions of the second kind, 343–345
Bessel inequality, 284
Bessel's equation, 327
Bianchi identity, 413
binomial coefficient, 141
Bloch's theorem, 105
bodies falling in air, 247
Boltzmann distribution, 54–55
Boltzmann's constant, 149
boundary conditions
  Dirichlet, 262
  natural, 260, 268
  Neumann, 260
Bravais lattice, 104
Bromwich integral, 129

calculus of variations, 443–447
  in nonrelativistic mechanics, 443–444
  in relativistic electrodynamics, 445
  in relativistic mechanics, 444–445
  particle in a gravitational field, 445–447
  strings, 643–645
Callan–Symanzik equation, 630–634
canonical commutation relations, 132, 233
Cartan subalgebras, 379
Casimir effect, 214–217
Cauchy's principal value, 199
  Feynman's propagator, 201–205

trick, 200
chaos, 635–642
attractors, 639–642
limit cycle, 639
of fractal dimension, 642
strange, 642
Bernoulli shift, 638
chaotic threshold, 635
Duffing's equation, 635
dynamical system, 636
autonomous, 636
fractals, 639–642
Cantor set, 640
fractal dimension, 640
Koch snowflake, 640
self-similar dimension, 641
Hénon's map, 638
invertible map, 636, 637
map, 636
period-two sequence, 635
Poincaré surface of section, 636
Rayleigh–Benard convection, 635
van der Pol's equation, 639
characteristic function, 119
and moments, 119
class C^k of functions, 85
Clifford algebra, 393
commutators, 353
compact, 351
complex arithmetic, 2
complex-variable theory, 160–222
Abel–Plana formula, 212–217
analytic continuation, 179–180
analyticity, 160–161
and string theory, 217–219
applications to string theory
radial order, 217
argument principle, 178–179
calculus of residues, 180–182
Cauchy's inequality, 173
Cauchy's integral formula, 165–169
Cauchy's integral theorem, 161–165
and Stoke's theorem, 169
Cauchy's principal value, 198–205
Cauchy–Riemann conditions, 169–170
conformal mapping, 197–198
contour integral with cut, 196
cuts, 193–197
dispersion relations, 205–208
essential singularity, 177
and Picard's theorem, 177
exercises, 219–222
fundamental theorem of algebra, 174
ghost contours, 182–191
harmonic functions, 170–171
isolated singularity, 177
Laurent series, 174–179
Liouville's theorem, 173–174
logarithms, 193–197
method of steepest descent, 210–212
phase and group velocities, 208–210
pole, 177
residue, 176
roots, 194
simple pole, 177
singularities, 177–179
Taylor series, 171–173
conformal mapping, 197–198
contractions, 416
contravariant vector field, 401
contravariant vectors, 401
convergence
of functions, 85
uniform, 85
uniform
and term-by-term integration, 85
convergence in the mean, 138–139
convex function, 560
convolutions, 121–124, 134
and Gauss's law, 121–123
and Green's functions, 121–123
and translational invariance, 123
coordinates, 400–401
correlation functions
Glauber and Titulaer, 69–71
cosmology, 289–291, 457–469
Ω, 462
time evolution of, 461–463
comoving coordinates, 458
cosmic microwave background radiation (CMBR), 458, 469
critical energy density, 458, 462
dark matter, 457
era of dark energy, 458
era of matter, 458
era of radiation, 457
first three minutes, 457
homogeneous and isotropic
line element, 458
Hubble constant, 458
Hubble rate, 461
inflation, 457
models, 463–469
dark-energy dominated, 468
equation of state, 464
inflation dominated, 465
matter dominated, 466–468
radiation dominated, 465–466
transparency, 468
without acceleration, 464–465
recombination, 468
redshift, 463
Robertson–Walker metric, 459–463
energy–momentum tensor of, 461
Friedmann equations, 461
transparency, 458
covariant derivatives
in Yang–Mills theory, 365
covariant vector field, 402
covariant vectors, 402

decuplet of baryon resonances, 646–647
degenerate eigenvalue, 40
delta function, 97–101, 112–115, 120, 130, 131, 281–283, 287–288
  and Green's functions, 122
  Dirac comb, 98–101, 114–115
  eigenfunction expansion of, 281–283, 287–288
  for continuous square-integrable functions, 112–115
  for real periodic functions, 101, 103
  for twice differentiable functions on [−1, 1], 310
  of a function, 113
density operators, 54–55
determinants, 27–33
  and antisymmetry, 28, 29
  and Levi–Civita symbol, 28
  and linear dependence, 29
  and linear independence, 29
  and permutations, 30
  and the inverse of a matrix, 31
  cofactors, 28
  invariances of, 28
  Laplace expansion, 28
  minors, 28
  product rule for, 32
  3×3, 27
  2×2, 27
dielectrics, 154
differential equations, 223–295
  exercises, 293–295
  terminal velocity
    of mice, men, falcons, and bullets, 247
diffusion, 133–134
  Fick's law, 133
dimensional regularization, 180
Dirac mass term, 395
Dirac notation, 19–27, 96–101
  and change of basis, 22
  and inner-product rules, 20
  and self-adjoint linear operators, 23
  and the adjoint of an operator, 22–23
  bra, 19
  bracket, 19
  examples, 22
  ket, 19
  outer products, 21
  tricks, 51
Dirac's delta function, 285
Dirac's gamma matrices, 393
direct product, 377
  adding spins, 68
  and hydrogen atom, 68
dispersion relations, 205–208
  and causality, 205
  Kramers and Kronig, 206–208
divergence, 228–230
division algebra, 380, 384
double-factorials, 143

eigenfunctions, 267–283
  complete, 274, 277–283
  orthonormal
    Gram–Schmidt procedure, 274
eigenvalues, 267–283
  algebraic multiplicity, 40
  degenerate, 39
  geometric multiplicity, 40
  nondegenerate, 39
  simple, 40
  unbounded, 275–283
eigenvectors, 36–55
Einstein's summation convention, 404–405
electric constant, 154, 413
electric displacement, 413
electric susceptibility, 154
electrodynamics, 411–414
electrostatic energy, 154
electrostatic potential
  multipole expansion, 286
electrostatics
  dielectrics, 153–157
emission rate from a fluorophore, 247
energy–momentum 4-vector, 410
entropy, 54–55
euclidean coordinates, 402–404
euclidean space, 402–404
Ewald summation, 115
expected value, 117
exterior derivative, 417–419

factorials, 141–145
  double, 143, 145
  Mermin's approximation, 141
  Mermin's infinite-product formula, 141
  Ramanujan's approximation, 141
  Stirling's approximation, 141
Faraday's law, 154, 412
Feynman's propagator, 120–121, 201–205
  as a Green's function, 201
field of charge near a membrane, 155–157
field of charge near dielectric interface, 154–157
forms, 416–419, 427–431, 479–501
  closed, 496–498
  differential forms, 416–419, 481–501
    p-forms, 417
    1-forms, 416
    2-forms, 416
    and exterior derivative, 417
    and gauge theory, 471
    and Stokes's theorem, 418
    closed, 418
    complex, 498
    curl, 436
    exact, 418
    Hodge star, 439–440, 442
    invariance of, 416
    wedge product, 416
  exact, 496–498
  exercises, 500–501
  exterior derivatives, 486–491

exterior forms, 479–481
Frobenius's theorem, 498–500
integration, 491–496
Stokes's theorem, 495
Fourier series, 75–107
and scalar fields, 132
better convergence of integrated series, 88
complex, 75–79, 83–89, 96–107
for real functions, 76
of nonperiodic functions, 78–89
convergence, 84–89
convergence theorem, 85
exercises, 105–107
Gibbs overshoot, 81–82
nonrelativistic strings, 103
of a C^1 function, 87
of nonperiodic functions, 78–89
Parseval's identity, 100
periodic boundary conditions, 103–105
Born–von Karman, 105
poorer convergence of differentiated series, 89
quantum-mechanical examples, 89–96
real functions, 79–82
Gibbs overshoot, 81
several variables, 84
stretched intervals, 83–84
the interval, 77
where to put the 2πs, 77
Fourier transforms, 108–125, 129–134
and Ampère's law, 123
and characteristic functions, 119
and convolutions, 121–124
and differential equations, 129–134
and diffusion equation, 133–134
and Fourier series, 110, 120
and Green's functions, 121–125
and momentum space, 116–119
and Parseval's relation, 113
and scalar wave equation, 131
and the delta function, 112–115
and the Feynman propagator, 120–121
and the uncertainty principle, 117–119
derivatives of, 115–119
exercises, 134–135
in several dimensions, 119–121
integrals of, 115–119
inverse of, 109
of a gaussian, 110–111, 183
of real functions, 111–112
Fourier–Legendre expansion, 310
Fourier–Mellin integral, 129
function
$f(x \pm 0)$, 85
continuous, 85
piecewise continuous, 85
functional derivatives, 578–585
and delta functions, 579–580
and variational methods, 579–582
exercises, 585
functional differential equation for ground state of a field theory, 583–585
functional differential equations, 583–585
notation used in physics, 579–580
of higher order, 581–582
Taylor series of, 582–583
functionals, 578
functions
analytic, 160–161
differentiable, 160
fundamental theorem of algebra, 174

gamma function, 141–145, 179–180
Gauss's law, 154, 284, 412
gaussian integrals, 586–588
Gell-Mann's $SU(3)$ matrices, 377
general relativity, 289–291, 445–469
black holes, 456–457
cosmological constant, 454
cosmology, 457–469
dark energy, 454
Einstein's equations, 453–456
Einstein–Hilbert action, 454–455
model cosmologies, 463–469
dark-energy dominated, 468
equation of state, 464
inflation dominated, 465
matter dominated, 466–468
radiation dominated, 465–466
transparency, 468
without acceleration, 464–465
Schwarzschild's solution, 456–457
static and isotropic gravitational field, 455–457
Schwarzschild's solution, 456–457
standard form, 455–456
geometric series, 139–140
gradient, 228–230
grand unification, 377
Grassmann numbers, 2, 6–7
Grassmann polynomials, 2
Grassmann variables, 613–619
Green's function
for Helmholtz's equation, 286
for Helmholtz's modified equation, 286
for Laplacian, 123
for Poisson's equation
and Legendre functions, 287
Green's functions, 284–289
and eigenfunctions, 287–289
Feynman's propagator, 287
for Helmholtz's equation, 285–286
for Poisson's equation, 284–287
of a self-adjoint operator, 288
Poisson's equation, 285
group index of refraction, 209
groups, 348–399
$O(n)$, 349
$SO(3)$
adjoint representation of, 366–367
$SO(n)$, 349

$SU(2)$, 368–376
  defining representation of, 371–374
  spin and statistics, 370
  tensor products of representations, 369
$SU(3)$, 377–379
$SU(3)$ structure constants, 378
Z_2, 358
Z_3, 358
Z_n, 358
abelian, 349
and symmetries in quantum mechanics, 351
and Yang–Mills gauge theory, 365
automorphism, 355
  inner, 355
  outer, 355
block-diagonal representations of, 351
centers of, 353
characters, 356–357
compact, 350–351
compact Lie groups
  real structure constants of, 364
  totally antisymmetric structure constants of, 364
completely reducible representations of, 351
conjugacy classes of, 353
continuous, 348–350
definition, 348
direct sum of representations of, 351
equivalent representations of, 351
exercises, 396–399
factor groups of by subgroups, 354
finite, 350, 358–361
  multiplication table, 358
  regular representation of, 359
further reading, 396
Gell-Mann matrices, 377–379
generators of adjoint representations of, 374–375
invariant integration, 384–385
irreducible representations of, 351
isomorphism, 354
Lie algebras, 361–399
  $SU(2)$, 368–374
Lie groups, 348–350, 361–399
  $SO(3)$, 366–367
  $SU(3)$, 377–379
  $SU(3)$ structure constants, 378
  adjoint representations of, 374–375
  antisymmetry of structure constants, 363
  Cartan subalgebra of $SU(3)$, 378
  Cartan subalgebras of, 379
  Cartan's list of compact simple Lie groups, 383–384
  Casimir operators of, 369, 375
  compact, 361–385
  defining representation of $SU(2)$, 371–374
  definition of, 361
  exponential parametrization, 362
  generators of, 362
  generators of adjoint representation, 374
  hermitian generators of, 363
  Jacobi identity, 374
  noncompact, 361–364
  nonhermitian generators of, 363
  of rotations, 366–374
  simple and semisimple, 376–377
  structure constants of, 363, 374–375
  $SU(2)$ tensor product, 372
  $SU(3)$ generators, 377
  symplectic group $Sp(2n,R)$, 382
  symplectic groups, 381–383
Lie groups have structure constants that are independent of the representation, 365
Lorentz, 349
Lorentz group, 386–396
  Dirac representation of, 393–395
  Lie algebra of, 386–389
  two-dimensional representations of, 389–393
matrix, 349
morphism, 354
noncompact, 350
nonabelian, 349
of matrices, 349–350
of orthogonal matrices, 349–350
of permutations, 360–361
of rotations, 366–374
  adjoint representation of, 366–367
  explicit 3×3 representation of, 367
  generators of, 366–367
  spin and statistics, 370
  tensor operators of, 376
of transformations, 348–350
  Lorentz, 348
  Poincaré, 348
  rotations and reflections, 348
  translations, 348
of unitary matrices, 349–350
order of, 349, 358
Poincaré, 349
Poincaré group, 395–396
  Lie algebra of, 395–396
reducible representations of, 351
representations of, 350–352
  dimensions of, 350
  in Hilbert space, 351–352
rotations
  representations of, 352
Schur's lemma, 355–356
semisimple, 353
similarity transformations, 351
simple, 353
simple and semisimple Lie algebras, 376–377
subgroups, 353–355
  cosets of, 354
  invariant, 353
  left cosets of, 354
  normal, 353
  quotient coset space, 354
  right cosets of, 354
  trivial, 353

symmetry
antilinear, antiunitary representations of, 352
linear, unitary representations of, 352
tensor product
addition of angular momenta, 357
tensor products, 357–358
translation, 348
unitary representations of, 351

harmonic function, 170
harmonic oscillator, 101–102, 272–273
Heaviside step function, 106
helicity
positive, 393
right-handed, 393
Helmholtz's equation
in cylindrical coordinates
and Bessel functions, 328–335
in spherical coordinates
and spherical Bessel functions, 335–341
in three dimensions, 231–232
in two dimensions, 230–231
rectangular coordinates, 231
spherical coordinates, 232
and associated Legendre functions, 317
and spherical Bessel functions, 317
with azimuthal symmetry, 315–316
and Legendre polynomials, 315–316
and spherical Bessel functions, 315–316
Hermite functions, 264
Hermite's system, 264
hermitian differential operators, 261
Hilbert spaces, 13–14, 25–26
homogeneous functions, 243–245
Euler's theorem, 243
virial theorem, 243–244

index of refraction, 207
infinite products, 157–158
infinite series, 136–159
absolute convergence, 136
asymptotic, 152–153
WKB & Dyson, 153
Bernoulli numbers and polynomials, 151–152
binomial series, 148
binomial theorem, 147, 148
Cauchy's criterion, 137
Cauchy's root test, 137
comparison test, 137
conditional convergence, 136
convergence, 136–140
d'Alembert's ratio test, 138
divergence of, 136
exercises, 158–159
Intel test, 138
logarithmic series, 148–149
of functions
convergence, 138–139
convergence in the mean, 138–139
Dirichlet series, 149–151
Fourier series, 146
geometric series, 139–140
power series, 139–140, 146
Taylor series, 145–149
uniform convergence, 139
Riemann zeta function, 149
uniform convergence, 138
and term-by-term integration, 139
inner products, 3, 11–14
and distance, 12
and norm, 12
degenerate, 12
hermitian, 12
indefinite, 12, 14
Minkowski, 14
nondegenerate, 12
of functions, 13
positive definite, 11
Schwarz, 11
inner-product spaces, 13–14
integral equations, 296–304
exercises, 304
Fredholm, 297–301
eigenfunctions, 297–301
eigenvalues, 297–301
first kind, 297
homogeneous, 297
inhomogeneous, 297
second kind, 297
implications of linearity, 298–304
integral transformations, 301–304
and Bessel functions, 302–304
Fourier, Laplace, and Euler kernels, 302
kernel, 297
numerical solutions, 299–301
Volterra, 297–301
eigenfunctions, 297–301
eigenvalues, 297–301
first kind, 297
homogeneous, 297
inhomogeneous, 297
second kind, 297
integral transformations, 301–304
and Bessel functions, 302–304
Fourier, Laplace, and Euler kernels, 302
invariant distance, 409
invariant subspace, 37

Jacobi identity, 374

kernel of a matrix, 355
Kramers–Kronig relations, 206–208
Kronecker delta, 5, 415

Lagrange multipliers, 35–36, 54–55, 267–283
Lapack, 33, 66
Laplace transforms, 125–134
and convolutions, 134
and differential equations, 128–134
derivatives of, 127–128

examples, 125
integrals of, 127
inversion of, 129
Laplace's equation
in two dimensions, 316–317
Laplacian, 228–230
Legendre functions, 263–264, 287, 305–324
exercises, 323–324
second kind, 266
Legendre polynomials, 305–313
addition theorem for, 321
generating function, 307–309
Helmholtz's equation
with azimuthal symmetry, 315–316
Legendre's differential equation, 309–311
normalization, 305
recurrence relations, 311–312
Rodrigues's formula, 306–307
Schlaefli's integral for, 312–313
special values of, 312
Legendre's system, 263–264
Leibniz's rule, 141
Lerch transcendent, 151
Levi-Civita symbol, 366
Lie algebras
ranks of, 379
roots of, 379
weight vector, 379
weights of, 379
light
slow, fast, and backwards, 209–210
linear algebra, 1–74
exercises, 71–74
linear dependence, 15–16, 224
and determinants, 29
linear independence, 15–16, 224
and completeness, 15
and determinants, 29
linear least squares, 34–35
linear operators, 9–11
and matrices, 9
density operator, 69
domain, 9
hermitian, 23
range, 9
real, symmetric, 23–24
self-adjoint, 23
unitary, 24–25
Lorentz force, 414
Lorentz transformations, 405–411
boost, 406
invariance under, 406

magnetic constant, 413
magnetic field, 413
magnetic induction, 411
Majorana field, 235
Majorana mass term, 391, 393
Maple, 33, 66
Mathematica, 33, 66
Matlab, 33, 61, 66
matrices, 4–7
adjoint, 4
and linear operators, 9
change of basis, 10
characteristic equation of, 38, 41–42
CKM, 63
and *CP* violation, 63
congruency transformation, 49
defective, 40
density operator, 69
diagonal form of square nondefective matrix, 41
functions of, 43–45
gamma, 393
hermitian, 5, 45–49
and diagonalization by a unitary transformation, 48
complete and orthonormal eigenvectors, 47
degenerate eigenvalues, 46
eigenvalues of, 45
eigenvectors and eigenvalues of, 48
eigenvectors of, 46
identity, 5
imaginary and antisymmetric, 48
inverse, 5
inverses of, 31
nonnegative, 6
nonsingular, 40
normal, 50–55
compatible, 52–55
diagonalization by a unitary transformation, 50
orthogonal, 5, 25
Pauli, 5, 68, 371
positive, 6
positive definite, 6
rank of, 65
example, 65
rank-nullity theorem, 58
real and symmetric, 48
similarity transformation, 10, 41
singular-value decomposition, 55–63
example, 62
quark mass matrix, 62
square, 38–42
eigenvalues of, 38–42
eigenvectors of, 38–42
trace, 4
cyclic, 4
unitary, 5
upper triangular, 33
Maxwell's equations, 412
in vacuum, 413
Maxwell–Ampère law, 412
mean value, 117
method of steepest descent, 210–212
metric spaces, 13–14
Minkowski space, 405–407
Monte Carlo methods, 563–577

and evolution, 576–577
and lattice gauge theory, 574, 577
detailed balance, 574
exercises, 577
genetic algorithms, 577
in statistical mechanics, 572–575
Metropolis step, 572
Metropolis's algorithm, 572–575
more general applications, 575–577
of data analysis, 566–572
example, 566–572
of numerical integration, 563–566
partition function, 574
smart schemes, 575
sweeps, 573
thermalization, 573
Moore–Penrose pseudoinverse, 63–65

natural units, 121
nearest-integer function, 108
nonlinear differential equations, 289–293
general relativity, 289–291
solitons, 291–293
notation for derivatives, 226–228
null space of a matrix, 355
numbers
complex, 1, 7
atan2, 2
phase of, 2
irrational, 1
natural, 1
rational, 1
real, 1

Octave, 33, 66
octet of baryons, 379
octet of pseudo-scalar mesons, 378
octonians, 384
open, 351
operators
adjoint of, 22–23
antilinear, 26–27
antiunitary, 26–27
compatible, 352
complete, 352
orthogonal, 25
real, symmetric, 23–24
unitary, 24–25
optical theorem, 207
ordinary differential equations, 223–225
and variational problems, 259–260
boundary conditions, 258–260
differential operators of definite parity, 255
even and odd differential operators, 254–255
exact, 238–242
Boyle's law, 239
condition of integrability, 239
Einstein's law, 239
human population growth, 239
integrating factors, 242
integration, 240–242
van der Waals's equation, 239
first-order, 235–248
exact, 238–242
separable, 235–238
self-adjoint, 266–267
Frobenius's series solutions, 251–254
Fuch's theorem, 253–254
indicial equation, 252
recurrence relations, 252
Green's formula, 260
hermitian operators, 267
homogeneous first-order, 245
homogeneous functions, 243–245
Lagrange's identity, 260
linear, 223–225
general solution, 224, 225
homogeneous, 224
inhomogeneous, 224, 225
order of, 223
linear dependence of solutions, 224
linear independence of solutions, 224
linear, first-order, 246–248
exact, 246
integrating factor, 246
meaning of exactness, 240–242
nonlinear, 225
second-order
eigenfunctions, 273–275
eigenvalues, 273–275
essential singularity of, 251
Green's functions, 288
irregular singular point of, 251
making operators self adjoint, 264–265
nonessential singular point of, 251
regular singular point of, 251
second solution, 255–257
self-adjoint, 260–283
self-adjoint form, 223
singular points at infinity, 251
singular points of, 250–251
weight function, 273
why not three solutions?, 257–258
wronskians of self-adjoint operators, 265–266
self-adjoint, 260–283
self-adjoint operators, 265
separable, 235–238
general integral, 236
hidden separability, 238
logistic equation, 236
Zipf's law, 236
separated, 235
singular points of Legendre's equation, 251
Sturm–Liouville problem, 265, 267–283
systems of, 248–250, 289–293
Friedmann's equations, 289–291
Lagrange's equations, 248–250
Wronski's determinant, 255–258
orthogonal coordinates, 228–230

divergence in, 228–230
gradient in, 228–230
Laplacian in, 228–230
orthogonal polynomials, 313–315
Hermite's, 314
Jacobi's, 313–314
Laguerre's, 314–315
outer products, 18–22
example, 18
in Dirac's notation, 21

partial differential equations, 225–235
Dirac equation, 234–235
general solution, 226
homogeneous, 226
inhomogeneous, 226
Klein–Gordon equation, 233
linear, 225–235
separable, 230–235
Helmholtz's equation, 230–232
wave equations, 233–235
photon, 233–234
spin-one-half fields, 234–235
spinless bosons, 233
path integrals, 586–625
and gaussian integrals, 586–588
and lattice gauge theories, 622
and nonabelian gauge theories, 619–624
ghosts, 622–624
the method of Faddeev and Popov, 620–624
and perturbative field theory, 605–624
and quantum electrodynamics, 609–613
and Schrödinger's equation, 592–593
and the Bohm–Aharonov effect, 594–595
and the principle of stationary action, 591–592
euclidean, 588–590
euclidean correlation functions, 599–600
exercises, 624–625
fermionic, 613–619
finite temperature, 588–590
for a free particle in imaginary time, 595
for a free particle in real time, 593–595
for harmonic oscillator in imaginary time, 597–598
for harmonic oscillator in real time, 595–597
in field theory, 603–624
in finite-temperature field theory, 600–603
in imaginary time, 588–590
in real time, 590–593
Minkowski, 590–593
of fields, 600–624
of fields in euclidean space, 600–603
of fields in imaginary time, 600–603
ratios of and time-ordered products, 604–624
Pauli matrices, 371
permittivity, 154, 413
permutations, 360–361
and determinants, 30
cycles, 360
phase and group velocities, 208–210
slow, fast, and backwards light, 209
Planck's constant, 149
Planck's distribution, 149–151
points, 400–401
Poisson summation, 114–115
power series, 139–140
pre-Hilbert spaces, 13–14
principle of least action, 259–260
principle of stationary action, 248–250, 267–273, 443–447
in nonrelativistic mechanics, 328, 443–444
in quantum mechanics, 267–273
in relativistic electrodynamics, 445
in relativistic mechanics, 444–445
particle in a gravitational field, 445–447
probability and statistics, 502–562
Bayes's theorem, 502–505
Bernoulli's distribution, 508
binomial distribution, 508–511
brownian motion, 520–527
Einstein–Nernst relation, 520–524
Langevin's theory of, 520–527
Cauchy distributions, 532
central limit theorem, 532–543
illustrations of, 535–543
central moments, 505–508
centroid method, 514
characteristic functions, 527–530
chi-squared distribution, 531
chi-squared statistic, 551–554
convergence in probability, 535
correlation coefficient, 507
covariance, 507
lower bound of Cramér and Rao, 546–550
cumulants, 529
diffusion, 519–527
diffusion constant, 523
direct stochastic optical reconstruction microscopy, 515
Einstein–Nernst relation, 520–524
Einstein's relation, 523
ensemble average, 521
error function, 515–518
estimators, 543–550
Bessel's correction, 545
bias, 543
consistent, 543
standard deviation, 546
standard error, 546
exercises, 560–562
expectation, 505–508
expected value, 505–508
exponential distribution, 531
fat tails, 530–532
Fisher's information matrix, 546–550
fluctuation and dissipation, 524–527
gaussian distribution, 512–519
Gosset's distribution, 530
Heisenberg's uncertainty principle, 506–507
information, 546–550

information matrix, 546–550
Kolmogorov's function, 556
Kolmogorov's test, 554–560
kurtosis, 530
Lévy distributions, 532
Lindeberg's condition, 535
log-normal distribution, 531
Lorentz distributions, 532
maximum likelihood, 550–551
Maxwell–Boltzmann distribution, 518–519
mean, 505–508
moment-generating functions, 527–530
moments, 505–508
normal distribution, 514
Pearson's distribution, 531
Poisson distribution, 511–512
coherent states, 512
power-law tails, 530
probability density, 505–519
probability distribution, 505–519
random-number generators, 537–538
skewness, 530
Student's t-distribution, 530
variance, 505–508
lower bound of Cramér and Rao, 546–550
viscous-friction coefficient, 523
proper time, 409
and time dilation, 410–411
pseudoinverse, 63–65, 551

quantum mechanics, 267–283
quaternions, 379–383
and the Pauli matrices, 380

R-C circuit, 247
regular and self-adjoint differential system, 262
relative permittivity, 154
renormalization group, 237, 626–634
exercises, 634
in condensed-matter physics, 632–634
in lattice field theory, 630–632
in quantum field theory, 626–634
rotations, 366–374

scalar fields, 131, 401
scalars, 401
Schwarz inequality, 14–15, 284
examples, 14
Schwarz inner products, 69–71
seesaw mechanism, 49
self-adjoint differential operators, 260–283
self-adjoint differential systems, 262–283
sequence of functions
convergence in the mean, 88
simple and semisimple Lie algebras, 376–377
simple and semisimple Lie groups, 376–377
simple eigenvalue, 40
simply connected, 164
slow, fast, and backwards light, 209
and Kramers–Kronig relations, 210
small oscillations, 250
solitons, 291–293
special relativity, 408–414
4-vector force, 410
and time dilation, 410–411
electrodynamics, 411–414
energy–momentum 4-vector, 410
kinematics, 410–411
spherical harmonics, 319–323
spin and statistics, 370
standard model of particle physics, 377
states
coherent, 71
Stefan's constant, 150
Stokes's theorem, 170
strings, 643–650
and infinities of quantum field theory, 643
Dirichlet boundary condition, 645
free-endpoint boundary condition, 645
Nambu–Goto action, 643–647
quantized, 647–650
D-branes, 647–648
Regge trajectories, 646–647
Riemann surfaces and moduli, 649–650
scattering of, 648
Sturm–Liouville equation, 265, 267–283
$SU(3)$ and quarks, 378–379
subspace, 351
invariant, 351
proper, 351
summation convention, 404–405
symmetric differential operators, 261
symmetry
in quantum mechanics, 26
systems of linear equations, 34–35

Taylor series, 145–149
tensor products, 377
adding angular momenta, 357–358, 371–374
adding spins, 68, 371–374
and hydrogen atom, 68
tensors, 400–478
affine connections, 431–433
and metric tensor, 436–437
and general relativity, 445–469
antisymmetric, 415
basic axiom of relativity, 422
basis vectors, 421
Bianchi identity, 436
Christoffel symbols, 431–433
connections, 431–433
contractions, 416
contravariant metric tensor, 422
covariant curl, 434–436
covariant derivatives, 431–434
and antisymmetry, 436
metric tensor, 437–438
covariant metric tensor, 422
curvature, 451–453
curvature of a sphere, 451–453

curvature scalar, 451
cylindrical coordinates, 425
divergence of a contravariant vector, 438–443
and Hodge star, 439–440
and Laplacian, 441–443
Einstein's equations, 453–456
exercises, 475–478
gauge theory, 469–475
geometry of, 474–475
role of vectors, 471–473
standard model, 469–473
gradient, 426–427
Hodge star, 428–431
and divergence, 428
and Laplacian, 428
and Maxwell's equations, 430–431
Laplacian, 441–443
and Hodge star, 442
Levi-Civita's symbol, 427–431
Levi-Civita's tensor, 427–431
metric of sphere, 421
metric tensor, 420–427
moving frame, 421
notation for derivatives, 433
orthogonal coordinates, 423–426
parallel transport, 433
particle in a weak, static gravitational field, 448–451
gravitational redshift, 450
gravitational time dilation, 449–450
perfect fluid, 453
polar coordinates, 424
principle of equivalence, 447–448
geodesic equation, 448
quotient theorem, 420
raising, lowering indices, 423
Ricci tensor, 451
Riemann tensor, 451
second-rank, 414–416
spherical coordinates, 425–426
symmetric, 415
tensor equations, 419–420
torsion tensor, 433
third-harmonic microscopy, 184
time dependence of Heisenberg operators, 604
time dilation, 409–410
in muon decay, 409–410
time-ordered product, 121, 604
total cross-section, 207

uncertainty principle, 117–119, 244

variance
of an operator, 118–119
variational methods, 248–250, 259–260, 267–273, 443–447
in nonrelativistic mechanics, 328, 443–444
in quantum mechanics, 267–273
in relativistic electrodynamics, 445
in relativistic mechanics, 444–445
particle in a gravitational field, 445–447
strings, 643–645
vector space, 16
dimension of, 16
vectors, 7–8
basis, 7, 16, 17
complete, 16
components, 7
direct product, 66–68
example, 68
eigenvalues, 37–55
example, 37
eigenvectors, 37–55
example, 37
eigenvectors of square matrix, 40
functions as, 8
orthonormal, 16–17
Gram–Schmidt method, 16–17
partial derivatives as, 8
span, 16
span a space, 16
states as, 8
tensor product, 66–68
Virasoro's algebra, 219, 222
virial theorem, 243–244

wedge product, 416
Weyl spinor
left-handed, 390
right-handed, 393
Wigner–Eckart theorem
special case of, 355–356
wronskian, 255–258

Yang–Mills theory, 469–475
geometry of, 474–475
role of vectors, 471–473
standard model, 469–473
Yukawa potential, 125, 286

zeta function, 149–151